IONIC LIQUID-BASED SURFACTANT SCIENCE

Wiley Series on
Surface and Interfacial Chemistry

Series Editors:
Ponisseril Somasundaran
Nissim Garti

Multiple Emulsion: Technology and Applications
By A. Aserin
November 2007

Colloidal Nanoparticles in Biotechnology
Edited by Abdelhamid Elaissari
April 2008

Self-Assembled Supramolecular Architectures: Lyotropic Liquid Crystals
Edited by Nissim Garti, Ponisseril Somasundaran, Raffaele Mezzenga
September 2012

Proteins in Solution and at Interfaces: Methods and Applications in Biotechnology and Materials Science
Edited by Juan M. Ruso and Ángel Piñeiro
March 2013

Ionic Liquid-Based Surfactant Science: Formulation, Characterization and Applications
Edited by Bidyut K. Paul and Satya P. Moulik
September 2015

IONIC LIQUID-BASED SURFACTANT SCIENCE

FORMULATION, CHARACTERIZATION, AND APPLICATIONS

Edited by

Bidyut K. Paul
Satya P. Moulik

Wiley Series on Surface and Interfacial Chemistry

WILEY

Published by John Wiley & Sons, Inc., Hoboken, New Jersey
Published simultaneously in Canada

Library of Congress Cataloging-in-Publication Data:

Ionic liquid-based surfactant science : formulation, characterization and applications / edited by Bidyut K. Paul, Satya P. Moulik.
pages cm. – (Wiley series on surface and interfacial chemistry)
Includes bibliographical references and index.
ISBN 978-1-118-83419-0 (cloth)
1. Surface chemistry. 2. Self-assembly (Chemistry) I. Paul, Bidyut K. (Bidyut Kumar), editor. II. Moulik, Satya P. (Satya Priya), editor. III. Series: Wiley series on surface and interfacial chemistry.
QD506.I585 2015
541′.33–dc23

2015010255

Cover image courtesy of Luca Jovine

Set in 9.5/11.5pt Times by SPi Global, Pondicherry, India

Printed in the United States of America

10 9 8 7 6 5 4 3 2 1

1 2015

CONTENTS

Contributors vii

Foreword xiii

Preface xvii

1. Ionic Liquids Modify the AOT Interfacial Curvature and Self-Assembly 1
Sergio Murgia, Sandrina Lampis, Marianna Mamusa, and Gerardo Palazzo

2. Characterization of Self-Assembled Amphiphiles in Ionic Liquids 23
Lang G. Chen, Stephen H. Strassburg, and Harry Bermudez

3. Self-Assembly of Nonionic Surfactants in Room-Temperature Ionic Liquids 47
Kenichi Sakai, Takeshi Misono, Masahiko Abe, and Hideki Sakai

4. Ionic Liquid-Based Surfactants: Synthesis, Molecular Structure, Micellar Properties and Applications 63
Paula D. Galgano and Omar A. El Seoud

5. Ionic Liquids in Bulk and at an Interface: Self-Aggregation, Interfacial Tension, and Adsorption 101
Mohammad Tariq, Karina Shimizu, José N. Canongia Lopes, Benilde Saramago, and Luís Paulo N. Rebelo

6. Aggregation Behavior of Ionic Liquid-Based Gemini Surfactants and Their Interaction with Biomacromolecules 127
Ting Zhou and Guiying Xu

7. Fluorescence Studies of the Microenvironments of the Morpholinium Room-Temperature Ionic Liquids 151
Dinesh Chandra Khara, Kotni Santhosh, and Anunay Samanta

8. Self-Assembly of Surface-Active Ionic Liquids in Aqueous Medium 175
K. Srinivasa Rao, Pankaj Bharmoria, Tushar J. Trivedi, and Arvind Kumar

9. Effect of a Surface-Active Lonic Liquid on Calixarenes 193
Shubha Pandey, Shruti Trivedi, Pramod S. Pandey, Siddharth Pandey, and Sandeep K. Mishra

10. **Ionic Liquids in Colloidal Regime** **207**
Indrajyoti Mukherjee and Satya P. Moulik

11. **Nanostructures of Amphiphiles and Microemulsions in Room-Temperature Ionic Liquids** **239**
Ahmed Mourchid

12. **Microemulsions with Ionic Liquids** **253**
Joachim Koetz

13. **Properties of Ionic Liquid-Based Microemulsions** **261**
Maria Figueira-González, Luis García-Río, Mercedes Parajó, and Pedro Rodríguez-Dafonte

14. **Ionic Liquids in Soft Confinement: Effect of Reverse Micelle Interfaces on the Entrapped Ionic Liquid Structure** **283**
Ruben Dario Falcone, N. Mariano Correa, Juana J. Silber, and Nancy E. Levinger

15. **Designing a New Strategy for the Formation of IL-In-Oil Microemulsions Containing Double Chain Surface-Active Ionic Liquid** **303**
Vishal Govind Rao, Chiranjib Banerjee, Surajit Ghosh, Sarthak Mandal, and Nilmoni Sarkar

16. **Ionic Liquid-Based Microemulsions** **325**
Jianling Zhang

17. **Ionic Liquid-Based Nonaqueous Microemulsion** **343**
Qilong Ren, Qiwei Yang, Baogen Su, Zhiguo Zhang, Zongbi Bao, and Huabin Xing

18. **Ionic Liquid Microemulsions and Applications** **359**
Xue Qin An and Jun Shen

19. **Ionic Liquid-In-Oil Microemulsions** **375**
Debostuti Ghosh Dastidar and Sanjib Senapati

20. **Recent Advances in Bioionic Liquids and Biocompatible Ionic Liquid-Based Microemulsions** **397**
Kaushik Kundu, Bidyut K. Paul, Soumik Bardhan, and Swapan K. Saha

21. **Density Prediction of Ternary Mixtures of Ethanol + Water + Ionic Liquid Using Backpropagation Artificial Neural Networks** **447**
J. Morales, O. A. Moldes, M. A. Iglesias-Otero, J. C. Mejuto, G. Astray, and A. Cid

22. **Effect of Ionic Liquids on Catalytic Properties and Structure of Biocatalysts** **459**
Maria H. Katsoura, Athena A. Papadopoulou, Angeliki C. Polydera, and Haralampos Stamatis

23. **Analytical Applications of Ionic Liquid-Based Surfactants in Separation Science** **475**
María J. Trujillo-Rodríguez, Providencia González-Hernández, and Verónica Pino

24. **Ionic Liquids: Surfactant Agents for Layered Silicates** **503**
Sébastien Livi, Jean-François Gérard, and Jannick Duchet-Rumeau

25. **Deep Eutectic Solvents as a New Reaction Medium for Biotransformations** **517**
Zhen Yang and Qing Wen

Index **533**

CONTRIBUTORS

Masahiko Abe, Department of Pure and Applied Chemistry, Faculty of Science and Technology, Research Institute for Science and Technology, Tokyo University of Science, Noda, Chiba, Japan

Xue Qin An, School of Chemistry and Molecular Engineering, East China University of Science and Technology, Shanghai, China

G. Astray, Department of Physical Chemistry, Faculty of Science, University of Vigo at Ourense, Ourense, Spain; Department of Geological Sciences, College of Arts and Sciences, Ohio University, Athens, OH, USA; Faculty of Law, International University of La Rioja, Logroño, Spain

Chiranjib Banerjee, Department of Chemistry, Indian Institute of Technology, Kharagpur, India

Zongbi Bao, Key Laboratory of Biomass Chemical Engineering of Ministry of Education, Department of Chemical and Biological Engineering, Zhejiang University, Hangzhou, China

Soumik Bardhan, Department of Chemistry, University of North Bengal, Darjeeling, India

Harry Bermudez, Department of Polymer Science and Engineering, University of Massachusetts Amherst, Amherst, MA, USA

Pankaj Bharmoria, AcSIR-Central Salt and Marine Chemicals Research Institute, Bhavnagar, India

José N. Canongia Lopes, Instituto de Tecnologia Química e Biológica, Universidade Nova de Lisboa, Oeiras, Portugal; Centro de Química Estrutural, Instituto Superior Técnico, Universidade de Lisboa, Lisboa, Portugal

Dinesh Chandra Khara, School of Chemistry, University of Hyderabad, Hyderabad, India; Department of Chemistry, Ben-Gurion University of the Negev, Beer-Sheva 8410501 Israel

Lang G. Chen, Department of Polymer Science and Engineering, University of Massachusetts Amherst, Amherst, MA, USA

A. Cid, Chemistry Department, REQUIMTE-CQFB, Faculty of Science and Technology, New University of Lisbon, Caparica, Portugal

Ruben Dario Falcone, Departamento de Química, Universidad Nacional de Río Cuarto, Río Cuarto, Argentina

Debostuti Ghosh Dastidar, Department of Biotechnology, Indian Institute of Technology Madras, Chennai, India

Jannick Duchet-Rumeau, Université de Lyon, Lyon, France; INSA Lyon, Villeurbanne, France; Laboratoire d'Ingénierie des Matériaux Polymères, CNRS, UMR 5223, Villeurbanne, France

Omar A. El Seoud, Institute of Chemistry, University of São Paulo, São Paulo, Brazil

Maria Figueira-González, Departamento de Química Física, Centro Singular de Investigación en Química Biológica y Materiales Moleculares (CiQUS), Universidade de Santiago de Compostela, Santiago, Spain

Paula D. Galgano, Nitro Química, São Paulo, Brazil

Luis García-Río, Departamento de Química Física, Centro Singular de Investigación en Química Biológica y Materiales Moleculares (CiQUS), Universidade de Santiago de Compostela, Santiago, Spain

Jean-François Gérard, Université de Lyon, Lyon, France; INSA Lyon, Villeurbanne, France; Laboratoire d'Ingénierie des Matériaux Polymères, CNRS, UMR 5223, Villeurbanne, France

Surajit Ghosh, Department of Chemistry, Indian Institute of Technology, Kharagpur, India

Providencia González-Hernández, Departamento de Química Analítica, Nutrición y Bromatología, Universidad de La Laguna (ULL), La Laguna, Tenerife, Spain

M. A. Iglesias-Otero, Department of Physical Chemistry, Faculty of Science, University of Vigo at Ourense, Ourense, Spain

Maria H. Katsoura, Laboratory of Biotechnology, Department of Biological Applications and Technologies, School of Health Sciences, University of Ioannina, University Campus, Ioannina, Greece

Joachim Koetz, Institut für Chemie, Universität Potsdam, Potsdam, Germany

Arvind Kumar, AcSIR-Central Salt and Marine Chemicals Research Institute, Bhavnagar, India; CSIR-Central Salt and Marine Chemicals Research Institute, Bhavnagar, India

Kaushik Kundu, Surface and Colloid Science Laboratory, Geological Studies Unit, Indian Statistical Institute, Kolkata, India

Sandrina Lampis, Dipartimento di Scienze Chimiche e Geologiche, Università di Cagliari, Monserrato, Italy

Nancy E. Levinger, Department of Chemistry, Colorado State University, Fort Collins, CO, USA

Sébastien Livi, Université de Lyon, Lyon, France; INSA Lyon, Villeurbanne, France; Laboratoire d'Ingénierie des Matériaux Polymères, CNRS, UMR 5223, Villeurbanne, France

Marianna Mamusa, Dipartimento di Chimica "Ugo Schiff," Università degli Studi di Firenze, Sesto Fiorentino, Italy

Sarthak Mandal, Department of Chemistry, Indian Institute of Technology, Kharagpur, India

N. Mariano Correa, Departamento de Química, Universidad Nacional de Río Cuarto, Río Cuarto, Argentina

J. C. Mejuto, Department of Physical Chemistry, Faculty of Science, University of Vigo at Ourense, Ourense, Spain

Sandeep K. Mishra, Delhi Pollution Control Committee, Delhi, India

Takeshi Misono, Department of Pure and Applied Chemistry, Faculty of Science and Technology, Research Institute for Science and Technology, Tokyo University of Science, Noda, Chiba, Japan

O. A. Moldes, Department of Physical Chemistry, Faculty of Science, University of Vigo at Ourense, Ourense, Spain

J. Morales, Department of Physical Chemistry, Faculty of Science, University of Vigo at Ourense, Ourense, Spain

Satya P. Moulik, Centre for Surface Science, Department of Chemistry, Jadavpur University, Kolkata, India

Ahmed Mourchid, Matière et Systèmes Complexes, UMR 7057 CNRS-Université Paris Diderot, Paris, France

Indrajyoti Mukherjee, Centre for Surface Science, Department of Chemistry, Jadavpur University, Kolkata, India

Sergio Murgia, Dipartimento di Scienze Chimiche e Geologiche, Università di Cagliari, Monserrato, Italy

Gerardo Palazzo, Dipartimento di Chimica, Università di Bari, Bari, Italy

Pramod S. Pandey, Department of Chemistry, Indian Institute of Technology Delhi, Hauz Khas, New Delhi, India

Shubha Pandey, Department of Science and Technology, Technology Bhawan, New Delhi, India; Indian Institute of Technology Delhi, Hauz Khas, New Delhi, India

Siddharth Pandey, Department of Chemistry, Indian Institute of Technology Delhi, Hauz Khas, New Delhi, India

Athena A. Papadopoulou, Laboratory of Biotechnology, Department of Biological Applications and Technologies, School of Health Sciences, University of Ioannina, University Campus, Ioannina, Greece

Mercedes Parajó, Departamento de Química Física, Centro Singular de Investigación en Química Biológica y Materiales Moleculares (CiQUS), Universidade de Santiago de Compostela, Santiago, Spain

Bidyut K. Paul, Surface and Colloid Science Laboratory, Geological Studies Unit, Indian Statistical Institute, Kolkata, India

Verónica Pino, Departamento de Química Analítica, Nutrición y Bromatología, Universidad de La Laguna (ULL), La Laguna, Tenerife, Spain

Angeliki C. Polydera, Laboratory of Biotechnology, Department of Biological Applications and Technologies, School of Health Sciences, University of Ioannina, University Campus, Ioannina, Greece

Vishal Govind Rao, Department of Chemistry, Indian Institute of Technology, Kharagpur, India

Luís Paulo N. Rebelo, Instituto de Tecnologia Química e Biológica, Universidade Nova de Lisboa, Oeiras, Portugal

Qilong Ren, Key Laboratory of Biomass Chemical Engineering of Ministry of Education, Department of Chemical and Biological Engineering, Zhejiang University, Hangzhou, China

Pedro Rodríguez-Dafonte, Departamento de Química Física, Centro Singular de Investigación en Química Biológica y Materiales Moleculares (CiQUS), Universidade de Santiago de Compostela, Santiago, Spain

Swapan K. Saha, Department of Chemistry, University of North Bengal, Darjeeling, India

Hideki Sakai, Department of Pure and Applied Chemistry, Faculty of Science and Technology, Research Institute for Science and Technology, Tokyo University of Science, Noda, Chiba, Japan

Kenichi Sakai, Department of Pure and Applied Chemistry, Faculty of Science and Technology, Research Institute for Science and Technology, Tokyo University of Science, Noda, Chiba, Japan

Anunay Samanta, School of Chemistry, University of Hyderabad, Hyderabad, India

Kotni Santhosh, School of Chemistry, University of Hyderabad, Hyderabad, India; Department of Chemical Physics, Weizmann Institute of Science, Rehovot, Israel

Benilde Saramago, Centro de Química Estrutural, Instituto Superior Técnico, Universidade de Lisboa, Lisboa, Portugal

Nilmoni Sarkar, Department of Chemistry, Indian Institute of Technology, Kharagpur, India

Sanjib Senapati, Department of Biotechnology, Indian Institute of Technology Madras, Chennai, India

Jun Shen, College of Continuing Education, East China University of Science and Technology, Shanghai, China

Karina Shimizu, Instituto de Tecnologia Química e Biológica, Universidade Nova de Lisboa, Oeiras, Portugal; Centro de Química Estrutural, Instituto Superior Técnico, Universidade de Lisboa, Lisboa, Portugal

Juana J. Silber, Departamento de Química, Universidad Nacional de Río Cuarto, Río Cuarto, Argentina

K. Srinivasa Rao, AcSIR-Central Salt and Marine Chemicals Research Institute, Bhavnagar, India

Haralampos Stamatis, Laboratory of Biotechnology, Department of Biological Applications and Technologies, School of Health Sciences, University of Ioannina, University Campus, Ioannina, Greece

Stephen H. Strassburg, Department of Polymer Science and Engineering, University of Massachusetts Amherst, Amherst, MA, USA

Baogen Su, Key Laboratory of Biomass Chemical Engineering of Ministry of Education, Department of Chemical and Biological Engineering, Zhejiang University, Hangzhou, China

Mohammad Tariq, Instituto de Tecnologia Química e Biológica, Universidade Nova de Lisboa, Oeiras, Portugal; Department of Chemical Engineering, College of Engineering, Qatar University, Doha, Qatar

Shruti Trivedi, Department of Chemistry, Indian Institute of Technology Delhi, Hauz Khas, New Delhi, India

Tushar J. Trivedi, AcSIR-Central Salt and Marine Chemicals Research Institute, Bhavnagar, India

María J. Trujillo-Rodríguez, Departamento de Química Analítica, Nutrición y Bromatología, Universidad de La Laguna (ULL), La Laguna, Tenerife, Spain

Qing Wen, College of Life Sciences, Shenzhen University, Shenzhen, Guangdong, China

Huabin Xing, Key Laboratory of Biomass Chemical Engineering of Ministry of Education, Department of Chemical and Biological Engineering, Zhejiang University, Hangzhou, China

Guiying Xu, Key Laboratory of Colloid and Interface Chemistry, Ministry of Education, Shandong University, Jinan, P. R. China

Qiwei Yang, Key Laboratory of Biomass Chemical Engineering of Ministry of Education, Department of Chemical and Biological Engineering, Zhejiang University, Hangzhou, China

Zhen Yang, College of Life Sciences, Shenzhen University, Shenzhen, Guangdong, China

Jianling Zhang, Institute of Chemistry, Chinese Academy of Sciences, Beijing, China

Zhiguo Zhang, Key Laboratory of Biomass Chemical Engineering of Ministry of Education, Department of Chemical and Biological Engineering, Zhejiang University, Hangzhou, China

Ting Zhou, Key Laboratory of Colloid and Interface Chemistry, Ministry of Education, Shandong University, Jinan, P. R. China; College of Science, China University of Petroleum (East China), Qingdao, P. R. China

FOREWORD

Over the last 15 or 20 years, the study of ionic liquids (ILs) has become very fashionable, and the published literature about these systems continues to grow exponentially. At the beginning of this voyage, it was argued that they are green, because their vapour pressure is negligible. However, it turned out that many of them are not green at all. This was recently discussed by Jessop in a landmark paper on the real needs in the field of green solvents [1]. He even concludes that the capacities of the ILs are overestimated, and too much efforts have been invested on their research.

Since the beginning of the widespread research on ILs, there has been another argument that they are "designer solvents," and can possibly be tuned to deliver any desired property. In the light of possible cation–anion combinations, there are possibilities of numerous products with melting points below 100°C, which is advantageous, especially for the scientists. We can publish thousands of articles, and since the community is ever growing, most of these papers will be highly cited. In return, the scientists have scope that in some decades all present unfavourable liquids will be replaced by the new ILs.

Of course, this is not realistic, and perhaps a big danger. It reminds me of all the promises made by electrochemists in the 1970s and 1980s. According to their views, conventional biofuel-driven cars should have been replaced by electro cars since many years. Everybody knows that they promised too much and that even a simple mobile phone must still be charged once a day. Therefore, we should be modest when speaking about so many "potential" applications of ILs. It is known now that most of them are not green, they are difficult to synthesize (and even more to purify), they often show high viscosities, and so forth.

So, is all bad with the ILs, and is the present monograph superfluous? Of course it is not. There is indeed a promising future for ILs, provided they are carefully chosen, and their properties and potential applications are compared with the existing systems so that their superiority can be proven. Or even better, that they fulfil tasks that could not be envisaged with current conventional liquids.

For the moment, I see the following advantages of ILs:

1. ILs are liquid over a considerable temperature range. For several applications that can be a significant advantage, for example in the field of nanoparticle synthesis at high or—even more—at very low temperature.

2. ILs contain a lot of charges. This is a disadvantage for the use of many enzymes, but it can be an advantage for electrochemical applications.
3. ILs can also be considerably surface active. And this is one of the main topics of the present monograph. Charged liquid surfactants are scarce, although not completely unknown in literature (indeed the so-called extended surfactants are liquid at room temperature [2]). But most of the classical charged surfactants are solid or delivered as dilute aqueous solutions. For applications, however, it is always better to work with liquids—they are easier to handle and it is not necessary to overcome the energy of crystallisation during mixing or dissolution. And there is a second advantage that is often overlooked. IL surfactants bear special charged headgroups, often with delocalised charges. This chemical specificity is interesting by itself. Not only is the special headgroup (like imidazolium, etc.) the main reason for the liquid state of the whole surfactant, but it can also deliver very special interfacial properties to the systems.

As shown and discussed in details in the present monograph, ILs can significantly modify the behavior of classical surfactants both in the bulk and at the interface. The obtained structures such as liquid crystalline ones can be similar to aqueous systems, but often the detected phases are significantly shifted to higher surfactant concentrations or, are even very different from their aqueous counterparts. The IL can be surface active on its own or in combination with classical surfactants. Surface-active ILs can undergo specific interactions with polymers and biomolecules, as well as with special chemicals, such as calixarenes. They may tune the activity of enzymes and can be used in separation science. Of special interest is their interaction with surfaces such as layered silicates. All these aspects—and some more—are treated in the first part of the book.

As far as colloidal chemistry is concerned, manifold applications could be interesting in this field, and several of them are discussed in this monograph. Within the liquid state, particularly microemulsions are considered, and this is the main topic of the second part of the book. Several scenarios are imaginable: the ILs serve either as base of the polar (pseudo-)phase or as apolar one or as surfactant interphase or as two of these phases or pseudo-phases [3, 4]. Until now, as far as I know, nobody mixed a polar, an apolar and a surface-active IL to get a solely IL-containing microemulsion. Water-free microemulsions are not new [5], but the variety is quite limited before ILs become available. This approach has a great promise, and could lead to a real fine-tuning of liquid nanoscopic environments leading to specially designed nano-reactors in chemical synthesis. Of course, it is mandatory to precisely know the structures of such microemulsions, and therefore several papers consider this topic from different angles. An attempt to calculate thermodynamic properties (densities) of ternary systems containing ILs is also presented.

As mentioned before, toxicity aspects are increasingly important. Not only in Europe and the United States it is very difficult to introduce new chemical substances to the market; if their toxicity (and non biodegradability) is significant, there is no future for them. Therefore, it is interesting to consider bio-based or at least biocompatible systems. They may be based on "drinkable" ILs or, as a similar type of systems, like, the deep eutectic solvents that have many points (positive and negative) in common with ILs, such as a high ionic strength and usually high viscosity.

This monograph has only covered selected aspects of the huge area of ILs. But I think that liquid IL-based surfactant systems and in particular IL-based microemulsions are a promising field where ILs could (finally) show their power so that "potential" applications may finally turn to real ones. But even without applications, fascinating new aspects of the liquid state of matter are found in this subject, and this, by its own, justifies the present efforts.

REFERENCES

[1] Philip G. Jessop, Searching for green solvents, *Green Chemistry* **3** (2011) 1391–1398.

[2] Angelika Klaus, Gordon Tiddy, Didier Touraud, Anette Schramm, Georgine Stühler, and Werner Kunz, Phase behavior of an extended surfactant in water and a detailed characterization of the concentrated phases, *Langmuir* **26** (22) (2010) 16871–16883.

[3] Werner Kunz, Thomas Zemb, and Agnes Harrar, Using ionic liquids to formulate microemulsions: current state of affairs, *Current Opinion in Colloid & Interface Science* **17** (2012) 205–211.

[4] Oliver Zech and Werner Kunz, Conditions for and characteristics of nonaqueous micellar solutions and microemulsions with ionic liquids, *Soft Matter* **7** (2011) 5507–5513.

[5] Armand Lattes and Isabelle Rico, Aggregation in formamide solution: reactivity and structure of non-aqueous microemulsions, *Colloids and Surfaces* **35** (1989) 221–235.

Professor Werner Kunz
Institute of Physical and Theoretical Chemistry,
University of Regensburg, D-93040 Regensburg, Germany
Email: Werner.Kunz@ur.de

PREFACE

Factually, ionic liquids (ILs) are both old and new. Although ethylammonium nitrate (EAN), an organic liquid of mp ≈ 14°C, is known since 1914 from the work of P. Walden, in recent years, ILs have received much attention as a class of neoteric nonaqueous solvents, because of their unusual properties, amply mentioned in this monograph. Functionalization of ILs by designing different cations and anions makes considerable room for flexibility in their properties, which qualify them to be termed "designer solvents." Studies on self-assemblies of conventional surfactants into micelles, vesicles, liquid crystals, and microemulsions in a variety of ILs have become an attractive field for both theoretical and applied research. Although significant literature (original papers and reviews) in this particular field are available in this decade to our knowledge, a comprehensive literature in the form of a book or monograph is yet to be published on IL-based self-assembled systems. Our endeavor is to fill this gap. In this book, we have attempted to provide a comprehensive presentation of the topics on the performance of IL-assisted micelles and microemulsions, discussing their fundamental characteristics and theories, and development of bio-ILs or greener biodegradable, non eco-toxic solvents. We comprehend that the book will be useful for advanced postgraduate and undergraduate students, researchers in institutes, universities, and industries. The landscape looks encouraging. Therefore, good-quality critical advancements in this field comprising prospective environment benign or greener IL-based self-assembled systems are expected to emerge in the coming years.

In Chapter 1, Murgia, Palazzo, and coworkers investigated the physicochemical behaviors of a binary IL $bmimBF_4$ and water, and the ternary NaAOT, water and $bmimBF_4$ mixtures essentially through the evaluation of the self-diffusion coefficients of the various chemical species in solution by PGSTE-NMR experiments. The diffusion of water molecules and $bmimBF_4$ ions were found to be within different domains, which suggested that the systems were nanostructured with formation of micelles having positive curvature and a bicontinuous micellar solution for the former and the later systems, respectively. The remarkable differences between the two systems are attributed to the specific counterion effect between the aforementioned ILs and the anionic surfactant. In Chapter 2, Bermudez and coworkers focused on the characterization of small (conventional surfactants) and polymeric amphiphiles (block copolymers) in different types of ILs (imidazolium, ammonium, phosphonium, etc.) with special reference to the interfacial and bulk behaviors, and compared them with aqueous systems to highlight similarities and dissimilarities between ILs and water as self-assembly media employing traditional techniques. Ultra-high vacuum (UHV) methods were also employed in the measurements. Possible applications and future directions of the studies on the fundamental behavior of amphiphiles at the interface and in the bulk have also been presented. In the Chapter 3, the self-assembly of nonionic surfactants (analogues polyoxyethylene alkyl ethers) in room-temperature ILs (RT-ILs) under varied physicochemical conditions emphasizing on different aspects, viz., thermodynamics of micellization, characterization of binary (surfactant-RT-ILs) phase behaviors, and adsorption characteristics at solid/RT–IL interfaces has been presented by Sakai and coworkers. In addition, the knowledge of the interfacial properties of RTILs with water in the absence and presence of non-ionic surfactants has been presented for a better understanding of the preparation mechanism of metal oxide particles in RT-ILs. A futuristic view concerning RT-ILs from the standpoint of colloid and interface chemistry has been addressed. In chapter 4, El Seoud and Galgano have made a detailed presentation on imidazole-derived IL-based surfactants (ILBSs; ILs with long-chain "tails"), including syntheses, determination of the properties of their solutions, comparison between their micellar properties and those of "conventional" cationic surfactants, for example, pyridine-based cationics, and their main applications. The authors have suggested that a single factor that distinguishes ILBSs from other conventional surfactants is their structural versatility. The most frequently employed schemes for the synthesis and purification of ILBSs are specified; in addition, the micellar properties (viz. the critical micelles concentration, counter-ion dissociation constant, surfactant aggregation number, thermodynamic parameters of aggregation) are also presented. The applications of the ILBSs are briefly discussed. The impact of ILs in terms of characterization of different types of interactions, they experience in the bulk and at the interface, has been addressed by Lopes and coworkers in Chapter 5, by taking into account three types of research work: self-aggregation behavior of dialkylpyrrolidinium bromide ILs in the bulk phase using isothermal titration calorimetry, energetics at the IL–air interface (using 1-alkyl-3-methylimidzolium bistriflamide homologous series of ILs over a wide temperature range) from surface tension measurements, and finally, characterization of the adsorption of ILs on solid substrates (viz., gold and glass) using quartz crystal microbalance with dissipation (QCM-D), and atomic force microscopy (AFM). The results yielded a fascinating picture of the complex surface behaviors of ILs at the solid/liquid interface. In Chapter 6, Xu and Zhou summarized the aggregation behavior of aqueous solution of IL-based gemini surfactants and their interactions with biomacromolecules (e.g., BSA, Gelatin, and DNA). These surfactants possess unique aggregation behaviors which have significant promise in industrial applications. Further, prospective applications, such as drug entrapment and release, gene transfection of IL-based gemini surfactants have been presented. Future directions of research on different aspects of IL-based gemini surfactants, including synthesis with new structure, understanding of the mechanism underlying interaction between these surfactants or with other substances, for example, polymers and biomacromolecules to develop their functional efficiency and application have been focused. In Chapter 7, Samanta and coworkers have presented the development of morpholinium ion-based ILs along with their physicochemical studies: these ILs have promise as potential benign (environment friendly) alternatives to the volatile organic compounds. The microheterogeneous nature of these ILs (morpholinium cations) has been established from steady-state and time-resolved fluorescence measurements. These *N*-alkyl-*N*-methylmorpholinium ILs are much more structured compared to the extensively studied imidazolium ILs. The dynamics of the solvent and the rotational relaxation in these media are also presented. Kumar and coworkers presented the formation and characterization of self-assembling nature of surface-active ILs (SAILs) of different ionic surfactant types (cationic, anionic, and catanionic) in aqueous medium in Chapter 8. The surface activity of SAILs has been found to be greater than their analogous surfactants in aqueous medium, and attributed to cation and anion used. The role of cation and anion in determining the surface activity of SAILs has been presented. The conspicuous phase behavior of SAILs-based mixed systems in aqueous medium has been reported and compared with analogous surfactant-based mixed systems. The conventional anionic surfactant, SDS converted to amino acid-based IL (AAIL) surfactants, has been found to be prospective for different applications such as, nanomaterial preparation and mitigation of harmful algal blooms. Mixed cationic and anionic type of SAILs form higher self-assembled structures such as vesicles like conventional surfactants. Catanionic surface-active ILs (CASAILs) have shown versatile solubility in different solvents, and form vesicles in aqueous medium and reverse micelles (RMs) in nonpolar organic solvents. Amino acid-derived ASAILs (AAILSs) have shown promising ability in the synthesis of CeO_2 nanoparticles and in the mitigation of harmful algal blooms. Pandey and coworkers reported in Chapter 9 that the interaction behaviors of common calixarenes with a SAIL [1-decyl-3-methylimidazolium chloride ([C_{10}mim][Cl])] depend on the functionalities present on the molecular architecture of the calixarene. UV-Vis absorbance and fluorescence measurements using pyrene as probe have been the methodologies used in the study. SAIL seems to effectively control the solubilization sites, and thus the properties of calixarenes in solution. In Chapter 10, Moulik and Mukherjee have presented physicochemical and interfacial behaviors of different IL (polar, apolar, amphiphilic)-based systems, establishing their versatility in the domain of colloids, in addition to the enhancement of their applicability as polar and nonpolar solvents as well as amphiphilic entities. Moreover, self-micellization, mixed micellization of ILs, and their influence on the micellization of conventional surfactants (ionic, nonionic, zwitterionic, etc.) have been evidenced with specific citations. Phase forming behaviors of IL-based microemulsions (of IL/O, O/IL, and bicontinuous types) are discussed along with the effects of additives on them. Formation of large single-phase (clear) zones in the pseudo ternary and ternary phase diagrams with ILs has been emphasized along with their application potentials as templates for nanomaterial synthesis, enzyme catalysis, and drug delivery. Antibacterial and anticancer activities of ILs and IL-derived systems like microemulsions, etc., are also discussed to elucidate the broad domain of ILs in the field of colloid and interface science. The experimental evidences of self-assembled structures of different surfactant molecules in ILs have been presented by Mourchid in Chapter 11. The analogies between the self-assembly and mesophase morphologies of the amphiphilic nano-aggregates in ILs compared to those usually found in conventional polar and non-polar media have been discussed. Some important differences between the RTILs and the molecular solvents, in respect of ability to promote self-assembly through solvophobic interactions of surfactant hydrocarbon chains, are pointed out. Finally, the data obtained from the phase behavior, properties, and microstructure of microemulsions for ternary water—RTIL—amphiphile systems have been discussed in the light of studies on conventional water-in-oil microemulsions, although the efficiency of the former remains low. This is a consequence of weak solvophobic interactions in ILs, which is also reflected in the measured short nanometric repeat distance and correlation length in RTILs microemulsions.

In Chapter 12, Koetz has discussed on recent developments on microemulsions containing ILs with special accentuation of their interfacial activities. Because of the flexibility in the properties of ILs, they can be polar (partial or total completely replacing water), or nonpolar as well as the surfactant component causing spontaneous formation of different new types of microemulsion. By combination of anionic surfactant and polar ILs, formation of water-free microemulsions has been reported. Further, IL-oil-IL microemulsions with tuneable properties have conveyed a novel direction to the surface chemical research with prospective applications. Formation and characterization of IL-based microemulsions comprising nonionic (TX-100; Brij and Tween) and cationic (polymeric and long chain imidazolium-based ILs) surfactants, and imidazole-based tetrafluoroborate and hexafluorophosphate ILs as substitutes for water and oil, respectively, have been presented by Rodríguez-Dafonte, García-Río, and coworkers in Chapter 13. In addition, new amphiphilic ILs categorized as (i) ILs with long alkyl chains incorporated into the imidazolium cation, and (ii) ILs with the counter-ions containing a long alkyl chain have been used to improve the properties of the aggregates. The interactions between different constituents, especially of surfactants and ILs, have been considered for the stabilization of the microemulsions. Their significant applications are reported commensurate with the development of tuned ILs with the desirable properties. Chapter 14 of Falcone, Silber, and coworkers highlights on the development and characterization of RMs comprising surfactant of different charge types, viz. anionic, nonionic and cationic formed with imidazole-based ILs with different anions viz. $[BF_4]$, $[Tf_2N]$, [TfO], and [TfA] as polar phases, and benzene and chlorobenzene as nonpolar solvents using dynamic light scattering (DLS), as well as absorption, multinuclear NMR and FT-IR spectroscopy. It was demonstrated that RMs comprise discrete spherical and non-interacting droplets of IL stabilized by the surfactants. The properties of the encapsulated IL appreciably depend on the nature of the interface present in the organized system. They showed important structural differences between the ILs entrapped in the cationic RMs and the neat ILs or the ILs entrapped in nonionic or anionic RMs, and suggested that confinement substantially modified the ionic interactions of both the surfactants and the ILs. It was concluded that these media could be useful nanoreactors with modulation of the microenvironment by simply changing the RMs components and the IL content. In Chapter 15, Sarkar and coworkers have discussed the possibility of creating large number of IL-in-oil microemulsions, simply by replacing the inorganic cation, Na^+ of NaAOT by any organic cation, and using different ILs (imidazolium-based ILs with different anions) as the polar core. In this sequel, formation and characterization of different IL-in-oil microemulsions containing an anionic surface-active IL (SAIL), $[C_4mim][AOT]$ were demonstrated. The results indicated that depending on IL used, amount of IL within the core of microemulsions can be easily manipulated to directly affect the size of aggregates in microemulsions. Further, the effect of water addition on microemulsions-containing hydrophobic ILs and compare it with microemulsions containing hydrophilic IL have been discussed. Different ways to tune the structure of microemulsions, which in turn can provide different routes to alter the size of the prepared nanoparticles/polymers and to afford environment for performing organic reactions have been proposed. In Chapter 16, Zhang reviewed the formation of microemulsions with ILs, which are very attractive owing to their unusual solvent properties with special reference to tunable and designable solvents with essentially zero volatility, wide electrochemical window, nonflammability, high thermal stability, and wide liquid range. This chapter delineated formation of various kinds of microemulsions containing ILs such as, IL-in-oil and oil-in-IL microemulsions, IL-in-water and water-in-IL microemulsions, IL-in-IL microemulsion, and IL-in-supercritical CO_2 and CO_2-in-IL microemulsions. The applications of these microemulsions in different fields, such as protein delivery, drug release, catalysis, and nanomaterial synthesis are presented. Future direction of research on these novel IL-based microemulsions with prospective applications has been suggested. The recent progress in the formation and characterization of IL-based nonaqueous microemulsions has been summarized by Ren and coworkers in Chapter 17. Their phase behavior, properties, microstructure, and intermolecular interactions among the constituents in microenvironment have been discussed in comparison with microemulsion systems comprising conventional surfactants. Further, this chapter outlines the applications of IL-based nonaqueous microemulsions in drug dissolution, material preparation, organic synthesis and polymerization. Future studies are warranted to resolve the issue of formation of nonaqueous microemulsions comprising ILs, which can be regarded as purely "green" solvents. In Chapter 18, An and Shen have reviewed water-in-IL microemulsions (wherein, IL substitutes the oil component) as well as IL-in-oil microemulsions (where, IL substitutes the polar or water component). Apart from morphology, physicochemical property, microstructure, phase equilibria, and critical phenomena, emphasis has also been given to their applications viz. prospective reaction media and drug carrier templates etc. Suggestion for the use of green or bio-compatible ILs as pharmaceutical solvents; alternative media for reactions; and functional solvents for nanoparticle synthesis, extraction, and separation has been made; efforts for synthesizing such ILs by combination of green properties of ILs with their unique tailor-made physicochemical properties have been proposed. In this perspective, future direction of research on exploring newer biocompatible IL microemulsions to achieve such applications has been emphasized. Senapati and Ghosh Dastidar have reviewed some of the recent advancements in the field of ILs-in-oil microemulsions, with a special emphasis on the structural characteristics and solvation dynamics of the confined IL pool in Chapter 19. The effects of added water and temperature on the stability of these microemulsions have also been critically surveyed. Recent applications in various areas such as, material chemistry, biotechnology, and sustainable synthesis of polymers, using these novel templates have been discussed. The authors have expressed possibilities of designing the "greener" isotypes, consisting of hydrophobic ILs or supercritical CO_2 ($scCO_2$) as the apolar phase. Several studies in this direction have already been reported and their applications are being tested. In view of this development, formation of more complex IL-in-CO_2 microemulsions could be critically examined. In addition, it is hoped that molecular dynamics simulations could play an important role in deciphering atomic-level understanding of these systems to unravel the formation of defined structures of these systems. In Chapter 20, Paul and coworkers have presented the achievements and current status of environmental risk assessment of different types of bio-ILs (BILs) with special reference to their synthetic strategies, physicochemical properties, antimicrobial activity, (eco) toxicological aspect, and biodegradability. The role of BILs in the fields of enzyme activity, biotransformation, and surfactant self-assembly formation with special reference to microemulsion systems has been summarized. It is envisaged that these systematic studies will be addressed to producers, developers, and downstream users of ILs in different fields of application, to facilitate the selection of (eco) toxicologically favorable structural elements and thus to contribute to the design of inherently safer BILs. Formation and characterization of novel IL-assisted nonaqueous microemulsions with pharmaceutically acceptable components, which could be effectively used in solubilizing many drug molecules (insoluble or poorly soluble in water and in most organic solvents) have been also reviewed. A new approach on the formation of aqueous nanometer-sized domains for carrying out enzymatic reactions in ILs has also been reported. Application of the combination of green properties of ILs with their unique tailor-made physicochemical properties should in near future generate biocompatible ILs for uses as pharmaceutical solvents and reagents. In Chapter 21, new insights in the prediction of density in ternary mixtures of ethanol + water + IL using back propagation artificial neural network (ANN) have been presented by Mejuto and coworkers. These predictions are compared with the corresponding ones with another model, that is, multiple linear regression (MLR) model, and the advantages of neural modeling than the traditional modeling MLR have been presented. The scope of Chapter 22 presented by Stamatis and coworkers has been to cover the effect of selected properties of ILs on the activity, stability, as well as the structure of enzymes, pointing out the main principles governing the aforesaid effects. Several parameters such as polarity, hydrogen-bonding capacity, viscosity, kosmotropicity/chaotropicity, and hydrophobicity were investigated, and various spectroscopic and scattering studies were used in order to explore the structural and conformational dynamics of enzymes in these media and also to understand how ILs affect the stability and activity of enzymes. Enzyme-catalyzed reactions in ILs have been reviewed; the use of ILs in various applications, including their uses as solvents for biocatalysis, has been addressed. In view of the bio-incompatibility of many ILs, the authors have drawn attention to the development of green and biodegradable ILs formulated with compounds derived from renewable resources, which may further stimulate their uses in industrial biocatalytic processes taking into account of both ecological and economic requirements. The research on the development and application of the third-generation ILs and deep-eutectic solvents (DESs), as media for enzymatic reactions are aimed as their future perspectives. In Chapter 23, Pino and coworkers have demonstrated successful employment of more than 50 ILBs as substitutes for conventional organic solvents in extraction schemes, or as modifiers of chemical structure of conventional sorbents, which has been a promising and developing field in separation science. The analytical performances of these novel ILBs have been shown to be better than conventional organic solvents and also the cationic surfactants. Low CMC values and higher interaction affinities for a variety of compounds compared to conventional cationic surfactant analogues are responsible for diverse analytical applications based on ILBSs. In Chapter 24, Livi and coworkers have presented an overview of the potential of ILs as surface active agents towards polymer materials. The preparation, characterization, and properties of different nanocomposites using

ILs based on pyridinium, imidazolium, or phosphonium cations to modify layered silicates (fillers) according to the nature of the polymer matrices have been reported. Recently, these types of ILs are emerging as new alternatives for the design of thermally stable organically modified clays.

In Chapter 25, Yang and Wen discussed the physicochemical properties of DESs and reviewed their uses as new reaction media for biocatalytic transformation, either as such or as a co-solvent with water. They have introduced a new type of DESs, natural DESs (NADESs), which possess an enormous potential for applications due to their non-toxicity, sustainability, and friendliness to the environment. The advantages of using DESs over the conventional ILs are low cost, easy preparation with high purity and biodegradability, and low toxicity. More studies of DES on biocatalysis with the following perspectives have been suggested, (i) correlation between the structure and composition of a DES and its physicochemical properties; (ii) correlation between the structure of a DES and its interaction with an enzyme; and (iii) correlation between the DES structure and enzyme function.

In the end, we wish to acknowledge a number of people who helped in various ways to bring the endeavor in reality. Our foremost thanks go to the chapter authors of the book for their willingness despite their busy schedules. In total, we received contributions of 82 individuals from 14 countries. Without their timely response, professionalism, and excellent updated information, the publication of the book would not have been possible. We express our sincere thanks and gratitude to Professor Dr. Werner Kunz for writing an excellent Foreword of this book. Our special thanks are due to the reviewers for their helps as peer-review is a requirement to preserve a high standard of a publication. Our appreciation goes to Ms. Anita Lekhwani, Senior Acquisitions Editor at John Wiley & Sons, Inc. for her unwavering interest and constant encouragement and assistance in this work. We are indebted to Ms. Cecilia Tsai, Senior Editorial Assistant for her cooperation and patience; she has worked very hard at the final editing stage for the production of the book. Dr. Kaushik Kundu (a senior research fellow) has rendered extensive help in the ratification and scrutiny of the chapters as per statutory requirements of the publisher; his help is acknowledged with thanks and appreciation. Finally, our appreciation and sincere thanks are for Professor P. Somasundaran for his genuine interest in the publication of this book.

Bidyut Kumar Paul
Satya Priya Moulik

CHAPTER 1

Ionic Liquids Modify the AOT Interfacial Curvature and Self-Assembly

SERGIO MURGIA and SANDRINA LAMPIS

Dipartimento di Scienze Chimiche e Geologiche, Università di Cagliari, Monserrato, Italy

MARIANNA MAMUSA

Dipartimento di Chimica "Ugo Schiff", Università degli Studi di Firenze, Sesto Fiorentino, Italy

GERARDO PALAZZO

Dipartimento di Chimica, Università di Bari, Bari, Italy

1.1 INTRODUCTION

Surfactants are amphiphilic molecules, that is, they simultaneously possess a portion that loves water and another that loves oil. This dual characteristic underpins the formation of nanoscale structures from biological cells to micelles, microemulsions, and liquid crystals.

The structure of surfactants systems can be idealized as a set of interfaces dividing polar and apolar domains. A peculiar and unifying feature of all surfactant systems is that the polar and apolar domains can arrange itself in a variety of shapes (e.g., lamellae, cylinders, spheres, and so on) depending on the intensive variables of the systems.

An interesting application of ionic liquids (ILs) concerns their use in combination with classical surfactants [1, 2]. Indeed, they can suitably replace each of the microemulsion components (aqueous phase, apolar phase, and surfactants) conferring peculiar features to self-assembled systems. Indeed, ILs are salts and as such have affinity for water, but they also typically possess a lipophilic moiety, and this means affinity for oils. Depending on their chemical structure, ILs can act as cosolvent either for water or for oil. In addition, when their hydrophilic and hydrophobic nature are both strong enough, a fraction of ILs will reside preferentially at the interface formed by the surfactant, and this can impact dramatically the interfacial physics, drastically changing the microemulsion structure and dynamics.

Ionic Liquid-Based Surfactant Science: Formulation, Characterization, and Applications, First Edition. Edited by Bidyut K. Paul and Satya P. Moulik.

In the following, the focus will be on the ability of two imidazolium-based ILs in modifying the polar–apolar curvature of the anionic, double-tailed surfactant sodium bis(2-ethylhexyl) sulfosuccinate (NaAOT). At first, the reader will be introduced to the NMR technique used to investigate these systems. Then, the microstructure of water/IL solutions will be discussed. The basic of surfactant systems thermodynamics will be subsequently recalled and the NaAOT behavior in water reviewed. Finally, the nanostructure of the micellar phases originated by loading aqueous solutions of imidazolium-based ILs with NaAOT will be discussed.

1.2 HOW TO INVESTIGATE SURFACTANT SYSTEMS: PGSE-NMR

The microstructure of complex fluids such as ILs, surfactant systems, and liquid crystals can be profitably investigated by means of pulsed gradient spin-echo nuclear magnetic resonance (PGSE-NMR) experiments, a technique that allows the determination of the self-diffusion coefficients.

PGSE-NMR has several advantages: (i) it gives a true self-diffusion coefficient that is easily associated to a chemical species through its NMR signal; (ii) it is unaffected by the optical appearance of the sample, and thus it is insensitive to critical phenomena; (iii) besides the sizing, it can give information on the partition of components; and (iv) interesting pieces of information can be obtained also on systems where the molecular diffusion is dramatically far from the unrestricted Brownian diffusion as in emulsions, liquid crystals, and even on porous solids.

The mechanism underlying PGSE-NMR is described in several reviews [3–6], and here, only the basic concepts will be recalled. The application of a suitable sequence of a radiofrequency pulse and of a magnetic field gradient (of magnitude $\boldsymbol{G}$ and duration δ) forces the transverse nuclear magnetization (i.e., the experimental observable in the NMR spectroscopy) along a well-defined spatial helix within the NMR tube. The helix axis is along the gradient direction, and it is characterized by the space vector $\boldsymbol{q}$:

$$\boldsymbol{q} = \frac{\gamma\delta\boldsymbol{G}}{2\pi} \tag{1.1}$$

where γ is the gyromagnetic ratio of the observed nucleus and the helix pitch is $\boldsymbol{q}^{-1}$. Then, after a time lapse, Δ, the process is reversed by another magnetic gradient pulse, and the spins refocalize giving an NMR signal (the so-called spin echo). However, such a refocusing is not complete because of spin diffusion during the interpulse interval (Δ). The experimental observable in the PGSE-NMR is the echo attenuation $E(q,\Delta)$, a function of both q and Δ. It is defined as $E(q,\Delta) = I(q,\Delta)/I(0,\Delta)$, that is, as the ratio between the NMR signal intensity $I(q,\Delta)$ after application of the pulse gradient and the signal intensity $I(0,\Delta)$ in absence of gradient. $E(q,\Delta)$ can be thought as the autocorrelation function of the spin phase changes induced by the first gradient pulse, and it coincides with the Fourier transform of the diffusion propagator. In the case of particle undergoing free Brownian motion, the diffusion propagator is Gaussian in the spatial displacement, and the echo attenuation decays exponentially with q^2, $E(q,\Delta) = \exp(-q^2 D\Delta)$, being D the self-diffusion coefficient.

The displacements accessible to PGSE-NMR investigation are bracketed by two length scales: The minimum observable displacement depends on the maximum q-value attainable (q_{max}) being equal to q_{max}^{-1} (in the 10–100 nm range depending on the gradient unit), while the maximum diffusional length probed corresponds to the $\text{RMSD} = \sqrt{2D\Delta}$ experienced during the observation time Δ.

Since each NMR signal gives rise to a distinct echo attenuation, using PGSE-NMR, it is possible to measure the diffusion coefficients of different components at the same system thus allowing an easy analysis of binding or association phenomena: When two species (having a different size and/or shape) share the same self-diffusion coefficient, it means that they are moving together. This is a powerful tool to discriminate the topological nature of the microemulsions. If surfactant and oil share the same diffusion coefficients ($D_s \approx D_{oil} << D_W$), the system is constituted by oil-swollen micelles dispersed in a continuous aqueous phase; if surfactant and water share the same diffusion coefficients ($D_s \approx D_W << D_{oil}$), the system is made by reverse micelles (a surfactant shell secluding a water core) dispersed in a continuous oil phase; finally, in the case of bicontinuous systems, the diffusion coefficients of the three components are uncorrelated, but the water and the oil have diffusion coefficients close to those of pure components and usually much higher than that of the surfactant self-diffusion. The diffusion within the continuous phase is influenced by the presence of barriers and thus reflects the size and shape of particles or interfaces. On the other hand, the self-diffusion coefficients of the disconnected particles permit the evaluation of the hydrodynamic radius R_h via the Stokes–Einstein equation

$$D = \frac{k_B T}{6\pi\eta R_h} \tag{1.2}$$

where η represents the viscosity of the continuous phase, k_B is the Boltzmann constant, and T is the temperature. The Stokes–Einstein relation has been demonstrated to hold for a plethora of systems as long as the size of the diffusing particle is larger than that of the solvent molecules.

Typical PGSE-NMR experiments use Δ-values of the order of several tens of milliseconds. This is a relatively long time with respect to molecular exchange. Therefore, when fast molecular exchange between sites characterized by different diffusion coefficients occurs, the observed self-diffusion coefficient D_{obs} is an average value. With regard to a two-site system, such as a ligand in fast exchange between free and bound forms (e.g., free to move in the solvent and bound to a much larger particle) with diffusion D_f and D_b, respectively, the observed diffusion coefficient is

$$D_{obs} = P_b D_b + (1 - P_b) D_f \tag{1.3}$$

where P_b represents the fraction of bound molecules, D_b is the diffusion coefficient of the particle (measured in the same experiment if the particle diffusion result unaltered by the presence of the bound ligand), and D_f is the diffusion coefficient of the ligand (measured in a separate experiment in the absence of particles). Once P_b is known, the partition equilibrium can be evaluated.

As a final remark, it should be noticed that, when dealing with systems where the nucleus investigated via NMR undergoes fast spin–spin relaxation (as in the cases described in this chapter), experiments for the determination of self-diffusion coefficients are usually performed using the pulsed gradient stimulated echo (PGSTE-NMR) rather than the PGSE-NMR sequence to allow for an increased Δ [3–6].

1.3 STRUCTURE OF THE WATER/BMIMBF$_4$ BINARY SYSTEM

One of the most peculiar features of ILs is the distinct degree of mesoscopic order they possess. Importantly, the latter is taken into account to explain at least part of the unique properties of ILs, such as their complex solvation dynamics. Loading IL with water has a twofold effect: (i) hydration of ions is likely to disrupt the ion pairs and (ii) the hydrophobic effect pushes toward the self-assembly of the organic cations.

The structural heterogeneities in water (W)/1-butyl-3-methylimidazolium (bmim$^+$) tetrafluoroborate (BF_4^-) mixtures were recently investigated, and at low water loading, the formation of water cluster and the IL organization into a polar network with a nanosegregation of the hydrophobic tails were inferred [7]. Upon increasing the water content, ion-pair interactions are gradually broken up, thus provoking the weakening of such a structural organization [8]. Moreover, the presence of a sharp diffraction peak at low frequency often found in X-ray or neutron scattering diffractograms of imidazolium-based room-temperature ILs was interpreted as indicative of mesoscopic organization. However, some recent neutron scattering and computational investigations evidenced that this peculiar spectroscopic feature of ILs could be accounted for without calling into play clustering or nanoscale structuring [9, 10]. Therefore, new experimental contributes are necessary to shed some more light on the nanoscopic organization of ILs. In this context, the entire W/bmimBF$_4$ phase diagram is here reinvestigated by means of diffusion NMR techniques [11].

The self-diffusion coefficients of W, bmim$^+$, and BF_4^- were obtained by ^{1}H and ^{19}F PGSTE-NMR experiments. For all components, the self-diffusion coefficients increase upon water loading. The dependence of D_{BF4} and D_{bmim} on the water content is of particular interest. At low water content, anions and cations share the same self-diffusion coefficient, but above a critical water concentration, the anion begins to diffuse faster than the cation. Such a threshold composition can be easily determined with the help of Figure 1.1 where the dependence of the difference $D_{BF4} - D_{bmim}$ on the water loading is shown. Clearly, only above $X_W = 0.2$ such a difference deviates significantly from zero.

As stated in Section 1.2, the measured diffusion is an average self-diffusion coefficient D_{obs}. According to Equation 1.3, it is strongly biased by fast diffusing species. In other words, a small fraction of free molecules can dominate the measured diffusion as long as D_f is much higher than D_b. On the basis of these arguments, the concentration $X_W \sim 0.2$ should be intended as the composition at which the ion pairs *start* to dissociate. Of course, further water addition drives the equilibrium toward dissociated ions until, above a certain water concentration, the system will behave as a classical electrolyte solution.

Since the response of PGSE-NMR (as well as PGSTE-NMR) measurements is insensitive to critical fluctuations, the diffusion data can profitably be used to detect the presence of micelle-like aggregates in the water-rich region. According to the Stokes–Einstein equation (Eq. 1.2), the self-diffusion coefficient is expected to scale as the reciprocal of the viscosity ($D \propto \eta^{-1}$). However, as shown in Figure 1.2, this

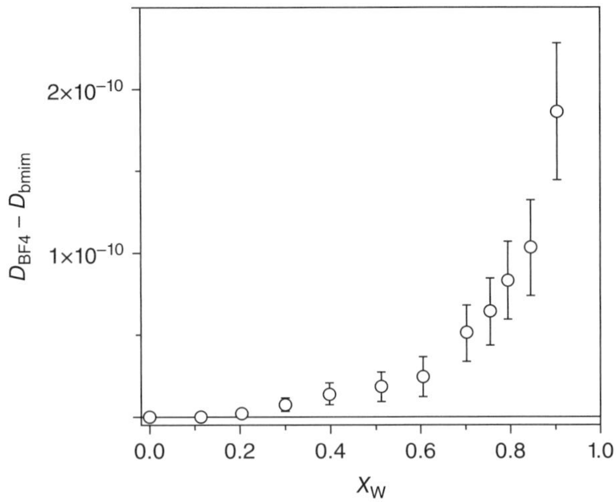

Figure 1.1 Difference between the self-diffusion coefficients of BF_4^- and $bmim^+$ ionic species at various water/ionic liquid mixtures (X_W = water molar fraction). Reproduced from Murgia et al. [11] with permission from Springer Science and Business Media.

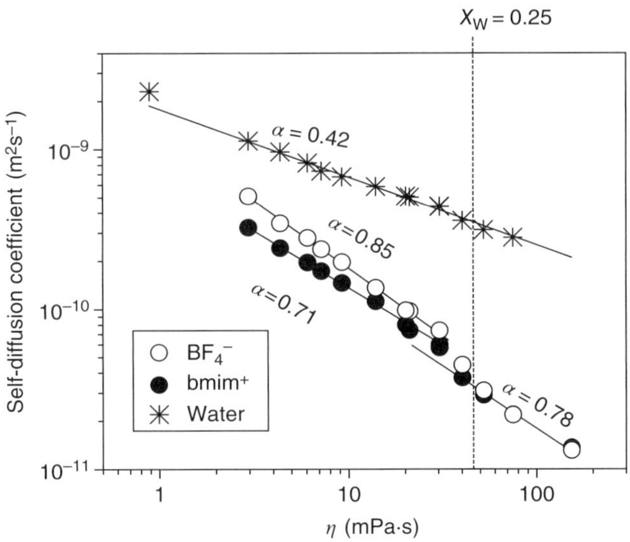

Figure 1.2 Double logarithmic plot of water, $bmim^+$, and BF_4^- self-diffusion coefficients versus viscosity. Reproduced from Murgia et al. [11] with permission from Springer Science and Business Media.

prediction is not fulfilled either by ions ($bmim^+$ and BF_4^-) or by water. Instead, all the components obey the power law

$$D = \text{const} \times \eta^{-\alpha} \tag{1.4}$$

with an exponent $\alpha < 1$. The above equation can be thought as a fractional formulation of the Stokes–Einstein equation. The emergence of fractional forms of Equation 1.2 was observed in a large number of systems when the size of the tagged particle is less than few nanometers and the fluid viscosity increases over orders of magnitude and is believed to take place when the particle size is comparable to that of the solvent molecules.

Inspection of Figure 1.2 reveals that the exponent α differs from unit and depends on the nature of the spin-bearing molecules. (Please note that, for the system under investigation, the dependence of viscosity on composition at 25°C was reported in literature [12]). Importantly, at low water content (high viscosity), the $bmim^+$ and BF_4^- ions share the same α value (0.78), while upon dilution, they show a different α value (0.71 and 0.85, respectively). According to the previously discussed self-diffusion results (Fig. 1.1), this finding evidenced once again the ion-pair dissolution that starts already at low water content.

Dimensional arguments require that, as long as Equation 1.4 holds, the self-diffusion coefficient must be related to the molecular size and mass (m) according to [13]

$$D = C \frac{(k_B T)^{\frac{1+\alpha}{2}}}{m^{\frac{1-\alpha}{2}} R_h^{2\alpha-1} \eta^{\alpha}} \tag{1.5}$$

where C is a dimensionless constant. For $\alpha = 1$, Equation 1.5 reduces to the classical Stokes–Einstein (Eq. 1.2), thus $C = 1/6\pi$. Applying Equation 1.5 to the diffusion data of Figure 1.2 allows estimating the ionic hydrodynamic radii for different water concentrations (see Fig. 1.3).

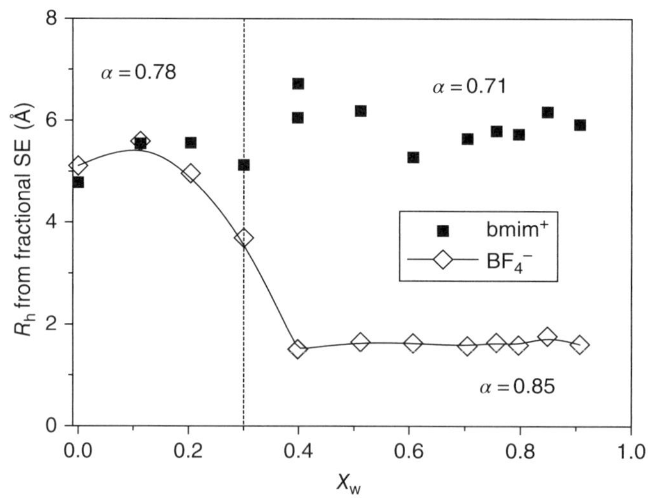

Figure 1.3 Hydrodynamic radius of $bmim^+$ and BF_4^- calculated from the fractional Stokes–Einstein equation (Eq. 1.5) at different water molar fractions. For $X_W < 0.3$, the same coefficient $\alpha = 0.78$ has been used for both the ions, while above that water content $\alpha = 0.71$ and $\alpha = 0.85$ for $bmim^+$ and BF_4^-, respectively, have been used in Equation 1.5. The correlation between X_W and η has been obtained from literature data as described in Murgia et al. [11]. Reproduced from Murgia et al. [11] with permission from Springer Science and Business Media.

The ion sizes evaluated according to this procedure are reasonable although the bmim$^+$ radius is systematically much higher than the van der Waals radius. This evidence strongly suggests that the cations experience some form of association. The bmim$^+$ size is essentially unaffected by the water content. On the contrary, the anion size undergoes to a dramatic drop passing from about 5 Å in pure IL to less than 2 Å for water contents larger than $X_W \sim 0.4$. For $X_W < 0.2$, bmimBF$_4$ diffuses as a whole *large* entity, but for further water loading, the measured D_{BF4} is dominated by the diffusion of *small* free BF_4^- ions, and thus the hydrodynamic size becomes essentially the van der Waals size of BF_4^- (1.95 Å). Since the contribution of BF_4^- to the volume of the ion pair is very small, the hydrodynamic size of bmim$^+$ remains essentially unchanged.

To summarize, the existence of mesoscopic domains in the W/bmimBF$_4$ binary system can be inferred from the analysis of the self-diffusion coefficients of the various molecular species in solution, since they were found to obey in a different way to a fractional Stokes–Einstein equation. In addition, bmim$^+$ and BF_4^- self-diffusion measurements, although suggesting some form of association of the cations, clearly evidenced that micellar aggregates did not form at any composition.

1.4 THERMODYNAMICS OF SURFACTANT SYSTEMS

Before discussing the consequences of adding NaAOT into a mixture of water and IL, it is useful to review its behavior in water and in ternary (oil containing) systems.

In surfactant systems, polar and apolar domains are separated by a dense self-assembled surfactant monolayer, with an area density of

$$\frac{A}{V} = \frac{\Phi_s}{l_s} \tag{1.6}$$

Here, Φ_s is the surfactant volume fraction and $l_s \equiv v_s/\alpha$ is the surfactant length, which is defined as the surfactant molecular volume v_s divided by the average area α that the surfactant molecule occupies at the water–oil interface.

Phases made up of flexible surfactant films can be understood in terms of the Helfrich curvature free energy:

$$g_c = 2\kappa(H - H_0)^2 + \bar{\kappa}K \tag{1.7}$$

Here, H_0 is the spontaneous curvature of the surfactant film and $H = (c_1 + c_2)/2$ and $K = c_1c_2$ are the mean and the Gaussian curvature, with c_1 and c_2 being the two principal curvatures. $\kappa > 0$ is the bending rigidity of the film and $\bar{\kappa}$ is the saddle-splay modulus that reports on the preferred topology of the film, which can take either positive or negative values. If $\bar{\kappa} < 0$, a spherically bent film is preferred (c_1 and c_2 having equal sign) favoring the formation of closed surfaces, that is, disconnected particles. If $\bar{\kappa} > 0$, a locally saddle-shaped surface is preferred favoring the formation of bicontinuous structures. κ is typically a few times k_BT, low enough to be flexible, but high enough so that $H \approx H_0$. $\bar{\kappa}$ is typically of the same magnitude as κ, but negative. Note that while H_0 is a property of the interfacial film, H depends on the volume/surface ratio and thus on the overall micelle composition. The total curvature free energy, G_c, for a given surface configuration is then formally obtained by integrating g_c over the total interfacial area.

Microstructure and phase behavior of surfactant systems strongly depend on H_0. Counting curvature toward apolar domains as positive, direct micelles are found when $H_0 >> 0$, and reverse micelles when $H_0 << 0$. For $H_0 \approx 0$, the ternary water–surfactant–oil phase diagram is dominated by a lamellar phase, because with planar layers, $H \approx 0$ irrespective of the composition. However, $H \approx 0$ can also be satisfied by bicontinuous structures, which are typically found at lower surfactant concentrations.

Sodium bis(2-ethylhexyl) sulfosuccinate (NaAOT) is the archetype of surfactant that forms monolayer with negative or null spontaneous curvature. The NaAOT surfactant in water has a quite low critical aggregation concentration (CAC = 2.2×10^{-3} mol l^{-1}) [14], and the binary NaAOT/W phase diagram is characterized by an extended lamellar (Lα) phase that forms at low concentration. Upon further surfactant additions, the Lα evolves toward an isotropic bicontinuous cubic gyroid (C_G) phase followed by a reverse hexagonal phase (H_2) [15]. Remarkably, AOT micelles having positive interfacial curvature (H) cannot form in water. Such a peculiarity is ascribed to the AOT geometrical parameters. Indeed, with a chain length (l) of 8.5 Å, a head group area (a_0) equal to 60 Å^2, and a hydrophobic chain volume (ν) of 480 Å^3, AOT possesses a packing parameter ($p = \nu / a_0 l$) of approximately 0.9, fully incompatible with an efficient packing into spheroidal aggregates with $H > 0$. On the other hand, the same molecular characteristics are called into play to justify the AOT well-known ability to form reverse micelles ($H < 0$) upon addition of oil [16, 17]. Actually, water-in-oil (W/O) spherical droplets with a hard-sphere behavior possibly occur only in a very limited region of the microemulsion (L_2) phases, namely, at low volume fraction of the disperse phase, that is, close to the oil corner. Conversely, the wide literature available concerning the microstructural features of L_2 phases formed by AOT provided evidence of anomalous behaviors with respect to a hard-sphere model. Discrepancies have been discussed within the context of the percolation theory [18–21], and both conductivity and water self-diffusion experiments demonstrated that transient fusion–fission processes among the droplets provoke huge modifications of the W/O droplet organization. Particularly, the lifetime of the particles' encounters is prolonged by attractive interactions thus originating clusters of droplets that allow the establishment of water networks all over the L_2 phase. This clustering, associated with changes of various macroscopic parameters such as viscosity and electrical conductivity, can be understood in terms of percolation [20–22]. Moreover, below the static percolation threshold, dynamic percolation can also take place. In this case, water channels form when the surfactant interface, separating adjacent water cores, breaks down during collisions or through the transient merging of droplets. Of course, the observation of the microstructural transitions in terms of static or dynamic percolation is strictly dependent on the timescale of the experimental technique used, ranging the kinetic constants regulating the lifetime of the clustering interaction over a wide timescale [23]. Finally, it deserves noticing that various factors may affect the percolation, including the type of counterion, as observed in the case of the ternary microemulsions formed by CaAOT, where the percolation threshold occurs at a much lower volume fraction of the dispersed phase than in the corresponding NaAOT system [24].

1.5 THE TERNARY SYSTEMS

The different phase regions originated when NaAOT is added to either $bmimBF_4$ or bmimBr aqueous solutions are shown in the NaAOT/W/$bmimBF_4$ and NaAOT/W/bmimBr phase diagrams reported in Figure 1.4 [25].

The remarkable differences observed represent a strong evidence of the alteration in the NaAOT interfacial packing induced by the ILs. If compared with the binary NaAOT/W diagram, the most striking difference is certainly the existence of the large liquid isotropic micellar phase found in both ternary diagrams. Moreover,

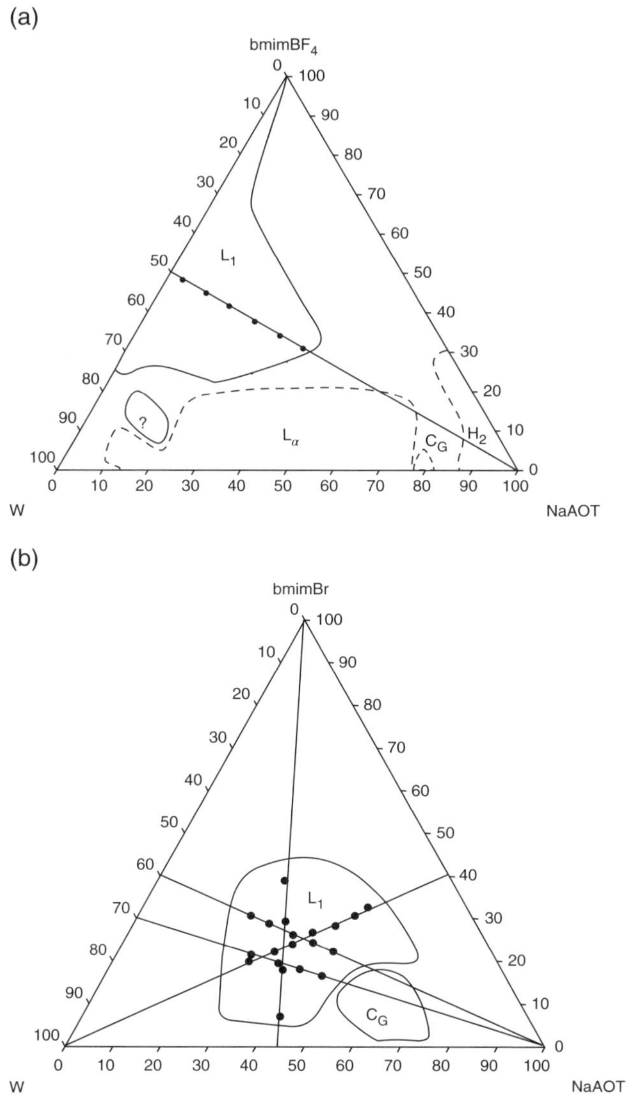

Figure 1.4 (a) NaAOT/W/$bmimBF_4$ and (b) NaAOT/W/bmimBr phase diagrams at 25°C. Dilution lines and samples analyzed are also shown. Reproduced from Murgia et al. [25] with permission from Royal Society of Chemistry.

exchanging the BF_4^- with Br^- caused dramatic modifications in the phase diagrams: The micellar region, originally connected to the W/IL binary axis, now approximately occupies the center of the phase diagram. Furthermore, the lamellar and hexagonal phases collapse, while the C_G phase appears greatly enlarged. While the nanostructures of the various liquid crystalline regions were successfully investigated by small-angle X-ray diffraction (SAXRD), the nanostructure of the micellar regions was inferred analyzing at 25°C the self-diffusion coefficients of W, $bmim^+$, AOT^-, Na^+, and BF_4^- obtained via PGSTE-NMR experiments.

Starting with the description of the NaAOT/W/bmimBF_4 ternary system, a decrease of all the measured self-diffusion coefficients upon surfactant loading was generally observed [26]. Particularly, the AOT^- self-diffusion coefficient (D_{AOT}) was found systematically lower than the diffusion of W, BF_4^-, and Na^+ by at least one order of magnitude. Important indication of the NaAOT self-assembling into micellar aggregates came from the comparison between the hydrodynamic radius calculated in the more diluted sample analyzed ($\Phi_{AOT}=0.03$, water/bmimBF_4=50/50 mixture) using Equation 1.2 (where $D_{AOT}=3.54\times10^{-10}\,m^2\,s^{-1}$ and $\eta=1.88\,mPa{\cdot}s$, interpolated from data reported in Liu et al. [12]) and that obtained for the NaAOT monomer, deduced from $D_{AOT}(3.54\times10^{-10}\,m^2\,s^{-1})$ measured in deuterated water below the NaAOT cmc. Indeed, values of 15 Å and of 5 Å were, respectively, found in the former and in the latter system. Moreover, being the water diffusion (D_W) always much higher than D_{AOT}, the presence of reverse aggregates can be excluded (because in that case, D_W should match D_{AOT}).

Figure 1.5 shows the evolution of the reduced self-diffusion coefficients (D/D^0) of water, BF_4^-, $bmim^+$, and Na^+ upon loading with NaAOT (the reference self-diffusion coefficient D^0 was taken as the diffusion measured in the binary solution water/bmimBF_4=50/50 for W and IL ions, while for Na^+, it was extrapolated at null NaAOT concentration). See Murgia et al. [11] for a comprehensive discussion on the reduced

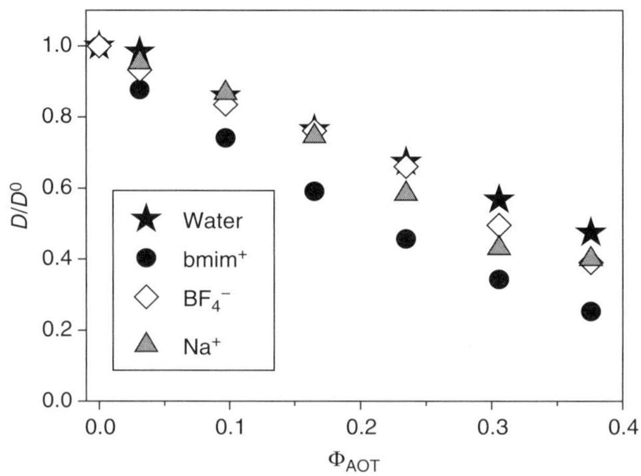

Figure 1.5 Reduced self-diffusion coefficients (D/D^0) of the components of the NaAOT/W/bmimBF_4 system as a function of the NaAOT volume fraction (Φ_{AOT}). Samples in the L_1 phase are made of equal mass of water and ionic liquid and different AOT mass. Reproduced from Murgia et al. [26] with permission from American Chemical Society.

diffusion coefficients. In the case of water, the reduced diffusion decreases only weakly while increasing the NaAOT concentration, a strong evidence that aggregates are disconnected, with O/W type curvature ($H > 0$). Similar trends were found for the reduced coefficients of BF_4^- and Na^+, while $bmim^+$ clearly deviates from this common trend.

Obstruction effects and specific interactions with the micellar wall can both be called into play to justify the observed D/D^0 trends. Specifically, the fact that W, BF_4^-, and Na^+ share the same D/D^0 values is a clear indication that their self-diffusion is mainly affected by obstruction effects (since they can hardly share the same interactions with AOT^-). Conversely, the systematically low D/D^0 values observed for $bmim^+$ denote a strong binding of a significant fraction of this cation to the AOT^- micelles. Obstruction and binding effects can be both treated according to Equation 1.3, slightly modified and rewritten as follows:

$$D^{obs} = b(1-P)D^0 + PD^{mic} \tag{1.8}$$

where P represents the fraction of bound molecules (moving along with the micelles). Accordingly, the observed self-diffusion coefficient, D^{obs}, is the population average of the self-diffusion coefficients in the two sites: the micelle (D^{mic}) and the continuous bulk. This last quantity is expressed in Equation 1.8 as the product of the self-diffusion coefficient in the absence of micelles, D^0, times the obstruction factor b. Hypothesizing a spherical shape of the aggregates, the equation $b = (1 + \Phi_{eff}/2)^{-1}$ describing the obstruction factor for spheres [27] can be used, and Equation 1.8 is rewritten explicitly for the $bmim^+$ as

$$P = \frac{\left(1 + \frac{\Phi_{eff}}{2}\right)^{-1} - \frac{D_{bmim}}{D^0_{bmim}}}{\left(1 + \frac{\Phi_{eff}}{2}\right)^{-1} - \frac{D_{AOT}}{D^0_{bmim}}} \tag{1.9}$$

where we have assumed that $D^{mic} = D_{AOT}$. Of course, the effective volume fraction of the micelles depends on the fraction of $bmim^+$ bound to the micelles themselves:

$$\frac{\Phi_{AOT} + \Phi^{bound}_{bmim}}{\Phi_{AOT}} = \frac{\Phi_{eff}}{\Phi_{AOT}} = 1 + \frac{\Phi_{bmim}}{\Phi_{AOT}} P \tag{1.10}$$

Equations 1.9 and 1.10 were used in an iterative procedure to evaluate the effective volume fraction from the self-diffusion coefficients of $bmim^+$ and AOT^-. In the initial step, Φ_{eff} in Equation 1.9 was assumed to be equal to the NaAOT volume fraction, and a first value of P was evaluated. Using such a value, Equation 1.10 allows obtaining a new value of Φ_{eff} that can be inserted in Equation 1.9 to obtain a new P-value and so on. Within three to five iterations, the previously described procedure converges giving the values of Φ_{eff}/Φ_{AOT} shown in Figure 1.6 as closed stars (this figure also contains an estimation of the volume of $bmim^+$ secluded in the micelles obtained from the analysis of the H_2 reverse hexagonal phase lattice parameter measured via SAXRD—for details in these calculations, the reader is referred to the original paper [26]).

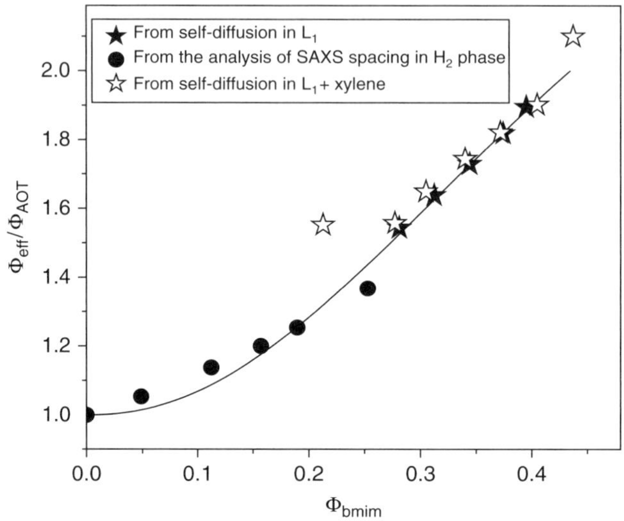

Figure 1.6 Effective volume of the interfacial film (normalized to the NaAOT volume) as a function of the bmimBF_4 concentration (volume fraction): (closed circles) data from the lattice parameter of H_2 phase obtained from SAXRD measurements, (closed stars) data from the self-diffusion coefficients of AOT^- and $bmim^+$ measured in the L_1 phase calculated from Equations 1.9 and 1.10 (see text), and (open stars) data obtained for the L_1 phase doped with *p*-xylene (see text). The curve represents the best fit according to the Hill's cooperative binding (Eq. 1.13). Reproduced from Murgia et al. [26] with permission from American Chemical Society.

Plainly, the $bmim^+$ cation accounts for a considerable fraction of the micellar volume. The plot in Figure 1.6 is peculiar because it does not contain any indication of saturation, while the steepness of the plot increases with Φ_{bmim}, a feature that excludes a description of $bmim^+$ binding to AOT^- according to a classical Langmuir's isotherm. Differently, the steepness of the plot Φ_{eff}/Φ_{AOT} versus Φ_{bmim} indicates a cooperative binding process that can be treated through the Hill's binding equilibrium [27, 28]

$$n\text{bmim}^+ + \text{AOT}^- = \text{AOT}^- - \text{bmim}_n^+ \quad (1.11)$$

with an equilibrium constant

$$K_{eq} = \frac{\Phi_{\text{AOT-bmim}}}{\Phi_{\text{AOT}}\Phi_{\text{bmim}}^n} \quad (1.12)$$

The ratio Φ_{eff}/Φ_{AOT} can be expressed according to the above equilibrium as

$$\frac{\Phi_{eff}}{\Phi_{AOT}} = 1 + \frac{nfK_{eq}\Phi_{bmim}^n}{1+K_{eq}\Phi_{bmim}^n} \quad (1.13)$$

where $f = 0.66$ is the ratio between the $bmim^+$ ($v = 316\,Å^3$) [29] and AOT^- molecular volumes.

Equation 1.13 was successfully fitted to the Φ_{eff}/Φ_{AOT} versus Φ_{bmim} data of Figure 1.6 (continuous curve) furnishing as best-fit parameters $K_{eq} = 5 \pm 1$ and $n = 2.23 \pm 0.07$.

When strong binding to the micelles occurs, the Lindman law (Eq. 1.3 or 1.8) represents an effective treatment for molecular diffusion affected by obstruction and/or interactions with surfactant aggregates. However, often the interactions are weak and cannot be described by a simple binding process. For example, the micellar surface can locally alter the diffusive properties causing a change in the measured self-diffusion coefficient also without any special affinity for the micelle. When dealing with colloidal particles, a proper theoretical description of the system can be obtained through the so-called effective cell model (ECM) [30]. The concept that underpins this model is the division of the system into small subsystems (cells) in order that they may represent the macroscopic properties of the whole system. The reader is referred to Jönsson et al. [30] for a complete description of the model. In this model, the effective self-diffusion coefficient for a component i will depend on both the diffusion of the cell and the diffusion within the cell. The equation for the total effective self-diffusion coefficient D_i of component i in a micellar system can be written as [31]

$$D_i = D_i^{\text{cell}}\left(1 - \frac{D^{\text{mic}}}{D_i^0}\right) + D^{\text{mic}} \tag{1.14}$$

where D_i^{cell} is the effective self-diffusion coefficient in a cell centered around the micelle, D_i^0 is the self-diffusion coefficient of component i in the bulk solution, and D^{mic} is the self-diffusion coefficient of the micelle. For water self-diffusion, being D^{mic} in the system under investigation extremely low, D_i^{cell} is identical to the collective self-diffusion coefficient (D_W). The key parameter of the model is the local variation of the product of the self-diffusion coefficient and the concentration of the component (C_iD_i). Simple cases are those where the cell is divided into two subvolumes. One subvolume is close to the micelle and is characterized by concentration C_1 and self-diffusion coefficient D_1, the rest of the cell having bulk concentration C_2 and self-diffusion coefficient D_2. The general equation has the form [30]

$$\frac{D^{\text{cell}}}{D_2} = \frac{U}{1 - \left(1 - \frac{C_1}{C_2}\right)\Phi_{\text{eff}}} \tag{1.15}$$

where the function U depends on the C_1D_1 and C_2D_2 products and on the symmetry (shape of the micelles). For spherical micelles, Equation 1.15 assumes the simple form [30]

$$\frac{D^{\text{cell}}}{D_2} = \frac{1}{1 - \left(1 - \frac{C_1}{C_2}\right)\Phi_{\text{eff}}} \, \frac{1 - \beta\Phi_{\text{eff}}}{1 + \beta\Phi_{\text{eff}}/2}$$

$$\beta = \frac{1 - C_1D_1/C_2D_2}{1 + C_1D_1/2C_2D_2} \tag{1.16}$$

In the case of weak binding ($C_1 < C_2$) and taking into account that D_2 is the bulk diffusion ($D_2 = D^0$), the following equation holds:

$$\frac{D^{\text{cell}}}{D^0}(1-\Phi_{\text{eff}}) = U = \frac{1-\beta\Phi_{\text{eff}}}{1+\beta\Phi_{\text{eff}}/2} \tag{1.17}$$

Note that in the absence of any adsorption ($C_1 = 0 \rightarrow \beta = 1$), Equation 1.17 describes only the obstruction effect and reduces to the definition of b used in Equation 1.9. Further rearrangement gives [31]

$$\frac{1-U}{1+U/2} = \beta\Phi_{\text{eff}} \tag{1.18}$$

Equation 1.18 can be used to discriminate between the possible micelle shapes. Indeed, although numerically indistinguishable from the case of prolate or cylindrical micelles, it is the closed solution for spherical micelles [30]. Conversely, oblate and discoid micelles show a very different behavior [27–30]. Thus, the function U was evaluated according to Equation 1.17, using the measured D_W values. The ratio $(1-U)/(1+U/2)$ was plotted as a function of Φ_{eff} in Figure 1.7; according to the prediction of Equation 1.18, it is a linear function of the effective volume fraction with null intercept. The slope of the linear regression gives $\beta = 1.24$, indicative of a low affinity of water for the micellar surface. It deserves noticing that for oblate micelles (having axial ratio larger than 4–5), the plot is expected markedly nonlinear [31].

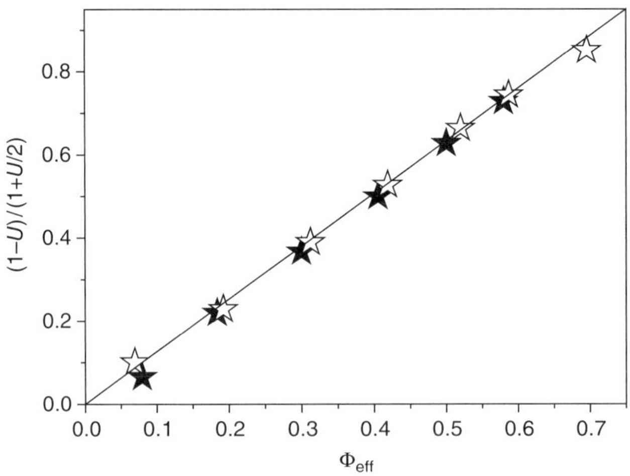

Figure 1.7 Analysis of the water diffusion according to the ECM (Eq. 1.18). Dependence of the function $(1-U)/(1+U/2)$ on the effective micelle volume fraction (Φ_{eff}). U was calculated from the water diffusion using Equation 1.17 and the micellar volume fraction from the AOT^- and $bmim^+$ diffusion using Equations 1.9 and 1.10. Closed symbols refer to the NaAOT/W/$bmimBF_4$ system; open symbols refer to the NaAOT/W/$bmimBF_4$/*p*-xylene system (in this last case, the micellar volume fraction also accounts for the *p*-xylene volume fraction). Reproduced from Murgia et al. [26] with permission from American Chemical Society.

This evidence safely excludes the presence of disk-like micelles in the present system. Such a conclusion is somehow surprising given the well-known preference of NaAOT to self-assemble into bilayers in water. On the other hand, the presence of spherical or cylindrical micelles cannot be excluded on the basis of the plot reported in Figure 1.7.

This goal can be achieved using the prediction of the ECM in the case of a strongly adsorbing component able to efficiently diffuse along the micellar contour, that is, a component with high C_1 and D_1. The matter of the micelles' shape at high surfactant volume fraction was unraveled by investigating the self-diffusion of a strongly adsorbed component that efficiently diffuses along the micellar contour. It has been demonstrated that, for Φ_{eff} large enough, the molecular diffusion within aggregates may significantly contribute to the macroscopic transport if exchanges among micelles occur [30]. In this study, *p*-xylene (*p*-xyl) was chosen as oil candidate. It is scarcely soluble (around 3 wt%) in the $bmimBF_4$/W = 50/50 solvent mixture, while it can be added up to 20.4 wt% in the NaAOT/W/$bmimBF_4$ = 38.4/30.8/30.8 ternary system without any phase separation, indicating that *p*-xylene is basically segregated within the micelles. The diffusion of *p*-xylene was investigated in a quaternary sample with composition NaAOT/$bmimBF_4$/W/*p*-xyl = 31.4/24.6/25.2/18.8. As confirmed by the high and mutually close values of D_W and $D_{p\text{-xyl}}$, the topology of the system is bicontinuous [32]. In addition, the surfactant self-diffusion ($D_{AOT} = 3.9 \times 10^{-11}\,m^2 s^{-1}$) is almost doubled with respect to the value measured in the absence of oil ($D_{AOT} = 2.1 \times 10^{-11}\,m^2 s^{-1}$). After this successful demonstration of the oil diffusion within the surfactant aggregates, a tiny amount of *p*-xylene was added as a molecular probe (NaAOT/*p*-xyl molar ratio equal to 4/1) to a series of samples having composition equivalent to that previously described: The measured self-diffusion coefficients of the various chemical species were found similar to those observed in the absence of *p*-xylene.

The binding of $bmim^+$ to the surfactant aggregates in the presence of *p*-xylene was iteratively calculated using Equations 1.9 and 1.10 as described in the previous section with two important modifications: (i) the *p*-xylene volume fraction is now taken into account (thus in the first iteration, we used $\Phi_{eff} = \Phi_{AOT} + \Phi_{p\text{-xyl}}$), and (ii) for the same reason, Equation 1.10 furnishes $\left(\Phi_{AOT} + \Phi_{bmim}^{bound}\right)/\Phi_{AOT} \neq \Phi_{eff}/\Phi_{AOT}$. The result of this calculation is shown in Figure 1.6 as open stars. It is clear that there is a full agreement with the results obtained in the absence of *p*-xylene. The analysis of the water diffusion according to Equations 1.17 and 1.18 reveals that, when the contribution of the oil to the micellar volume fraction is considered, *p*-xylene does not induce any change in the obstruction effect probed by water (see open stars in Fig. 1.7). It can be therefore inferred that the oil loading leaves the systems' microstructure unchanged. The self-diffusion coefficients of *p*-xylene are shown as a function of the effective volume fraction in Figure 1.8; $D_{p\text{-xyl}}$ drops upon increasing Φ_{eff} to 0.3 and then remains essentially steady around $1.3 \times 10^{-10}\,m^2 s^{-1}$. This value is several times greater than the corresponding D_{AOT}. In other words, the *p*-xylene diffusion is slower than expected for continuous diffusion paths but systematically higher than the diffusion of the micelles. A possible explanation is that, above $\Phi_{eff} = 0.3$, dynamic percolation takes place. Accordingly, the diffusion within the aggregates as well as the reduction of the distance between the aggregates themselves contributes sufficiently to the macroscopic diffusion to steady it. The ECM implicitly accounts for this mechanism since the reduction of the interaggregate distance corresponds to a reduction of the distance between the micelle and the cell

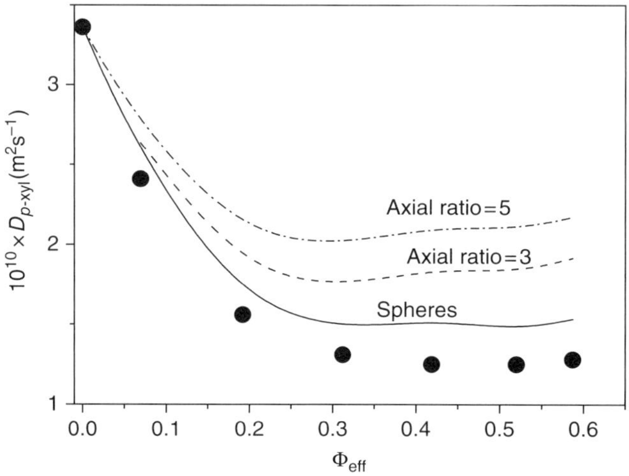

Figure 1.8 Self-diffusion coefficients of *p*-xylene as a function of the micellar volume fraction. Solid curve is the prediction for spherical micelles (Eqs. 1.14 and 1.16); dashed curves are the prediction for prolate spheroids with axial ratio = 3 or 5; for all the simulations, the parameters are $C_2 = 0.003$, $D_2 = 3.36 \times 10^{-10}\,m^2 s^{-1}$, $C_1 = \Phi_{xyl} / \Phi_{eff}$, and $D_1 = 5 \times 10^{-10}\,m^2 s^{-1}$. Reproduced from Murgia et al. [26] with permission from American Chemical Society.

boundary [30]. Therefore, this hypothesis can be quantitatively tested through Equations 1.14 and 1.16. The terms C_2 and D_2, respectively, correspond to the solubility (0.003 volume fraction) and the self-diffusion coefficient ($3.36 \times 10^{-10}\,m^2 s^{-1}$) of *p*-xylene in the absence of NaAOT; the concentration within the micelle can be set as $C_1 = \Phi_{p\text{-xyl}} / \Phi_{eff}$, and an estimate of the lateral diffusion is given by that measured in the bicontinuous phase $D_1 = 5 \times 10^{-10}\,m^2 s^{-1}$.

The result of this calculation, shown in Figure 1.8 (solid curve), is in good agreement with the experimental data. We have applied the ECM to calculate the diffusion expected in the case of prolate micelles of different axial ratios using equation 49 of Jönsson et al. [30] and the same parameters (C_1, C_2, D_1, D_2) described previously. As shown in Figure 1.8, already for axial ratios of 3 and 5, the agreement between the ECM prediction for anisotropic shapes and the experimental data is much worse than that obtained in the case of spherical micelles. The straightforward conclusion is that the aggregates made by NaAOT in water/bmimBF_4 solutions are likely to be spherical in shape.

Turning the attention to the NaAOT/W/bmimBr ternary system, the ^1H PGSTE-NMR experiments conducted for the determination of W, $bmim^+$, and AOT^- self-diffusion coefficients, first of all, evidenced a decrease of D_W upon NaAOT addition along the W/bmimBr = 60/40 dilution line, with values ($5.5 - 4.3 \times 10^{-10}\,m^2 s^{-1}$) always much lower than the self-diffusion coefficient of pure water ($D_W^0 = 2.3 \times 10^{-9}\,m^2 s^{-1}$) [25]. However, these values were found significantly higher than D_{AOT} ($1.6 - 1.2 \times 10^{-11}\,m^2 s^{-1}$). These findings indicate the formation of NaAOT micelles having positive curvature of the interface ($H > 0$). Indeed, a confinement of the water molecules within the NaAOT aggregates would imply the measurement of comparable D_W and D_{AOT}. Therefore, the obstruction effect due to the presence of the micelles can be accounted for the D_W decrease. The positive curvature of the NaAOT aggregates was unambiguously confirmed performing a conductometric titration of the W/bmimBr = 60/40 binary system upon addition of NaAOT. Although the conductivity (χ) decreases linearly while increasing the NaAOT content, its value remains high (around tenths of mS/cm) clearly indicating that water is the continuous medium, while the recorded reduction of χ can be entirely ascribed to the increased volume of the dispersed phase (the micelles) that determines a greater obstruction to the motion of the ions. Moreover, self-diffusion experiments performed in aqueous solution of bmimBr in the presence and in the absence of NaAOT in a wide range of compositions reveal a huge (around one order of magnitude) decrease of D_{bmim} in the presence of the surfactant. The reasons for the remarkable decrease of the D_{bmim} values recorded in the presence of NaAOT can be twofold. On the one hand, bmim$^+$ will certainly experience the obstruction effect caused by the micelles. On the other hand, if bmim$^+$ is adsorbed at the interface, the observed D_{bmim} would be a weighted value according to Equation 1.3. As will be demonstrated in the following, obstructed diffusion can only be a concurrent cause in the decrease of D_{bmim}, while the effect due to the bmim$^+$ adsorption at the interface prevails.

In Figure 1.9a, b, the W and AOT self-diffusion coefficients for different dilution lines as a function of the dispersed phase volume fraction (Φ_{AOT}) are reported.

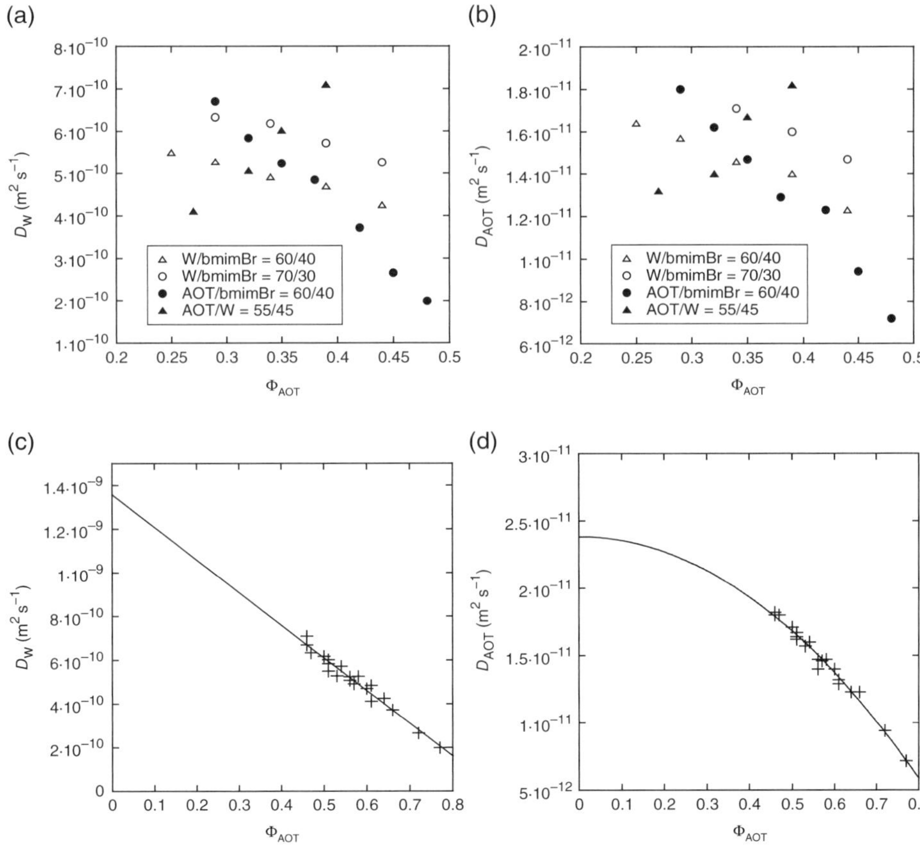

Figure 1.9 Water and AOT$^-$ self-diffusion coefficients measured at 25°C in the L_1 region of the NaAOT/W/bmimBr system as function of both the NaAOT (Φ_{AOT}) and the NaAOT + bmimBr ($1 - \Phi_W$) volume fractions in different dilution lines. Reproduced from Murgia et al. [25] with permission from Royal Society of Chemistry.

Evidently, data are rather scattered in the diagrams and do not show a common trend. The situation radically changes if the dispersed phase volume fraction is calculated summing the contributions of NaAOT and bmimBr $(1-\Phi_W)$. In systems having a bicontinuous nanostructure, while increasing Φ, theory predicts a D_W and D_{AOT} evolution according to the following linear [33] and quadratic [34] equations:

$$\frac{D_W}{D_W^0} = \beta - \mu(1-\Phi_W) \tag{1.19}$$

$$\frac{D_{AOT}}{D_{AOT}^0} = \beta - \mu(1-\Phi_W)^2 \tag{1.20}$$

where D_W/D_W^0 and D_{AOT}^0, respectively, represent the reduced self-diffusion coefficient of water and the lateral self-diffusion coefficient of the surfactant, $\beta=2/3$ is a topologically independent constant, and μ is a constant that depends on the topology of the dispersed phase [33]. Experimentally, by fitting the water diffusion data to Equation 1.19, one obtains an apparent β-value of 0.6 instead of 0.67, but this is because the water molecules necessary to hydrate the ions are not taken into account resulting in a β-value slightly underestimated [26].

As can be noticed in Figure 1.9c, d, data definitely follow the linear and quadratic trends predicted by these equations, giving a strong indication that the system has a bicontinuous nanostructure and that the bmim$^+$ cation is adsorbed at the interface.

A further confirmation that the bmim$^+$ cation is really adsorbed at the interface comes from linear interpolation of data described in Figure 1.10, where the D_{bmim} is

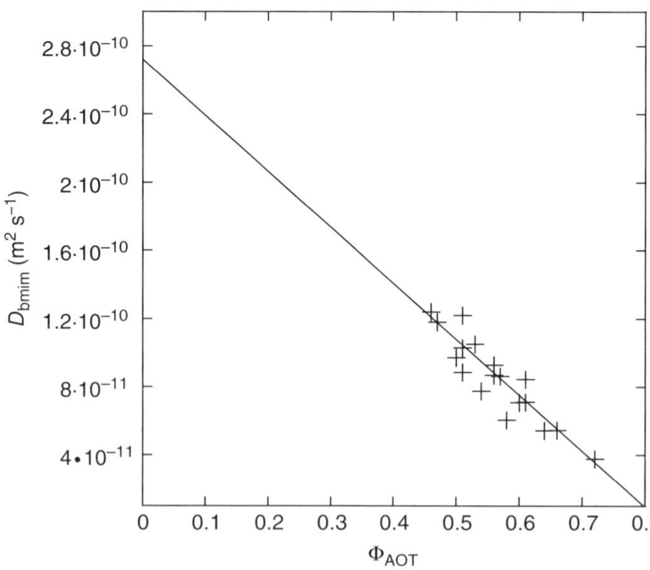

Figure 1.10 bmim$^+$ self-diffusion coefficients measured in the L_1 region of the NaAOT/W/bmimBr system in all the investigated dilution lines at 25°C as function of the NaAOT + bmimBr $(1-\Phi_W)$ volume fraction. Reproduced from Murgia et al. [25] with permission from Royal Society of Chemistry.

reported as a function of $(1-\Phi_W)$. Actually, the self-diffusion coefficient extracted at infinite dilution $(2.7\times10^{-10}\,m^2\,s^{-1})$ is too low if compared with that obtained from the D_{bmim} values measured in the W/bmimBr binary system $(8.3\times10^{-10}\,m^2\,s^{-1})$. Such a low value can be explained taking into account that the $bmim^+$ cation is partially bound to the micelle and partially free to move in the bulk according to a two-site fast-exchange model (Eq. 1.3).

To understand the causes at the basis of the remarkable differences observed in the NaAOT/W/bmimBF_4 and the NaAOT/W/bmimBr phase diagrams, the $bmim^+$ adsorption isotherms in both systems were compared. For a suitable description of the results, the effective volume fraction of the interface $(\Phi_{eff}=\Phi_{AOT}+\Phi_{bmim}^{mic})$ normalized to the NaAOT volume fraction (Φ_{AOT}) and the $bmim^+$ volume fraction (Φ_{bmim}) is, respectively, reported in ordinate and in abscissa in Figure 1.11. The volume fraction of the micellized $bmim^+$ (Φ_{bmim}^{mic}) was calculated under the assumption that the self-diffusion coefficient of the micellized $bmim^+$ is much lower than that of free $bmim^+$ in aqueous solution. When such a condition holds, Equation 1.3 simplifies in

$$D_{bmim}^{obs}=D_{bmim}^{free}p_{bmim}^{free} \tag{1.21}$$

Moreover, from Equation 1.19

$$D_{bmim}^{free}=D_{bmim}^{0}\left[0.6-0.65\left(1-\Phi_W\right)\right] \tag{1.22}$$

In Equation 1.22, the D_{bmim}^{0} value previously extracted from the binary system $(8.3\times10^{-10}\,m^2\,s^{-1})$ was used. In addition, since the domain over which water and $bmim^+$ diffuse is the same, the constants β and μ were assessed treating the water data. Once D_{bmim}^{obs} and D_{bmim}^{free} are known, the amount $(1-p_{bmim}^{free})$, that is the micellized $bmim^+$ fraction, is easily extracted through Equation 1.21. Finally, Φ_{bmim}^{mic} is obtained

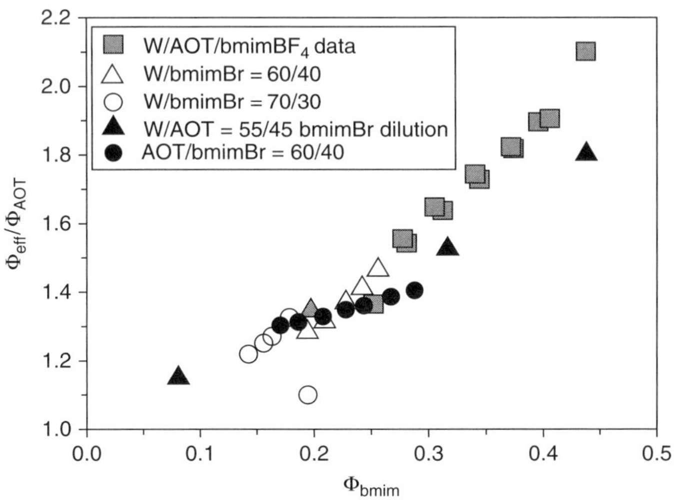

Figure 1.11 Comparison between the $bmim^+$ adsorption isotherms in the micellar region of both NaAOT/W/bmimBF_4 and NaAOT/W/bmimBr (all the investigated dilution lines) systems. Reproduced from Murgia et al. [25] with permission from Royal Society of Chemistry.

multiplying this amount to Φ_{bmim}. In Figure 1.11, the result of such data processing is compared with the adsorption isotherm from the NaAOT/W/bmimBF$_4$ ternary system. Clearly, the amount of adsorbed bmim$^+$ cation is the same in both systems.

The picture emerging from the experiments conducted on the NaAOT/W/bmimBr system describes an L$_1$ region characterized by highly branched bicontinuous micelles. Both AOT$^-$ and bmim$^+$ participate to the formation of a nanostructure having a positive (null on average) interfacial curvature and dispersed in a bulk phase composed by water and bmim$^+$ nonadsorbed at the interface. It is here also worth noticing that an inversion of the locally positive interfacial curvature adjacent to the branches is a stringent requirement in a bicontinuous micellar system. This condition is certainly favored by a nonhomogeneous partition of bmim$^+$ at the interface. In other words, the original negative curvature of NaAOT-based aggregates can be reestablished by a depletion of bmim$^+$ close to the branches so that a local excess of AOT$^-$ is produced. This scenario represents an intermediate condition between that reported for the NaAOT/W/bmimBF$_4$ system (spherical micelles with positive curvature of the surfactant film) and that commonly observed in NaAOT-based systems (structures with reverse curvature or flat interfaces). Given the same adsorption at the interface of the bmim$^+$ cation, the bromide ion can be easily accounted for the significant differences found in the two ternary systems investigated. This fact deserves some notes.

While the increase of the polar head area originated by the bmim$^+$ adsorption at the interface may fully justifies the inversion of the ordinary (reverse) surfactant film curvature, the observed ion specificity may be rationalized taking into account the screening effect Br$^-$ and BF$_4^-$ cause at the interface. Indeed, the higher Br$^-$ basicity with respect to BF$_4^-$ should imply a higher hydration and a consequent bulkier character of the former anion. In other words, it is reasonable that BF$_4^-$ adsorb at the interface at a larger extent than Br$^-$. The same subject can be debated through a different approach. Ion specificity is ubiquitous in chemistry and biology. Effects connected to such specificity exhibit a recurrent trend, known as the Hofmeister series [35–38]. In conformity with current nomenclature, at the extremities of this series are located the kosmotropic and the chaotropic ions (in allusion to their hypothetical aptitude to enhance/reduce the structure of neighbor water molecules). Regarding the ionic species here under examination, Br$^-$ and BF$_4^-$, they are universally considered as chaotropic in the Hofmeister series, with the order Br$^-$ < BF$_4^-$. Chaotropic ions have the tendency of being poorly hydrated and more easily adsorb at the interface with respect to their counterpart, the kosmotropic ions. Nevertheless, within this theory, this order may be reversed depending on the system investigated because of a number of motivations that range from the salt concentration [39] to the interface charge and polarity [40]. Plainly, this could be particularly true for neighboring terms, just as Br$^-$ and BF$_4^-$. In the colloidal systems here investigated, the AOT effective packing parameter decreases (because of an increased polar head area) in the presence of bmimBr (that causes the establishment of a bicontinuous interface, $H \approx 0$) or, even more, in the presence of bmimBF$_4$ (that causes the formation of spherical aggregates with $H > 0$). Conversely, from the predicted binding order (Br$^-$ < BF$_4^-$), a higher decrease of the effective packing parameter should be expected in the presence of bmimBr rather than bmimBF$_4$. Indeed, a stronger binding to the micellar surface should result in a reduction of the effective polar head area because of screening effects. Therefore, the phase behavior detected is in contradiction with the predicted binding order.

However, the delicate equilibrium experienced by the molecules at the interface is ruled by a complex set of electrostatic and steric factors that, finally, result in the macroscopic phenomenology observed. Although further work is needed to clarify the mechanism that supports the observed phase behavior, actually, ion specificity in the present work emerges in ion ability of modulating the interface curvature.

1.6 CONCLUSIONS AND OUTLOOK

Although amphiphilic, the $bmim^+$ cation does not form micelles in water and thus is not a surfactant. However, it is adsorbed at the interface and affects the structure of the AOT micelles. Therefore, in the matter of micelle making, $bmim^+$ can be rather considered as a cosurfactant whose influence is tuned by the nature of the counterion.

Micelle and microemulsion systems find applications in human activities, spanning from the conservation of cultural heritage [41] to the preparation of a variety of nanocrystalline materials [42], and their possible uses in drug delivery [43], synthesis of nanomaterials, etc. Micelle and IL-based microemulsion systems are prospective to improve many features of their traditional counterparts [44, 45], and, in this context, the possibility of modifying both the topology and the morphology of these soft nanoparticles can be of great importance in view of their industrial applications.

REFERENCES

[1] Eastoe, J.; Gold, S.; Rogers, S. E.; Paul, A.; Welton, T.; Heenan, R. K.; Grillo, I., *J. Am. Chem. Soc.* 2005, **127**, 7302–7303.

[2] Hao, J.; Zemb, T., *Curr. Opin. Colloid Interface Sci.* 2007, **12**, 129–137.

[3] Johnson, C. S. J., *Prog. Nucl. Magn. Reson. Spectrosc.* 1999, **34**, 203–256.

[4] Price, W. S., *Concepts Magn. Reson.* 1997, **9**, 299–336.

[5] Price, W. S., *Concepts Magn. Reson.* 1998, **10**, 197–237.

[6] Stilbs, P., *Prog. Nucl. Magn. Reson. Spectrosc.* 1987, **19**, 1–45.

[7] Triolo, A.; Russina, O.; Bleif, H. J.; Di Cola, E., *J. Phys. Chem. B* 2007, **111**, 4641–4644.

[8] Fazio, B.; Triolo, A.; Di Marco, G., *J. Raman Spectrosc.* 2008, **39**, 233–237.

[9] Annapureddy, H. V. R.; Kashyap, H. K.; De Biase, P. M.; Margulis, C. J., *J. Phys. Chem. B* 2010, **114**, 16838–16846.

[10] Hardacre, C.; Holbrey, J. D.; Mullan, C. L.; Youngs, T. G. A.; Bowron, D. T., *J. Chem. Phys.* 2010, **133**, 074510.

[11] Murgia, S.; Monduzzi, M.; Lopez, F.; Palazzo, G., *J. Solut. Chem.* 2013, **42**, 1111–1122.

[12] Liu, W.; Zhao, T.; Zhang, Y.; Wang, H.; Yu, M., *J. Solut. Chem.* 2006, **35**, 1337–1346.

[13] Pollack, G. L.; Enyeart, J. J., *Phys. Rev. A* 1985, **31**, 980.

[14] Mukherjee, K.; Moulik, S. P.; Mukherjee, D. C., *Langmuir* 1993, **9**, 1727–1730.

[15] Rogers, J.; Winsor, P. A., *Nature* 1967, **216**, 477–479.

[16] De, T. K.; Maitra, A., *Adv. Colloid Interface Sci.* 1995, **59**, 95–193.

[17] Nave, S.; Eastoe, J.; Heenan, R. K.; Steytler, D.; Grillo, I., *Langmuir* 2000, **16**, 8741–8748.

[18] Bergenholtz, J.; Romagnoli, A.; Wagner, N., *Langmuir* 1995, **11**, 1559–1570.

[19] Cametti, C.; Codastefano, P.; Tartaglia, P.; Chen, S. H.; Rouch, J., *Phys. Rev. A* 1992, **45**, R5358–R5361.

[20] Feldman, Y.; Kozlovich, N.; Nir, I.; Garti, N., *Phys. Rev. E* 1995, **51**, 478–491.
[21] Kotlarchyk, M.; Chen, S. H.; Huang, J. S.; Kim, M. W., *Phys. Rev. Lett.* 1984, **53**, 941–944.
[22] Ponton, A.; Bose, T. K.; Delbos, G., *J. Chem. Phys.* 1991, **94**, 6879–6886.
[23] Fletcher, P. D. I.; Howe, A. M.; Robinson, B. H., *J. Chem. Soc. Faraday Trans.* 1987, **83**, 985–1006.
[24] Pitzalis, P.; Angelico, R.; Soderman, O.; Monduzzi, M., *Langmuir* 2000, **16**, 442–450.
[25] Murgia, S.; Palazzo, G.; Mamusa, M.; Lampis, S.; Monduzzi, M., *Phys. Chem. Chem. Phys.* 2011, **13**, 9238–9245.
[26] Murgia, S.; Palazzo, G.; Mamusa, M.; Lampis, S.; Monduzzi, M., *J. Phys. Chem. B* 2009, **113**, 9216–9225.
[27] Jóhannesson, H.; Halle, B., *J. Chem. Phys.* 1996, **104**, 6807–6817.
[28] Weiss, J. N., *FASEB J.* 1997, **11**, 835–841.
[29] Jin, H.; O'Hare, B.; Dong, J.; Arzhantsev, S.; Baker, G. A.; Wishart, J. F.; Benesi, A. J.; Maroncelli, M., *J. Phys. Chem. B* 2008, **112**, 81–92.
[30] Jönsson, B.; Wennerström, H.; Nilsson, P. G.; Linse, P., *Colloid Polym. Sci.* 1986, **264**, 77–88.
[31] Jonströmer, M.; Jönsson, B.; Lindman, B., *J. Phys. Chem.* 1991, **95**, 3293–3300.
[32] Guering, P.; Lindman, B., *Langmuir* 1985, **1**, 464–468.
[33] Balinov, B.; Olsson, U.; Soederman, O., *J. Phys. Chem.* 1991, **95**, 5931–5936.
[34] Anderson, D. M.; Wennerstroem, H., *J. Phys. Chem.* 1990, **94**, 8683–8694.
[35] Murgia, S.; Monduzzi, M.; Ninham, B. W., *Curr. Opin. Colloid Interface Sci.* 2004, **9**, 102–106.
[36] Murgia, S.; Monduzzi, M.; Palazzo, G., *Langmuir* 2012, **28**, 1283–1289.
[37] Murgia, S.; Portesani, F.; Ninham, B. W.; Monduzzi, M., *Chem. Eur. J.* 2006, **12**, 7889–7898.
[38] Ninham, B. W.; Lo Nostro, P., *Molecular Forces and Self Assembly in Colloid, Nano Sciences and Biology*. Cambridge University Press, Cambridge, UK, 2010.
[39] Zhang, Y.; Cremer, P. S., *Proc. Natl. Acad. Sci. U. S. A.* 2009, **106**, 15249–15253.
[40] Schwierz, N.; Horinek, D.; Netz, R. R., *Langmuir* 2010, **26**, 7370–7379.
[41] Baglioni, P.; Berti, D.; Bonini, M.; Carretti, E.; Dei, L.; Fratini, E.; Giorgi, R., *Adv. Colloid Interface Sci.* 2013, **205**, 361–371.
[42] Ganguli, A. K.; Ganguly, A.; Vaidya, V., *Chem. Soc. Rev.* 2010, **39**, 474–485.
[43] Lawrence, M. J.; Rees, G. D., *Adv. Drug Deliv. Rev.* 2012, **64**, 175–193.
[44] Kunz, W.; Zemb, T.; Harrar, A., *Curr. Opin. Colloid Interface Sci.* 2012, **17**, 205–211.
[45] Qiu, Z.; Texter, J., *Curr. Opin. Colloid Interface Sci.* 2008, **13**, 252–262.

CHAPTER 2

Characterization of Self-Assembled Amphiphiles in Ionic Liquids

LANG G. CHEN, STEPHEN H. STRASSBURG, and HARRY BERMUDEZ

Department of Polymer Science and Engineering, University of Massachusetts Amherst, Amherst, MA, USA

2.1 INTRODUCTION

Ionic liquids (ILs) are a special class of salts composed solely of ions yet are liquids at room temperature. They often comprise organic cations paired with organic or inorganic anions (Fig. 2.1). The bulky and asymmetric shape of the cations destabilize crystalline ordering, leading to the much lower melting temperatures as compared to classical salts. ILs generally have low volatility, high thermal stability, and high ionic conductivity. In addition, ILs can dissolve an enormous range of inorganic, organic, and polymeric materials. Therefore, the study of ILs has been expanding and attracting interest for many potential applications such as solvent media for extraction and reaction [1–4]. Moreover, recent developments have been focused on the various self-assemblies of amphiphiles in ILs including both small surfactant molecules and macromolecules [5–9].

By combining amphiphiles and ILs, surface properties such as surface tension and surface charge can be modified and controlled. Furthermore, aggregates such as micelles, microemulsions, and liquid crystalline phases can be formed in bulk IL solution, just as in aqueous solution. Traditional amphiphiles include small molecular surfactants, polymeric amphiphiles, and more complex systems such as their mixtures (Table 2.1). Due to the liquid form of ILs, the surface adsorption and bulk aggregation of all types of amphiphiles in ILs have been studied by different techniques. Comprehensive reviews are published on the surface chemistry and interfacial properties of neat ILs [4, 10] and bulk solvation properties in ILs [11, 12]. Due to the negligible vapor pressure of ILs, a number of ultrahigh vacuum (UHV) techniques have been applied to IL surfaces, such as X-ray photoelectron spectroscopy (XPS), direct recoil spectrometry (DRS), time-of-flight secondary ion mass spectrometry (ToF-SIMS), neutral impact collision ion scattering spectroscopy (NICISS), metastable-induced electron spectroscopy (MIES), and ultraviolet photoelectron

Ionic Liquid-Based Surfactant Science: Formulation, Characterization, and Applications, First Edition. Edited by Bidyut K. Paul and Satya P. Moulik.

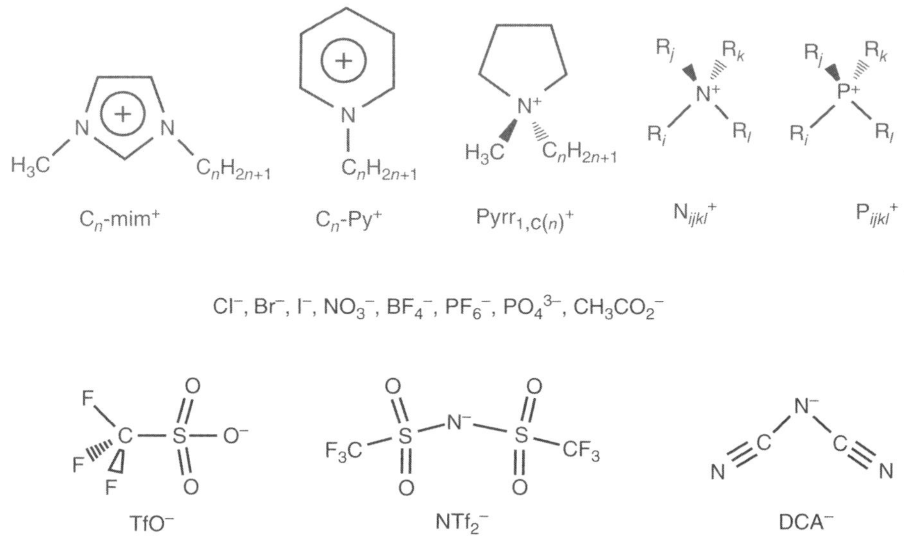

Figure 2.1 Chemical structures for common IL cations and anions. Source: Castner and Wishart [1] and reproduced with permission from American Institute of Physics. *See insert for color representation of the figure.*

TABLE 2.1 Scheme of the Structures of Amphiphiles

Amphiphiles	Schematic	Example
Surfactants	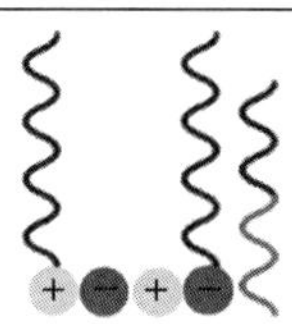	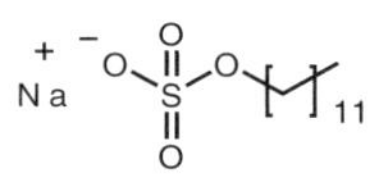 ($NaC_{12}S$) 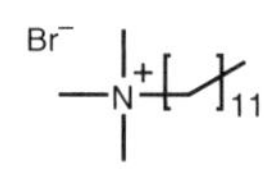 ($C_{12}TAB$) 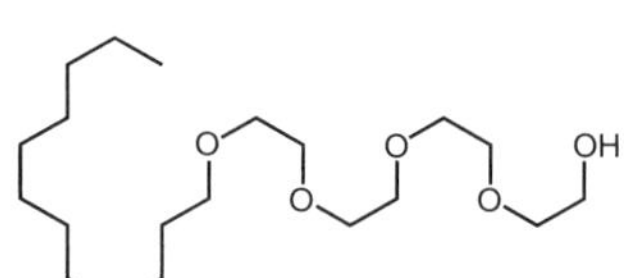 (Brij-30 or $C_{12}E_4$)
Block copolymers	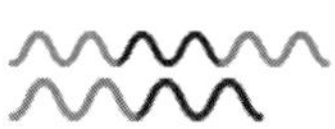	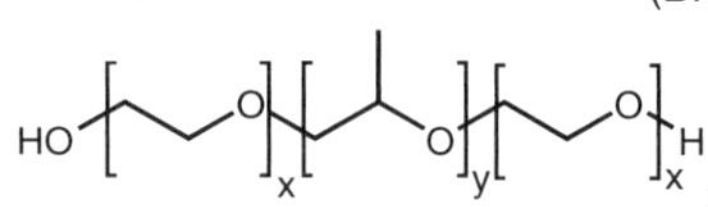 (Pluronic) 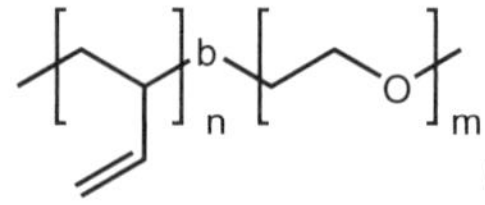 (PB-b-PEO)

spectroscopy (UPS). Other traditional surface techniques such as neutron reflectivity (NR), sum frequency generation spectroscopy (SFG), atomic force microscopy (AFM), and tensiometry have also been used to understand different interfaces with IL systems. On the other hand, self-assembly of amphiphiles in the bulk of ILs has also been characterized by numerous techniques such as fluorescence spectroscopy, polarized optical microscopy (POM), scattering (dynamic light scattering (DLS), small-angle X-ray scattering (SAXS), and small-angle neutron scattering (SANS)), nuclear magnetic resonance (NMR), and transmission electron microscopy (TEM).

Both interface and bulk properties are of vital importance in the understanding of their novel solvent behavior and future applications. To the best of our knowledge, a review focusing on the *techniques* applied to surfactant-IL systems has not been published yet. In this article, we review techniques applied to the interfaces of surfactant-IL systems, as well as in the bulk. Due to space limitation, we have chosen to focus on selected ILs and amphiphiles. Hence, an exhaustive literature review was not possible. Instead, we illustrate with examples that range from the common to the esoteric, with an intention that the reader will be motivated to explore new techniques and materials in this topic.

2.2 CHARACTERIZATION TECHNIQUES

2.2.1 Interface Techniques

Unlike the homogeneous bulk, the interface is unique in the break of symmetry, which may induce order or disorder. Numerous techniques have been used to acquire structure and composition information at the interface of ILs, many of which require UHV. The negligible vapor pressure of ILs makes those techniques possible to interrogate their air–liquid interface. Other surface-sensitive methods with ambient conditions including NR, SFG, AFM, and tensiometry are also applicable to measure air–liquid, solid–liquid, and liquid–liquid interfaces.

2.2.1.1 XPS Among all the surface-sensitive techniques, XPS is arguably the most common and prominent UHV-based tool to provide unique information on chemical composition, chemical state identification, and even composition depth profiles of the near-surface region. Recently, Lovelock et al. [13] published a comprehensive review article on photoelectron spectroscopy applied to IL interfaces. The first experimental investigation of the air–IL interface using XPS was reported by Smith et al. in 2005 [14]. This technique is based on the kinetic energy and number of photoelectrons that are irradiated by a beam of X-rays at the emission angle of θ (Fig. 2.2) [14]. The electron binding energy of each of the emitted electrons is given by

$$E_{\text{binding}} = E_{\text{X-ray}} - E_{\text{kinetic}} \tag{2.1}$$

where E_{binding} is the binding energy of the electron, $E_{\text{X-ray}}$ is the energy of X-ray being used, and E_{kinetic} is the kinetic energy of the electron measured. Different elements or the same element with different chemical environments will have different characteristic binding energy. The XPS signal originates from the top 1 to 10 nm of the sample depending on the emission angle θ.

Figure 2.3a is the XPS survey spectrum for the IL $[C_2C_1IM][EtSO_4]$ [15]. The presence of C, N, O, and S atoms confirms the formula and absence of other impurity at the air–IL interface. In contrast, silicone impurities were detected for

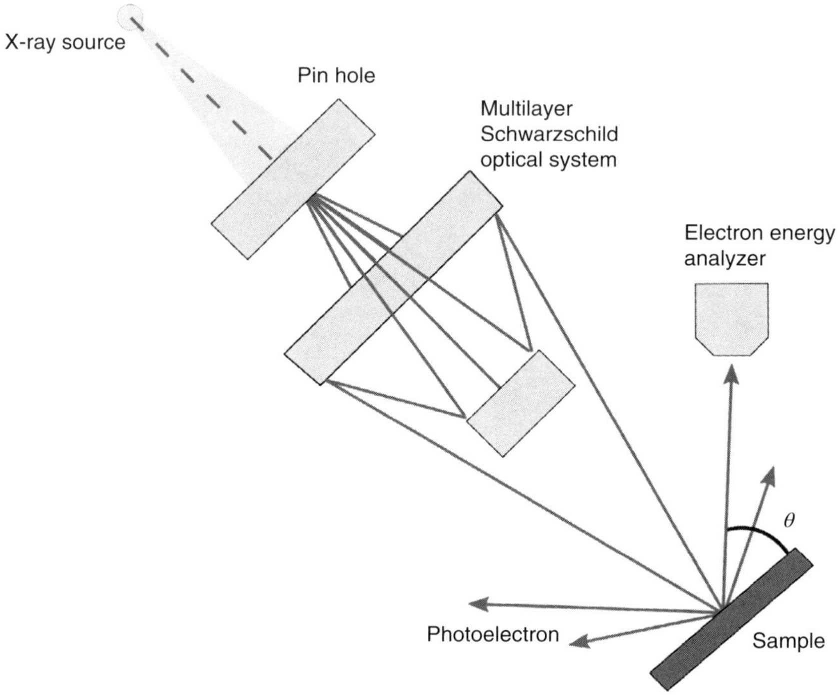

Figure 2.2 Schematic of X-ray photoelectron spectroscopy.

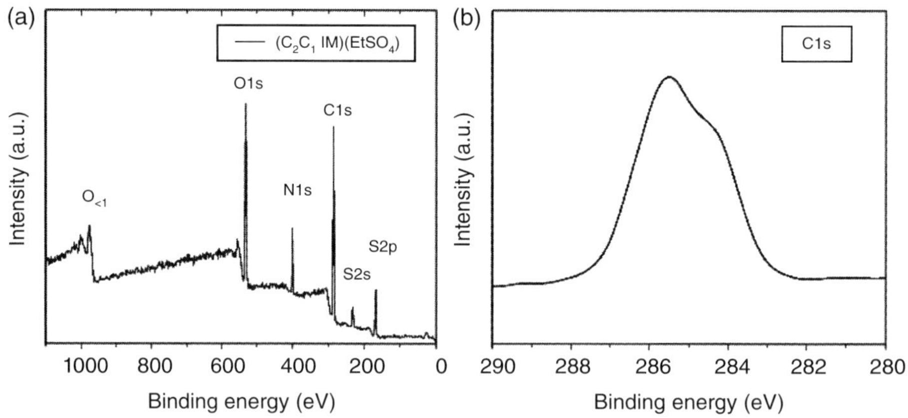

Figure 2.3 (a) XPS spectra of $[C_2C_1IM][EtSO_4]$, recorded at $\theta = 45°$ emission angle. (b) XPS C1s regional spectra of (C_8TAB) on $[C_2C_1IM][EtSO_4]$. Source: Chen et al. [15] and reproduced with permission from American Chemical Society.

contaminated commercial $[C_2C_1IM][EtSO_4]$ [16]. A series of XPS studies on the influence of anions and substituents for neat ILs have been reported by several groups [17–19].

As expected, most XPS studies have focused exclusively on neat ILs. However, the interfacial self-assembly of surfactants in IL has also been investigated by

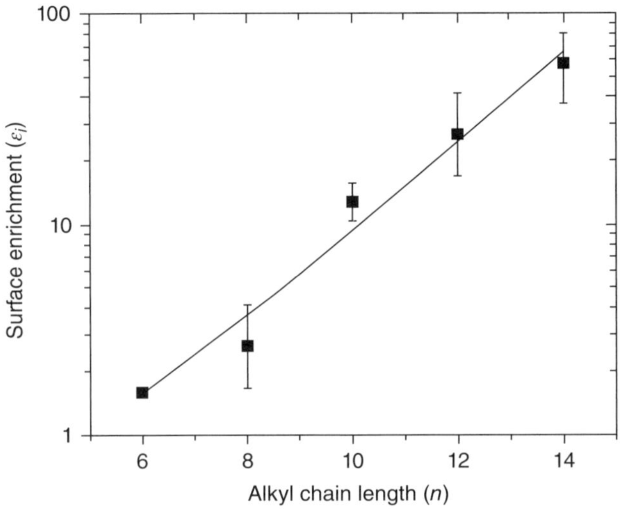

Figure 2.4 Surface enrichment of C_nTAB on $[C_2C_1IM][EtSO_4]$. Surface enrichment is defined as the ratio $\varepsilon_i = x_i^{surf}/x_i^{bulk}$, where x_i^{surf} are taken over the XPS information depth $d \approx 3.2$ nm. Source: Visser [20] and reproduced with permission from American Chemical Society.

using XPS. For example, our group examined the influence of positively charged surfactants alkyltrimethylammonium bromide (C_nTAB) on the $[C_2C_1IM][EtSO_4]$ interface by XPS [20]. Figure 2.3b shows the C1s regional XPS scan of C_8TAB on $[C_2C_1IM][EtSO_4]$. Due to the numerous species present, decoupling of the peaks into contributions from particular elemental types is somewhat complicated. Therefore, we analyzed the compositional data by means of atomic mass balance, instead of attempting to fit multiple Gaussian functions to the spectra. The mass balance approach relies on the chemical formulas of each species to determine the molecular compositions at the interface [15]. Furthermore, the role of surfactant alkyl chain length, concentration, and information depth on interfacial properties was determined. For example, in Figure 2.4, we defined the "surface enrichment" as the ratio of surface fraction to bulk fraction, $\varepsilon_i = x_i^{surf}/x_i^{bulk}$ [20]. This ratio provides a measure of the relative tendency of a species to segregate to the interface (i.e., for an ideal mixture $\varepsilon_i = 1$). We found that all surfactants exhibit surface enrichment at $\varepsilon_i \gg 1$, confirming their surface activity at the IL interface. Therefore, XPS can be used as a direct measure of the surface activity in IL systems, which may prove advantageous in situations where other techniques are either impossible or inconvenient (e.g., high temperature, low pressure). The increase of surface enrichment with chain length is attributed to solvophobicity, analogous to the hydrophobic effect in water.

2.2.1.2 NR In NR, a beam of neutrons is directed at the interface at an incident angle. The reflected beam intensity, R, is measured as a function of the scattering vector $\boldsymbol{Q}$:

$$\boldsymbol{Q} = \frac{4\pi \sin\theta}{\lambda} \tag{2.2}$$

where λ is the wavelength [21]. NR is attractive primarily due to its nondestructive nature, the contrast provided by deuterium labeling, and its subnanometer resolution.

The adsorption of nonionic polyoxyethylene *n*-alkyl ether surfactants C_nE_m ($n=12$–16; $m=4$–8) at the ethylammonium nitrate (EAN) surface has been determined using NR by Wakeham et al. [21, 22]. NR profiles for $C_{12}E_4$, $C_{14}E_4$, and $C_{16}E_4$ at the D-EAN/air and H-EAN/air interface are shown in Figure 2.5. Because the bulk concentrations are kept higher than the respective critical micelle concentration (CMC) values, the surfactant films are saturated and in equilibrium with micelles in the subphase. To interpret the profiles, the data were fit using a two-layer model where the surfactant alkyl chains are oriented away from the bulk liquid and the ethoxy groups toward the bulk liquid. From the results, morphology of the adsorbed surfactant monolayer and surface properties such as surface layer thickness, surface roughness, and the molecular areas at the EAN–air interface are all determined with precision. The thickness of the tail layer increases by approximately 1.2 Å for each CH_2 unit, and the presence of gauche defects indicates a high degree of conformational disorder. Surface roughness increases the effective thickness of the interfacial layers, and the roughness values were between 4 and 6 Å. Similar to aqueous systems, the molecular areas at the EAN–air interface are decreasing with increasing alkyl chain length due to solvophobicity, resulting in higher saturated adsorption densities and lower molecular area. In addition, the effect of surfactant concentration, temperature, and solvent mixture composition on surfactant adsorption have also been studied.

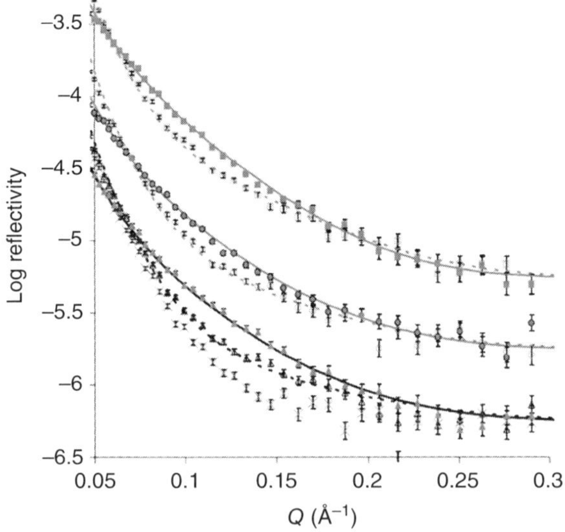

Figure 2.5 Neutron reflectivity profiles for $C_{12}E_4$ in (▲) H-EAN and (△) D-EAN, $C_{14}E_4$ in (●) H-EAN and (○) D-EAN, and $C_{16}E_4$ in (■) H-EAN and (□) D-EAN at room temperature. The $C_{14}E_4$ and $C_{16}E_4$ data have been offset vertically by 0.5 and 1.0 for clarity. Source: Wakeham et al. [21] and reproduced with permission from American Chemical Society.

2.2.1.3 Sum Frequency Generation SFG spectroscopy is a nonlinear optical technique used to investigate vibrational resonances of molecules at an interface. Two laser beams, one pulsed tunable infrared beam (ω_{IR}) and one fixed visible beam (ω_{Vis}), are overlapped on the surface of interest, producing an output beam at the sum frequency ($\omega_{SFG} = \omega_{IR} + \omega_{Vis}$). This technique can probe the interface for a few molecular layers and provide information such as chemical structure and orientation. By using the vibrational frequencies and intensities, the type and amount of the absorbed species can be determined. By analyzing different polarization combinations of the incident beams, SFG can also determine the orientation of molecules at the interface. SFG has been used in many surface studies including air–IL interface [23–25], liquid–IL interface [26], and even solid–IL interface [27, 28]. Besides neat ILs, surfactant adsorption at air–IL interface has been investigated using SFG [21]. Figure 2.6 shows the SFG spectra for the air/EAN and air/d$C_{14}E_4$-EAN interfaces, in the CH region. The use of tail-deuterated d$C_{14}E_4$ surfactant can highlight EA^+ alkyl chain contribution to the spectrum. The air/EAN interface shows that the EAN methyl peak intensities are decreased by almost an order of magnitude in the presence of added $C_{14}E_4$. This signal decrease may be due to several factors including redistribution of species at the interface, an orientational change of interfacial species, or a combination thereof. Because the ratio of two methyl peak intensity (Fermi resonance methyl symmetric peak (2945 cm^{-1}) to the methyl symmetric peak (2890 cm^{-1})) is a constant before and after surfactant adsorption, it suggests that the orientation of the interfacial EAN is not significantly affected by surfactant adsorption. Therefore, the decrease in signal strength is primarily due to interfacial surfactant displacing the EAN. This conclusion is consistent with the ion exchange between charged surfactant and IL at air–IL interface determined by XPS [15, 20].

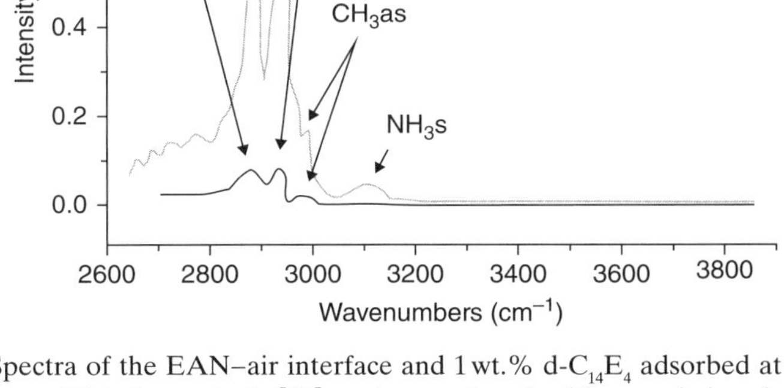

Figure 2.6 Spectra of the EAN–air interface and 1 wt.% d-$C_{14}E_4$ adsorbed at the EAN–air interface. Source: Wakeham et al. [21] and reproduced with permission from American Chemical Society.

2.2.1.4 AFM AFM is a type of scanning probe microscopy, with demonstrated resolution on the order of fractions of a nanometer. AFM carries the advantage of probing nanoscopic areas with characterizable electrode roughness and crystallographic orientation. The information is gathered by "feeling" the surface with a mechanical probe (Fig. 2.7) [28]. AFM force measurements may be divided into normal force and lateral force measurements depending on the movement direction of the cantilever probe. Atkin and coworkers [29–32] used AFM to study neat ILs at mica surfaces. Depending on the IL, they found several solvation layers between the AFM tip and the mica. The self-assembled structures at the EAN–silica interface when three Pluronic surfactants (P65, L81, L121) were introduced into ILs were investigated in 2009 [33]. An AFM image of P65–EAN–silica at room temperature is shown in Figure 2.8. A lateral surface structure or surface aggregation is revealed, and the layer has an amorphous character: globular aggregates of roughly 10–20 nm. This was the first report of adsorption of aggregates at a charged surface in an IL. In addition, EAN was also demonstrated to support nonionic surfactant aggregate formation at the graphite/IL interface [34]. These observations suggest that ILs may find usefulness as media for the suspension of hydrophobic particles.

2.2.1.5 Tensiometry The balance of forces at the free boundary of liquids reveals a net inward force toward the bulk, known as surface tension γ. Thus, surface tension is a property intimately related to both the bulk and the interface. To modify the surface tension of a given liquid, surfactants are often used as additives, primarily because the bulk properties of the liquid remain relatively unchanged. Other approaches to changing surface tension (e.g., temperature, solvent mixing) have the undesired effect that they alter both interfacial and bulk properties simultaneously.

With respect to neat ILs, their surface tensions span the range of organic solvents and in some cases approach the value of water [35, 36]. Such a wide range is

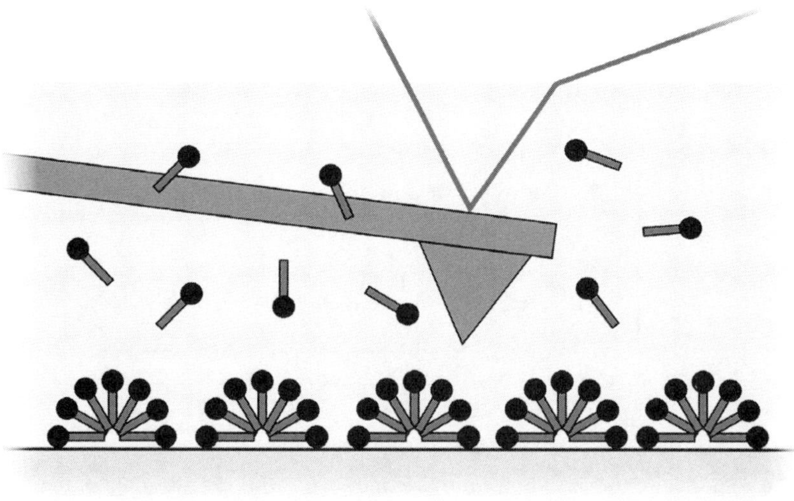

Figure 2.7 Schematic of AFM probing liquid–solid interface.

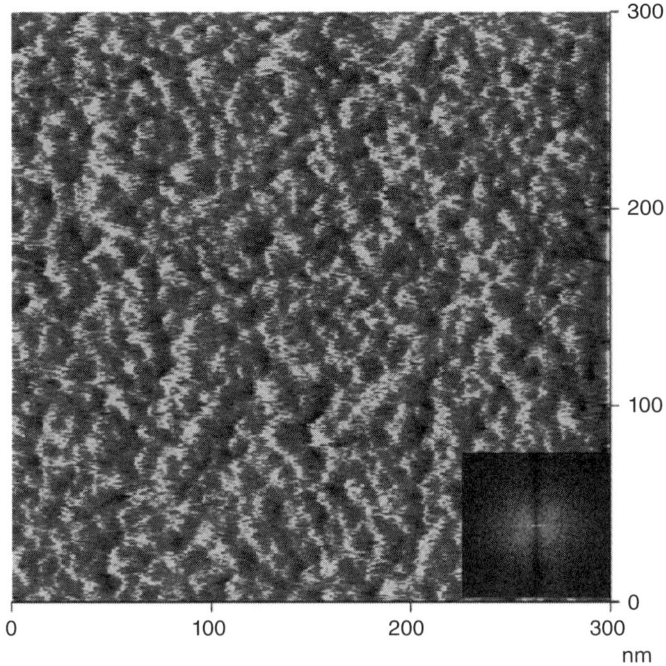

Figure 2.8 AFM image of 5 wt.% P65–EAN–silica at 21°C. The inset shows the 2-D Fourier transform. Source: Atkin et al. [33] and reproduced with permission from American Chemical Society.

expected, given the organic character of IL ions and potential for hydrogen-bonding interactions.

Several methods are available to measure surface tension, differing in sensitivity and ease of use. The most basic (and crude) is the capillary rise method, where a capillary of known radius, *r*, is partially immersed into the liquid. The height, *h*, of the liquid column inside the capillary is related to the surface tension γ by: $(2\gamma(\cos\theta))/r = gh(\Delta\rho)$, where θ is the contact angle, *g* is the gravitational constant, and $\Delta\rho$ is the density difference between the inner and the surrounding fluid. In the case of ideal wetting and with air as the surrounding fluid, the above equation can be approximated as $\gamma = \rho ghr/2$. Other approaches include the bubble pressure, pendant drop, and Wilhelmy plate methods. The Wilhelmy plate method is perhaps the most sensitive and relies on the downward force applied to a probe by surface tension and the force due to buoyancy.

Monitoring the surface tension as a function of surfactant concentration at constant temperature yields a so-called "isotherm" and can be used to calculate interfacial properties as well as the onset of aggregation. The CMC is identified by the sharp transition between a window gradual decrease and a plateau in the surface tension. During the decrease of surface tension, the Gibbs equation and thermodynamic analysis yield characteristics such as maximum surface excess concentration Γ, molecular area *A*, and free energies of micellization and adsorption. Zheng and coworkers used a novel fluorinated surfactant FC-4 ($C_{20}H_{20}F_{23}N_2O_4I$) to study

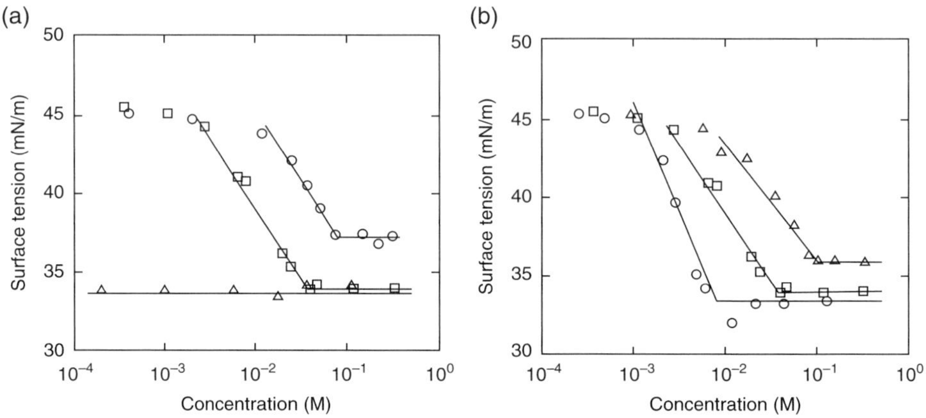

Figure 2.9 (a) Surface tension of $C_{14}E_8$ in $[C_4C_1IM][BF_4]$ (squares), $[C_4C_1IM][PF_6]$ (circles), and $[C_4C_1IM][NTf_2]$ (triangles) at 25°C. (b) Surface tension of solutions of $C_{12}E_8$ (triangles), $C_{14}E_8$ (squares), and $C_{16}E_8$ (circles) in $[C_4C_1IM][BF_4]$ at 25°C. Source: Patrascu et al. [7] and reproduced with permission from Wiley.

interfacial and aggregation properties in $[C_4C_1IM][BF_4]$ and $[C_4C_1IM][PF_6]$ [6] and found the expected result that this surfactant was more efficient than traditional alkyl chain surfactants in reducing surface tension. However, this fluorinated surfactant was shown to form micelles in $[C_4C_1IM][NTf_2]$ [37], something that no traditional surfactant can achieve.

Other groups have also examined the role of surfactant properties in ILs. For example, Patrascu et al. studied various C_nE_m surfactants in $[C_4C_1IM][BF_4]$, $[C_4C_1IM][PF_6]$, and $[C_4C_1IM][NTf_2]$ ILs [7]. The C_nE_m surfactants reduced surface tension in a concentration-dependent manner, displaying a CMC for $[C_4C_1IM][BF_4]$ and $[C_4C_1IM][PF_6]$, whereas no change in surface tension was observed for the most hydrophobic IL $[C_4C_1IM][NTf_2]$ (Fig. 2.9a). With increasing surfactant alkyl chain length, the CMC in $[C_4C_1IM][BF_4]$ was reduced (Fig. 2.9b), consistent with the greater surface activity of longer-chain surfactants.

Our group [38] measured isotherms for the cationic C_nTAB surfactant in various ILs toward understanding the role of the IL. The ILs studied included those with imidazolium, pyrrolidinium, and ammonium cations with varied anions. This study identified CMC values by solubility measurements in addition to isotherms, revealing the key importance of temperature on micellization in ILs. Further, we also confirmed the logarithmic dependence of the CMC with surfactant alkyl chain length (Fig. 2.10). The relative CMC values for the same C_nTAB in different ILs was adequately described by a mean-field model, lending support for a dominant role of interfacial energy between surfactant and IL and a minimal role of electrostatic effects [38].

There is a great opportunity to use other interfacial techniques to study surfactant-IL systems, for example, ToF-SIMS, DRS, and X-ray reflectivity (XRR). All of these techniques have already been used to study neat IL interfaces [39–41] and should be easily extended to surfactant-IL systems.

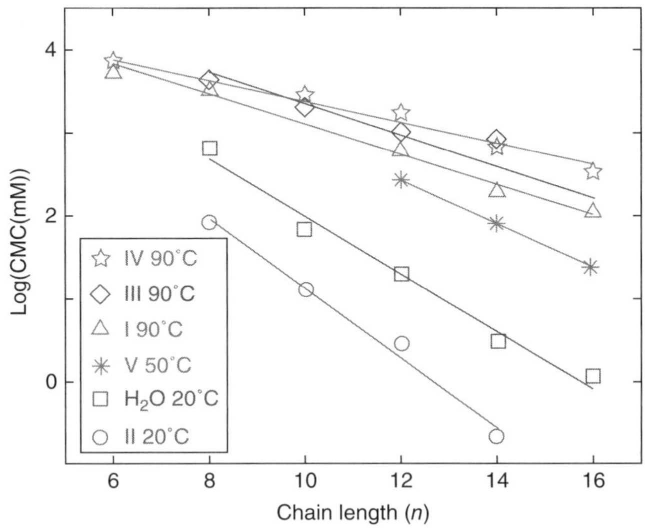

Figure 2.10 Dependence of CMC on C_nTAB alkyl chain length in different ILs: [C_2C_1IM][$EtSO_4$], I; [BHEDMA][$MeSO_3$], II; [C_4C_1IM][BF_4], III; [C_4C_1Pyr][DCA], IV; EAN, V. Source: Chen and Bermudez [38] and reproduced with permission from American Chemical Society. *See insert for color representation of the figure.*

2.2.2 Bulk Techniques

2.2.2.1 Fluorescence Spectroscopy The fluorescence emission of light from dyes under suitable excitation has wide usage in biochemistry and analytical chemistry to discriminate and/or detect specific molecules. Depending on the context, spectrometers are used to record either a range of wavelengths or a specific wavelength of interest.

In a few cases, ILs have been synthesized that possess inherent fluorescence properties [42, 43]. However, a more common approach is the use of a trace dye as a reporter for the aggregation state of amphiphiles in ILs. This approach relies on dyes that are sparingly soluble in the IL but which preferentially partition into the interior space of aggregates. Due to the close proximity of the dyes within aggregates, the resulting changes in spectral properties can be used to monitor the onset of aggregation.

One of the earliest studies reported the use of three dyes: pyrene, Reichardt's betaine, and 1,3-bis-(1-pyrenyl)propane (BPP), in order to investigate several surfactants in [C_2C_1IM][NTf_2] [5, 44]. Although discussed later, a typical pyrene emission spectrum is shown in Figure 2.11a. With increasing concentration in the [C_2C_1IM][NTf_2], nonionic surfactants (C_nE_m, Triton X-100, and Tween-20) all induced a decrease in the pyrene signal I_1/I_3 with a corresponding increase in the BPP signal I_M/I_E. From the relatively abrupt change in either the pyrene or BPP signal with concentration (Fig. 2.11b), the CMC can be estimated. For example, Brij-35 ($C_{12}E_{23}$) has a CMC in [C_2C_1IM][NTf_2] of about 0.05 M. The longer Brij-700 ($C_{18}E_{100}$) has a CMC in [C_2C_1IM][NTf_2] of about 0.01 M, which is presumably due to its alkyl portion outweighing the effect of its ethoxy portion. Interestingly, the response of Reichardt's dye was

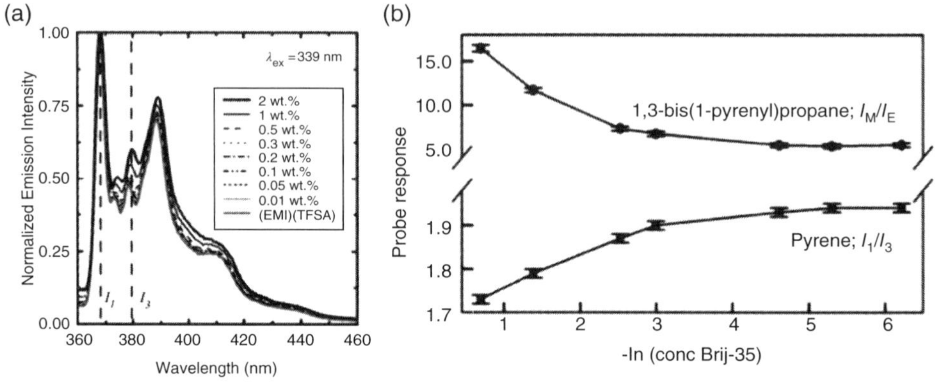

Figure 2.11 (a) Normalized emission spectra at λ_{ex} = 339 nm of pyrene in PS-*b*-PEO block copolymer concentration series in $[C_2C_1IM][NTf_2]$. Pyrene response is characterized from the ratio of I_1/I_3, indicated with dashed line. Source: Mok et al. [44] and reproduced with permission from American Chemical Society. (b) Response of pyrene (I_1/I_3) and BPP (I_M/I_E) solubilized in $[C_2C_1IM][NTf_2]$ with increasing $[C_{12}E_{23}]$. Source: Fletcher and Pandey [5] and reproduced with permission from American Chemical Society.

unaffected by all of the surfactants studied, indicating that this dye most likely lacks the needed partitioning behavior between IL and surfactant aggregates.

A more recent study used pyrene to monitor the aggregation of amphiphilic block copolymers in $[EMIM][NTf_2]$ [44]. Specifically, polystyrene-*b*-poly(ethylene oxide) (PS-*b*-PEO) and polystyrene-*b*-poly(methyl methacrylate) (PS-*b*-PMMA) copolymers were used to investigate the effect of molecular weight on the CMC. As expected, increasing either the relative or absolute length of the hydrophobic PS block caused a decrease in the CMC for both the copolymers. The decrease in CMC with polymer length is rather modest (nearly all the CMCs are roughly 0.1 wt.%), but this effect has been previously observed for long amphiphilic polymers in water.

2.2.2.2 POM Optical microscopy combined with a pair of cross-polarizers is a relatively simple method for the identification of sample birefringence. With regard to ILs, they are typically optically homogeneous and do not display any birefringence on their own. However, the addition of amphiphiles often leads to the formation of mesophases, in particular at higher concentrations. These mesophases are responsible for anisotropy within the IL, which is easily seen as a characteristic birefringent texture (Fig. 2.12a). However, because such textures are qualitative, POM is often followed by scattering techniques to confirm the type of mesophase (e.g., cylindrical, lamellar) and its lattice spacing.

POM was first used to identify mesophases of nonionic surfactants C_nE_m in EAN [46]. This study revealed the similarity of the mesophases in EAN with those in water (Fig. 2.12b), even following the same trends with alkyl and ethoxy chain lengths. The key difference from water is that longer alkyl chains are needed to form mesophases in EAN, an effect attributed to solvophobicity. The potential hydrogen-bonding between C_nE_m and EAN is not necessary for the formation of mesophases, as demonstrated by C_{16}TAB mesophases in EAN [47]. Indeed, the positively charged C_{16}TAB surfactant is able to form mesophases in a wide variety of (protic) ammonium ILs.

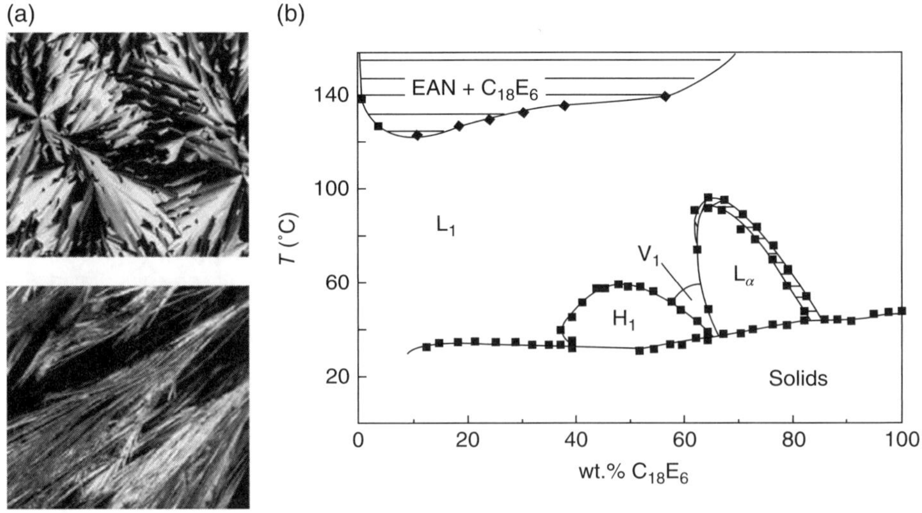

Figure 2.12 (a) Change in birefringence during phase transitions for a 52 wt.% C_{16}TAB/EAN solution. Top: at 55°C well-defined conical fanlike birefringence texture were observed; bottom: at 40°C, birefringence textures disappeared and replaced by needlelike solid crystals. Source: Lopez-Barron and Wagner [45] and reproduced with permission from American Chemical Society. (b) Phase diagram obtained for POM of the $C_{18}E_6$/EAN system, showing a large single-phase isotropic (L_1) region, together with hexagonal, bicontinuous cubic, and lamellar phases. Dashed lines indicate approximate phase boundaries determined for the smaller lyotropic phases, and horizontal hatching denote tie lines for coexistence of two phases. Source: Araos and Warr [46] and reproduced with permission from American Chemical Society.

More recently, thermal transitions between mesophases of C_{12}TAB in EAN were studied by POM and other techniques [45].

POM has also been used to study the phase behavior of the amphiphilic triblock copolymer Pluronic P123 in [C_4C_1IM][BF_6] [48] and in EAN [49]. Unlike the similarity observed between C_nE_m in EAN and in water [46], different mesophases were obtained for P123 in ILs as compared to water. For example, in contrast to P123 in water, P123 in [C_4C_1IM][PF_6] lacks a cubic micellar I_1 mesophase. P123 in EAN also differs from water in that it displays a reverse bicontinuous cubic V_2 mesophase. Even when all the systems share the same mesophase, there are differences: for the hexagonal H_1 and lamellar $L\alpha$ mesophases, [C_4C_1IM][PF_6] and EAN lattice spacings were smaller than those in water, with [C_4C_1IM][PF_6] being the smallest. Clearly, further work is needed to resolve the role of molecular weight in the assembly of nonionic amphiphiles in ILs.

2.2.2.3 Scattering All scattering follows the same basic premise, which is illustrated in Figure 2.13. An incident beam of light for light scattering (LS), X-rays for SAXS, or neutrons for SANS is passed through a sample, and the scattered beam is measured at different scattering angles (θ). Each of the three previously mentioned scattering techniques has different wavelengths (λ), which correspond to their emission sources. Scattering is a robust approach for determining d-spacing (SAXS and SANS), diffusion rate (DLS), and size.

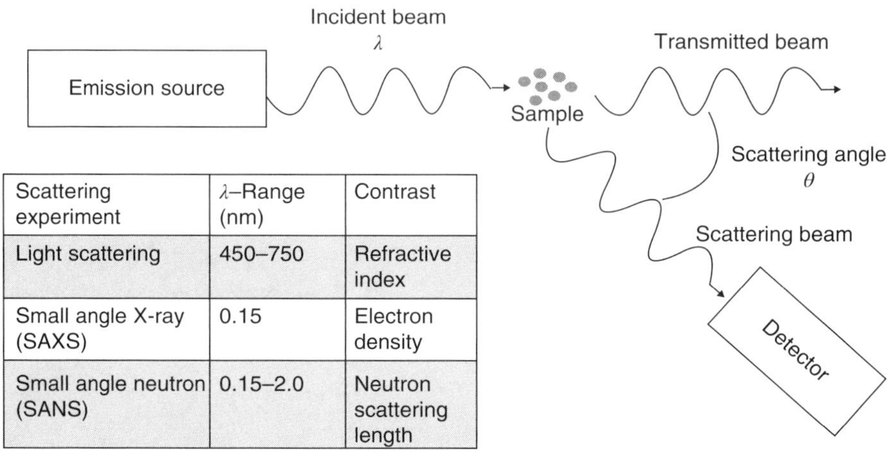

Scattering experiment	λ–Range (nm)	Contrast
Light scattering	450–750	Refractive index
Small angle X-ray (SAXS)	0.15	Electron density
Small angle neutron (SANS)	0.15–2.0	Neutron scattering length

Figure 2.13 General schematic of scattering. Differences in wavelength and contrast between the different scattering experiments are shown in the table inset.

DLS In DLS, as molecules move in and out of the scattering volume, the intensity fluctuations are measured in real time. These fluctuations can be related to a relaxation time, τ, and subsequently the diffusion coefficient, D. Once D is known, the hydrodynamic radius, R_h, may be determined from the Stokes–Einstein equation:

$$R_h = \frac{kT}{6\pi\eta D} \tag{2.3}$$

where k is the Boltzmann constant, T is the temperature, and η is the viscosity. DLS is especially useful for surfactant systems since both the size of the micelle and the CMC can be determined. Furthermore, only a single scattering vector, $\boldsymbol{Q}$ (see Eq. 2.2), is needed to determine the hydrodynamic radius (R_h).

The Inoue group has performed a great deal of size and CMC characterization of nonionic surfactants in ILs using DLS and NMR [50–53]. A plot of average hydrodynamic diameter, $\langle D \rangle$, versus surfactant wt% (Fig. 2.14) shows a sharp increase in size at the onset of CMC [53]. This increase corresponds to the initial formation of micelles (CMC) of the surfactant $C_{12}E_5$ in $[C_4C_1IM][PF_6]$. The plateau at higher concentrations not only reveals the size of fully formed micelles but also is useful in showing the weak size dependence on concentration.

DLS is also useful for determining phase behavior and micelle formation in block copolymers dissolved in ILs. The work by Lee et al. with poly(ethylene oxide)-*b*-poly(*N*-isopropylacrylamide) (PEO-*b*-PNIPAM) in $[C_2C_1IM][BF_4]$ and $[C_4C_1IM][BF_4]$ shows how the lower critical solution temperature (LCST) behavior of PEO and the upper critical solution temperature (UCST) behavior of PNIPAM can be exploited to form PNIPAM-core micelles at low temperature and PEO-core micelles at high temperature. As can be seen in Figure 2.15, this system is thermally reversible allowing for potential applications as thermosensitive materials [8].

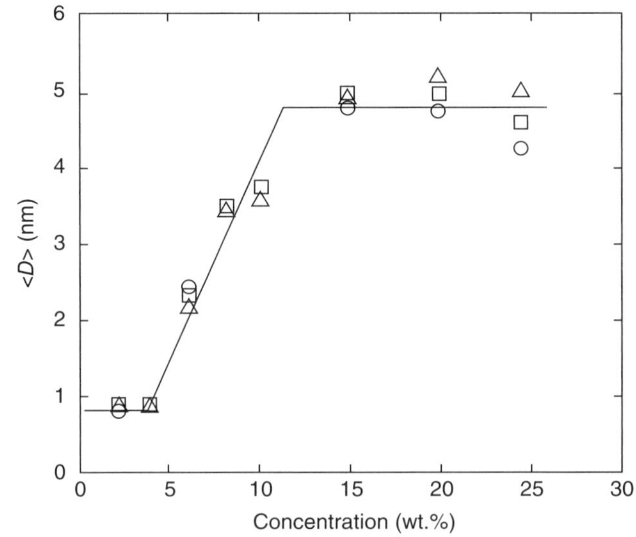

Figure 2.14 Mean hydrodynamic diameter <D> of $C_{12}E_5$ in $[C_4C_1IM][PF_6]$ at 25°C (circles), 30°C (squares), and 35°C (triangles). Source: Misono et al. [53] and reproduced with permission from Elsevier.

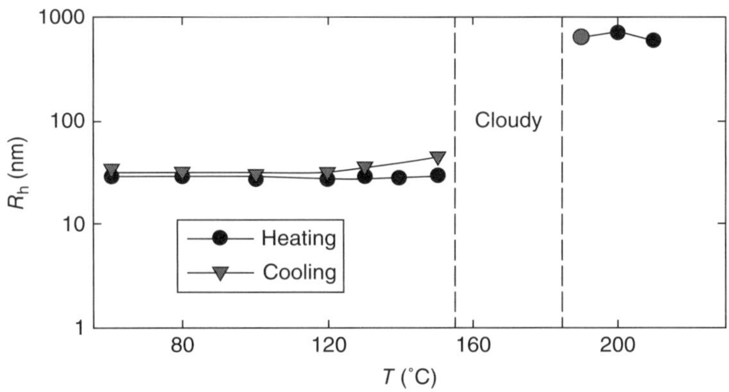

Figure 2.15 Hydrodynamic radius of 1 wt.% PEO-PNIPAM in a 50/50 (w/w) mixture of $[C_2C_1IM][BF_4]/[C_4C_1IM][BF_4]$ after heating and cooling. Source: Lee et al. [8] and reproduced with permission from American Chemical Society.

SAXS SAXS is a scattering technique that measures elastically scattered X-rays. Contrast arises from differing electron densities in a sample, whereby the heavier the element, the larger the scattering intensity. Two-dimensional scattering data can be converted to one-dimensional intensity (I) versus scattering vector ($\boldsymbol{Q}$) plots. A plot of ln(I) versus $\boldsymbol{Q}$ (like one shown in Fig. 2.16) can be used to determine the d-spacing and lattice orientation of a material. The d-spacing is related to $\boldsymbol{Q}$ through Bragg's law, $\lambda = 2d\sin\theta$. Combining Bragg's law with Equation 2.2, $Q = (4\pi\sin\theta)/\lambda = (2\pi)/d$. Subsequently, the $\boldsymbol{Q}$ values for multiple peaks can be used to find the lattice orientation.

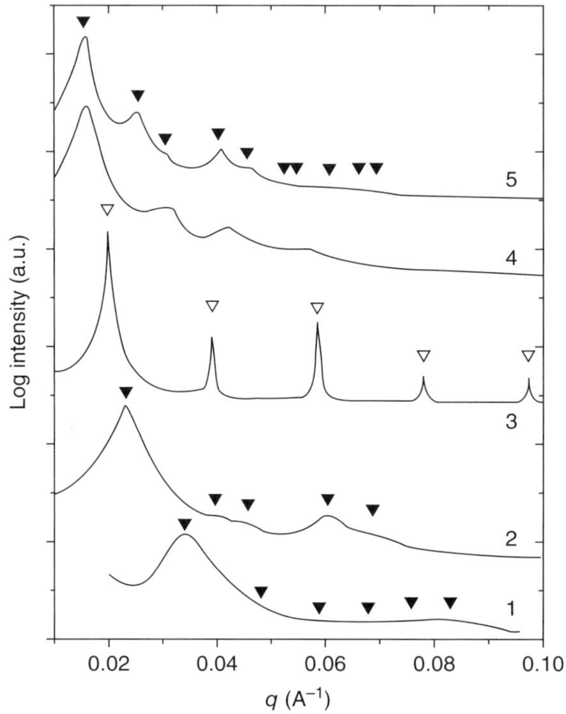

Figure 2.16 1D scattering plots for PB-*b*-PEO blocks in $[C_2C_1IM][NTf_2]$ where 1, (9:3) bulk; 2, (9:3) 81 wt.%; 3, (9:3) 60 wt.%; 4, (9:6) 40 wt.%; and 5, (9:6) 30 wt.%. All ratios are monomers of PB:PEO. Source: Simone and Lodge [9] and reproduced with permission from American Chemical Society.

The initial ***Q*** peak is known as the ***Q****** peak, and each subsequent peak can be taken as a ratio from the ***Q*/*Q****. These ratios obtained correspond to well-defined morphologies. For example, an integer ratio of 1:2:3:4 is representative of a lamellar orientation.

The lyotropic phase behavior of polybutadiene-*b*-poly(ethylene oxide) (PB-*b*-PEO) copolymers in $[C_2C_1IM][NTf_2]$ and $[C_4C_1IM][PF_6]$ has been studied using SAXS by Simone and Lodge [9]. These systems show multiphase behavior dependence on the polymer diblock ratios and concentration. As can be seen by Figure 2.16, there are distinct morphologies depending on the ratio of blocks of PB to PEO. While the morphologies vary slightly between the two ILs, $[C_2C_1IM][NTf_2]$ and $[C_4C_1IM][PF_6]$, the multiphase behavior is similar to that seen in PB-*b*-PEO block copolymers in water.

SANS SANS is a scattering technique that measures the deflection of elastically scattered neutrons. Similar to SAXS, SANS primarily reveals d-spacing and lattice structure. However, where SAXS is dependent on electromagnetic interactions for contrast, SANS derives its contrast from nuclear interactions. Nuclear interactions offer two unique advantages over other scattering techniques. Firstly, the neutron is noninvasive meaning it does not influence the sample during scanning. Secondly,

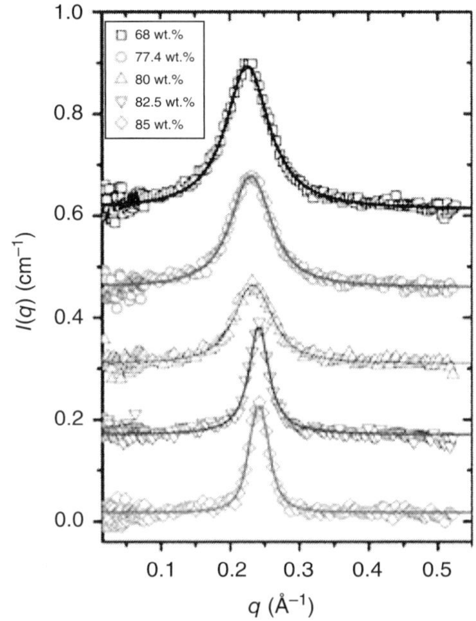

Figure 2.17 SANS profile at 25°C for DDAB/D-EAN solutions at various concentrations. Narrowing and shifting in peak from 80 to 82.5% correspond to the shift from L_3 to $L\alpha$ region. Source: Lopez-Barron et al. [55] and reproduced with permission from American Chemical Society. *See insert for color representation of the figure.*

due to scattering occurring from nuclear interactions, there is no molecular density (Z) dependence upon scattering/absorption, making SANS a bulk technique that is compatible with a very wide range of materials. However, it does suffer from a few drawbacks, the primary one being the large cost of operation and requirement of a nuclear energy source to generate neutrons [54].

Lopez-Barron et al. [55] studied a mixture of didodecyldimethylammonium bromide (DDAB) in a protic IL EAN, which shows an L_3–$L\alpha$ transition as determined by SANS. From the SANS data (Fig. 2.17), there is a slight peak shift and narrowing between 80 and 82.5 wt.%. This shift corresponds to the transitions from the spongelike L_3 phase to the more ordered lamellar ($L\alpha$) phase at 25°C. In the case of the DDAB/EAN mixture, there is a stable spongelike phase over a much wider range of compositions compared to that of DDAB/water presumably due to the lower solvophobic effect of the IL. The DDAB/EAN mixture is therefore a more robust system for applications of the sponge phase due to the relatively narrow composition region that exists in the aqueous system.

2.2.2.4 NMR

^{1}H-NMR By measuring the chemical shifts of different concentrations of surfactant in IL, the CMC can be determined. For example, Figure 2.18 shows the chemical shift of $C_{14}E_6$ in $[C_4C_1IM][BF_4]$ [52]. The peak used for analysis corresponds to the chemical shift of the hydrogen attached to the alkyl carbon adjacent to the ether

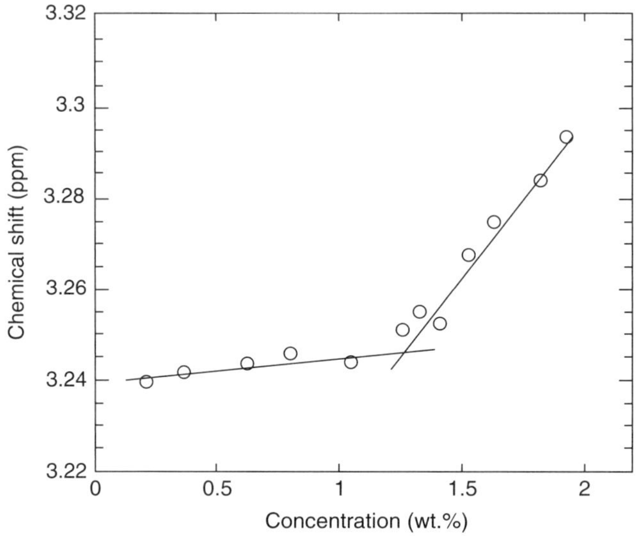

Figure 2.18 Chemical shift (δ) of surfactant proton versus concentration of $C_{14}E_6$ in $[C_4C_1IM][BF_4]$. Source: Inoue and Yamakawa [52] and reproduced with permission from Elsevier.

oxygen. However, a hydrogen on the IL cation works equally well. Usually, the best resolved proton peak is used regardless of its origin [53].

Pulsed-Field Gradient Spin-Echo NMR Diffusion NMR experiments resolve different compounds in a mixture based on their diffusion coefficients, depending on physical parameters such as size and shape of the molecules, temperature, and viscosity. The diffusion NMR technique is often referred to as diffusion-ordered spectroscopy (DOSY) or pulsed-field gradient spin-echo (PGSE) NMR. A series of NMR diffusion spectra are acquired as a function of the gradient strength G (Fig. 2.19) [56], and the slope of the peak decay is used to obtain the diffusion coefficient D. Furthermore, the hydrodynamic radius can be obtained from the Stokes–Einstein equation (Eq. 2.3).

Diffusion NMR can indicate the formation of aggregates (e.g., micelles, emulsions). In the case of 1-alkyl-3-methylimidazolium ibuprofenate ($[C_nC_1][Ibu]$) in water, the diffusion coefficients of cation and anion were found to be different, yet the CMC values obtained from the diffusion measurements for the cation and anion were identical, thus confirming that they were both participating in aggregate formation (Fig. 2.20) [57]. Moreover, the CMC values were in agreement with those obtained by other techniques (surface tension, conductivity). Because of the dynamic equilibrium between monomer and micelle, the observed diffusion coefficient is the mean value of the two states. If the exchange kinetics between the two states is fast on the diffusion time scale, the measured diffusion coefficient D_{obs} can be expressed as

$$D_{obs} = \begin{cases} \dfrac{CMC}{C_t} D_{mon} + \dfrac{C_t - CMC}{C_t} D_{mic}, & C_t > CMC \\ D_{mon}, & C_t \le CMC \end{cases}$$

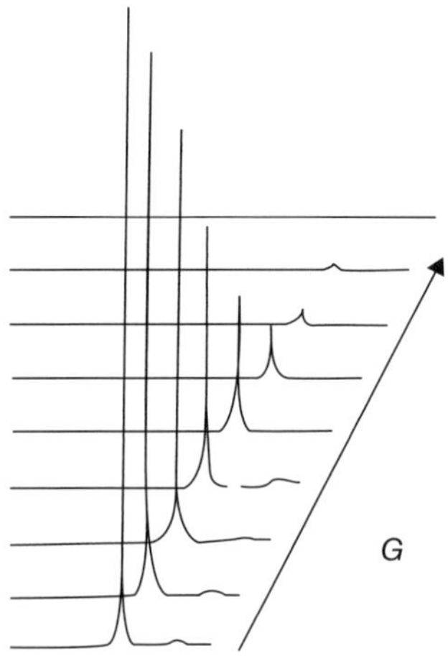

Figure 2.19 Schematic of diffusion NMR spectroscopy. Source: Cohen et al. [56] and reproduced with permission from Wiley.

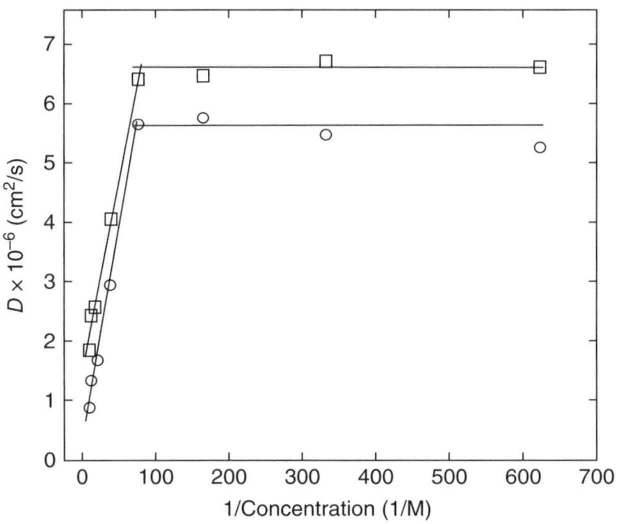

Figure 2.20 Variation of self-diffusion coefficients of $[C_8C_1IM][Ibu]$ (D_{obs}^{Ibu} circle and D_{obs}^{Im} square) as a function of the reciprocal concentration at 25°C. Source: Tourne-Peteilh et al. [57] and reproduced with permission from Royal Society of Chemistry.

where C_t is the total concentration of surfactants.

Diffusion NMR studies have also been performed on the self-aggregation of neat IL with or without the presence of salt [58, 59]. To our knowledge, surfactant aggregation in IL has not been studied yet by using PGSE-NMR although the analysis in aqueous solution can be also applied in surfactant-IL systems. Our group is presently investigating the diffusion and size of surfactant aggregates in $[C_2C_1IM]$ $[EtSO_4]$ by PGSE-NMR.

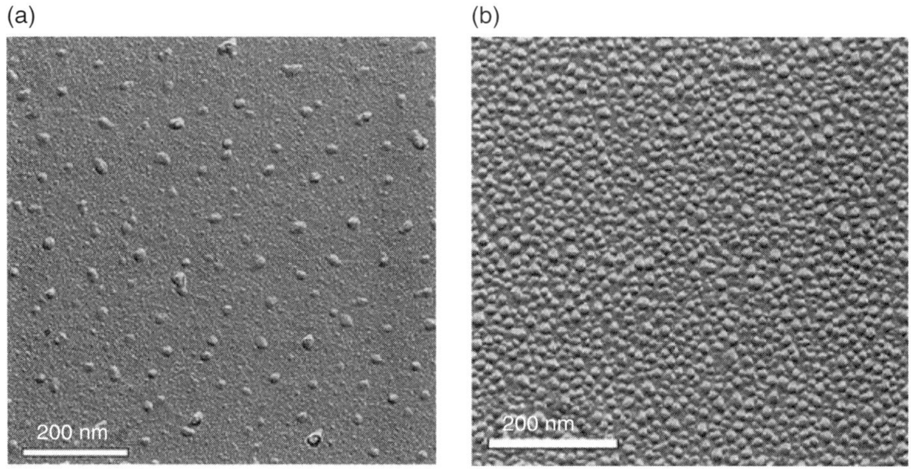

Figure 2.21 FF-TEM images of FC-4 surfactant in $[C_4C_1IM][NTf_2]$ at concentrations (a) 0.280 mol/l, (b) 0.344 mol/l. The CMC for FC-4 surfactant in neat $[C_4C_1IM][NTf_2]$ is 0.212 mol/l at 25°C. Source: Li et al. [37] and reproduced with permission from American Chemical Society.

2.2.2.5 TEM TEM allows for direct observation of micelle and vesicle in ILs. In TEM, a beam of electrons is sent through a thin sample, and the inelastically scattered electrons are measured by a detector producing an image. The primary method of contrast for TEM images of ILs comes from mass thickness contrast. Thicker sections or areas with higher molecular density (Z) appear as darker regions due to more scattering events, while areas with lower Z values appear brighter.

TEM is advantageous over traditional optical microscopy because it allows for much higher resolution than would otherwise be attainable in light microscopy. However, TEM does suffer from the disadvantage of operating in UHV. But as already mentioned, UHV conditions are not a concern for ILs due to their negligible vapor pressure.

Two main techniques have been used in imaging micelles in ILs, freeze-fracture TEM (FF-TEM) and cryo-TEM. Both techniques were developed for biological systems but have proven robust for studying nonaqueous IL systems. In FF-TEM and cryo-TEM, the sample is quenched (approximately −200°C) to vitrify the solvent. For FF-TEM, the sample is then fractured and shadow coated with a metal alloy to improve contrast and to prevent evaporation of solvent [60]. Because fracturing the material is required, FF-TEM samples typically are thicker than cryo-TEM. Cryo-TEM does not require fracturing but needs very thin samples as well as a cryogenic stage and preparation area. Such equipment costs can make cryo-TEM prohibitively expensive. However, a major advantage of cryo-TEM is that it allows for in situ observation of micelle and vesicles that cannot otherwise be seen with FF-TEM [61].

In the work by Li et al. [37], FF-TEM was used to analyze fluorinated cationic surfactant FC-4 ($C_{20}H_{20}F_{23}N_2O_4I$) micelles in $[C_4C_1IM][NTf_2]$ (Fig. 2.21). Below the CMC, there were no aggregates seen in the TEM images. However, once above the

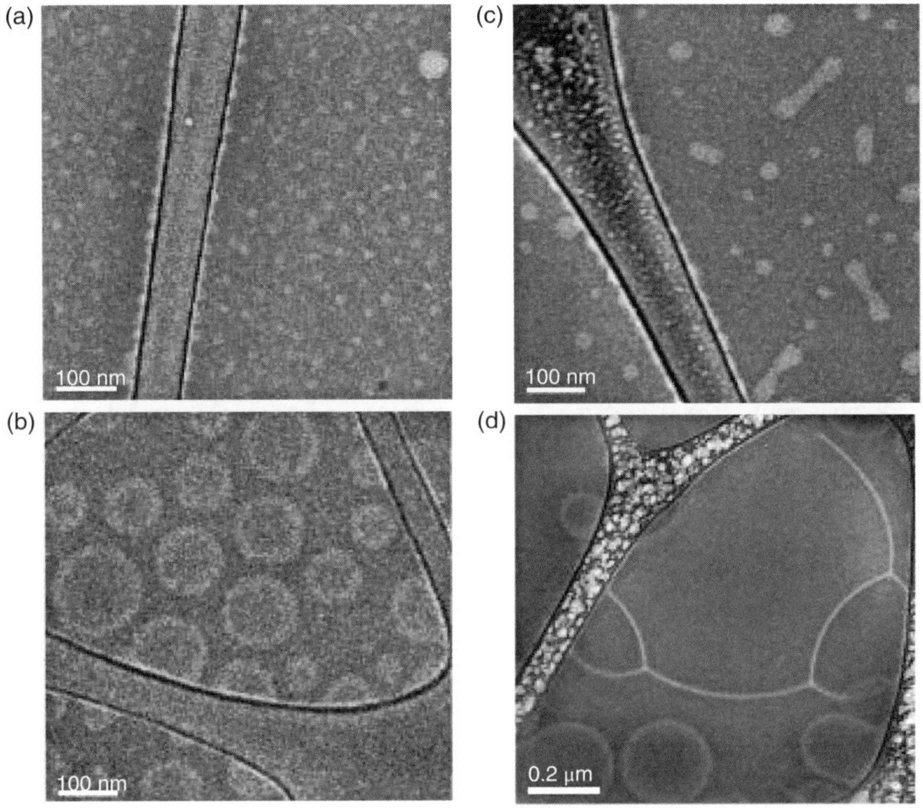

Figure 2.22 Cryo-TEM images of PB-*b*-PEO copolymers dissolved in $[C_2C_1IM][NTf_2]$ (a) and (b) in $[C_4C_1IM][PF_6]$ (c) and (d) at 1.0 wt.%. Ratios of monomers of PB and PEO are 9:6 for a and c and 9:3 for b and d. Source: Simone and Lodge [9] and reproduced with permission from American Chemical Society.

measured CMC values, micelle aggregation was observed. As the concentration increased, the number of micelles increased, but their size remained constant.

In the work by Simone and Lodge [9], the phase behavior of two block copolymers comprised of PB-*b*-PEO was studied in the ILs $[C_2C_1IM][NTf_2]$ and $[C_4C_1IM][PF_6]$. As can be seen by Figure 2.22, the PB blocks of the copolymer form the inner layer of micelles and vesicles (seen as the lighter region compared to the dark solvent regions). The concentration of block copolymer was held constant at 1 wt.%, while the ratio of PB:PEO block length was manipulated. In $[C_2C_1IM][NTf_2]$ as the PB block was increased, spherical micelles increased in size until large vesicles were formed (Fig. 2.22a and b). However, in $[C_4C_1IM][PF_6]$ at the same 1 wt.% concentration, PB blocks formed both spherical and rodlike micelles. As the PB block was increased, large vesicles were formed including vesicles that were fused together (Fig. 2.22c and d). Simone et al. argued that the primary driving force was a decrease in the interfacial tension of the core/corona [9]. Thus, increasing the interfacial tension of the solvent led to morphologies with lower interfacial curvature.

2.3 CONCLUSIONS AND OUTLOOK

The unique character of ILs allows for numerous techniques to be applied to their characterization, in ways not possible with traditional aqueous or molecular solvents. Thus, ILs provides an opportunity to more closely examine the fundamental behavior of amphiphiles at the interface and in the bulk. Current studies have already established that much of the behavior seen in water (i.e., surface activity, micelle assembly, mesophase formation) can also be obtained in ILs, although the conditions required can vary substantially. The specific role of the IL in determining amphiphile behavior is manifested at a simple level in terms of solvophobicity, analogous to hydrophobicity in water, and at a more complex level in terms of hydrogen-bonding and ion-pairing interactions. Nevertheless, intriguing areas remain to be explored with amphiphiles and ILs, such as the stabilization of nanoscale clusters within ILs, improving control over layering at IL interfaces, and the development of these colloidal systems as media for emulsion polymerization or catalysis.

ACKNOWLEDGMENT

The authors thank the National Science Foundation (DMR-0847558) for financial support.

REFERENCES

[1] Castner, E. W.; Wishart, J. F. *J. Chem. Phys.* 2010, **132**, 120901.
[2] Greaves, T. L.; Drummond, C. J. *Chem. Rev.* 2008, **108**, 206–237.
[3] Greaves, T. L.; Drummond, C. J. *Chem. Soc. Rev.* 2008, **37**, 1709–1726.
[4] Aliaga, C.; Santos, C. S.; Baldelli, S. *Phys. Chem. Chem. Phys.* 2007, **9**, 3683–3700.
[5] Fletcher, K. A.; Pandey, S. *Langmuir* 2004, **20**, 33–36.
[6] Li, N.; Zhang, S. H.; Zheng, L. Q.; Wu, J. P.; Li, X. W.; Yu, L. *J. Phys. Chem. B* 2008, **112**, 12453–12460.
[7] Patrascu, C.; Gauffre, F.; Nallet, F.; Bordes, R.; Oberdisse, J.; de Lauth-Viguerie, N.; Mingotaud, C. *ChemPhysChem* 2006, **7**, 99–101.
[8] Lee, H. N.; Bai, Z. F.; Newell, N.; Lodge, T. P. *Macromolecules* 2010, **43**, 9522–9528.
[9] Simone, P. M.; Lodge, T. P. *Macromolecules* 2008, **41**, 1753–1759.
[10] Santos, C. S.; Baldelli, S. *Chem. Soc. Rev.* 2010, **39**, 2136–2145.
[11] Wasserscheid, P.; Keim, W. *Angew. Chem. Int. Ed. Engl.* 2000, **39**, 3772–3789.
[12] Rogers, R. D.; Seddon, K. R. *Science* 2003, **302**, 792–793.
[13] Lovelock, K. R. J.; Villar-Garcia, I. J.; Maier, F.; Steinruck, H. P.; Licence, P. *Chem. Rev.* 2010, **110**, 5158–5190.
[14] Smith, E. F.; Villar Garcia, I. J.; Briggs, D.; Licence, P. *Chem. Commun.* 2005, 5633–5635.
[15] Chen, L. G.; Lerum, R. V.; Aranda-Espinoza, H.; Bermudez, H. *J. Phys. Chem. B* 2010, **114**, 11502–11508.
[16] Gottfried, J. M.; Maier, F.; Rossa, J.; Gerhard, D.; Schulz, P. S.; Wasserscheid, P.; Steinruck, H. P. *Z. Phys. Chem.* 2006, **220**, 1439–1453.

[17] Kolbeck, C.; Cremer, T.; Lovelock, K. R. J.; Paape, N.; Schulz, P. S.; Wasserscheid, P.; Maier, F.; Steinruck, H. P. *J. Phys. Chem. B* 2009, **113**, 8682–8688.

[18] Lovelock, K. R. J.; Kolbeck, C.; Cremer, T.; Paape, N.; Schulz, P. S.; Wasserscheid, P.; Maier, F.; Steinruck, H. P. *J. Phys. Chem. B* 2009, **113**, 2854–2864.

[19] Lockett, V.; Sedev, R.; Bassell, C.; Ralston, J. *Phys. Chem. Chem. Phys.* 2008, **10**, 1330–1335.

[20] Visser, A. E., ed., *Ionic Liquids: Science and Applications*, ACS Symposium Series **1117**; American Chemical Society, Washington, DC, 2012, pp. 289–302.

[21] Wakeham, D.; Niga, P.; Warr, G. G.; Rutland, M. W.; Atkin, R. *Langmuir* 2010, **26**, 8313–8318.

[22] Wakeham, D.; Warr, G. G.; Atkin, R. *Langmuir* 2012, **28**, 13224–13231.

[23] Baldelli, S. *J. Phys. Chem. B* 2003, **107**, 6148–6152.

[24] Santos, C. S.; Baldelli, S. *J. Phys. Chem. B* 2007, **111**, 4715–4723.

[25] Jeon, Y.; Sung, J.; Bu, W.; Vaknin, D.; Ouchi, Y.; Kim, D. *J. Phys. Chem. C* 2008, **112**, 19649–19654.

[26] Iwahashi, T.; Sakai, Y.; Kanai, K.; Kim, D.; Ouchi, Y. *Phys. Chem. Chem. Phys.* 2010, **12**, 12943–12946.

[27] Romero, C.; Baldelli, S. *J. Phys. Chem. B* 2006, **110**, 6213–6223.

[28] Baldelli, S. *J. Phys. Chem. Lett.* 2013, **4**, 244–252.

[29] Hayes, R.; El Abedin, S. Z.; Atkin, R. *J. Phys. Chem. B* 2009, **113**, 7049–7052.

[30] Atkin, R.; El Abedin, S. Z.; Hayes, R.; Gasparotto, L. H. S.; Borisenko, N.; Endres, F. *J. Phys. Chem. C* 2009, **113**, 13266–13272.

[31] Wakeham, D.; Hayes, R.; Warr, G. G.; Atkin, R. *J. Phys. Chem. B* 2009, **113**, 5961–5966.

[32] Segura, J. J.; Elbourne, A.; Wanless, E. J.; Warr, G. G.; Voitchovsky, K.; Atkin, R. *Phys. Chem. Chem. Phys.* 2013, **15**, 3320–3328.

[33] Atkin, R.; De Fina, L. M.; Kiederling, U.; Warr, G. G. *J. Phys. Chem. B* 2009, **113**, 12201–12213.

[34] Atkin, R.; Warr, G. G. *J. Am. Chem. Soc.* 2005, **127**, 11940–11941.

[35] Sedev, R. *Curr. Opin. Colloid Interface Sci.* 2011, **16**, 310–316.

[36] Tariq, M.; Freire, M. G.; Saramago, B.; Coutinho, J. A. P.; Lopes, J. N. C.; Rebelo, L. P. N. *Chem. Soc. Rev.* 2012, **41**, 829–868.

[37] Li, N.; Zhang, S.; Zheng, L.; Inoue, T. *Langmuir* 2009, **25**, 10473–10482.

[38] Chen, L. G.; Bermudez, H. *Langmuir* 2012, **28**, 1157–1162.

[39] Gunster, J.; Hofft, O.; Krischok, S.; Souda, R. *Surf. Sci.* 2008, **602**, 3403–3407.

[40] Law, G.; Watson, P. R.; Carmichael, A. J.; Seddon, K. R.; Seddon, B. *Phys. Chem. Chem. Phys.* 2001, **3**, 2879–2885.

[41] Sloutskin, E.; Ocko, B. M.; Taman, L.; Kuzmenko, I.; Gog, T.; Deutsch, M. *J. Am. Chem. Soc.* 2005, **127**, 7796–7804.

[42] Shi, L. J.; Li, N.; Zheng, L. Q. *J. Phys. Chem. C* 2011, **115**, 18295–18301.

[43] Marwani, H. M. *J. Fluoresc.* 2013, **23**, 251–257.

[44] Mok, M. M.; Thiagarajan, R.; Flores, M.; Morse, D. C.; Lodge, T. P. *Macromolecules* 2012, **45**, 4818–4829.

[45] Lopez-Barron, C. R.; Wagner, N. J. *Langmuir* 2012, **28**, 12722–12730.

[46] Araos, M. U.; Warr, G. G. *J. Phys. Chem. B* 2005, **109**, 14275–14277.

[47] Greaves, T. L.; Weerawardena, A.; Fong, C.; Drummond, C. J. *Langmuir* 2007, **23**, 402–404.

[48] Wang, L. Y.; Chen, X.; Chai, Y. C.; Hao, J. C.; Sui, Z. M.; Zhuang, W. C.; Sun, Z. W. *Chem. Commun.* 2004, 2840–2841.

[49] Zhang, G. D.; Chen, X.; Zhao, Y. R.; Ma, F. M.; Jing, B.; Qiu, H. Y. *J. Phys. Chem. B* 2008, **112**, 6578–6584.

[50] Inoue, T.; Misono, T. *J. Colloid Interface Sci.* 2009, **337**, 247–253.

[51] Inoue, T. *J. Colloid Interface Sci.* 2009, **337**, 240–246.

[52] Inoue, T.; Yamakawa, H. *J. Colloid Interface Sci.* 2011, **356**, 798–802.

[53] Misono, T.; Sakai, H.; Sakai, K.; Abe, M.; Inoue, T. *J. Colloid Interface Sci.* 2011, **358**, 527–533.

[54] http://www.ncnr.nist.gov/staff/hammouda/the_SANS_toolbox.pdf (accessed March 14, 2015).

[55] Lopez-Barron, C. R.; Basavaraj, M. G.; DeRita, L.; Wagner, N. J. *J. Phys. Chem. B* 2012, **116**, 813–822.

[56] Cohen, Y.; Liat, A.; Frish, L. *Angew. Chem. Int. Ed.* 2005, **44**, 520–554.

[57] Tourne-Peteilh, C.; Devoisselle, J. M.; Vioux, A.; Judeinstein, P.; In, M.; Viau, L. *Phys. Chem. Chem. Phys.* 2011, **13**, 15523–15529.

[58] Kunze, M.; Jeong, S.; Paillard, E.; Schonhoff, M.; Winter, M.; Passerini, S. *Adv. Energy Mater.* 2011, **1**, 274–281.

[59] Pramanik, R.; Sarkar, S.; Ghatak, C.; Rao, V. G.; Sarkar, N. *J. Phys. Chem. B* 2011, **115**, 2322–2330.

[60] Mondain-Monval, O. *Curr. Opin. Colloid Interface Sci.* 2005, **10**, 250–255.

[61] Glaeser, R. M. *Phys. Today* 2008, **61**, 48–54.

CHAPTER 3

Self-Assembly of Nonionic Surfactants in Room-Temperature Ionic Liquids

KENICHI SAKAI, TAKESHI MISONO, MASAHIKO ABE, and HIDEKI SAKAI

Department of Pure and Applied Chemistry, Faculty of Science and Technology, Research Institute for Science and Technology, Tokyo University of Science, Noda, Chiba, Japan

3.1 INTRODUCTION

Room-temperature ionic liquids (RTILs) are a relatively new class of organic solvents. They exhibit unique physicochemical properties such as excellent thermal stability up to 300°C or more, negligible volatility, and nonflammability. These physicochemical properties qualify them to be "green solvents" in many industrially important reactions. RTILs have also gained importance in the fields of colloid and interface chemistry [1, 2]. In a recent review presented by Inoue, prior studies on RTILs were classified into four categories from the standpoint of surfactant chemistry [3]. These are (i) RTILs as a new type of surfactant, (ii) effects of RTILs on the aqueous solution properties of surfactants, (iii) the formation of microemulsions using RTILs as polar solvents, and (iv) the self-assembly of surfactants in RTILs. In this chapter, we review our recent research findings classified into the fourth category.

It is known that surfactants form molecular assemblies in aqueous solution. In a manner similar to aqueous surfactant systems, surfactants can form molecular assemblies in RTILs as well. However, there may be significant differences between the two systems. The aim of this chapter is to demonstrate similarities and differences in the self-assemblies of nonionic surfactants that occur in aqueous and RTIL media. This chapter deals with the following aspects given in the subsequent sections:

Section 3.2: Micellization of nonionic surfactants in RTILs
Section 3.3: Binary phase behavior of nonionic surfactants with RTILs
Section 3.4: Adsorption of nonionic surfactants at solid/RTIL interfaces

A future direction of studies concerning RTILs from the standpoint of colloid and interface chemistry is also addressed.

Ionic Liquid-Based Surfactant Science: Formulation, Characterization, and Applications, First Edition. Edited by Bidyut K. Paul and Satya P. Moulik.

3.2 MICELLIZATION OF NONIONIC SURFACTANTS IN RTILs

The micellization of nonionic surfactants in RTILs has been reported in the literature [4–24]. The combinations of RTILs and nonionic surfactants employed in these papers are summarized in Table 3.1. In this table, the RTILs are classified by their protic/aprotic nature: that is, protic RTILs can form a hydrogen-bonding network structure between ion pairs [25] leading to a sponge-like phase in the bulk as a result of their self-assembly. In contrast, aprotic RTILs do not form such a network structure, and hence, their self-assembly in the bulk is absent or weak. One of the key conclusions regarding the micellization of nonionic surfactants in RTILs is that the micellization is only observed when surfactant molecules are associated with each other in appropriate RTILs. In general, nonionic surfactants with excellent solubility in RTILs cannot form micelles because of their insufficient intermolecular solvophobic interaction and those with poor solubility in RTILs cannot do that either [6].

We have studied the dilute solution properties of nonionic surfactants in an aprotic RTIL for better understanding of their self-assemblies in bulk solutions (i.e., micellization) as well as at air/solution interfaces (i.e., adsorption) [11]. The aprotic RTIL used in this study was 1-butyl-3-methylimidazolium hexafluorophosphate, $BmimPF_6$. Four analogues of polyoxyethylene alkyl ether surfactants (abbreviated as C_nEO_m; n is the hydrocarbon chain length and m is the number of oxyethylene units, $C_{12}EO_5$, $C_{12}EO_6$, $C_{12}EO_7$ and $C_{14}EO_6$) were used in this study in order to see effects of the hydrocarbon and oxyethylene chain lengths on their interfacial properties. Figure 3.1 shows the static surface tension (γ) of C_nEO_m in $BmimPF_6$ as a function of their concentrations (C). For all of the four surfactants, the surface tension decreases sharply with increasing concentration, attains a break point, and then becomes almost constant at concentrations above the break point. This behavior is generally seen in aqueous surfactant systems, suggesting that the C_nEO_m surfactants start to form molecular assemblies (micelles) at the concentration corresponding to this break point (cmc).

The surface tension data allow calculation of several physicochemical parameters (see Table 3.2). The surface excess concentration estimated at cmc (Γ_{cmc}) and the occupied area per surfactant molecule adsorbed at the air/solution interface (A_{cmc}) were calculated from the Gibbs adsorption equation:

$$\Gamma_{cmc} = -\frac{1}{2.303iRT}\left(\frac{d\gamma}{d\log C}\right)_T \tag{3.1}$$

$$A_{cmc} = \frac{1}{N_A\Gamma_{cmc}} \tag{3.2}$$

where T is the absolute temperature, N_A is Avogadro's constant, and R is the gas constant. The number of adsorption species (i in Eq. 3.1) is assumed to be 1. It is clear that, at a given oxyethylene chain length ($m = 6$), the longer hydrocarbon surfactant ($C_{14}EO_6$) gives lower cmc, lower γ_{cmc} (surface tension measured at the cmc), larger Γ_{cmc}, and smaller A_{cmc} values when compared with the shorter hydrocarbon surfactant ($C_{12}EO_6$). In addition, we can also see that at a given hydrocarbon chain

TABLE 3.1 Summary of Studies Focusing on Micellization of Nonionic Surfactants in RTILs

RTILs	Surfactants	Major Analytical Methods	References
Protic RTILs			
EAN	C_nEO_m	ST, SANS	4, 5
	C_nEO_m	ST, SAXS	6
	BPS-*m*	ST	7
PAN	C_nEO_m	ST, SAXS	6
DEAF	C_nEO_m	ST, SAXS	6
TEAF	C_nEO_m	ST, SAXS	6
Aprotic RTILs			
BmimCl	C_nEO_m (Brij 35, Brij 700)	ST, IGC	8
$BmimPF_6$	C_nEO_m (Brij 35, Brij 700)	ST, IGC	8
	C_nEO_m (Brij 35)	Flu, Rhe	9
	C_nEO_m	ST	10
	C_nEO_m	ST, Flu, DLS, NMR	11
	C_nEO_m (Brij 35, Brij 700), TX-100, OBG	NIR	12
	TX-100	ST, NMR	13
	Tween-20	ST, NMR	14
	BPS-*m*	ST, TEM	15, 16
	PB-PEO	TEM, DLS	17
$BmimBF_4$	C_nEO_m	ST	10
	C_nEO_m	Flu	18
	C_nEO_m	DLS, NMR	19, 20
	TX-100	ST, NMR	13, 21
	Tween-20	ST, NMR	14
	BPS-*m*	ST	7
BmimTFSI	C_nEO_m	ST	10
$Bmim(CF_3SO_3)$	C_nEO_m	ST, DLS, NMR	22
$EmimBF_4$	C_nEO_m	DLS, NMR	23
EmimTFSI	C_nEO_m (Brij 35, Brij 700), TX-100, Tween-20	Flu	24
	C_nEO_m (Brij 35, Brij 700), TX-100, OBG	NIR	12
$HmimBF_4$	C_nEO_m	DLS, NMR	23

$BmimBF_4$, 1-butyl-3-methylimidazolium tetrafluoroborate; $Bmim(CF_3SO_3)$, 1-butyl-3-methylimidazolium trifluoromethanesulfonate; BmimCl, 1-butyl-3-methylimidazolium chloride; $BmimPF_6$, 1-butyl-3-methylimidazolium hexafluorophosphate; BmimTFSI, 1-butyl-3-methylimidazolium bis(trifluoromethylsulfonyl) imide; BPS-*m*, phytosterol ethoxylate (*m* is the number of oxyethylene units); Brij 35, polyoxyethylene (23) dodecyl ether; Brij 700, polyoxyethylene (100) octadecyl ether; C_nEO_m, polyoxyethylene alkyl ether (*n* is the hydrocarbon chain length and *m* is the number of oxyethylene units); DEAF, diethylammonium formate; DLS, dynamic light scattering; EAN, ethylammonium nitrate; $EmimBF_4$, 1-ethyl-3-methylimidazolium tetrafluoroborate; EmimTFSI, 1-ethyl-3-methylimidazolium bis(trifluoromethylsulfonyl)imide; Flu, fluorescence spectroscopy; $HmimBF_4$, 1-hexyl-3-methylimidazolium tetrafluoroborate; IGC, inverse gas chromatography; NIR, near-infrared spectroscopy; NMR, nuclear magnetic resonance spectroscopy; OBG, *n*-octyl-β-D-glucopyranoside; PAN, propylammonium nitrate; PB-PEO, poly((1,2-butadiene)-ethylene oxide) diblock copolymer; Rhe, rheology; SANS, small-angle neutron scattering; SAXS, small-angle X-ray scattering; ST, surface tensiometry; TEAF, triethylammonium formate; TEM, transmission electron microscopy; Tween-20, polyoxyethylene (20) sorbitan monolaurate; TX-100, polyoxyethylene (10) octylphenyl ether.

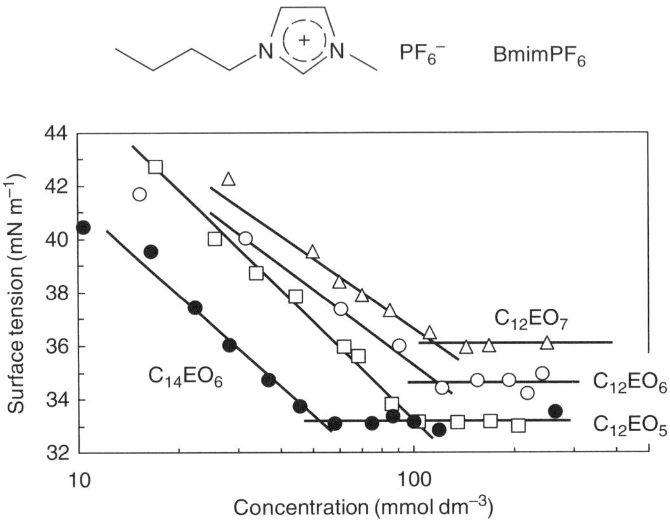

Figure 3.1 Surface tension of C_nEO_m in $BmimPF_6$ as a function of surfactant concentration (in mmol dm^{-3}). These measurements were performed at 25°C. The chemical structure of $BmimPF_6$ is also shown in the figure. Reproduced from Misono et al. [11] with permission from Elsevier.

TABLE 3.2 Physicochemical Parameters Estimated for the C_nEO_m Systems at 25°C

	cmc (wt.%)	cmc (mmol dm^{-3})	γ_{cmc} (mN m^{-1})	Γ_{cmc} (μmol m^{-2})	A_{cmc} (nm^2)
$C_{12}EO_5/BmimPF_6$[a]	2.98	101	33.2	2.1	0.78
$C_{12}EO_6/BmimPF_6$[a]	3.99	121	34.6	1.7	1.0
$C_{12}EO_7/BmimPF_6$[a]	5.04	140	36.1	1.2	1.4
$C_{14}EO_6/BmimPF_6$[a]	1.88	53.7	33.2	1.9	0.88
$C_{12}EO_5$/water[b]		0.064		3.3	0.50
$C_{12}EO_7$/water[b]		0.082		2.9	0.57

[a]From Misono et al. [11].
[b]From Rosen [26].

length ($n = 12$), an increased oxyethylene chain length results in an increased cmc, an increased γ_{cmc}, a decreased Γ_{cmc}, and an increased A_{cmc}. These results are consistent with a general behavior of nonionic surfactants in aqueous media and suggest that the hydrocarbon chain acts as a solvophobic group whereas the oxyethylene units act as a solvophilic group in $BmimPF_6$. This situation is qualitatively similar to the general behavior of nonionic surfactants dissolved in aqueous solutions. Such a similarity between RTIL and aqueous systems has also been reported in the study focusing on the surface active properties of nonionic phytosterol ethoxylates (BPS-*m*; *m* is the number of oxyethylene units, 5, 10, 20, and 30) in $BmimPF_6$ [15].

It is interesting to compare the physicochemical parameters estimated for the $C_nEO_m/BmimPF_6$ systems with the corresponding data obtained for the C_nEO_m/water

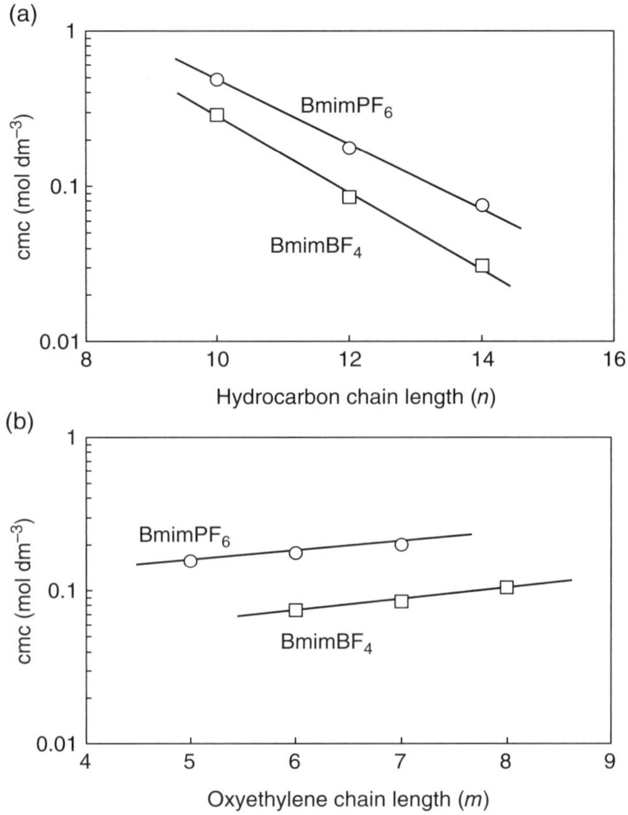

Figure 3.2 Plots of cmc of C_nEO_m in $BmimPF_6$ and $BmimBF_4$ as a function of (a) hydrocarbon chain length (n) at a fixed number (6) of oxyethylene unit and (b) oxyethylene chain length (m) at a fixed hydrocarbon chain length (12). The cmc values were determined by ^{1}H-NMR spectroscopy at 25°C. Reproduced from Misono et al. [11] with permission from Elsevier.

systems [26]. As shown in Table 3.2, we see higher cmc, smaller Γ_{cmc}, and larger A_{cmc} values of the $C_nEO_m/BmimPF_6$ systems compared with the corresponding C_nEO_m/water systems. This indicates that the surface activity of C_nEO_m is lower in $BmimPF_6$ than in water, being reflective of weaker solvophobicity of C_nEO_m in $BmimPF_6$ than in water.

The dependence of cmc on the hydrocarbon and oxyethylene chain lengths has been further investigated by ^{1}H-NMR measurements [11, 19, 20]. The solvents used in these studies were $BmimPF_6$, 1-butyl-3-methylimidazolium tetrafluoroborate, $BmimBF_4$, and water. These two RTILs have the same cation ($Bmim^+$) with different anions (PF_6^- and BF_4^-), and this difference leads to a different hydrophobicity; that is, $BmimPF_6$ is more hydrophobic than $BmimBF_4$. The resulting cmc data are given in Figure 3.2 as functions of hydrocarbon chain length (n) and oxyethylene chain length (m). For both of the two RTIL systems, the log cmc values are decreased linearly with increasing hydrocarbon chain length of C_nEO_6 (Fig. 3.2a). This behavior is similarly seen in the aqueous C_nEO_6 systems [26]. The slope values calculated for the log cmc against n plots are as follows: −0.21 for $BmimPF_6$, −0.24 for $BmimBF_4$, and ca. −0.5 for water [26]. This means that the addition of two methylene units to the surfactant hydrocarbon chain reduces the cmc by a factor of 1/2.6 in $BmimPF_6$ and of 1/3.0 in $BmimBF_4$, whereas the cmc is reduced by a factor of 1/10 in water. That is, the effect of hydrocarbon chain length on cmc is much weaker in the RTILs compared with that in water. Again, this reflects weaker solvophobic interaction of C_nEO_m in the two RTILs than in water. A greater solvophobic interaction between surfactant hydrocarbon chains (in other words, greater hydrophilicity of the solvents) leads to a lower cmc at a given hydrocarbon chain length, and hence, we see a change in cmc in the following order: $BmimPF_6$ > $BmimBF_4$ > water.

The effect of oxyethylene chain length on cmc is rather weak as can be seen in Figure 3.2b. The slope of the linear increase is calculated as 0.054 for $BmimPF_6$, 0.074 for $BmimBF_4$, and 0.058 for water [27]. This corresponds to the cmc increase per oxyethylene unit by a factor of 1.13 for $BmimPF_6$, 1.18 for $BmimBF_4$, and 1.14 for water. This suggests no significant differences in the affinity of the oxyethylene units with the three solvents.

Thermodynamics of micellization of C_nEO_m in RTILs have been studied by measuring cmc at various temperatures [11]. As a typical example, the cmc values estimated for $C_{12}EO_5$ in $BmimPF_6$ are shown in Figure 3.3a as a function of temperature. The cmc values are gradually increased with increasing temperature. The standard Gibbs free energy change associated with the micelle formation (ΔG^0_m) is relevant to the cmc. In the case of nonionic surfactants, the relationship between ΔG^0_m and cmc is given by the following equation [28]:

$$\Delta G^0_{\mathrm{m}} = RT \ln X_{\mathrm{cmc}} \tag{3.3}$$

where X_{cmc} is the cmc expressed in mole fraction unit. The Gibbs–Helmholtz equation gives the standard enthalpy of micelle formation (ΔH^0_m) by plotting the $\Delta G^0_m / T$ values as a function of $1/T$:

$$\left[\frac{\partial\left(\Delta G^0_{\mathrm{m}} / T\right)}{\partial(1/T)}\right]_{\mathrm{P}} = \Delta H^0_{\mathrm{m}} \tag{3.4}$$

A second-order polynomial is fitted to the data points, and the values of ΔH^0_m are calculated from the slopes of the tangential lines at the temperatures corresponding to the experimental data points. Finally, the standard entropy change associated with the micelle formation (ΔS^0_m) is calculated using the following equation:

$$\Delta S^0_{\mathrm{m}} = \frac{\Delta H^0_{\mathrm{m}} - \Delta G^0_{\mathrm{m}}}{T} \tag{3.5}$$

The thermodynamic parameters (ΔG^0_m, ΔH^0_m, and $-T\Delta S^0_m$) estimated by this procedure are plotted in Figure 3.3b as a function of temperature. The contribution of entropy term ($-T\Delta S^0_m$) to negative ΔG^0_m is superior to that of enthalpy term (ΔH^0_m) at low temperatures, whereas this relationship becomes opposite at high temperatures. This indicates that the micelle formation results from an entropy-driven process at low temperatures, whereas from an enthalpy-driven process at high

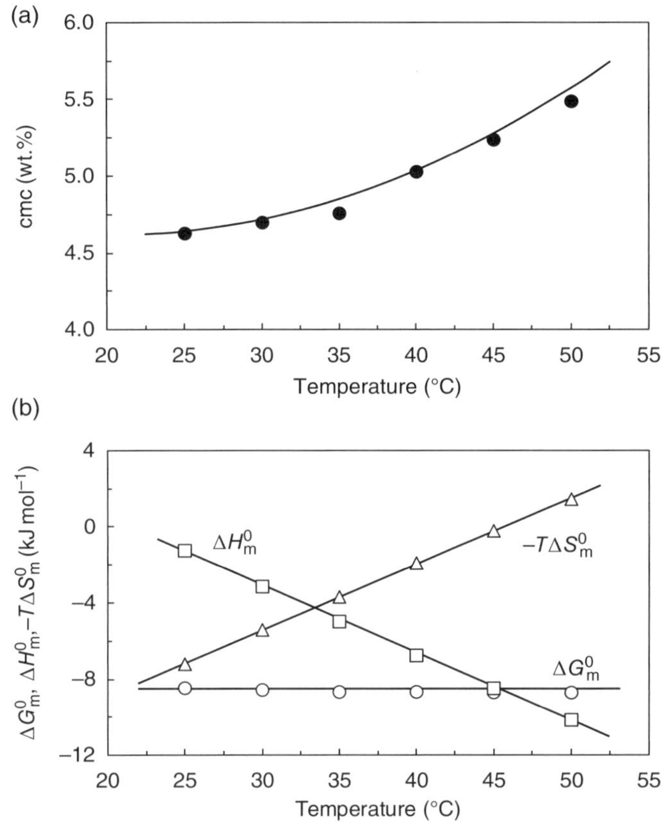

Figure 3.3 (a) Plots of cmc (wt.%) against temperature for $C_{12}EO_5$ in $BmimPF_6$. (b) Plots of thermodynamic parameters of micellization (ΔG^0_m, ΔH^0_m, and $-T\Delta S^0_m$) against temperature for $C_{12}EO_5$ in $BmimPF_6$. Reproduced from Misono et al. [11] with permission from Elsevier.

temperatures. This behavior is similar to that observed for aqueous surfactant systems. The increased entropy is attributed to the release of highly ordered solvent molecules existing around the surfactant hydrocarbon chain. In conclusion, the driving force for the micelle formation in $BmimPF_6$ is a solvophobic interaction between hydrocarbon chains of the surfactant molecules just like a hydrophobic interaction in aqueous solutions.

3.3 BINARY PHASE BEHAVIOR OF NONIONIC SURFACTANTS WITH RTILs

The binary phase behavior of nonionic surfactants in RTILs has been reviewed by Hao and Zemb [1], Greaves and Drummond [2], Inoue [3], and Dong and Hao [29]. The primary focus of these studies has been on the self-assembly of nonionic surfactants in protic RTILs such as ethylammonium nitrate (EAN) [30–33]. Araos and Warr [30] have studied the phase behavior of C_nEO_m in EAN and demonstrated that the overall phase behavior in EAN is very similar to that in water, although longer hydrocarbon chains are required in EAN to obtain the same phases as those obtained in water due to the higher solubility of the surfactants in EAN than in water. In comparison with the phase studies using protic RTILs, there have been fewer systematic studies focusing on the phase behavior that occurs in aprotic RTILs [34–36]. In addition, no mesophases have been observed for C_nEO_m in aprotic, imidazolium-type RTILs at any temperature and concentration [29]. In order to fill the lacunae of data and deep understanding of the phase behavior of nonionic surfactants in aprotic RTILs, we studied the temperature-concentration phase diagram of phytosterol-type nonionic surfactants, phytosterol ethoxylate (m is the number of oxyethylene units), and BPS-m (m = 5, 10, 20, and 30) in $BmimPF_6$ [37]. It has been reported that, in the mixed system comprising BPS-m with $BmimPF_6$, the oxyethylene unit of BPS-m acts as a solvophilic group and the phytosterol unit acts as a solvophobic group [15].

Figure 3.4 shows the resultant temperature-concentration phase diagrams of BPS-m with $BmimPF_6$ [37]. Several phase states including a micellar phase (L_1), discontinuous cubic phase (I_1), hexagonal phase (H_1), lamellar phase (L_α), lamellar gel phase (L_β), and reverse micellar phase (L_2) are observed along with coexistence of these phases in certain regions. One can see in these diagrams that lesser-curved molecular assemblies tend to be formed at higher surfactant concentrations, at lower temperatures, and for shorter oxyethylene chain analogues. For example, the increased surfactant concentration results in phase changes as L_1-I_1-H_1-L_β for BPS-20 and as L_1-I_1-L_β for BPS-30 at relatively low temperatures. In addition, the mixtures of BPS-5 and BPS-10 with $BmimPF_6$ yield a two-phase coexistence region at high temperatures. The region consists of an immiscible BPS-m phase and a $BmimPF_6$ phase. It appears plausible that the two-phase separation results from the cloud point phenomenon [38]; that is, the increased temperature results in decreased solvophilicity of the oxyethylene units with $BmimPF_6$. In contrast, no corresponding two-phase region appears for the longer chain analogues (BPS-20 and BPS-30), reflecting the greater solvophilicity of the longer chain analogues with $BmimPF_6$. In the mixed systems of BPS-20 and BPS-30 with $BmimPF_6$, an increased temperature induces a phase transition from the I_1, H_1, and L_β phases to the L_1 phase. This indicates that the increased temperature results in increased thermal motion of the surfactant molecules in the systems, leading to the destabilization of the well-defined molecular assemblies.

The phase behaviors shown in Figure 3.4 are similar to those observed in the BPS-m + water systems [39], and hence, the overall phase behavior in the BPS-m + $BmimPF_6$ systems is largely similar to that in the BPS-m + water systems. However, notable differences in the phase states are observed in each of the two liquid systems. First, fewer phases are observed in the $BmimPF_6$ system than in the water system. For example, the H_1 phase is observed in the BPS-10 + water and BPS-30 + water systems [39], whereas this phase does not appear in the corresponding $BmimPF_6$ systems. Similarly, the L_2 phase is observed in the BPS-10, BPS-20, and BPS-30 aqueous systems (but the observed region is limited at high temperatures and at high surfactant concentrations) [39], whereas this phase is not identified in the corresponding $BmimPF_6$ systems. Second, the temperature-concentration region in which a phase that is common to both systems is observed is somewhat different for the $BmimPF_6$ versus the aqueous system. This is exemplified by the experimental observation that at the fixed temperature of 25°C, the I_1–L_1 phase transition occurs

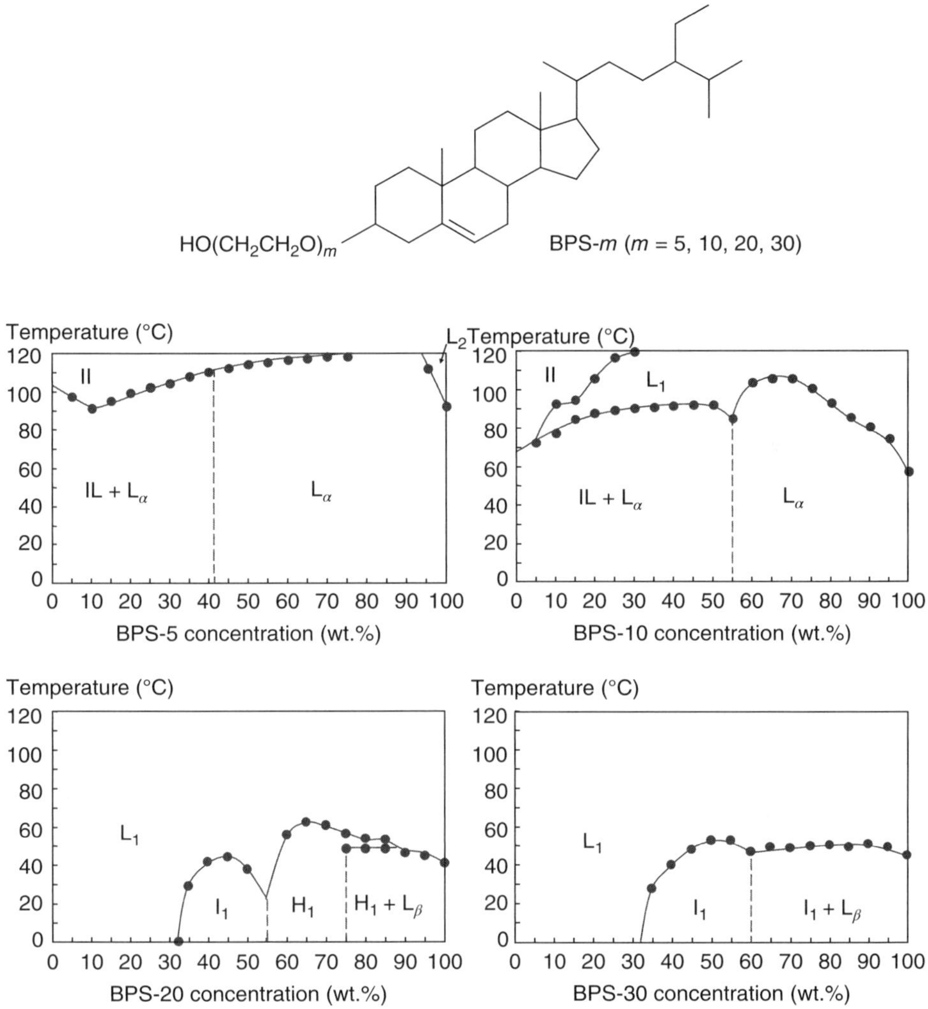

Figure 3.4 The temperature-concentration phase diagrams of BPS-*m* ($m = 5$, 10, 20, and 30) with $BmimPF_6$. The phase abbreviations are as follows: micellar phase (L_1), discontinuous cubic liquid crystal phase (I_1), hexagonal liquid crystal phase (H_1), lamellar liquid crystal phase (L_α), lamellar gel phase (L_β), reverse micellar phase (L_2), ionic liquid phase (IL), and two-phase separation (II). The chemical structure of β-sitosterol ethoxylates as a typical example of BPS-*m* is also shown in this figure. Reproduced from Sakai et al. [37] with permission from Japan Oil Chemists' Society.

at a BPS-20 concentration of 34–35 wt.% in the BPS-20 + $BmimPF_6$ system, whereas this transition occurs at a BPS-20 concentration of ca. 26 wt.% in the BPS-20 + water system [39]. The occurrence of the phase transition at higher BPS-20 concentrations in the $BmimPF_6$ system relative to the water system results from the greater solvophilic nature of BPS-*m* with $BmimPF_6$ than with water, as was similarly suggested for the C_nEO_m + EAN/water systems [30].

3.4 ADSORPTION OF NONIONIC SURFACTANTS AT SOLID/RTIL INTERFACES

The self-assembly (adsorption) of surfactants that occurs at solid/solution interfaces is an interesting subject not only from academia but also from industry. For example, adsorption characteristics of surfactants on solid surfaces from RTILs will be fundamentally useful in developing lubrications, dye-sensitized solar cells, electrodepositions, etc. Pioneering work relevant to the adsorption characteristics of nonionic surfactants in RTILs has been carried out by Atkin and Warr [25, 40–46]. The key findings reported by these authors are summarized as follows. Solvation layers consisting of ion pairs (and, in some cases, cations in the innermost solvation layer) are formed at solid/RTIL interfaces. The number of detectable layers depends not only on the nature of the solid substrates (such as surface charge density and surface roughness) but also on the physicochemical properties of RTILs (such as their protic–aprotic nature and molecular flexibility) [25, 40–42]. Recently, the lateral structure of solvation layers formed on a solid surface (mica) has been studied by means of soft-contact and amplitude-modulation atomic force microscopy (AFM) [43]. The protic RTIL (EAN) forms an interfacial innermost layer in a slightly disordered wormlike morphology, and an upper layer formed on the innermost layer gives less ordered and more undulating appearance. In the case of the aprotic RTIL (EmimTFSI), the imidazolium cations adsorb onto mica in a more isolated fashion. This difference is attributed to the hydrogen-bonding network of the protic RTIL to promote self-assembly at the solid/liquid interface as well as in the bulk liquid [43].

The addition of nonionic surfactants results in the formation of a surfactant layer at solid/RTIL interfaces [44–46]. For example, $C_{16}EO_6$ spontaneously forms hemicylindrical surface aggregates on graphite in EAN [44]. The overall morphology of the surface aggregates observed in the RTIL is similar to that in aqueous media. In contrast, the adsorption of the nonionic surfactant has not been detected at a hydrophilic silica/EAN interface [44]. Instead, the force-distance data measured by AFM demonstrate force oscillations resulting from $C_{16}EO_6$ micelles that are nonadsorbing on silica from EAN [45]. This indicates that the EAN solvation layers originally formed onto the silica surface prevent surfactant adsorption at the silica/EAN interface. Interestingly, polymeric amphiphiles consisting of polyethylene oxide (PEO) and polypropylene oxide (PPO) blocks (PEO*n*-PPO*m*-PEO*n*, Pluronics) can adsorb at the silica/EAN interface due to the large number of EO units (i.e., multiple attachment points to the silica surface) in a molecule [46].

In order to understand the effects of physicochemical properties of RTILs on surfactant adsorption, we studied the adsorption of BPS-20 on a hydrophilic silica surface in an aprotic RTIL (1-ethyl-3-methylimidazolium bis(trifluoromethylsulfonyl)imide, EmimTFSI) [47]. Figure 3.5 shows the resulting force-distance data. In the absence of BPS-20 (i.e., pure EmimTFSI), oscillating solvation forces are clearly observed. The number of detectable layers is (at least) four in this case. The layer thickness suggests that $Emim^+$ ions form the innermost (first) solvation layer directly contacting with the silica surface, whereas the second, third, and fourth solvation layers are composed of EmimTFSI ion pairs.

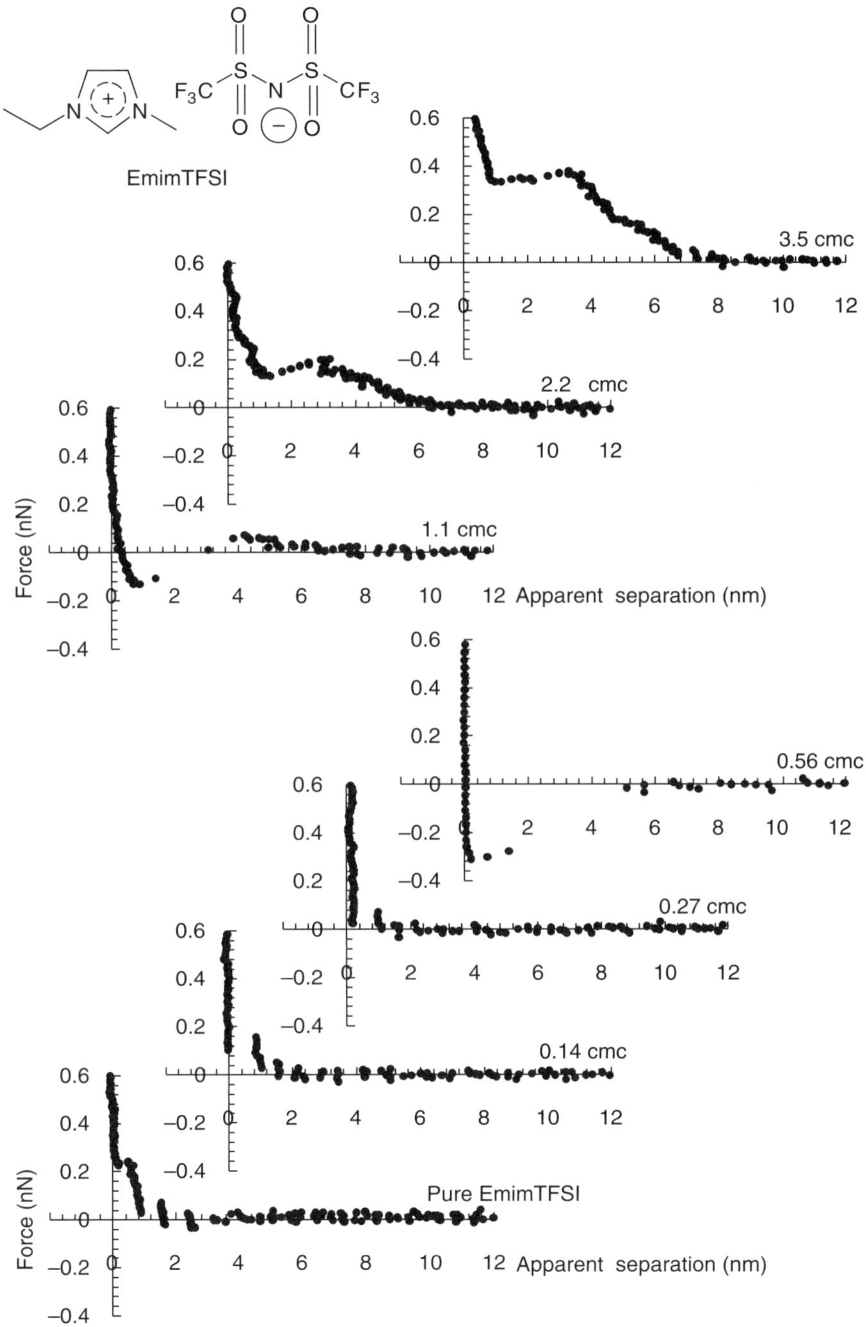

Figure 3.5 Force curve data obtained for the BPS-20/EmimTFSI systems. The BPS-20 concentrations were varied from 0 to 3.5 cmc in EmimTFSI. The measurements were taken at room temperature (25°C). The chemical structure of EmimTFSI is also shown in the figure. Reproduced from Sakai et al. [47] with permission from the American Chemical Society.

The addition of BPS-20 dramatically affects the force-distance data. At concentrations well below the cmc (at 0.14 and 0.27 cmc), one can still see oscillating solvation forces, but the force magnitude is gradually decreased with increasing surfactant concentration. A further increase in the BPS-20 concentration (at 0.56 cmc) results in the disappearance of the solvation forces, and instead one can see only an attractive jump to contact from an apparent separation of 5 nm. This attractive interaction results from either the solvophobic or van der Waals interaction, induced by a surfactant monolayer adsorption. When the BPS-20 concentration exceeds the cmc, a weak repulsive interaction is detected, although the attractive interaction is still observed from ca. 4 nm. The magnitude of the repulsive interaction increases with the BPS-20 concentration in the range of 1.1–3.5 cmc. In this concentration range, force instabilities are observed at short separations. Such instabilities are generally observed in the system of surfactant adsorption that occurs at solid/aqueous solution interfaces and arise from the "squeezed out" of the surfactant layer adsorbed on the solid surface during the tip approach [48]. The force-distance data shown in Figure 3.5 suggest, therefore, that the adsorption of BPS-20 occurs at the silica/EmimTFSI interface and the adsorbed BPS-20 layer grows with increasing BPS-20 concentration. This is a totally different behavior reported for the $C_{16}EO_6$/EAN/silica system, where no surfactant adsorption is evidenced [44, 45].

We proposed the following adsorption mechanisms [47]. In the first mechanism, the solvation layers consisting of $Emim^+$ ions and EmimTFSI ion pairs are replaced by a BPS-20 monolayer adsorbed on silica. A possible driving force for this adsorption is the hydrogen bonding between the oxyethylene headgroups and the surface silanol groups. As a result, solvophobic moieties of BPS-20 will face the EmimTFSI solution phase. In the second step, further adsorption takes place as a result of the solvophobic interaction between the solvophobic moieties of BPS-20, and finally, a lamellar bilayer is spontaneously formed. This adsorption mechanism suggests that the interaction between EmimTFSI and the silica surface is weaker than that between BPS-20 and the silica surface, and hence, the EmimTFSI solvation layers originally formed on the silica surface can be replaced by BPS-20.

The other adsorption mechanism is that the solvation layers (at least the innermost $Emim^+$ cation layer) remain on the silica surface but the disruption of the solvation layer(s) occurs as a result of the surfactant adsorption onto the solvation layer(s). If this is true, then the force-distance data shown in Figure 3.5 may result from the "screening" of the solvation layers by surfactant adsorption. The only difference between the two adsorption mechanisms is whether the RTIL solvation layers remain on the silica surface. It is suggested that the surfactant adsorption disrupts the structured solvation layers whatever the presence or absence of $Emim^+$ ions (and EmimTFSI ion pairs) adsorbed on the silica surface. This depends on the adsorption affinity for the two species ($Emim^+$ cation and BPS-20) with silica surface sites. The sum frequency vibrational spectroscopic (SFVS) data have demonstrated that the water molecules associated with EmimTFSI at the silica surface form hydrogen bonds with the IL anions and the structure of the imidazolium cations bound to the surface does not change significantly with water content [49]. This indicates the adherence to the imidazolium cations adsorbed onto the silica surface against the replacement by water (and even by the oxyethylene-type nonionic surfactant).

3.5 CONCLUSION AND OUTLOOK

We have discussed the self-assembly of nonionic surfactants that occurs in RTILs. Overall, the self-assembly properties in RTILs are largely similar to the aqueous medium. Notable differences between the aqueous and nonaqueous systems are sometimes seen when nonionic surfactants form micelles or lyotropic liquid crystals at certain compositions and temperatures, and this mainly results from the different affinity of the nonionic surfactants with the liquids. In other words, it may be possible to expect the formation of micelles or lyotropic liquid crystals to a certain degree by considering the solvophobic or solvophilic nature of the nonionic surfactants in the RTILs. An interesting feature of RTILs is their self-assembly in bulk liquids and at interfaces. This feature also makes a significant impact on the self-assembly of nonionic surfactants in RTILs. Particularly, we have demonstrated the importance of this feature when nonionic surfactants adsorb at solid/RTIL interfaces. We believe that the self-assembled structures of amphiphilic molecules with RTILs are of great interest not only from academic but also from industrial standpoints. One of the potential applications based on such self-assembled structures should be high-performance ion-conductive electrolytes as a new device system with nanolevel order [50].

Our research in this area will be directed toward understanding the effects of water on the interfacial properties of RTILs. There are a few papers studying the ternary phase diagram of surfactant/water/RTIL systems [51–54]. For example, Wang et al. [51] have studied the ternary phase diagrams of Brij-97 [polyoxyethylene (10) oleoyl ether]/water/BmimPF$_6$ and Brij-97/water/BmimBF$_4$, and demonstrated that both the two systems yield an H$_1$ phase at certain compositions. However, the molecular packing arrangement is quantitatively different between the two systems as a result of the different hydrophilicity of the two RTILs; that is, the more hydrophobic RTIL (BmimPF$_6$) is dominantly penetrated between the oxyethylene chains of the surfactant molecules, whereas the more hydrophilic RTIL (BmimBF$_4$) is localized into the water region in the H$_1$ phase. In addition, surfactant/water/RTIL or surfactant/water/RTIL/oil mixtures have been frequently used in order to prepare microemulsions [13, 55–58]. These studies focus on the bulk properties of RTILs in the presence of water and surfactants; however, the interfacial properties of such ternary systems have not been understood clearly yet. We note that, as far as we are aware, only a few research groups have reported the effects of water on air/liquid [5, 59], solid/liquid [49, 60], or liquid/liquid [61] interfacial properties in RTILs, but the number of studies focusing on the adsorption of surfactants at the interfaces is very limited.

In the field of inorganic material chemistry, the preparation of metal oxide particles in RTILs has attracted much attention because of their unique morphologies and crystalline structures [62–66]. The metal oxide particles are synthesized through the reactions of hydrolysis and polycondensation of metal oxide precursors with water dissolved in RTILs. We expect, therefore, that in these reactions the interaction between water and metal oxide surfaces plays a key role in determining the unique morphologies and crystalline structures of the particles. In fact, our preliminary data suggest that the morphology of titania particles prepared in RTILs is strongly dependent on the water concentration in the system [67]. In addition, surfactants usually act as a "structure-directing agent" when preparing metal oxide particles, and hence, the shape and size of the products are expected to be changed by the presence of surfactant assemblies formed in RTILs. From this point of view, the knowledge regarding the interfacial properties of RTILs with water in the absence and presence of surfactants will contribute to a better understanding of the preparation mechanism of metal oxide particles in RTILs.

REFERENCES

[1] Hao, J.; Zemb, T. *Curr. Opin. Colloid Interface Sci.*, 2007, **12**, 129–137 (and references therein).

[2] Greaves, T. L.; Drummond, C. J. *Chem. Soc. Rev.*, 2008, **37**, 1709–1726 (and references therein).

[3] Inoue, T. *J. Oleo Sci.*, 2010, **10**, 179–185 (and references therein).

[4] Araos M. U.; Warr, G. G. *Langmuir*, 2008, **24**, 9354–9360.

[5] Wakeham, D.; Warr, G. G.; Atkin, R. *Langmuir*, 2012, **28**, 13224–13231.

[6] Greaves, T. L.; Mudie, S. T.; Drummond, C. J. *Phys. Chem. Chem. Phys.*, 2011, **13**, 20441–20452.

[7] Yue, X.; Chen, X.; Li, Q. *J. Phys. Chem. B*, 2012, **116**, 9439–9444.

[8] Anderson, J. L.; Pino, V.; Hagberg, E. C.; Sheares, V. V.; Armstrong, D. W. *Chem. Commun.*, 2003, 2444–2445.

[9] Chakraborty, A.; Seth, D.; Chakrabarty, D.; Setua, P.; Sarkar, N. *J. Phys. Chem. A*, 2005, **109**, 11110–11116.

[10] Patrascu, C.; Gauffre, F.; Nallet, F.; Bordes, R.; Oberdisse, J.; de Lauth-Viguerie, N.; Mingotaud, C. *ChemPhysChem*, 2006, **7**, 99–101.

[11] Misono, T.; Sakai, H.; Sakai, K.; Abe, M.; Inoue, T. *J. Colloid Interface Sci.*, 2011, **358**, 527–533.

[12] Tran, C. D.; Yu, S. *J. Colloid Interface Sci.*, 2005, **283**, 613–618.

[13] Li, N.; Zhang, S.; Ma, H.; Zheng, L. *Langmuir*, 2010, **26**, 9315–9320.

[14] Wu, J.; Li, N.; Zheng, L.; Li, X.; Gao, Y.; Inoue, T. *Langmuir*, 2008, **24**, 9314–9322.

[15] Sakai, H.; Saitoh, T.; Endo, T.; Tsuchiya, K.; Sakai, K.; Abe, M. *Langmuir*, 2009, **25**, 2601–2603.

[16] Sakai, H.; Saitoh, T.; Misono, T.; Tsuchiya, K.; Sakai, K.; Abe, M. *J. Oleo Sci.*, 2011, **60**, 563–567.

[17] He, Y.; Li, Z.; Simone, P.; Lodge, T. P. *J. Am. Chem. Soc.*, 2006, **128**, 2745–2750.

[18] Seth, D.; Chakraborty, A.; Setua, P.; Sarkar, N. *J. Phys. Chem. B*, 2007, **111**, 4781–4787.

[19] Inoue, T. *J. Colloid Interface Sci.*, 2009, **337**, 240–246.

[20] Inoue, T.; Yamakawa, H. *J. Colloid Interface Sci.*, 2011, **356**, 798–802.

[21] Gao, Y.; Li, N.; Li, X.; Zhang, S.; Zheng, L.; Bai, X.; Yu, L. *J. Phys. Chem. B*, 2009, **113**, 123–130.

[22] Inoue, T.; Maema, K. *Colloid Polym. Sci.*, 2011, **289**, 1167–1175.

[23] Inoue, T.; Kawashima, K.; Miyagawa, Y. *J. Colloid Interface Sci.*, 2011, **363**, 295–300.

[24] Fletcher, K. A.; Pandey, S. *Langmuir*, 2004, **20**, 33–36.

[25] Hayes, R.; Warr, G. G.; Atkin, R. *Phys. Chem. Chem. Phys.*, 2010, **12**, 1709–1723.

[26] Rosen, M. J. Surfactants and Interfacial Phenomena (2nd Edition), John Wiley & Sons, Inc., New York, 1989.

[27] Rosen, M. J.; Cohen, A. W.; Dahanayake, M.; Hua, X. Y. *J. Phys. Chem.*, 1982, **86**, 541–545.

[28] Heimenz, P. C.; Rajagopalan, R. Principles of Colloid and Surface Chemistry (3rd Edition), Marcel Dekker, New York, 1997.

[29] Dong, R.; Hao, J. *Chem. Rev.*, 2010, **110**, 4978–5022.

[30] Araos, M. U.; Warr, G. G. *J. Phys. Chem. B*, 2005, **109**, 14275–14277.

[31] Greaves, T. L.; Weerawardena, A.; Fong, C.; Drummond, C. J. *Langmuir*, 2007, **23**, 402–404.

[32] Greaves, T. L.; Weerawardena, A.; Fong, C.; Drummond, C. J. *J. Phys. Chem. B*, 2007, **111**, 4082–4088.

[33] Greaves, T. L.; Weerawardena, A.; Krodkiewska, I.; Drummond, C. J. *J. Phys. Chem. B*, 2008, **112**, 896–905.

[34] Wang, L.; Chen, X.; Chai, Y.; Hao, J.; Sui, Z.; Zhuang, W.; Sun, Z. *Chem. Commun.*, 2004, 2840–2841.

[35] Tang, J.; Li, D.; Sun, C.; Zheng, L.; Li, J. *Colloids Surf. A*, 2006, **273**, 24–28.

[36] Simone, P. M.; Lodge, T. P. *Macromolecules*, 2008, **41**, 1753–1759.

[37] Sakai, H.; Saitoh, T.; Misono, T.; Tsuchiya, K.; Sakai, K.; Abe, M. *J. Oleo Sci.*, 2012, **61**, 135–141.

[38] Inoue, T.; Misono, T. *J. Colloid Interface Sci.*, 2009, **337**, 247–253.

[39] Folmer, B. M.; Svensson, M.; Holmberg, K.; Brown, W. *J. Colloid Interface Sci.*, 1999, **213**, 112–120.

[40] Atkin, R.; Warr, G. G. *J. Phys. Chem. C*, 2007, **111**, 5162–5168.

[41] Wakeham, D.; Hayes, R.; Warr, G. G.; Atkin, R. *J. Phys. Chem. B*, 2009, **113**, 5961–5966.

[42] Hayes, R.; Abedin, S. Z. E.; Atkin, R. *J. Phys. Chem. B*, 2009, **113**, 7049–7052.

[43] Segura, J. J.; Elbourne, A.; Wanless, E. J.; Warr, G. G.; Voïtchovsky, K.; Atkin, R. *Phys. Chem. Chem. Phys.*, 2013, **15**, 3320–3328.

[44] Atkin, R.; Warr, G. G. *J. Am. Chem. Soc.*, 2005, **127**, 11940–11941.

[45] Atkin, R.; Fina, L.-M. D.; Kiederling, U.; Warr, G. G. *J. Phys. Chem. B*, 2009, **113**, 12201–12213.

[46] Werzer, O.; Warr, G. G.; Atkin, R. *Langmuir*, 2011, **27**, 3541–3549.

[47] Sakai, K.; Onuma, Y.; Torigoe, K.; Biggs, S.; Sakai, H.; Abe, M. *Langmuir*, 2011, **27**, 3244–3248.

[48] Wanless, E. J.; Ducker, W. A. *J. Phys. Chem.*, 1996, **100**, 3207–3214.

[49] Fitchett, B. D.; Conboy, J. C. *J. Phys. Chem. B*, 2004, **108**, 20255–20262.

[50] Yoshio, M.; Mukai, T.; Kanie, K.; Yoshizawa, M.; Ohno, H.; Kato, T. *Adv. Mater.*, 2002, **14**, 351–354.

[51] Wang, Z.; Liu, F.; Gao, Y.; Zhuang, W.; Xu, L.; Han, B.; Li, G.; Zhang, G. *Langmuir*, 2005, **21**, 4931–4937.

[52] Zhuang, W.; Chen, X.; Cai, J.; Zhang, G.; Qiu, H. *Colloids Surf. A*, 2008, **318**, 175–183.

[53] Wang, Z.; Zhou, W.; Li, G. *J. Colloid Interface Sci.*, 2008, **318**, 405–410.

[54] Wu, J.; Zhang, J.; Zheng, L.; Zhao, X.; Li, N.; Dong, B. *Colloids Surf. A*, 2009, **336**, 18–22.

[55] Mehta, S. K.; Kaur, K. *Indian J. Chem.*, 2010, **49A**, 662–684 (and references therein).

[56] Gao, Y.; Han, S.; Han, B.; Li, G.; Shen, D.; Li, Z.; Du, J.; Hou, W.; Zhang, G. *Langmuir*, 2005, **21**, 5681–5684.

[57] Gao, Y.; Li, N.; Zheng, L.; Zhao, X.; Zhang, S.; Han, B.; Hou, W.; Li, G. *Green Chem.*, 2006, **8**, 43–49.

[58] Gao, Y.; Li, N.; Zheng, L.; Zhao, X.; Zhang, J.; Cao, Q.; Zhao, M.; Li, Z.; Zhang, G. *Chem. Eur. J.*, 2007, **13**, 2661–2670.

[59] Misono, T.; Aburai, K.; Endo, T.; Sakai, K.; Abe, M.; Sakai, H. *J. Oleo Sci.*, 2013, **62**, 363–370.

[60] Smith, J. A.; Werzer, O.; Webber, G. B.; Warr, G. G.; Atkin, R. *J. Phys. Chem. Lett.*, 2010, **1**, 64–68.

[61] Iwahashi, T.; Sakai, Y.; Kim, D.; Ishiyama, T.; Morita, A.; Ouchi, Y. *Faraday Discuss.*, 2012, **154**, 289–301.

[62] Nakashima, T.; Kimizuka, N. *J. Am. Chem. Soc.*, 2003, **125**, 6386–6387.
[63] Zhou, Y.; Antonietti, M. *J. Am. Chem. Soc.*, 2003, **125**, 14960–14961.
[64] Li, Z.; Zhang, J.; Du, J.; Han, B.; Wang, J. *Colloids Surf. A*, 2006, **286**, 117–120.
[65] Zhao, M.; Zheng, L.; Bai, X.; Li, N.; Yu, L. *Colloids Surf. A*, 2009, **346**, 229–236.
[66] Ding, K.; Miao, Z.; Hu, B.; An, G.; Su, Z.; Han, B.; Liu, Z. *Langmuir*, 2010, **26**, 5129–5134.
[67] Kinoshita, Y.; Saito, T.; Endo, T.; Sakai, K.; Sakai, H.; Abe, M. *Abstract of Annual Meeting of Division of Colloid and Interface Chemistry*, Chemical Society of Japan, Tokyo, 2011.

CHAPTER 4

Ionic Liquid-Based Surfactants: Synthesis, Molecular Structure, Micellar Properties and Applications

PAULA D. GALGANO
Nitro Química, São Paulo, Brazil
OMAR A. EL SEOUD
Institute of Chemistry, University of São Paulo, São Paulo, Brazil

4.1 INTRODUCTION

A simple and valid approach for appreciating the importance of ionic liquids (ILs) and the structurally related ionic liquid-based surfactants (ILBSs) is to examine the evolution of publications on both subjects during the past 25 years, based on a search in SciFinder data base and depicted in Figure 4.1. As shown in Figure 4.1a, at the start of the current millennium, the versatile and extremely useful properties of ILs have been recognized; this led to an explosive growth in the number of basic research papers on ILs, accompanied by a sustained increase in the number of patents. Shortly afterward, it was recognized that these properties should also apply to compounds that carry long hydrocarbon chains, that is, compounds that exhibit surface activity [1, 2]. This was verified and led to a parallel increase of publications of basic research and patents on ILBSs, as shown in Figure 4.1b.

We dwell on certain aspects of ILBSs that are important to their applications. This includes their synthesis and aggregation in water and nonaqueous solvents and the properties of the aggregates formed. This is followed by showing representative applications that exploit the surface-active nature of the ILBSs.

We advance several reasons for discussing the deceptively simple synthesis of ILBSs and their parent ILs: a search in the catalog of a major chemical reagents supplier has shown that only a very few ILBSs are commercialized. If the purities in these surfactants are similar to those of the structurally related ILs, commercialized by the same supplier, then this purity is not satisfactory, because 81% have purity less than or equal to 98% (Fig. 4.2). Note that "surface-active purity"

Ionic Liquid-Based Surfactant Science: Formulation, Characterization, and Applications, First Edition. Edited by Bidyut K. Paul and Satya P. Moulik.

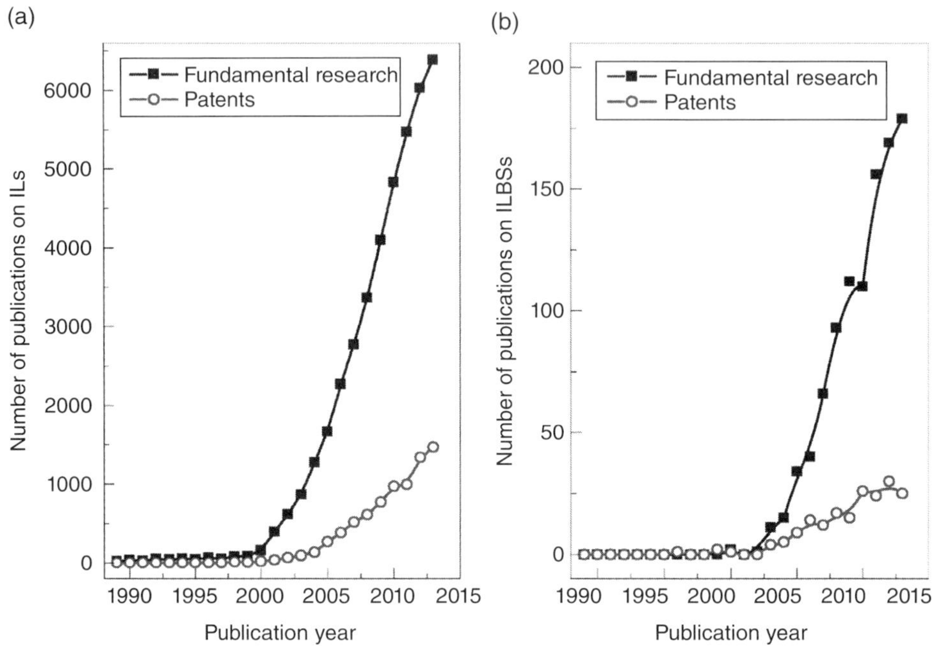

Figure 4.1 Evolution of publications on ILs and ILBSs during the period 1989–2013. Basic research papers and patents on ILs are shown in (a), whereas the corresponding plots for ILBSs are depicted in (b).

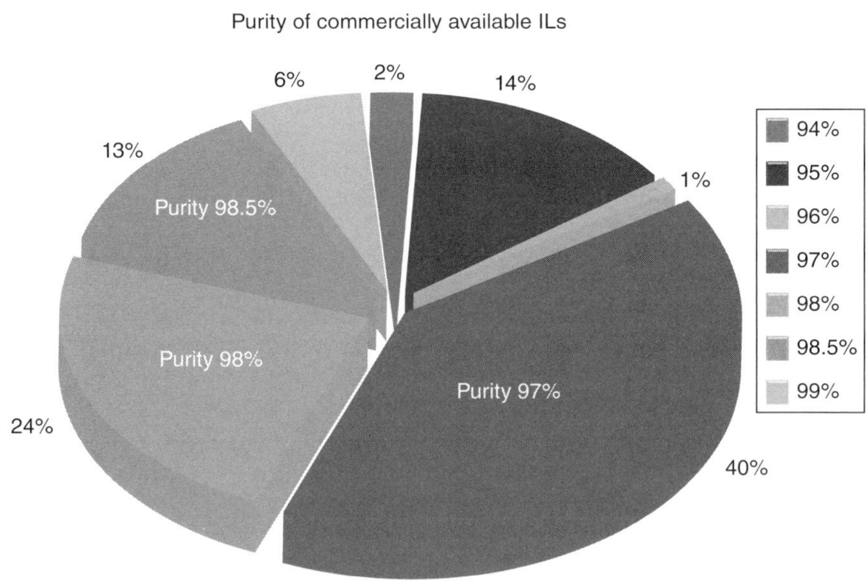

Figure 4.2 Distribution of purity of commercially available ILs from a major chemicals supplier.

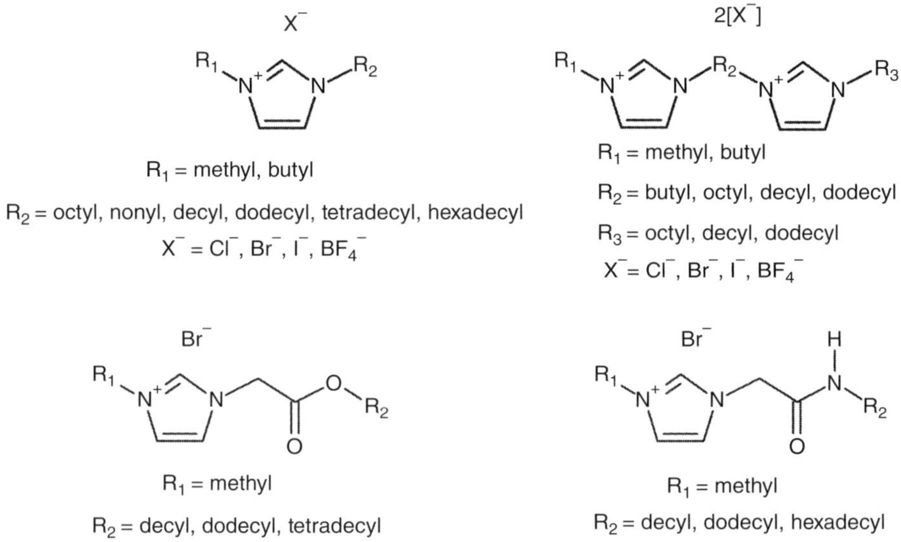

Figure 4.3 Molecular structure of the ILBSs that we cover in this review.

is important for determination of micellar properties, for example, the area per surfactant molecule at the interface (Γ) and degree of counterion dissociation (α_{mic}) [3]. Note also that surfactant purification from surface-active impurities is a tedious and laborious task [4]; synthesizing pure ILBSs from pure reactants is a superior approach.

The micellar parameters that we will discuss include critical micelle concentration (cmc), α_{mic}, the micellar average aggregation number (N_{agg}), and the dependence of thermodynamic parameters of micellization on the molecular structure of the surfactant, including the length of the hydrophobic "tail" and the nature of the counterion. Of particular interest are cationic ILBSs with counterions that are biodegradable, zwitterionic and gemini ILBSs (see Fig. 4.3), and their aggregates. In gemini ILBSs, two head ions, usually cationic, are joined by a spacer, whose length determines the properties of the aggregates. Finally, we show some applications where the surface-active nature of the compounds is exploited, for example, as antibacterial agents and templates for the synthesis of nanoparticles of controlled geometry.

4.2 OUTLINE OF SYNTHETIC STRATEGIES

The most employed synthetic routes for ILs or ILBSs are summarized in Scheme 4.1. Their synthesis involves one or two distinct steps: an S_N reaction between a tertiary amine, for example, 1-methylimidazole and an alkyl halide or alkyl methane sulfonate, to yield halides or methane sulfonates. The second step consists, in most cases, of metathesis (from the Greek words "meta" for change and "tithemi" for place) of the halide or methane sulfonate counterion, leading to ILBSs containing complex

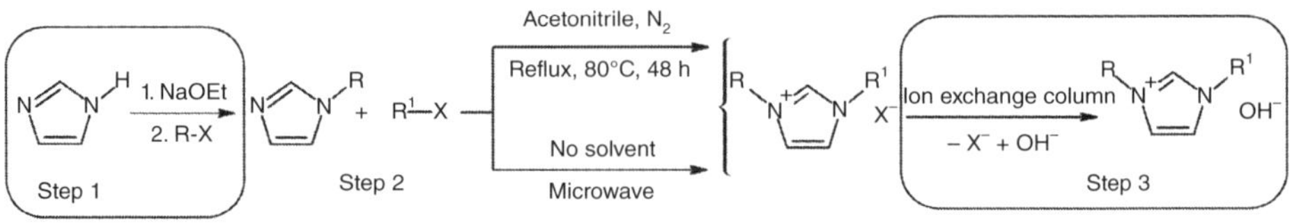

Scheme 4.1 Summary of synthetic paths to obtain ILBSs.

anions, for example, BF_4^-, PF_6^-, $C_6H_5CO_2^-$, and $(F_3CSO_2)_2N^-$, having interesting properties such as lower melting points, variable solubilities in water and organic solvents, etc. [5, 6]. Second-generation ILs have also been synthesized directly by reacting MeIm, alkyl halide in the presence of an inorganic electrolyte that carries the required final anion, for example, KPF_6 or KBF_4, under the combined effect of ultrasound (cavitation) and microwave (MW) irradiation (dielectric dissipation) [7].

The most frequently employed procedures for the S_N reaction include reflux in an appropriate molecular solvent, for example, acetonitrile, usually for many hours [8–14]. Alternatively, ILBSs have been synthesized in the absence of solvents, in a MW reactor [15], or by using MW and ultrasound irradiation [7]. The metathesis step is usually carried out in one- or two-phase systems [16, 17]. One convenient route to convert first- to second-generation IL is to exchange, for example, the halide into hydroxide on macro porous ion exchange column, followed by careful neutralization of the IL-OH by an appropriate acid.

New ILBSs structures have been recently introduced, for example, compounds that are zwitterionic, anionic, and cationic with functionalized side chain (e.g., containing ester or ether group) and gemini ILBSs. Their syntheses deserve separate comments. The synthesis of gemini-type ILBSs has been carried out by two successive S_N reactions by imidazole. This requires protecting one of the nitrogen atoms of the heterocycle, for example, by nucleophilic addition to acrylonitrile. After the first alkylation, the protecting group is removed by E1cB-type elimination, followed by the second alkylation. This protection scheme has been employed by Balthazar for preparing ILs with two different alkyl groups on the nitrogen atoms (part A), as well as gemini ILBSs (part B), and is shown in Scheme 4.2 [12].

Alternatively, gemini ILBSs have been synthesized by reacting imidazole with a di-haloalkane, followed by alkylation of the two "outer" nitrogen atoms (see Scheme 4.3).

Note that the spacer may contain ether or thioether linkage, as shown in Schemes 4.4 and 4.5.

A new type of gemini ILBSs that carries no imidazolium rings has been synthesized by Li and coworkers. These novel compounds were found to be ILs at room temperature—apart from $(C_{16}–EO_2–C_{16})Br_2$ and $(C_n–EO_3–C_n)Br_2$ ($n = 14$ or 16), which were yellow waxy solids at room temperature but melt at low temperatures.

ILBSs that carry ester, amide, or ether groups are interesting especially because these groups render the ILBSs more biodegradable [20]. ILBSs with functionalized side chain have been synthesized by the S_N reaction of substituted imidazole with α-bromoester or α-bromoamide (Scheme 4.6).

Zwitterionic ILBSs have been synthesized by reacting bromoacetic acid with *N*-alkylimidazole and then neutralizing with NaOH (Scheme 4.7). Alternatively, the

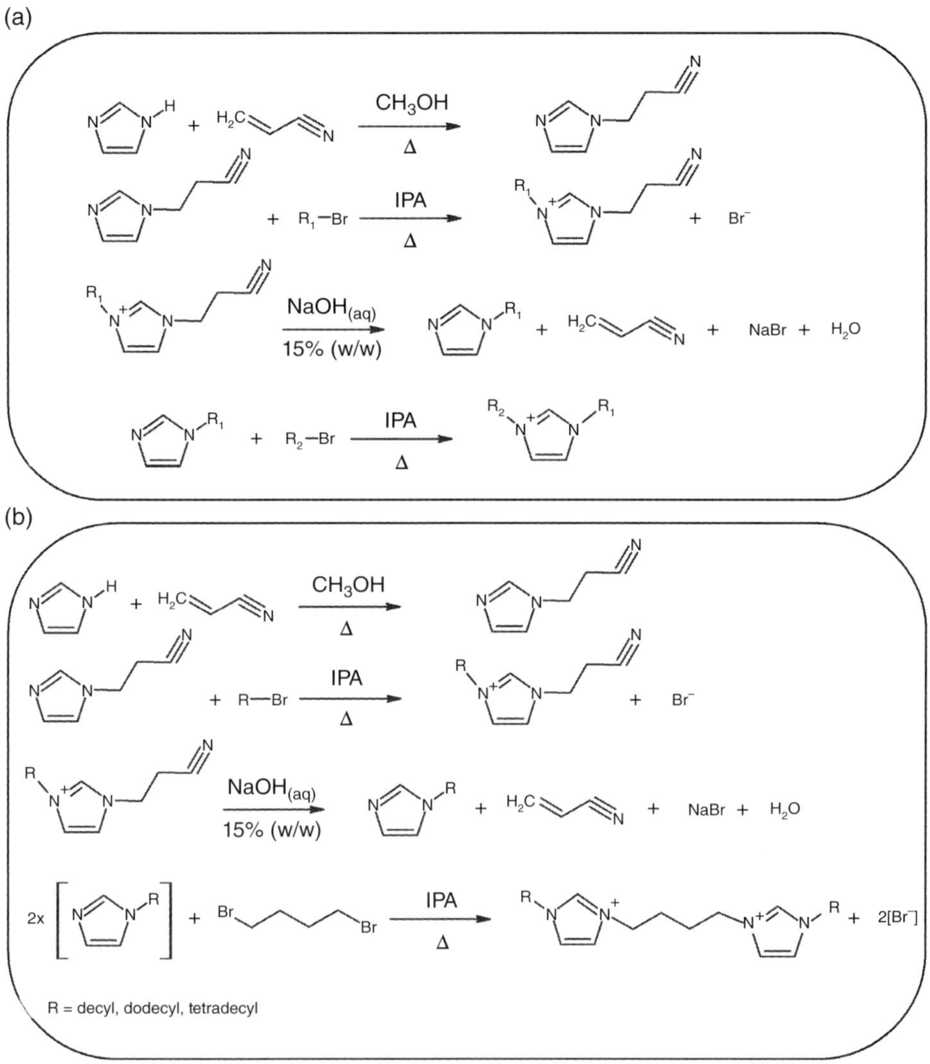

Scheme 4.2 Reaction scheme illustrating the synthesis of 1,3-dilakylimidazolium salt, and gemini cationic ILBS, via the use of a protecting group. Redrawn based on the data taken from Baltazar et al. [12] with permission from Elsevier.

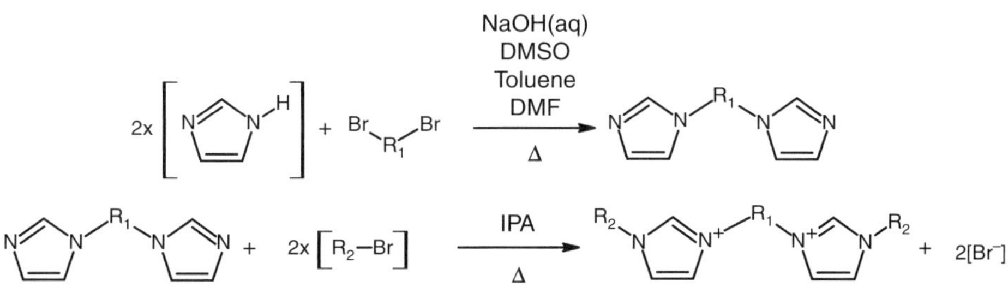

Scheme 4.3 Synthetic schemes illustrating the synthesis of dicationic (*gemini*) ionic liquids from dihaloalkanes. Redrawn based on the data taken from Baltazar et al. [12] with permission from Elsevier.

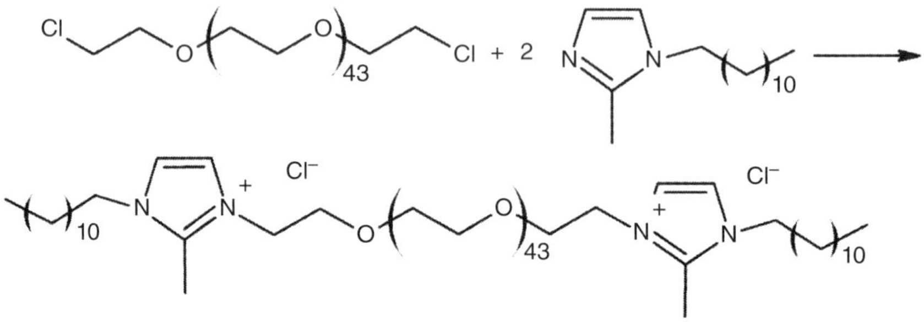

Scheme 4.4 Ether-containing gemini ILBS. Source: Zhu et al. [18] with permission from RSC Publishing.

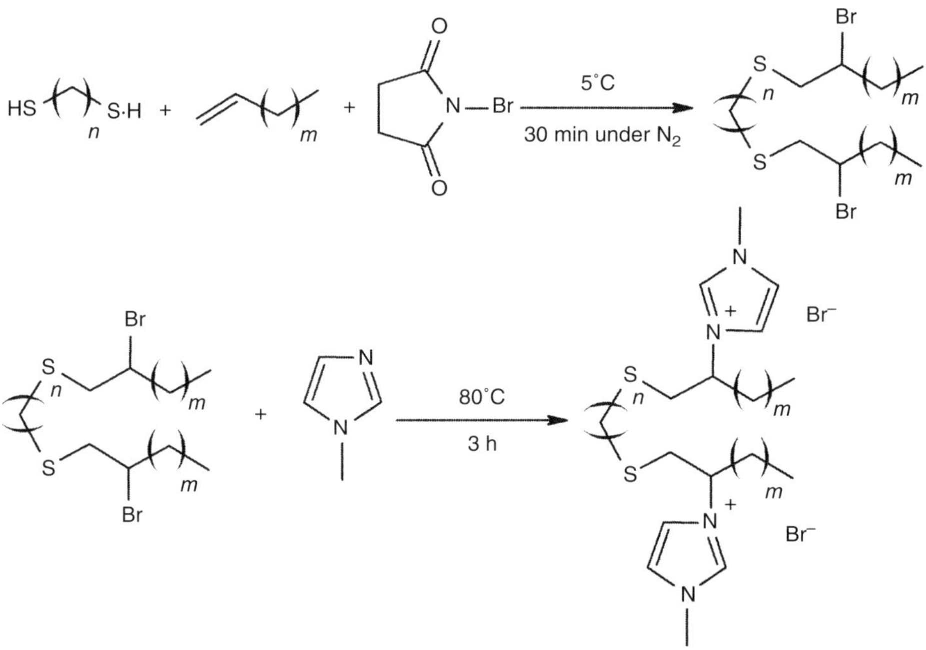

Scheme 4.5 Thioether-containing gemini ILBS. Redrawn based on the data taken from Bhadani et al. [19] with copyright permission from ACS Publications.

nitrogen of imidazole has been protected with acrylonitrile (see Scheme 4.5), followed by reaction with 1-bromoalkane to form the imidazolium salt, deprotection, and reaction with sodium chloroacetate (Scheme 4.8) [23].

4.3 SOLUTION PROPERTIES OF MICELLIZED ILBS

We set to compile literature data that show the effects of structure of ILBSs on the properties of their solutions, hence on their applications. Because the imidazole ring carries two nitrogen atoms that can be substituted independently, most of these

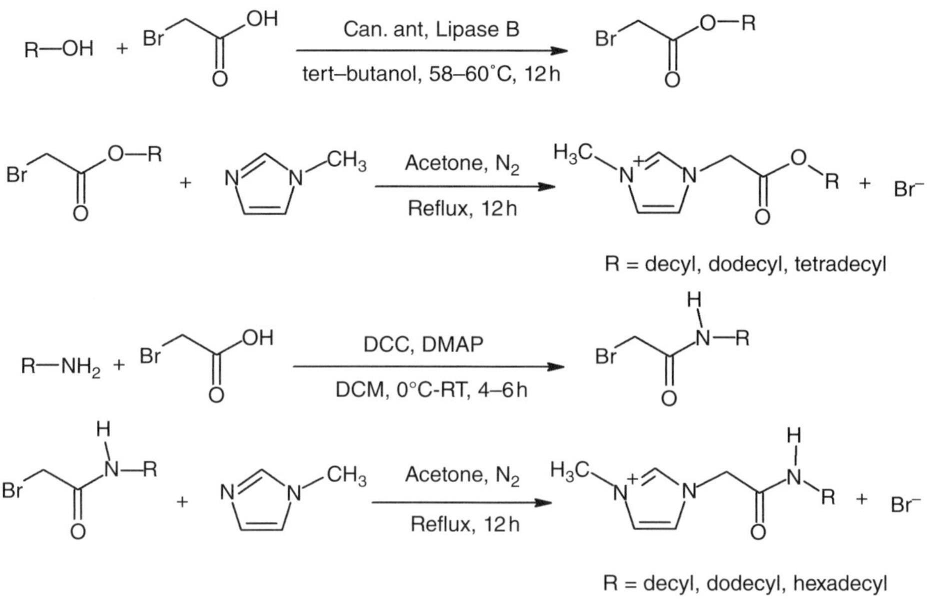

Scheme 4.6 Synthetic route to ILBSs with side ester or amide group side chain. Source: Kanjilal et al. [13] and reproduced with permission from Wiley-VCH.

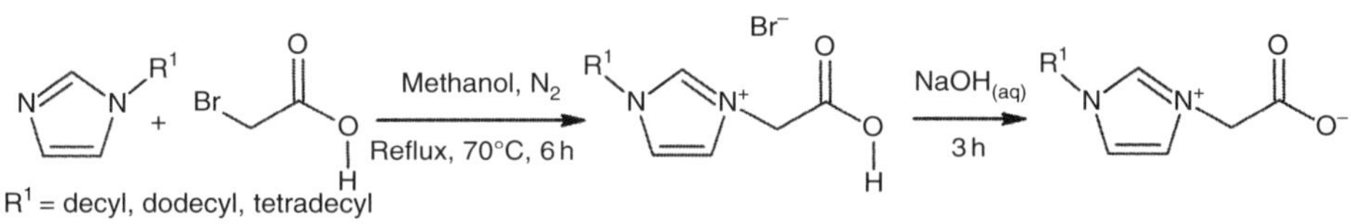

Scheme 4.7 Synthesis of zwitterionic ILBSs, from *N*-alkylimidazole and bromoacetic acid. Source: Wang et al. [21, 22] and reproduced with permission from Elsevier and ACS, respectively.

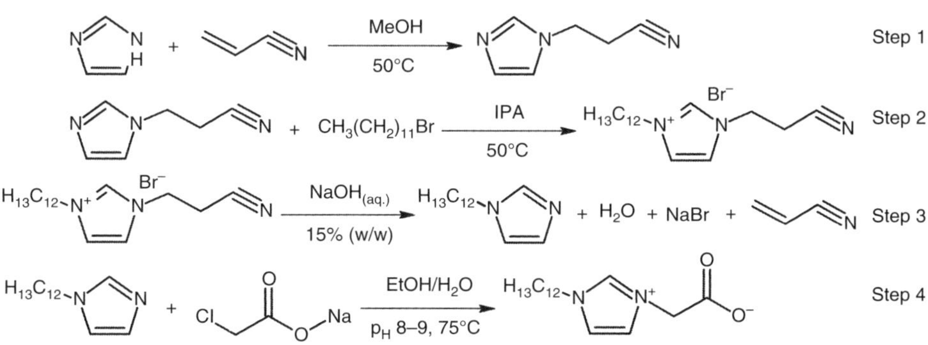

Scheme 4.8 Synthesis of zwitterionic ILBS by a sequence of N-protection, SN2 reaction, deprotection, and a second SN2 reaction with sodium chloroacetate. Source: Liu et al. [23] and reproduced with permission from Springer.

surfactants are based on the 1,3-dialkylimidazolium ion, usually 1-R-3-methylimidazolium, with R ranging from C_8 to C_{16}; the counterions are mostly halides. Most published work has been on single-chain, cationic ILBSs; representative examples of these will be discussed, along with the more recently introduced surfactants.

The progress of application of ILBSs in diverse fields, for example, organic and polymer synthesis and nanotechnology [5, 6, 24–27] rests on a clear understanding of the molecular structure/property/performance relationships of both bulk compounds and their solutions in water and organic solvents.

Attempt of correlating the molecular structures and experimental data, for example, cmc, α_{mic}, and the thermodynamic parameters of micellization (enthalpy, entropy, and free energy), rests on the assumption that they have been calculated by a consistent procedure; this point needs further consideration. At the outset, it should be noticed that there are systematic differences between the results, for example, the cmc, obtained by using distinct experimental techniques. The reason is that the function plotted (absorbance of micelle-solubilized dye, conductivity, surface tension, light scattering intensity, etc.) versus [surfactant] measures different averages of the various species in solution. Examples are surface tension that primarily depends on monomer concentration and solubilization of (water-insoluble) dye that depends mainly on the total amount of micelles present. The consequence is that cmc measured from surface tension will always be lower than cmc measured by dye solubilization [28]. In fact, values of the cmc of the same surfactant, determined by different groups, by the same technique show differences. For example, fifty-four cmc's *determined by the same technique* for $C_{16}NMe_3Br$ (measurements at 25°C) differ by 22% [29].

Two equations have been employed for calculating α_{mic}, one introduced by Frahm et al. (see Eq. 4.1) [30] and the other by Evans (see Eq. 4.2) [31]; for simplicity, both equations refer to a quaternary ammonium chloride surfactant. The basic difference between the two approaches is that solution conductivity at [surfactant] > cmc is attributed either solely to the free, that is, dissociated counterion (Frahm), or to the latter plus the micelle (a macroion; Evans):

$$\alpha_{mic} = \frac{S_2}{S_1} \tag{4.1}$$

$$1000S_2 = \frac{\alpha_{mic}^2}{(N_{agg})^{-2/3}}(1000S_1 - \Lambda_{Cl^-}) + \alpha_{mic}\Lambda_{Cl^-} \tag{4.2}$$

where S_1 and S_2 are the slopes of the pre- and post-cmc regions of conductivity versus [Surf] plot and Λ_{Cl-} refers to the equivalent conductivity of the surfactant counterion at infinite dilution. Models that are more sophisticated are also available for calculating (α_{mic}) from conductivity data at various (T) and ionic strengths; these are based on the mass action micellization thermodynamics and the Debye–Hückel–Onsager conductivity theory [32].

Frahm stressed that α_{mic} calculated by (Eq. 4.1) "may be taken as an *approximation* if no theoretical calculations on the bases of surface potential measurements are available." This important proviso was largely ignored; the method has become popular because of its simplicity. The problem, however, is that it overestimates α_{mic}, because *it ignores the contribution of the micelle to solution conductivity above the cmc*; this contribution is far from being negligible [32, 33]. Overestimation of α_{mic} leads to smaller $|\Delta G^0_{mic}|$ (see Eq. 4.3 below) [34, 35]; the error is then carried over to ΔS^0_{mic}:

$$\Delta G^0_{mic} = RT\left[(2-\alpha_{mic})\ln \chi_{cmc}\right] \tag{4.3}$$

Evans equation requires that N_{agg} is available; this is not always the case. However, because of the form of Equation 4.2, an uncertainty of 12% in N_{agg} (which is more than the uncertainty of most methods for calculating N_{agg}) leads to a change of 0.01 in α_{mic}; this leads to uncertainty of only 0.16 kcal mol^{-1} in $|\Delta G^0_{mic}|$, (calculation for C_{16}MeImCl and C_{16}PyCl at 25°C) [36]. That is, the use of Evans equation is superior; N_{agg} can be calculated from the simple Tanford equation [37] or by more sophisticated geometry optimization software [36].

The enthalpies of micellization, ΔH^0_{mic}, can be calculated indirectly by use of the van't Hoff treatment or directly by isothermal titration calorimetry (ITC). Except for few cases (e.g., some nonionic surfactants), the results of these methods do not agree [38]. The main reason is that there is no provision in van't Hoff equation for factors that are important for micelle formation of ionic surfactants, in particular, the dependence of micellar geometry, surface-charge density, and extent of hydration on temperature (T) [38]. On the other hand, the effects of (T) on the aforementioned micellar parameters are included in the direct (i.e., calorimetric) determination of ΔH^0_{mic}. From Gibbs free energy relationship, any uncertainty introduced in the calculation of ΔH^0_{mic} will be carried over to ΔS^0_{mic}, so that $\Delta S^0_{mic,ITC} > \Delta S^0_{mic,Cond}$. Where available, therefore, we compare the thermodynamic quantities of micellization, based on experimental data of the same technique.

The following micellar properties of aqueous solution of ILBSs are listed in Table 4.1: the cmc, frequently determined by surface tension and conductivity (Cond); degree of counterion dissociation (α_{mic}), usually determined from conductivity measurements by using Frahm or Evans equation; and aggregation number (N_{agg}), determined by static light scattering (SLS), conductance, or fluorescence (Fl). The thermodynamic parameters of micellization, namely, Gibbs free energy (ΔG^0_{mic}), enthalpy (ΔH^0_{mic}), entropy (ΔS^0_{mic}), and heat capacity ($\Delta C^0_{p\ mic}$), are presented in Table 4.2. Because of article length limitation, we dwell here on ILBSs with the halide counterion and gemini ILBSs.

4.4 DISCUSSION OF THE MICELLAR PROPERTIES OF ILBSs

At the outset, the data listed in Tables 4.1 and 4.2 are not exhaustive, because of article length limitation. On request, we can supply more extended versions of these tables, with 229 and 167 entries, respectively. A comparative discussion of the effects of structure on the micellar properties of a large number of surfactants, as those reported in Tables 4.1 and 4.2, is not trivial. These data have been generated by a large number of authors, using not only different experimental techniques but also different assumptions in order to calculate the micellar parameters, such as the examples that we discussed earlier (α_{mic}, and ΔH^0_{mic}). Using different techniques and assumptions in calculations will most certainly lead to larger uncertainties, as shown by the values of N_{agg} of C_8MeImBr at 25°C, namely, 24 [41] and 53 [44] for SANS and

TABLE 4.1 Published Data on the Micellar Properties of ILBSs[a]

Entry	ILBS; T, °C	$10^3 \times$ cmc, mol l^{-1}		α_{mic}, Equation Employed for α_{mic}[b]; N_{agg}, Technique; Employed for N_{agg}; Other Information Available	Reference
		Surface Tension	Conductivity		
1. Cationic ILBSs with hydrocarbon side chains					
1.1. 1,3-Dialkylimidazolium halides					
1.1.1. Bromide counterion					
1	C_8MeImBr; 25	121			12
2	C_8MeImBr; RT			α_{mic} = —; 22, Th; NMR-D	39
3	C_8MeImBr; 25	170	160	0.57, Frahm; SP, NMR-cmc, Fl-cmc, NMR-D	40
4	C_8MeImBr; 25	150	150	α_{mic} = —; 24, SANS; SANS-cmc	41
5	C_8MeImBr; 25		130	0.47, Frahm; Cal-cmc = 210	42, 43
6	C_8MeImBr; 25		160	0.37, Frahm; 53, Fl, 22, Th; Fl-cmc, Dens-cmc	44
7	C_8MeImBr; 25			α_{mic} = —; 25; NMR-cmc	45
8	C_8MeImBr; 25	120.0	150.0	0.630, Frahm; cmc-electrolyte, α-electrolyte	46
9	C_8MeImBr; 30	120.7	150.7	0.628, Frahm; cmc-electrolyte, α-electrolyte	46
10	C_8MeImBr; 40	122.1	152.1	α_{mic} = 0.624, Frahm; cmc-electrolyte; α-electrolyte	46
11	C_9MeImBr; 25		74	0.26, Frahm; 45, Fl	47
12	C_{10}MeImBr; 25	20			12
13	C_{10}MeImBr; 25	33	42	0.33, Frahm; SP, NMR-cmc, Fl-cmc, NMR-D, MIC	40
14	C_{10}MeImBr; 25	29.3	32.9	0.31, Frahm	48
15	C_{10}MeImBr; 25	40	30	α_{mic} = —; 45, SANS; SANS-cmc	41
16	C_{10}MeImBr; 25		38.2	0.40, Frahm; Cal-cmc = 39.7	42, 43
17	C_{10}MeImBr; 25		41	0.29, Frahm; 42, Fl	47
18	C_{10}MeImBr; RT	14.51		SP, MIC	13
19	C_{10}MeImBr; 25		39	0.34, Frahm; 35, Fl, 33, Th; Fl-cmc, Dens-cmc	44
20	C_{10}MeImBr; 25			α_{mic} = —; 27, NMR; NMR-cmc	45
21	C_{10}MeImBr; 25		37.4	α_{mic} = 0.47, Frahm; cmc-electrolyte; α-electrolyte; N_{agg}-electrolyte FL	49
21	C_{12}MeImBr; 5		10.30		50
22	C_{12}MeImBr; 15		10.1	0.23, Shanks; 45, Cond	11

23	C_{12}MeImBr; 15		9.84	0.23, Shanks	50
24	C_{12}MeImBr; 20		9.60	0.23, Shanks; 44 Cond	11
25	C_{12}MeImBr; RT	9.89		0.27, Frahm	51
26	C_{12}MeImBr; RT	8.98	4.3		12
27	C_{12}MeImBr; 25	9.0	10.0	0.26, Frahm; SP, NMR-cmc, Fl-cmc, NMR-D, MIC	40
28	C_{12}MeImBr; 25	10.9	10.1	0.34, Frahm, Cal-cmc; Cond; SP,	102
29	C_{12}MeImBr; 25	10.9	8.5	0.27, Frahm	48
30	C_{12}MeImBr; 25		9.0	0.22, Evans; Dens-V	52
31	C_{12}MeImBr; 25		9.68	0.24, Shanks; Cal-cmc	50
32	C_{12}MeImBr; 25		9.53	0.24, Shanks; 43, Cond	11
38	C_{12}MeImBr; 30		10.1	0.26, Shanks; 44, Cond	11
39	C_{12}MeImBr; 35	10.10		0.26, Shanks	50
40	C_{12}MeImBr; 35		10.4	0.26, Shanks; 43, Cond	11
47	C_{14}MeImBr; 25	1.9	2.5	0.26, Frahm; 61, Cond; SP, NMR-cmc, Fl-cmc, NMR-D, MIC	40
48	C_{14}MeImBr; 25	2.69		0.22, Shanks	50
49	C_{14}MeImBr; 25		2.61	0.22, Shanks; 61, Cond	11
50	C_{14}MeImBr; 25		2.68	0.33, Frahm; Cal-cmc = 2.78	42, 43
51	C_{14}MeImBr; 25		2.5	0.24, Frahm; 59, Fl	47
60	C_{16}MeImBr; RT	0.646		0.32, Frahm	51
61	C_{16}MeImBr; 25	0.8			12
62	C_{16}MeImBr; 25	0.51		α_{mic} = 0.23, Shanks; Cal-cmc	50
63	C_{16}MeImBr; 25		0.653	α_{mic} = 0.23, Shanks; 76, Cond	11
64	C_{16}MeImBr; 25	0.54		Cal-cmc; SAXS of solid	53
65	C_{16}MeImBr; RT	0.77		SP, MIC	13
66	C_{16}MeImBr; 25		0.59	0.33, Frahm; Cal-cmc = 0.61	42, 43
67	C_{16}MeImBr; 28		0.61	0.30, Frahm; 66, Fl	47
68	C_{16}MeImBr; 30		0.667	0.22, Shanks; 74, Cond	11
69	C_{16}MeImBr; 35	0.52		0.24, Shanks	50
70	C_{16}MeImBr; 35		0.716	0.24, Shanks; 73, Cond	11
71	C_{16}MeImBr; 40		0.762	0.26, Shanks; 74, Cond	11
72	C_{16}MeImBr; 35	0.54			50

(*Continued*)

TABLE 4.1 (*Continued*)

Entry	ILBS; T, °C	$10^3 \times$ cmc, mol l^{-1}		α_{mic}, Equation Employed for α_{mic} ^{b}; N_{agg}, Technique; Employed for N_{agg}; Other Information Available	Reference
		Surface Tension	Conductivity		
73	C_{16}MeImBr; 35	0.57			50
74	C_8BuImBr; 25	41			12
75	C_{10}BuImBr; 25	6.3			12
76	C_{12}BuImBr; 25	2.4			12
77	C_{16}BuImBr; 25	0.1			12
78	C_8OcImBr; 25	5.6			12
79	C_{12}DoImBr; 25	0.1			12
1.1.2. Chloride counterion					
86	C_8MeImCl; 25	101.7	101.2	0.41, Frahm; 24, SANS; cmc-electrolyte, α-electrolyte, N_{agg}-electrolyte	46
87	C_8MeImCl; 25	234	220	0.45, Frahm, ΔG^0_{mic}, micellar retention time	54
88	C_{10}MeImCl; 9.85		62	0.52, Frahm	55
89	C_{10}MeImCl; 25	39.90	40.47	0.26, Evans; 40, Fl, 48, Th; SP, Fl-cmc	9
90	C_{10}MeImCl; 25		56.42	0.25, Evans; 23, SLS, 32, Th; Cal-cmc	56
91	C_{10}MeImCl; 25			NMR-cmc, Fl-cmc, cmc-electrolyte	57
92	C_{10}MeImCl; 45		57.12	α_{mic} = 0.28, Evans; 18 SLS, 30, Th; Cal-cmc	56
93	C_{10}MeImCl; 55		56.83	0.30, Evans; 24, Th; Cal-cmc	56
94	C_{10}MeImCl; 65		59.79	0.32, Evans; 22, Th; Cal-cmc	56
95	C_{10}MeImCl; 75		69.02	0.33, Evans; 19, Th; Cal-cmc	56
96	C_{12}MeImCl; 15		14.57	0.22, Evans; 54 SLS, 57, Th; Cal-cmc	56
97	C_{12}MeImCl; 25	13.7	13.47	0.23, Evans; 58, Fl, 67, Th; EMF-cmc, Fl-cmc, EMF-α	9
98	C_{12}MeImCl; 25		14.14	0.24, Evans; 48 SLS, 50, Th; Cal-cmc	56
99	C_{12}MeImCl; 25			NMR-cmc, Fl-cmc, cmc-electrolyte	57
100	C_{12}MeImCl; 35		14.23	0.25, Evans; 42 SLS, 43, Th; Cal-cmc	56
101	C_{12}MeImCl; 35		14	0.44, Frahm; 37 SLS; Cal-cmc	58
102	C_{12}MeImCl; 45		15.00	0.27, Evans; 34 SLS, 37, Th; Cal-cmc	56
103	C_{12}MeImCl; 55		16.00	0.31, Evans; 30, Th; Cal-cmc	56
104	C_{12}MeImCl; 65		17.58	0.31, Evans; 24, Th; Cal-cmc	56

105	C_{12}MeImCl; 75		20.42	0.33, Evans; 17, Th; Cal-cmc	56
106	C_{14}MeImCl; 15		3.64	0.24, Evans; 79 SLS, 84, Th; Cal-cmc	56
107	C_{14}MeImCl; 25	2.98	3.68	0.20, Evans; 79, Fl, 89, Th; EMF-cmc, Fl-cmc, EMF-α_{mic}	9
108	C_{14}MeImCl; 25		3.32	0.26, Evans; 70 SLS, 76, Th; Cal-cmc	56
109	C_{14}MeImCl; 25	4		NMR-cmc, Fl-cmc	57
110	C_{14}MeImCl; 25	3.15	3.38	0.42, Frahm, ΔG^0_{mic}, micellar retention time	54
111	C_{14}MeImCl; 35		3.70	0.24, Evans; 61 SLS, 68, Th; Cal-cmc	56
112	C_{14}MeImCl; 35		3.8	0.41, Frahm; 46, SLS; cal-cmc	58
113	C_{14}MeImCl; 45		3.76	0.27, Evans; 55 SLS, 60, Th; Cal-cmc	56
114	C_{14}MeImCl; 55		3.83	0.27, Evans; 51, Th; Cal-cmc	56
115	C_{14}MeImCl; 65		4.38	0.29, Evans; 42, Th; Cal-cmc	56
116	C_{14}MeImCl; 75		5.53	0.29, Evans; 34, Th; Cal-cmc	56
117	C_{16}MeImCl; 15		0.85	0.19, Evans; 111 SLS, 109, Th; Cal-cmc	56
118	C_{16}MeImCl; 25	0.87	0.86	0.19, Evans; 99, Fl, 114, Th; EMF-cmc, Fl-cmc, EMF-α_{mic}	9
119	C_{16}MeImCl; 25		0.86	0.21, Evans; 97 SLS, 96, Th; Cal-cmc	56
120	C_{16}MeImCl; 25	1.14	1.26	0.55, Frahm, ΔG^0_{mic}, micellar retention time	54
121	C_{16}MeImCl; 25	1.1		Cal-cmc; SAXS of solid	53
122	C_{16}MeImCl; 35		0.91	0.23, Evans; 79 SLS, 83, Th; Cal-cmc	56
123	C_{16}MeImCl; 45		1.01	0.25, Evans; 73 SLS, 70, Th; Cal-cmc	56
124	C_{16}MeImCl; 55		1.17	0.27, Evans; 56, Th; Cal-cmc	56
125	C_{16}MeImCl; 65		1.40	0.29, Evans; 43, Th; Cal-cmc	56
126	C_{16}MeImCl; 75		1.75	0.31, Evans; 30, Th; Cal-cmc	56
127	C_{18}MeImCl; 25	0.45	0.43	0.60, Frahm, ΔG^0_{mic}, micellar retention time	54
1.1.3. Iodide counterion					
128	C_8MeImI; 25	94.9	149.8	0.644, Frahm; cmc-electrolyte, α-electrolyte	46
129	C_8MeImI; 30	95.4	150.3	0.644, Frahm; cmc-electrolyte, α-electrolyte	46
130	C_8MeImI; 40	96.5	151.4	0.640, Frahm; cmc-electrolyte, α-electrolyte	46
1.1.4. Miscellaneous counterions: Tetrafluoroborate and methanesulfonate					
130	C_{12}MeImBF$_4$; 25	9.2	7.6	0.27, Frahm	48
131	C_{16}MeImBF$_4$; 25			cmc-electrolyte, α-electrolyte	8
132	C_{16}MeImBF$_4$; 40	1.37			8

(*Continued*)

TABLE 4.1 (*Continued*)

Entry	ILBS; T, °C	$10^3 \times$ cmc, mol l^{-1}		α_{mic}, Equation Employed for α_{mic} [b]; N_{agg}, Technique; Employed for N_{agg}; Other Information Available	Reference
		Surface Tension	Conductivity		
133	$C_8MeImCH_3SO_3$; 23	220		SP	59
134	$C_{10}MeImCH_3SO_3$; 23	60		SP, Fl-cmc	59
135	$C_{12}MeImCH_3SO_3$; 23	14		SP, Fl-cmc	59
2. Cationic ILBSs with functionalized side chains: Amides and esters					
136	$C_{10}NHCOCH_2MeImBr$	4.11		SP, MIC	13
137	$C_{12}NHCOCH_2MeImBr$	0.34		SP, MIC	13
138	$C_{16}NHCOCH_2MeImBr$	0.058		SP, MIC	13
139	$C_{10}OCOCH_2MeImBr$	0.76		SP, MIC	13
140	$C_{12}OCOCH_2MeImBr$	0.23		SP, MIC	13
141	$C_{14}OCOCH_2MeImBr$	0.038		SP, MIC	13
3. Cationic ILBSs with chiral centers					
142	C_{16}(Hydroxy-propan-2-yl) ImBr; 25	0.577	0.563	25, Fl; SP, Fl-α, NMR-α	60
143	C_{16}(Hydroxy-propan-2-yl) ImBr; 30	0.558			60
144	C_{16}(Hydroxy-propan-2-yl) ImBr; 35	0.562			60
145	C_{16}(Hydroxy-propan-2-yl) ImBr; 40	0.576			60
146	C_{16}(Hydroxy-propan-2-yl) ImBr; 45	0.593			60
147	C_{16}(Hydroxy-propan-2-yl) ImBr; 45	0.608			60
4. Gemini ILBSs[d]					
148	$(C_1Im\text{-}10\text{-}C_1Im)Br_2$; 25	149			12
149	$(C_4Im\text{-}8\text{-}C_4Im)Br_2$; 25	323			12
150	$(C_4Im\text{-}10\text{-}C_4Im)Br_2$; 25	117			12

151	$(C_4Im\text{-}12\text{-}C_4Im)Br_2$; 25	72			12
152	$(C_{10}Im\text{-}4\text{-}C_{10}Im)Br_2$; 15		5.05	0.30, Frahm	61
153	$(C_{10}Im\text{-}4\text{-}C_{10}Im)Br_2$; 20		5.06	0.31, Frahm	61
154	$(C_{10}Im\text{-}4\text{-}C_{10}Im)Br_2$; 25	2.0			12
155	$(C_{12}Im\text{-}2\text{-}C_{12}Im)Br_2$; 25	0.55	0.64	0.23, Frahm; 21, Fl; 0.470 (1.0 mM NaCl), 0.125 (1.0 mM NaBr), 0.028 (1.0 mM NaI), cmc-electrolyte	10
156	$(C_{12}Im\text{-}2\text{-}C_{12}Im)Br_2$; 30		0.66	0.26, Frahm	10
157	$(C_{12}Im\text{-}2\text{-}C_{12}Im)Br_2$; 35		0.67	0.31, Frahm	10
158	$(C_{12}Im\text{-}4\text{-}C_{12}Im)Br_2$			$M_T = 55.0$, $\Delta H^0_M = 68.3\,kJ\,mol^{-1}$	62
159	$(C_{12}Im\text{-}4\text{-}C_{12}Im)Br_2$; 15		0.68	0.25, Frahm	61
160	$(C_{12}Im\text{-}4\text{-}C_{12}Im)Br_2$; 20		0.71	0.28, Frahm	61
161	$(C_{12}Im\text{-}4\text{-}C_{12}Im)Br_2$; 25	0.72	0.73	0.30, Frahm; 16, Fl; 0.608 (1.0 mM NaCl), 0.270 (1.0 mM NaBr), 0.043 (1.0 mM NaI), cmc-electrolyte	10, 61
162	$(C_{12}Im\text{-}4\text{-}C_{12}Im)Br_2$; 30		0.76	0.35, Frahm	61
163	$(C_{12}Im\text{-}4\text{-}C_{12}Im)Br_2$; 35		0.79	0.40, Frahm	61
164	$(C_{14}Im\text{-}4\text{-}C_{14}Im)Br_2$; 15		0.10	0.39, Frahm	61
165	$(C_{14}Im\text{-}4\text{-}C_{14}Im)Br_2$; 20		0.11	0.42, Frahm	61
166	$(C_{14}Im\text{-}4\text{-}C_{14}Im)Br_2$; 25	0.1	0.12	0.43, Frahm	61
167	$(C_{14}Im\text{-}4\text{-}C_{14}Im)Br_2$; 30		0.13	0.44, Frahm	61
168	$(C_{14}Im\text{-}4\text{-}C_{14}Im)Br_2$; 35		0.14	0.45, Frahm	61
169	$(C_{12}Im\text{-}6\text{-}C_{12}Im)Br_2$; 15		0.61	0.37, Frahm	10
170	$(C_{12}Im\text{-}6\text{-}C_{12}Im)Br_2$; 20		0.67	0.39, Frahm	10
171	$(C_{12}Im\text{-}6\text{-}C_{12}Im)Br_2$; 25	0.78	0.74	0.40, Frahm; 14, Fl; 4.0 nm, R_H-DLS; 0.707 (1.0 mM NaCl), 0.307 (1.0 mM NaBr), 0.050 (1.0 mM NaI), cmc-electrolyte	10
172	$(C_{12}Im\text{-}6\text{-}C_{12}Im)Br_2$; 30		0.80	0.43, Frahm; 2.5 nm, R_H-DLS	10
173	$(C_{12}Im\text{-}6\text{-}C_{12}Im)Br_2$; 35		0.86	0.45, Frahm	10
174	$(C_8\text{-}EO_2\text{-}C_8)Br_2$; 20	1.822			63
175	$(C_{10}\text{-}EO_2\text{-}C_{10})Br_2$; 20	1.70			63
176	$(C_{12}\text{-}EO_2\text{-}C_{12})Br_2$; 20	1.239			63
177	$(C_{14}\text{-}EO_2\text{-}C_{14})Br_2$; 20	0.333			63

(Continued)

TABLE 4.1 *(Continued)*

Entry	ILBS; T, °C	$10^3 \times$ cmc, mol l^{-1}		α_{mic}, Equation Employed for α_{mic}[b]; N_{agg}, Technique; Employed for N_{agg}; Other Information Available	Reference
		Surface Tension	Conductivity		
178	$(C_{16}\text{-}EO_2\text{-}C_{16})Br_2$; 20	0.389			63
179	$(C_8\text{-}EO_3\text{-}C_8)Br_2$; 20	0.701			63
180	$(C_{10}\text{-}EO_3\text{-}C_{10})Br_2$; 20	0.649			63
181	$(C_{12}\text{-}EO_3\text{-}C_{12})Br_2$; 20	0.607			63
182	$(C_{14}\text{-}EO_3\text{-}C_{14})Br_2$; 20	0.502			63
183	$(C_{16}\text{-}EO_3\text{-}C_{16})Br_2$; 20	0.398			63
184	$(C_{12}\text{-}S\text{-}2\text{-}S\text{-}C_{12})Br_2$; 25	0.32	0.40	0.37, Frahm; $A_{min} = 0.63\,nm^2$	19
185	$(C_{12}\text{-}S\text{-}3\text{-}S\text{-}C_{12})Br_2$; 25	0.26	0.35	0.37, Frahm; $A_{min} = 0.77\,nm^2$	19
186	$(C_{12}\text{-}S\text{-}4\text{-}S\text{-}C_{12})Br_2$; 25	0.22	0.31	0.42, Frahm; $A_{min} = 0.80\,nm^2$	19
187	$(C_{14}\text{-}S\text{-}2\text{-}S\text{-}C_{14})Br_2$; 25	0.072	0.10	0.46, Frahm; $A_{min} = 0.78\,nm^2$	19
188	$(C_{14}\text{-}S\text{-}3\text{-}S\text{-}C_{14})Br_2$; 25	0.063	0.085	0.48, Frahm; $A_{min} = 0.53\,nm^2$	19
189	$(C_{14}\text{-}S\text{-}4\text{-}S\text{-}C_{14})Br_2$; 25	0.058	0.071	0.47, Frahm; $A_{min} = 0.53\,nm^2$	19
190	$(C_{16}\text{-}S\text{-}2\text{-}S\text{-}C_{16})Br_2$; 25		0.022	0.33, Frahm	19
191	$(C_{16}\text{-}S\text{-}3\text{-}S\text{-}C_{16})Br_2$; 25		0.021	0.34, Frahm	19
192	$(C_{16}\text{-}S\text{-}4\text{-}S\text{-}C_{16})Br_2$; 25		0.020	0.35, Frahm	19

[a] *Abbreviations*: A_{min}, minimum area of the surfactant monomer at the water/air interface; AOT, bis(2-ethylhexyl) sulfosuccinate; C_4MePoDS, *N*-(1-butyl),*N*-methylpyrrolidinium; Cal-cmc, cmc determined by calorimetry measurements; cmc-electrolyte, cmc determined in the presence of an additional electrolyte; CMe, carboxymethyl group; Dens-cmc, cmc calculated from density measurements; DS, dodecyl sulfate; Fc, ferrocene; Fc^+, ferrocenium ion; Fl, fluorescence quenching of micelle-incorporated probe; Fl-cmc, cmc calculated from fluorescence quenching of micelle-incorporated probe; MIC, minimum inhibitory concentration; NMR-cmc, cmc calculated from NMR measurements; NMR-D, micellar diffusion coefficient calculated from NMR data; Po, pyrrolidinonium; R_H-DLS, hydrodynamic radius measured by dynamic light scattering; RT, room temperature; SANS, small-angle neutron scattering; SANS-cmc, cmc, calculated from small-angle neutron scattering measurements; SLS, static light scattering; SP, surfactant properties at the solution/air interface, calculated from surface tension measurements; TC, aerosol-OT trichain; $T_{cmc,min}$, temperature of minimum value of cmc; Th, theoretically calculated N_{agg}; α_{mic}-electrolyte, counterion dissociation degree determined in the presence of an additional electrolyte; γ_{cmc}, surface tension at the cmc.

[b] The equations for determination of α_{mic} are those of Frahm et al. [30], Evans [31], and Shanks and Franses [32].

[c] An example of the nomenclature employed for gemini surfactants is $(C_4Im\text{-}8\text{-}C_4Im)Br_2$. This means that two *N*-(1-butyl)imidazolium rings are joined at the second *N* (of the heterocyclic ring) by eight (CH_2) groups. This surfactant is synthesized by the reaction of *N*-(1-butyl)imidazole with 1,8-dibromooctane.

TABLE 4.2 Thermodynamic Parameters of Micelle Formation of ILBSs[a]

Entry	ILBS; T, °C	ΔG^0_{mic}	ΔH^0_{mic}, Technique[b]	$-T\Delta S^0_{mic}$	$\Delta C^0_{p\,mic}$; Other Information Available	Reference
1,3-Dialkylimidazolium chlorides						
1	C_{10}MeImCl; 15	–28.69	4.75, ITC	–33.44	–316	56
2	C_{10}MeImCl; 25	–31.14				9
3	C_{10}MeImCl; 25	–29.45	2.85, ITC	–32.30	–303	56
4	C_{10}MeImCl; 35	–30.20	0.16, ITC	–30.36	–289	56
5	C_{10}MeImCl; 35	–24.8	–1.3, ITC	–23.5		58
6	C_{10}MeImCl; 45	–30.86	–2.85, ITC	–28.01	–275	56
7	C_{10}MeImCl; 55	–31.04	–4.67, ITC	–26.37	–262	56
8	C_{10}MeImCl; 65	–31.69	–6.33, ITC	–25.36	–248	56
9	C_{10}MeImCl; 75	–32.26	–7.43, ITC	–24.83	–234	56
10	C_{12}MeImCl; 15	–34.92	6.78, ITC	–41.70	–579	56
11	C_{12}MeImCl; 25	–36.51				9
12	C_{12}MeImCl; 25		3.913			64
13	C_{12}MeImCl; 25	–36.15	2.04, ITC	–38.19	–551	56
14	C_{12}MeImCl; 32		1.206			64
15	C_{12}MeImCl; 35	–36.88	–3.10, ITC	–33.78	–524	56
16	C_{12}MeImCl; 35	–33.2	–2.9, ITC	–30.3		58
17	C_{12}MeImCl; 40		–1.253			64
18	C_{12}MeImCl; 45	–37.78	–7.00, ITC	–30.78	–496	56
19	C_{12}MeImCl; 55	–37.52	–10.14, ITC	–27.38	–469	56
20	C_{12}MeImCl; 65	–38.23	–12.77, ITC	–25.46	–441	56
21	C_{12}MeImCl; 75	–38.39	–15.68, ITC	–22.71	–413	56
22	C_{14}MeImCl; 15	–40.56	5.51, ITC	–46.07	–752	56
23	C_{14}MeImCl; 25	–42.91				9
24	C_{14}MeImCl; 25	–41.82	–0.54, ITC	–41.28	–707	56
25	C_{14}MeImCl; 35	–43.39	–6.32, ITC	–37.07	–662	56
26	C_{14}MeImCl; 35	–39.6	–6.3, ITC	–32.5		58

(Continued)

TABLE 4.2 (*Continued*)

Entry	ILBS; T, °C	ΔG^0_{mic}	ΔH^0_{mic}, Technique[b]	$-T\Delta S^0_{mic}$	$\Delta C^0_{p\,mic}$; Other Information Available	Reference
27	C_{14}MeImCl; 45	−43.62	−10.41, ITC	−33.21	−617	56
28	C_{14}MeImCl; 55	−44.815	−15.05, ITC	−29.77	−572	56
29	C_{14}MeImCl; 65	−45.057	−17.31, ITC	−27.75	−527	56
30	C_{14}MeImCl; 75	−45.661	−19.11, ITC	−26.55	−482	56
31	C_{16}MeImCl; 10		5.198			64
32	C_{16}MeImCl; 15	−47.42	4.25, ITC	−51.67	−1016	36
33	C_{16}MeImCl; 18		2.239			64
34	C_{16}MeImCl; 25	−49.58				9
35	C_{16}MeImCl; 25		−2.176		ΔH^0_{mic} - electrolyte	64
36	C_{16}MeImCl; 25	−48.63	−2.75, ITC	−45.88	−948	36
37	C_{16}MeImCl; 25		−5.9, ITC			53
38	C_{16}MeImCl; 32		−5.323		ΔH^0_{mic} - electrolyte	64
39	C_{16}MeImCl; 35	−49.64	−10.08, ITC	−39.56	−879	36
40	C_{16}MeImCl; 40		−8.281		ΔH^0_{mic} - electrolyte	64
41	C_{16}MeImCl; 45	−50.41	−16.5, ITC	−33.91	−811	36
42	C_{16}MeImCl; 50		−11.899			64
43	C_{16}MeImCl; 55	−50.67	−21.12, ITC	−29.55	−742	36
44	C_{16}MeImCl; 65	−50.75	−24.12, ITC	−26.63	−673	36
45	C_{16}MeImCl; 75	−50.71	−24.64, ITC	−26.07	−605	36
Gemini ILBSs						
46	$(C_{10}Im\text{-}4\text{-}C_{10}Im)Br_2$; 25	−26.96	−2.84	−24.12		61
47	$(C_{10}Im\text{-}4\text{-}C_{10}Im)Br_2$; 25	−26.96			$\Delta G^0_{solv} = -29.44\,kJ\,mol^{-1}$	65
48	$(C_{12}Im\text{-}2\text{-}C_{12}Im)Br_2$; 25	−34.41	−5.21	−29.2		10
49	$(C_{12}Im\text{-}2\text{-}C_{12}Im)Br_2$; 25	−34.41			$\Delta G^0_{solv} = -36.05\,kJ\,mol^{-1}$	65
50	$(C_{12}Im\text{-}4\text{-}C_{12}Im)Br_2$; 25	−33.41	−5.77	−27.64		61
51	$(C_{12}Im\text{-}4\text{-}C_{12}Im)Br_2$; 25	−33.41			$\Delta G^0_{solv} = -34.51\,kJ\,mol^{-1}$	65
52	$(C_{12}Im\text{-}6\text{-}C_{12}Im)Br_2$; 25	−30.59	−15.98	−14.61		10
53	$(C_{12}Im\text{-}8\text{-}C_{12}Im)Br_2$; 25				$\Delta G^0_{solv} = -33.44\,kJ\,mol^{-1}$	65

54	$(C_{14}Im\text{-}4\text{-}C_{14}Im)Br_2$; 15	−35.15	−15.49	19.66		61
55	$(C_{14}Im\text{-}4\text{-}C_{14}Im)Br_2$; 15	−35.15	−10.20	−25.66	$\Delta G^0_{solv} = -32.86\,kJ\,mol^{-1}$	65
56	$(C_{14}Im\text{-}4\text{-}C_{14}Im)Br_2$; 20	−34.54	−14.52	20.02		61
57	$(C_{14}Im\text{-}4\text{-}C_{14}Im)Br_2$; 20	−34.54	−10.92	−25.60	$\Delta G^0_{solv} = -33.70\,kJ\,mol^{-1}$	65
58	$(C_{14}Im\text{-}4\text{-}C_{14}Im)Br_2$; 25	−34.58	−13.78	20.80		61
59	$(C_{14}Im\text{-}4\text{-}C_{14}Im)Br_2$; 25	−34.58	−11.67	−25.53	$\Delta G^0_{solv} = -34.51\,kJ\,mol^{-1}$	65
60	$(C_{12}\text{-}S\text{-}2\text{-}S\text{-}C_{12})Br_2$; 25	−33.14			$\Delta G^0_{ads} = -45.57\,kJ\,mol^{-1}$	19
61	$(C_{12}\text{-}S\text{-}3\text{-}S\text{-}C_{12})Br_2$	−33.52			$\Delta G^0_{ads} = -48.21\,kJ\,mol^{-1}$	19
62	$(C_{12}\text{-}S\text{-}4\text{-}S\text{-}C_{12})Br_2$; 25	−32.36			$\Delta G^0_{ads} = -47.51\,kJ\,mol^{-1}$	19
63	$(C_{14}\text{-}S\text{-}2\text{-}S\text{-}C_{14})Br_2$; 25	−34.08			$\Delta G^0_{ads} = -47.80\,kJ\,mol^{-1}$	19
64	$(C_{14}\text{-}S\text{-}3\text{-}S\text{-}C_{14})Br_2$; 25	−33.83			$\Delta G^0_{ads} = -42.31\,kJ\,mol^{-1}$	19
65	$(C_{14}\text{-}S\text{-}4\text{-}S\text{-}C_{14})Br_2$; 25	−34.62			$\Delta G^0_{ads} = -42.82\,kJ\,mol^{-1}$	19
66	$(C_{16}\text{-}S\text{-}2\text{-}S\text{-}C_{16})Br_2$; 25	−42.72				19
67	$(C_{16}\text{-}S\text{-}3\text{-}S\text{-}C_{16})Br_2$; 25	−42.49				19
68	$(C_{16}\text{-}S\text{-}4\text{-}S\text{-}C_{16})Br_2$; 25	−42.26				19

[a] For a list of abbreviations and an example of the nomenclature applied for gemini surfactants, see footnotes on Table 4.1.
[b] ITC, isothermal titration calorimetry; vH, van't Hoff equation.

fluorescence quenching, respectively, a 220% difference! Faced with this difficulty, we took a conservative approach in order to extract the dependence of the micellar parameters on surfactant molecular structure. Therefore, we plotted the data and considered those which "cluster" around a certain value *and whose use in surfactant structure/micellar property correlations agrees with chemistry*. For example, the value of the free energy of transfer of a CH_2 group from bulk aqueous pseudophase to the interior of the micellar aggregate, $\Delta G^0_{\mathrm{Trans,\,CH_2}}$, is known to be independent of the head group (HG) structure and the charge of the surfactant [66]. Consequently, the cmc values that we selected for the same cation and different counterions were those whose use led to HG-*independent* $\Delta G^0_{\mathrm{Trans,\,CH_2}}$.

A final point is the effect of HG structure on the micellar properties. For the same counterion, these effects are usually small, because they refer to the transfer of the HG from aqueous pseudophase to the (highly aqueous) interfacial region. The exception is when the structure of the head ion is changed substantially, for example, in going from an alkyltrimethyl- to the corresponding alkyl-tri(1-butyl)ammonium group [67]. The reason is that this change is associated with noticeable decrease in the "effective" water concentration in the interfacial region, hence produces a larger $\Delta G^0_{\mathrm{Trans}}$ [68]. We compare the results of imidazolium and pyridinium surfactants in order to show the important effect on micellization of the relatively acidic C2-H of the former series. Because the change of the HG structure is not massive, this comparison has to be done with care. Therefore, we limited the discussion to our own data, because they were obtained by using the same techniques and calculation methods to obtain the micellar properties.

Before showing data correlations, we discuss briefly the appropriateness of the techniques and data treatment employed. With regard to the former, the two most employed techniques to calculate $\Delta H^0_{\mathrm{mic}}$ are conductivity and ITC. Whereas use of the latter allows accurate determination of cmc and $\Delta H^0_{\mathrm{mic}}$, conductivity is appropriate to determine cmc and α_{mic}. The deficiency of the van't Hoff treatment for the determination of $\Delta H^0_{\mathrm{mic}}$ and the subsequent effect of the other micellar thermodynamic parameters have been discussed previously [36, 38].

4.4.1 Dependence of cmc and $\Delta G^0_{\mathrm{mic}}$ on the Number of Carbon Atoms of the Hydrophobic Tail, nC

As expected (see Fig. 4.4), the relationship between ln (cmc) and the number of carbon atoms (nC) is linear; the slope is fairly independent of the counterion and of the cationic HG structure (see Table 4.2). As discussed earlier, this is a coherent result because $\Delta G^0_{\mathrm{Trans,\,CH_2}}$ from the aqueous pseudophase to the micellar interior should be independent of the nature of the surfactant head ions. For the C_nMeImBr and C_nMeImCl series, all data were recorded at 25°C, and as there is a large body of data available, we used a mean value. The trend for gemini surfactants is similar, and the cmc values are lower than the single long-chain counterpart due to the presence of two hydrophobic chains and higher energetic gain from aggregation.

As discussed previously, the cmc values for gemini surfactants are always lower than their single long-chain counterpart due to more favorable ($\Delta G^0_{\mathrm{mic}}$). As can be seen in Figure 4.5, there is also a decrease in cmc as a function of increasing *n*C (slope = −0.89 ± 0.03). Concerning the influence of the number of carbons in the

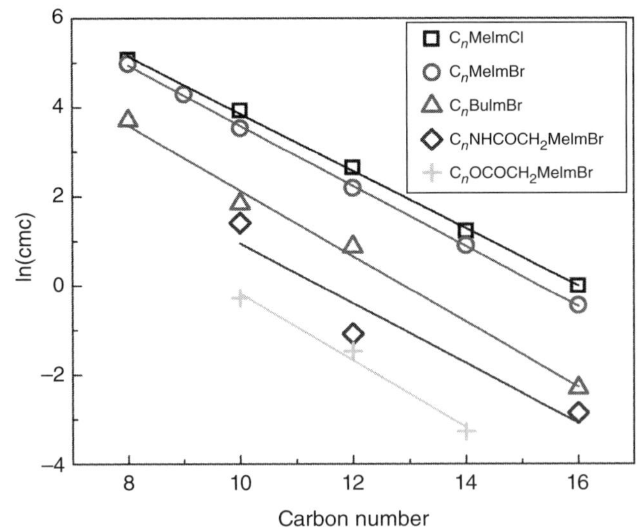

Figure 4.4 Dependence of ln(cmc) on the number of carbon atoms in the surfactants hydrophobic chain.

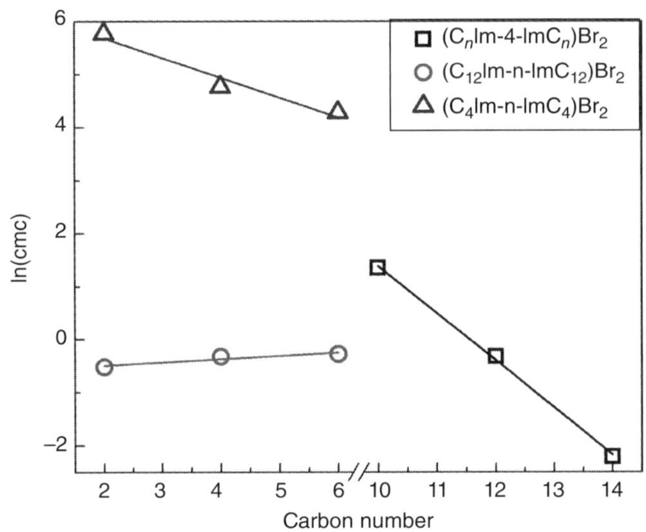

Figure 4.5 Dependence of ln(cmc) on the number of carbon atoms in the surfactants hydrophobic chain and the spacer between the head groups.

spacer between head ions, it can either increase or decrease cmc ((C_{12}Im-*n*-ImC_{12})Br_2 slope = 0.06 ± 0.02; (C_4Im-*n*-ImC_{12})Br_2 slope = 0.38 ± 0.08). The small number of gemini surfactants examined does not warrant further discussion of this effect of molecular structure on micellar property.

The influence of the length on hydrophobic chain on the energetics of micellization can be analyzed by considering the effects of the surfactant discrete groups on the thermodynamic parameters of micelle formation. The general equation is (Eq. 4.4)

$$\Delta G^0_{mic} = \Delta G^0_{mic}(HG + CH_3) + N_{CH_2}\,\Delta G^0_{mic}(CH_2) \quad (4.4)$$

where $\Delta G^0_{mic}(HG + CH_3)$, $\Delta G^0_{mic}(CH_2)$, and N_{CH_2} refer to the contributions of the HG plus the terminal methyl group of the hydrophobic chain and each methylene group, respectively. The analysis is simplified because the contribution of the terminal CH_3 is independent of the chain length; hence, it is a constant along the surfactants series. Consequently, the intercept of Equation 4.4 reflects essentially the effects of the transfer of the HGs from bulk solution to the interfacial region, and the slope reflects the transfer for one methylene group from bulk solution to the micellar interior [34, 35]. As can be seen from Figure 4.6 and Equations 4.5 and 4.6, ΔG^0_{mic} decreases as a function of increasing N_{CH_2}; $\Delta G^0_{Trans,CH_2}$ is practically 3.3 kJ mol^{-1}, in agreement with other "conventional" surfactants [34]. The general trend is followed by gemini surfactants; this will not be presented here because the assumptions used in Equation 4.4 for these surfactants are not straightforward. The increase in number of methylene groups at the spacer decreases $|\Delta G^0_{mic}|$ (see entries 48, 50, and 52 in Table 4.2):

$$\Delta G^0_{mic}(Br^-) = 2.02(\pm 1.04) - N_{CH_2}\,3.56(\pm 0.09),\ r^2 > 0.997 \quad (4.5)$$

$$\Delta G^0_{mic}(Cl^-) = 0.44(\pm 1.31) - N_{CH_2}\,3.31(\pm 0.11),\ r^2 > 0.995 \quad (4.6)$$

Figures 4.4 and 4.6 show that the values of cmc, hence ΔG^0_{mic}, is slightly lower for the surfactants whose counterion is bromide. Relative to the chloride ion, Br^- is more polarizable and less hydrated; this leads to stronger Br^--micellar cation interactions, hence lower cmc [66, 69, 70]. Iodide counterion lowers further the cmc as

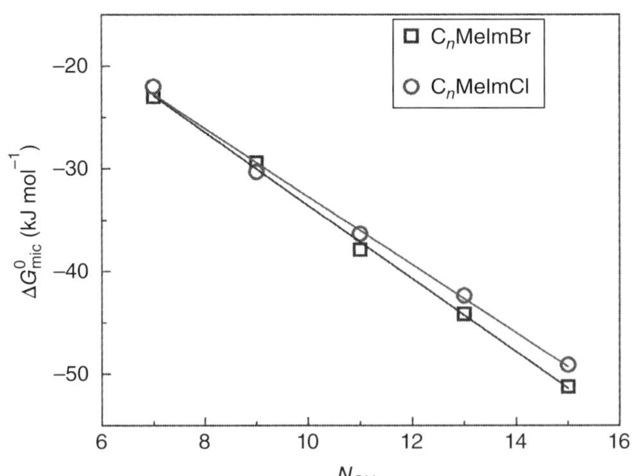

Figure 4.6 Dependence of ΔG^0_{mic} on the number of methylene groups in the surfactant hydrophobic chain, N_{CH_2}.

compared to bromide (see entries 128 (C_8MeImI, 25°C) and 1–8 (C_8MeImBr, 25°C)). Keeping the same counterion (Br^-) as a reference (entries 1–8, 12–21, and 26–32), other counterions either increase the cmc, for example, methanesulfonate (entries 133–135), or decrease the cmc, for example, tetrafluoroborate (entry 130). Surfactants with functionalized side chains at the HG have lower cmc values than their C_nMeIm counterparts (see entries 136–141 in Table 4.1).

4.5 TYPICAL APPLICATIONS OF ILBSs

As shown earlier, the structural versatility of ILs and the derived ILBSs results in concomitant variation in the properties of the solvent proper or the micellar solution of ILBS; these are important for applications. Examples that are worth mentioning include solvents for organic, organometallic, and inorganic compounds [71] and biopolymers [25]; nonaqueous, polar solvents whose miscibility with water and organic solvents can be employed for carrying out reactions in two-phase systems [72]; and templates for specific nanostructure [24, 26, 27].

The list of ILBS-based applications shown below is not exhaustive; it serves to show their present and potential fields of applications. We dwell first on bio-related applications and then their use in nanotechnology and in the preparation of nanoparticles of controlled properties.

4.5.1 Bio-related Applications

Many "conventional" cationic surfactants have antibacterial properties, in particular the 1-alkylpyridinium halides and the benzalkonium chlorides (*N*-benzyl-*N,N*-dimethyl-*N*-R chloride, R = C_8 to C_{18}). It is, therefore, only natural that ILBSs were tested for the same property, especially because this bears on their impact in the environment, for example, their toxicological effects on aquatic organisms [73–76]. The emphasis in these studies has been on the types of microorganism sensitive to ILBSs and the structure/activity relationships (SAR), in particular with regard to the length of the alkyl group, the heterocycle employed, and the counterion.

Microbial biofilms are ubiquitous in nature and represent the predominant mode of growth of microorganisms. Generally, they tend to exhibit significant tolerance to antimicrobial challenge compared with planktonic bacteria of the same species. The antibiofilm activity of a series of 1-alkyl-3-methylimidazolium chloride ILs and ILBSs (RMeImCl, R = C_6, C_8, C_{10}, C_{12}, and C_{14}) has been evaluated against a panel of clinically significant microbial pathogens, cocci, rods, and fungi, including *m*ethicillin-*r*esistant *S*taphylococcus *a*ureus (MRSA, a bacterium responsible for several difficult-to-treat infections in human). RMeImCl with R = C_4 was found to be ineffective, whereas the C_{16} and C_{18} ILBSs showed water solubility problems, hence were not investigated further. The biofilms were grown and the 24 h biofilm viable counts were performed. ILBSs with R ≤ C_{10} were ineffective. The minimum biofilm eradication concentration (MBEC) of the remaining compounds showed an inverse dependence of log (MBEC) on the length of R. Therefore, ILBSs are potentially effective toward development of novel disinfectants/antiseptics for improved infection control within the hospital environment [77].

The same dependence of antimicrobial efficiency on the length of (R) was observed for ILBSs with bromide counterion. Thus, 1-R-3-methylimidazolium bromide and 1-R-pyridinium bromide series were synthesized, and the properties of their micellar solutions examined by tensiometry, conductometry, spectrofluorimetry, and PGSE-NMR. The ILBSs investigated displayed the characteristic chain length dependence of the micellar properties. The antimicrobial activity was evaluated against Gram-negative and Gram-positive bacteria and fungi (*Micrococcus luteus*, *Staphylococcus epidermis*, *S. aureus*, *Escherichia coli*, *Klebsiella pneumoniae*, *Pseudomonas aeruginosa*, *Candida albicans*, and *Bacillus subtilis*). As shown in Figure 4.7, there is a linear correlation between the antimicrobial activity (MIC) and log(cmc); the C_{14} members are the most efficient. That is, log(cmc) seems to be an appropriate index to estimate the efficiency of imidazolium and pyridinium ILBSs as antimicrobial agents. On the other hand, the nature of the hydrophilic group, imidazolium or pyridinium cation, has no significant effect on their biological activity. This indicates that the hydrophilic properties of both cationic HGs, 1-methylimidazolium and pyridinium, are similar [40].

A number of ILBSs were synthesized, and their antibacterial and antifungal activities have been evaluated. Three surfactant groups were synthesized: series (A) contained bromides and chlorides of 1-R-3-methylimidazolium (R = C_8, C_{10}, C_{12}, C_{14}, C_{16}), series B had 1-R-2-methyl-3-hydroxyethylimidazolium chlorides with R = C_{14}, C_{16}, and series C contained *N*-C_{12}-*N*-hydroxyethylpyrrolidinonium chloride. The microorganisms employed were *E. coli*, *S. typhimurium*, *S. aureus*, *B. subtilis*, *C. albicans*, and *Chlorella regularis*. The antimicrobial efficiency was inferred from the observed (MIC) values. Series A and B imidazolium salts showed very good antimicrobial activity against the examined Gram-negative and Gram-positive bacteria and fungi. Also, the pyrrolidinonium salt was found to have low MIC for some of the tested microorganisms. The antibacterial and antifungal active properties of these

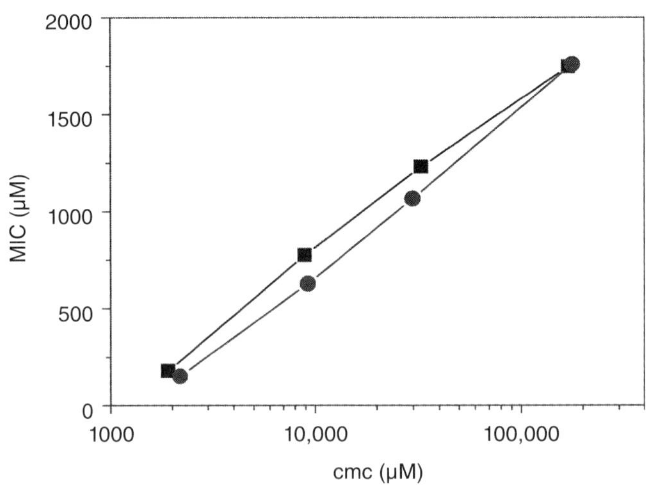

Figure 4.7 Dependence of the antimicrobial activity (MIC) of the ILBSs RMeImBr and RPyBr on the corresponding cmc. Source: Cornellas et al. [40] with copyright permission from Elsevier.

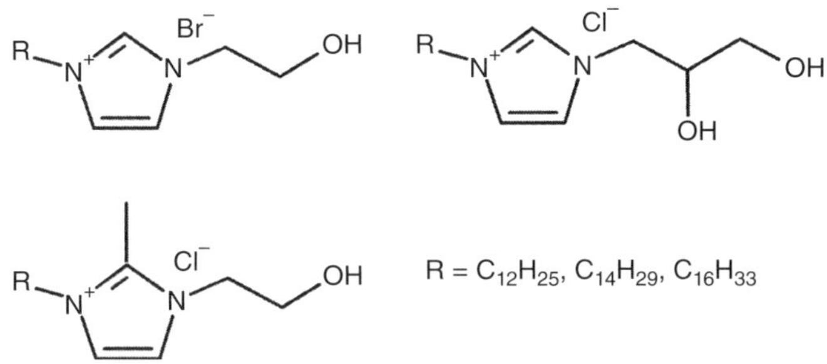

Figure 4.8 Molecular structures of the synthesized functionalized ILBSs.

salts depend upon the structure of the functional groups and the alkyl chain length. Among the synthesized imidazolium and pyrrolidinonium salts, the ones containing a long alkyl chain and a hydroxyethyl and methyl group showed broad-spectrum antimicrobial activity (bacteriostatic and bactericide). The results obtained are comparable or better than those of classic antibacterial agents benzalkonium chloride and *N*-C_{16}PyCl [78].

The beneficial effect of the hydroxyethyl side chain on the antimicrobial properties of ILBSs was investigated further by synthesizing silver nanoparticles stabilized by two surfactant series: 1-C_{14}-3-hydroxyethylimidazolium bromide, 1-C_{14}-3-(2′,3′-dihydroxy)propylimidazolium chloride, and 1-C_{16}-3-hydroxyethyl-2-methylimidazolium chloride (see Fig. 4.8 [79]).

The microorganisms investigated included *B. subtilis* and *S. aureus*, among Gram-positive bacteria; *E. coli* and *S. typhimurium*, among Gram-negative bacteria; and *C. albicans* as a representative fungus. The efficiency of Ag nanoparticles stabilized by these salts was evaluated from the value of MIC. The silver nanoparticles were the most efficient salts among those tested. The two series of quaternary salts stabilized Ag nanoparticles showed the broadest bactericidal and fungicidal activity. C_{14}-3-hydroxyethylimidazolium bromide was found to be more effective against both types of bacteria than benzalkonium chloride and *N*-C_{16}-pyridinium chloride [79].

Knowledge of SAR for several series of ILBSs allows the design and synthesis of new cationic amphiphiles with optimized antimicrobial activities for future development of new disinfectants, sanitizers, or preservatives. This rests on the understanding of the role of the IL cation (more studied) and anion (much less studied). The dependence of the antimicrobial activity on chemical structure, chain length, and anion type of 30 compounds was determined. The introduction of a longer substituent on the imidazolium cation results in a lower MIC. Thus, antifungal and antibacterial activities of 1-R-3MeIMCl were found to increase with chain length, very often up to a point exhibiting a cutoff effect at chain lengths of C_{16} or C_{18}, as shown in Figure 4.10. The SAR indicated an antimicrobial mode of action dependent on the surface activity of the molecule, inferring that surface activity may contribute to the cutoff effect in the biological activity of ILBSs.

The SAR was taken a step further by studying the series RMeImX, R = C_2, C_4, C_6, C_8, C_{10}, C_{14}, C_{16}, and C_{18}, with the following counterions (X): Cl^-, BF_4^-, $MeSO_4^-$ and $C_8SO_4^-$, Tf_2N^-, $TFMS^-$, and TS^-. The following microorganisms have been investigated: *C. albicans*, *Candida glabrata*, *Candida tropicalis*, *Saccharomyces cerevisiae*,

Saccharomyces cerevisiae JG, *Saccharomyces cerevisiae JG* CDR1, *Rhodotorula mucilaginosa*, *Geotrichum candidum*, *E. coli*, *S. aureus*, *Enterococcus faecium*, and *Enterococcus hirae* [80]. For the same anion, an inverse linear correlation has been observed between log(MIC) and the number of carbon atoms in the long chain, from C_8 to C_{14}, followed by leveling off for C_{16} and C_{18} (see Figs. 4.9 and 4.10). This "cutoff" effect has been observed for the biological activity of other ILBSs [81, 82]. Another way to look at these results is to correlate log [MIC] with pC20. For several ILBSs and bacteria, this correlation was found to be linear, as shown in Figure 4.10, indicating the intimate SAR between the surface activity and the efficiency of ILBSs as antimicrobial agents.

As shown in Figure 4.11, the antimicrobial efficiency also depends on the nature of the counterion. For example, for *C. albicans*, the antifungal activity order is $Tf_2N^- > C_8SO_4^- \approx BF_4^- \approx TS^- > MeSO_4^- > Cl^- > TFMS^-$. For the same cation, the most hydrophobic counterion caused the most significant reduction pC_{20} and thereby exhibited higher antimicrobial activity. Thus, the antibacterial and antifungal activities of this series of ILBs are directly related to the effects of the hydrophobicity of the cation and anion on their surface activity.

A study on ILBSs containing amide or ester side chains 1-($RNHCOCH_2$)-3-MeImBr and 1-($ROCOCH_2$)-3-MeImBr ($R = C_{10}, C_{12}, C_{14}, C_{16}$) was carried out [13]. These surfactants are interesting because their biodegradation is faster than the corresponding RMeImBr, due to the presence of (hydrolysable) side chain. These authors studied the surface-active properties of the surfactants synthesized and assayed their antimicrobial activities against Gram-positive bacteria, *B. subtilis*, *S. aureus, and Staphylococcus epidermidis*, and Gram-negative ones, *E. coli*, *P. aeruginosa*, and *K. pneumoniae.* They found that they possess good activities against selected strains. They pointed out that the influence of the ester or the amide

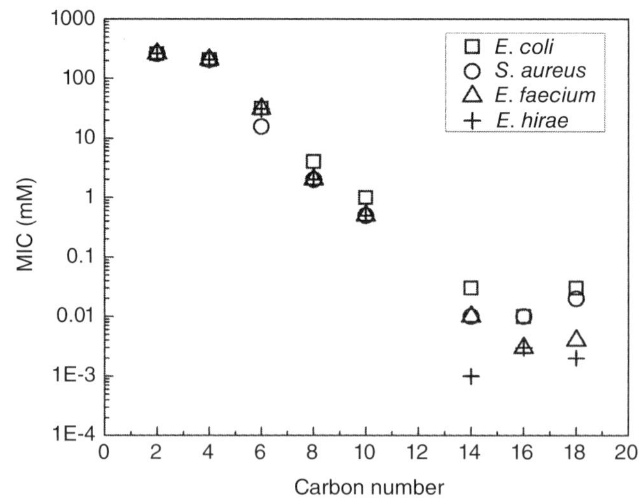

Figure 4.9 Relationship between the activities of RMeImCl against several strains of bacteria and the number of carbon atoms in the long side chain for (□) *E. coli*, (○) *S. aureus*, (°) *E. faecium*, and (+) *E. hirae*. Source: Łuczak et al. [80] and reproduced with permission from RSC Publishing.

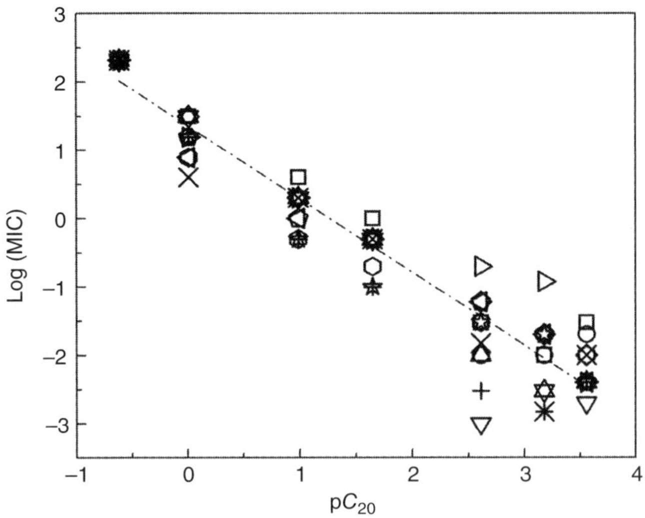

Figure 4.10 Correlation between the efficiency of ILBSs as antimicrobial agents (log [MIC]) with their surface activity (pC20). Source: Łuczak et al. [80] and reproduced with permission from RSC Publishing.

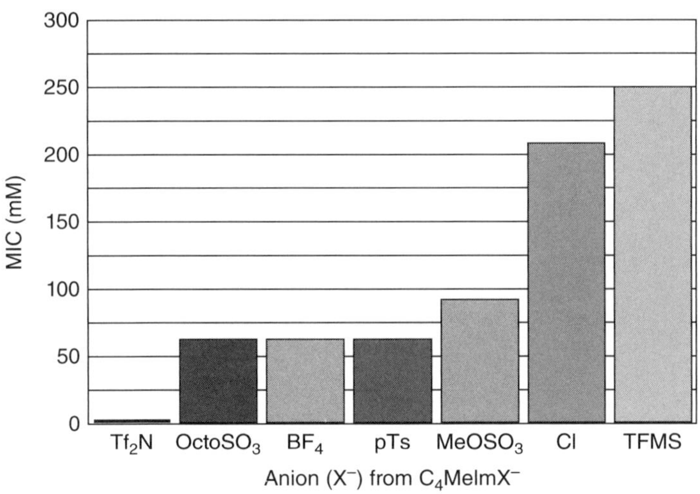

Figure 4.11 Dependence of the antifungal activity of C4MeImX on the nature of the counterion, determined for *C. albicans*. Source: Łuczak et al. [80] and reproduced with permission from RSC Publishing.

functional group on the antibacterial activity was not pronounced. These functional groups, however, caused a significant increase in the antifungal activity. This indicated that the compounds were more specific toward fungi in influencing membrane physiology compared to bacteria. At present, the reason for this difference is unclear.

4.5.2 Antitumor Activity of ILs and ILBSs

The volume of work on the antitumor activity and cytotoxicity of ILs and ILBSs is limited, with more publications on the effects ILs than their surface-active counterparts. Although discussion on the former class of compounds is outside the scope of the present article, we note that imidazolium-, phosphonium-, and ammonium-based ILs have been employed against 60 human tumor cell lines, including brain tumor cells, [83], human hepatocellular liver carcinoma [84], HeLa cancer cells [85], and human melanoma cell lines [86].

The effects of the molecular structure of phosphonium- and ammonium-based ILs and ILBSs (C_{14} side chain; counterion BF_4^-, PF_6^-, Tf_2N^-) on the antitumor activity and cytotoxicity were evaluated *in vitro* for the first time for the NCI-60 human tumor cell lines. All surface-active compounds (see Fig. 4.12) were found to be active against leukemia, melanoma, lung, colon, kidney, ovary, breast, prostate, and central nervous system cancer cells; the tetra(1-butyl)ammonium Tf_2N^- was not [87].

The same research group studied on the antitumor activity and cytotoxicity of imidazolium-based ILs and ILBSs [88]. The ILs and ILBSs studied included 1-R-3-MeImX, R = C_3, C_5, C_7, C_{11}, C_{15}, and C_{17}, and the counterions included Cl^-, BF_4^-, PF_6^-, and Tf_2N^-. As observed for the antibacterial effect of ILBSs, the length of (R) played a crucial role toward the antitumor activity and cytotoxicity. Thus, C_{12}MeImX was found to be effective against all 60 tumor cell lines and showed very low cytotoxicity in most cases. Further increase in chain length resulted in enhanced growth inhibition of tumor cell lines as well as high cytotoxicity. Interestingly, the ILBSs C_{12}MeImCl, C_{12}MeImBF$_4$, C_{16}MeImCl, C_{18}MeImCl, C_{18}EtImPF$_6$, and C_{18}MeIm Tf_2N

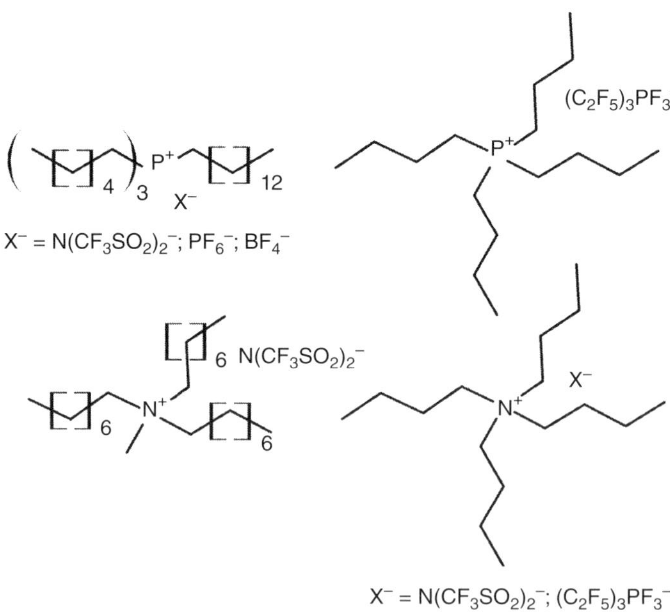

Figure 4.12 Molecular structures of the compounds tested against NCI-60 cancer cells. Compounds showing 60% tumor growth inhibition in eight or more cells were considered to be active. Source: Kumar and Malhotra [87] and reproduced with permission from Elsevier.

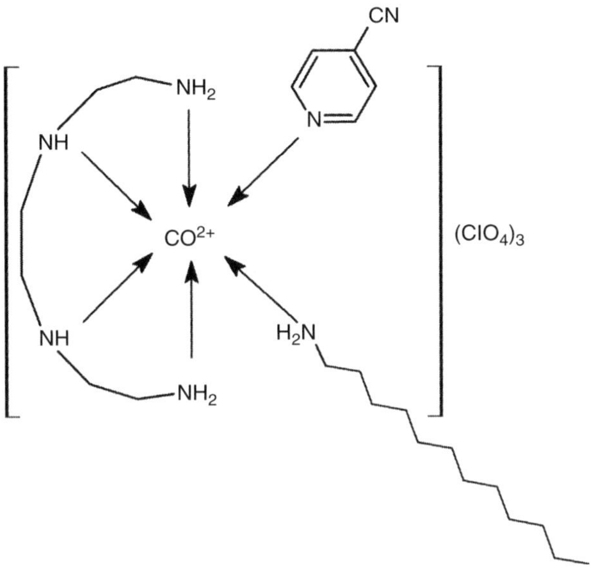

Figure 4.13 The molecular structure of the complex surfactant cis-[Co(trien)(4-CNP)(DA)] $(ClO_4)_3$. Source: Nagaraj and Arunachalam [84] and reproduced with permission from RSC Publishing.

were highly active against leukemia cell lines, with very low cytotoxicity, $LC_{50} > 100\,\mu M$ in all six leukemia cell lines tested.

The surfactant–cobalt(III) complex, *cis*-[Co(trien)(4CNP)(DA)]$(ClO_4)_3$ (trien = triethylene tetramine, 4CNP = 4-cyanopyridine, and DA = dodecylamine), was synthesized (Fig. 4.13) and characterized by various spectroscopic and physicochemical techniques. The complex was tested *in vitro* on human hepatocellular liver carcinoma tumor cell lines and found to be active [84].

The compounds triisobutyl (methyl) phosphonium tosylate (a) and trihexyl (tetradecyl) phosphonium bis 2,4,4-(trimethylpentyl)phosphinate (b) were synthesized (Fig. 4.14), and their surface-active properties studied. The polar compound (a) is water soluble and surface active, does not form micelles, but affects the micellization properties of ionic, nonionic, and zwitterionic surfactants more strongly than conventional electrolytes. The less polar compound (b) forms micelles and has very low aqueous solubility. Both compounds form mixed micelles with Triton X-100 nonionic surfactant in aqueous solution. Compound (a) replaces water to form microemulsions with isopropyl myristate as oil, stabilized by (b). Compound (a) showed a clear antitumor activity, for example, 5 mg (a) ml^{-1} in 0.9% NaCl solution caused 100% killing of Sarcoma-180 cell line in 1 h. More diluted solutions were still active: 2.5 and 1 mg (a) ml^{-1} caused 81 and 53% killing of the same cells, respectively. On the other hand, compound (b) was less active than (a): 10 mg (b) ml^{-1} in 0.9% NaCl solution caused 89% killing of Sarcoma-180 cell line in 2 h. Note that the concentration of (b) employed was 33 times higher than its cmc ($0.03 \times 10^{-3}\,mol\,l^{-1}$). The efficiency of (a) with respect to (b) may be due to the fact that the former does not form micellar aggregates [89].

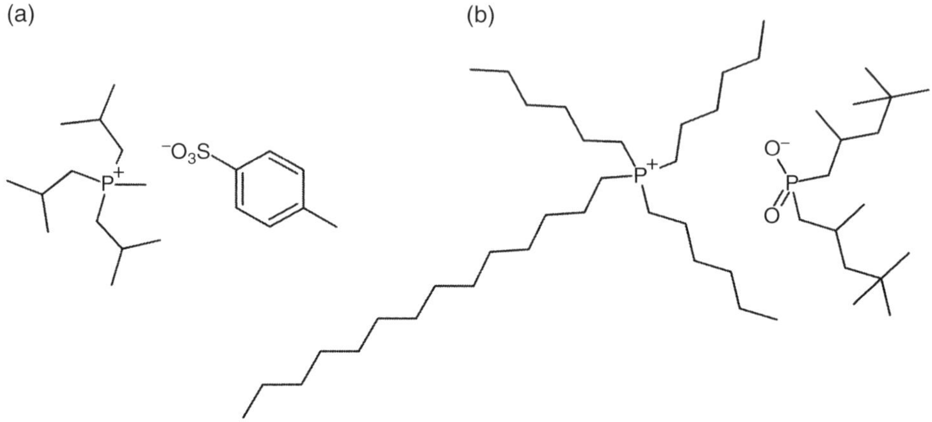

Figure 4.14 Molecular structures of compounds (a) and (b). Source: Mukherjee et al. [89] and reproduced with permission from Elsevier.

4.5.3 Interactions with Biomacromolecules

To shed more light on the potential hazard of ILs to humans and the environment, an acetyl cholinesterase (AchE) inhibition screening assay was used to identify toxicophore substructures and interaction potentials mediating enzyme inhibition. Qualitative and quantitative SAR for the inhibitory effects of cationic HGs, functionalized side chains, and anions of ILs on AchE was carried out [90]. The quaternary nitrogen atoms and the lipophilic side chain were identified as the key structural elements in binding ILBSs to the enzyme active site. Introduction of polar hydroxyl, ether, or nitrile functions into the alkyl side chain reduced the inhibitory potential. Additionally, a broad set of commonly employed counterions (halides, BF_4^-, PF_6^-, acetate, trifluoroacetate, and complex borate anions) were tested; the majority exhibited no effect on AchE. Only the fluoride- and fluoride-containing anionic species that readily undergo hydrolytic cleavage (e.g., BF_4^-) were identified as AchE inhibitors.

Ester-functionalized ILBSs with surface-active anion, 1-(ethoxycarbonylmethyl)-3-methylimidazolium dodecylsulfate (a) and 1-(ethoxycarbonylmethyl)-3-methylpyrrolidinium dodecylsulfate (b), were synthesized, and their surface-active properties investigated by tensiometry, fluorimetry, circular dichroism, and ITC. Their interactions with BSA were investigated. The results showed that the heterocycle (imidazolium or pyrrolidinium) had a mild effect on the BSA–ILBS interaction. The binding isotherms of BSA with the studied ILBSs exhibited four characteristic regions with increasing [ILBS]. The unfolding of BSA occurred in the third region. Fluorescence spectra indicated that the studied ILBSs caused Trp residues to expose to a hydrophobic environment. Compound (a) reduced more effectively the fluorescence intensity of BSA at lower [ILBS] as compared with (b). The denaturation extent of BSA followed the same order, that is, compound (a) > (b) [91].

4.5.4 Applications in Nanotechnology

Particle geometry and size distribution is of prime importance in nanotechnology; ILBSs play an important role in controlling the growth of NPs, for example, those of polymer latex, gold, silica, and iron oxides. Other applications include their use as

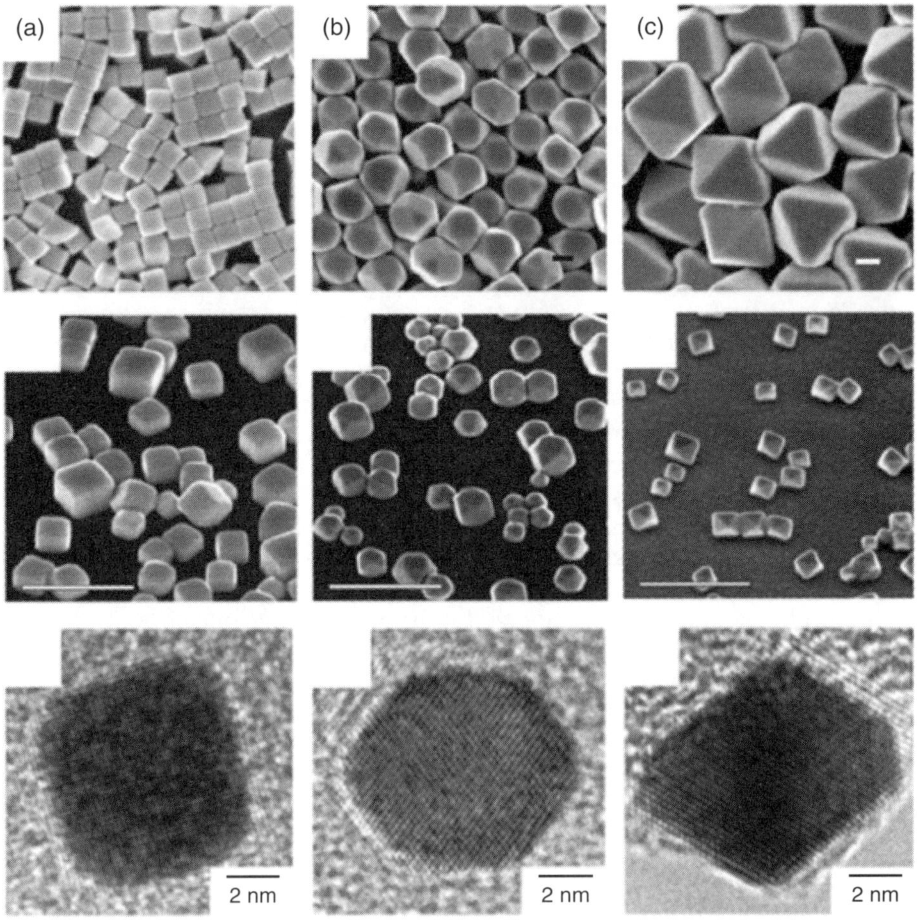

Figure 4.15 Possible morphologies of nanocrystals, obtained by controlling the process of their synthesis. Cubes, cuboctahedra, and octahedra have been obtained for (a) silver (scale bar = 100 nm), (b) gold (scale bar = 1 mm), and (c) platinum (scale bar = 2 nm). The similarity between these noble metals suggests a general nucleation and growth mechanism where stabilization of the {111} facets relative to the {100} facets can be used to tune final nanocrystal shape. Source: Tao et al. [92] and reproduced with permission from Wiley.

ligands or for particle surface modification. Figure 4.15 shows that the geometry of NPs can be controlled by their method of synthesis.

Template-directed synthetic methods have been widely employed to create ordered mesoporous materials, in particular SiO_2 and TiO_2. Mesoporous materials have unique features, for example, controllable pore size, high surface area, and regularly arranged channel systems [93].

Emulsion polymerization has been extensively studied in order to prepare quasi monodisperse polymer latex particles. Ding and coworkers employed C_{16}MeImCl as a surfactant in the emulsion polymerization of PS [24]. Monodisperse PS latex particles with average diameter of 126 nm and extremely low polydispersity (index of 0.002) were obtained by using this process. When the concentration of C_{16}MeImCl was increased while maintaining constant the other experimental variables, the average diameter of PS-C_{16} latex particles decreased, and the polydispersity index increased. The zeta potential value of PS-C_{16} indicated the adsorption of C_{16}MeImCl on the particle surface. Subsequently, $AuCl_4$ was immobilized on the surface of PS particles via the imidazolium cation; the adsorbent acted as oxidant for aniline polymerization, leading to core–shell structure formation. This is the first example in which an oxidant is immobilized on the surface of PS particles to direct the PS–polyaniline core–shell structure.

The synthesis of Au NPs with controlled architectures using supramolecular methods has remained a major obstacle. Control of the self-assembly of the aggregates formed requires special techniques for the control of the relevant parameters such as interparticle and surfactant interactions. A comprehensive understanding of the mechanisms involved in those interactions permits some level of control so that, in principle, the desired structures would become accessible. The controlled assembly of monodispersed Au NPs into multidimensional 1D, 2D, and 3D structures by using ILBSs and conventional cationic and nonionic surfactants was investigated [94]. A ligand-driven self-assembly strategy was employed, in which charged ligands interacted with and partially or fully replace the initial stabilizer molecules leading to assembly of gold NPs. This approach represents a general and versatile route for the fabrication of multidimensional assemblies with excellent control over morphology. It provides a method for the manipulation of nanoscale materials into dimension-controlled nano- or microstructured assemblies.

In another work, ultrathin BiOCl nanoflakes, nanoplate arrays, and curved nanoplates were successfully synthesized via an ionothermal synthetic route by using C_{16}MeImCl as "all-in-one" solvent, by simply adjusting reaction temperature [27].

Due to the increase in the number of catalytic active sites, silica material containing both ordered mesoporous channels and various hybrid atoms in porous matrix has enhanced properties relative to pure mesoporous silica. ILBSs displayed a significantly unique templating behavior, based on the distinct polarizability of the HGs—large hydrophilic groups and well-distributed positive charges on its micelles, which resulted in highly ordered pore systems [95]. Zhou and Antonietti demonstrated that C_{16}MeImCl displayed significantly stronger tendency toward self-aggregation and supramolecular templating in the preparation of supermicroporous lamellar silica by nanocasting. This has been traced to the polarizability of the HG [96, 97]. Wang et al. [98] and Adams et al. [99] have found that ILBSs can be used to generate mesoporous silica with a 2D hexagonal structure of cylindrical mesopores using hydrothermal synthetic method.

The synthesis of porous silica with ordered cubic and hexagonal mesophases using C_{16}MeImCl as template and a hydrothermal process has been carried out [100]. In their work, Zhang et al. [101] employed IL-coated magnetic Fe_3O_4 nanoparticles for the preconcentration of polycyclic aromatic hydrocarbons (PAHs) from environmental samples. Due to the high surface area and excellent adsorption capacity of the Fe_3O_4 NPs after modification with ILBSs, satisfactory extraction recoveries have been achieved. A comprehensive study of the adsorption conditions such as the amount of Fe_3O_4 NPs and ILs, solution pH, ionic strength, and contact time, breakthrough volume, and desorption solvents was presented. The extraction ability of different ILBSs as coating agents, for example, C_{16}MeImBr and C_{10}MeImBr, as compared to conventional cationic surfactant CTABr was compared. The extraction efficiencies obtained with ILBSs were much higher, presumably due to the polarizability of the heterocyclic HG.

4.6 CONCLUSIONS

If we were to select a single factor that distinguishes ILBSs from other conventional surfactants, it is their structural versatility. A wide variety of physical properties, including solubility in-/miscibility with molecular solvents, polarity, electrochemical behavior, etc., can be "fine-tuned" by a judicious combination of the structure of the cation and the basicity and polarizability of the counterion. The ease with which gemini ILBSs can be synthesized with the possibility of including a heteroatom in the "spacer" between the HGs (in particular oxygen and sulfur) adds to the structural variation possibilities. Therefore, the fields of applications of ILBSs are extensive and diverse. The possibility of fine-tuning miscibility renders biphasic reactions feasible and facilitates recycling into the process, an important aspect of green chemistry. The most extensively studied ILBSs are the 1-R-3-MeImX, with counterions halides, BF_4^- and PF_6^-. The dependence on counterion is important, as they affect the properties of their interfacial regions, akin to conventional surfactants. Furthermore, the introduction of a functional group close to the interfacial region, as shown by entries 136–141 in Table 4.1, is another structural possibility that has interesting consequence, not only for the micellar properties but also for the environmental impact of these compounds. The study of SAR of ILBSs is still in its infancy; this may change due to their proven antitumor activity. We hope that the applications shown in this account will interest more workers toward this attractive and highly promising field.

ACKNOWLEDGMENTS

O. A. El Seoud thanks FAPESP (State of São Paulo Research Foundation) for financial support and the CNPq (National Council for Scientific and Technological Research) for a research productivity fellowship.

NOTE ON THE ACRONYMS

The acronyms employed are as follows: we refer to each of the discrete structural moieties by using two letters. For example, Me, Et, Pr, Bu, Hx, Oc, Py, and Im refer to methyl, ethyl, 1-propyl, 1-hexyl, 1-octyl, pyridine, and imidazole, respectively. Unless specified otherwise, the alkyl groups are 1-alkyl. Usually, one of the two groups attached to the imidazolium ring is a long-chain one. We use C_n to denote this group, where (**n**) is the number of carbon atoms present. Unless specified otherwise, C_{10}, C_{12}, C_{14}, etc., refer to 1-decyl, 1-dodecyl, and 1-tetradecyl alkyl chain, respectively. The acronyms for the cation and the anion of the ILBSs are placed next to each other, without showing the corresponding charges; the substitution is on the two nitrogen atoms of the imidazole located at the 1 and 3 positions of the heterocycle, unless specified otherwise. Therefore, C_{16}MeImCl, C_{12}MeImBF$_4$, and C_{10}PyAc refer to 1-(1-hexadecyl)-3-methylimidazolium chloride, 1-(1-dodecyl)-3-methylimidazolium tetrafluoroborate, and 1-(1-decyl)pyridinium acetate, respectively.

Gemini ILBSs have two HGs joined by a "spacer" and two long hydrophobic chains. The spacer is composed either of methylene groups only or contains a hetero-atom (O or S). Thus, $(C_{10}Im\text{-}4\text{-}C_{10}Im)Br_2$, $(C_{16}\text{-}Im\text{-}EO_1\text{-}ImC_{16})Br_2$, $(C_{12}\text{-}Im\text{-}EO_2\text{-}ImC_{12})Br_2$, and $(C_{12}\text{-}S\text{-}4\text{-}S\text{-}C_{12})Br_2$ refer to the following structures: $(C_{10}Im\text{-}(CH_2)_4\text{-}C_{10}Im)Br_2$, $(C_{16}\text{-}Im\text{-}CH_2OCH_2\text{-}ImC_{16})Br_2$, $(C_{12}\text{-}Im\text{-}CH_2OCH_2\text{-}CH_2OCH_2\text{-}ImC_{12})Br_2$, and $(C_{12}\text{-}S\text{-}(CH_2)_4\text{-}S\text{-}C_{12})Br_2$, respectively.

REFERENCES

[1] Gaillon L., Sirieix-Plenet J., Letellier P. *J Solut Chem* 2004, **33**, 1333–1347.

[2] Galgano P. D., El Seoud O. A. Surface-Active Ionic Liquids: Syntheses, Solution Properties, and Applications. In: Mun J., Sim H., editors. *Handbook of Ionic Liquids: Properties, Applications and Hazards*, 2012, Nova Publisher, Hauppauge, pp. 521–548.

[3] Chiu Y. C., Wang S. J. *Colloids Surf* 1990, **48**, 297–309.

[4] Lunkenheimer K., Wienskol G., Prosser A. J. *Langmuir* 2004, **20**, 5738–5744.

[5] Wasserscheid P., Welton T., editors. *Ionic Liquids in Synthesis (Green Chemistry)*, 2007, 2nd ed. Wiley-VCH Verlag GmbH, Weinheim.

[6] Welton T. *Chem Rev* 1999, **99**, 2071–2084.

[7] Cravotto G., Gaudino E. C., Boffa L., Lévêque J.-M., Estager J., Bonrath W. *Molecules* 2008, **13**, 149–156.

[8] Thomaier S., Kunz W. *J Mol Liq* 2007, **130**, 104–107.

[9] El Seoud O. A., Pires P. A. R., Abdel-Moghny T. *J Colloid Interface Sci* 2007, **313**, 296–304.

[10] Ao M., Huang P., Xu G., Yang X., Wang Y. *Colloid Polym Sci* 2009, **287**, 395–402.

[11] Inoue T., Ebina H., Dong B., Zheng L. *J Colloid Interface Sci* 2007, **314**, 236–241.

[12] Baltazar Q. Q., Chandawalla J., Sawyer K., Anderson J. L. I. *Colloids Surf A Physicochem Eng Asp* 2007, **302**, 150–156.

[13] Kanjilal S., Sunitha S., Reddy P. S., Kumar K. P., Murty U. S. N., Prasad R. B. N. *Eur J Lipid Sci Technol* 2009, **111**, 941–948.

[14] Gordon C. M., Holbrey J. D., Kennedy A. R., Seddon K. R. *J Mater Chem* 1998, **8**, 2627–2636.

[15] Aupoix A., Pégot B., Vo-Thanh G. *Tetrahedron* 2010, **66**, 1352–1356.

[16] Dupont J., Consorti C. S., Suarez P. A. Z., De Souza R. F. P. *Org Synth* 2003, **79**, 236–243.

[17] Cook R., Dispenziere N., Mehnert C. Method for Preparing High-Purity Ionic Liquids. US20040074842 A1, 2004.

[18] Zhu W., Yang H., Yu Y., Hua L., Li H., Feng B., et al. *Phys Chem Chem Phys* 2011, **13**, 13492–13500.

[19] Bhadani A., Singh S. *Langmuir* 2011, **27**, 14033–14044.

[20] Deng Y., Morrissey S., Gathergood N., Delort A.-M., Husson P., Costa Gomes M. F. *ChemSusChem* 2010, **3**, 377–385.

[21] Wang X., Liu J., Yu L., Jiao J., Wang R., Sun L. *J Colloid Interface Sci* 2013, **391**, 103–110.

[22] Wang X., Wang R., Zheng Y., Sun L., Yu L., Jiao J., et al. *J Phys Chem B* 2013, **117**, 1886–1895.

[23] Liu X., Dong L., Fang Y. *J Surfactant Deterg* 2011, **14**, 497–504.

[24] Ding K., Miao Z., Liu Z., An G., Xie Y., Tao R., et al. *J Mater Chem* 2008, **18**, 5406–5411.

[25] El Seoud O. A., Koschella A., Fidale L. C., Dorn S., Heinze T. *Biomacromolecules* 2007, **8**, 2629–2647.

[26] Kaper H., Sallard S., Djerdj I., Antonietti M., Smarsly B. M. *Chem Mater* 2010, **22**, 3502–3510.

[27] Ma J., Liu X., Lian J., Duan X., Zheng W. *Cryst Growth Des* 2010, **10**, 2522–2527.

[28] Mukerjee P. *Adv Colloid Interface Sci* 1967, **1**, 242–275.

[29] Mukerjee P., Mysels K. J. *Critical Micelle Concentrations of Aqueous Surfactant Systems*, 1971, U.S. National Bureau of Standards, U.S. Government Printing Office, Washington, DC.

[30] Frahm J., Diekmann S., Haase A. *Ber Bunsen Ges* 1980, **84**, 566–571.

[31] Evans H. C. *J Chem Soc* 1956, 579–586.

[32] Shanks P. C., Franses E. I. *J Phys Chem* 1992, **96**, 1794–1805.

[33] Schulz P. C., Hernandez-Vargas M. E., Puig J. E. *Lat Am Appl Res* 1995, **25**, 153–159.

[34] Hiemenz P. C., Rajagopalan R. *Principles of Colloid and Surface Chemistry*, 1997, 3rd ed. Marcel Deker, New York.

[35] Rosen M. J. *Surfactants and Interfacial Phenomena*, 2004, 3rd ed. Wiley-Interscience, Hoboken.

[36] Galgano P. D., El Seoud O. A. *J Colloid Interface Sci* 2010, **345**, 1–11.

[37] Tanford C. *The Hydrophobic Effect: Formation of Micelles and Biological Membranes*, 1980, Wiley-Interscience, New York.

[38] Moulik S. P., Mitra D. *J Colloid Interface Sci* 2009, **8**, 569–578.

[39] Bhargava B. L., Klein M. L. *J Phys Chem B* 2009, **113**, 9499–9505.

[40] Cornellas A., Perez L., Comelles F., Ribosa I., Manresa A., Garcia M. T. *J Colloid Interface Sci* 2011, **355**, 164–171.

[41] Goodchild I., Collier L., Millar S. L., Prokes I., Lord J. C. D., Butts C. P., et al. *J Colloid Interface Sci* 2007, **307**, 455–468.

[42] Liu J., Zhang Q., Huo Y., Zhao M., Sun D., Wei X., et al. *Colloid Polym Sci* 2012, **290**, 1721–1730.

[43] Liu J., Zhao M., Zhang Q., Sun D., Wei X., Zheng L. *Colloid Polym Sci* 2011, **289**, 1711–1718.

[44] Wang J., Wang H., Zhang S., Zhang H., Zhao Y. *J Phys Chem B* 2007, **111**, 6181–6188.

[45] Zhao Y., Gao S., Wang J., Tang J. *J Phys Chem B* 2008, **112**, 2031–2039.

[46] Vaghela N. M., Sastry N. V., Aswal V. K. *Colloids Surf A Physicochem Eng Asp* 2011, **373**, 101–109.

[47] Vanyúr R., Biczók L., Miskolczy Z. *Colloids Surf A Physicochem Eng Asp* 2007, **299**, 256–261.

[48] Dong B., Li N., Zheng L., Yu L., Inoue T. *Langmuir* 2007, **23**, 4178–4182.

[49] Wang H., Feng Q., Wang J., Zhang H. *J Phys Chem B* 2010, **114**, 1380–1387.

[50] Geng F., Liu J., Zheng L., Yu L., Li Z., Li G., et al. *J Chem Eng Data* 2010, **55**, 147–151.

[51] Brinchi L., Germani R., Braccalenti E., Spreti N., Tiecco M., Savelli G. *J Colloid Interface Sci* 2010, **348**, 137–145.

[52] Feng Q., Wang H., Zhang S., Wang J. *Colloids Surf A Physicochem Eng Asp* 2010, **367**, 7–11.

[53] Kaper H., Smarsly B. *Z Phys Chem (Muenchen, Ger)* 2006, **220**, 1455–1471.

[54] Jungnickel C., Luczak J., Ranke J., Fernández J. F., Müller A., Thöming J. *Colloids Surf A Physicochem Eng Asp* 2008, **316**, 278–284.

[55] Łuczak J., Jungnickel C., Joskowska M., Thöming J., Hupka J. *J Colloid Interface Sci* 2009, **336**, 111–116.

[56] Galgano P. D., El Seoud O. A. *J Colloid Interface Sci* 2011, **361**, 186–194.

[57] Blesic M., Marques M. H., Plechkova N. V., Seddon K. R., Rebelo L. P. N., Lopes A. *Green Chem* 2007, **9**, 481–490.

[58] Bai G., Lopes A., Bastos M. *J Chem Thermodyn* 2008, **40**, 1509–1516.

[59] Blesic M., Swadźba-Kwaśny M., Holbrey J. D., Canongia Lopes J. N., Seddon K. R., Rebelo L. P. N. *Phys Chem Chem Phys* 2009, **11**, 4260–4268.

[60] Li X.-W., Gao Y.-A., Liu J., Zheng L.-Q., Chen B., Wu L.-Z., et al. *J Colloid Interface Sci* 2010, **343**, 94–101.

[61] Ao M., Xu G., Zhu Y., Bai Y. *J Colloid Interface Sci* 2008, **326**, 490–495.

[62] Zhang S., Yuan J., Ma H., Li N., Zheng L., Inoue T. *Colloid Polym Sci* 2011, **289**, 213–218.

[63] Li H., Yu C., Chen R., Li J., Li J. *Colloids Surf A Physicochem Eng Asp* 2012, **395**, 116–124.

[64] Heintz A., Lehmann J. K., Kozlova S. A., Balantseva E. V., Bazyleva A. B., Ondo D. *Fluid Phase Equilib* 2010, **294**, 187–196.

[65] Liu G., Gu D., Liu H., Ding W., Li Z. *J Colloid Interface Sci* 2011, **358**, 521–526.

[66] Myers D. *Surfaces, Interfaces, and Colloids: Principles and Applications*, 1999, 2nd ed., Wiley-VCH Verlag GmbH, New York.

[67] Bazito R. C., El Seoud O. A. *Berichte Bunsenges Phys Chem* 1997, **101**, 1933–1941.

[68] Tada E. B., Novaki L. P., El Seoud O. A. *Langmuir* 2001, **17**, 652–658.

[69] Impey R. W., Madden P. A., McDonald I. R. *J Phys Chem* 1983, **87**, 5071–5083.

[70] Sepulveda L., Cortes J. I. *J Phys Chem* 1985, **89**, 5322–5324.

[71] Reichardt C., Welton T. *Solvents and Solvent Effects in Organic Chemistry*, 2011, Wiley-VCH Verlag GmbH, Weinheim.

[72] Hallett J. P., Welton T. *Chem Rev* 2011, **111**, 3508–3576.

[73] Bernot R. J., Brueseke M. A., Evans-White M. A., Lamberti G. A. *Environ Toxicol Chem* 2005, **24**, 87–92.

[74] Bernot R. J., Kennedy E. E., Lamberti G. A. *Environ Toxicol Chem* 2005, **24**, 1759–1765.

[75] Docherty K. M., Charles F., Kulpa J. *Green Chem* 2005, **7**, 185–189.

[76] Pretti C., Chiappe C., Pieraccini D., Gregori M., Abramo F., Monni G., et al. *Green Chem* 2006, **8**, 238–240.

[77] Carson L., Chau P. K. W., Earle M. J., Gilea M. A., Gilmore B. F., Gorman S. P., et al. *Green Chem* 2009, **11**, 492–497.

[78] Demberelnyamba D., Kim K.-S., Choi S., Park S.-Y., Lee H., Kim C.-J., et al. *Bioorg Med Chem* 2004, **12**, 853–857.

[79] Dorjnamjin D., Ariunaa M., Shim Y. K. *Int J Mol Sci* 2008, **9**, 807–820.

[80] Łuczak J., Jungnickel C., Łącka I., Stolte S., Hupka J. *Green Chem* 2010, **12**, 593–601.

[81] Birnie C. R., Malamud D., Schnaare R. L. *Antimicrob Agents Chemother* 2000, **44**, 2514–2517.

[82] Pernak J., Sobaszkiewicz K., Mirska I. *Green Chem* 2003, **5**, 52–56.

[83] Kaushik N. K., Attri P., Kaushik N., Choi E. H. *Molecules* 2012, **17**, 13727–13739.

[84] Nagaraj K., Arunachalam S. *New J Chem* 2013, **38**, 366–375.

[85] Hu X., Deng J., Wang L., Yuan Y. *Med Chem* 2013, **9**, 1058–1062.

[86] Rao V. K., Tiwari R., Chhikara B. S., Shirazi A. N., Parang K., Kumar A. *J Phys Chem B* 2013, **3**, 15396–15403.

[87] Kumar V., Malhotra S. V. *Bioorg Med Chem Lett* 2009, **19**, 4643–4646.

[88] Malhotra S. V., Kumar V. *Bioorg Med Chem Lett* 2010, **20**, 581–585.

[89] Mukherjee I., Mukherjee S., Naskar B., Ghosh S., Moulik S. P. *J Colloid Interface Sci* 2013, **395**, 135–144.

[90] Arning J., Stolte S., Böschen A., Stock F., Pitner W.-R., Welz-Biermann U., et al. *Green Chem* 2008, **10**, 47–58.

[91] Wang X., Liu J., Sun L., Yu L., Jiao J., Wang R. *J Phys Chem B* 2012, **116**, 12479–12488.

[92] Tao A. R., Habas S., Yang P. *Small* 2008, **4**, 310–325.

[93] Hoffmann F., Cornelius M., Morell J., Fröba M. *Angew Chem* 2006, **118**, 3290–3328.

[94] Ji Q., Acharya S., Hill J. P., Richards G. J., Ariga K. *Adv Mater* 2008, **20**, 4027–4032.

[95] Antonietti M., Kuang D., Smarsly B., Zhou Y. *Angew Chem Int Ed* 2004, **43**, 4988–4992.

[96] Zhou Y., Antonietti M. *Adv Mater* 2003, **15**, 1452–1455.

[97] Zhou Y., Antonietti M. *Chem Mater* 2004, **16**, 544–550.

[98] Wang T., Kaper H., Antonietti M., Smarsly B. *Langmuir* 2007, **23**, 1489–1495.

[99] Adams C. J., Bradley A. E., Seddon K. R. *Aust J Chem* 2001, **54**, 679–681.

[100] Yue H., Zhao H., Liu L., Wang S., Ruan Q., Wang T. *Front Chem Eng China* 2008, **2**, 135–139.

[101] Zhang Q., Yang F., Tang F., Zeng K., Wu K., Cai Q., et al. *Analyst* 2010, **135**, 2426–2433.

[102] Liu J., Sun D., Wei X., Wang S., Yu L., Zheng L. J. Disper. *Sci. Technol.*, 2012, **33**, 5–14.

CHAPTER 5

Ionic Liquids in Bulk and at an Interface: Self-Aggregation, Interfacial Tension, and Adsorption

MOHAMMAD TARIQ

Instituto de Tecnologia Química e Biológica, Universidade Nova de Lisboa, Oeiras, Portugal
Department of Chemical Engineering, College of Engineering, Qatar University, Doha, Qatar

KARINA SHIMIZU and JOSÉ N. CANONGIA LOPES

Instituto de Tecnologia Química e Biológica, Universidade Nova de Lisboa, Oeiras, Portugal
Centro de Química Estrutural, Instituto Superior Técnico, Universidade de Lisboa, Lisboa, Portugal

BENILDE SARAMAGO

Centro de Química Estrutural, Instituto Superior Técnico, Universidade de Lisboa, Lisboa, Portugal

LUÍS PAULO N. REBELO

Instituto de Tecnologia Química e Biológica, Universidade Nova de Lisboa, Oeiras, Portugal

5.1 INTRODUCTION

Ionic liquids (ILs) are a new class of salts with melting point below 100°C, so that they are liquid at or near room temperature. Due to their remarkable properties, like negligible vapor pressure, high thermal stability, nonflammability, wide electrochemical window, enhanced solvent quality, and, above all, easy recycling procedures, they are considered to be potential substitutes for many traditional organic solvents. There has been an explosion of academic research, and their introduction in industrial applications is well underway (Plechkova and Seddon [1]). Much of this interest has centered on their designation as "designer solvents"; that is, they can be fine-tuned by the independent selection of cations and anions to be the optimum solvent with a desired set of properties for a given application.

Despite the recent widespread interest in room-temperature ILs for numerous applications, especially of interfacial relevance, they have not yet been fully subjected to the molecular-level scrutiny that other liquid surfaces and interfaces have. For instance, ILs are finding use as gas-capture media. This process involves collision between the

Ionic Liquid-Based Surfactant Science: Formulation, Characterization, and Applications, First Edition. Edited by Bidyut K. Paul and Satya P. Moulik.

gas-phase molecules and the liquid surface, which is the first step in establishing equilibrium. Upon collision, gas-phase molecules can be scattered instantaneously or can be accommodated by the liquid-phase molecules present at the surface. This elementary process is highly dependent on the chemical nature of the liquid surface (Lovelock [2]).

The arrangement of ions at the interface is a topic of fundamental importance. Identification of the ions potentially present at a variety of interfaces and of their arrangement there is often not obvious. At present, there are several descriptions available for the gas–liquid interface. In general, there are three possibilities (Fig. 5.1): the anions are able to approach the gas interface closer, on average, than the cations; cations are closer to the interface than anions; and both ions are evenly distributed at the interface. These limiting conditions will have an impact on the uptake and accommodation of the gas-phase molecules into the IL (Santos and Baldelli [3]). The variety of ions that compose ILs and their inbuilt complexity will also have an obvious impact on such superficial structure.

In a classical system like an electrolyte solution, ions generally avoid the gas–liquid boundary mainly due to the existence of weaker interactions at the surface. Therefore, ions prefer to remain in the bulk. This is the traditional interpretation based on the Gibbs equation and surface tension (ST) measurements. It should be noted that some recent molecular dynamics (MD) simulations and spectroscopy results challenge this interpretation (Jungwirth and Tobias [4] and Ghosal et al. [5]). Most electrolyte anions typically approach closer to the gas boundary than cations. Again, this is interpreted in two ways: many of those (molecular) anions are less strongly solvated than cations by molecular solvents, and/or those anions are more polarizable and thus experience less self-charge repulsions. In essence, the overall structure and energetics of the ions at the surface are intimately influenced by the solvent, often water (Santos and Baldelli [3]).

ILs are salts—substances composed essentially of cations and anions. This fact differentiates them from simple ionic solutions, where ions are dissolved in a molecular medium. The lack of a molecular solvent has a profound impact in the structure of the IL near a surface boundary. This chapter encompasses our most recent efforts in the investigation of such behavior.

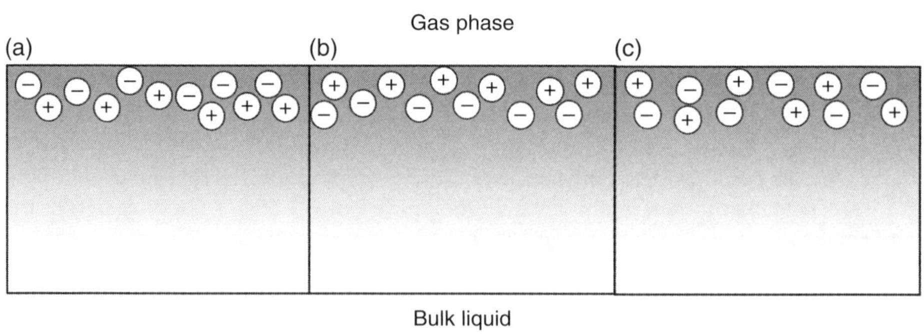

Figure 5.1 Three limiting cases for ions at the gas–liquid boundary: (a) anions closer than cations, (b) cations closer than anions, or (c) mixed cations and anions.

5.2 ILs IN BULK: AGGREGATE FORMATION

Diluted aqueous solutions of ILs containing ions with long alkyl side chains (typically larger than octyl chains) can exhibit self-aggregation phenomena (Blesic et al. [6]) similar to that of conventional surfactants (Tariq et al. [7], Greaves and Drummond [8], Lee et al. [9], and Ali et al. [10]). In these cases, some of the unique properties of ILs, such as their low melting point or their rather sophisticated solvation behavior, become minor or of no consequence (Angell et al. [11]). Nevertheless, aqueous solutions of ILs have some unique and important characteristics: on the one hand, demixing leads to liquid–liquid equilibria instead of solid precipitation (like in the case of traditional inorganic salts); on the other hand, hydrophilic ILs that are totally miscible with water can form electrolyte solutions ranging in concentration from pure IL to pure water. The mixture of an IL with a molecular compound such as water can exhibit different kinds of ionic association or aggregation over a wide range of compositions (Bernardes et al. [12]). Within this framework, we have selected a few IL systems in order to enhance the present understanding about the aggregation behavior of ILs in aqueous solutions and to compare their behavior with that of traditional surfactants. The studies started with diluted regimes that lead to the existence of either isolated ionic species, ion pairs, or micellar states; later, they were progressively extended to more concentrated IL solutions, where the appearance of other aggregation patterns are expected.

The first step was to use isothermal titration calorimetry (ITC), a powerful technique for the quantitative investigation of the aggregation properties of surfactants, to study the temperature-dependent aggregation behavior of three dialkylpyrrolidinium bromide ILs in aqueous solutions, namely, *N*-dodecyl-*N*-methylpyrrolidinium bromide, $[C_1C_{12}Pyrr]Br$; *N*-butyl-*N*-dodecyl-pyrrolidinium bromide, $[C_4C_{12}Pyrr]Br$; and *N*-butyl-*N*-octylpyrrolidinium bromide, $[C_4C_8Pyrr]Br$. These systems are structurally very close to tetraalkylammonium salts that are used as traditional ionic surfactants, and the similarities or differences to be encountered can contribute to elucidate the mechanisms and the structures responsible for the aggregation of ILs in aqueous solutions.

The values of critical micelle concentration (CMC) and micellization enthalpy, ΔH_{mic}, determined using ITC for the aqueous solutions of $[C_1C_{12}Pyrr]Br$, $[C_4C_{12}Pyrr]Br$, and $[C_4C_8Pyrr]Br$ at three different temperatures are listed in Table 5.1. For all systems and temperatures, except for $[C_4C_8Pyrr]Br$ at 323.15 K,

TABLE 5.1 Critical Micellar Concentration (CMC) and Micellization Enthalpy, ΔH_{mic}, Data as a Function of Temperature for Three Pyrrolidinium-Based Ionic Liquids

Ionic Liquid	T (K)	CMC (mM)	ΔH_{mic} (kJ.mol^{-1})
$[C_1C_{12}Pyrr]Br$	288.15	13.6 ± 0.3	2.5 ± 0.3
	298.15	15.3 ± 0.5	−1.8 ± 0.4
	323.15	16.1 ± 0.5	−11.4 ± 0.6
$[C_4C_{12}Pyrr]Br$	288.15	7.2 ± 0.5	7.4 ± 0.4
	298.15	6.1 ± 0.4	2.6 ± 0.1
	323.15	8.7 ± 0.3	−8.8 ± 0.2
$[C_4C_8Pyrr]Br$	288.15	169.3 ± 0.6	10.20 ± 0.06
	298.15	144.7 ± 0.8	7.2 ± 0.2
	323.15	148 ± 3	−0.3 ± 0.7

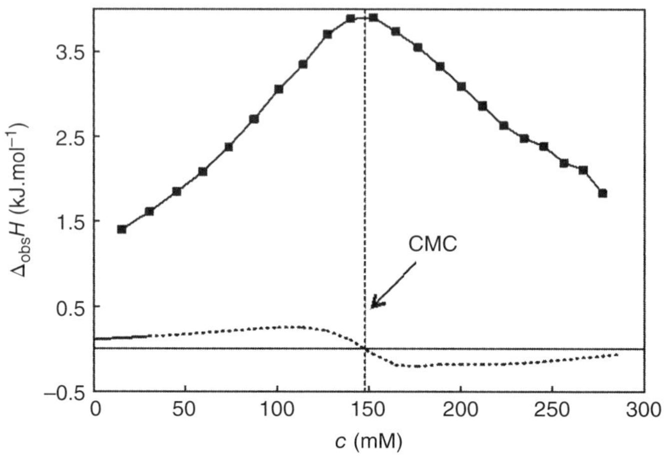

Figure 5.2 Enthalpy change as a function of concentration for titration of the aqueous solution of $[C_4C_8Pyrr]Br$ (1.6 M) into water at 323.15 K. The data (squares) were fitted in a sigmoidal curve (filled curve), and the CMC was obtained as the zero of the second derivative (dotted curve, multiplied by 10).

the determination of CMC and ΔH_{mic} from titration curves was not difficult because the aggregation process is complete and heat flow differences before and after the CMC are sharp. With $[C_4C_8Pyrr]Br$ at 323.15 K, the shape of the titration curve is more complex due to a less abrupt transition in the heat flow trends, ΔH_{obs}, before and after CMC (cf. Fig. 5.2). Here, contrary to $[C_1C_{12}Pyrr]Br$ and $[C_4C_{12}Pyrr]Br$, the dilution effect in the premicellar region is much more important than the effect due to demicellization. The slope after passing the CMC reflects the equilibrium between smaller, less structured aggregates and a solution containing a higher concentration of free monomers. Both effects contribute to gradual, rather than steep, changes in the enthalpy in the CMC region. This behavior was attributed to shorter alkyl chains being less hydrophobic, leading to smaller aggregation numbers and to lower cooperativity in the micellization process (Olofsson and Loh [13] and Bijma et al. [14]).

By analyzing the calorimetric data, further insights could be gained into the energetics of the self-organization phenomena. The CMC values for $[C_1C_{12}Pyrr]Br$ and $[C_4C_{12}Pyrr]Br$ are one order of magnitude lower than those for $[C_4C_8Pyrr]Br$, an observation compatible with easier micellization processes—and thus lower CMCs—when longer alkyl chains are present. For instance, it is known that in aqueous solutions of ILs composed of 1-alkyl-3-methylimidazolium cations with chloride or bromide ions, the lower limit of alkyl side chain length leading to the formation of micelles is around C6 or C7 (Blesic et al. [6, 15, 16]). The smaller CMC values of $[C_4C_{12}Pyrr]Br$ compared to those of $[C_1C_{12}Pyrr]Br$ also indicate a more favorable micellization when two longer chains are present, C4 and C12, because of more favorable interactions between the hydrophobic groups.

Table 5.1 also shows typical nonmonotonous dependences of CMC on temperature for $[C_4C_{12}Pyrr]Br$ and $[C_4C_8Pyrr]Br$, which can be interpreted as the interplay between the two driving forces that concur to micellization in aqueous solutions: enthalpy- versus entropy-driven processes (Attwood and Fluorence [17]). The gradual breakdown of the hydrogen-bonded network of liquid water as temperature increases (and the resulting increase of the entropy of liquid water) leads to enthalpy-driven micellization processes at higher temperatures, whereas entropy-driven processes usually dominate at lower temperatures. The CMC of [C_1C_{12}Pyrr]Br increases monotonically with temperature in the range studied, indicating that any CMC minima are outside this range, probably below 288.15 K. This is the IL with lower ΔH_{mic} values, which means that a large contribution from the entropy term is not necessary to achieve micellization (negative ΔG_{mic} values), a situation that occurs at relatively low temperatures.

In the temperature range studied, the enthalpy of micellization of [C_1C_{12}Pyrr]Br and [C_4C_{12}Pyrr]Br is positive at lower temperatures, becoming negative at higher temperatures (Table 5.1), which is consistent with the behavior observed for tetraalkylammonium bromide surfactants (Bashford and Woolley [18]). Higher values of ΔH_{mic} for [C_4C_{12}Pyrr]Br in comparison with [C_1C_{12}Pyrr]Br indicate stronger interactions between the alkyl chains in the former case, compatible with the lower values of CMC obtained. On the other hand, for [C_4C_8Pyrr]Br, the enthalpy of micellization decreases and approaches zero around 323.15 K, meaning that both contributions to the enthalpy are equally important: these are the disruption of the structural organization of water molecules around hydrophilic and hydrophobic domains of the IL surfactant and the restoring of the hydrogen-bond structure of water when the micelle is formed. This can also explain the peculiar shape of the calorimetric titration curve observed in that case.

In order to prove that diluted aqueous solutions of IL-based surfactants have similar characteristics to those of traditional ionic surfactants, we have conducted a set of comparisons and empirical correlations between the properties of solutions of the two classes of surfactants (Tariq et al. [7]). Data concerning the aggregation process are compared for tetraalkylammonium bromide and *N,N*-pyrrolidinium bromide surfactants with either one or two long alkyl chains attached to the cationic head group. The four series—tetraalkylammonium or *N,N*-pyrrolidinium head groups with either one or two alkyl chains—yield an interesting and broad base for comparison and correlation. We note that both head groups contain a nitrogen atom connected to four aliphatic chains, two of which can "grow" to any desired length, while the two others retain their size (either two methyl groups or a pyrrolidinium ring). The main difference between the tetraalkylammonium bromide and *N,N*-pyrrolidinium bromide surfactants is that the former are traditional ionic surfactants (when pure, they melt only above 200°C, near their decomposition temperature), while the latter belong to the IL class, with melting points below or around 100°C when pure.

The most assessable data, obtained experimentally for all compounds, are the values of CMC. These are, of course, directly related to the tendency for self-aggregation—the lower the CMC, the higher the tendency—and many authors have correlated CMC values (or their logarithms) with the size of the alkyl chains of different surfactants (Zana [19]). In the four series under discussion, such comparisons are made in Figure 5.3, plots a and b, where ln(CMC) are plotted as a function of n, the number of carbons in the longest chain of the surfactant cation (plot a), or $n+m$, the total number of carbon atoms in the two "variable" chains of the cations (plot b). Both plots show distinct trends for each series, although both tetraalkylammonium and *N,N*-dialkylpyrrolidinium cations with a single long chain show almost superimposed trends.

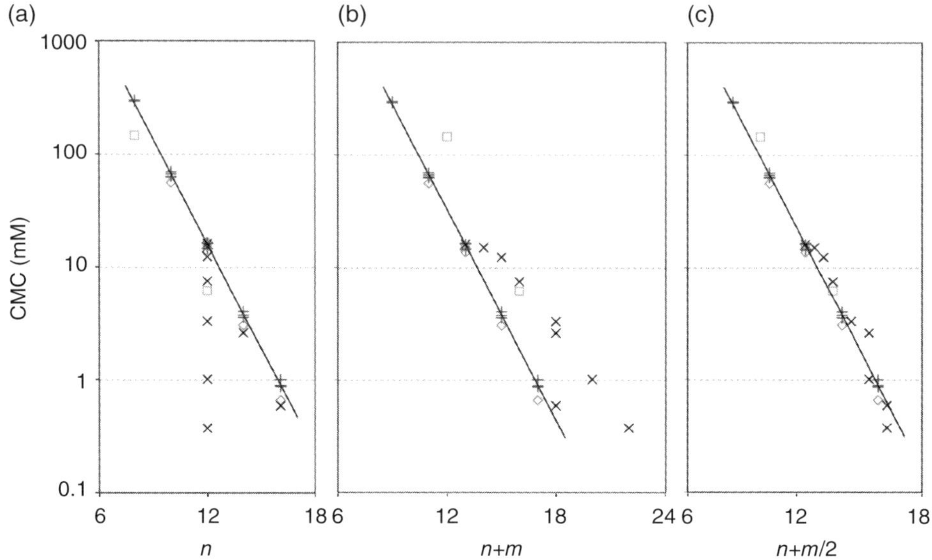

Figure 5.3 Critical micelle concentration (CMC) at 298 K as a function of the number of carbons in the first (n) and second (m) chains of C_nTAB (cross), C_nC_mDAB (x-cross), [C_nC_1Pyrr]Br (rhomb), and [C_nC_mPyrr]Br (square). (a) As a function of n only; (b) as a function of n plus m; (c) as a function of n plus m/2.

In plot 3a, the surfactants belonging to series with two long side chains show, for a given size of the longest chain, lower CMC values than the analogous compound with just one chain, for instance, CMC ($C_{12}C_4$DAB) < CMC ($C_{12}C_1$DAB) ≈ CMC(C_{12}TAB), with DAB = dialkylammonium bromide and TAB = trialkylammonium bromide. The vertically aligned x-crossed markers in the plot refer to different $C_{12}C_n$DAB surfactants with different n values. The n value in each case can be calculated by confronting the CMC values between plots 3a and 3b. The second chain probably assists the self-aggregation process, as observed for the [C_nC_mPyrr] cations studied in this work. However, if the length of both chains is added in terms of the ln(CMC) correlation (as in plot 3b), we see that it is more advantageous in terms of self-aggregation to have one very long and one short chain (a high ratio) instead of a more even structure (lower ratio): CMC ($C_{16}C_2$DAB) < CMC ($C_{12}C_8$DAB). Plots 3a and 3b suggest a way to make all series fit in a single master line: the shorter chain plays an important part (plot 3a) but apparently not as significant as that of the longer one (plot 3b). By adjusting the relative weight of both alkyl chains, we can easily arrive at plot 3c, where ln(CMC) is plotted against (n + m/2) and all series fit into a single trend. In other words, in terms of self-aggregation, the shorter hydrophobic chain is just half as important as the longest one. Curiously, the nature of the "polar" head group plays a rather modest role, with all four series following quite well the (n + m/2) rule. It must be stressed that the present and following correlations were performed taking into account only these four series of bromide-based surfactants with mono-/dialkyl substituted quaternary ammonium or pyrrolidinium salts. Generalizations to ammonium salts substituted with more than two long alkyl chains or to gemini surfactants were not considered, and the conclusions here obtained will not necessarily apply. The focus of the present study was the comparison of traditional ionic surfactants with the ones based on ILs, represented here by the pyrrolidinium cation that can only have a maximum of two long alkyl chains substituted at the quaternary nitrogen atom of the ring.

Although CMC correlations offer a straightforward way to relate the structure of a given surfactant with its tendency to self-aggregate, they do not provide a direct quantitative assessment of the energetics involved in the micellization process. The equation that is sometimes used to relate the free energy of micellization (ΔG^0_{mic}) with the CMC value at a given temperature, $\Delta G^0_{\text{mic}} = \text{R}T\ln(\text{CMC})$, should be used for nonionic surfactants only, yielding incorrect ΔG^0_{mic} results for uni-univalent ionic surfactants (Zana [20]). The correct expression for uni-univalent ionic surfactants is

$$\Delta G^0_{\text{mic}} = RT(1+\beta)\ln(\text{CMC}) \tag{5.1}$$

where β is the fraction of counterions (bromide in this study) that are bound to the aggregate. The values of β can be obtained experimentally by determining the slope of the conductivity versus concentration curves below and above CMC. Those are also represented in Figure 5.4, plot a, as a function of n.

For surfactants substituted with a single long chain, the values of β increase with the number of carbon atoms in the alkyl chain—this increase has always been

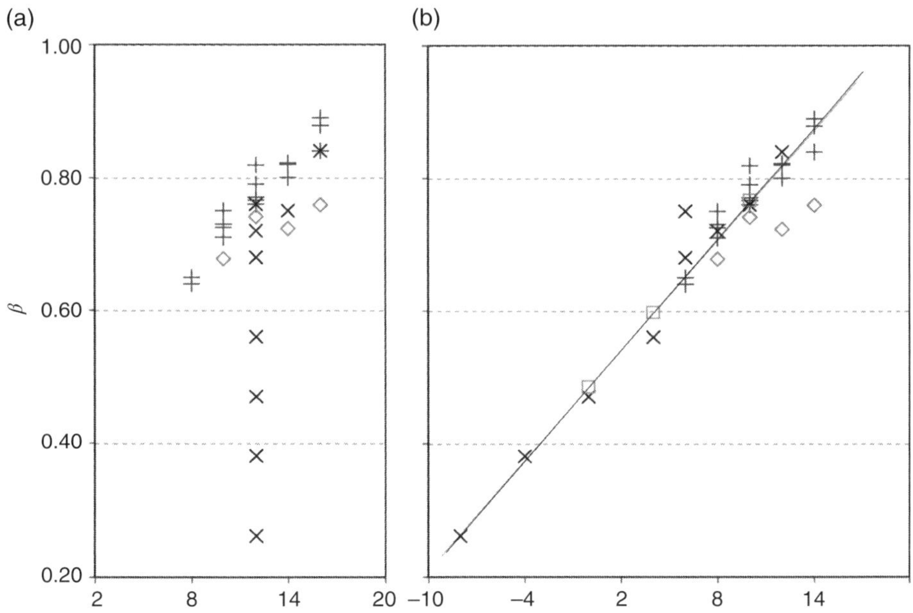

Figure 5.4 Mole fraction of bound counterions (β) at 298 K as a function of the number of carbons in the first (n) and second (m) chains of C_nTAB (cross), C_nC_mDAB (x-cross), $[C_nC_1\text{Pyrr}]$Br (rhomb), and $[C_nC_m\text{Pyrr}]$Br (square). The values for $[C_nC_m\text{Pyrr}]$Br were obtained by interpolation of a linear regression of all other points. (a) As a function of n only; (b) as a function of n minus $2m$.

attributed to larger charge densities at the micelle surface (Zana [19]). In other words, longer-chain surfactants tend to aggregate in bulkier structures where the surface-to-volume ratio is smaller, in which both the volume per surfactant ion and the number of ions per micelle are larger. This means that the polar head groups will be packed more closely and will be bound by a larger fraction of counterions. However, when a second alkyl side chain (of length m) is present, the values of β decrease with increasing m, showing that as the second chain becomes longer, the charge density at the surface of the micelle decreases, the head group (which now has two hydrophobic moieties) probably penetrating more in the core of the aggregate. A general correlation for β, which is able to predict the behavior of the different surfactants under discussion, must take into account the opposite trends caused by the growth of the first and second hydrophobic chains. Figure 5.4b shows such a correlation, where β is plotted as a function of $(n-2m)$. Since β was not determined experimentally in the present work, we have used the linear correlation implicit in Figure 5.4b to estimate the corresponding values for the IL-based surfactants studied here.

The values of CMC and β can now be used in Equation 5.1 to calculate correct ΔG^0_{mic} values for all surfactants. ΔG^0_{mic} is a measure of the free energy involved in transferring one mole of surfactant from the aqueous solution into the micellar pseudophase (Zana [20]). For the tetraalkylammonium- and pyrrolidinium-based surfactants with a single long chain of n carbon atoms, ΔG^0_{mic} as a function of n yields a linear plot (Fig. 5.5a). This situation is the norm for many conventional ionic and nonionic surfactants with a single long chain attached to a polar group (Zana [20]). The slope of the line represents the free energy change associated with the transfer of one methylene group from the aqueous phase to the micellar pseudophase, $\Delta G^0_{\text{tr}}(-CH_2-)$.

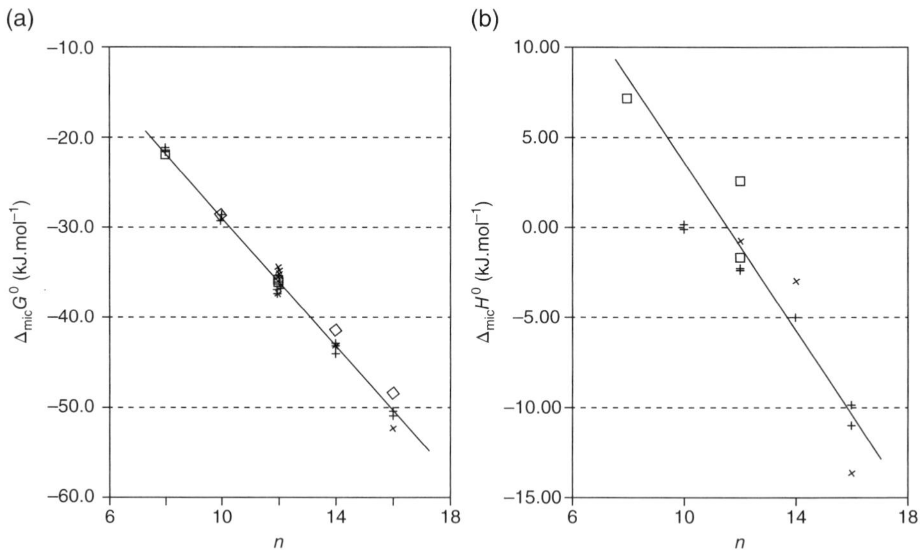

Figure 5.5 (a) Free energies (ΔG^0_{mic}) and (b) enthalpies (ΔH^0_{mic}) of micellization at 298 K as a function of the number of carbons in the longest (n) chain of C_nTAB (cross), C_nC_mDAB (x-cross), [C_nC_1Pyrr]Br (rhomb), and [C_nC_mPyrr]Br (square).

This is clearly independent of the nature of the head group of the surfactant, which is the reason why such slopes are similar for many different single-chained conventional surfactants. The first conclusion here is that [C_nC_1Pyrr]Br surfactants act exactly like the corresponding conventional single-chained surfactants, represented by C_nTAB. Moreover, not only the two plots show a similar slope, but they lie almost superimposed to each other, indicating that the structural similarity of the two head groups leads to similar free energies of transfer of one surfactant molecule from the aqueous to the micellar environment.

However, the most striking feature of Figure 5.5a is that the plots of ΔG^0_{mic} versus n for double-chained surfactants [C_nC_mPyrr]Br and C_nC_mDAB can also be superimposed on the plots obtained for their single-chained counterparts. Apparently, the presence of a second long chain (and the m carbon atoms it contains) does not affect much the value of ΔG^0_{mic}: as an example, at 298 K, 35.0 kJ.mol^{-1} < ΔG^0_{mic} (C_{12}TAB) ~ ΔG^0_{mic}($C_{12}C_m$DAB) ~ ΔG^0_{mic}([$C_{12}C_1$Pyrr]Br) ~ ΔG^0_{mic} ([$C_{12}C_m$Pyrr] Br) < 37.3 kJ.mol^{-1}. It seems as ΔG^0_{tr} (—CH_2—) refers only to methylene groups in the first chain. This surprising result—the second chain plays an important role in the self-aggregation process, as can be attested by the change of CMC values, *but* does not affect the values of ΔG^0_{mic}—can be explained by taking into account the effects of the second chain both in terms of ln(CMC) and β. Longer second chains mean at the same time lower CMC values (more tendency to aggregate as micelles) but also lower β values (less neutralized, less stable micelles). The empirical correlations found above imply that the two factors together may cancel out and thus the contribution of the atoms of the second chain to ΔG^0_{mic} is negligible when compared to that of the atoms of the first chain. The first correlation is in terms of $(n+m/2)$, while the second is in terms of $(n-2m)$, indicating the opposing nature of the contribution of the second chain in terms of ln(CMC) and β. In other words, the gain in free energy of removing second-chain methylene groups from the aqueous to the micellar phase—which would eventually lead to more negative ΔG^0_{mic} values—is canceled out because the presence of the second chain also leads to a slightly less stable micellar phase with smaller β values.

The free energy of micellization can be decomposed in its enthalpic and entropic contributions ($\Delta G^0_{mic} = \Delta H^0_{mic} - T\Delta S^0_{mic}$), and the latter can be estimated in cases where ΔH^0_{mic} has been measured experimentally. Figure 5.5b shows the dependence of the enthalpy of micellization with the number of carbons in the longest chain. Although a similar trend between ΔH^0_{mic} and ΔG^0_{mic} is observable when one compares the two panels of Figure 5.5, the correlation between ΔH^0_{mic} and n is not so clear. The rather scarce number of experimental data (and possible errors) does not allow a better definition of the trend and is not sufficient for further investigation of possible correlations with both n and m.

We have seen that ΔG^0_{mic} is almost independent of m due to the cancelation of two effects (micelle aggregation vs. micelle stability). Such cancelation does not necessarily take place in the two terms, ΔH^0_{mic} and ΔS^0_{mic}, that compose ΔG^0_{mic}. It seems that the dependence of ΔH^0_{mic} with n and m follows a trend similar to that of β (n and m have opposing effects), while ΔS^0_{mic} follows the trend of ln(CMC) (n and m leading to concurring effects). Although speculative, this makes sense from a molecular point of view: in aqueous solutions at 298 K, it is known that the micellization process is entropically driven by gains in the entropy of water, when hydrophobic solvation of individual surfactant molecules is compared with the favorable solvation of micelles. This means that both long chains, which are responsible for those effects, must contribute to the value of ΔS^0_{mic}. On the other hand, ΔH^0_{mic} will depend not only on the interactions that the surfactant ions are experiencing when individually solvated by water but also on the interactions in the micellar media. The latter type of interactions will depend heavily on the values of β, hence the possible relation between ΔH^0_{mic} and β.

Aqueous solutions of ILs have some unique and important characteristics: on the one hand, demixing leads to liquid–liquid equilibria instead of solid precipitation (like in the case of traditional inorganic salts); on the other hand, hydrophilic ILs that are totally miscible with water can form electrolyte solutions ranging in concentration from pure IL to pure water. In this context, we have performed a study on the determination of volumetric or volume-related (compressibility) data that will allow a more direct understanding of the relations between the self-aggregation phenomena (micelle formation in this case), the concomitant structural changes that occur in the aqueous solutions, and the corresponding consequences at a volumetric level.

The three ILs selected for this study (Tariq et al. [21]) are members of the 1-alkyl-3-methylimidazolium chloride homologous series—$[C_8C_1im]Cl$, $[C_{10}C_1im]Cl$, and $[C_{12}C_1im]Cl$—and are hydrophilic ILs that form micelles in aqueous solutions at quite distinct concentration ranges. The choice of these ILs also reflects the emphasis given to the measurement of volumetric properties: their specific densities are all very close to that of pure water—at 298 K, $[C_8C_1im]Cl$ is slightly more dense than water, whereas the other two are slightly less dense than water.

These two sets of properties (different CMCs, similar densities) will impose a rather stringent test to the capability of evaluating micelle formation using volumetric data in the corresponding diluted aqueous solutions.

The global volumetric behavior at 298 K of the three aqueous solutions containing the three ILs under analysis is depicted in Figure 5.6. Panel a shows the expected density profiles of each solution if all mixtures exhibited null excess molar volumes (ideal volumetric behavior, i.e., $\rho^{id} = (x_1M_1 + x_2M_2)/(x_1V_1 + x_2V_2)$, with M_i and V_i the molar mass and molar volume of the ith component of the mixture). The panel also shows that whereas $[C_8C_1im]Cl$ is around 1% more dense than pure water, $[C_{10}C_1im]Cl$ and $[C_{12}C_1im]Cl$ are less dense than pure water by around 1 and 3%, respectively. Panels b–d show the experimental density data (also at 298 K) for very diluted IL solutions superimposed on the corresponding ideal behavior lines. Some general comments are possible at this stage: (i) the experimental density data are always larger than the corresponding ideal values; (ii) for the most diluted concentrations in each case, the slopes of the experimental data are almost constant and always larger than the slope of the corresponding ideal lines; (iii) at a certain point and for slightly more concentrated solutions, the experimental data slopes start to change and approach the corresponding values (slopes) of the ideal lines. As we shall see in the following discussion, these last trend shifts can be correlated with the attainment of the CMC in each case. It must also be noted that in the case of $[C_{10}C_1im]Cl$, the positive deviation between the experimental and ideal values implies that aqueous solutions between 0 and 0.001 mole fraction in the IL exhibit densities that are larger than the densities of both pure water and the pure IL: in that case, the density measurements are directly giving a measure of the excess volumetric properties of the mixtures.

One straightforward way to predict the occurrence of a self-aggregation process using volumetric data is to calculate the apparent molar volumes of the solute, $V\phi$, as a function of the concentration of the solutions

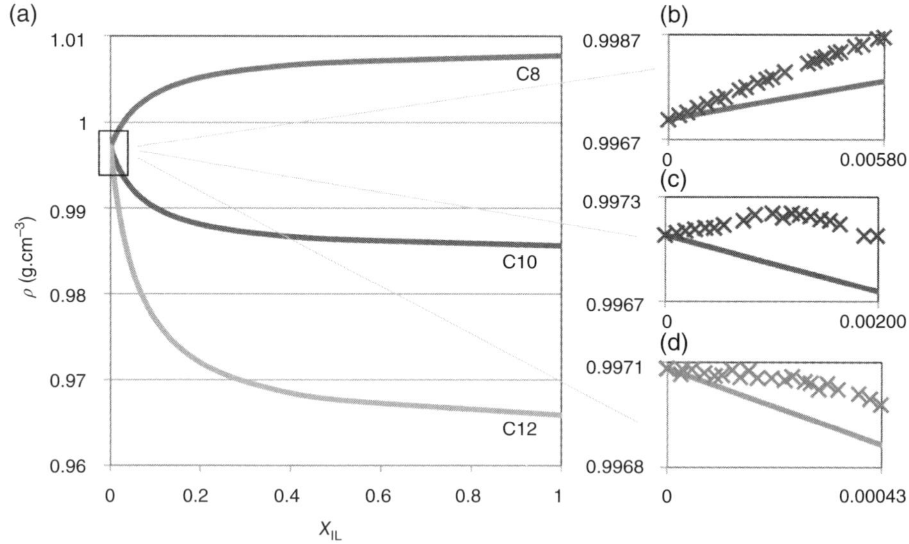

Figure 5.6 Global volumetric behavior at 298 K of aqueous solutions of $[C_8C_1im]Cl$ (top), $[C_{10}C_1im]Cl$ (middle), and $[C_{12}C_1im]Cl$ (bottom). (a) Density, ρ, versus mole fraction considering ideal volumetric behavior. (b–d) Zoom of the diluted regime comparing the ideal density lines with the experimental density data (x-cross).

$$V\phi\left(\text{cm}^3.\text{mol}^{-1}\right) = \frac{M}{\rho} + \frac{10^3(\rho - \rho_0)}{m\rho} \tag{5.2}$$

where M and ρ_0 are the molar mass (in g.mol^{-1}) of the solute and density (in g.cm^{-3}) of the pure solvent, respectively, and m and ρ are the molality and density (in g.cm^{-3}) of the solutions, respectively (Gonzalez-Perez et al. [22]). The main approximation behind this simple calculation lies in the assumption that the contraction of the solution is wholly attributed to the solute (which, in this case, will occupy an apparent molar volume smaller than that of the pure IL).

The top panels of Figure 5.7 show the apparent molar volumes of the three IL solutes as a function of solution composition for all studied temperatures. As expected, at a given temperature, the apparent volumes are always lower than the corresponding values for the pure IL, showing in this way the more dense nature of the real solutions relative to an ideal case. The lower panels of Figure 5.7 show, for each temperature, the deviations of the apparent molar volumes at each composition from the average apparent volume at the corresponding temperature (a way to magnify the trends observed in the curves depicted in the upper panels for each temperature). The rectangles depicted in the bottom panels show the positions of the corresponding CMCs determined at 298 K by ST, fluorescence (FL), or NMR methods (Blesic et al. [6]). In all cases, it is obvious that there is a trend shift in the apparent volumes when the CMC region is crossed—in most cases, the different apparent volume isotherms reach minimum values inside (or near) the CMC regions. Such trend shifts are even more

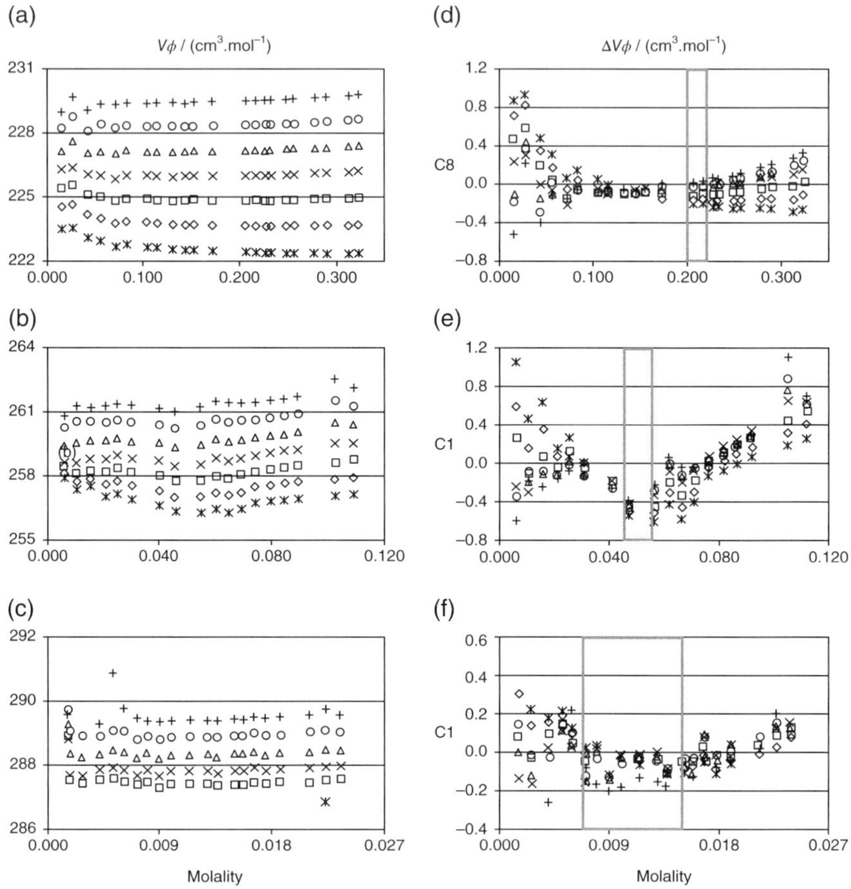

Figure 5.7 (a–c) Apparent molar volumes, $V\phi$, as a function of molality of the different IL solutions at the seven studied temperatures. (d–f) Absolute deviations, $\Delta V\phi$, of the apparent molar volumes at each composition and temperature from the average apparent molar volumes at each temperature. The rectangles represent the range of CMC values determined using other methods at 298 K [6]. The other studied temperatures are 283 (asterisk, ✳), 288 (rhomb, ◇), 293 (square, □), 303 (triangle, Δ), 308 (circle, ○), and 313 K (cross, +).

noticeable if the apparent solute molar volume is represented as a function of the inverse of the concentration (as is usually the case in CMC studies).

The apparent volume data show that is indeed possible to estimate the position of the CMC based solely on volumetric data. Moreover, compared to other methods (ST, FL, NMR) that are frequently used to estimate CMCs, such analysis also provides some insights concerning the behavior of the solute in the midst of the water molecules. As was mentioned before, the increase in the solutions density is caused by the partial breakup of the hydrogen-bonded network of water molecules due to the intrusion of the solute ions, specially the aliphatic moieties that cause the existence of hydrophobic effects. The small apparent molar volumes of the solutes at the most diluted solutions (around 226 $cm^3.mol^{-1}$ for $[C_8C_1im]Cl$ solutions below CMC at 298 K (cf. Fig. 5.7), as compared to 229 $cm^3.mol^{-1}$ for pure $[C_8C_1im]Cl$ at the same temperature)

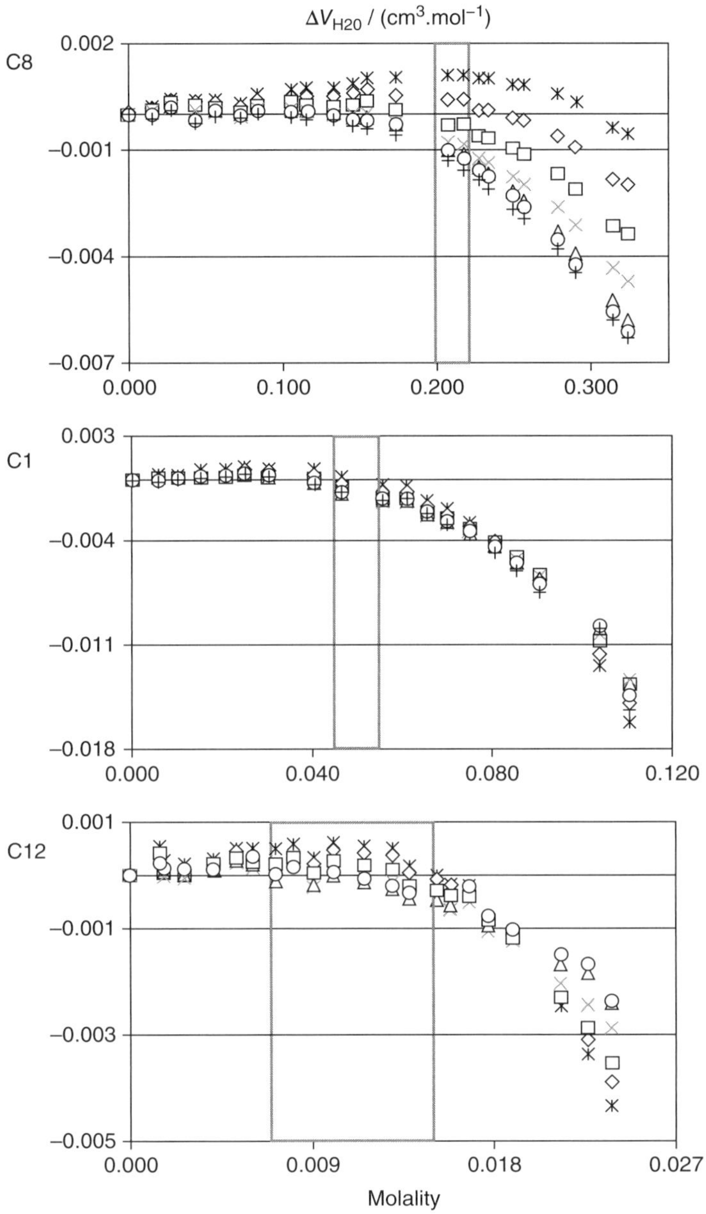

Figure 5.8 Absolute difference, ΔV_{H2O}, between the partial molar volume of water at a given temperature and molal composition, and the molar volume of pure water at the same temperature (asterisk, ✳), 288 (rhomb, ◇), 293 (square, □), 303 (triangle, Δ), 308 (circle, ○), and 313 K (cross, +).

reflect the collapse of the open water structure by the aliphatic chains of the IL cations. From the CMC region onward, most of the chains will be "hidden" from the water molecules due to micelle formation, which means that the collapse of the water structure will be less pronounced and the apparent molar volume of the solute will increase.

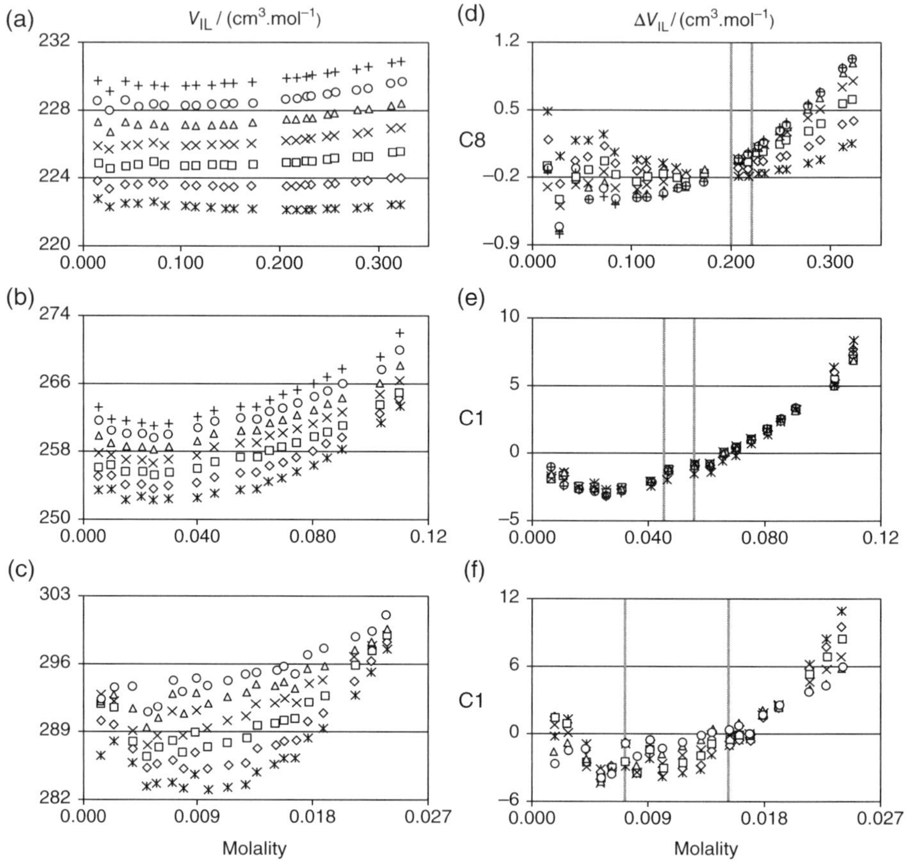

Figure 5.9 (a–c) Partial molar volumes of each IL, V_{IL}, as a function of molality of the different IL solutions at the seven studied temperatures. (d–f) Absolute deviations, ΔV_{IL}, of the partial molar volumes at each composition and temperature from the average partial molar volumes at each temperature.

Instead of doing the calculations based on the apparent molar volumes of the solute (and the corresponding approximation of attributing all volumetric changes of the solution to changes in the volume of the solute), one can also do the calculations based on the estimation of partial molar volumes for both components. This is in fact a much better option since it does not contain the aforementioned approximation. In order to do so, the molar volumes of the solutions as a function of the solute mole fractions were fitted to third-order polynomial functions, and the corresponding derivatives were calculated in order to yield (by extrapolation to the pure component compositions) the corresponding partial molar volumes.

The estimated partial molar volumes are shown in Figures 5.8 and 5.9 (data for water and IL solutes, respectively). In Figure 5.8, the absolute deviation of the partial molar volume of water from the molar volume of pure water at each studied temperature is shown as a function of concentration for the set of isotherms studied for each solute. In most cases, the partial molar volume of water remains constant below the corresponding CMCs and then starts to drop for concentrations larger than the CMCs. Furthermore, the partial molar volumes of the solutes only start to increase markedly after the CMCs are attained (Fig. 5.9, lower panels). This state of affairs is consistent with the apparent volume results and provides a more complete analysis of the volumetric behavior of the solutions: when the solutes are added to pure water in concentrations below the CMC, the contraction of the solution is directly and linearly proportional to the amount of solute ions added (which will exist in solution either as isolated ions or ion pairs). This behavior can already be seen in panels b–d of Figure 5.6 where, in the left side of each panel, the density of the solution diverges in a linear fashion from the density of pure water. This means that the corresponding partial molar volume of water in any of these pre-CMC solutions will remain close to the value for the molar volume of pure water at that temperature (it also validates the approximation used for the calculation of the apparent molar volumes of the solutes). From a physical perspective, the increase in the overall density observed in the pre-CMC solutions due to the collapse of the open structure of water around the isolated solute ions can be mainly attributed to the fact that the solute will occupy cavities (solvent cages) that, per mole of solute, are smaller than those found for the pure ILs. The water that constitutes the walls of such solvent cages still remains at adequate distances from other water molecules in order to continue to interact via hydrogen bonds (in many cases, those "cage" water molecules are even more structured than those found in bulk water), which means that the partial molar volume of water will remain quite close to that of pure water. In other words, from the perspective of the water molecules, the collapse of the three-dimensional structure of water is partially offset by the hydrophobic effect (formation of highly structured solvent cages) caused by the interactions with the aliphatic residues of the IL cations.

For concentrations higher than the CMC values, two phenomena occur (and it is here that the main departures from the apparent molar volume scheme are noticeable): the IL species start to occupy larger molar volumes (and the increase is more rapid in terms of partial molar volumes (Fig. 5.9) than in terms of apparent molar volumes (Fig. 5.7)), and the partial molar volume of water starts to decrease (Fig. 5.8). From a physical perspective, the formation of micelles promotes the interactions between alkyl chains from different cations, which means that those residues will occupy volumes more similar to those found in pure or concentrated solutions of ILs. Moreover, since a part of the alkyl residues are now hidden inside the micelles, the water molecules will start interacting mainly with the charged parts of the ions (chloride anions and the exterior of the micelles, which also include some chloride counterions), which will cause a structure collapse without the opposing hydrophobic effect. The decrease in the partial molar volume of water implies a more pronounced increase in the partial molar volume of the ILs, which in turn justifies the much clearer trend shifts found in the lower panels of Figure 5.9 as compared to those of the lower panels of Figure 5.7. The CMC windows determined by other experimental methods depicted in Figures 5.7, 5.8, and 5.9 show that volumetric studies can yield physical information about the processes involved around the CMC concentrations. However, they should not supersede those other types of experiments since most of the trend shifts are not as pronounced as in the determination of other properties, for instance, ST.

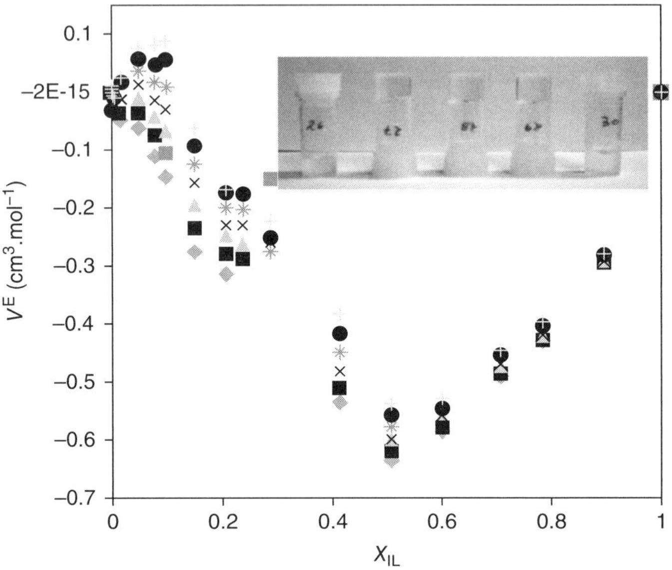

Figure 5.10 Excess volumes, V^E, of H_2O–[C_8C_1im]Cl system as a function of mole fraction of ionic liquid, [C_8C_1im]Cl, x_{IL} and temperature (the order of markers, for instance, at $x_{IL}=0.4$ indicates temperatures from 283 K (◆) to 313 K (+)). Vials in the inset numbered from 26 to 30 correspond to a concentration of $x_{IL}=0.15, 0.21, 0.24, 0.29$, and 0.41, respectively. Those upside down indicate gel-like solutions.

The last part of this section deals with volumetric measurements performed at more concentrated regimes of the solutions under discussion. In the case of [C_8C_1im] Cl solutions, we have observed gel-like structures in the $0.2>x_{IL}>0.4$ concentration range at room temperature. At higher temperatures and IL concentrations, the solutions became clear again. Figure 5.10 presents the corresponding volumetric data, expressed as the excess volume of the solutions. For [$C_{10}C_1$im]Cl, above $x_{IL}=0.06$ concentrations, the solutions become very viscous (solid-like), thereby restricting our volumetric measurements in the whole composition range. The presence of gel or solid-like phases may indicate the formation of different mesophases in these water–IL systems and needs further investigation.

Nevertheless, these experiments also indicate that it is possible to cover the entire concentration range from almost pure water containing IL ions at infinite dilution to pure ILs containing monomeric water molecules dispersed in their nanostructured liquid phases.

5.3 ILs AT AIR–LIQUID INTERFACES: HIGH-TEMPERATURE ST

Among the many unique properties of ILs, their ST plays a special role for different reasons. Even at conditions well above room temperature, ILs typically exhibit extremely low vapor pressure values (Earle et al. [23]). This fact is a consequence of their ionic nature and makes it difficult to accurately determine the cohesive forces present in the liquid. The use of corresponding state principles and other powerful correlation techniques commonly used for molecular fluids is also difficult to implement. ST values are a way to circumvent an issue that is hard to solve for the bulk fluid, but it is possible to access at the liquid–gas boundary—ST is a measure of cohesive forces between liquid molecules present at the surface, and it represents the quantification of force per unit length of free energy per unit area. Thus, the measurement of ST of ILs is one of the most effective ways to (indirectly) access the intrinsic energetics that are involved in the interactions between the ions. Such measurements are also vital to anchor theoretical models that are able to describe realistically, at a molecular level, the fluid properties and structural nature of ILs.

Another distinctive characteristic of ILs is their nanosegregated nature, that is, ILs are complex, highly nonisotropic fluids from a structural point of view. This fact is a consequence of the interplay between two types of important interactions present in the fluid (electrostatic and dispersive) acting at different length scales between distinct atoms that constitute the ions. The lack of fluid isotropy already present in the bulk is also present (and sometimes enhanced) at the surface (Santos and Baldelli [3]). This means that ST data are also a powerful means to explore the different types of segregation/orientation that occur at an ionic/molecular level and how these influence the interactions with other molecule (Lockett et al. [24]).

For liquids composed of spherical molecules, for example, liquid argon or methane, the surface ordering effects are negligible, and the ST correlates well with the intermolecular interactions in the bulk. In contrast, for liquids composed of nonspherical molecules with polar and nonpolar groups, the molecules at the interface may be strongly oriented. According to the Langmuir's principle of independent surface action, each part of a molecule that is actually present at the interface contributes with its own surface free energy to the ST of the liquid. In the case of spherical molecules, the liquid surface should represent a random truncation of the bulk; in contrast to nonspherical molecules, the interface is different from the bulk, and the ST results from the contributions of the groups that form the outer layer of the interface. Langmuir's principle was successfully applied to the ST of ILs (Restolho et al. [25], Law and Watson [26], and Kolbeck et al. [27]), but its application demands the knowledge of the ST of the separated ions and their moieties. Analysis of a large set of experimental values of ST of ILs led Kolbeck et al. [27] to conclude that the "observed behavior can be explained in a consistent way considering Langmuir's principle, namely that the surface composition is determined by the interplay of cohesive energy and surface orientation." This means, for instance, that ILs with long alkyl side chains on either the cation or the anion have the surface region dominated by these tails, which are oriented away from the bulk. However, if the alkyl chains are short or functionalized, different effects may occur.

As stated in the previous paragraphs, the ST of pure liquids is an important physical property that, unlike other bulk properties, is not easily correlated with the characteristics of the substances at a molecular level: the structure and molecular distribution at the surface are, in many cases, quite different from that present in the bulk. In the case of ILs, this situation is particularly significant due to the fact that these are substances composed by two ionic species that generally possess a high degree of structural and charge anisotropy, which often leads to the existence of nanosegregated domains, even in the bulk phase (Law and Watson [26] and Kolbeck et al. [27]).

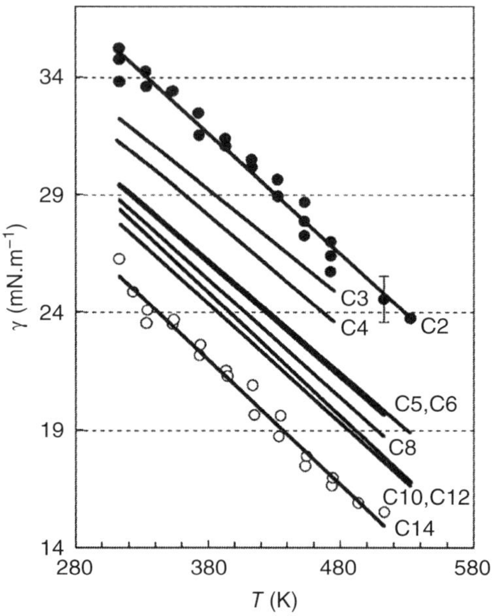

Figure 5.11 Surface tension as a function of temperature for the $[C_nC_1im][Ntf_2]$ ionic liquid series. Experimental points are plotted only for the $[C_2C_1im][Ntf_2]$ and $[C_{14}C_1im][Ntf_2]$ series. The scatter for the other series is similar. The error bar corresponds to the estimated overall uncertainty of the surface tension measurements.

We have measured liquid density and ST data in temperature ranges up to 473 and 532 K, respectively, for 1-alkyl-3-methylimidazolium bis(trifluoromethylsulfonyl) amide ILs, $[C_nC_1im][Ntf_2]$ with $n = 2–10$, 12, and 14 (Tariq et al. [28]). The aim was to test the validity of the aforementioned theories as well as the soundness of the approximations previously performed (Rebelo et al. [29]) by confronting the current results with recently obtained vapor pressure and enthalpy of vaporization data (Esperança et al. [30]).

Figure 5.11 shows that the ST decreases with increasing temperature and with the increase of the alkyl side chain length along the $[C_nC_1im][Ntf_2]$ family. The decrease is more significant from $[C_2C_1im][Ntf_2]$ to $[C_5C_1im][Ntf_2]$ and less so in the $[C_6C_1im][Ntf_2]$ to $[C_{12}C_1im][Ntf_2]$ range and shows a noticeable drop from $[C_{12}C_1im][Ntf_2]$ to $[C_{14}C_1im][Ntf_2]$. Part of this irregular behavior can be attributed to the overall uncertainty of the data, including the fact that the ST is a property sensitive to the presence of small amounts of impurities. Nevertheless, one can always speculate that part of the observed trends are a manifestation of different types of arrangements at the surface as the ratio of the nonpolar to polar moieties of the ILs is altered by the increase in the length of the alkyl side chains. Interestingly, this could also be a reflection of what is happening in the bulk phase: MD simulations and X-ray diffraction studies (Shimizu et al. [31]) have shown that in the case of the $[C_nC_1im][Ntf_2]$ family, there is a transition in the bulk structure of the ILs around $[C_6C_1im][Ntf_2]$ (formation of nonpolar continuous domains). On the other hand, Waring et al. [32] have shown a similar trend (albeit around C4) for the same family of ILs at the liquid–air interface. It is also known that with alkyl side chains above 12 carbon atoms long, there is the possibility of a transition from an isotropic (*albeit* nanostructured) fluid to a partially layered (more smectic-like) fluid (Blesic et al. [16]), although in the case of $[C_nC_1im][Ntf_2]$ ($n = 12$–18) ILs, no evidence of liquid crystal formation was ever found (Hardacre et al. [33]).

Several other conclusions can be drawn from the analysis of Figure 5.11: (i) there is an inherent lack of precision in the determination of the slope of the $\gamma(T)$ functions, a fact that will always partially undermine attempts to extrapolate such data; (ii) previously reported literature values (Rebelo et al. [29]) are even less reliable in terms of defining the temperature dependence of γ due to their short measuring ranges; (iii) nevertheless, the $\gamma(T)$ functions exhibit very similar slopes for the different members of the $[C_nC_1im][Ntf_2]$ family, specially if one takes into consideration the large uncertainty of the data.

The Eötvos equation, $\gamma \cdot V^{(2/3)} = m \cdot (T - T_c)$, states that the product $\gamma \cdot V^{(2/3)}$ at a given temperature should be directly proportional to the interval between that temperature and the critical point temperature, T_c. On the other hand, the Guggenheim equation establishes a linear correlation between the function $\gamma^{(9/11)}$ and the temperature $\gamma^{(9/11)} = \mathrm{b} - (\mathrm{b}/T_c) \cdot T$. Both equations were used with density and ST data in the extended temperature range (~300 – 500 K) to yield new estimates for the (hypothetical) critical point temperatures for all studied members of the $[C_nC_1im][Ntf_2]$ family. The results were then converted to estimates of normal boiling point temperatures using the empirical relation $T_b = 0.7 \cdot T_c$. In contrast to the preliminary estimates performed before Rebelo et al. [29], where an averaged (over many types of substances) value of $0.6 \cdot T_c$ was employed, we are now convinced that $0.7 \cdot T_c$ should better describe the critical-to-boiling point scaling of many ILs because this is the factor experimentally known for *n*-alkanols, substances with many similarities to these commonly used aprotic liquid salts (Guedes et al. [34]).

The results are presented in Figure 5.12 along with the previously published T_b values calculated using data in a much more limited temperature range (Rebelo et al. [29]) and values obtained using vapor pressure data (Zaitsau et al. [35], Paulechka et al. [36], and Emel'yanenko et al. [37]) extrapolated to atmospheric pressure using the Clausius–Clapeyron relation. The rather large uncertainty associated with the estimated T_b values reflects the lack of precision of the ST data, the large extrapolations involved in the use of the Eötvos and Guggenheim empirical correlations, and the (also empirical) conversion from T_c to T_b values using the 0.7 factor. On the other hand, the extrapolation of vapor pressure data (from a few points five to six orders of magnitude below atmospheric pressure) also entails a significant loss of precision.

Nevertheless, all predictions indicate normal boiling point temperatures well above the decomposition temperatures of the ILs, which means that the results may never be corroborated experimentally by direct measurements. Moreover, the different sets of data show that there is a tendency for a decrease in the normal boiling point temperature as the length of the alkyl side chain is increased, thus qualitatively corroborating preliminary estimates (Rebelo et al. [29]). Nonetheless, the observed tendency is not so strong if one uses these new γ and V data in an extended temperature range instead of the old data taken from a much narrower temperature interval. This subdued tendency is also in better agreement with the data calculated from the vapor pressure measurements.

The fact that cations with larger alkyl side chains seem to exhibit lower normal boiling point temperatures may seem in contradiction with experimental fractional distillation results (Earle et al. [23] and Widegren et al. [38]) that indicate that in a

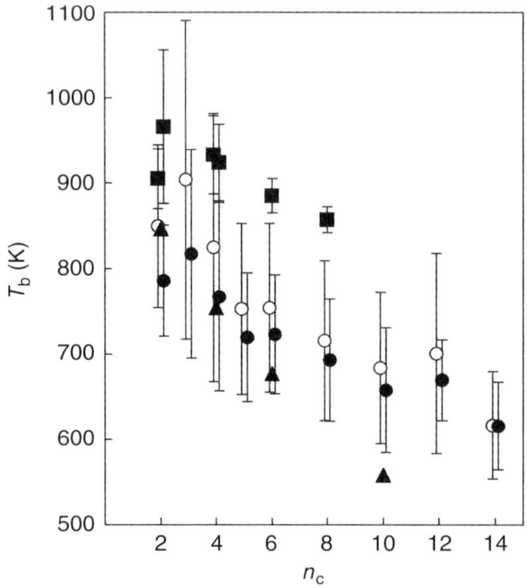

Figure 5.12 Estimated normal boiling point temperatures, T_b, of $[C_nC_1im][Ntf_2]$ ionic liquids as a function of alkyl side chain length, n. ○, T_b using the Eötvos equation with the data under discussion [26]; ●, T_b using the Guggenheim equation with the data under discussion [26]; ▲, T_b using the Guggenheim equation with data from [1]; ■, T_b using the Clausius–Clapeyron relation with experimental vapor pressure values from [33–35].

($[C_nC_1im][Ntf_2]+[C_mC_1im][Ntf_2]$, $4 \leq n < m$) binary mixture, the first fractions of distillate (enriched in the most volatile component) will contain predominately the IL with a shorter alkyl side chain, $[C_nC_1im][Ntf_2]$. The analysis of the vapor above such binary mixtures (accessed using ion cyclotron resonance mass spectroscopy techniques) also showed the same trend (Leal et al. [39]). However, the two situations can be reconciled if one takes into account the fact that such distillation experiments (and also vapor pressure experiments using the Knudsen effusion method; Zaitsau et al. [35]) must be performed at temperatures well below the decomposition temperature of the ILs and, consequently, at vapor pressures well below one atmosphere. Under those conditions, the heavier members of the family are the less volatile (lower vapor pressures). However, due to their larger values of molar enthalpy of vaporization (Zaitsau et al. [35] and Santos et al. [40]), their vapor pressure will increase more rapidly with temperature, and if a vapor pressure crossover intervenes before atmospheric pressure is reached, then they will indeed exhibit lower boiling point temperatures and will be the more volatile members of the family in that higher temperature range, as predicted by the density–ST correlations presented in this work.

5.4 ILs AT SOLID SURFACES: ADSORPTION AND MORPHOLOGY

The adsorption of surfactants on solid surfaces has been the subject of numerous studies. This phenomenon has widespread technological applications in wetting, adhesion, nanofabrication, and cleaning and degreasing of metal surfaces. The adsorbed layers (typically of nanometric size) are known to play a key role in the development of new electronic devices, sensor materials, and surfaces with special wetting properties. Several attempts have been made to understand the nanoscopic features of the adsorbed layers, which are required for the design of interfaces with desired characteristics, such as wetting, agglomeration, and friction properties (Paria and Khilar [41]).

The research on this front has started in 1955, when the first proposal of hemimicelle formation at the solid–liquid interface was forwarded by Gaudin and Fuerstenau [42]. Various techniques, especially FL probe studies, optical reflectometry, NMR, and zeta potential, were employed to understand the microscopic nature of those interfaces. In 1995, a breakthrough occurred in this area, when Manne and Gaub [43] reported the direct observation of organized adsorbed structures using AFM. Since then, there was an upsurge in the studies related to the self-organization of amphiphilic molecules (anionic, cationic, zwitterionic, and mixed micellar systems) and the factors responsible for the alteration of the corresponding structures. The resulting shapes of the adsorbed aggregates are considered to be a compromise between packing effects and specific constraints due to molecule–substrate interactions. The micelles of conventional surfactants adsorbed at the solid–liquid interface are generally considered to be stable only in the presence of the solution—the ordered structure may be lost when the solvent is taken out from the system. However, some cases have been reported where the regular packing of two-dimensional (2D) aggregates formed by amphiphilic molecules at the solid–liquid interface has been successfully preserved at the solid substrate after solvent removal (Atkin et al. [44]).

After the first reports on the self-aggregation behavior of some ILs, the research was quickly extended in different directions, for example, the aggregation behavior of ILs in aqueous media, nonaqueous solutions, and mixed solvents and at solid surfaces. The behavior at solid surfaces, which has been scarcely investigated (Tariq et al. [45]), is the focus of our work.

The aim of the work was (i) to compare the adsorption behavior of a family of ILs at a gold surface and to check whether the dissipation factor measured with a quartz crystal microbalance (QCM) can be conclusive as regards the viscoelasticity of the adsorbed layer, (ii) to observe the morphology and structural patterns formed by diluted IL surfactant solutions dried on a gold surface, and (iii) to find possible correlations between the adsorption of IL surfactants and the corresponding frictional properties and, most importantly, between the alkyl side chain length of the ILs and all those properties. The selection of the present family of surfactants, which belongs to one of the most commonly studied groups of ILs, is an extension of our previous studies on self-aggregation phenomena with 1-alkyl-3-methylimidazolium chloride ILs (Blesic et al. [6]).

The average values of the frequency shift from the third overtone after adsorption of the ILs were determined and are presented in Figure 5.13. The relationship between the absorbed mass and the corresponding frequency change for viscoelastic films may be described through a modified version of the Sauerbrey equation (Gispert et al. [46]), which requires the knowledge of the density, viscosity, and elasticity of the adsorbed film. Unfortunately, we do not have a way of estimating these parameters for the ILs under study. However, the decrease of frequency change with the increase of the chain length (Fig. 5.13) indicates that the amount of IL adsorbed at the gold surface decreases with the length of the IL chains.

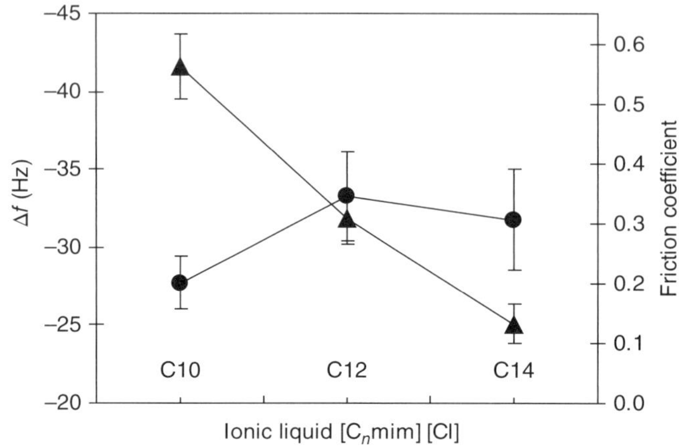

Figure 5.13 Changes in the third overtone frequency (▲) and average friction coefficient (●) for [$C_{10}C_1$im]Cl, [$C_{12}C_1$im]Cl, and [$C_{14}C_1$im]Cl. The error bars show the precision of the underlying experimental data.

The friction and adsorption results of the ILs can be related with the morphology and structure of the adsorbed layers. According to Figure 5.13, the adsorbed masses are lower for C_{12} and C_{14}, and from the AFM images (Fig. 5.14), a multilayer structure should be present on both cases. So, it is quite probable that the film is adsorbed in islands, larger than the dimensions of the images obtained by AFM. Recently, Ao et al. [47] claimed that a single-chain IL, [$C_{12}C_1$im]Br, forms multilayers on the silicon surface, in contrast with its gemini/dimeric counterpart [C_{12}-4-$C_{12}C_1$im][Br_2], which forms only a bilayer, if the concentration is a multiple of the CMC. They suggested that the [$C_{12}C_1$im]Br ILs start to cover the gold surface from 0 to almost 50% if the concentration of the solutions is raised from 1.0 × CMC to 2.0 × CMC, but for greater ratios, the fraction of surface coverage remains constant, even if the concentration is raised to 5.0 × CMC. According to the same authors, after the initial nucleation and growth process, layered island-like structures start to merge. After the formation of a bilayer structure, the added surfactant starts to accumulate on the surface, layer by layer, due to electrostatic and hydrophobic interactions. Thus, a similar phenomenon seems to occur in the present cases for C_{12} and C_{14}.

For C_{10}, the higher adsorbed mass should be due to the formation of an extensive homogeneous layer, which gives rise to a low friction, whereas the multilayer structures of C_{12} and C_{14}, besides being loosely packed, should present bare zones, which led to high frictional behavior.

Knowledge of the parameters that characterize the aggregates formed by the ILs in aqueous solutions may help us to understand their behavior when adsorbed on gold surfaces. However, it should be stressed that direct extrapolations to interpret the AFM observations may not be valid because the effect of the solvent removal upon the shape of the adsorbed agglomerates is not known. The chosen concentration of the IL solutions (0.07 M) corresponds to ~1.3 × CMC of C_{10}, ~4.7 × CMC of C_{12}, and ~17.5 × CMC of C_{14}.

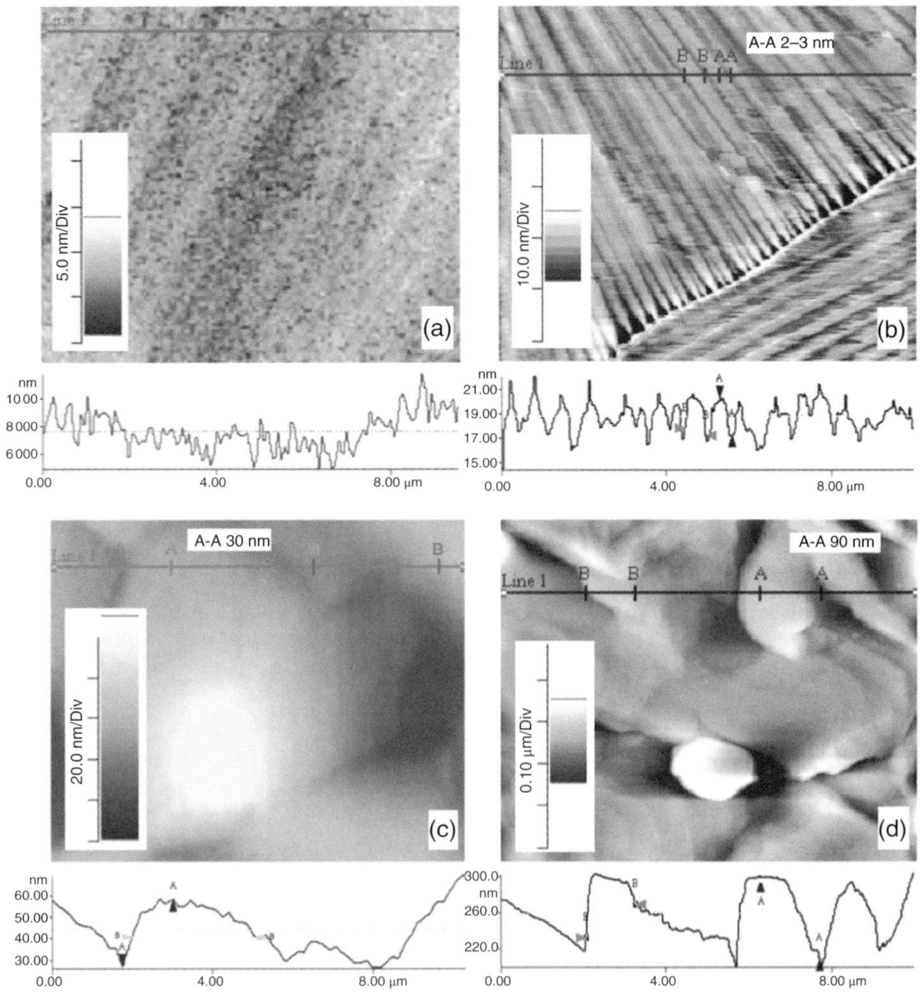

Figure 5.14 AFM images (10 × 10 μm) of pure gold substrates. (a) Bare gold surface. (b–d) Gold surfaces treated with a 0.07 M concentration of different ionic liquids: (b) $[C_{10}C_1im]Cl$, (c) $[C_{12}C_1im]Cl$, and (d) $[C_{14}C_1im]Cl$. Cross-sectional profiles are depicted below each AFM image.

Comparison of the height profile of the $[C_{10}C_1im]Cl$ layer (2–3 nm; Fig. 5.14b) with the hydrodynamic diameter (1.4 nm; Bai et al. [48]) for this IL suggests that the uppermost part of this layer reflects the structure of cylindrical micelles. We should stress that the AFM images only provide information about the topography of the surface and do not allow estimating the thickness of the adsorbed layers.

The images for $[C_{12}C_1im]Cl$ and $[C_{14}C_1im]Cl$ show much higher heights compared to $[C_{10}C_1im]Cl$, which could be a result of their lower CMC. In those cases, the larger and more abundant micelles in solution do not retain their shape when adsorbed onto the surface and form loosely bound structures, which partially cover the surface and lead to higher friction coefficients.

5.5 CONCLUSION

The impact of ILs in terms of surface science was addressed taking into account three research areas: (i) IL self-aggregation behavior in the bulk phase, (ii) energetics at the IL–air interface, and (iii) characterization of the adsorption of ILs at solid substrates.

Two characteristics of ILs are specially relevant in the context of surface studies:

(1) As pure bulk phases, ILs exhibit intermediate-range (nanoscale) ordering that reflects on the one hand the need to establish local electroneutrality conditions among their ions and on the other hand the need to allow for the presence of non-polar moieties attached to the ions. Such nanoscale ordering has to adapt to the presence of 2D boundaries such as solid or gas interfaces.
(2) As components of liquid mixtures, many ILs exhibit surfactant properties that promote self-aggregation phenomena. The main asset of ILs in this context is that such aggregation studies can be conducted in extended concentration ranges without the occurrence of precipitation, as in the case of traditional surfactants.

REFERENCES

[1] Plechkova, N.V.; Seddon, K.R. *Chem. Soc. Rev.*, 2008, **37**, 123–150.
[2] Lovelock, K.R.J. *Phys. Chem. Chem. Phys.*, 2012, **14**, 5071–5089.
[3] Santos, C.S.; Baldelli, S. *Chem. Soc. Rev.*, 2010, **39**, 2136–2145.
[4] Jungwirth, P.; Tobias, D.J. *Chem. Rev.*, 2006, **106**, 1259–1281.
[5] Ghosal, S.; Brown, M.A.; Bluhm, H.; Krisch, M.J.; Salmeron, M.; Jungwirth, P.; Hemminger, J.C. *J. Phys. Chem. A*, 2008, **112**, 12378–12384.
[6] Blesic, M.; Marques, M.H.; Plechkova, N.V.; Seddon, K.R.; Rebelo, L.P.N.; Lopes, A. *Green Chem.*, 2007, **9**, 481–490.
[7] Tariq, M.; Podgorsek, A.; Ferguson, J.L.; Lopes, A.; Costa Gomes, M.F.; Pádua, A.A.H.; Rebelo, L.P.N.; Canongia Lopes, J.N. *J. Colloid Interface Sci.*, 2011, **360**, 606–616.
[8] Greaves, T.L.; Drummond, C.J. *Chem. Rev.*, 2008, **108**, 206–237.
[9] Lee, C.K.; Huang, H.W.; Lin, I.J.B. *Chem. Commun.*, 2000, **19**, 1911–1912.
[10] Ali, M.; Sarkar, A.; Tariq, M.; Ali, A.; Pandey, S. *Green Chem.*, 2007, **9**, 1252–1258.
[11] Angell, C.A.; Byrne, N.; Belieres, J.P. *Acc. Chem. Res.*, 2007, **40**, 1228–1236.
[12] Bernardes, C.E.S.; Minas da Piedade, M.E.; Canongia Lopes, J.N. *J. Phys. Chem. B*, 2011, **115**, 2067–2074.
[13] Olofsson, G.; Loh, W. *J. Braz. Chem. Soc.*, 2009, **20**, 577–593.
[14] Bijma, K.; Engberts, J.; Blandamer, M.J.; Cullis, P.M.; Last, P.M.; Irlam, K.D.; Soldi, L.G. *J. Chem. Soc. Faraday Trans.*, 1997, **93**, 1579–1584.
[15] Blesic, M.; Lopes, A.; Melo, E.; Petrovski, Z.; Plechkova, N.V.; Canongia Lopes, J.N.; Seddon, K.R.; Rebelo, L.P.N. *J. Phys. Chem. B*, 2008, **112**, 8645–8650.

[16] Blesic, M.; Swadzba-Kwasny, M.; Holbrey, J.D.; Canongia Lopes, J.N.; Seddon, K.R.; Rebelo, L.P.N. *Phys. Chem. Chem. Phys.*, 2009, **11**, 4260–4268.

[17] Attwood, D.; Fluorence, A.T. *Surfactant Systems, Their Chemistry, Pharmacy and Biology*, Chapman and Hall, London, 1983, pp. 72–129.

[18] Bashford, M.T.; Woolley, E.M. *J. Phys. Chem.*, 1985, **89**, 3173–3178.

[19] Zana, R. *J. Colloid Interface Sci.*, 1980, **78**, 330–337.

[20] Zana, R. *Langmuir*, 1996, **12**, 1208–1211.

[21] Tariq, M.; Moscoso, F.; Deive, F.J.; Rodriguez, A.; Sanromán, M.A.; Esperança, J.M.S.S.; Canongia Lopes, J.N.; Rebelo, L.P.N. *J. Chem. Thermodyn.*, 2013, **59**, 43–48.

[22] Gonzalez-Perez, A.; del Castillo, J.L.; Czapkiewicz, J.; Rodriguez, J.R. *J. Phys. Chem. B*, 2001, **105**, 1720–1724.

[23] Earle, M.J.; Esperanca, J.M.S.S.; Gilea, M.A.; Canongia Lopes, J.N.; Rebelo, L.P.N.; Magee, J.W.; Seddon, K.R.; Widegren, J.A. *Nature*, 2006, **439**, 831–834.

[24] Lockett, V.; Sedev, R.; Harmer, S.; Ralston, J.; Horne, M.; Rodopoulos, T. *Phys. Chem. Chem. Phys.*, 2010, **12**, 13816–13827.

[25] Restolho, J.; Serro, A.P.; Mata, J.L.; Saramago, B. *J. Chem. Eng. Data*, 2009, **54**, 950–955.

[26] Law, G.; Watson, P.R. *Langmuir*, 2001, **17**, 6138–6141.

[27] Kolbeck, C.; Lehmann, J.; Lovelock, K.R.J.; Cremer, T.; Paape, N.; Wasserscheid, P.; Fröba, A.P.; Maier, F.; Steinrück, H.-P. *J. Phys. Chem. B*, 2010, **114**, 17025–17036.

[28] Tariq, M.; Serro, A.P.; Mata, J.L.; Saramago, B.; Esperanca, J.M.S.S.; Canongia Lopes, J.N.; Rebelo, L.P.N. *Fluid Phase Equilib.*, 2010, **294**, 131–138.

[29] Rebelo, L.P.N.; Canongia Lopes, J.N.; Esperança, J.M.S.S.; Filipe, E. *J. Phys. Chem. B*, 2005, **109**, 6040–6043.

[30] Esperança, J.M.S.S.; Canongia Lopes, J.N.; Tariq, M.; Santos, L.M.N.B.F.; Magee, J.W.; Rebelo, L.P.N. *J. Chem. Eng. Data*, 2010, **55**, 3–12.

[31] Shimizu, K.; Bernardes, C.E.S.; Canongia Lopes, J.N. *J. Phys. Chem. B*, 2014, **118**, 567–576.

[32] Waring, C.; Bagot, P.A.J.; Slattery, J.M.; Costen, M.L.; McKndrick, K.G. *J. Phys. Chem. A*, 2010, **114**, 4896–4904.

[33] Hardacre, C.; Holbrey, J.D.; McMath, S.E.J.; Nieuwenhuyzen, M. in: R.D. Rogers, K.R. Seddon (Eds.), *Small-Angle Scattering from Long-Chain Alkylimidazolium-Based Ionic Liquids*, ACS Symposium Series, American Chemical Society, Washington, DC, 2002, pp. 400–412.

[34] Guedes, H.J.R.; Cordeiro, T.C.; Canongia Lopes, J.N.; Esperança, J.M.S.S.; Rebelo, L.P.N.; Huq, S.; Seddon, K.R. *J. Phys. Chem. B*, 2005, **109**, 3519–3525.

[35] Zaitsau, D.H.; Kabo, G.J.; Strechan, A.A.; Paulechka, Y.U.; Tschersich, A.; Verevkin, S.P.; Heintz, A. *J. Phys. Chem. A*, 2006, **110**, 7303–7306.

[36] Paulechka, Y.U.; Zaitsau, D.H.; Kabo, G.J.; Strechan, A.A. *Thermochim. Acta*, 2005, **439**, 158–160.

[37] Emel'yanenko, V.N.; Verevkin, S.P.; Heintz, A. *J. Am. Chem. Soc.*, 2007, **129**, 3930–3937.

[38] Widegren, J.A.; Wang, Y.-M.; Henderson, W.A.; Magee, J.W. *J. Phys. Chem. B*, 2007, **111**, 8959–8964.

[39] Leal, J.P.; Esperança, J.M.S.S.; Minas da Piedade, M.E.; Canongia Lopes, J.N.; Rebelo, L.P.N.; Seddon, K.R. *J. Phys. Chem. A*, 2007, **111**, 6176–6182.

[40] Santos, L.M.N.B.F.; Canongia Lopes, J.N.; Coutinho, J.A.P.; Esperança, J.M.S.S.; Gomes, L.R.; Marrucho, I.M.; Rebelo, L.P.N. *J. Am. Chem. Soc.*, 2007, **129**, 284–285.

[41] Paria, S.; Khilar, K.C. *Adv. Colloid Interface Sci.*, 2004, **110**, 75–95.

[42] Gaudin, A.M.; Fuerstenau, D.W. *Trans. AIME*, 1955, **202**, 958–962.

[43] Manne, S.; Gaub, H.E. *Science*, 1995, **270**, 1480–1483.

[44] Atkin, R.; Craig, V.S.J.; Biggs, S. *Langmuir*, 2000, **16**, 9374–9380.

[45] Tariq, M.; Serro, A.P.; Colaço, R.; Saramago, B.; Canongia Lopes, J.N.; Rebelo, L.P.N. *Colloid Surf. A*, 2011, **377**, 361–366.

[46] Gispert, M.P.; Serro, A.P.; Colaço, R.; Saramago, B. *Surf. Interface Anal.*, 2008, **40**, 1529–1537.

[47] Ao, M.; Xu, G.; Pang, J.; Zhao, T. *Langmuir*, 2009, **25**, 9721–9727.

[48] Bai, G.; Lopes, A.; Bastos, M. *J. Chem. Thermodyn.*, 2008, **40**, 1509–1516.

CHAPTER 6

Aggregation Behavior of Ionic Liquid-Based Gemini Surfactants and Their Interaction with Biomacromolecules

TING ZHOU

Key Laboratory of Colloid and Interface Chemistry, Ministry of Education, Shandong University, Jinan, P. R. China
College of Science, China University of Petroleum (East China), Qingdao, P. R. China

GUIYING XU

Key Laboratory of Colloid and Interface Chemistry, Ministry of Education, Shandong University, Jinan, P. R. China

6.1 INTRODUCTION

Gemini surfactants are novel surfactants that are attracting considerable interest in the academic and industrial fields, such as drug entrapment and release, gene transfection, phase transfer catalysts, oil recovery, and polymerizations [1–10]. The "gemini surfactants" term was named by Menger in 1991 to describe a family of surfactants that are constructed by two hydrophobic chains and two polar/ionic head groups covalently connected by a spacer group at the level of or in close vicinity to the head groups [7], as shown in Figure 6.1. The terminology has been extended to include all dimeric surfactants [1, 2, 6, 8–10] and the variation in the length of the hydrophobic tails, the head groups, and the rigid or flexible spacer groups. Compared to conventional monochain surfactants, the interfacial activity of gemini surfactants in solution is excellent, such as lower critical micelle concentration (CMC), higher adsorption efficiency, and better soluble wetting, foaming, and limber-soap dispersing properties [3–5]. The ionic liquid-based (IL-based) gemini surfactants are a new generation of amphiphilic molecules, which have been drawn attention to researchers, because they not only own the properties of traditional gemini surfactant but also sustain the inherent ionic nature of IL [11–14]. According to the recent literature, three types of IL-based gemini surfactants, namely, imidazolium [11–15], pyridinium [13], and pyrrolidinium based [11, 13, 15], are reported. Owing to the specific properties of ILs and structures of gemini surfactants [16–26], IL-based gemini surfactants show several advantages, for example, formation of compact

Ionic Liquid-Based Surfactant Science: Formulation, Characterization, and Applications, First Edition. Edited by Bidyut K. Paul and Satya P. Moulik.

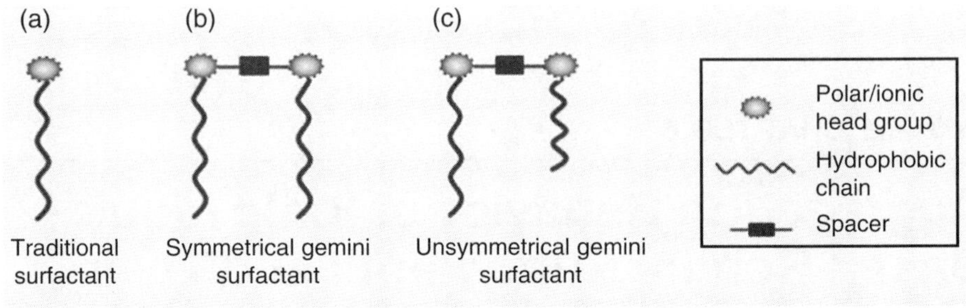

Figure 6.1 Schematic representation of a traditional surfactant (a), symmetrical gemini surfactant (b), and unsymmetrical gemini surfactant (c).

membranes and potential applications in biology due to the strong attraction between IL head groups and aromatic rings through π–π interaction [27–33].

In other aspects, the possibilities of aggregate formation of IL-based gemini surfactants have concern in the synthesis and purification of bulk ILs, the solubility properties of the IL molecules by simple solutes, and the formation of phase-separated systems. As well known, ILs have been extensively studied solvents for synthesis and catalysis, which are considered as environment-friendly compounds. Additionally, enhanced reaction rates obtained in ILs allow the reduction of solvent volumes in a given technological process. Other important properties are their ability to dissolve an enormous number of organic, inorganic, and polymeric materials in a wide temperature range [16, 26]. The structural tunability and versatility of IL-based surfactants allow for the synthesis of compounds with unique micellization properties. IL-based surfactants have been successfully used in analytical chemistry and separation science, such as used as substitutes in extraction processes (dissolved in water or linked to a stationary phase), or as mobile or stationary phase modifiers in high-performance liquid chromatography (HPLC), or as the micellar component in capillary electrophoresis [34]. Owing to the distinct polarizability of imidazolium head groups, IL-based surfactants are reported to be used as supramolecular templates for the preparation of functional materials [35], or modifier of various types of chemical reactions [28]. In recent years, IL-based gemini surfactants (imidazolium or pyridinium based in particular) have generated efficient gene delivery vectors. Their positive charge (due to delocalization on an aromatic ring) allows a strong association of nucleic acid and generation of surfactant/nucleic acid complexes [32, 33]. These special amphiphilic properties and optimum DNA association/release profile of the IL-based polar head, together with the high charge/mass ratio (loading capacity) of the gemini surfactant design, prompted an active research in the field of IL-based gemini surfactants.

In view of overwhelming reports on IL-based gemini surfactants, we mainly summarize the recent developments in imidazolium, pyridinium, and pyrrolidinium-based gemini surfactants over the past few years in this chapter. The first part reviews the progress of a series of IL-based gemini surfactants with respect to synthesis and the aggregation behavior in the aqueous solution and at the air/water interface and related issues. The second part reviews the interactions between IL-based gemini surfactants and biomacromolecules (proteins—gelatin and BSA;

DNA). Finally, conclusions and future prospects for IL-based gemini surfactants, in brief, are summarized.

6.2 IL-BASED GEMINI SURFACTANT

The IL-based gemini surfactants are comprised of two imidazolium (or pyridinium or pyrrolidinium) cations and two aliphatic chains connected by a spacer [36–40]. The first works involving the amphiphilic association structures of imidazolium-, pyridinium-, and pyrrolidinium-based gemini surfactants were carried out in 2005 by Anderson et al. [11], 2000 by Meekel et al. [32], and 2004 by Baker et al. [39], respectively. Recently, many researchers have directed their efforts toward synthesis and studying new IL-based gemini surfactants by using different techniques [41–43]. In this section, we summarize the synthesis methods of IL-based gemini surfactants and the aggregation behavior in the aqueous solution and at the air/water interface and related issues.

6.2.1 Synthesis of IL-Based Gemini Surfactants

The methods used to synthesize IL-based gemini surfactants can be found in the literature [12, 13, 15, 32, 36–38, 42, 43]. In general, cationic IL-based gemini surfactants can be synthesized by heating, condensation, salifying, or coupling reactions. There are three main approaches for the preparation of IL-based gemini surfactants as follows:

(1) Connect two hydrophobic alkyl chains directly to hydrophilic head group through chemical bonds.

This method is attractive when the dibromide is commercially available, including extremely simple steps. For example, a one-step process for the synthesis of the IL-based gemini surfactants has been used in Baltazar et al. [36] and Ding et al. [12] researches, mixing either 1-butylimidazole or 1-methylimidazole with the appropriate dibromoalkane. However, if the dibromide or monomeric surfactants are not purchasable, researchers have to synthesize the intermediates by several steps. Our group has synthesized imidazolium gemini surfactants using four steps [37], including salifying and coupling, and the schematic diagram of synthesis route is presented in Figure 6.2. Another example is that Patial et al. [44] synthesize by reaction of epichlorohydrin with long-chain fatty alcohol products (2-(alkoxymethyl)oxirane), followed by their subsequent treatment with imidazole, resulting in the formation of 1-(1H-imidazol-1-yl-3 alkoxy)propane-2-ol, which on subsequent treatment with 1,2-dibromoethane and 1,3-dibromopropane resulted in the formation of targeted IL-based gemini surfactants.

(2) Connect the hydrophilic head groups with a spacer, followed by linking the hydrophobic chains.

Here are two typical examples. The synthesis of C_4 and $C_{12}(Dim)_2$-2Br ILs (i.e., ILs possess butyl and dodecyl linkage chains separating the dications with decyl groups on the three positions of the imidazolium cation) was a

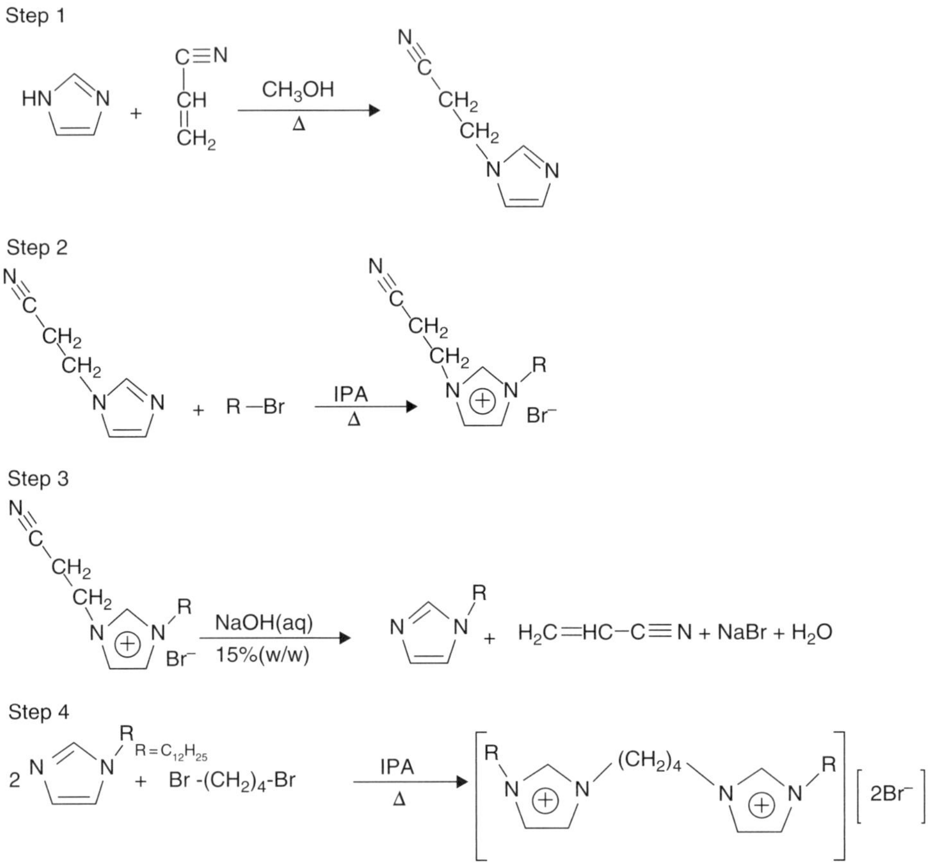

Figure 6.2 The synthesis of IL-based gemini surfactants [C_{12}-4-C_{12}im]Br_2. Reproduced from Ao et al. [37] with permission from Elsevier.

two-step process by Baltazar et al. [36]. Either 1,4-dibromobutane or 1,12-dibromododecane was added to interact with imidazole. The subsequent quaternization reaction was accomplished by the reaction of this product with 1-bromodecane, and the resulting products were recrystallized. Jin et al. [45] prepared 1,11-diimidazolium-3,6,9-trioxaundecane. Subsequently, 1,11-diimidazolium-3,6,9-trioxaundecane was quaternized with excess alkyl halide or polyfluoroalkyl halide under neat reaction conditions, and the halide derivatives underwent metathesis reactions with a slight excess of $LiN(SO_2CF_3)_2$ to form the corresponding gemini surfactants that contain polyether linkage chains.

(3) Connect two hydrophobic chains by a spacer; then connect the hydrophilic head groups.

New pyridinium gemini amphiphiles with ethane-1,2-dithiol spacer have been synthesized by regioselective electrophilic cobromination of R-olefins [46]. Firstly, ethane-1,2-dithiol and *N*-bromosuccinimide on reaction with R-olefins (dodecene, tetradecene, hexadecene, and octadecene) gave the respective 1,2-bis(2-bromoalkylthio)ethane. Subsequently, the bromoalkylthio ethers when reacted with pyridine gave the respective gemini bispyridinium bromide. Another example is that Kouwer and Swager [47] have synthesized a series of ionic liquid crystals (ILCs) that bridges the classes of ILs and LCs. Alkoxy tails were introduced at the far end of the mesogen under standard Williamson etherification conditions. The key step in the synthesis process was an Ullmann-type coupling of an aryl iodide or bromide with imidazole and 2-substituted imidazole derivatives.

In summary, no matter which methods were used to synthesize IL-based gemini surfactants, researchers should try to simplify preparation and reaction process according to the actual suitable situation of synthetic route. Also, it must be ensured that the certain active sites participate in the reaction intermediate, so that the hydrophilic groups, hydrophobic chains, and spacer can be connected at the active point in the following step. On the other hand, the reaction processes of gemini surfactants generally involve some subsidiary reactions, and they must be separated and purified from the reaction product by recrystallization and chromatography. In addition to the effort to synthesize new IL-based gemini surfactants, additional experimental efforts has focused on studying properties of these IL-based gemini surfactants. The self-assembly and aggregation behavior of IL-based gemini surfactants was reviewed in the following section.

6.2.2 The Aggregation Behavior of IL-Based Gemini Surfactants in Aqueous Solution

Considering the growing number of IL-based gemini surfactant investigations [48–52], the possible directions of the development and applications can be identified: the amphiphilic nature of surfactants [53, 54], for example, leading to aggregation phenomena in aqueous solutions. With the possibility of the fine-tuning hydrophobicity by changing the alkyl chain length and/or the nature and size of the counterion (anion), one can affect both the structure and the dynamics of aggregates. Various experimental methods are used in the surfactant aggregation behavior characterization, such as calorimetry, conductometry, light scattering, NMR spectroscopy, neutron scattering, spectrophotometry, and tensiometry [37, 53, 55, 56].

6.2.2.1 Micellization (CMC) Micellization research has been the intense subject for many years [5–8, 27, 36, 56]. When the surface becomes saturated with surfactant monomers, micelles begin to form in the bulk solution. The threshold concentration at which micellization begins is named the CMC (Zana [6]). Table 6.1 summarizes the CMC values of IL-based gemini surfactants according to the literature. Owing to the two hydrophobic alkyl chains, the CMC values of IL-based gemini surfactants are much lower than those of the corresponding monomes. Since surfactants at the air/solution interface project their hydrophobic tails toward the air, the surface tension of the solution is decreased when surfactant is added. The CMC can be easily estimated using surface tension measurements [36, 58]. The CMC values, which were determined by conductivity method, have been found to be much higher than those obtained from surface tension measurements, due to the existence of nonsurface-active premicellar aggregates [15]. The CMC can also been determined using pyrene

TABLE 6.1 Surface Activity Parameters for IL-Based Gemini Surfactants

Surfactants	CMC ($mmol \cdot l^{-1}$)	γ_{cmc} ($mN \cdot m^{-1}$)	Γ_{max} ($\mu mol \cdot m^{-2}$)	A_{min} (nm^2)	pc_{20}	CMC (c_{20})	Reference
$[C_{12}\text{-2-}C_{12}im]Br_2$	0.55	33.6	1.23	1.35	4.54	18.97	55
$[C_{12}\text{-4-}C_{12}im]Br_2$	0.72	35.7	1.19	1.4	3.94	6.32	
$[C_{12}\text{-6-}C_{12}im]Br_2$	0.78	39.5	1.16	1.43	3.73	4.13	
$[C_{10}\text{-4-}C_{10}im]Br_2$	4.5	35.2	1.25	1.33	3.14	6.16	37
$[C_{14}\text{-4-}C_{14}im]Br_2$	0.1	37.2	0.88	1.88	5.04	10.85	
[12-(S-2-S)-12]im	0.4	39.7	2.6	0.63	0.93		48
[12-(S-4-S)-12]im	0.31	40.8	2.06	0.8	1.12		
[14-(S-2-S)-14]im	0.1	42.9	2.12	0.78	1.52		
[14-(S-4-S)-14]im	0.071	46.6	3.1	0.53	1.4		
$C_6[C_{10}ArOCH_2Im]_2 \cdot 2Br$	3.02×10^{-3}	39.8	7.3	0.23			57
$C_6[C_{12}ArOCH_2Im]_2 \cdot 2Br$	0.175×10^{-3}	42.8	17.2	0.09			
$C_6[C_{14}ArOCH_2Im]_2 \cdot 2Br$	68.6×10^{-3}	48.7	5.8	0.29			
$[C_{12}im]_2$	0.12	35.5	1.38	1.21			41
$[C_{14}im]_2$	0.089	36.3	1.19	1.39			
$[C_{16}im]_2$	0.029	49.3	3.56	0.47			
1,2-bis(1(3-alkoxy-2-hydroxypropyl)-1H-imidazol-3-ium)ethane bromide	0.021	30.04	4.53	0.036	2	2.21	44
Bromo-*N*-(2-hydroxyalkyl)-*N*-(2-hydroxyethyl) imidazolium surfactant	0.44	36.85	4.04	0.41	0.7	2.2	15
Chloro-*N*-(2-hydroxyalkyl)-*N*-(2-hydroxyethyl) imidazolium surfactant	1.02	37.38	3.02	0.55	0.39	2.5	
Py-10-4-10	2.69	38.5	2.76	0.6	0.11		52
Py-12-4-12	0.56	40.5	1.77	0.94	0.64		
$C_{10}\text{-4-}_{10}PB$	3.36	39.1		1.6	3.22		40
$C_{16}\text{-4-}_{16}PB$	1.5	37.6		1.88	5.64		
$C_{12}\text{-3-}C_{12}PB$	0.41	35.4		1.47			14
$C_{12}\text{-12-}C_{12}PB$	0.19	38.2		2.64			

as an extrinsic fluorescence probe, by the comparison of the intensities of the first, I_1, and third, I_3 (I_3/I_1) [40].

The structure of IL-based gemini surfactants (the alkyl chains, head groups, spacer) has great effect on the CMC values. New series of *N*-(2-hydroxyalkyl)-*N′*-(2-hydroxyethyl)imidazolium surfactants were investigated for their aggregation properties by Kamboj et al. [15], and the CMC values of imidazolium surfactants decreased with an increase in alkyl chain length. In Baltazar et al. [36] work, imidazolium gemini surfactants exhibited nearly twofold lower CMC values than monocationic surfactants. The spacer can be used to manipulate the hydrophobic interactions as well as to constrain the electrostatic repulsion between charged head groups and hence their CMC [2]. The nature and structure of spacers between the two head groups of IL-based gemini surfactant are important factors. An increase in the spacer chain length will reduce the CMC. However, with the progressive increase in spacer chain length, a stage will come when the spacer itself starts behaving like hydrophobic chain. At that stage, the spacers penetrate into the hydrophobic core of the micelles, which increase the intramolecular hydrophobic interaction, resulting in decrease of the CMC values. Both in C_{10} and C_{12} homologues of the gemini series $[C_nH_{2n}(ROArCH_2Im)_2][2Br]$, about 1.4- to 2.0-fold decrease in the CMC was observed with the increase of spacer length from C_3 to C_6. However, an increasing trend (0.6-fold) was observed in C_{14} gemini series with the increase in spacer length from C_3 to C_6 [57]. This may be due to the increased intermolecular hydrophobic interaction at the cost of intramolecule. Pal et al. [56] obtained that the CMC values of 16-Im-*n*-Im-16 ($n=2, 3, 4, 5, 6, 8, 10, 12$) pass through a minimum around $n=3$, which again increase with the further increase of the *n*-values and plateau off at higher *n*-value. It is slightly different from Zana [2, 6] research about the CMC value change trends as a function of the spacer lengths, that is, the CMC values of ammonium-based gemini surfactants go through a maximum at $n=6$, and with higher spacer length, the CMC goes down. It can be explained as that rigid and planar imidazolium ions might interfere with the packing of the long hydrophobic spacer, and hence, both the individual chains in surfactants start to behave as independently beyond $n=3$. The head groups also have the dramatic effect on the CMC values. The CMC values of gemini pyridinium surfactants (long chain contains phenoxy ring) were found to be lower as compared to gemini imidazolium congeners in Bhadani et al. [13, 43] study. Researchers have paid attention to the effect of counterion [43, 55]. A comparison of bromo and chloro derivatives of homologous series shows that $[Br^-]$ is more effective than $[Cl^-]$ for the screening of the electrostatic repulsion between the polar head groups, since $[Br^-]$ is less hydrated than $[Cl^-]$ [43].

IL-based gemini surfactants present much lower CMC values than their monocationic analogues, as well as a strong efficacy in decreasing the surface tension. As a general trend, the CMC values decrease with increasing the length of the alkyl chain, independent of the position in which the alkyl chain is linked to the ring. This aspect is quite interesting for using limited amounts of surfactants by designing the proper surfactants.

6.2.2.2 *Surface Activity Parameters*

The surface activity parameters of surfactants are mainly obtained via surface tension measurement [36, 37, 40, 42, 44, 46]. From the surface tension, isotherms of the saturation surface excess, Γ_{max}, can be calculated using the Gibbs adsorption equation, which is used as a measure of maximum extent of adsorption of surfactant at the air/solution or solution/solution interface.

The minimum area per surfactant molecule at the interface (A_{min}) can also be obtained using the Gibbs adsorption equation [6]. Moreover, the "efficiency" and "effectiveness" of surfactant are necessary to consider (Rosen [8]). The former is the concentration of surfactant in the bulk phase required to produce a given surface tension reduction, commonly 20 mN·m^{-1}, and the latter is the maximum reduction that can be obtained by the addition of surfactant. The surfactant efficiency can be defined as the negative logarithm of the bulk phase concentration (C_{20}) at 20 mN·m^{-1}, that is, $-\log c_{20} = pc_{20}$. The effectiveness of surface tension reduction can be described by the value of γ_{cmc} (the surface tension at the CMC) or π_{cmc} ($\pi_{cmc} = \gamma_0 - \gamma_{cmc,}$ γ_0 is the surface tension of pure solvent) [6], which is surface tension or surface pressure at the CMC. Table 6.1 summarizes the surface activity parameters obtained for the IL-based gemini surfactants characterized up to date.

In Anderson et al. [11] study, the surface tension values of IL-based gemini surfactants, which contain nonpolar butyl groups in the three positions of the imidazolium rings, are significantly smaller than those containing the 3-methylimidazolium dications separated by a nonane or dodecane linkage chain. These data indicate that the alkyl substituent located on the three positions of the imidazolium ring plays a more important role in lowering the surface tension than the alkyl linkage chain that separates the geminal dications. We have investigated the surface activity of imidazolium-based gemini surfactants with different alkyl chain and spacer lengths [C_n-*s*-$C_n$im]Br_2 (n = 10, 12, 14; s = 2, 4, 6). The value changes of A_{min}, π_{cmc}, pc_{20}, and CMC/C_{20} indicate that the shorter the spacer, the higher the surface activity of [C_{12}-*s*-C_{12}im]Br_2, and the adsorption at the air/solution surface for [C_n-4-$C_n$im]Br_2 with longer hydrophobic chain is easier [37]. The similar result is also evident for quaternary ammonium gemini surfactants [5, 6]. In contrast to these results, a continuous decrease in the value of A_{min} for the series of gemini pyridinium surfactants is obtained [59]. The values of A_{min} of imidazolium gemini surfactants with different spacers (i.e., -S-$(CH_2)_n$-S-) have been found to be exceptionally low, due to the tendency of imidazolium gemini cationics to form premicellar aggregates at sufficiently low concentrations below their CMC values [48].

The difference in surface properties of IL-based gemini surfactants is based on surface activity parameters. However, no regular patterns or trends in surface activity parameters were observed in case of different structural IL-based gemini surfactants. It is true that these parameters depend on the state of surfactant adsorption at the air/water interface.

6.2.2.3 Thermodynamic Parameters The CMC value dependence on temperature is used for the determination of thermodynamic parameters applying models [60]. The mass action model, apparent and partial model, and phase separation model [14, 59, 60] are applied to estimate the thermodynamic parameters: Gibbs energy, enthalpy, and entropy of micelle formation. The enthalpy and entropy change for the micellization can be determined using the Gibbs–Helmholtz equation [55].

Table 6.2 summarizes the thermodynamic parameters of IL-based gemini surfactants. The thermodynamic parameters of surfactants with pyrrolidinium head groups 1,1′-(alkane-1,*s*-diyl)bis(1-dodecylpyrrolidinium)bromide (C_{12}-C_S-C_{12}PB) suggest that the standard Gibbs free energy changes ΔG^0_m were negative, showing that the micelle formation is spontaneous process; the negative standard enthalpy changes ΔH^0_m

TABLE 6.2 Thermodynamic Parameters for IL-Based Gemini Surfactants

Surfactants	Temperature (K)	ΔG^0_m (kJ·mol^{-1})	ΔH^0_m (kJ·mol^{-1})	$T\Delta S^0_m$ (kJ·mol^{-1})	β	Reference
[C_{12}-2-C_{12}im]Br_2	298	−34.41	−5.21	29.2	0.77	
[C_{12}-4-C_{12}im]Br_2	298	−33.41	−5.77	27.64	0.7	55
[C_{12}-6-C_{12}im]Br_2	298	−30.59	−15.98	14.61	0.6	
[C_{10}-4-C_{10}im]Br_2	298	−26.96	−2.84	24.12	0.67	37
[C_{14}-4-C_{14}im]Br_2	298	−34.65	−12.12	22.53	0.55	
ILGIS (C_{14}-4-C_{14})	298	−36.52	−10.92	25.6		61
[12-(S-2-S)-12]im	298	−33.14			0.63	
[12-(S-4-S)-12]im	298	−32.36			0.58	48
[14-(S-2-S)-14]im	298	−34.08			0.54	
[14-(S-4-S)-14]im	298	−34.62			0.53	
[C_{12}im]$_2$	298	−34.6			0.64	
[C_{14}im]$_2$	298	−38.6			0.7	41
[C_{16}im]$_2$	298	−41.9			0.7	
C_{12}-5(OH)$_3$-C_{12}	298	−41.7	−16.32	25.33		62
C_{12}-5-C_{12}	298	−33.24	−9.11	24.13		
[PPC_{14}im][Cl]	298	−67.29	11.23	78.52		27
1,2-bis(1(3-alkoxy-2-hydroxypropyl)-1H-imidazol-3-ium)ethane bromide	298	−36.49				44
Bromo-*N*-(2-hydroxyalkyl)-*N*-(2-hydroxyethyl) imidazolium surfactant	298	−48.37				15
Chloro-*N*-(2-hydroxyalkyl)-*N*-(2-hydroxyethyl) imidazolium surfactant	298	−45.06				
12-Py(2)-2-(2)Py-12 MS	298		−3.9		74	
12-Py(2)-4-(2)Py-12 MS	298		−4.4		40	59
12-Py(2)-12-(2)Py-12 MS	298		17.6		57	
C_{10}-4-$_{10}$PB	298	−27.45	−0.77	26.68	0.71	
C_{12}-4-$_{12}$PB	298	−33.75	−5.79	27.96	0.73	40
C_{14}-4-$_{14}$PB	298	−35.7	−8.69	27.01	0.57	
C_{16}-4-$_{16}$PB	298	−39.08	−23.94	15.14	0.59	
C_{12}-3-C_{12}PB	298	−36.29	−12.41	23.88	0.79	
C_{12}-6-C_{12}PB	298	−32.34	−6.04	26.3	0.67	14
C_{12}-12-C_{12}PB	298	−31.17	−10.57	20.6	0.56	

suggest that the micellization is an exothermic process; the micellization of C_{12}-C_S-C_{12}PB is an entropy-driven process regardless of the spacer length *s* (using electrical conductivity measurement). The micellization is entropy-driven process with shorter hydrophobic chain lengths, but the process is enthalpy driven for longer hydrophobic chain lengths of (C_n-4-C_nPB) [40]. Besides electrical conductivity methods, molecular dynamic simulations and free energy perturbation (FEP) were performed for studying the enthalpy–entropy compensation. The predicted ΔG^0_m, ΔH^0_m, and ΔS^0_m demonstrate that the micellization of imidazolium gemini surfactants is entropy driven and enthalpy–entropy compensated [61]. The enthalpy–entropy compensation plots exhibit an excellent linearity, and the compensation temperature was 307 ± 2 K. The compensation temperature of novel branched surfactants with hydroxyl group is lower than that of alky chain spacer ones [62]. With the aim of verifying the additive rule and finding out the contribution of each constituting group, apparent and partial molar enthalpies of the homologous series of 1,1-didodecyl-2,2′-alkylenebispyridinium dimethanesulfonate are obtained by means of a pseudophase transition [59]. A molecule conformational change is caused by stacking interactions between the two pyridinium rings, mediated by the counterion and appearing at an optimum length of the spacer.

Thermodynamic parameters of IL-based gemini surfactants are reported in quite a systematic way. Finding out the contribution of each constituting group and predicting the thermodynamic behavior are especially important. In case of enthalpies, the effect of a methylene group when added to extend the hydrophobic tail in gemini surfactants seems to be greater than that in the case of monomers, and many more work is needed to confirm and explain this trend. Moreover, it is not completely understood whether the contribution of the head groups in the IL-based Gemini surfactant aggregates.

6.2.2.4 Degree of Counterion Binding (β) Discussion has concentrated on the degree of counterion binding (*β*) [38, 55, 59]. The ratio of the slopes of the conductivity versus concentration curve above and below the CMC gives the degree of counterion dissociation *α* (i.e., $\alpha = S_{micellar}/S_{premicellar}$), and $(1-\alpha)$ gives the degree of counterion binding *β*. *β* is an important parameter because it manifests the counterions that are contained in the Stern layer to counterbalance the electrostatic force that opposes micelle formation. *β* values of typical IL-based gemini surfactants were summarized in Table 6.2. *β* values of imidazolium gemini surfactant [C_n-4-$C_n$im]Br_2 decrease with the increase of temperature, suggesting that the binding of counterions to micellar surface is an exothermic process (the counterion binding is caused by an electrostatic interaction between opposite charges) [55]. The hydrophilic and huge head groups in gemini surfactants cannot be closer to each other than that of monomer, resulting in a lower effective positive charge density on the micellar surface and thus decreasing the attraction with counterions. Patial et al. [44] found that the *β* value decreases with increasing alkyl chain or spacer length. In Bhadani and Singh [46] study, *β* values of bispyridinium surfactants decrease with an increase in chain length. The smaller degree of hydration of the bromide ion makes the hydrated chloride ion bigger than that of the bromide, which interacts strongly with the imidazolium, as indicated by higher degree of counterion binding [63]. The *β* values of *N*-(2-hydroxyalkyl)-*N*′-(2-hydroxyethyl) imidazolium surfactants with bromide counterions are higher than the compounds with chloride counterion [15].

6.2.2.5 *Structures of Aggregate* Structure of surfactant is worth discussing because it affects rheological and solubilization properties, which are important issues from a practical application. Best known structures are normal micelle that spontaneously form surfactant aggregates (with the hydrophobic tails toward the center and hydrophilic head groups at the outer surface) in water [6]. The spherical micelle is the most common structure of surfactant aggregate. However, amphiphiles can associate into a variety of structures in aqueous solutions: ellipsoidal, cylindrical, or disk-like micelles, bilayers, and vesicles [1, 7]. Noninvasive techniques such as small-angle neutron scattering (SANS) allow the structural elucidation of the assemblies formed by surfactants [6, 24]. Pal et al. [56] have reported detailed measurements of the SANS cross sections from different imidazolium-based gemini surfactant ([16-Im-*n*-Im-16] [2Br$^-$]) (where $n = 2, 3, 4, 5, 6, 8, 10$, and 12) micelles in D_2O. Derived from gemini with relatively longer spacer chains, the micelles become progressively smaller in size, and the micellar morphologies of imidazolium surfactants were sensitive to spacer length, which mediated ellipsoid-to-spheroid transitions. Cryo-transmission electron microscopy (cryo-TEM) and dynamic light scattering (DLS) have produced the following morphologies and sizes: (i) besides the dominated spherical micelles, a cylindrical shape with the length of about 100 nm for C_{12}-C_{14}-C_{12}PB (cryo-TEM), resulting in two peaks at 2.4 and 165 nm (DLS), and (ii) unilamellar and multilamellar vesicles for C_{12}-C_{16}-C_{12}PB (cryo-TEM) [14]. Fiber-like aggregates were observed for imidazolium gemini surfactant by optical microscopy, and a dynamic micelle–vesicle equilibrium exists in solution for ester-containing gemini imidazolium surfactants using DOSY. Considering the fact that nuclear magnetic resonance (NMR) is a sensitive technique for the study of the aggregation behavior of surfactants [64], ^{1}H NMR and 2D ^{1}H-^{1}H NOESY (nuclear overhauser effect spectroscopy) are used to determine possible microscopic structures of IL-based gemini surfactant. This possible aggregation structure resembles the structure of stairs. There are π–π stacking interactions between the imidazolium rings, and the alkyl chain extends to the middle of the other two imidazolium rings.

Although previous studies have explored some important aspects of the aggregation of surfactants in aqueous solutions, the microscopic structure of aggregates has not been well defined. To date, little is known about the exact structure of IL-based gemini surfactants. It is unclear about how the formation of micelles, emulsions, or dissociated ions is affected by temperature and solution conditions.

6.2.2.6 *ILCs* The self-organizing systems of ILCs offer great advantages over isotropic systems. For instance, ion conduction was enhanced in the smectic A phase (the lowest symmetry layered phase, SmA) and columnar phases compared to that in the isotropic phase. There has been considerable interest in the field of nonconventional low molar mass LCs, especially in liquid crystal dimmers because of their interesting mesomorphic properties due to restricted molecular motions [29, 47, 65, 66]. Early papers focused on the mesomorphic properties of small aromatic systems (imidazolium, pyridinium, etc.), substituted by one or multiple long aliphatic tails. The SmA phase displayed by such systems often requires a minimum length of the tail (usually C_{12}), and the liquid crystalline properties originate from their strong amphiphilic character Kouwer and Swager [47] conveniently integrated imidazolium groups in the core of the mesogen, which has the shape of a conventional rod-shaped mesogen. Mesogens substituted with ethylene glycol tails show deviant behavior (reduced mesophase stability), due to the interaction of the tail with the imidazolium cation. Considerable research effort is currently focused on the interplay between ionic conduction properties of ILs and mesomorphic behavior of LCs [29]. Some liquid crystalline gemini surfactants have been displayed as lyotropic and thermotropic mesophases. Symmetrical gemini surfactants based on imidazole and triphenylene moieties exhibit columnar mesophase over a wide temperature range and display ionic conductivity in the range of 10^{-6}–10^{-5} S/m [65]. Yoshio et al. have successfully designed and prepared 1D anisotropically ion-conductive polymeric films by photopolymerization of aligned columnar liquid crystals of a fan-shaped imidazolium salt [29], and these compounds have imidazolium and tris(alkoxy) phenyl moieties containing acrylate polymerizable groups at their periphery. The ILCs and mixtures have many potential applications, such as used as an anisotropic medium for electrochemistry and applied in dye-sensitized solar cells.

6.2.2.7 Krafft Temperatures and Melting Points of IL-Based Gemini Surfactants Although the solubility of surfactant generally increases with increasing temperature, it typically increases dramatically at a point known as its Krafft temperature (T_k). Aggregation of surfactants can occur only above T_k value [6]. The new 2,2′(α,ω alkanediyl)bis(1-alkyl pyridinium) series have Krafft point below 0°C [52], which can only be used in cold water. T_k values of gemini surfactants are 18°C for $[C_{14}im]_2$ and 37°C for $[C_{16}im]_2$ by Zhuang et al. [41], and T_k values increase with an increase of the alkyl chain length. In our research, we found that the shorter the spacer of ($[C_{12}$-s-$C_{12}im]Br_2$, $s = 2, 4, 6$), the higher the T_k values [55]. The T_k values of pyrophosphate surfactants, which contain imidazolium counterions, are uniformly <23°C. The substitution of sodium for imidazolium counterions results in higher T_k values. The lesser T_k value of branched chain compared to straight chain reflects the effect of branching on T_k values. The organic salts whose melting temperature (T_m) is lower than 100°C are called IL [19, 37]. The melting temperatures determined by a microbinocular loupe melting point apparatus were 47.5 ± 0.5, 55.0 ± 0.5, and 64.5 ± 0.5 °C for $[C_{10}$-4-$C_{10}im]Br_2$, $[C_{12}$-4-$C_{12}im]Br_2$, and $[C_{14}$-4-$C_{14}im]Br_2$, respectively, which one can also call them as gemini surfactant "ILs" [37]. Zhang and Liu [49] have observed that the hydrated solid with the composition of $[C_{12}im$-4-$C_{12}im]Br_2 \cdot 2H_2O$ is formed, and it exhibits melting point (33°C) at which the hydrated solid decomposes into H_1 phase and unhydrated surfactant solid. Differential scanning calorimetry (DSC) heating and cooling curves for 1,4-bis(3-teradecylimidazolium-1-yl)butane bromide and 1-*n*-tetradecyl-3-methylimidazolium bromide are illustrated by Ding et al. [12], and the melting points of the former (about 58°C) is higher than that of the later (about 54 °C). When longer linkage chains are used to separate the geminal dications (e.g., imidazolium vs. pyrrolidinium), a decrease in the melting point was observed for most ILs [11]. This may be related to an increase in the number of closely related configurational states and possibly crystalline polymorphs. Kouwer and Swager have found that mesogens with a large core display higher melting temperatures, which can be reduced by introducing small laterally attached substituent [47]. The anions also have great effect on the T_m, and the trend of T_m (the anions) in the capability of the crystallization is $[NO_3]^- > [BF_4]^- > [NPf_2]^- > [NTf_2]^-$ [53]. Generally, low symmetry, weak ion interactions (such as suppressing hydrogen bonding), and effective charge distribution over the cation and/or anion tend to reduce the crystal lattice energy of the salts, thus resulting in low melting temperature. Therefore, imidazolium gemini surfactant of more symmetrical structure has higher melting point.

6.2.3 The Behavior of IL-Based Gemini Surfactants at the Interface

IL-based gemini surfactants possess foaming, emulsification, and dispersion. These functions are related to the surfactant adsorbing at the interface [67–70]. Surface properties of IL-based gemini surfactants have effect on many unit operations and processes. For example, the compound transfer through the interface is an important part of the extraction process [34].

6.2.3.1 *Air/Solution Interface* When molecule adsorption occurs at the air/solution interface, surfactants arrange at the interface orienting their hydrophilic head groups toward water and hydrophobic groups toward air to avail favorable thermodynamic interactions [6]. Due to the electrostatic interaction of two ion head groups in IL-based gemini molecules, the distance of two ion head groups in molecules is narrow. The hydrophobic chains are arranged closely, and the curvature of hydrophobic chains is reduced. The surfactant molecules arranged vertically at the air/solution interface and the surface tension are decreased effectively. $d\gamma/d\log c$ of pyrrolidinium gemini surfactants is the slope of the linear part in the surface tension isotherms, where the surfactant concentration is below the CMC. According to the classic theory of Gibbs analysis, the interface is considered to be saturated with surfactant molecules, and the continued reduction in the surface tension is mainly due to the increased activity of surfactant in the bulk phase rather than at the interface. Mahajan et al. [51] have obtained the smaller A_{min} value of pyridinium gemini surfactants [12-(S-2-S)-12] (59.30 Å^2) than [14-(S-2-S)-14] (136.09 Å^2), suggesting that the former has higher packing density at the air/water interface than the latter, which is due to the fact that the longer hydrophobic chains are more prone to bend. Fisicaro et al. [59] have found a peculiar behavior as a function of the spacer length. A_{min} taken at the air/water interface is significantly lower for 12-Py(2)-4-(2)Py-12 MS than that of the other homologous compounds with different spacer length. The mechanism of traditional dimeric surfactant adsorption at the air/water interface can be investigated by dynamic surface tension measurements, which have been found to strongly depend on the nature of the surfactants [6, 7]. Even though the IL-based gemini surfactants with various head groups, spacer, and chains have been designed and characterized, the mechanism of IL-based gemini surfactant aggregate behavior has not been investigated in these literatures.

6.2.3.2 *Solid/Liquid Interface* The aggregation behaviors of surfactants on the solid/liquid interface have been investigated, which is important in surfactant application, such as mineral flotation and lubrication [45, 70]. Compared with conventional ammonium surfactants, IL-based surfactants have novel properties on solid surface due to the existence of IL head groups [67, 69]. The mechanism of IL-based gemini surfactant adsorption mainly involves three steps. The first step occurs at very low concentration, corresponding to the binding of individual gemini surfactants on charge sites at the silica surface by an ion-exchange mechanism. The second step occurs at a concentration slightly below the CMC, corresponding to the formation of surface aggregates. The third step occurs above the critical surface aggregation concentrations (CSAC), corresponding to the formation of layer structures. We have investigated $[C_{12}\text{-}4\text{-}C_{12}im]Br_2$ aggregates on a silicon wafer by AFM, and aggregation experiences an evolving course from no aggregates to circular islands, then to semi-continuous islands, and finally bilayer with increasing concentrations [67], as shown in Figure 6.3. Surfactant forms a continuous flat bilayer rather than the multilayer. The probable causes are as follows. The spacer between the hydrophilic head groups limits the freedom of the monomer parts, which is energetically unfavorable to form the multilayer film. Both the spacer length and the counterion (Br^-) binding degree reflect the weaker degree of Br^- of $[C_{12}\text{-}4\text{-}C_{12}im]Br_2$ binding on the aggregate surface, leading to the lower surface negative charge density for the additional $[C_{12}\text{-}4\text{-}C_{12}im]Br_2$ cations adsorbed on the surface. Zheng groups [68, 69] have used IL-based gemini surfactant (imidazolium $[C_nim\text{-}4\text{-}C_nim]Br_2$; pyrrolidinium $[C_npy\text{-}4\text{-}C_npy][Br_2]$; $n = 12, 14$) as dispersed solvents, and they all can disperse multiwalled carbon nanotubes (MWCNTs) effectively even at low concentrations. The hydrophobic tails of $[C_{14}py\text{-}4\text{-}C_{14}py][2Br]$ tend to lie flat on the graphitic surface of MWCNTs by the hydrophobic interaction, while the hydrophilic heads are inclined to extend into the water. Since the surface of MWCNTs adsorbs some molecules of $[C_{14}py\text{-}4\text{-}C_{14}py][Br_2]$, it becomes positively charged. Thus, MWCNTs can be dispersed effectively in aqueous solutions due to the electrostatic repulsion forces between them [69]. The adsorption of IL-based surfactants on solid surface could produce two-dimensional supramolecular nanostructures, which have the potential to be used as lateral nanomaterials or templates of biomimetic or biomineralization processes [66, 71]. On the other aspect, the imidazolium surfactants have good antimicrobial activities, and the modified materials may also be used as antimicrobial materials [72]. Thus, the investigation of aggregation behavior at solid/liquid surface would have certain theoretical guidance of the modification of solid surface.

6.3 THE INTERACTIONS BETWEEN IL-BASED GEMINI SURFACTANTS AND BIOMACROMOLECULES

The unique physicochemical properties of IL-based gemini surfactants have attracted increasing interests due to their potential applications in various areas [73]. Especially in life science and biological areas, the interactions between IL-based gemini surfactants and biomacromolecules are of interest, for example, the potential application in gene delivery and transfection [74–77]. In this section, we mainly summarize the complexation and thermodynamics of interactions between IL-based gemini surfactants and biomacromolecules (DNA, proteins).

6.3.1 IL-Based Gemini Surfactants/DNA Complexes

Gene therapy has been extensively pursued as an important therapeutic strategy by substitution of defective gene at transcriptional or translational level, with the aim of finding efficient and safe delivery systems [78, 79]. *Nonviral* gene delivery is the use of chemical agents that interact with DNA and transport genes to the cell [80–82], such as amphiphiles [9]. However, gene transfection efficiency must be improved by adopting the structure of these self-assembled delivery systems. Therefore, designed and characterized surfactants with unique structures become interesting. As typical novel surfactants, IL-based gemini surfactants have received particular intensive attention in their interactions with DNA molecules.

Pyridinium gemini surfactants possess two positive charges, a high charge/mass ratio (loading capacity) and a high molecular flexibility. In Ilies et al. research [76],

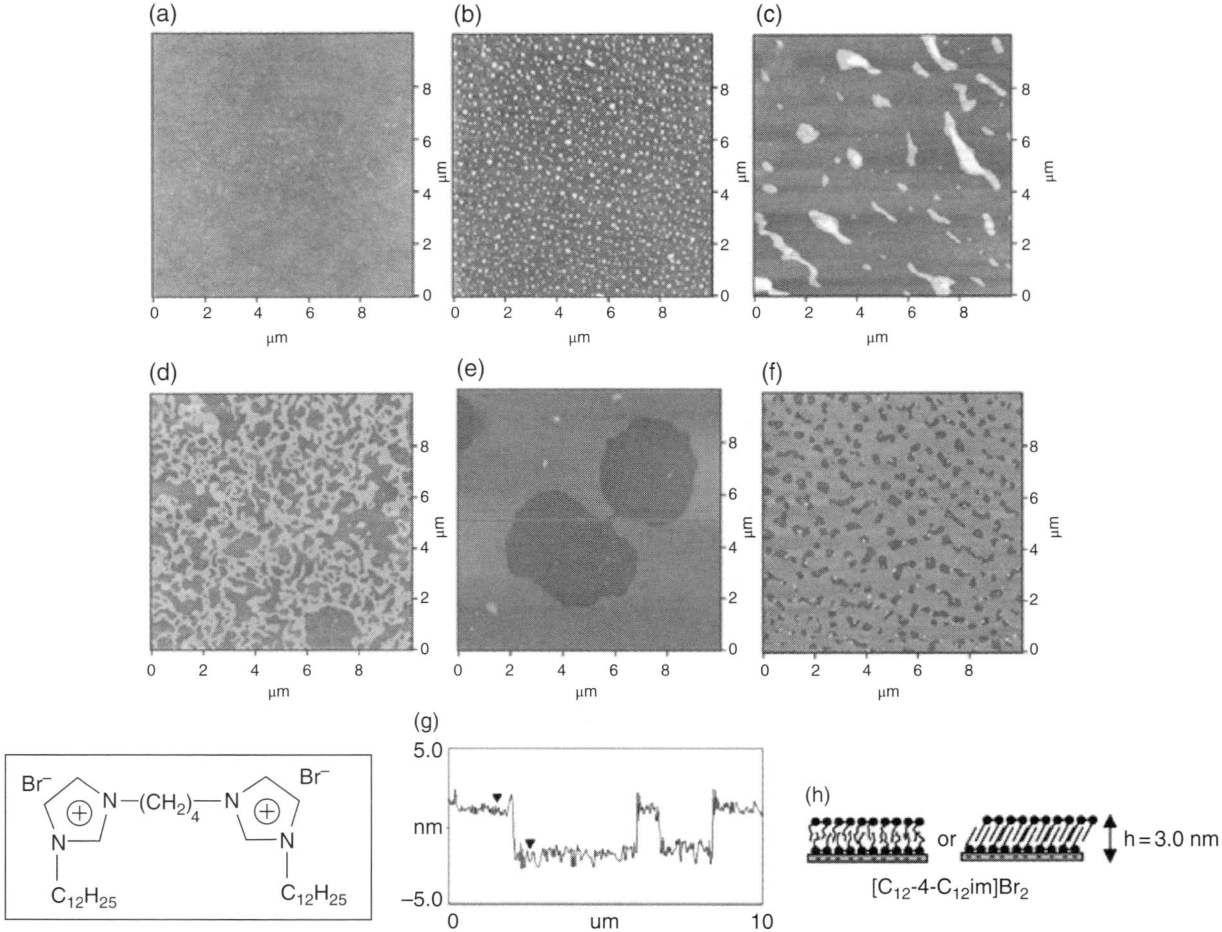

Figure 6.3 AFM height images of silica substrates treated with different $[C_{12}\text{-}4\text{-}C_{12}im]Br_2$ concentrations: (a) 0.50, (b) 0.75, (c)1.0, (d) 2.0, (e) 3.0, and (f) 5.0 cmc. (g) Corresponding section analyses for e. (h) Schematic model of $[C_{12}\text{-}4\text{-}C_{12}im]Br_2$ aggregates on the silica surface. Reproduced from Ao et al. [67] with permission from the American Chemical Society.

they found that pyridinium gemini surfactants with procationic hydrophilic linker are less biologically efficient than that of bearing hydrophobic linkers, and a polar but neutral linker could guarantee a good transfection profile. Bhadani and Singh [46] obtained that pyridinium gemini surfactants with ethane-1,2-dithiol spacer have low cytotoxicity by MTT (3-(4,5-dimethylthiazol-2-yl)-2,5-diphenyltetrazolium bromide) assay, and agarose gel electrophoresis indicated chain length-dependent DNA binding abilities. The imidazolium gemini surfactants display stronger self-aggregation tendency and strong π–π interaction with aromatic rings (Fig. 6.4) [33]. Meanwhile,

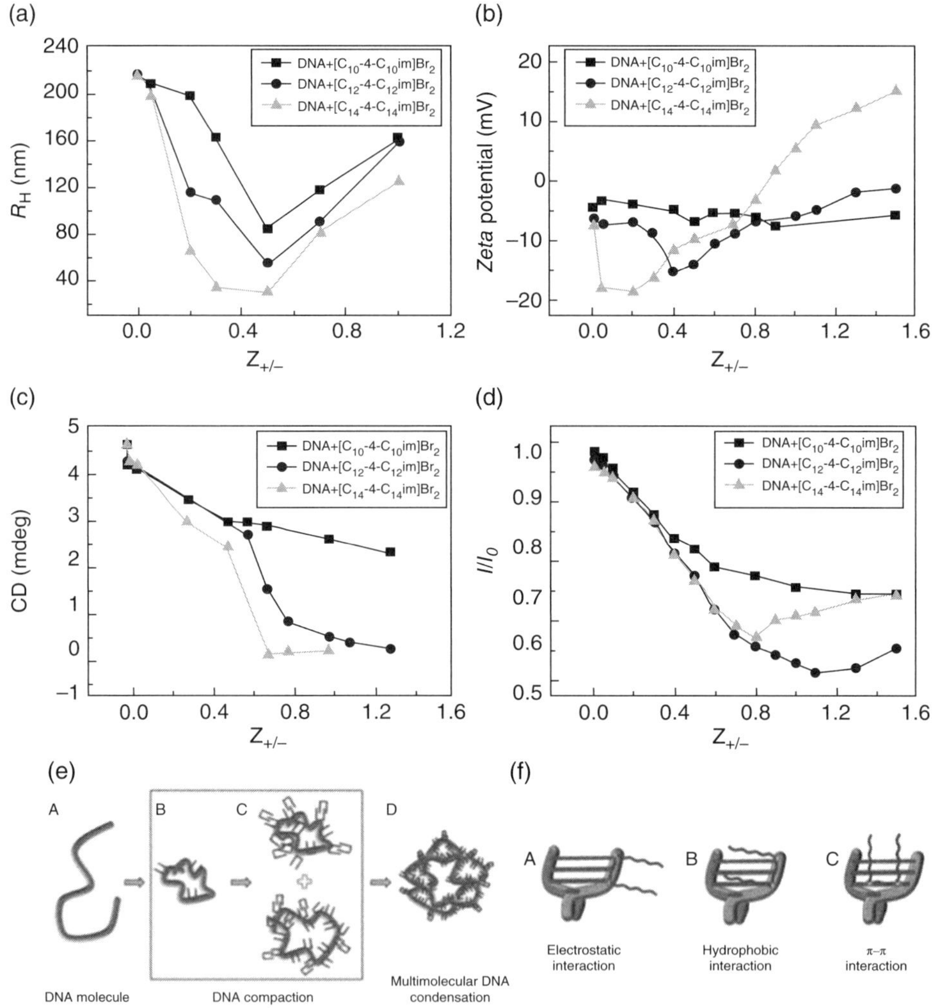

Figure 6.4 Hydrodynamic radius (a), *zeta* potential (b), CD intensities at 276 nm (c), and relative fluorescence intensity I/I_0 (d) of DNA/[C_n-4-$C_n$im]Br_2 versus $Z_{+/-}$. (e) Schematic illustration of DNA compaction to multimolecular DNA condensation progress with the increasing of [C_n-4-$C_n$im]Br_2. (f) Schematic illustration of the three types of DNA and [C_n-4-$C_n$im]Br_2 interactions. Reproduced from Zhou et al. [33] with permission from Elsevier.

imidazolium gemini surfactants have low critical aggregation concentrations (CAC), which would result in the decrease of agent amount needed for biological applications (highly desirable for optimization of the safety profiles *in vivo*). Moreover, the introduction of histidine residues in DNA condensing agents was reported to promote gene transfection through the enhancement of "proton-sponge" effect. Upon addition of $[C_n\text{-}4\text{-}C_n\text{im}]Br_2$ ($n = 10, 12, 14$), DNA molecules undergo the process from compaction to multimolecular condensation accompanied by conformation change, and the comparison between DNA compaction and condensation by surfactants with different tail lengths demonstrated the important contribution of the hydrophobic interaction [33]. We have also found that $[C_{12}\text{-}4\text{-}C_{12}\text{im}]Br_2$ could condense DNA into spherical particles with size of approximately 100 nm in the proper charge ratio. Highly efficient gene transfection *in vitro* is demonstrated in human embryonic kidney 293 (HEK293) and HeLa cells, which could be attributed to the effective DNA condensation into uniform nanoparticles induced by $[C_{12}\text{-}4\text{-}C_{12}\text{im}]Br_2$. In addition, the low cytotoxicity of $[C_{12}\text{-}4\text{-}C_{12}\text{im}]Br_2$ at transfection concentration region verified by cell viability assay (MTT assay) also supports $[C_{12}\text{-}4\text{-}C_{12}\text{im}]Br_2$ as an effective gene vector [83, 84]. Kamboj et al. [63] also have found that imidazolium gemini surfactants are much low toxic *in vivo*, which has a greater probability of application as a synthetic vector. A detailed structure–property relationship study was investigated by Sharma et al. [75], which was carried out within a series of pyridinium gemini surfactants bearing a 3-oxapentane-1,5-diyl hydrophilic spacer with different chains and counterions. The surfactant aggregates with optimum charge density, which can be achieved by tuning the self-assembling process at the level of counterion, hydrophobic chain, added co-lipids, and so on, may be effectively used for DNA delivery.

6.3.2 IL-Based Gemini Surfactants/Protein Complexes

Studying of protein/surfactant interaction is an active field of interest, ever since surfactants are known to interact with proteins thereby inducing polarity and stability changes of proteins [85–88]. Bovine serum albumin (BSA) is often selected as a model globular protein for studying, because its structure and physicochemical properties have been well characterized. The work investigated by Wang et al. [89] offered a whole view of the interaction of BSA with IL-based gemini surfactant, glycol bis-*N*-tetradecyl nicotinate dibromide (P14-1-14). The molecular docking manifested P14-1-14 embedded into subdomain of BSA, which made possible the π–π stacking between the electron-deficit pyridinium rings in P14-1-14 and the electron-abundant pyrrole ring in tryptophan (Trp) residues of BSA, and that hydrogen bonding and hydrophobic interaction also took place during the binding. The binding process of BSA with 1,4-bis(3-(dodecyloxylacyl)pyridinium)butane dibromide (NAE12-4-12) was exothermic, spontaneous, and enthalpy driven, and the distance between them decreased with incremental concentration of NAE12-4-12, which is helpful for a further understanding of gemini surfactants binding to BSA at molecular level [90]. Our previous studies have shown that quaternary ammonium surfactants can protect the BSA structure against the thermal denaturation or directly interact with BSA by binding, which leads to substantial changes in BSA conformation [86, 87]. Using the spectroscopy methods, we have obtained that imidazolium gemini surfactants with a shorter spacer or with a longer chain have a larger effect on BSA unfolding, and the imidazolium gemini surfactant interacts with BSA more strongly than its corresponding monomer and the quaternary ammonium gemini surfactant. Moreover, fluorescence quenching of BSA went probably through the static quenching procedure (that is initiated by ground-state complex formation rather than by dynamic collision) through obtaining the Stern–Volmer quenching constants [91]. A fundamental understanding on the effect of surfactant architecture in the interaction between surfactant and protein may provide guidance in selecting appropriate surfactants in pharmaceutical, cosmetic, and biological systems.

Gelatin is a polydisperse polypeptide derived from the structural protein collagen. When used as a material for photographic coatings, surfactants are commonly added to gelatin solutions to promote emulsification and control surface in multilayer coatings and to ensure the formation of uniformly thick films. In recent years, there has been great interest in the surface rheology because it is not only more sensitive to the interactions between the molecules in the adsorbed layer and the formation of networks on the surface (provide more accurate information about the structure of adsorption layer) but also useful for probing dynamic adsorption/desorption processes (understand stability of the emulsion and foam), which can include macromolecule conformational changes/segment redistribution on the surface. Both surface shear and dilational rheology are widely used to investigate the adsorbed layer of protein [88, 92, 93]. We investigated the interactions of imidazolium gemini surfactants with gelatin by surface rheology and surface tension methods. Both the surface shear and dilational rheological properties show that the strength of gelatin film is enhanced in the presence of 0.005 mmol l^{-1} [C_{12}-4-C_{12}im]Br_2, and the storage modulus is larger than loss modulus for gelatin/[C_{12}-4-C_{12}im]Br_2, indicating that gelatin/[C_{12}-4-C_{12}im]Br_2 is an elastic film. With increasing the concentration of imidazolium gemini surfactant, the dilational modulus increases firstly and then passes through a maximum value, due to the electrostatic and hydrophobic interactions between surfactant molecules and gelatin chains, as shown in Figure 6.5 [94]. Although the surface tension of gelatin/surfactant mixed solution is identical to that of the pure surfactant solution at high surfactant concentration, the dilational modulus suggests that the surfactant molecules are unable to fully replace gelatin molecules from the air/water surface. The structure of the gemini surfactant has great effect on the surface viscoelastic modulus of gelatin, and it is found that the gelatin film in the presence of imidazolium gemini surfactant with longer hydrophobic chain or longer spacer reveals the larger strength. Understanding effect of the molecular structure of IL-based gemini surfactants on the surface properties of gelatin could provide more useful information for their practical applications.

6.4 SUMMARY AND FUTURE DIRECTIONS

The present review mainly summarizes the aggregation behavior of IL-based gemini surfactants (in the aqueous solution and at the interface) and the interactions of IL-based gemini surfactants with biomacromolecules (e.g., protein, DNA). It reveals from these investigations that IL-based gemini surfactants possess unique aggregation behaviors, which have significant promise in industrial applications. Further, the interaction between IL-based gemini surfactants and biomacromolecules (DNA, BSA, gelatin) reveals predominance in the fields of life and biological sciences. However, there are still a number of areas that require further investigation

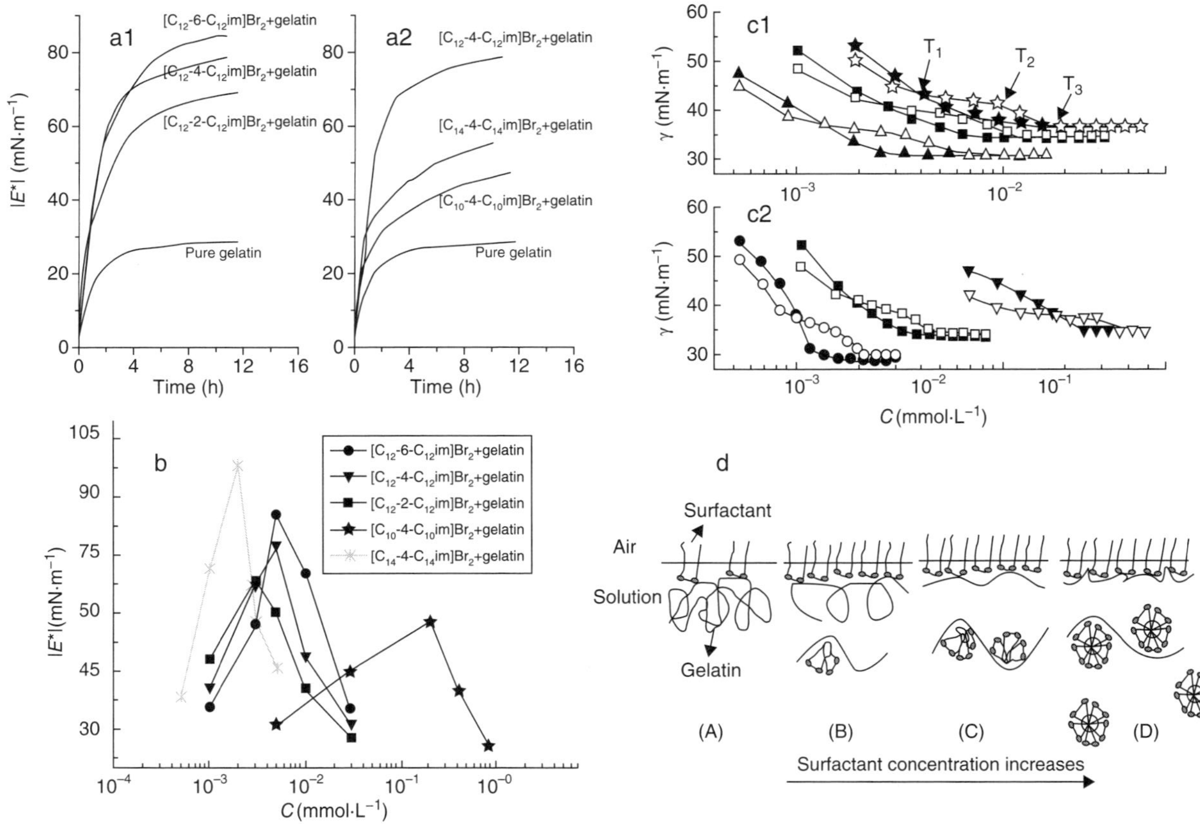

Figure 6.5 (a) Dynamic surface dilational moduli for pure gelatin and gelatin/[C_n-*s*-$C_n$im]Br_2 systems. (b) Static surface dilational viscoelasticity as a function of [C_n-*s*-$C_n$im]Br_2 concentrations. (c) Surface tension as a function of surfactant concentration for pure [C_n-*s*-$C_n$im]Br_2 and gelatin/[C_n-*s*-$C_n$im]Br_2 systems. (d) A schematic illustration of the interaction between gelatin and imidazolium gemini surfactant on the air/solution surface. *The surface dilational modulus. Reproduced from Ao et al. [94] with permission from the Royal Society of Chemistry.

and mention why on each point. First, a comprehensive analysis and further understanding of the dependence of interfacial activities on structural aspects of gemini surfactants are needed. Secondly, more complicated systems (including surfactants itself and surrounding conditions) might be investigated. Thirdly, it is still utterly necessary to have a full understanding of the mechanism underlying the interaction between IL-based gemini surfactants and various substances, such as polymers, biomacromolecules, and other surfactants. Fourthly, the novel IL-based gemini surfactants with new structure may lead to the development of efficient functions and also to exploiting practical applications.

REFERENCES

[1] Zana, R.; Talmon, Y.; *Nature*, 1993, **362**, 228–230.

[2] Zana, R.; *Curr. Opin. Colloid Interface Sci.*, 1996, **1**, 566–571.

[3] Shukla, D.; Tyagi, V. K.; *J. Oleo Sci.*, 2006, **55**, 381–390.

[4] Zhang, X.; Wang, C.; *Chem. Soc. Rev.*, 2011, **40**, 94–101.

[5] Han, Y. C.; Wang, Y. L.; *Phys. Chem. Chem. Phys.*, 2011, **13**, 1939–1956.

[6] Zana, R.; *Adv. Colloid Interface Sci.*, 2002, **97**, 205–253.

[7] Menger, F. M.; Littau, C. A.; *J. Am. Chem. Soc.*, 1991, **113**, 1451–1452.

[8] Rosen, M. J.; *Surfactants and Interfacial Phenomena*; Wiley-Interscience, New York, 2004.

[9] Kirby, A. J.; Camilleri, P.; Engberts, J. B. F. N.; Feiters, M. C.; Nolte, R. J. M.; Söderman, O.; Bergsma, M.; Bell, P. C.; Fielden, M. L.; Rodríguez, C. L. G.; et al., *Angew. Chem. Int. Ed.*, 2003, **42**, 1448–1457.

[10] Wettig, S. D.; Verrall, R. E.; Foldvari, M.; *Curr. Gene Ther.*, 2008, **8**, 9–23.

[11] Anderson, J. L.; Ding, R. F.; Ellern, A.; Armstrong, D. W.; *J. Am. Chem. Soc.*, 2005, **127**, 593–604.

[12] Ding, Y. S.; Zha, M.; Zhang, J.; Wang, S. S.; *Colloids Surf. A*, 2007, **298**, 201–205.

[13] Bhadani, A.; Kataria, H.; Singh, S.; *J. Colloid Interface Sci.*, 2011, **361**, 33–41.

[14] Cai, B.; Dong, J. F.; Cheng, L.; Jiang, Z.; Yang, Y.; Li, X. F.; *Soft Matter*, 2013, **9**, 7637–7646.

[15] Kamboj, R.; Singh, S.; Chauhan, V.; *Colloids Surf. A*, 2014, **441**, 233–241.

[16] Welton, T.; *Chem. Rev.*, 1999, **99**, 2071–2083.

[17] Huddleston, J. G.; Visser, A. E.; Reichert, W. M.; Willauer, H. D.; Broker, G. A.; Rogers, R. D.; *Green Chem.*, 2001, **3**, 156–164.

[18] Dupont, J.; *J. Braz. Chem. Soc.*, 2004, **15**, 341–350.

[19] Lee, S. G.; *Chem. Commun.*, 2006, **42**, 1049–1063.

[20] Dupont, J.; Suarez, P. A. Z.; *Phys. Chem. Chem. Phys.*, 2006, **8**, 2441–2452.

[21] Wang, Y.; Jiang, W.; Yan, T.; Voth, G. A.; *Acc. Chem. Res.*, 2007, **40**, 1193–1199.

[22] Seoud, O. A. E.; Koschella, A.; Fidale, L. C.; Dorn, S.; Heinze, T.; *Biomacromolecules*, 2007, **8**, 2629–2647.

[23] Plechkova, N. V.; Seddon, K. R.; *Chem. Soc. Rev.*, 2008, **37**, 123–150.

[24] Łuczak, J.; Hupka, J.; Thöming, J.; Jungnickel, C.; *Colloids Surf. A*, 2008, **329**, 125–133.

[25] Leclercq, L.; Schmitzer, A. R.; *Supramol. Chem.*, 2009, **21**, 245–263.

[26] Neouze, M. A.; *J. Mater. Chem.*, 2010, **20**, 9593–9607.

[27] Ding, W.; Liu, K.; Liu, X. J.; Luan, H. X.; Lv, C. F.; Yu, T.; Qu, G. M.; *J. Appl. Polym. Sci.*, 2013, 2057–2062.

[28] Han, X. X.; Armstrong, D. W.; *Org. Lett.*, 2005, **7**, 4205–4208.

[29] Yoshio, M.; Kagata, T.; Hoshino, K.; Mujai, T.; Ohno, H.; Kato, T.; *J. Am. Chem. Soc.*, 2006, **128**, 5570–5577.

[30] Wu, F. G.; Wang, N. N.; Zhang, Q. G.; Sun, S. F.; Yu, Z. W.; *J. Colloid Interface Sci.*, 2012, **374**, 197–205.

[31] Liu, L. H.; Wu, H.; Riduan, S. N.; Ying, J. Y.; Zhang, Y. G.; *Biomaterials*, 2013, **34**, 1018–1023.

[32] Meekel, A. A. P.; Wagenaar, A.; Šmisterová, J.; Kroeze, J. E.; Haadsma, P.; Bosgraaf, B.; Stuart, M. C. A.; Brisson, A.; Ruiters, M. H. J.; Hoekstra, D.; et al.; *Eur. J. Org. Chem.*, 2000, 665–673.

[33] Zhou, T.; Xu, G. Y.; Ao, M. Q.; Yang, Y. L.; Wang, C.; *Colloids Surf. A*, 2012, **414**, 33–40.

[34] Pino, V.; Hernández, M. G.; Pérez, A. M.; Anderson, J. L.; *Sep. Sci. Technol.*, 2012, **47**, 264–276.

[35] Safavi, A.; Zeinali, S.; *Colloids Surf. A*, 2010, **362**, 121–126.

[36] Baltazar, Q. Q.; Chandawalla, J.; Sawyer, K.; Anderson, J. L.; *Colloids Surf. A*, 2007, **302**, 150–156.

[37] Ao, M. Q.; Xu, G. Y.; Zhu, Y. Y.; Bai, Y.; *J. Colloid Interface Sci.*, 2008, **326**, 490–495.

[38] Quagliotto, P.; Viscardi, G.; Barolo, C.; Barni, E.; Bellinvia, S.; Fisicaro, E.; Compari, C.; *J. Org. Chem.*, 2003, **68**, 7651–7660.

[39] Baker, G. A.; Pandey, S.; Pandey, S.; Baker, S. N.; *Analyst*, 2004, **129**, 890–892.

[40] Cai, B.; Li, X. F.; Yang, Y.; Dong, J. F.; *J. Colloid Interface Sci.*, 2012, **370**, 111–116.

[41] Zhuang, L. H.; Yu, K. H.; Wang, G. W.; Yao, C.; *J. Colloid Interface Sci.*, 2013, **408**, 94–100.

[42] Patial, P.; Shaheen, A.; Ahmad, I.; *J. Surfactant Deterg.*, 2013, **16**, 49–56.

[43] Bhadani, A.; Singh, S.; Kamboj, R.; Chauhan, V.; *Colloid Polym. Sci.*, 2013, **291**, 2289–2297.

[44] Patial, P.; Shaheen, A.; Ahmad, I.; *J. Surfactant Deterg.*, 2014, **17**, 929–935.

[45] Jin, C. M.; Ye, C. F.; Phillips, B. S.; Zabinski, J. S.; Liu, X. Q.; Liu, W. M.; Shreeve, J. M.; *J. Mater. Chem.*, 2006, **16**, 1529–1535.

[46] Bhadani, A.; Singh, S.; *Langmuir*, 2009, **25**, 11703–11712.

[47 Kouwer, P. H. J.; Swager, T. M.; *J. Am. Chem. Soc.*, 2007, **129**, 14042–14052.

[48] Bhadani, A.; Singh, S.; *Langmuir*, 2011, **27**, 14033–14044.

[49] Zhang, G. C.; Liu, M. H.; *J. Phys. Chem. B*, 2008, **112**, 7430–7437.

[50] Akbay, C.; Hoyos, Y.; Hooper, E.; Arslan, H.; Rizvi, S. A. A.; *J. Chromatogr. A*, 2010, **1217**, 5279–5287.

[51] Mahajan, R. K.; Mahajan, S.; Bhadani, A.; Singh, S.; *Phys. Chem. Chem. Phys.*, 2012, **14**, 887–898.

[52] Zhou, L. M.; Jiang, X. H.; Li, Y. T.; Chen, Z.; Hu, X. Q.; *Langmuir*, 2007, **23**, 11404–11408.

[53] Shirota, H.; Mandai, T.; Fukazawa, H.; Kato, T.; *J. Chem. Eng. Data*, 2011, **56**, 2453–2459.

[54] Anouti, M.; Jones, J.; Boisset, A.; Jacquemin, J.; Caravanier, M. C.; Lemordant, D.; *J. Colloid Interface Sci.*, 2009, **340**, 104–111.

[55] Ao, M. Q.; Huang, P. P.; Xu, G. Y.; Yang, X. D.; Wang, Y. J.; *Colloid Polym. Sci.*, 2009, **287**, 395–402.

[56] Pal, A.; Datta, S.; Aswal, V. K.; Bhattacharya, S.; *J. Phys. Chem. B*, 2012, **116**, 13239–13247.

[57] Sunitha, S.; Reddy, P. S.; Prasad, R. B. N.; Kanjilal, S.; *Eur. J. Lipid Sci. Technol.*, 2011, **113**, 756–762.

[58] Li, H. Q.; Yu, C. C.; Chen, R.; Li, J.; Li, J. X.; *Colloids Surf. A*, 2012, **395**, 116–124.

[59] Fisicaro, E.; Compari, C.; Biemmi, M.; Duce, E.; Peroni, M.; Barbero, N.; Viscardi, G.; Quagliotto, P.; *J. Phys. Chem. B*, 2008, **112**, 12312–12317.

[60] Borse, M. S.; Devi, S.; *Adv. Colloid Interface Sci.*, 2006, **123**, 387–399.

[61] Liu, G. Y.; Gu, D. M.; Liu, H. Y.; Ding, W.; Li, Z.; *J. Colloid Interface Sci.*, 2011, **358**, 521–526.

[62] Liu, H. Y.; Gu, D. M.; Liu, G. Y.; Ding, W.; *J. Surfactant Deterg.*, 2013, **16**, 495–502.

[63] Kamboj, R.; Singh, S.; Bhadani, A.; Kataria, H.; Kaur, G.; *Langmuir*, 2012, **28**, 11969–11978.

[64] Gao, H. J.; Shi, L. J.; Zhang, S. H.; Li, J.; Wang, X. D.; Zheng, L. Q.; *Colloid Polym. Sci.*, 2013, **291**, 1077–1084.

[65] Kumar, S.; Gupta, S. K.; *Tetrahedron Lett.*, 2010, **51**, 5459–5462.

[66] Li, C. X.; Zhu, W.; Li, J.; Liu, X. Y.; Li, W. N.; Jiang, Y.; Yang, H. W.; Li, G. T.; *Chem. Eur. J.*, 2013, **19**, 10146–10149.

[67] Ao, M. Q.; Xu, G. Y.; Pang, J. Y.; Zhao, T. T.; *Langmuir*, 2009, **25**, 9721–9727.

[68] Liu, Y. H.; Yu, L.; Zhang, S. H.; Yuan, J.; Shi, L. J.; Zheng, L. Q.; *Colloids Surf. A*, 2010, **359**, 66–77.

[69] Zhang, S. H.; Lu, F.; Zheng, L. Q.; *Colloid Polym. Sci.*, 2011, **289**, 1815–1819.

[70] Hegazy, M. A.; Tabei, A. S. E.; *J. Surfactant Deterg.*, 2013, **16**, 221–232.

[71] Dujat, L. C.; Rodrigues, M.; Yagüe, A.; Calpena, A. C.; Amabilino, D. B.; Linares, J. G.; Borràs, M.; García, L. P.; *Langmuir*, 2012, **28**, 2368–2381.

[72] Docherty, K. M.; Kulpa, C. F.; *Green Chem.*, 2005, **7**, 185–189.

[73] Wang, C. J.; Jiang, X. H.; Zhou, L. M.; Xia, G. Q.; Chen, Z. J.; Duan, M.; Jiang, X. M.; *Chem. Eng. J.*, 2013, **219**, 469–477.

[74] Ilies, M. A.; Seitz, W. A.; Ghiviriga, I.; Johnson, B. H.; Miller, A.; Thompson, E. B.; Balaban, A. T.; *J. Med. Chem.*, 2004, **47**, 3744–3754.

[75] Sharma, V. D.; Aifuwa, E. O.; Heiney, P. A.; Ilies, M. A.; *Biomaterials*, 2013, **34**, 6906–6921.

[76] Ilies, M. A.; Seitz, W. A.; Johnson, B. H.; Ezell, E. L.; Miller, A. L.; Thompson, E. B.; Balaban, A. T.; *J. Med. Chem.*, 2006, **49**, 3872–3887.

[77] Chen, T.; Xiong, H. Y.; Wen, W.; Zhang, X. H.; Wang, S. F.; *Bioelectrochemistry*, 2013, **91**, 8–14.

[78] Qasba, P. K.; Aposhian, H. V.; *Proc. Natl. Acad. Sci. U. S. A.*, 1971, **68**, 2345–2349.

[79] Robbins, P. D.; Ghivizzani, S. C.; *Pharmacol. Ther.*, 1998, **80**, 35–47.

[80] Luo, D.; Saltzman, W. M.; *Nat. Biotechnol.*, 2000, **18**, 33–37.

[81] Guo, X.; Huang, L.; *Acc. Chem. Res.*, 2012, **45**, 971–979.

[82] He, Y. F.; Shang, Y. Z.; Liu, Z. H.; Shao, S.; Liu, H. L.; Hu, Y.; *Colloids Surf. B*, 2013, **101**, 398–404.

[83] Zhou, T.; Llizo, A.; Li, P.; Wang, C. X.; Guo, Y. Y.; Ao, M. Q.; Bai, L. L.; Wang, C.; Yang, Y. L.; Xu, G. Y.; *J. Phys. Chem. C*, 2013, **117**, 26573–26581.

[84] Zhou, T.; Llizo, A.; Wang, C.; Xu, G. Y.; Yang, Y. L.; *Nanoscale*, 2013, **5**, 8288–8306.

[85] Liu, Y.; Yang, L.; Guo, R.; *Soft Matter*, 2013, **9**, 3671–3680.

[86] Wu, D.; Xu, G. Y.; Feng, Y. J.; Wang, Y J.; Zhu, Y. Y.; *Colloid Polym. Sci.*, 2009, **287**, 225–230.

[87] Wu, D.; Xu, G. Y.; Sun, Y. H.; Zhang, H. X.; Mao, H. Z.; Feng, Y. J.; *Biomacromolecules*, 2007, **8**, 708–712.

[88] Wu, D.; Xu, G. Y.; Feng, Y. J.; Li, Y. M.; *Int. J. Biol. Macromol.*, 2007, **40**, 345–350.

[89] Wang, Y.; Jiang, X. H.; Zhou, L. M.; Yang, L. J.; Xia, G. Q.; Chen, Z. J.; Duan, M.; *Colloids Surf. A*, 2013, **436**, 1159–1169.

[90] Tu, S.; Jiang, X. H.; Zhou, L. M.; Yin, W. M.; Wang, H. C.; Duan, M.; Liu, P. L.; Jiang, X. M.; *J. Lumin.*, 2012, **132**, 381–385.

[91] Zhou, T.; Ao, M. Q.; Xu, G. Y.; Liu, T.; Zhang, J.; *J. Colloid Interface Sci.*, 2013, **389**, 175–181.

[92] Gong, H. J.; Xu, G. Y.; Liu, T.; Xu, L.; Zhai, X. R.; *Langmuir*, 2012, **28**, 13590–13600.

[93] Liu, T.; Xu, G. Y.; Gong, H. J.; Pang, J. Y.; He, F.; *Langmuir*, 2011, **27**, 9253–9260.

[94] Ao, M. Q.; Xu, G. Y.; Kang, W. L.; Meng, W.; Gong, H. J.; Zhou, T.; *Soft Matter*, 2011, **7**, 1199–1206.

CHAPTER 7

Fluorescence Studies of the Microenvironments of the Morpholinium Room-Temperature Ionic Liquids

DINESH CHANDRA KHARA

School of Chemistry, University of Hyderabad, Hyderabad, India
Department of Chemistry, Ben-Gurion University of the Negev, Beer-Sheva 8410501 Israel

KOTNI SANTHOSH

School of Chemistry, University of Hyderabad, Hyderabad, India
Department of Chemical Physics, Weizmann Institute of Science, Rehovot, Israel

ANUNAY SAMANTA

School of Chemistry, University of Hyderabad, Hyderabad, India

7.1 INTRODUCTION

The concern over environmental pollution by volatile organic compounds (VOCs), which are widely used in chemical transformation in research laboratories and industries, has led to the realization of the importance of "green" alternatives such as solvent-free synthesis [1, 2], use of water as solvent [3], supercritical carbon dioxide [4–6], and room-temperature ionic liquids (RTILs) as reaction media [7–10]. Even though extensive investigations have shown that it may not be possible to replace the VOCs with the ionic liquids (ILs), the studies have revealed the importance and some interesting and unusual properties of these substances generating a great deal of attention worldwide [11–15].

ILs are salts, composed entirely of ions having melting points below 100°C. They are termed as RTILs as they exist as liquids at ambient temperature (20–30°C) and pressure (1 bar). Weak electrostatic interactions between the counterions and lack of a symmetric structure of the cation or anion or both contribute to the low melting points of the ILs. The chemical formulae of some of the constituent ions of the ILs are shown in Chart 7.1. Properties like liquidity, low volatility, high thermal stability, good electrical conductivity, and ability to dissolve a wide range of organic and inorganic compounds have drawn the attention of these ILs as possible alternatives to VOCs in chemical reactions, mass spectrometry, catalysis, separation processes, etc. [7, 16–21]. Biphasic Acid Scavenging by Ionic Liquids (BASIL) by BASF was the first

Ionic Liquid-Based Surfactant Science: Formulation, Characterization, and Applications, First Edition. Edited by Bidyut K. Paul and Satya P. Moulik.

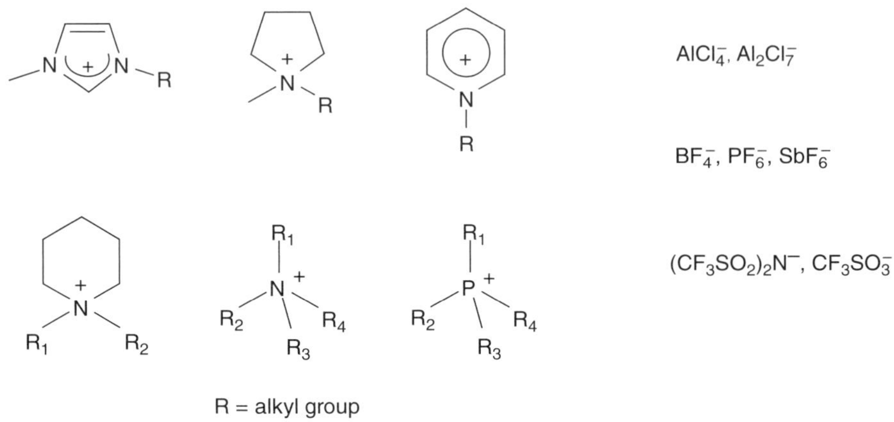

Chart 7.1 Some commonly used constituent ions of the ILs.

successful and major industrial application of the ILs, where 1-alkylimidazole was used to scavenge the acid produced during the chemical reactions [22, 23]. Due to good electrical conductivity and low reactivity, ILs serve as electrolytes in electrochemical applications such as in lithium-ion batteries, fuel cells, and capacitors [18–20, 24, 25]. Synthesis of biocompatible ILs allows biocatalysis, enzyme-based chemical reactions, and protein folding studies in ILs [26, 27]. These materials have also been explored as solvent for the synthesis of nanoparticles showing high stability and good catalytic activity [28]. These are some of the key areas, where application of the ILs has been realized to minimize the use and generation of hazardous chemical substances.

Even though the ILs were originally considered to be environmentally benign primarily due to their low vapor pressure, it has been revealed that many of the ILs are moderately or highly toxic, particularly to aquatic organisms, and their biodegradability depends on their structural design [29, 30]. Though ILs are often termed as nonvolatile because of their extremely low vapor pressure, it is shown that commonly used imidazolium ILs can be distilled under high vacuum and temperature conditions in pure form [31–34]. Though the ILs are nonflammable media, highly energetic ILs rich in oxygen and nitrogen ignite under oxygen environment [35]. As ILs are "designer solvents" [36] and a combination of various cations and anions allows generation of new ILs, it is still possible to think of new ILs with low vapor pressure and low toxicity and that are economically viable to serve the purpose of green chemistry. Consequently, the design and development of new ILs and investigation of their physicochemical properties remain as an active area of research.

Even though the ILs appear homogeneous in the naked eye, recent experimental, theoretical, and simulation studies have revealed organized structures of these liquids [37–44]. As most of these studies have dealt mainly with most frequently used imidazolium ILs, it is not possible to comment whether the organized structure and properties of the imidazolium ILs represent the generic features of the other ILs or not unless one measures the physiochemical properties of the other ILs and compares them with those of the imidazolium ILs. This requires physicochemical characterization of each family of new ILs. ILs based on morpholinium cation have been developed recently [45–48], and these are found to be cost-effective compared to the imidazolium and pyrrolidinium ILs. However, as the physicochemical properties of the morpholinium ILs have not been explored carefully, we recently embarked on the study of the fluorescence response of some carefully chosen probe molecules in these media (cf. Chart 7.2) to understand their physicochemical and structural properties [49, 50]. It was revealed that the morpholinium ILs are more structured/organized than the imidazolium ILs.

7.2 POLARITY

Static dielectric constant and refractive index are the common parameters that represent the macroscopic polarity of the solvent medium. The measured static dielectric constants of several ILs are low ($\varepsilon = 11.4$–12.3 at 298 K) [51, 52] and indicate that these media are much less polar than acetonitrile ($\varepsilon = 35.9$ at 298 K) [53]. However, the experimental observations with ILs cannot be explained on the basis of the measured dielectric constants [54]. To understand the interaction of the solvent molecules with the solute molecules at the microscopic level in the solution, the solvent polarity is often expressed in terms of the microscopic solvent polarity parameter, $E_T(30)$ [54], which is defined as

$$E_T(30)/\text{kcal mol}^{-1} = 28{,}591/\lambda_{\text{max}}(\text{nm})$$

and is most frequently estimated by measuring the molar transition energies of a solvent-sensitive pyridinium-N-phenoxide betaine dye. The λ_{max} in the above expression represents the wavelength corresponding to the maximum of the long-wavelength, solvatochromic, intramolecular charge transfer absorption band of the standard betaine dye.

The $E_T(30)$ values of solvents are often estimated using solvent polarity-sensitive fluorescence probes. In the present case, we have chosen coumarin-153 (C-153; Chart 7.3) for estimation of the $E_T(30)$ values of the ILs. The excited state dipole moment of a molecule like C-153 is higher than that in the ground state. Consequently,

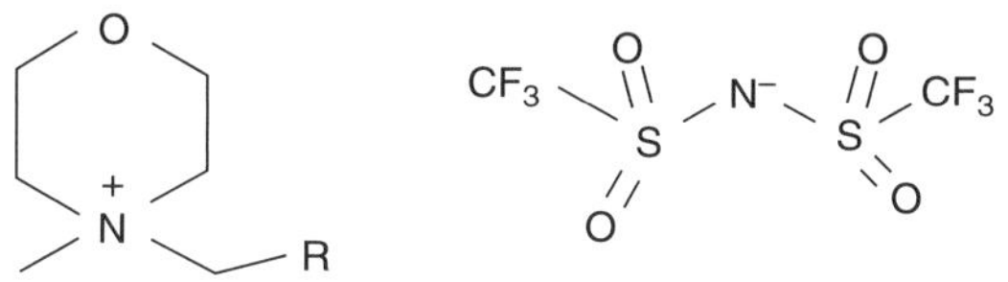

R = CH_3, $[Mor_{1,2}](Tf_2N)$
= C_3H_7, $[Mor_{1,4}](Tf_2N)$
= C_5H_{11}, $[Mor_{1,6}](Tf_2N)$
= C_7H_{15}, $[Mor_{1,8}](Tf_2N)$

Chart 7.2 Chemical formulae of the morpholinium ILs.

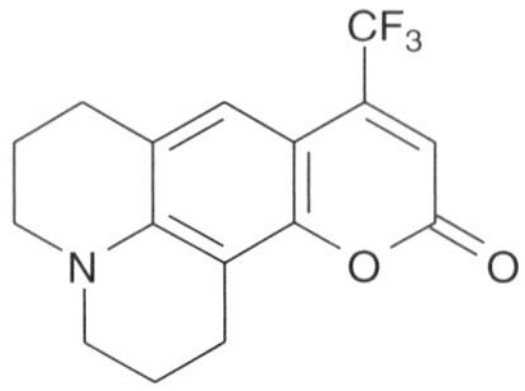

C153

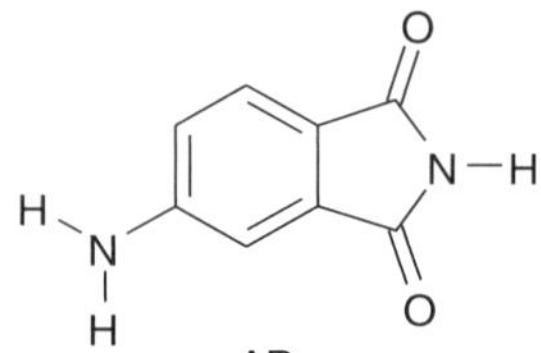

AP

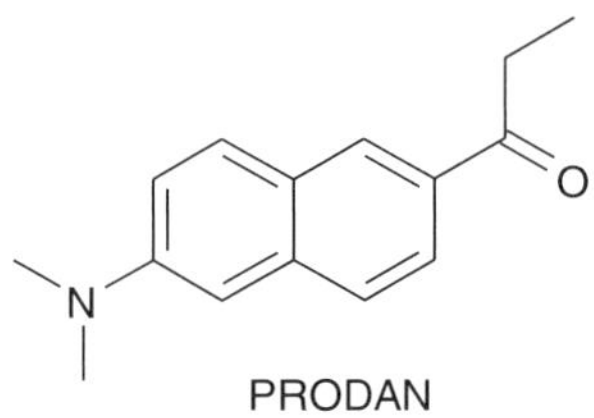

PRODAN

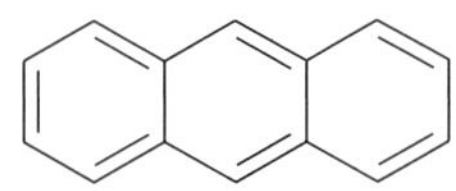

Anthracene

Chart 7.3 Fluorescence probes used in this study.

with increase in polarity of the medium, the excited state is stabilized more compared to the ground state leading to a Stokes shift in the emission maximum. As the extent of shift depends on the solute–solvent interactions, the frequency corresponding to the emission maximum ($\bar{\nu}_{max}^{em}$) of a dipolar probe can be considered as a measure of the microscopic polarity of the surrounding medium. Once the $\bar{\nu}_{max}^{em}$ values of C-153 are measured in different solvents of known polarities, the unknown microscopic solvent polarity of an IL can be estimated using the linear relationship between $\bar{\nu}_{max}^{em}$ and $E_T(30)$ values and the measured $\bar{\nu}_{max}^{em}$ value of C-153 in the ILs. Additional details on the method of estimation of the microscopic solvent polarity parameter using solvent polarity-sensitive probes can be found elsewhere [55, 56].

The absorption and emission spectra of C-153 (Fig. 7.1) in four ILs show that the absorption maximum of C-153 is independent of the length of the alkyl chain attached to the morpholinium ion, but the emission maximum shifts approximately 15 nm toward the shorter wavelength when the IL is changed from $[Mor_{1,2}][Tf_2N]$ to $[Mor_{1,8}][Tf_2N]$, thus indicating that the polarity of the morpholinium ILs decreases with increase in the alkyl chain length. The shift is more prominent for transition from $[Mor_{1,2}][Tf_2N]$ to $[Mor_{1,4}][Tf_2N]$. Only a small change is observed as the IL is changed from $[Mor_{1,4}][Tf_2N]$ to $[Mor_{1,8}][Tf_2N]$. The effect of temperature on the $\bar{\nu}_{max}^{em}$ value of C-153 is studied over a modest temperature range of 298–348 K. It is observed that the $\bar{\nu}_{max}^{em}$ values decrease with increase in temperature (Table 7.1) indicating an increase of the polarity of these ILs. The extent of change is found to be dependent on the IL, and the effect is most prominent in ILs comprising a long alkyl chain length. Considering the fact that an increase in temperature usually results in a small increase in the $\bar{\nu}_{max}^{em}$ value of C-153 in conventional solvents (due to a small decrease in polarity of these media), the present result suggests that the temperature-induced small decrease in the polarity of these media is masked by another factor, which produces a greater effect of an opposite kind. It is known that the viscosity of a solvent decreases at a higher temperature. However, this small change in viscosity does not influence the $\bar{\nu}_{max}^{em}$ value of C-153 in conventional solvents. Interestingly, the situation is very different in ILs. The viscosity of the ILs

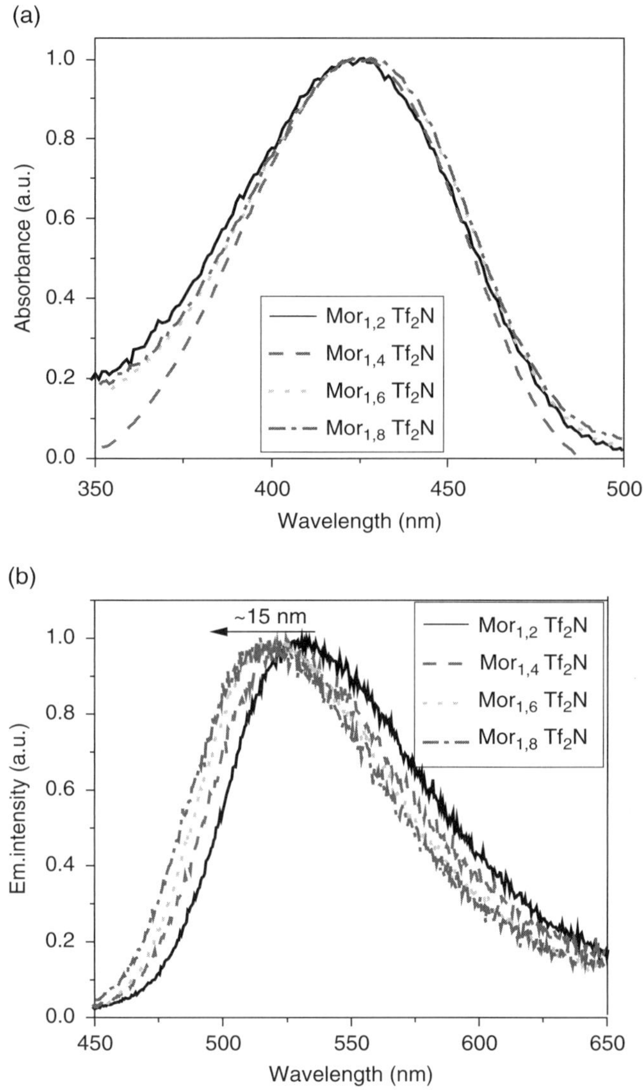

Figure 7.1 Normalized steady-state absorption (a) and fluorescence spectra (b) of C-153 in morpholinium ILs of different alkyl chain lengths, at 298 K. λ_{exc} for fluorescence measurements is 405 nm. Reproduced from Khara and Samanta [49] with permission from the American Chemical Society. *See insert for color representation of the figure.*

decreases sharply with increase in temperature (vide the following section), and this drastic decrease in viscosity of the ILs at higher temperature has a profound influence on the $\bar{\nu}_{max}^{em}$ values of C-153. This is because a lowering of viscosity allows a much faster and greater stabilization of the emitting state of dipolar molecules like C-153, thus accounting for the temperature dependence of $\bar{\nu}_{max}^{em}$ and/or the polarity of the ILs, as presented in Table 7.1.

TABLE 7.1 Temperature Dependence of the $\bar{\nu}_{max}^{em}$ Values of C-153 and the Viscosity and Estimated Polarities of the ILs

	$[Mor_{1,2}][Tf_2N]$			$[Mor_{1,4}][Tf_2N]$			$[Mor_{1,6}][Tf_2N]$			$[Mor_{1,8}][Tf_2N]$		
Temp (K)	η	$\bar{\nu}_{max}^{em}$	$E_T(30)$	η	$\bar{\nu}_{max}^{em}$	$E_T(30)$	η	$\bar{\nu}_{max}^{em}$	$E_T(30)$	η	$\bar{\nu}_{max}^{em}$	$E_T(30)$
298	—	18,830	48.5	503	19,080	47.0	578	19,230	45.8	673	19,340	45.1
308	147	18,760	—	236	18,975	—	269	19,160	—	327	19,230	—
318	85	18,725	—	126	18,900	—	142	19,050	—	183	19,160	—
328	52	18,725	—	72	18,900	—	80	19,010	—	113	19,050	—
338	34	18,725		45	18,900		48	19,010		76	19,050	
348	—	18,725	49.3	—	18,900	48.1	—	19,010	47.3	—	19,050	47.1

Reproduced from Khara and Samanta [49] with permission from the American Chemical Society.

A look at the $E_T(30)$ values estimated from the measured $\bar{\nu}_{max}^{em}$ values of C-153 at 298 K suggests that these media are slightly more polar than phosphonium ILs and less polar compared to the imidazolium and pyridinium ILs with identical alkyl chain lengths [56]. However, a comparison of the data at 338 K indicates that the polarities of the morpholinium ILs are similar to those of imidazolium ILs of the similar alkyl chain lengths [56]. These contradictory findings at two different temperatures can be rationalized considering that at 298 K, the viscosity of these ILs is sufficiently high and, consequently, complete stabilization (solvation) of the emitting state does not take place during the lifetime of the excited state and the emission originates from a partially relaxed state of this molecule. Hence, the $E_T(30)$ value estimated from the measured $\bar{\nu}_{max}^{em}$ under high viscous condition does not correctly represent the microscopic polarity of the medium.

7.3 VISCOSITY

The ILs are known to be viscous liquids, and the presence of impurities like halides and water severely influences the viscosity of these substances [57, 58]. The least viscous IL is almost 30 times more viscous than commonly used conventional solvents. Apart from the electrostatic forces operating between the ionic components, the size and symmetry of the constituent ions and H-bonding interactions also determine the viscosity of the ILs [59–61]. The measured viscosities of the four morpholinium ILs at various temperatures are presented in Table 7.1. As can be seen, at any given temperature, the viscosity increases with increase in the alkyl chain length attached to the morpholinium cation, and interestingly, the measured viscosities of the morpholinium-based ILs are higher compared to ILs based on imidazolium ions of the identical alkyl chain lengths [62].

That the viscosities of the ILs are highly sensitive to the temperature is evident from the data presented in Table 7.1. The temperature dependence of the viscosity of morpholinium ILs, depicted in Figure 7.2, is found to be better represented by the Vogel–Fulcher–Tammann (VFT) equation [63], an equation widely used to describe the temperature dependence of the viscosity of glass-forming substances compared to the Arrhenius equation.

7.4 STRUCTURAL HETEROGENEITY OF THE MORPHOLINIUM ILs

7.4.1 Excitation Wavelength-Dependent Emission Spectra

According to Kasha's rule, the emission behavior of a molecule in the condensed phase is independent of the excitation energy, and fluorescence generally originates from the lowest vibrational level of the first excited state of same multiplicity [64]. However, excitation wavelength-dependent emission behavior of molecular systems is reported in many cases such as in proteins [65], polymers [66], micelles [67], phospholipid vesicles [68], and, recently, in some ILs as well [37–43, 63]. This unusual excitation wavelength-dependent emission behavior is termed as red edge effect (REE) or red edge excitation shift (REES) or edge excitation shift (EES) or edge excitation red shift (EERS) [69, 70].

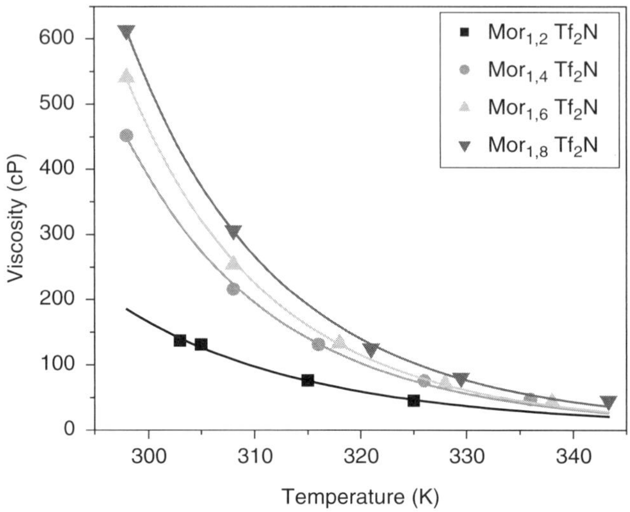

Figure 7.2 Viscosities of the morpholinium ILs measured at different temperatures, and the solid lines represent VFT fits to the experimental data points. Reproduced from Khara and Samanta [49] with permission from the American Chemical Society.

It is now understood that the excitation wavelength dependence of the emission behavior is the result of (i) the existence of a distribution of energetically different molecules in the ground state (a heterogeneity) and (ii) a slower rate of excited state relaxation process compared to the fluorescence decay rate of the molecule [69, 70]. Inhomogeneous broadening of the absorption spectrum toward the red side is suggestive of the presence of molecules with differing energies, and this is why REE is identified through excitation on the red side of the absorption spectrum (Scheme 7.1).

In the case of the morpholinium ILs, the emission behavior of C-153 is found to be dependent on the excitation wavelength. This dependence is depicted in Figure 7.3. With increase in excitation wavelength (along the red edge of the absorption/excitation spectrum), the emission maximum shifts toward red, and this effect is more pronounced in ILs having a higher alkyl chain length. While REE has been observed previously in ILs, the present observation is surprising for C153 did not exhibit excitation wavelength dependence in imidazolium ILs [37, 38, 71] as its fluorescence lifetimes are significantly larger than the solvation times. As the fluorescence lifetime of C-153 ($\tau_f = 5.0 \pm 0.2$ ns) is much higher than its solvation time in the morpholinium ILs as well [49], the observation of REE of C-153 in these media is unexpected. This surprising observation can be understood considering structural heterogeneity of these ILs. The heterogeneity originates from the isolation of the ionic constituents from the alkyl chains, thus forming polar and nonpolar domains into which C-153 is distributed. As the molecules belonging to these regions are energetically different, a change in the excitation wavelength along the red side of the absorption spectrum changes the proportion of molecules excited from the two regions, and in the absence of any communication between the molecules from the two regions, the emission behavior is determined by the population of the excited molecules from the two regions. As no excitation wavelength dependence of C-153

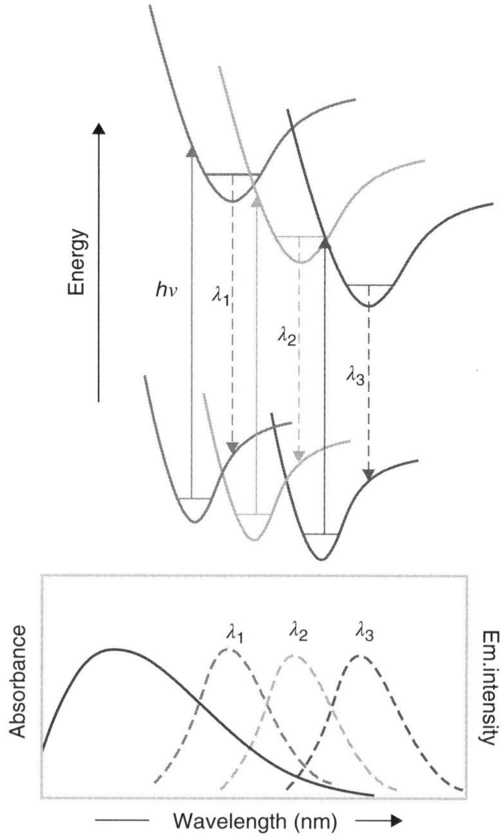

Scheme 7.1 Pictorial representation of molecular basis of REE.

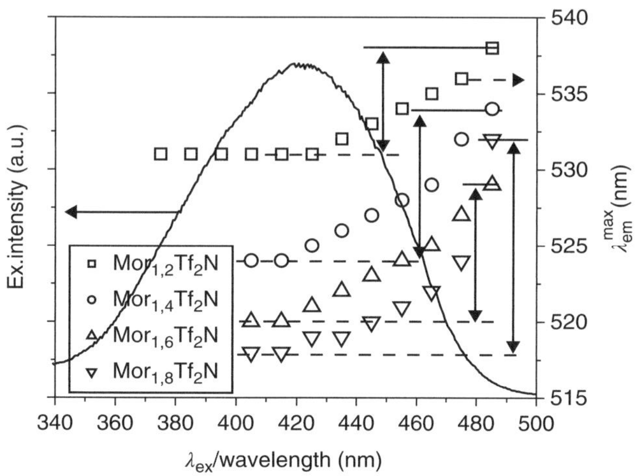

Figure 7.3 Excitation wavelength dependence of the emission maxima of C-153 in four different morpholinium ILs, at 298 K. A typical fluorescence excitation spectrum of C-153 in these ILs is also shown. Reproduced from Khara and Samanta [49] with permission from the American Chemical Society. *See insert for color representation of the figure.*

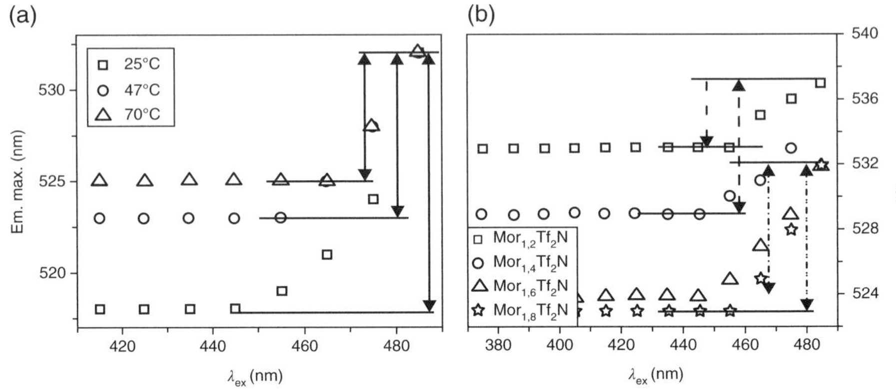

Figure 7.4 Excitation wavelength dependence of the emission maximum of C-153 in [$Mor_{1,8}$][Tf_2N] at different temperatures (a) and in four different morpholinium ILs under isoviscous conditions ($\eta = 130$ cP) (b). Reproduced from Khara and Samanta [49] with permission from the American Chemical Society. *See insert for color representation of the figure.*

is observed in imidazolium ILs with similar alkyl chain lengths, it can be argued that morpholinium ILs are structurally more heterogeneous than the imidazolium ILs.

The structural heterogeneity of the morpholinium ILs is further established by the results obtained at different temperatures and under isoviscous conditions. Significantly lower excitation wavelength dependence at higher temperature (Fig. 7.4) is a reflection of partially destroyed domain structure of the IL. Under isoviscous conditions, a more pronounced excitation wavelength dependence of C-153 in morpholinium ILs with higher alkyl chain length (Fig. 7.4) is also a reflection of the domain-like structures in them. Dynamic fluorescence Stokes shift studies and time-resolved fluorescence anisotropy experiments further support the organized structures of the morpholinium ILs.

7.4.2 Solvation Dynamics

The time taken for reorganization of the solvent molecules around an instantly created dipole is termed as solvation time or solvent relaxation time (τ_{solv}) [72]. As the ILs are polar, time-resolved fluorescence studies on dipolar fluorescent molecules provide valuable information on the timescales of reorganization of the constituents of the ILs around a photoexcited molecule. The timescale of solvation depends on the viscosity, temperature, and molecular structure of the surrounding solvent [72]. As ILs are highly viscous, the solvation in ILs is a much slow process compared with that in less viscous conventional solvents. The dynamics of solvation is commonly studied by monitoring the time-dependent fluorescence Stokes shift of a dipolar molecule following its excitation by a short pulse (Scheme 7.2). This phenomenon is called dynamic fluorescence Stokes shift [73], and the solvation dynamics in several ILs has been studied by this method using various fluorescent probes.

As dynamic fluorescence Stokes shift studies can also provide information on the structural heterogeneity of a medium, the solvation dynamics of C-153 is studied under isoviscous ($\eta = 130$ cP) conditions in several morpholinium ILs. The time-resolved

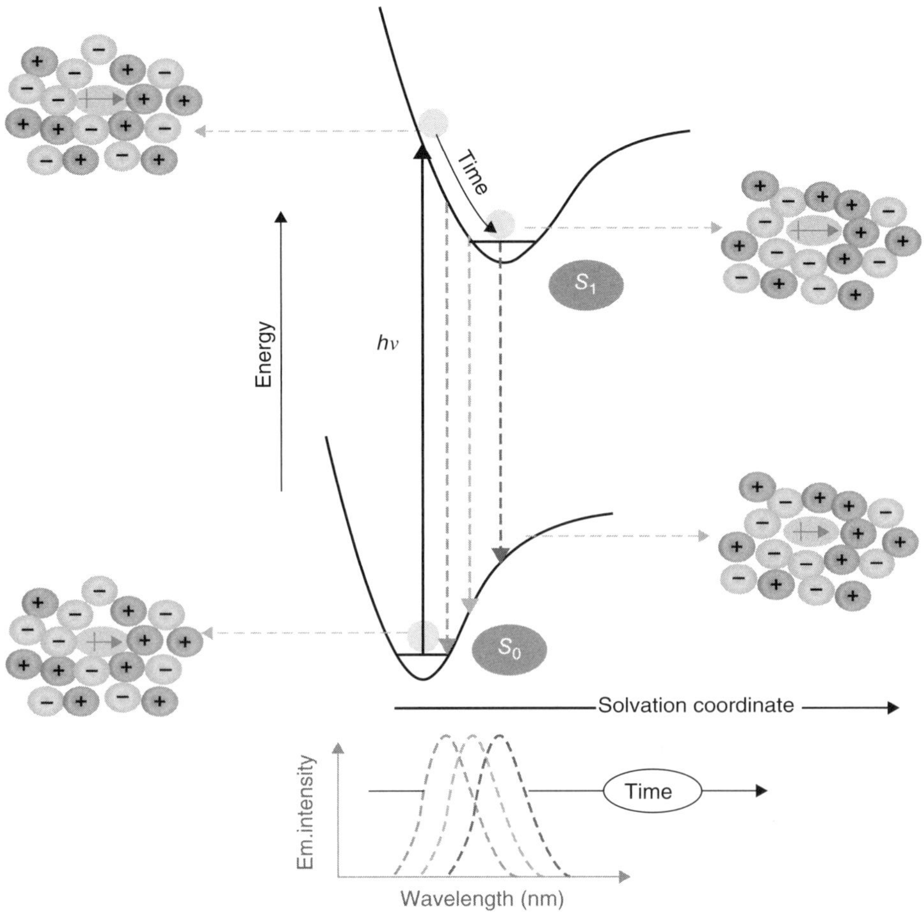

Scheme 7.2 Pictorial illustration of dynamic fluorescence Stokes shift in ILs.

emission spectra (TRES) of the system have been constructed using the fitting parameters of the fluorescence decay profiles measured at different wavelengths in the four ILs under isoviscous conditions [72]. Figure 7.5 shows typical fluorescence decay profiles at a few selected wavelengths and the TRES of C-153 in $[Mor_{1,2}][Tf_2N]$ obtained from these data at $\eta = 130$ cP.

A gradual shift of the emission maximum from a higher to a lower energy in the nanosecond timescale is a manifestation of slow relaxation/reorientation of the solvent molecules around photoexcited C-153. A similar situation is encountered in other three ILs. The solvation time is estimated by first measuring the peak frequencies ($\bar{\nu}$) at various times from the TRES and then following the time dependence of the solvent shift correlation function $C(t)$, defined as

$$C(t) = \frac{[\bar{\nu}(t) - \bar{\nu}(\infty)]}{[\bar{\nu}(0) - \bar{\nu}(\infty)]}$$

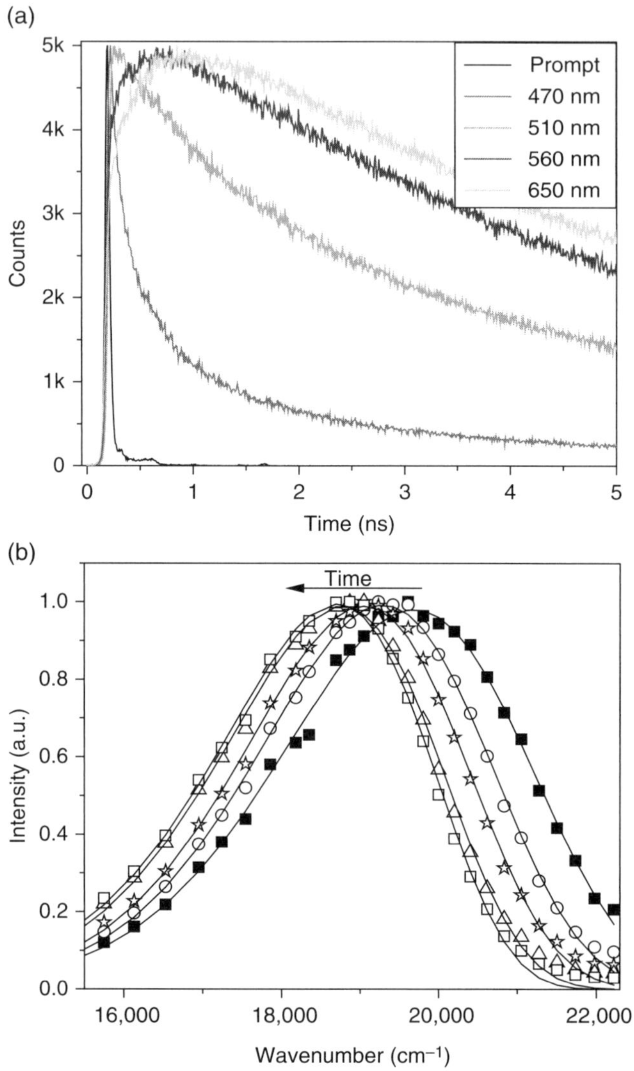

Figure 7.5 Wavelength-dependent fluorescence decay profiles (a) and time-dependent dynamic fluorescence Stokes shift (b) of dipolar probe C-153 in $[Mor_{1,2}][Tf_2N]$ IL at $\lambda_{exc} = 405$nm and $\eta = 130$cP. Reproduced from Khara and Samanta [49] with permission from the American Chemical Society. *See insert for color representation of the figure.*

The plot of $C(t)$ versus t (Fig. 7.6) is best described by a biexponential decay function, $C(t) = \alpha_1 \exp(-t/\tau_1) + \alpha_2 \exp(-t/\tau_2)$, where α_1 and α_2 represent normalized preexponential factors and τ_1 and τ_2 are the two components of the solvent relaxation times. The measured individual solvation time components and the average solvation times are presented in Table 7.2.

Due to finite time resolution of the instrumental setup (typically 50ps), complete solvent relaxation process of C-153 could not be observed. Hence, employing the

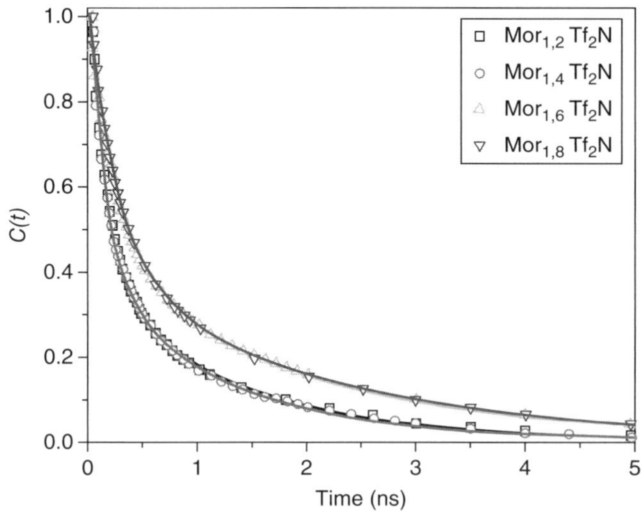

Figure 7.6 Decays of spectral shift correlation functions, $C(t)$, of C-153 in four morpholinium ILs under isoviscous conditions. Solid lines represent biexponential fit to the data. Reproduced from Khara and Samanta [49] with permission from the American Chemical Society. *See insert for color representation of the figure.*

actual time-zero maximum, which is estimated by a standard procedure [74], the missing components of the solvation dynamics are estimated and presented in Table 7.2. It can be seen that the missing component is almost approximately 50% in all ILs.

Generally, the solvent relaxation time is dependent on the viscosity and increases with increase in viscosity of the surrounding solvent medium. Interestingly, under isoviscous conditions, the solvent relaxation times of C-153 in four ILs are not the same, but they increase with increase in the alkyl chain length of the morpholinium cation. Specifically, of the two time constants, the long time component (τ_2) increases considerably as the IL is changed from $[Mor_{1,2}][Tf_2N]$ to $[Mor_{1,8}][Tf_2N]$. The average solvent relaxation time estimated from the individual time components increases approximately twofold as one moves from $[Mor_{1,2}][Tf_2N]$ to $[Mor_{1,8}][Tf_2N]$.

It is reported that the solvent relaxation time, $<\tau_{solv}>$, is dependent on the size of the constituent cations as $<\tau_{solv}> \alpha\, (\eta/T)R_+^4$ [75]. However, under isoviscous condition, a plot of $<\tau_{solv}>$ versus $(\eta/T)R_+^4$ exhibits a poor linear dependence. This is possible when the solvent is more structured and the bulk viscosity of the medium and the size of the large ion cannot determine the solvation time of a probe molecule.

7.4.3 Time-Resolved Fluorescence Anisotropy

Fluorescence anisotropy is the measure of emission depolarization of a fluorescent molecule excited by a polarized light [72]. Rotation of the molecule is commonly responsible for the emission depolarization. Several factors such as viscosity, temperature, structure of the surrounding solvent, and specific solute–solvent

TABLE 7.2 Solvent Relaxation Parameters Estimated from the TRES of C-153 in Four Morpholinium ILs under the Isoviscous ($\eta = 130$ cP) Conditions

		Solvent Relaxation Data			
ILs (T/K)[a]	η (cP)	τ_1 (α_1)	τ_2 (α_2)	$<\tau_{solv}>$[b]	Missing Comp. (%)
$[Mor_{1,2}][Tf_2N]$ (305)	~130	0.17 (0.64)	1.39 (0.36)	0.58	53
$[Mor_{1,4}][Tf_2N]$ (316)		0.15 (0.59)	1.12 (0.41)	0.55	51
$[Mor_{1,6}][Tf_2N]$ (318)		0.22 (0.55)	1.88 (0.45)	0.97	55
$[Mor_{1,8}][Tf_2N]$ (320)		0.25 (0.58)	2.04 (0.42)	1.00	49

Reproduced from Khara and Samanta [49] with permission from the American Chemical Society.
[a] Temperatures at which the experiments are performed.
[b] Average solvation time $<\tau_{solv}> = a_1\tau_1 + a_2\tau_2$, where $a_1 + a_2 = 1, \pm 5\%$.

interactions determine the timescale of rotational relaxation of a solute molecule [72]. Hence, in addition to the dynamic fluorescence Stokes shift studies, rotational dynamics of probe molecules is also studied by measuring time-resolved fluorescence anisotropy of molecular systems in morpholinium ILs to obtain information on the structural heterogeneity of these media. The molecules chosen for this purpose are two dipolar molecules, 4-aminophthalimide (AP) and 6-propionyl-2-dimethylaminonaphthalene (PRODAN), and one nonpolar probe molecule, anthracene (Chart 7.3).

The time-resolved fluorescence anisotropy values, $r(t)$, are calculated using [72]

$$r(t) = \frac{I_{II}(t) - GI_{\perp}(t)}{I_{II}(t) + 2GI_{\perp}(t)}$$

where $I_{II}(t)$ and $I_{\perp}(t)$ are, respectively, the fluorescence intensities polarized parallel and perpendicular to the polarized excitation light and G is a quantity that corrects for the difference in the detector sensitivity toward vertically and horizontally polarized light. Fluorescence anisotropy decay profiles of the three fluorescent probes in $[Mor_{1,8}][Tf_2N]$ IL at 298 K are shown in Figure 7.7. In the same manner, time-resolved fluorescence anisotropy decay profiles of the probes in rest of the ILs at different temperatures and for different excitation wavelengths are recorded. The average rotational relaxation times (τ_r) of the probe molecules measured at different temperatures and for different excitation wavelengths are obtained by fitting the data to a single or biexponential function, and the measured values are collected in Table 7.3.

We would like to highlight a few interesting observations here. As viscosity of the IL increases with increase in alkyl chain length of the constituent cation, the τ_r values are expected to increase. While this is indeed the case for AP and PRODAN, near constancy of the τ_r values of anthracene with change of viscosity of the medium is quite surprising (Table 7.3). Another point to note here is that while AP and PRODAN exhibit a similar trend, the increase is more appreciable in case of AP. The reorientation times of all the three probes decrease with increase in temperature of the system irrespective of the ILs. This probe dependence of the fluorescence anisotropy is explained in the following discussion considering the domain structure of the morpholinium ILs and specific solute–solvent interactions.

Stokes–Einstein–Debye hydrodynamic theory is the most commonly used model for understanding the rotational diffusion of molecular systems [76, 77].

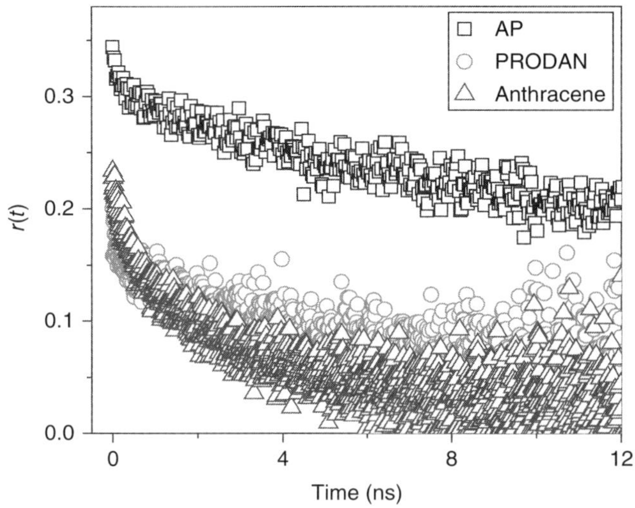

Figure 7.7 Time-resolved fluorescence anisotropy decay profiles of AP, PRODAN, and anthracene in $[Mor_{1,8}][Tf_2N]$ IL at 298 K. $\lambda_{exc} = 375\text{nm}$. *See insert for color representation of the figure.*

According to this model, the rotational relaxation time $\left(\tau_r^{SED}\right)$ of a solute noninteracting with the solvent continuum of viscosity η is given by

$$\tau_r^{SED} = \frac{V_h \eta}{k_B T}$$

where V_h is the solute hydrodynamic volume, a product of van der Waals volume (V) of the molecule, its shape factor (f), and boundary condition parameter C; k_B is the Boltzmann constant; and T is the temperature of the system. The shape factor, f, which accounts for nonspherical shape of the molecule, is a function of the ratio of semimajor and semiminor axis. For spherical molecule, this value is unity, and for asymmetric ellipsoids, this value is greater than unity. The slip and stick boundary conditions can be computed with the knowledge of molecular dimensions and diffusion coefficients along the three axes. The shape factor and the slip boundary condition parameter (C_{slip}) in turn can be estimated from stick and slip rotational relaxation times. The procedure for computation of the stick (τ_{stick}) and slip (τ_{slip}) rotational relaxation times is described elsewhere [78–81].

Figure 7.8 shows the dependence of experimentally measured rotational relaxation times (τ_r) of AP, PRODAN, and anthracene on η/T in four morpholinium ILs. The calculated τ_{stick} (upper solid line) and τ_{slip} (lower dashed line) values from the SED theory are also shown.

It can be seen that with increase in the chain length, AP changes from a stick to a superstick behavior. PRODAN rotational times lie mostly between the stick and slip boundary conditions, and a gradual shift from slip to subslip condition is observed in higher alkyl chain length ILs at lower temperature. The rotational times of anthracene are closer to the slip values at higher temperature, but they shift to a subslip regime in high viscous condition, except in $[Mor_{1,8}][Tf_2N]$, where the behavior is far

TABLE 7.3 Rotational Relaxation Times[a] of the Probe Molecules in Four Morpholinium ILs as a Function of Temperature and Excitation Wavelength

ILs	Temp. (K)	Rotational Relaxation Time (ns) AP $\lambda_{exc}=375$ nm	AP $\lambda_{exc}=439$ nm	PRODAN $\lambda_{exc}=375$ nm	PRODAN $\lambda_{exc}=439$ nm	Anthracene $\lambda_{exc}=375$ nm
$[Mor_{1,2}][Tf_2N]$	308	6.35	7.02	5.62	6.10	1.54
	318	3.76	4.25	4.20	4.26	0.90
	328	2.33	2.73	2.78	2.88	0.55
	338	1.52	1.71	1.94	2.05	0.36
$[Mor_{1,4}][Tf_2N]$	298	24.49	28.53	8.77	8.96	4.25
	308	12.11	12.80	8.17	8.87	2.27
	318	6.44	7.92	5.83	6.27	1.44
	328	3.73	4.45	3.81	4.17	0.98
	338	2.29	2.41	2.35	2.62	0.74
$[Mor_{1,6}][Tf_2N]$	298	32.07	36.30	8.70	9.26	3.31
	308	16.50	20.94	7.35	8.05	2.40
	318	9.50	10.46	6.21	6.91	1.59
	328	5.35	6.00	3.70	4.26	1.11
	338	3.26	3.77	2.49	2.84	0.78
$[Mor_{1,8}][Tf_2N]$	298	36.0	41.40	14.18	19.90	3.25
	308	21.10	21.98	8.56	12.83	2.03
	318	12.31	12.50	6.06	9.80	1.36
	328	7.03	8.03	3.94	5.74	0.86
	338	4.37	4.64	2.84	3.80	0.62

Reproduced from Khara et al. [50] with permission from the American Chemical Society.

[a] ±10%.

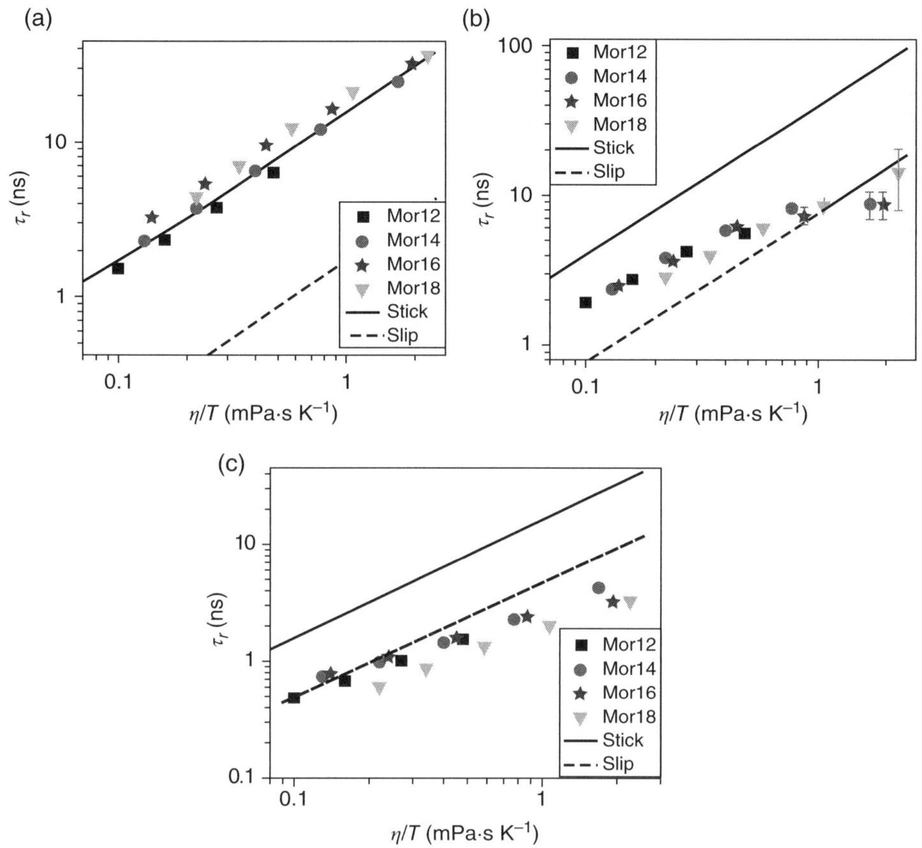

Figure 7.8 Plots of τ_r versus η/T for AP (a), PRODAN (b), and anthracene (c) in four different morpholinium ILs. The points indicate the experimentally measured rotational times. $\lambda_{exc} = 375\text{nm}$. Reproduced from Khara et al. [50] with permission from the American Chemical Society. *See insert for color representation of the figure.*

away from the slip boundary condition. Considering the fact that a superstick behavior is suggestive of a close association or specific interaction between solute and solvent molecules and a subslip behavior is indicative of nearly free rotation of the solute molecules, the observation clearly indicates that the solvent environments/interactions experienced by the three probe molecules are very different.

AP is known to exhibit stick to superstick behavior in conventional solvents and other ILs as it interacts with the solvent molecules through H-bonding interaction [77, 82, 83]. In the present case, the stick to superstick behavior of AP is perhaps a reflection of the H-bonding interaction of the N—H proton of AP with the O atom on the morpholinium cations and/or the F atom of $[Tf_2N]^-$ anion. The specific interaction of AP with the solvent is also evident from the increase of the coupling constant $\left(C_{rot} = \tau_r^{expt}/\tau_r^{stick}\right)$ with increase in the alkyl chain length (or van der Waals volume) of the morpholinium ILs (Fig. 7.9). When the τ_r values are fitted to (η/T) for each IL, the p values lie between 0.82 and 0.92, indicating almost linear dependence and confirming the fact that AP is associated with one environment of the IL through strong H-bonding interaction.

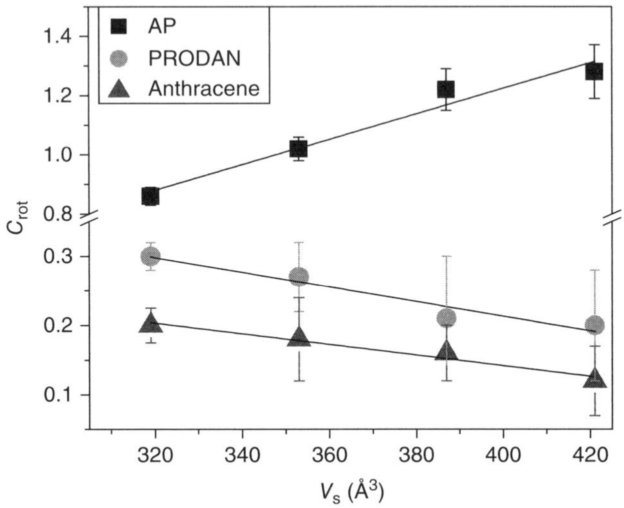

Figure 7.9 Plot of rotational coupling constant (C_{rot}, 308 K) versus van der Waals volume (V_s) of the ILs. λ_{exc} = 375 nm. The solid lines represent linear fit to the data points. Reproduced from Khara et al. [50] with permission from the American Chemical Society.

The rotational behavior of PRODAN, in particular the τ_r values gradually shifting toward slip and subslip boundary conditions with increase in chain length of the cation at lower temperatures, can only be explained if PRODAN is distributed in both hydrophilic and hydrophobic domains of the ILs. At higher temperature, an increase in the chain length does not have any effect on the rotational relaxation time, which lies between the stick and slip boundary condition. This is suggestive of the loss/perturbation of the domain structure of the IL at higher temperature. The effect of the alkyl chain length on the distribution of PRODAN between the polar and nonpolar domains is also evident from the gradual decrease of C_{rot} with increase in chain length (or van der Waals volume) of the morpholinium ILs (Fig. 7.9). The shift of the rotational behavior of PRODAN toward slip and subslip condition is also evident from the *p* values of the dependence of τ_r on $(\eta/T)^p$ of the ILs.

The *p* values, which lie between 0.40 and 0.66, suggest a high degree of nonlinearity in the case of PRODAN. As this behavior is similar to rotational dynamics of many probe molecules in higher *n*-alkanes [84], it is reasonable to assume that PRODAN is mostly surrounded by the alkyl groups, particularly in morpholinium ILs with higher chain lengths. As PRODAN is located in the hydrophobic domains of the ILs and can rotate freely in the voids, it shows gradual shift from the slip to subslip behavior with increase in alkyl chain length of the cation.

As the rotational relaxation time of anthracene lies between the slip and subslip condition in most cases but shifts completely toward the subslip regime in IL with a higher alkyl chain length, it can be considered that this nonpolar probe resides typically in the hydrophobic domains formed by the alkyl groups of the IL. A steady decrease of the C_{rot} values of anthracene with increase in volume (or alkyl chain length) of the IL suggests that the interaction between the nonpolar solute and the solvent changes with increase in the alkyl chain length of the cation.

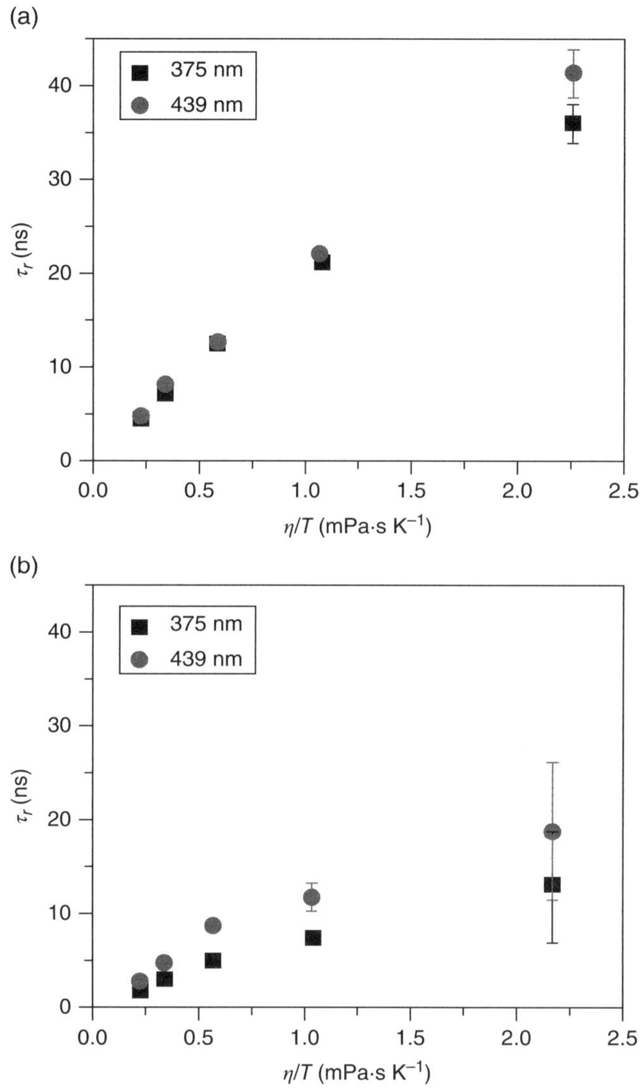

Figure 7.10 Excitation wavelength-dependent rotational relaxation times (τ_r) of AP (a) and PRODAN (b) on η/T in $[Mor_{1,8}][Tf_2N]$ IL. Reproduced from Khara et al. [50] with permission from the American Chemical Society.

Excitation wavelength dependence of the rotational relaxation times of AP and PRODAN also confirm the microheterogeneous nature of the morpholinium ILs. At 298 K, when the excitation wavelength is changed from 375 to 439 nm, the τ_r value increases by 15% for AP and 40% for PRODAN (Table 7.3). It is evident from Figure 7.10 that excitation wavelength dependency of AP in $[Mor_{1,8}][Tf_2N]$ is observable mostly at 298 K. However, PRODAN exhibits this effect even at higher temperature. A more prominent effect in the case of PRODAN suggests that this molecule is distributed more widely in different regions of the IL compared to the other dipolar molecule AP. The excitation wavelength dependence of the rotational relaxation time can be explained as follows. Excitation on red edge of the absorption spectrum excites preferentially the highly solvated molecules that reside in a more polar region. Since these molecules experience more hindrance or friction compared to the ones located in the nonpolar region, excitation at longer wavelength (439 nm) results in longer rotational times than the excitation at 375 nm, where the molecules occupying the nonpolar region are preferentially excited.

A few studies on the effect of alkyl chain length on the rotational dynamics have been carried out [75, 82, 85–87]. These results show no clear indication of hydrophilic and hydrophobic domain structures of the imidazolium ILs. However, our studies on morpholinium ILs clearly point to the effect of alkyl chain length on the rotational dynamics of polar and nonpolar solute molecules, suggesting their location in the hydrophilic and hydrophobic domains of the structured media.

7.5 CONCLUDING REMARKS

In essence, microheterogeneous nature of the ILs comprising a series of morpholinium cations is evident from the results of steady-state and time-resolved fluorescence measurements on selected probe molecules. The excitation wavelength dependence of the emission spectra, dynamic fluorescence Stokes shift, and time-resolved fluorescence anisotropy experiments not only substantiate an organized structure of the N-alkyl-N-methylmorpholinium ILs but also show that these are much more structured compared to the extensively studied imidazolium ILs. Even though our fluorescence studies are the first of its kind to indicate the organized environments of these little-explored ILs, we recognize that no detailed structural information on the spatial or temporal dimensions of this heterogeneity emerges from this study. Low-angle X-ray diffraction and molecular dynamics simulation studies could provide more insight into the microscopic structural details of these ILs.

ACKNOWLEDGMENTS

This work is supported by the J.C. Bose National Fellowship (to AS) of the Department of Science and Technology (DST), Government of India. The Senior Research Fellowship received by DCK and KS from the Council of Scientific and Industrial Research (CSIR) is gratefully acknowledged.

REFERENCES

[1] Li, M., Lv, X.-L., Wen, L.-R., Hu, Z.-Q., *Org. Lett.*, 2013, **15**, 1262–1265.

[2] Tanaka, K., Toda, F., *Chem. Rev.*, 2000, **100**, 1025–1074.

[3] Hang, C.T., Lianhai, L., Yang, Y., Wenshuo, L., Developing Green Chemistry: Organometallic Reactions in Aqueous Media. In *Clean Solvents: Alternative Media for Chemical Reactions and Processing*. Eds. Abraham, M., Moens, L., ACS Symposium Series 819; American Chemical Society, Washington, DC, 2002, pp. 166.

[4] Wai, C.M., Hunt, F., Ji, M., Chen, X., *J. Chem. Edu.*, 1998, **75**, 1641–1645.

[5] Licence, P., Ke, J., Sokolova, M., Ross, S.K., Poliakoff, M., *Green Chem.*, 2003, **5**, 99–104.

[6] Mayadevi, S., *Indian J. Chem.*, 2012, **51A**, 1298–1305.

[7] Hallett, J.P., Welton, T., *Chem. Rev.*, 2011, **111**, 3508–3576.

[8] Rogers, R.D., Voth, G., Special issue on ionic liquids, *Acc. Chem. Res.*, 2007, **40** 11, p. 1077.

[9] Plechkova, N.V., Seddon, K.R., *Chem. Soc. Rev.*, 2008, **37**, 123–150.

[10] Hough, W.L., Smiglak, M., Rodriguez, H., Swatloski, R.P., Spear, S.K., Daly, D.T., Pernak, J., Grisel, J.E., Carliss, R.D., Soutullo, M.D., Davis Jr., J.H., Rogers, R.D., *New J. Chem.*, 2007, **31**, 1429–1436.

[11] Wasserscheid, P., Welton, T., *Ionic Liquids in Synthesis*, Wiley-VCH Verlag GmbH, Weinheim, 2003.

[12] Holbrey, J.D., Reichert, W.M., Swatloski, R.P., Broker, G.A., Pritner, W.R., Seddon, K.R., Rogers, R.D., *Green Chem.*, 2002, **4**, 407–413.

[13] Brinchi, L., Germani, R., Savelli, G., *Tetrahedron Lett.*, 2003, **44**, 2027–2029.

[14] Pringle, J.M., Golding, J., Baranyai, K., Forsyth, C.M., Deacon, G.B., Scott, J.L., McFarlane, D.R., *New J. Chem.*, 2003, **27**, 1504–1510.

[15] Golding, J.J., Macfarlane, D.R., Spiccia, L., Forsyth, G.B., Skelton, B.W., White, A.H., *Chem. Commun.*, 1998, 1593–1594.

[16] Seddon, K.R., *Ionic Liquids, Industrial Applications for Green Chemistry*, American Chemical Society, Washington, DC, 2002.

[17] Dubreuil, J.F., Bourahla, K., Rahmouni, M., Bazureau, J.P., Hamelin, J., *Catal. Commun.*, 2002, **3**, 185–190.

[18] Enomoto, T., Nakamori, Y., Matsumoto, K., Hagiwara, R., *J. Phys. Chem. C*, 2011, **115**, 4324–4332.

[19] Ngo, H.L., LeCompte, K., Hargens, L., McEwen, A., *Thermochim. Acta*, 2000, **357**, 97–102.

[20] Lewandowski, A., Swiderska-Mocek, A., *J. Power Sources*, 2009, **194**, 601–609.

[21] Armstrong, D.W., Zhang, L.K., He, L., Gross, M.L., *Anal. Chem.*, **73**, 2001, 3679–3686.

[22] Rogers, R.D., Seddon, K.R., *Science*, 2003, **302**, 792–793.

[23] Seddon, K.R., *Nat. Mater.*, 2003, **2**, 363–365.

[24] Armand, M., Endres, F., MacFarlane, D.R., Ohno, H., Scrosati, B., *Nat. Mater.*, 2009, **8**, 621–629.

[25] Ding, J., Zhou, D., Spinks, G., Wallace, G., Forsyth, S., Forsyth, M., MacFarlane, D., *Chem. Mater.*, 2003, **15**, 2392–2398.

[26] Sankaranarayanan, K., Sathyaraj, G., Nair, B.U., Dhathathreyan, A., *J. Phys. Chem. B*, 2012, **116**, 4175–4180.

[27] Fujita, K., Ohno, H., *Biopolymers*, 2010, **93**, 1093–1099.

[28] Migowski, P., Dupont, J., *Chem. Eur. J.*, 2007, **13**, 32–39.

[29] Romero, A., Santos, A., Tojo, J., Rodríguez, A., *J. Hazard. Mater.*, 2008, **151**, 268–273.

[30] Coleman, D., Gathergood, N., *Chem. Soc. Rev.*, 2010, **39**, 600–637.

[31] Wasserscheid, P., *Nature*, 2006, **439**, 797.

[32] Earle, M.J., Esperanca, J.M.S.S., Gilea, M.A., Lopes, J.N.C., Rebelo, L.P.N., Magee, J.W., Seddon, K.R., Widegren, J.A., *Nature*, 2006, **439**, 831–834.

[33] Taylor, A.W., Lovelock, K.R.J., Deyko, A., Licence, P., Jones, R.G., *Phys. Chem. Chem. Phys.*, 2010, **12**, 1772–1783.

[34] Chambreau, S.D., Vaghjiani, G.L., Koh, C.J., Golan, A., Leone, S.R., *J. Phys. Chem. Lett.*, 2012, **3**, 2910–2914.

[35] Fox, D.M., Gilman, J.W., Morgan, A.B., Shields, J.R., Maupin, P.H., Lyon, R.E., Long, H.C.D., Trulove, P.C., *Ind. Eng. Chem. Res.*, 2008, **47**, 6327–6332.

[36] Freemantle, M., *Chem. Eng. News*, 1998, **76**, 32–37.

[37] Mandal, P.K., Sarkar, M., Samanta, A., *J. Phys. Chem. A*, 2004, **108**, 9048–9053.

[38] Samanta, A., *J. Phys. Chem. B*, 2006, **110**, 13704–13716.

[39] Paul, A., Samanta, A., *J. Phys. Chem. B*, 2008, **112**, 16626–16632.

[40] Hu, Z., Margulis, C.J., *Proc. Natl. Acad. Sci. U. S. A.*, 2006, **103**, 831–836.

[41] Jin, H., Li, X., Maroncelli, M., *J. Phys. Chem. B*, 2007, **111**, 13473–13478.

[42] Wang, Y., Voth, G.A., *J. Am. Chem. Soc.*, 2005, **127**, 12192–12193.

[43] Nishiyama, Y., Fukuda, M., Terazima, M., Kimura, Y., *J. Chem. Phys.*, 2008, **128**, 164514.

[44] Hardacre, C., Holbrey, J.D., Mullan, C.L., Youngs, T.G.A., Bowron, D.T., *J. Chem. Phys.*, 2010, **133**, 074510.

[45] Brigouleix, C., Anouti, M., Jacquemin, J., Caillon-Caravanier, M., Galiano, H., Lemordant, D., *J. Phys. Chem. B*, 2010, **114**, 1757–1766.

[46] Yu, W., Peng, H., Zhang, H., Zhou, X., *Chin. J. Chem.*, 2009, **27**, 1471–1475.

[47] Lava, K., Binnemans, K., Cardinaels, T., *J. Phys. Chem. B*, 2009, **113**, 9506–9511.

[48] Kim, K.S., Choi, S., Demberelnyamba, D., Lee, H., Oh, J., Lee, B.B., Mun, S.J., *Chem. Commun.*, 2004, 828–829.

[49] Khara, D.C., Samanta, A., *J. Phys. Chem. B*, 2012, **116**, 13430–13438.

[50] Khara, D.C., Kumar, J.P., Mondal, N., Samanta, A., *J. Phys. Chem. B*, 2013, **117**, 5156–5164.

[51] Wakai, C., Oleinikova, A., Ott, M., Weingartner, H.J., *J. Phys. Chem. B*, 2005, **109**, 17028–17030.

[52] Daguenet, C., Dyson, P.J., Krossing, I., Oleinikova, A., Slattery, J., Wakai, C., Weingartner, H.J., *J. Phys. Chem. B*, 2006, **110**, 12682–12688.

[53] Reichardt, C., *Solvents and Solvent Effects in Organic Chemistry*, Wiley-VCH Verlag GmbH, Weinheim, 1988.

[54] Paul, A., Samanta, A., *J. Phys. Chem. B*, 2007, **111**, 1957–1962.

[55] Aki, S.N.V.K., Brennecke, J.F., Samanta, A., *Chem. Commun.*, 2001, 413–414.

[56] Reichardt, C., *Green. Chem.*, 2005, **7**, 339–351.

[57] Seddon, K.R., Stark, A., Torres, M., *Pure Appl. Chem.*, 2000, **72**, 2275–2287.

[58] Widegren, J.A., Laesecke, A., Magee, J.W., *Chem. Commun.*, 2005, 1610–1612.

[59] Xu, W., Wang, L.-M., Nieman, R.A., Angell, C.A., *J. Phys. Chem. B*, 2003, **107**, 11749–11756.

[60] Dzyuba, S., Bartsch, R.A., *ChemPhysChem*, 2002, **3**, 161–166.

[61] Carda-Broch, S., Berthold, A., Armstrong, D.W., *Anal. Bioanal. Chem.*, 2003, **375**, 191–199.

[62] Patra, S., Samanta, A., *J. Phys. Chem. B*, 2012, **116**, 12275–12283.

[63] Santhosh, K., Banerjee, S., Rangaraj, N., Samanta, A., *J. Phys. Chem. B*, 2010, **114**, 1967–1974.

[64] Birks, J.B., *Photophysics of Aromatic Molecules*, Wiley-Interscience, London, 1970.

[65] Demchenko, A.P., *Ukr. Biochim. Zh.*, 1981, **53**, 22–27.

[66] Al-Hassan, K.A., El-Bayoumi, M.A., *J. Polym. Sci. B*, 1987, **25**, 495–500.

[67] Rawat, S., Mukherjee, S., Chattopadhyay, A., *J. Phys. Chem. B*, 1997, **101**, 1922–1929.

[68] Lakowicz, J.R., Nakamoto, S.K., *Biochemistry*, 1984, **23**, 3013–3021.

[69] Demchenko, A.P., *Luminescence*, 2002, **17**, 19–42.

[70] Demchenko, A.P., *Topics in Fluorescence Spectroscopy*, Plenum Press, New York, 1991.

[71] Mandal, P.K., Paul, A., Samanta, A., *J. Photochem. Photobiol. A Chem.*, 2006, **182**, 113–120.

[72] Lakowicz, J.R., *Principles of Fluorescence Spectroscopy*, Kluwer Academic/Plenum Publishers, New York, 1999.

[73] Samanta, A., *J. Phys. Chem. Lett.*, 2010, **1**, 1557–1562.

[74] Fee, R.S., Maroncelli, M., *Chem. Phys.*, 1994, **183**, 235–247.

[75] Jin, H., Baker, G.A., Arzhantsev, S., Dong, J., Maroncelli, M., *J. Phys. Chem. B*, 2007, **111**, 7291–7302.

[76] Fleming, G.R., *Chemical Applications of Ultrafast Spectroscopy*, Oxford University Press, Oxford, 1986.

[77] Ingram, J.A., Moog, R.S., Ito, N., Biswas, R., Maroncelli, M., *J. Phys. Chem. B*, 2003, **107**, 5926–5932.

[78] Sension, R.J., Hochstrasser, R.M., *J. Chem. Phys.*, 1993, **98**, 2490.

[79] Small, E.W., Isenberg, I., *Biopolymers*, 1977, **16**, 1907–1928.

[80] Hartman, R.S., Alavi, D.S., Waldeck, D.H., *J. Phys. Chem.*, 1991, **95**, 7872–7880.

[81] Khara, D.C., Samanta, A., *Phys. Chem. Chem. Phys.*, 2010, **12**, 7671–7677.

[82] Das, S.K., Sarkar, M., *J. Phys. Chem. B*, 2012, **116**, 194–202.

[83] Paul, A., Samanta, A., *J. Phys. Chem. B*, 2007, **111**, 4724–4731.

[84] Horng, M.L., Gardecki, J.A., Maroncelli, M., *J. Phys. Chem. A*, 1997, **101**, 1030–1047.

[85] Fruchey, K., Fayer, M.D., *J. Phys. Chem. B*, 2010, **114**, 2840–2845.

[86] Dutt, G.B., *J. Phys. Chem. B*, 2010, **114**, 8971–8977.

[87] Gangamallaiah, V., Dutt, G.B., *J. Phys. Chem. B*, 2012, **116**, 12819–12825.

CHAPTER 8

Self-Assembly of Surface-Active Ionic Liquids in Aqueous Medium

K. SRINIVASA RAO, PANKAJ BHARMORIA, and TUSHAR J. TRIVEDI

AcSIR-Central Salt and Marine Chemicals Research Institute, Bhavnagar, India

ARVIND KUMAR

AcSIR-Central Salt and Marine Chemicals Research Institute, Bhavnagar, India
CSIR-Central Salt and Marine Chemicals Research Institute, Bhavnagar, India

8.1 INTRODUCTION

Ionic liquids (ILs) are compounds comprising entirely of ions that melt below 100°C [1]. The physicochemical properties of ILs depend on the cation, anion, and alkyl chain substitutions [2, 3] and can be tuned by changing the nature of constituent components; these properties brand ILs as designer solvents [4]. ILs due to their unique physicochemical properties offer many advantages and are widely used for various applications [5–8]. Some early investigations made on molecular level revealed the inherent amphiphilicity of 1-methyl-3-alkylimidazolium salts with smaller counterions such as Cl, Br, and BF_4 showing the behavior similar to short-chain cationic surfactants in aqueous medium [9]. The surfactant properties of ILs, *N*-methyl-*N*-alkyl pyrrolidines, have also confirmed the amphiphilicity in ILs [10]. Based on these earlier findings, different classes of SAILs such as cationic surfactant-like ILs (CSAILs), anionic surfactant-like ILs (ASAILs), catanionic surfactant-like ILs (CASAILs) have been developed and investigated for their self-assembling behavior.

Many changes or modifications have been further made on the molecular structure of CSAILs by changing the head group [11–14], the substituent hydrophobicity [15–17], and the counterion [12], whose aggregation properties in aqueous medium have shown dependence on all the three parameters [18]. The CSAILs have shown improved surface-active properties compared to the conventional surfactants of similar alkyl chain length. Substituents have specific effects on the aggregation properties of imidazolium-based CSAILs, for example, hydroxyl substituent [15] or larger hydrophobic substituents [16] have improved the surface activity, and carbazole substituted on the tail have additionally induced the fluorescent property into IL [17]. Dicationic surface-active ILs

Ionic Liquid-Based Surfactant Science: Formulation, Characterization, and Applications, First Edition. Edited by Bidyut K. Paul and Satya P. Moulik.

(DCSAILs) and gemini cationic surface-active ILs (GCSAILs) are the further developments made on CSAILs. The molecular structures of DCSAILs and GCSAILs have many similarities but have shown distinct behavior in aqueous medium. The DCSAILs are SAILs with long-chain spacer and short-chain substituents, whereas GCSAILs are SAILs with short-chain spacer with long-chain substituents. The DCSAILs and GCSAILs are found to be higher surface active than CSAILs and structurally similar to gemini surfactants with quaternary ammonium head groups [19–22]. The ILs with medium to longer alkyl chain anions behave similar to anionic surfactants [23]. The structural difference between conventional anionic surfactants and ASAILs is the modified larger counterion that brings the melting point down to less than 100°C [23–28]. The counterions explored till date are imidazolium, quaternary ammonium, pyridinium, and various amino acid-derived cations in combination with single-, double-, or triple-chain alkyl sulfonates or alkylsulfates as anions. The advantages of these compounds over the existing SAILs are facile synthesis and tunable nature. They can also be readily generated from commercially available, cheap starting materials, with lower environmental hazards (halogen free) and toxicity (short imidazolium alkyl chain) than normal imidazolium-based SAILs. The green credentials of these ASAILs have been further improved tremendously by the development of bio-based AAILSs [25]. Protic ASAILs have also been developed by taking ethanolammonium as cation and oleate as anion [29]. The ASAILs have shown improved surface-active properties in comparison with their conventional anionic surfactant analogues. ASAILs have shown several promising advantageous properties such as improved solubility in hydrophobic ILs and promising emulsifying properties for molecular and IL solvents [30, 31] in nanomaterial synthesis and in the mitigation of algal blooms [25].

Mixtures of surfactants offer several advantageous properties in terms of improved surface activity [32], vesicle formation [33, 34], and rich phase behavior [35], where synergistic interactions play a crucial role. Different IL-based mixed surfactant systems have been developed by taking SAIL as one or both the components [36–41]. The ILs have shown synergistic interaction with the mixed surfactant; because of which, these are found to be more surface active than the individual surfactant and have formed vesicles in aqueous medium at some particular mole fractions similar to the conventional mixed surfactant systems [36, 39–41]. IL-based mixed surfactant systems have shown rich phase behavior in aqueous medium at varying mole fractions with some new phases, which are uncommon with conventional surfactant systems in aqueous medium [37, 38]. The vesicular structures formed by the mixed system [C_{12}mim]Br–SDS have been used as soft templates in the synthesis of hollow silica nanospheres [40]. The aqueous solutions of mixed IL–surfactants at three different mole fractions have been used as a medium for the photochemical synthesis of gold nanoparticles [41]. Another important class of surfactants, that is, CASAILs, has also been synthesized by taking medium or long alkyl chain attached cation and anion [42–44]. These CASAILs are found to be more surface active than the other class of SAILs and versatile. These CASAILs self-assembled into vesicles in aqueous medium [42, 44] and reverse micelles in benzene [42]. Aqueous solutions of imidazolium-based CASAILs have shown promising ability for the synthesis of hexagonal gold nanoplates through photochemical reduction of $HAuCl_4$ [44].

8.2 RESULTS AND DISCUSSION

The aggregation behavior of the different types of SAILs has been discussed in the following sections.

8.2.1 Aggregation Behavior of Cationic Surfactant-Like ILs

ILs with medium- or long-chain substituent on the cation (≥8) with smaller counterions behave similar to the cationic surfactants and form micelles in aqueous medium. The major structural difference between the conventional cationic surfactants and CSAILs is the head group that brings the melting point of corresponding surfactant below 100°C. The majority of cationic surfactants have quaternary ammonium head groups, whereas CSAILs have head groups such as imidazolium, pyridinium, piperidinium, and pyrrolidinium [11–14]. Molecular structures of various types of CSAILs are shown in Figure 8.1. The type of head group, substituent chain length, and counterion significantly affect the aggregation behavior of CSAILs. The CSAILs with all different head groups have shown improved surface-active properties compared to the conventional surfactants of similar alkyl chain length [45]. The surface activity of CSAILs was further increased with the increase of the substituent hydrophobicity, that is, larger chain length or bigger aromatic substituents on the imidazolium group have further decreased the critical micelle concentration (CMC) of the corresponding IL in aqueous medium [16]. The inclusion of carbazole moiety into the alkyl chain of IL induced the fluorescent activity in addition to the significant improvement in the surface activity [17]. The counterion with lower hydration ability binds more number with the head group at the interface, which increases the surface activity of the corresponding IL by effectively decreasing the same charge repulsion between the head groups [12]. The second type of CSAILs, that is, DCSAILs of varying side chain and linkage chain length, was developed and investigated for their aggregation behavior in aqueous medium. DCSAILs have shown lower CMC values than CSAILs of similar alkyl chain length, and among the DCSAILs, those with long linkage chains have shown much lower CMC values than those with long-chain substituents with small spacer [19]. Another class of CSAILs, that is, GCSAILs, is developed by attaching two CSAILs with a small spacer such as butyl chain, thioether, etc. The surface activity of GCSAILs has been found to be increased with the increase in either substituent or spacer chain length [20–22]. The thioether spacer containing imidazolium-based GCSAILs has been found to be more surface active than their analogues with the quaternary ammonium head groups of similar alkyl length [22]. The surface activity of CSAILs was estimated with the help of surface parameters derived from the surface tension data obtained by measuring the surface tension of water at increasing concentrations of CSAIL. The parameters adsorption efficiency pC_{20} and effectiveness of surface tension reduction Π_{CMC} were obtained by using Equations 8.1 and 8.2:

$$pC_{20} = -\log C_{20} \tag{8.1}$$

where C_{20} is the molar concentration of surfactant needed to reduce the surface tension of water by 20 units and pC_{20} is a measure of surface efficiency of a surfactant

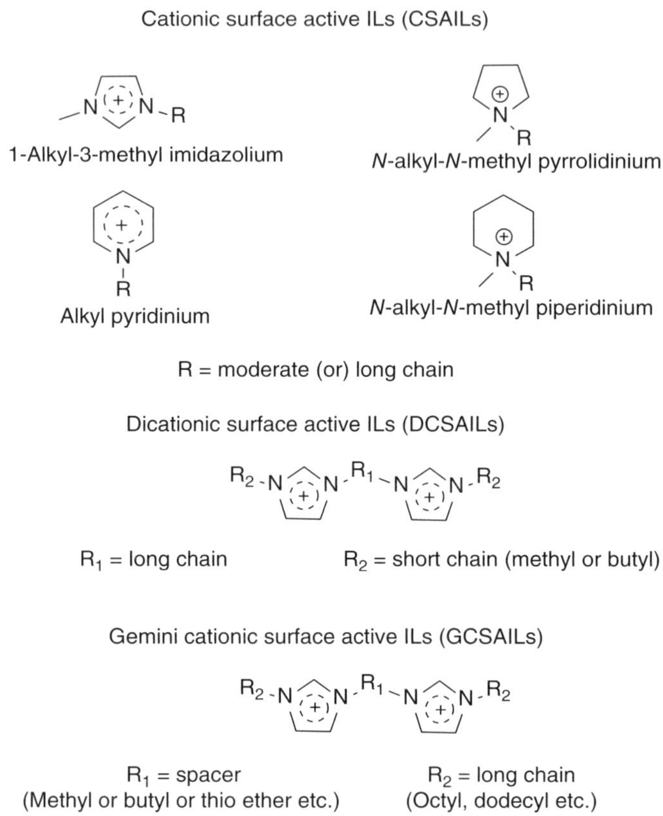

Figure 8.1 Molecular structures of different types of CSAILs.

or SAIL; the higher pC_{20} values indicate the higher adsorption efficiency of the corresponding surfactant or SAIL:

$$\Pi_{CMC} = \gamma_0 - \gamma_{CMC} \tag{8.2}$$

where γ_0 is the surface tension of pure water and γ_{CMC} is the surface tension at CMC.

The lower γ_{CMC} values or higher Π_{CMC} values indicate the efficiency of the respective surfactant in reducing the surface tension. The pC_{20} and Π_{CMC} values obtained for CSAILs are found to be higher than the values obtained for cationic surfactant of similar alkyl chain length indicating the higher adsorption efficiency of CSAILs than conventional cationic surfactants of same alkyl chain length. The other two surface parameters, surface excess concentration (Γ_{max}) and minimum area per molecule at air–water interface (A_{min}), are derived using Equations 8.3 and 8.4 by applying Gibbs adsorption isotherm to the surface tension data:

$$\Gamma_{max} = -\frac{1}{nRT}\frac{\partial \gamma}{\partial \ln C} \tag{8.3}$$

$$A_{min} = \frac{1}{\Gamma_{max} N_A} \tag{8.4}$$

TABLE 8.1 Critical Micelle Concentration (CMC), Surface Tension at CMC (γ_{CMC}), Effectiveness of Surface Tension Reduction (Π_{CMC}), Adsorption Efficiency (pC_{20}), Surface Excess Concentration (Γ_{max}), and Minimum Area per Molecule at Air–Water Interface (A_{min}) for Various CSAILs in Aqueous Medium at 298.15 K

	CMC (mM)	γ_{CMC} (mN.m^{-1})	Π_{CMC} (mN.m^{-1})	pC_{20}	Γ_{max} (μmol.m^{-2})	A_{min} (Å^2)
Cationic Surface-Active Ionic Liquid (CSAIL)						
[C_4mim]BF_4 [9]	800[a]	—	—	—	—	63
[C_4mim]Cl [18]	935[b]	—	—	—	—	—
[C_4mpy]Cl [18]	609[b]	—	—	—	—	—
[C_6mim][Cl] [46, 47]	902 [47] (390 [48])	39.2	32.6	0.9	1.94 [48], 1.3 [47]	130 [47]
[C_6mim][Br] [47]	470[a]	33.6	38.2	1.1	2.0	83
[C_6mim][I] [47]	250[a]	34	38.3	1.8	1.3	124
[CarbazoleC$_6$mim]Br [17]	5.07[a]	60.1	—	—	0.79	211
[C_8mim]Cl [9, 47, 49]	100,[a] (102[a] [47])	28.3	43.3	1.8	1.6	104 [47]
[C_8mim]I [9, 47]	100 (95 [47])	28.2	44.4	2.0	1.4	117
[C_8mim]Br [47]	120 [47]	28.7 [47]	44.9 [47]	1.8 [47]	2.7 [47]	60 [47]
[C_8mim]BF_4 [50]	28[c]	—	—	—	—	—
[C_8mim][CH_3COO] [12]	220[c]	—	—	—	—	—
[C_8mim][NO_3] [12]	150[c]	—	—	—	—	—
[C_8mim][CF_3COO] [12]	120[c]	—	—	—	—	—
4m-[C_8Pyr]Br [12]	160[c]	—	—	—	—	—
[C_8py]Br [14]	180	41.9	—	—	—	66
[C_8mpyrr]Br [12]	200[c]	—	—	—	—	—
[C_4C_8Pyrr][Br] [13]	150[d]	—	—	—	—	—
[C_9mim]Br [51]	74	—	—	—	—	—
[C_{10}mim]Cl [52, 53]	39.9 [52], 54[a] [53]	—	—	1.65 [53]	—	85 [52]
[C_{10}mpy]Cl [11]	45[d] [11]	—	27.9	1.62 [53]	—	121
[C_{10}mim]Br [20]	29.3[a]	39.7	33.3	2.15	1.72	96.7
[C_{10}Pyr]Br [14]	30[a]	40.7	—	—	—	61
[C_{10}pim]Br [17]	9.30[a]	38.3	—	—	1.90	87.4
[carbazoleC$_{10}$mim]Br [17]	0.71	56.2	—	—	0.87	191
[C_{12}mim]Cl [52]	13.17[a]	—	—	—	—	72
[C_2OHC$_{12}$im]Cl [15]	2.31	38.6	33.4	2.36	1.82	91.2

(Continued)

TABLE 8.1 (*Continued*)

	CMC (mM)	γ_{CMC} (mN.m^{-1})	Π_{CMC} (mN.m^{-1})	pC_{20}	Γ_{max} (μmol.m^{-2})	A_{min} (Å^2)
[C_{12}mim]Br [54]	10.9	39.4	33.6	2.67	1.91	86.8
[C_{12}pim]Br [16]	2.34[a]	38.4	—	—	2.09	79.4
[CarbazoleC_{12}mim]Br [17]	0.19	56.3	—	—	0.86	193
[C_{12}mim]BF_4 [45]	9.2	38.2	34.8	2.74	2.16	76.7
[C_{12}mpy]Cl [11]	13[d]	—	27.9	2.22	—	108
[C_{12}mpy]Br [11]	10[d]	—	—	—	—	—
[C_{12}mpip]Br [11]	11[d]	—	—	—	—	—
[C_1C_{12}mpyrr]Br [13]	15[d]	—	—	—	—	—
[C_4C_{12}pyrr]Br [13]	6[d]	—	—	—	—	—
[C_{12}Pyr]Br [14]	9.3	39.3	—	—	—	71
[C_{12}MP]Br [46]	13.5	42.4	30.3	—	3.03	54.8
[C_{14}mim]Cl [52]	2.98[a] [52], 3.2[a] [53]			2.61 [53]		56 [52]
[C_{14}mim]Br [20, 54]	2.8[a] [54]	39.2 [54]	33.8 [54]	3.33 [54]	1.96 [54]	84.7 [54]
[C_{14}pim]Br [17]	0.61	38.3	—	—	2.21	75.1
[C_{14}mpy]Cl [11]	3.1[d]		28.0	2.82		92
[C_{14}MP]Br [46]	3.3	42.7	30.0	—	3.55	46.8
[C_{16}mim]Cl [52]	0.87[a] [52], 1.1 [53]	—	—	3.18 [53]	—	49 [52]
[C_{16}mpy]Cl [11]	0.8	—	27.8	3.87	—	76
[C_{14}pyr]Br [14]	2.2	38.0	—	—	—	86
[C_{16}mim]Br [54]	0.55[a]	39.1	33.9	3.78	2.03	81.6
[C_{16}hpim]Br [55]	0.56[a]	43.4	28.6	3.55	1.99	83.2
[C_{16}MP]Br [46]	0.86	41.2	31.5		3.67	45.2
[C_{18}mim]Cl [53]	0.45[a]	—	—	3.56[a]	—	—
[C_{18}mpy]Cl [11]	0.3[d]	—	27.8	3.87	—	76
[C_6bim]Br [19]	271	42	—	—	—	—
[C_8bim]Br [19]	41	40	—	—	—	—
[C_{10}bim]Br [19]	6.3	36	—	—	—	—
[C_{12}bim]Br [19]	2.4	38	—	—	—	—
[C_{16}bim]Br [19]	0.1	45	—	—	—	—
[$(O)_2$im]Br [19]	5.6	32	—	—	—	—
[$(DD)_2$im]Br [19]	0.1	28	—	—	—	—

Gemini Cationic Surface-Active ILs (GCSAILs)						
$C_8(bim)_2$-2Br [19]	323	47	—	—	—	—
$C_{10}(bim)_2$-2Br [19]	117	38	—	—	—	—
$C_{12}(bim)_2$-2Br [19]	72	46	—	—	—	—
$C_{10}(mim)_2$-2Br [19]	149	38	—	—	—	—
$C_4(dim)_2$-2Br [19]	2.0	35	—	—	—	—
$C_{12}(dim)_2$-2Br [19]	0.6	48	—	—	—	—
$[C_{10}$-4-$C_{10}im]Br_2$ [20]	4.5	35.2	—	3.14	1.25	133
$[C_{12}$-4-$C_{12}im]Br_2$ [20]	0.72	35.7	—	3.94	1.19	140
$[C_{14}$-4-$C_{14}im]Br_2$ [20]	0.10	37.2	—	5.04	0.88	188
$[C_{12}$–2–$C_{12}im]Br_2$ [21]	0.55	33.6	38.4	4.54	1.23	135
$[C_{12}$–4–$C_{12}im]Br_2$ [21]	0.72	35.7	36.3	3.94	1.19	140
$[C_{12}$–6–$C_{12}im]Br_2$ [21]	0.78	39.5	32.5	3.73	1.16	143
[12-(S-2-S)-12]im [22]	0.32	39.7	—	—	2.60	63
[12-(S-3-S)-12]im [22]	0.26	40.7	—	—	2.13	77
[12-(S-4-S)-12]im [22]	0.22	40.8	—	—	2.06	80
[14-(S-2-S)-14]im [22]	0.07	42.9	—	—	2.12	78
[14-(S-3-S)-14]im [22]	0.063	45.8	—	—	3.09	53
[14-(S-4-S)-14]im [22]	0.058	46.6	—	—	3.1	53
[16-(S-2-S)-16]im [22]	0.022[a]	—	—	—	—	—
[16-(S-3-S)-16]im [22]	0.021[a]	—	—	—	—	—
[16-(S-4-S)-16]im [22]	0.020[a]	—	—	—	—	—

[a] Surface tension.
[b] NMR.
[c] Conductivity.
[d] IFT.

The CMC and surface parameters derived for various CSAILs in aqueous medium are given in Table 8.1.

Similar to the conventional surfactants, CSAILs form micelles/vesicles in aqueous medium after saturation of the air–water interface (i.e., above CMC/*CVC*, critical vesicle concentration). Long-chain imidazolium-based CSAILs form micelles in aqueous medium. It was predicted from NMR experiments that the orientation of the imidazolium ring in the aggregated structure depends on the interaction of the ring protons with the medium, which in fact is controlled by the relative interaction of the anion (counterion) with the water (medium) and imidazolium ring. The head groups of CSAILs have shown specific behavior in packing or arrangement of the aggregated structures. In particular, long-chain imidazolium CSAILs ([$C_n$mim]Br; $n = 10, 12, 14$) have shown spherical micelle-to-rodlike micelle-to-vesicle transitions with increase in concentration in aqueous medium [56]. It has been speculated by fluorescence studies that with increase in the IL concentration, counterion binding increased, which decreased the head group repulsion between the surfactants, consequently allowing the micelles to grow further to rodlike micelles and vesicles [56]. Further, CSAILs with pyridinium head group have shown unique self-quenching behavior for selective fluorescent probes [11]. The long-chain imidazolium and pyridinium ILs have shown excellent antimicrobial properties that increased with the increase in the chain length [14, 57]. The thioether spacer containing GCSAILs interacts strongly with the DNA, forming DNA–surfactant complex, thus protecting it against enzymatic degradation by DNase I [22]. The imidazolium CSAILs can be used for the stabilization of nanoparticles by suitable functionalization, for example, PEG functionalized CSILs have been used for stabilizing Pd nanoparticles [57]. Overall, CSAILs are versatile surfactants with the improved surface-active properties; further, the head groups with the conjugated electrons are found promising for many applications; thus, many other such head groups can be explored for SAIL synthesis by targeting specific applications.

8.2.2 Self-Assembly of Anionic Surfactant-Like ILs (ASAILs) in Aqueous Medium

The anionic surfactants with the melting point less than 100°C are considered as ASAILs [23–28]. Melting point of these ILs depends upon the packing of the corresponding cation and anion. In the case of quaternary ammonium counterions of different alkyl chain length substituents such as tetramethylammonium (TMA), tetraethylammonium (TEA), tetrapropylammonium (TPA), and tetrabutylammonium (TBA), the combinations that form ASAILs are single-chain anion dodecylsulfate (DS) with the larger counterion TBA, double-chain dioctylsulfosuccinate anion (AOT) with TBA, TPA counterions, and trichain (TC) anion with all the four counterions TMA, TEA, TPA, and TBA [27]. In combination with the imidazolium counterion 1-butyl-3-methylimidazolium ([C_4mim]), all three single-chain LS, double-chain AOT, and triple-chain TC anions have been found to form ILs [28]. The melting point of the ASAIL [C_4mim]DS is higher than room temperature (although <100°C), which is further decreased to less than room temperature by amino acid-derived counterions [25, 26]. Thus, asymmetry of both the cation and the anion plays a crucial role in decreasing the melting point of the corresponding ASAIL. The molecular structures of various cations and anions used for the synthesis of ASAILs are given in Figure 8.2. All the ASAILs synthesized had decreased CMC values compared to

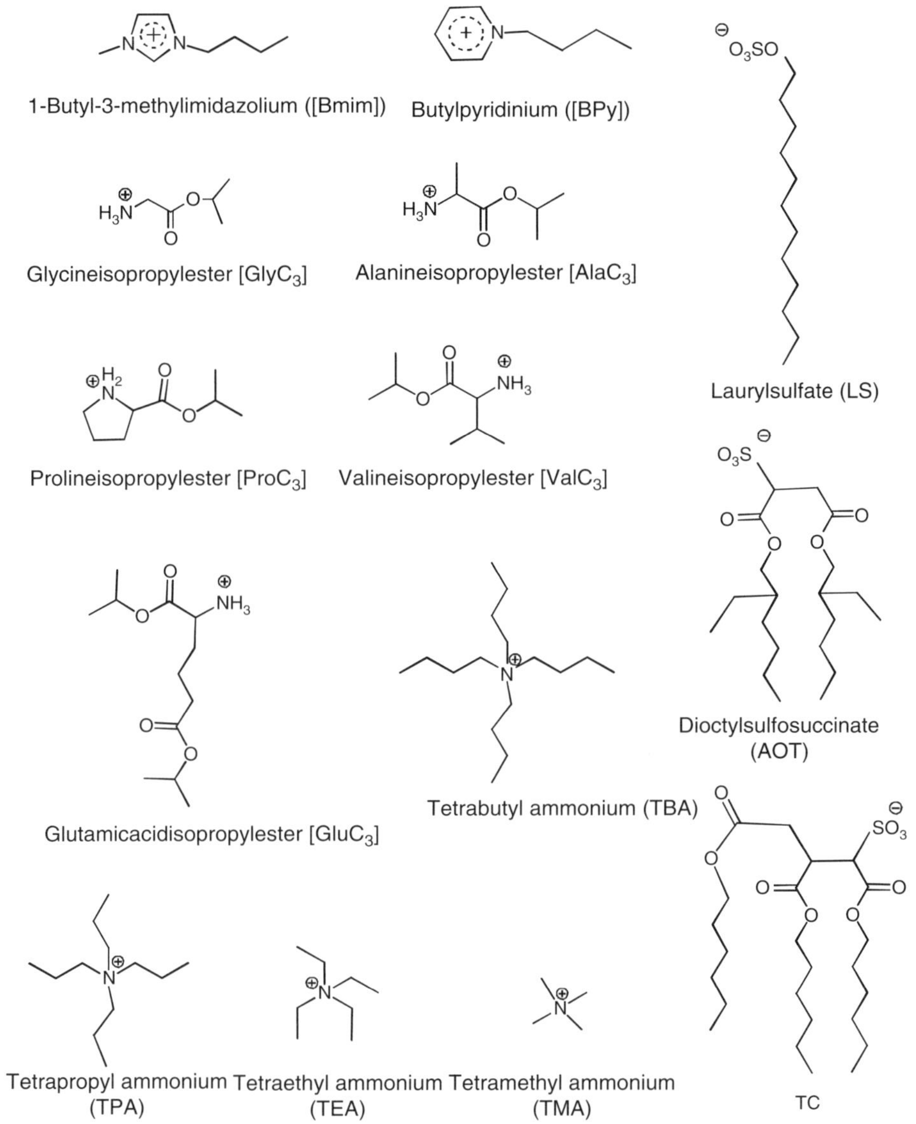

Figure 8.2 Molecular structures of various cations and anions used for the synthesis of ASAILs.

their analogue surfactants of the same alkyl chain length. This decreased CMC values are attributed to the higher counterion binding (β) of the ASAILs due to lower solvation of the bigger counterion thus effectively screening the same charge repulsion between the head groups of the anion. The surface tension plot at different concentrations of anionic surfactants, sodium dodecylsulfate (SDS), and ASAILs ([C_4mim]DS, [ProC_3]DS) in aqueous medium is shown in Figure 8.3. The plot shows the lower CMC values of ASAILs over its conventional analogue SDS. The CMC and various surface parameters derived from surface tension experiments for

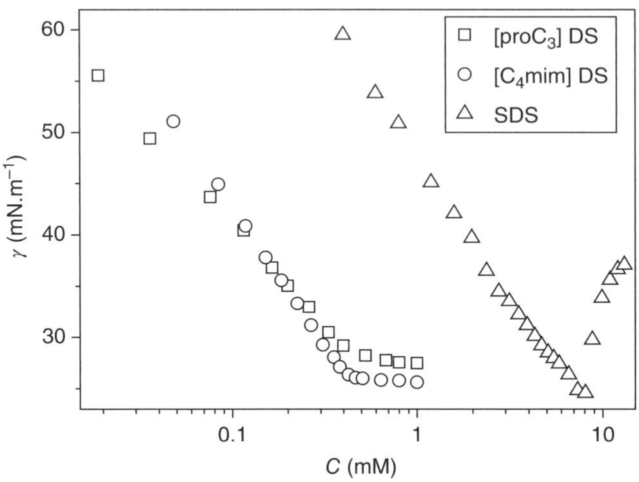

Figure 8.3 Decrease in surface tension of water with increase in concentration of anionic surfactant (SDS) and the ASAILs $[C_4mim]DS$ or $[ProC_3]DS$ in aqueous medium at 298.15 K.

ASAILs are given in Table 8.2. The ASAILs present lower limiting surface tension (γ_{CMC}) values than their conventional analogues. The large minimum of surface tension at CMC of SDS than $[C_4mim]DS$ and $[ProC_3]DS$ as shown in Figure 8.3 is due to less close packing at the interface compared to ASAILs, indicating the improved surface efficiency of ASAILs. The increase in surface tension in the case of SDS after reaching CMC is may be due to conformational changes in the arrangement of adsorbed surfactant at interface. Further, the effective screening by larger counterions increased the adsorption efficiency of corresponding ASAILs as estimated by the parameter pC_{20} derived from surface tension data, indicating dense packing of ASAILs at air–water interface. The surface excess concentration (Γ_{max}) and minimum area per molecule (A_{min}) calculated for various ASAILs and analogous surfactants by applying Equations 8.3 and 8.4 are given in Table 8.2. Similar to the conventional surfactants, ASAILs form aggregated structures in the bulk solution after saturation of the air–water interface. The type of aggregates formed by these ASAILs depends on chain length of the anion. The medium-chain ASAIL $[C_4mim]OS$ form micelles of 2.8 nm just above CMC (31 mM) [23, 40], which has shown a unique rearrangement in the micellar aggregates at higher concentrations (above 360 mM) due to the penetration of counterion ($[C_4mim]$) into the preformed micelles [58]. Long-chain $[C_4mim]DS$ exists as cluster of aggregated micelles of size (R_h, 261 nm) where the number of micelles are connected by the counterion bridging between them. As evidenced from SANS, the ASAIL $[C_4mim]DS$ at concentration 0.5 wt.% forms multilayer stacks where the bilayer thickness is approximately twice the length of the anion DS, unlike the homologue surfactants with quaternary ammonium (TAA ions) and Na^+ counterions, which exist as ellipsoidal micelles up to a large concentration range of 4 wt.% without much change in the aspect ratio [23]. Further increase in concentration of $[C_4mim]DS$ increased the bilayer rigidity and number of bilayers. The double-chain ASAIL $[C_4mim]AOT$ exhibits a mixed structure ($L_1 + L_\alpha$; L_1 (micellar phase), L_α (birefringent phase)) at 0.5 wt.% in water,

TABLE 8.2 Critical Micelle Concentration (CMC), Surface Tensions at CMC (γ_{CMC}), Effectiveness of Surface Tension Reduction (Π_{CMC}), Adsorption Efficiency (pC_{20}), Surface Excess Concentration (Γ_{max}), and Minimum Area per Molecule at Air–Water Interface (A_{min}) for Various ASAILs in Aqueous Medium at 298.15 K

Anionic Surface-Active ILs (ASAILs)	*CMC* (mM)	γ_{CMC} (mN.m^{-1})	Π_{CMC} (mN.m^{-1})	pC_{20}	Γ_{max} (μmol.m^{-2})	A_{min} (Å^2)
[C_4mim]OS [40]	35	25.2	46.8	2.52	2.17	77
Na^+ DS [27]	8.0	34.9	—	—	—	56
NH_4^+ DS [27]	5.52	30.6	—	—	—	48
TMA DS [27]	4.21	33.2	—	—	—	65
TEA DS [27]	2.76	33.7	—	—	—	78
TPA DS [27]	1.42	32.7	—	—	—	67
[C_4mim]DS [24]	1.8	31.9	40.3	3.4	2.53	66
[C_4MP]DS [24]	2.7	34.3	37.9	3.5	2.27	74
[GlyC_3]DS [26]	0.74	26.0	46.0	4.4	1.14	146
[ProC_3]DS [26]	0.53	27.8	44.2	4.6	1.74	96
[AlaC_3]DS [26]	0.48	25.7	46.3	4.8	1.24	134
[ValC_3]DS [26]	0.40	27.5	44.5	4.6	1.61	104
[GluC_3]DS [26]	0.35	30.6	41.4	4.7	1.75	95
Na^+ AOT [27]	2.88	30.6	—	—	—	70
NH_4^+ AOT [27]	3.01	28.3	—	—	—	109
TMA AOT [27]	2.48	29.4	—	—	—	96
TEA AOT [27]	2.07	28.7	—	—	—	101
TPA AOT [27]	1.27	26.1	—	—	—	96
[Bmim]AOT [28]	1.78	25.7	—	—	—	86
Na^+ TC [28]	0.11	24.9	—	—	—	136
NH_4^+ TC [27]	0.07	25.5	—	—	—	117
TMA TC [27]	0.12	24.7	—	—	—	129
TEA TC [27]	0.10	25.1	—	—	—	132
TPA TC [27]	0.07	25.4	—	—	—	137
[Bmim]TC [28]	0.76	24.8	—	—	—	108

similar to the analogues of the double-chain surfactants and ASAILs with quaternary ammonium counterions [22]. The SANS behavior of the trichain ASAIL [C_4mim]TC and its analogous ILs with TAA counterions is much similar to the corresponding AOT-based ASAILs, which have shown a mixed ($L_1 + L_\alpha$) phase at a concentration of 0.5 wt.% in water [23]. Further, it is observed that on increasing the anion size from straight chain to double and triple chain, the bilayer separation increased substantially. The ASAILs of single-chain anion DS, with amino acid-derived counterions (AAILSs), have shown good biodegradability and are found to be promising in the mitigation of algal blooms, which are toxic for the aquatic animals [20]. Also, these AAILSs are used for the preparation of shape-controlled CeO_2 nanomaterials, showing the versatility of these ASAILs [20]. The other parameter that is modified by converting conventional anionic surfactant into ASAIL is the solubility in a solvent. The ASAILs formed by double-chain AOT and triple-chain TC with the combination of TBA cation are found to be insoluble in water [27]; being liquid at room temperature, these ASAILs will be useful in the aqueous biphasic extraction processes. Most importantly, the ASAILs [C_4mim]DS and [ProC_3]DS have overcome the solubility limitation of their conventional analogue SDS in the hydrophobic IL [C_4mim]Tf_2N [24], which can be utilized in the preparation of IL-based microemulsions. Although the counterion has little role in the adsorption behavior at air–water interface, which is prominently driven by anion, it decreases the CMC significantly and incorporates additional advantages that can be tuned according to the requirement. Care should be taken in selecting the counterion, so that it should not drop the aqueous solubility of the IL; bigger counterions with the polar functional groups such as choline and amino acids are some of the good choices.

8.2.3 Mixed Surfactants with IL as One or Both the Components in Aqueous Medium

Closed bilayer structures such as vesicles have abundant applications in chemistry and biology. A convenient method of vesicle formation is by mixing two surfactants (oppositely charged in most cases). The mixed surfactant systems provide additional benefit in controlling the size of catanionic vesicles by varying the surfactant mixing ratio, chain length, and head group of the components. The CSAILs in combination with anionic surfactants or ASAILs have been investigated in aqueous medium. The mixed systems explored are CSAILs such as [C_4mim]PF_6, [C_6mim]BF_4, [C_6mim]Br, [C_{10}mim]Br, [C_6Py]Br, [C_6Py]BF_4, [C_{12}mim]Br, and [C_{16}mim]Br in combination with the conventional anionic surfactant SDS and ASAIL such as [C_4mim]OS in combination with the conventional cationic surfactants CTAB and CSAIL [C_8mim]Cl. The molecular structures of different types of SAIL-based mixed systems are given in Figure 8.4. Smirnova et al. investigated the self-assembling nature of SDS in combination with ILs of varying chain length ($n = 4, 6$ or 10) and counterion (Br, BF_4, or PF_6) in a wide mole fraction range in aqueous medium [36]. When these short-chain ILs are added to the aqueous solution of SDS, the CMC reduction of SDS is more efficient than inorganic salts due to the additional hydrophobic interaction provided by the ILs, which increased with increasing IL chain length. The mixture of the short-chain ILs [C_6Py]BF_4 or [C_6Py]Br with SDS leads to spontaneous vesicle

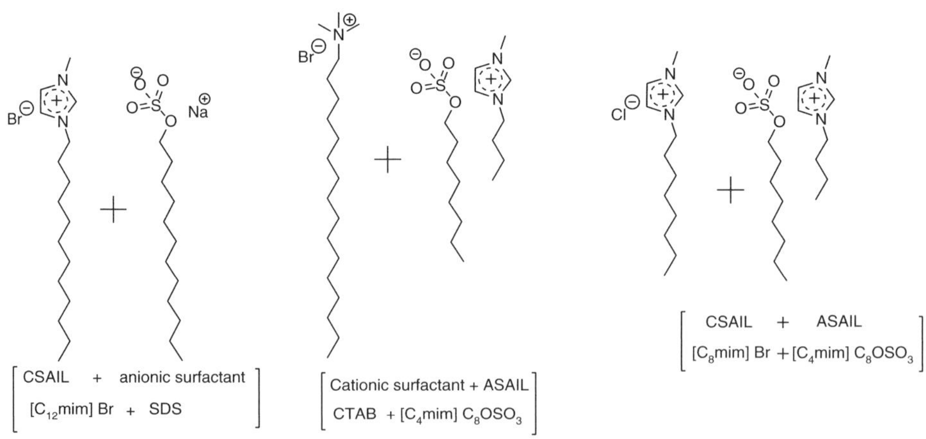

Figure 8.4 Molecular structures of IL-based mixed surfactant systems.

formation in aqueous medium [39]. The decrease in CMC of the mixture of IL, [C_{10}mim]Br with SDS, is exactly similar to the mixture of cationic–anionic surfactants in aqueous medium. For SDS–[C_{10}mim]Br solutions, the shape of the CMC versus X_{IL} curve ($CMC = CMC_{NaDS} + C_{IL}$) is typical of mixtures of cationic and anionic surfactants. Thus, it is believed that SDS in combination with ILs [$C_n$mim]Br with $n \geq 8$ behaves similar to cationic–anionic surfactant mixtures [36]. Different phases have been observed when SDS was mixed with the IL of same alkyl chain length with increase in the mole ratio (r) of SDS in aqueous medium at a total concentration of 100 mM. The sequence of the phase regions on the cationic-rich side with increasing SDS amounts is as follows: bluish micellar phase L_1 ($r = 0$–0.4); viscous and a slightly bluish L_1 phase (0.4–0.52); a two-phase region with an isotropic L_1 phase at the bottom and a birefringent L_α phase that has the appearance of creamy white floating precipitates at the top L_1/L_α ($r = 0.52$–0.58); a birefringent, turbid, and bluish L_α phase ($r = 0.58$–0.64); flow birefringence transparent and bluish L_α phase ($r = 0.64$–0.75); and a two-phase region with an L_1 phase at the top and thick precipitates at the bottom, L_1/thick precipitate ($r = 0.75$–1). The birefringent phases between $r = 0.58$–0.64 and between $r = 0.64$–0.75 contain vesicles proved by TEM [37]. Vesicle formation in the [C_{12}mim]Br/SDS aqueous solution is driven by both electrostatic and hydrophobic interactions. Differences in geometric packing and charge density between surfactant head groups induce different phase behavior for [C_{12}mim]Br/SDS than conventional DTAB/SDS aqueous solutions [37]. These catanionic vesicles formed by the mixtures can be used as soft templates for the synthesis of hollow silica spheres. When the IL of the even longer alkyl chain [C_{16}mim]Cl is mixed with SDS, all phases similar to the conventional catanionic systems are detected in addition to the two novel gel and lamellar (L_α) phases. The gel phase exists between the weight ratios of SDS to [C_{16}mim]Cl ranging from 3.4:6.6 to 4.5:5.5 with a total concentration not less than 2.7%, that is, the gel phase formed is of higher water content. The G, G″ values obtained from rheological measurements for gel phase are lower than the values reported earlier for the gel phase but higher than those for vesicle phases formed for conventional catanionic systems. This behavior is probably due to the presence of large amount of water in the gel phase, similar to vesicle phases [38]. When the parallel systems are investigated by replacing one of the sodium dodecylbenzenesulfonate (SDBS), sodium laurate (SL), sodium lauryl sulfonate (SLS), and CTAC as a replacement for one of the two components at a time, no gel formation is observed, which reveals the requirement of the presence of both the components. Experiments with variation in alkyl chain of both the components show that the minimum alkyl chain length required for gel formation is 14 for imidazolium cation and no gel formation in the case of sodium octylsulfate (SOS), reflecting the importance of hydrophobic interaction for the gel formation [38]. The existence of another new lamellar phase at SDS-rich region was detected by polarized optical microscope (POM). The total surfactant concentration needed to form this lamellar phase is lower than that required for CTAT–SDBS but is much similar to the system sodium laurate (SL)–dodecyltrimethylammonium bromide (DTAB) in water medium; the formed lamellar phases are minorly different from the known catanionic systems [28]. When the short anionic IL–surfactant [C_4mim]OS was mixed with the other medium-chain IL [C_8mim]Cl, micelle–vesicle–micelle transitions were observed at a mole fraction range of 0.05–0.1, 0.2–0.7, and 0.8–0.98, respectively, at a concentration 10 times CMC/*CVC* [30]. The advantages of the system over

TABLE 8.3 Self-Aggregate Formation of Mixed ILs and IL–Surfactant Systems in Aqueous Medium at Different Mole Fractions of the Component (1) at 298.15 K

[C_4mim]C_8OSO_3 (1) + [C_8mim]Cl (2) [40]						
X_1[a]	0–0.1	0.1–0.15	0.15–0.78			0.78–1
Aggregate	Micelle	Precipitate	Vesicle			Vesicle
[C_4mim]C_8OSO_3 (1) + Cetyltrimethylammonium Bromide (2) [41]						
X_1[a]	0–0.25	0.25–0.45	0.45–0.65	0.65–0.70	0.70–0.80	0.9–1
Aggregate	Micelle	Micelle + vesicle	Vesicle	Vesicle + precipitate	Elongated micelle	Micelle

[a] X_1, mole fraction of component (**1**).

conventional mixtures of IL are the minimum precipitating region and extended stable vesicle region (0.3–0.7) [40]. The same IL, [C_4mim]OS in combination with CTAB, shows similar transitions but at different mole fraction regions along with additional elongated micelles. The sequence of transitions observed with respect to the mole fraction of [C_4mim]OS is micelle, elongated micelle, vesicle + precipitate, vesicle, vesicle + micelle, and micelle at the mole fraction range of 0–0.1, 0.2–0.3, 0.3–0.35, 0.35–0.55, 0.55–0.75, and 0.75–1.0, respectively [41]. A comparison of the phase behavior of two systems is given in Table 8.3. The π-conjugation present in imidazolium head group of IL–surfactants is responsible for the previously discussed additional advantageous phase behavior of IL-based mixed surfactant systems unlike in conventional surfactant-based mixed systems. This π-conjugation present in the IL is very much useful in nanomaterial synthesis, with photochemical synthesis of gold nanoparticles in aqueous-mixed IL medium and silica hollow spheres prepared using IL–SDS vesicles as templates. These advantageous properties of IL-based mixed systems have prospects for different applications. For specific applications, newer systems can be tailored.

8.2.4 Catanionic Surfactant-Like ILs (CASAILs)

ILs with both the ions having surfactant properties, that is, CASAILs, have been developed by suitably selecting the constituent ions in a way that it decreases the symmetry of the resulting surfactant and bring its melting point down to less than 100°C. They are synthesized by mixing cationic and anionic precursors in 1:1 ratio by completely removing the counterions released from the precursor surfactants (in the form of inorganic salts). Several new CASAILs have been synthesized using different classes of cations and anions, such as imidazolium-based cation of moderate alkyl chain length ([C_8mim]) in combination with octylsulfate (OS) and dodecylsulfate (DS) anions, that is, [C_8mim]OS and [C_8mim]DS [44]; quaternary ammonium counterion of varying alkyl chain length in combination with double- or triple-chain anion [C_8TMA]AOT, [C_{10}TMA]AOT, and [C_4TMA]TC [43]; and benzenehexadecylammonium cation with AOT (AOT-BHC) [42]. The CASAIL AOT-BHC with melting point less than 100°C has formed large unilamellar vesicles (LUVs) in aqueous medium and has shown improved solubility in hydrocarbon solvents, overcoming the solubility limitation of its cationic precursor benzenehexadecylammonium chloride in hydrocarbon solvents. AOT-BHC has formed reverse micelles in hydrocarbon solvent benzene, in which aqueous confined domains have been prepared. The newly developed CASAIL, AOT-BHC, is found to be a versatile and efficient surfactant for both aqueous and hydrocarbon solvent media [42]. The imidazolium-based CASAILs [C_8mim]OS and [C_8mim]DS have formed unilamellar vesicles spontaneously in aqueous medium. Further, the colloidal solutions of CASAILs have shown promising ability in the synthesis of shape-controlled gold nanoparticles. CASAILs have also been synthesized using quaternary ammonium ions (C_nTMA; n=4, 6, 8, and 10) in combination with single-chain (DS), double-chain (AOT), or triple-chain (TC) surface-active anions. The melting point of the resultant CASAIL is guided by the chain length and branching of the constituent ions. No quaternary ammonium ion (among the aforementioned) can reduce the melting point of corresponding surfactant to below 100°C in combination with single-chain dodecylsulfate anion, whereas long-chain ammonium ions in combination with the double-chain anion (AOT) C_8TMA–AOT and C_{10}TMA–AOT can reduce the melting point to 18C and 1°C, respectively. The combination of all the quaternary ammonium cation (C_nTMA; n=4, 6, 8, and 10) with trichain surfactant anion is found to be liquid at room temperature; some of them are liquids even at very low temperature, which can have special interest for electrochemical applications [43]. The counterion has significant effect on the aggregated structures of CASAILs in aqueous medium. These ILs with long-chain counterions have formed single- and multilamellar sheets above 0.5 wt.% with the distance between the lamellar sheets increasing with the size of the counterion, whereas the smaller-chain counterions or sodium counterion formed ellipsoid-shaped aggregates whose aspect ratio increased with the size of the counterion. ILs prepared from the conventional solid surfactants significantly affect their solubility. The CASAILs with long-chain counterions decrease the solubility in water but increase it in the hydrocarbon solvent hexane. Overall, the CASAILs are very advantageous versatile molecules whose properties can be tuned by tuning cation, anion, and alkyl chain length for different applications in colloid and interface science.

8.3 CONCLUSIONS

Surfactants with melting point less than 100°C are considered as SAILs. Various types of SAILs such as cationic, anionic, and catanionic have been developed. The surface activity of SAILs is found to be higher than their analogous surfactants in aqueous medium. SAILs have also shown additional advantageous properties besides the improved surface activity based on its molecular structure. Some examples are, unlike conventional surfactants, CSAILs of imidazolium head group that have shown micelle-to-rodlike micelle-to-vesicle transitions with increase in concentration in the aqueous medium. CSAILs with pyridinium head groups have shown self-quenching behavior for selected fluorescent probes. Amino acid-derived ASAILs (AAILSs) have shown promising ability in the synthesis of CeO_2 nanoparticles and in the mitigation of harmful algal blooms. AAILSs have overcome the solubility limitation of their conventional analogue SDS in the hydrophobic IL [C_4mim]Tf_2N, thus making them suitable for construction of stable reverse aggregated structures in hydrophobic IL medium. CSAILs in combination with anionic surfactants have formed higher self-assembled structures such as vesicles. The CSAIL [C_{16}mim]Cl in combination with SDS has formed a gel phase with rich water content.

CASAILs have shown versatility in terms of solubility in a solvent medium. The CASAIL AOT-BHC has formed vesicles in aqueous medium and reverse micelles in benzene medium. Imidazolium-based CASAILs are found to be promising in the photochemical synthesis of gold nanoplates. The other most useful property of SAILs is their tunable nature; because of which, these are standing as versatile materials.

ACKNOWLEDGMENTS

The authors are thankful to Department of Science and Technology (DST), Government of India, for the financial support for this work (No. SR/S1/PC-04/2010). K. S. Rao is thankful to CSIR for the Senior Research Fellowship.

REFERENCES

[1] Hallett, J. P.; Welton, T. *Chem. Rev.* 2011, **111**, 3508–3576.

[2] Tokuda, H.; Hayamizu, K.; Ishii, K.; Susan, M. A. B. H.; Watanabe, M. *J. Phys. Chem. B* 2004, **108**, 16593–16600.

[3] Tokuda, H.; Hayamizu, K.; Ishii, K.; Susan, M. A. B. H.; Watanabe, M. *J. Phys. Chem. B* 2005, **109**, 6103–6110.

[4] Davis Jr., J. H. *Chem. Lett.* 2004, **33**, 1072–1077.

[5] MacFarlane, D. R.; Tachikawa, N.; Forsyth, M.; Pringle, J. M.; Howlett, P. C.; Elliott, G. D.; Davis, J. H.; Watanabe, M.; Simon, P.; Angell, C. A. *Energy Environ. Sci.* 2014, **7**, 232–250.

[6] Keskin, S.; Kayrak-Talay, D.; Akman, U.; Hortaçsu, Ö. *J. Supercrit. Fluids* 2007, **43**, 150–180.

[7] Olivier-Bourbigou, H.; Magna, L.; Morvan, D. *Appl. Catal. Gen.* 2010, **373**, 1–56.

[8] Rogers, R. D.; Seddon, K. R.; Volkov, S. *Green Industrial Applications of Ionic Liquids*, Nato Science Series; Kluwer Academic Publishers, Dordrecht, The Netherlands 2002.

[9] Bowers, J.; Butts, P.; Martin, J.; Vergara-Gutierrez, C.; Heenan, K. *Langmuir* 2004, **20**, 2191.

[10] Baker, G. A.; Pandey, S.; Pandey, S.; Baker, S. N. *Analyst* 2004, **129**, 890–892.

[11] Blesic, M.; Lopes, A.; Melo, E.; Petrovski, Z.; Plechkova, N. V.; Lopes, J. N. C.; Seddon, K. R.; Rebelo, L. P. N. *J. Phys. Chem. B* 2008, **112**, 8645–8650.

[12] Wang, H.; Wang, J.; Zhang, S.; Xuan, X. *J. Phys. Chem. B* 2008, **112**, 16682–16689.

[13] Tariq, M.; Podgoršek, A.; Ferguson, J. L.; Lopes, A.; Costa Gomes, M. F.; Pádua, A. A. H.; Rebelo, L.P.N.; Canongia Lopes, J. N. *J. Colloid Interface Sci.* 2011, **360**, 606–616.

[14] Cornellas, A.; Perez, L.; Comelles, F.; Ribosa, I.; Manresa, A.; Garcia, M. T. *J. Colloid Interface Sci.* 2011, **355**, 164–171.

[15] Liu, X.; Dong, L.; Fang, Y. *J. Surfactants Deterg.* 2011, **14**, 203–210.

[16] Shi, L.; Li, N.; Yan, H.; Gao, Y.; Zheng, L. *Langmuir* 2011, **27**, 1618–1625.

[17] Dong, B.; Gao, Y.; Su, Y.; Zheng, L.; Xu, J.; Inoue, T. *J. Phys. Chem. B* 2010, **114**, 340–348

[18] Singh, T.; Kumar, A. *J. Phys. Chem. B* 2007, **111**, 7843–7851.

[19] Baltazar, Q. Q.; Chandawalla, J.; Sawyer, K.; Anderson, J. L. *Colloids Surf. A* 2007, **302**, 150–156.

[20] Ao, M.; Xu, G.; Zhu, Y.; Bai, Y. *J. Colloid Interface Sci.* 2008, **326**, 490–495.

[21] Ao, M.; Huang, P.; Xu, G.; Yang, X.; Wang, Y. *Colloid Polym. Sci.* 2009, **287**, 395–402.

[22] Bhadani, A.; Singh, S. *Langmuir* 2011, **27**, 14033–14044.

[23] Miskolczy, Z.; Sebők-Nagy, K.; Biczók, L.; Göktürk, S. *Chem. Phys. Lett.* 2004, **400**, 296–300.

[24] Jiao, J.; Dong, B.; Zhang, H.; Zhao, Y.; Wang, X.; Wang, R.; Yu, L. *J. Phys. Chem. B* 2012, **116**, 958–965.

[25] Trivedi, T. J.; Rao, K. S.; Singh, T.; Mandal, S. K.; Sutradhar, N.; Panda, A. B.; Kumar, A. *ChemSusChem* 2011, **4**, 604–608.

[26] Rao, K. S.; Singh, T.; Trivedi, T. J.; Kumar, A. *J. Phys. Chem. B* 2011, **115**, 13847–13853.

[27] Brown, P.; Butts, C.; Dyer, R.; Eastoe, J.; Grillo, I.; Guittard, F.; Rogers, S.; Heenan, R. *Langmuir* 2011, **27**, 4563–4571.

[28] Brown, P.; Butts, C. P.; Eastoe, J.; Fermin, D.; Grillo, I.; Lee, H.; Parker, D.; Plana, D.; Richardson, R. M. *Langmuir* 2012, **28**, 2502–2509.

[29] Álvarez, V. H.; Mattedi, S.; Martin-Pastor, M.; Aznar, M.; Iglesias, M. *Fluid Phase Equilib.* 2010, **299**, 42–50.

[30] Rao, K. S.; So, S.; Kumar, A. *Chem. Commun.* 2013, **49**, 8111–8113.

[31] Rao, V. G.; Ghosh, S.; Ghatak, C.; Mandal, S.; Brahmachari, U.; Sarkar, N. *J. Phys. Chem. B* 2012, **116**, 2850–2855.

[32] Shiloach, A.; Blankschtein, D. *Langmuir* 1998, **14**, 1618–1636.

[33] Kaler, E. W.; Murthy, A. K.; Rodriguez, B. E.; Zasadainski, J. A. N. *Science* 1989, **245**, 1371–1374.

[34] Safran, S. A.; Pincus, P.; Andelman, D. *Science* 1990, **248**, 354–356.

[35] Kaler, E. W.; Herrington, K. L.; Murthy, A. K.; Zasadzinski, J. A. N. *J. Phys. Chem. B* 1992, **96**, 6698–6707.

[36] Smirnova, N. A.; Vanin, A. A.; Safonova, E. A.; Pukinsky, I. B.; Anufrikov, Y. A.; Makarov, A. L. *J. Colloid Interface Sci.* 2009, **336**, 793–802.

[37] Yuan, J.; Bai, X.; Zhao, M.; Zheng, L. *Langmuir* 2010, **26**, 11726–11731.

[38] Zhao, Y.; Chen, X.; Jing, B.; Wang, X.; Ma, F. *J. Phys. Chem. B* 2009, **113**, 983–988.

[39] Singh, K.; Marangoni, D. G.; Quinn, J. G.; Singer, R. D. *J. Colloid Interface Sci.* 2009, **335**, 105–111.

[40] Rao, K. S.; Singh, T.; Kumar, A. *Langmuir* 2011, **27**, 9261–9269.

[41] Ghosh, S.; Ghatak, C.; Banerjee, C.; Mandal, S.; Kuchlyan, J.; Sarkar, N. *Langmuir* 2013, **29**, 10066–10076.

[42] Villa, C. C.; Moyano, F.; Ceolin, M.; Silber, J. J.; Falcone, R. D.; Correa, N. M. *Chem. Eur. J.* 2012, **18**, 15598–15601.

[43] Brown, P.; Butts, C. P.; Eastoe, J.; Grillo, I.; James, C.; Khan, A. *J. Colloid Interface Sci.* 2013, **395**, 185–189.

[44] Rao, K. S.; Trivedi, T. J.; Kumar, A. *J. Phys. Chem. B* 2012, **116**, 14363–14374.

[45] Dong, B.; Li, N.; Zheng, L.; Yu, L.; Inoue, T. *Langmuir* 2007, **23**, 4178–4182.

[46] Zhao, M.; Zheng, L. *Phys. Chem. Chem. Phys.* 2011, **13**, 1332–1337.

[47] Vaghela, N. M.; Sastry, N. V.; Aswal, V. K. *Colloid Polym. Sci.* 2011, **289**, 309–322.

[48] Ghasemian, E.; Najafi, M.; Rafati, A. A.; Felegari, Z. *J. Chem. Thermodyn.* 2010, **42**, 962–966.

[49] Jungnickel, C.; Łuczak, J.; Ranke, J.; Fernández, J. F.; Müller, A.; Thöming, J. *Colloids Surf. A* 2008, **316**, 278–284.

[50] Singh, T.; Kumar, A. *Colloids Surf. A* 2008, **318**, 263–268.

[51] Vanyúr, R.; Biczók, L.; Miskolczy, Z. *Colloids Surf. A* 2007, **299**, 256–261.

[52] El Seoud, O. A.; Pires, P. A. R.; Abdel-Moghny, T.; Bastos, E. L. *J. Colloid Interface Sci.* 2007, **313**, 296–304.

[53] Łuczak, J.; Jungnickel, C.; Łącka, I.; Stolte, S.; Hupk, J. *Green Chem.* 2010, **12**, 593–601.

[54] Dong, B.; Zhao, X.; Zheng, L.; Zhang, J.; Li, N.; Inoue, T. *Colloids Surf. A* 2008, **317**, 666–672.

[55] Li, X.; Gao, Y.; Liu, J.; Zheng, L.; Chen, B.; Wub, L.; Tung, C. *J. Colloid Interface Sci.* 2010, **343**, 94–101.

[56] Wang, H.; Zhang, L.; Wang, J.; Li, Z.; Zhang, S. *Chem. Commun.* 2013, **49**, 5222–5224.

[57] Zhu, W.; Yang, H.; Yu, Y.; Hua, L.; Li, H.; Fenga, B.; Hou, Z. *Phys. Chem. Chem. Phys.* 2011, **13**, 13492–13500.

[58] Singh, T.; Drechsler, M.; Müller, A. H. E.; Mukhopadhyay, I.; Kumar, A. *Phys. Chem. Chem. Phys.* 2010, **12**, 11728–11735.

CHAPTER 9

Effect of a Surface-Active Lonic Liquid on Calixarenes

SHUBHA PANDEY

Department of Science and Technology, Technology Bhawan, New Delhi, India
Indian Institute of Technology Delhi, Hauz Khas, New Delhi, India

SHRUTI TRIVEDI, PRAMOD S. PANDEY, and SIDDHARTH PANDEY

Department of Chemistry, Indian Institute of Technology Delhi, Hauz Khas, New Delhi, India

SANDEEP K. MISHRA

Delhi Pollution Control Committee, Delhi, India

9.1 INTRODUCTION

Room temperature ionic liquids (RTILs) have generated increased attention as versatile substances in science and technology due to their potential applications in multidisciplinary areas, such as organic synthesis, electrochemistry, liquid phase extraction, catalysis, biotechnology, and polymerization process [1–16]. Their uniqueness derives from the physicochemical properties they offer, namely, negligible vapor pressure, wide liquidus range, high thermal stability, relatively high ionic conductivity, and wide electrochemical window [17–21]. RTILs are composed entirely of ions (cations and anions) that are liquid at ambient conditions, and their physicochemical properties can be readily tuned for a specific application by different combinations of anions and cations [18–20, 22, 23].

RTILs, where their anion and/or cation bear charged hydrophilic head group along with one or more long-chain hydrophobic tails, are expected to be surface active and are categorized as surface-active ILs (SAILs) [24, 25]. Owing to the existence of intrinsic amphiphilicity similar to traditional surfactants, these RTILs can self-assemble to form aggregates with specific structure, shape, and properties. One can modify the structure and morphology of these aggregates by simply changing the alkyl chain length, the type of cation/anion, and the counterion [26, 27]. The impetus for the increasing interest in studying SAILs is due to the fact that they possess favorable properties of both RTILs and surfactants and have exhibited improved surface-active properties as compared to their more traditional counterparts in several

Ionic Liquid-Based Surfactant Science: Formulation, Characterization, and Applications, First Edition. Edited by Bidyut K. Paul and Satya P. Moulik.

instances [28–30]. For example, it is established that chiral and long-chain RTILs (e.g., [C_{16}mim][Br]) can be used in micellar electrokinetic capillary chromatography (MEKCC) where they can compete effectively with classical surfactants [31, 32]. In another work, Wang et al. showed that long-chain imidazolium ILs (e.g., [C_{16}mim][Cl]) are superior in comparison with classical surfactants and were widely applied as templates in sol–gel technology [33].

Calixarenes and their derivatives have been of growing interest to the scientific community due to their widespread applications as unique and versatile hosts [34–42]. This class of compounds possesses a cavity with distinct hydrophobic and hydrophilic sites represented by upper and lower rims. By suitable modifications of these sites, one may obtain several different calixarenes for a wide variety of applications. It has been demonstrated that a variety of guests with different size, shape, properties, and characteristics can interact with calixarenes and their derivatives [35, 36].

Complexation study of RTILs with macrocycles, such as calixarenes, rotaxanes, cucurbiturils, cyclodextrins, etc., is an upcoming area of great interest mainly because of the application potentials of thus formed supramolecular assemblies in extraction and chromatographic methods and also in material synthesis. Research groups have started to report on the inclusion complex formation of RTILs with cyclodextrins (CDs), calixpyrroles, cucurbiturils, and calixarenes, among others, though the number of such reports is scarce. In all these studies, a complex association pattern of RTILs with macrocyclic compounds has been reported, which drastically changes with the structure of macrocyclic compounds and RTILs. Gao et al. have reported the formation of 1:1 (guest:host) inclusion complex between RTIL 1-butyl-3-methylimidazolium hexafluorophosphate ([bmim][PF_6]) and β-CD using NMR spectroscopy [43]. Scherman's group has highlighted the inclusion complexes of imidazolium-based RTILs with cucurbit[6]uril in aqueous environment where the dissolution ability was shown to be tuned by augmenting the imidazolium structure [44]. RTIL–calixpyrrole inclusion complexes were shown by Bates, Gale, and Light to form between pyridinium-based RTILs and meso-octamethylcalix[4]pyrrole where both the anion and the cation of the RTIL bind to the tetrapyrrole macrocycle [45]. In one of the recent works, it was demonstrated that the self-assembly of a dimeric host–guest complex is formed by the association of two pyrogallol[4]arene macrocyclic hosts and a geminal (dicationic) RTIL guest through noncovalent interactions [46]. Biczók and coworkers used a fluorescence probe, berberine, to study the encapsulation of imidazolium-type RTILs in 4-sulfonato-substituted calix[4]arene and calix[6]arene at pH 2 [47]. Further, Sueishi et al. showed the effect of pressure on inclusion complexation of phenothiazine dyes with *p*-sulfonatocalix[6]arene in a mixture of RTIL 1-butyl-3-methylimidazolium tetrafluoroborate and ethanol [48]. Later, Tran's group established the gas chromatographic separation of isotopic species using a cavitand-impregnated RTIL stationary phase [49]. Interaction of ions of various RTILs with α-, β-, and γ-CD were studied by Ondo et al. in a recent investigation [50].

In the light of the foregoing discussion, it is expected that SAIL can alter the properties of calixarenes and their derivatives as it possesses favorable properties of both RTILs and surfactants. Thus, understanding the effect of SAIL on the properties of calixarenes with different molecular architecture can open up new avenues in the field of host–guest chemistry. We report herein the outcomes of the investigation of the effect of a SAIL, 1-decyl-3-methylimidazolium chloride ([C_{10}mim][Cl]), on four structurally different calixarenes, (**1**), (**2**), (**3**), and (**4**), within basic aqueous media

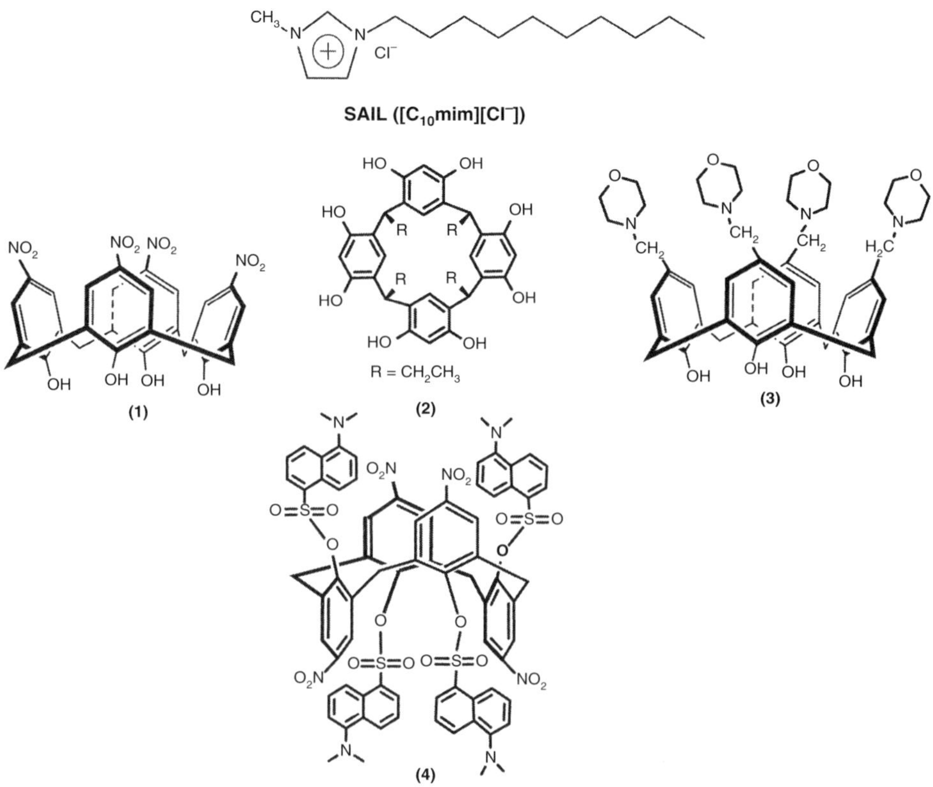

Figure 9.1 Structure of SAIL 1-decyl-3-methylimidazolium chloride ([C_{10}mim][Cl]) and the calixarenes used in this work.

using UV–Vis absorbance and fluorescence spectroscopy (*cf.* Fig. 9.1 for structures). While a common and popular fluorescence probe pyrene is used to study the interactions between each of the calixarenes, (**1**), (**2**), and (**3**), and the SAIL [C_{10}mim][Cl], fluorescence from fluorophore dansyl appended to a calixarene, (**4**), is used to obtain insight to the interactions present within the system composed of basic aqueous solution of SAIL [C_{10}mim][Cl]. Our investigation reveals site-specific interaction of calixarenes with aggregated and nonaggregated [C_{10}mim][Cl].

9.2 EXPERIMENTAL SECTION

9.2.1 Materials

The fluorescence probe pyrene was purchased from Sigma–Aldrich in the highest purity possible and was recrystallized thrice before use. Spectroscopic-grade high-purity 1-decyl-3-methylimidazolium chloride ([C_{10}mim][Cl], liquid at room temperature) and decyltrimethylammonium chloride (DeTACl, solid at room temperature) were purchased from Sigma–Aldrich in the highest purity possible and were stored under inert atmosphere. *p*-Nitrophenol and sodium hydroxide were purchased from

Sigma–Aldrich and Fisher Scientific, respectively, in the highest purity and were used as received. Ethanol (99.9%) and water (chromatography grade, for HPLC) were obtained from Merck Ltd. and were used as received.

9.2.1.1 Synthesis of Calixarenes (1)–(4) Compounds *p*-5,11,17,23-tetranitro-25,26,27,28-tetrahydroxy-calix[4]arene, (**1**), and 4,6,10,12,16,18,22,24-octahydroxy-2,8,14,20-tetraethyl-calix[4]resorcinarene, (**2**), were prepared in-house following procedures described elsewhere [37]. Compound 5,11,17,23-tetra-*kis*-(*N*-methylmorpholine)-25,26,27,28-tetrahydroxy-calix[4]arene, (**3**), was prepared by adapting the procedure described by Pandey [51]. Compound 5,11,17,23-tetranitro-25,26,27,28-tetra(*N*-(5-dimethylaminonaphthalene-1-sulfonyl))-calix[4]arene, (**4**), was synthesized according to the reported literature procedure [52].

9.2.2 Methods

A PerkinElmer Lambda 35 UV–Vis double-beam spectrophotometer with variable bandwidth was used for the acquisition of the UV–Vis absorbance data. Fluorescence spectra were acquired on a model FL 3–11 Fluorolog-3 modular spectrofluorometer (HORIBA Jobin Yvon, Inc.) with single Czerny–Turner grating excitation and emission monochromators as wavelength selection devices, a 450 W Xe arc lamp as the excitation source, and a photomultiplier tube (PMT) as the detector. All reported spectroscopic values were averages based on performing triplicate measurements on independently prepared samples. The spectral responses from appropriate blanks were subtracted before data analysis in each case. All absorbance and fluorescence data were acquired using 10.0 mm pathlength quartz cuvettes.

9.3 RESULTS AND DISCUSSION

9.3.1 Micelle Formation and Critical Micelle Concentration of [C_{10}mim][Cl]

Due to the fact that [C_{10}mim][Cl] possesses a long decyl chain appended to the polar imidazolium ring, the spontaneous self-aggregation of [C_{10}mim][Cl] into normal micelles in water is first investigated. Molecular fluorescence from an appropriate fluorophore is well suited to furnish information regarding self-aggregation of long-chain molecules owing to the higher sensitivity and orthogonality of information inherent to fluorescence techniques [53]. Pyrene is one of the most widely used neutral fluorescence probes for polarity studies [54–56]. The pyrene solvent polarity scale (Py I_1/I_3) is defined by its I_1/I_3 emission intensity ratio, where I_1 is the intensity of the solvent-sensitive band arising from the $S_1(v=0) \rightarrow S_0(v=0)$ transition and I_3 corresponds to the solvent-insensitive $S_1(v=0) \rightarrow S_0(v=1)$ transition. The I_1/I_3 ratio increases with increasing solvent dipolarity and is a function of both the solvent dielectric (ε) and the refractive index (n) via the dielectric cross term, $f(\varepsilon,n^2)$. In aqueous solution, in the absence of any self-assembling moieties, Py I_1/I_3 is fairly high. As self-assembling molecules are added to aqueous media, the I_1/I_3 starts to decrease rapidly till the critical micelle concentration (cmc) is reached after which the decrease in I_1/I_3 is not so rapid anymore.

As calixarenes (**1**)–(**4**) are soluble in 1 M aqueous NaOH, we collected fluorescence emission spectra of pyrene dissolved in 1 M NaOH solution of varying

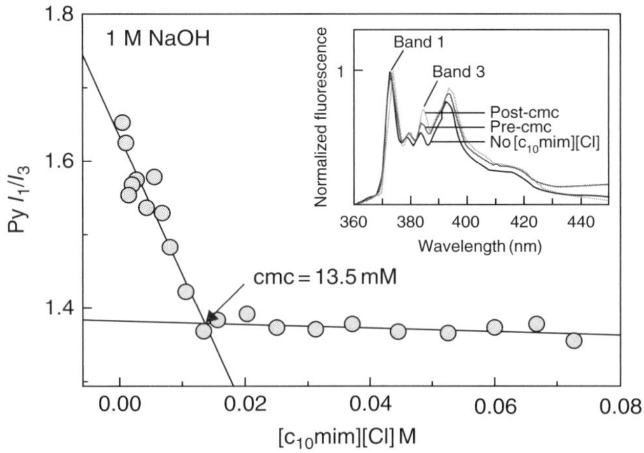

Figure 9.2 Pyrene (1 μM) band 1-to-band 3 emission intensity ratio (Py I_1/I_3) as a function of [C_{10}mim][Cl] concentration in 1 M aqueous NaOH. Inset, normalized fluorescence emission spectra of pyrene (λ_{ex} = 337 nm, slits = 1/1 nm) in 1 M aqueous NaOH in the absence of [C_{10}mim][Cl], in the presence of 1.1 mM [C_{10}mim][Cl] (pre-cmc region), and in the presence of 60 mM [C_{10}mim][Cl] (post-cmc region). *See insert for color representation of the figure.*

concentration of [C_{10}mim][Cl] (Fig. 9.2). Py I_1/I_3, obtained from pyrene fluorescence spectra, are plotted against [C_{10}mim][Cl] concentration. It is clear that after initial rapid decrease, Py I_1/I_3 becomes almost a constant as the concentration of [C_{10}mim][Cl] is increased further. The inflection point (please refer to Fig. 9.2) in Py I_1/I_3 versus concentration of [C_{10}mim][Cl] data is a clear indication of the onset of the micelle formation and is taken as the cmc. The cmc of [C_{10}mim][Cl] in 1 M aqueous NaOH is estimated to be 13.5 mM. It is clear that SAIL [C_{10}mim][Cl] readily self-aggregates in basic aqueous media and forms micelles. It is interesting to note that using the same procedure, the cmc of surfactant decyltrimethylammonium chloride (DeTACl), the conventional counterpart of SAIL [C_{10}mim][Cl], in 1 M aqueous NaOH is found to be 64 mM, which is significantly higher than the cmc of [C_{10}mim][Cl]. It is clear that SAIL [C_{10}mim][Cl] with 1,3-dialkylimidazolium head group self-aggregates into micelles much more efficiently than the conventional surfactant DeTACl. The SAIL, in this context, exhibits superior properties to the conventional surfactants.

Once the self-aggregation behavior and the cmc of [C_{10}mim][Cl] are established, we next investigated the interaction of SAIL [C_{10}mim][Cl] with various calixarenes dissolved in 1 M aqueous NaOH.

9.3.2 Molecular Absorbance of Calixarenes in the Presence of [C_{10}mim][Cl]

We first used molecular absorbance to investigate the effect of SAIL [C_{10}mim][Cl] on the properties of calixarenes (**1**), (**2**), and (**3**). UV–Vis molecular absorbance spectra of calixarenes (**1**), (**2**), and (**3**), respectively, in the presence of increasing concentration (from pre- to post-cmc) of [C_{10}mim][Cl] within 1 M NaOH are presented in Figure 9.3. A careful examination of the data presented in Figure 9.3 reveals that the absorbance behavior of (**1**) changes dramatically with increasing

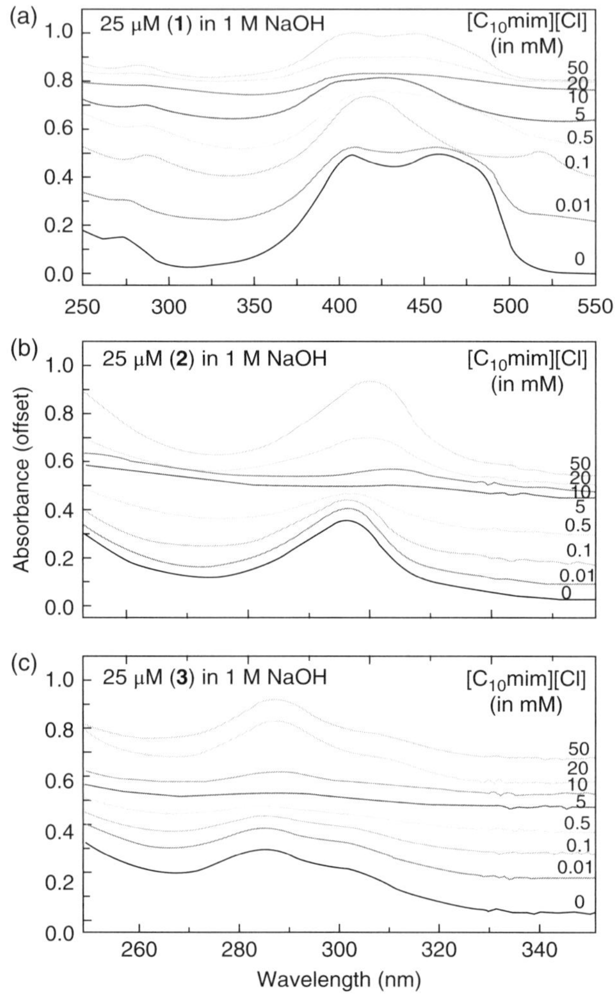

Figure 9.3 (a–c) UV–Vis absorbance spectra of 25 μM each of (**1**), (**2**), and (**3**) dissolved in 1 M aqueous NaOH at varying concentration of [C_{10}mim][Cl]. *See insert for color representation of the figure.*

concentration of [C_{10}mim][Cl] both below and above cmc. The structured absorbance spectral features of (**1**) highlighted by two absorbance bands with maxima centered at 405 and 450 nm, respectively, in 1 M aqueous NaOH appear to merge into one broad band as the concentration of [C_{10}mim][Cl] is increased to cmc and above (Fig. 9.3a). The changes are clearly not as dramatic for calixarenes (**2**) and (**3**) under identical conditions (Fig. 9.3b and c, respectively). The UV–Vis absorbance measurements imply strong interaction of (**1**) with both monomeric and micellar forms of [C_{10}mim][Cl]; interaction of calixarenes (**2**) and (**3**), respectively, with SAIL [C_{10}mim][Cl] is clearly not as dominant. A comparison of the structures of the calixarenes (**1**), (**2**), and (**3**) hints at the presence of electron-withdrawing –NO_2 functionality on (**1**) to be responsible for the strong interaction between this calixarene and [C_{10}mim][Cl].

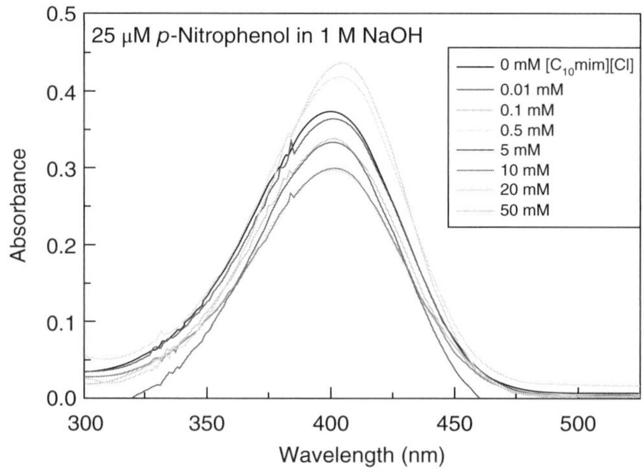

Figure 9.4 UV–Vis absorbance spectra of 25 μM of *p*-nitrophenol dissolved in 1 M aqueous NaOH with varying concentration of [C_{10}mim][Cl]. *See insert for color representation of the figure.*

In order to highlight the role of calixarene molecular framework in the interaction of (**1**) with [C_{10}mim][Cl], we collected UV–Vis absorbance spectra of *p*-nitrophenol under identical conditions as the control experiment (Fig. 9.4). It is clear from the minimal changes in the absorbance spectral features of *p*-nitrophenol as the concentration of [C_{10}mim][Cl] is increased through the pre- and the post-cmc regimes that the presence of calixarene framework is essential for the interactions.

9.3.3 Effect of Calixarenes on the Fluorescence of Probe Pyrene within Micellar [C_{10}mim][Cl]

In order to explore the interactions between calixarene and SAIL, we next used fluorescence from pyrene as the indicator. The concentration of the three calixarenes, (**1**), (**2**), and (**3**), respectively, were varied within the pyrene solution in 1 M aqueous NaOH in the presence of 50 mM (post-cmc) concentration of SAIL [C_{10}mim][Cl]. We extracted information from both pyrene polarity scale and the changes in the fluorescence signal of the pyrene as the concentration of the calixarene is changed.

9.3.3.1 Pyrene Polarity Scale I_1/I_3 As mentioned earlier, due to its sensitivity to the immediate microenvironment around pyrene, the polarity scale (Py I_1/I_3) could be a very effective parameter that can reveal information about the interactions present. Figure 9.5 presents Py I_1/I_3 as a function of calixarene concentration from the solution of 50 mM [C_{10}mim][Cl] in 1 M aqueous NaOH. The Py I_1/I_3 reveals interesting features that nicely complement the outcomes of UV–Vis absorbance data. It is abundantly clear from the data presented in Figure 9.5 that pyrene I_1/I_3 increases significantly as the concentration of (**1**) is increased within micellar [C_{10}mim][Cl] solution indicating the polarity of the pyrene cybotactic region to increase substantially with increasing concentration of (**1**). In complete contrast, instead of increase, a small decrease in pyrene I_1/I_3 is observed when concentrations of (**2**) and (**3**), respectively,

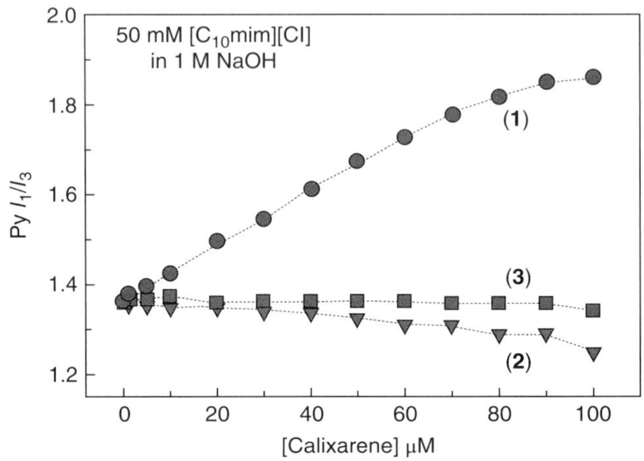

Figure 9.5 Pyrene (1 μM) band 1-to-band 3 emission intensity ratio (Py I_1/I_3) as function of concentration of (**1**), (**2**), and (**3**), respectively, at 50 mM [C_{10}mim][Cl] in 1 M aqueous NaOH. λ_{ex} = 337 nm, slits = 1/1 nm.

are increased. This observation strongly suggests the interaction of (**1**) with pyrene to perhaps take place via the $-NO_2$ functionality, whereas the interaction of (**2**) and (**3**) with pyrene to happen possibly through the organic framework of the calixarenes. Insignificant changes in pyrene I_1/I_3 are observed in the absence of [C_{10}mim][Cl], whereas similar increase in pyrene I_1/I_3 could be seen when *p*-nitrophenol instead of (**1**) is used (data not shown). The strong interaction of (**1**) through $-NO_2$ functionalities involving aggregated (micellar) SAIL [C_{10}mim][Cl] is amply established.

9.3.3.2 Quenching of Pyrene Fluorescence Nitro- and hydroxy-containing aromatic compounds are known to quench the fluorescence of pyrene [53, 57–59]. Each of the calixarenes (**1**), (**2**), and (**3**), in this respect, has the ability to quench the fluorescence of pyrene. Accordingly, we found that the above three calixarenes do quench the fluorescence of pyrene within micellar [C_{10}mim][Cl]. Plots of F_0/F versus the concentration of the calixarenes (where F_0 and F are the fluorescence intensities in the absence and the presence of the calixarene, respectively) are presented in Figure 9.6.

Linear regression analysis of the data reveals that the F_0/F varies linearly ($r^2 > 0.99$) with the concentration of the calixarenes for [calixarene] <100 μM. In accordance with the Stern–Volmer formulation for such cases, the intercept is close to unity, whereas the slope provides the Stern–Volmer quenching constant (K_{SV}) that represents the efficiency of quenching. The K_{SV} values for (**1**), (**2**), and (**3**) are recovered to be 1.6×10^4, 1.7×10^4, and $0.4 \times 10^4\,M^{-1}$, respectively, that, in general, represent high quenching efficiencies. The compartmentalization of pyrene along with the calixarene in the palisade layer of the micellar [C_{10}mim][Cl] provides close proximity to the fluorophore–quencher duo, which results in effective quenching of the pyrene fluorescence. However, the relative magnitudes of K_{SV} for the three calixarenes reveal important information. The appreciable quenching by (**1**) strongly hints at the involvement of $-NO_2$ functionalities as $-NO_2$ functionality is known to quench pyrene fluorescence by electron/charge accepting mechanism [37]. The quenching by (**2**) is perhaps due to the presence of

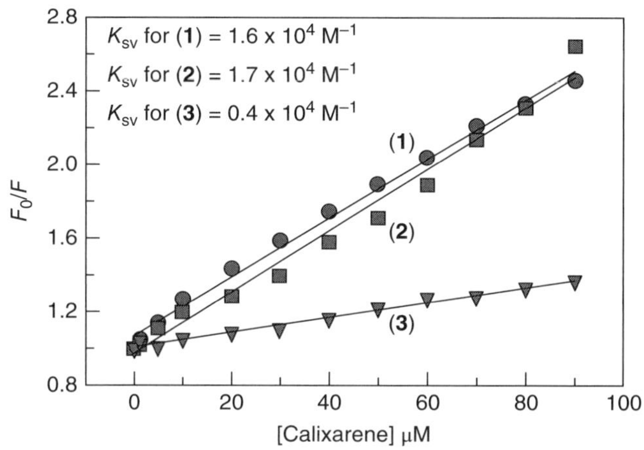

Figure 9.6 Plot of F_0/F (band 1 of pyrene) as function of different concentration of (**1**), (**2**), and (**3**), respectively, at 50 mM [C_{10}mim][Cl] in 1 M aqueous NaOH. Lines are best fitted by least squares method.

maximum number of –OH functionalities on this calixarene, the quenching by (**3**), which is the least among the three, is assigned to the fact that this calixarene locates itself around pyrene within micellar [C_{10}mim][Cl] in such a way as the organic framework of the calixarene is in close proximity with pyrene. Fluorescence quenching data reveals the orientation, and thus the interactions present, between calixarenes and [C_{10}mim][Cl] within the micellar solution of the SAIL.

9.3.4 Interaction of Dansylated Calixarene (4) with [C_{10}mim][Cl]

A calixarene-based host can be suitably modified by covalent attachment(s) of judiciously selected fluorophoric moiety(ies). This approach has generated considerable attention of chemists for its well-documented effectiveness and potential applications [52, 60, 61]. Several common and popular fluorophore moieties have been utilized to design and synthesize fluorophore-appended calixarenes [62–64]. Among others, dansylated calixarenes have received ever-increasing attention due to their recognition and sensing of heavy, transition, and alkali metals, such as Hg^{2+}, Pb^{2+}, Cd^{2+}, Cr^{3+}, Cu^{2+}, Na^{+}, etc. [41, 62–66]. Dansyl group, when covalently bound to a host molecule, offers many attractive and pertinent features due partly to (i) its strong fluorescence (absorption bands in the near-UV and intense fluorescence in the visible region) [67], (ii) relatively long emission wavelength [65], and (iii) its solvatochromic nature (i.e., sensitivity to the polarity of the medium) owing to the presence of twisted intramolecular charge transfer (TICT) excited state [52, 68, 69].

In order to investigate the effect of SAIL on a dansyl-appended calixarene, a series of experiments were carried out using compound (**4**). In the first set of experiments, fluorescence emission spectra of (**4**) are collected where concentration of (**4**) is increased from 5 to 25 μM with an increment of 5 μM within micellar solution of 50 mM SAIL [C_{10}mim][Cl] in 1 M aqueous NaOH (Fig. 9.7). A broad fluorescence band and location of the band maxima are characteristics of the dansyl functionality. A plot of fluorescence intensity at the band maxima versus the concentration of (**4**)

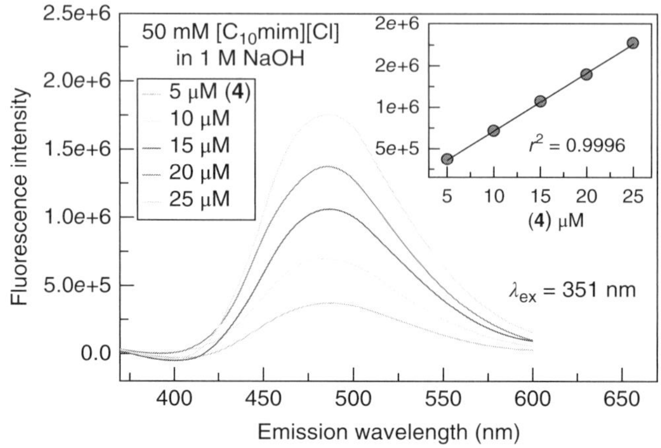

Figure 9.7 Fluorescence emission spectra of (**4**) at different concentrations dissolved in 50 mM $[C_{10}mim][Cl]$ in 1 M aqueous NaOH. λ_{ex} = 351 nm, slits = 3/3 nm. Inset, linear regression analysis for fluorescence intensity *versus* concentration of (**4**). *See insert for color representation of the figure.*

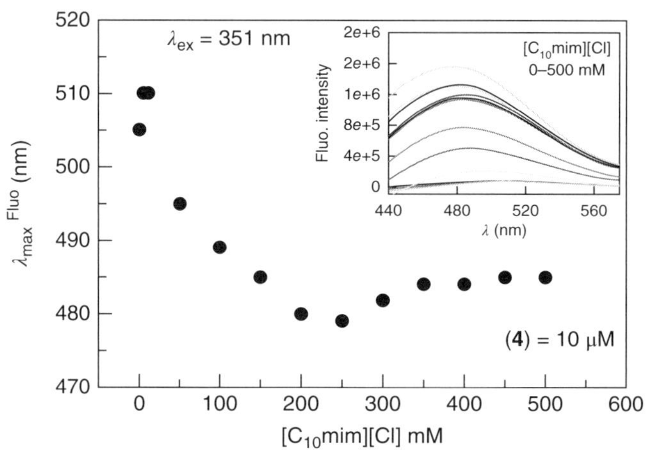

Figure 9.8 Variation in the lowest-energy fluorescence emission maxima (λ_{max}^{Fluo}) of (**4**) (10 μM) with the concentration of $[C_{10}mim][Cl]$ in 1 M aqueous NaOH. λ_{ex} = 351 nm, slits = 3/3 nm. Inset, fluorescence emission spectra of (**4**) at different concentrations of $[C_{10}mim][Cl]$ (increasing downward) in 1 M aqueous NaOH. *See insert for color representation of the figure.*

exhibits excellent linear behavior strongly suggesting dansyl-appended calix[4]arene to readily partition into the $[C_{10}mim][Cl]$ micellar pseudophase. Although (**4**) is readily soluble in 1 M aqueous NaOH, it is clear that it prefers micellar phase formed by the SAIL $[C_{10}mim][Cl]$ thus suggesting the affinity of calixarene for the SAIL.

In the second set of the experiments, the concentration of SAIL $[C_{10}mim][Cl]$ is varied from 0 to 500 mM within 1 M aqueous NaOH solution of fixed concentration (10 μM) of dansyl-appended calixarene (**4**). Fluorescence emission spectra of (**4**), acquired in the presence of varying concentrations of $[C_{10}mim][Cl]$, are presented in Figure 9.8.

It reveals that as the concentration of [C_{10}mim][Cl] is increased to cmc and above, more and more (**4**) partitions into the micellar pseudophase thus showing an increase in the fluorescence intensity as the fluorescence quantum yield of (**4**) increases for the hydrophobicity of the cybotactic region increases. Interestingly, the increase in the fluorescence intensity of (**4**) is also accompanied by a significant hypsochromic shift of its emission band. This arises due to the increase in hydrophobicity of the solubilizing microenvironment of (**4**) with the increasing concentration of [C_{10}mim][Cl] as the calixarene partitions into the micellar pseudophase of the SAIL.

9.4 CONCLUSIONS

A SAIL [C_{10}mim][Cl] is found to self-aggregate to form micelles in basic aqueous solution in a more efficient manner than its more conventional surfactant counterpart DeTACl. Interaction and behavior of common calixarenes with [C_{10}mim][Cl] depend on the functionalities present on the molecular architecture of the calixarene. Calixarene with $-NO_2$ (electron-withdrawing functionalities) appears to strongly interact with both unaggregated and micellar [C_{10}mim][Cl]. A common calixarene with $-NO_2$ functionality locates itself in such a manner to have its $-NO_2$ groups oriented toward the palisade layer, whereas other calixarenes have their organic frameworks oriented in the palisade layer. A tetradansylated calixarene readily partitions into the hydrophobic microenvironment of [C_{10}mim][Cl] micelles. SAIL appears to effectively control the solubilization sites and thus the properties of calixarenes in the solution.

ACKNOWLEDGMENTS

ShP thanks Department of Science and Technology, Government of India, for the Fast-Track Fellowship (grant no. SB/FT/CS-021/2012). SiP thanks Council of Scientific and Industrial Research (CSIR), Government of India (grant no. 01(2767)/13/EMR-II), for partially funding this work. ST thanks CSIR for her Research Assistantship.

REFERENCES

[1] Wasserscheid, P.; Welton, T. *Ionic Liquids in Synthesis*; Wiley-VCH Verlag GmbH, Weinheim: 2003.

[2] Rogers, R. D.; Seddon, K. R. *Science* 2003, **302**, 792–793.

[3] Malhotra, S. *Ionic Liquid Applications: Pharmaceuticals, Therapeutics, and Biotechnology*; ACS Symposium Series; American Chemical Society, Washington, DC: 2010.

[4] Freemantle, M. *An Introduction to Ionic Liquids*, 1st ed.; RSC Publishing, Cambridge, UK: 2009.

[5] Koel, M., ed. *Ionic Liquids in Chemical Analysis*; CRC Press, Boca Raton: 2009.

[6] Brennecke, J. F.; Rogers, R. D.; Seddon, K. R. *Ionic Liquids IV: Not Just Solvents Anymore*; ACS Symposium Series; American Chemical Society, Washington, DC: 2007.

[7] Rogers, R. D.; Seddon, K. R., eds. *Ionic Liquids as Green Solvents: Progress and Prospects*; ACS Symposium Series **856**; American Chemical Society, Washington, DC: 2003.

[8] Liu, J. F.; Jiang, G. B.; Chi, Y. G.; Cai, Y. Q.; Zhou, Q. X.; Hu, J. T. *Anal. Chem.* 2003, **75**, 5870–5876.

[9] Berthod, A.; Carda-Broch, S. *J. Liq. Chromatogr. Relat. Technol.* 2003, **26**, 1493–1508.

[10] Qin, W.; Li, S. F. Y. *Analyst* 2003, **128**, 37–41.

[11] Liang, C.; Yuan, C.-Y.; Warmack, R. J.; Barnes, C. E.; Dai, S. *Anal. Chem.* 2002, **74**, 2172–2176.

[12] Baker, G. A.; Baker, S. N.; McClesky, T. M. *Chem. Commun.* 2003, 2932–2933.

[13] Duchet, L.; Legeay, J. C.; Carrié, D.; Paquin, L.; Vanden Eynde, J. J.; Bazureau, J. P. *Tetrahedron* 2010, **66**, 986–994.

[14] Zanoni, M. V. B.; Rogers, E. I.; Hardacre, C.; Compton, R. G. *Anal. Chim. Acta* 2010, **659**, 115–121.

[15] Fontanals, N.; Ronka, S.; Borrull, F.; Trochimczuk, A. W.; Marcé, R. M. *Talanta* 2009, **80**, 250–256.

[16] Seddon, K. R. *J. Chem. Technol. Biotechnol.* 1997, **68**, 351–356.

[17] Sanghi, S.; Willett, E.; Versek, C.; Tuominen, M.; Coughlin, E. B. *RSC Adv.* 2012, **2**, 848–853.

[18] Chen, X.; Li, X.; Hu, A.; Wang, F. *Tetrahedron: Asymmetry* 2008, **19**, 1–14.

[19] Bonanni, M.; Manuelli, M.; Goti, A.; Faggi, C.; Cardona, F. *Arkivoc* 2014, **2014** (**iii**), 54–64.

[20] Heckel, T.; Winkel, A.; Wilhelm, R. *Tetrahedron: Asymmetry* 2013, **24**, 1127–1133.

[21] Plaquevent, J.-C.; Levillain, J.; Guillen, F.; Malhiac, C.; Gaumont, A.-C. *Chem. Rev.* 2008, **108**, 5035–5060.

[22] Pandey, S.; Baker, G. A.; Sze, L.; Pandey, S.; Kamath, G.; Zhao, H.; Baker, S. N. *New J. Chem.* 2013, **37**, 909–919.

[23] Pandey, S.; Baker, S. N.; Pandey, S.; Baker, G. A. *J. Fluoresc.* 2012, **22**, 1313–1343.

[24] El Seoud, O. A.; Pires, P. A. R.; Abdel-Moghny, T.; Bastos, E. L. *J. Colloid Interface Sci.* 2007, **313**, 296–304.

[25] Bharmoria, P.; Kumar, A. *RSC Adv.* 2013, **42**, 19600–19608.

[26] Bowers, J.; Butts, C. P.; Martin, P. J.; Gutierrez, M. C. V. *Langmuir* 2004, **20**, 2191–2198.

[27] Baker, G. A.; Pandey, S.; Pandey, S.; Baker, S. N. *Analyst* 2004, **129**, 890–892.

[28] Smirnova, J. N. A.; Safonova, E. A. *Russ. J. Phys. Chem. A.* 2010, **84**, 1695–1704.

[29] Adams, J.; Bradley, A. E.; Seddon, K. R. *Aust. J. Chem.* 2001, **54**, 679–681.

[30] Mahajan, S.; Sharma, R.; Mahajan, K. R. *Colloids Surf. A Physicochem. Eng. Asp.* 2013, **424**, 96–104.

[31] Niu, J.; Qiu, H.; Li, J.; Liu, X.; Jiang, S. *Chromatographia* 2009, **69**, 1093–1096.

[32] Rizvi, S. A. A.; Shamsi, S. A. *Anal. Chem.* 2006, **78**, 7061–7069.

[33] Wang, T.; Kaper, H.; Antonietti, M.; Smarsly, B. *Langmuir* 2007, **23**, 1489–1495.

[34] Aoyama, Y.; Tanaka, Y.; Sugahara, S. *J. Am. Chem. Soc.* 1989, **111**, 5397–5404.

[35] Gutsche, C. D. *Calixarenes: Monographs in Supramolecular Chemistry*; Royal Society of Chemistry, London: 1989.

[36] Gutsche, C. D. *Calixarene: An Introduction*, 2nd edn.; Royal Society of Chemistry, Cambridge, UK: 2008.

[37] Pandey, S.; Ali, M.; Bishnoi, A.; Azam, A.; Pandey, S.; Chawla, H. M. *J. Fluoresc.* 2008, **18**, 533–539.

[38] Ruth, L.; Johnson, H. B.; Redshaw, C.; Mathews, S. E.; Mueller, A. *J. Am. Chem. Soc.* 2008, **130**, 2892–2893.

[39] Sun, X. H.; Li, W.; Xia, P. F.; Luo, H. B.; Wei, Y.; Wong, M. S.; Cheng, Y. K.; Shuang, S. *J. Org. Chem.* 2007, **72**, 2419–2426.

[40] Chawla, H. M.; Srinivas, K. *J. Chem. Soc. Chem. Commun.* 1994, 2593–2594.

[41] Galina, G.; Talanova, G. G.; Roper, E. D.; Buie, N. M.; Gorbunova, M. G.; Bartsch, R. A.; Talanov, V. S. *Chem. Commun.* 2005, 5673–5675.

[42] Kim, J. S.; Kim, H. J.; Kim, H. M.; Kim, S. H.; Lee, J. W.; Kimand, S. K.; Cho, B. R. *J. Org. Chem.* 2006, **71**, 8016–8022.

[43] Gao, Y.-A.; Li, Z.-H.; Du, J.-M.; Han, B.-X.; Li, G.-Z.; Hou, W.-G.; Shen, D.; Zheng, L.-Q.; Zhang, G.-Y. *Chem. Eur. J.* 2005, **11**, 5875–5880.

[44] Jiao, D.; Biedermann, F.; Tian, F.; Scherman, O. A. *J. Am. Chem. Soc.* 2010, **132**, 15734–15743.

[45] Bates, G. W.; Gale, P. A.; Light, M. E. *CrystEngComm* 2006, **8**, 300–302.

[46] Fowler, D. A.; Atwood, J. L.; Baker, G. A. *Chem. Commun.* 2013, **49**, 1802–1804.

[47] Miskolczy, Z.; Biczók, L. *Chem. Phys. Lett.* 2009, **477**, 80–84.

[48] Inazumi, N.; Yamamoto, S.; Sueishi, Y. *J. Incl. Phenom. Macrocycl. Chem.* 2007, **59**, 33–39.

[49] Tran, C. D.; Mejac, I.; Rebek, Jr., J.; Hooley, R. J. *Anal. Chem.* 2009, **81**, 1244–1254.

[50] Ondo, D.; Tkadlecová, M.; Dohnal, V.; Rak, J.; Kvíčala, J.; Lehmann, J. K.; Heintz, A.; Ignatiev, N. *J. Phys. Chem. B* 2011, **115**, 10285–10297.

[51] Pandey, S. *Studies on Molecular Organization for Molecular and Ionic Recognition, Ph.D. Thesis*, Jamia Millia Islamia, New Delhi: 2010.

[52] Pandey, S.; Azam, A.; Pandey, S.; Chawla, H. M. *Org. Biomol. Chem.* 2008, **7**, 269–279.

[53] Lakowicz, J. R. *Principles of Fluorescence Spectroscopy*, 3rd edn.; Kluwer Academics/ Plenum Publishers, New York: 2006.

[54] Acree, Jr., W. E. *Absorption and Luminescence Probes in Encyclopedia of Analytical Chemistry: Theory and Instrumentation*; ed., Meyer, R. A.; John Wiley & Sons, Ltd., Chichester: 2000.

[55] Karpovich, D. S.; Blanchard, G. J. *J. Phys. Chem.* 1995, **99**, 3951–3958.

[56] Kalyanasundaram, K.; Thomas, J. K. *J. Phys. Chem.* 1977, **81**, 2176–2180.

[57] Acree, Jr., W. E., Pandey, S., Tucker, S. A., Fetzer, J. C. *Polycycl. Aromat. Compd.* 1997, **12**, 71–123.

[58] Rehm, D.; Weller, A. *Isr. J. Chem.* 1970, **8**, 259–271.

[59] Saik, V. O.; Goun, A. A.; Fayer, M. D. *J. Chem. Phys.* 2004, **120**, 9601–9611.

[60] Wosnick, J. H.; Swager, T. M. *Chem. Commun.* 2004, 2744–2745.

[61] Bugler, J.; Engbersen, J. F. J.; Reinhoudt, D. N. *J. Org. Chem.* 1998, **63**, 5339–5344.

[62] Kim, J. S.; Quang, T. Q. *Chem. Rev.* 2007, **107**, 3780–3799.

[63] Souchon, V.; Lerayand, I.; Valeur, B. *Chem. Commun.* 2006, 4224–4226.

[64] Othman, A. B.; Lee, J. W.; Wu, J. S.; Kim, J. S.; Abidi, R.; Thuéry, P.; Strub, J. M.; Dorsselaer, A. V.; Vicens, J. *J. Org. Chem.* 2007, **72**, 7634–7640.

[65] Bhalla, V.; Kumar, R.; Kumar, M.; Dhir, A. *Tetrahedron* 2007, **63**, 11153–11159.

[66] Talanova, G. G.; Elkarim, N. S. A.; Talanov, V. S.; Bartsch, R. A. *Anal. Chem.* 1999, **71**, 3106–3109.

[67] Beyeh, N. K.; Aumanen, J.; Åhman, A.; Luostarinen, M.; Mansikkamaki, H.; Nissinen, M.; Tommola, J. K.; Rissanen, K. *New J. Chem.* 2007, **31**, 370–376.

[68] Valeur, V.; Leray, I. *Inorg. Chim. Acta* 2007, **360**, 765–774.

[69] Cardona, C. M.; Alvarez, J.; Kaifer, A. E.; McCarley, T. D.; Pandey, S.; Baker, G. A.; Bonzagni, N. J.; Bright, F. V. *J. Am. Chem. Soc.* 2000, **122**, 6139–6144.

CHAPTER 10

Ionic Liquids in Colloidal Regime

INDRAJYOTI MUKHERJEE and SATYA P. MOULIK

Centre for Surface Science, Department of Chemistry, Jadavpur University, Kolkata, India

10.1 INTRODUCTION

Persistent research starting from 1914 with ethylammonium nitrate (EAN) (m.p. 12°C) by Walden opened up a new avenue in research on ionic liquids (ILs). Convenient physicochemical characteristics like nonflammability, low vapor pressure, good thermal and chemical stability, tunable solubility, etc. demand immense prospects of ILs in research and application. These convenient features have identified them as alternative solvents for the present-day research in the fields of synthesis, catalysis, and electrochemical applications [1, 2]. The ILs contain cations and anions with mismatched sizes and can solubilize both polar and nonpolar components without addition of other solvents. They are a special class of compounds comprising bulky asymmetric organic cations (that prevent crystal packing) and either inorganic or organic anions (with delocalized electron clouds that reduce interionic interactions) that results in low lattice energy with low melting points. Incorporation of typical functional groups changes their general physical properties like polarity, reactivity, viscosity, etc. [3–7]. Ammonium (N), pyridinium (py), pyrrolidinium (pyrr), imidazolium (im), phosphonium (P), etc. are common classes of cationic moieties present in the ILs, whereas halides, fluoroborates, and triflamides are typical examples of anions though recent researches have enhanced variations of ion combinations (depicted in Scheme 10.1).

ILs are categorized in three generations (Scheme 10.2) with respect to their inherent properties and applicability [7].

10.2 PHYSICOCHEMICAL PROPERTIES (BULK AND INTERFACIAL)

ILs are more viscous as compared with the polar solvents commonly employed. At ambient temperature, their viscosities (η) fall in the range of 10–500 cP [8], whereas viscosities of water, ethylene glycol (EG), and glycerol at room temperature are 0.89,

Ionic Liquid-Based Surfactant Science: Formulation, Characterization, and Applications, First Edition. Edited by Bidyut K. Paul and Satya P. Moulik.

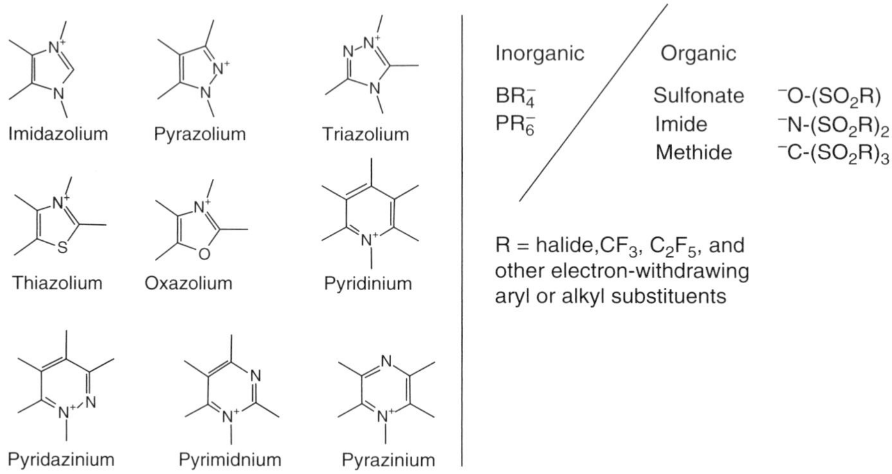

Scheme 10.1 Common cations and anions of ionic liquids. They are categorized in three generations with respect to their inherent properties and applications.

16.1 and 943 cP, respectively [8]. The typical viscous nature of ILs is governed by van der Waals interactions and their inherent tendency to form hydrogen bonds [9]. Imidazolium cation-based ILs with varied anions exhibit variations in their viscous behavior as follows: $[(CF_3SO_2)_2N^-] \leq [BF_4^-] \leq [CF_3CO_2^-] \leq [CF_3SO_3^-] \leq [(C_2H_5SO_2)_2N^-] < [C_3F_7CO_2^-] < [CH_3CO_2^-] \leq [CH_3SO_3^-] < [C_4F_9SO_3^-]$ [10]. ILs with unaltered anions manifest increase in viscosity with increasing chain length of the alkyl substituent [10]. Some phosphonium cation-based ILs (triisobutyl(methyl)phosphonium tosylate (called IL-1) and trihexyl(tetradecyl)phosphoniumbis(2,4,4-trimethylpentyl)phosphinate (called IL-2)) have produced fascinating viscous nature, like a non-Newtonian flow type under shear and temperature [11, 12]. At 298 K and under zero shear, viscosity of IL-1 ($\eta_0^{\text{IL}-1} = 88\text{cP}$) < viscosity of IL-2 ($\eta_0^{\text{IL}-2} = 300\text{cP}$). Temperature dependence of IL's viscosity obeys the empirical Vogel–Fulcher–Tammann equation (VFT equation) [13]

$$\ln \eta = \ln A + \frac{B}{T - T_0}$$

where A and B are fitting parameters and the latter (T_0) is called the VFT temperature.

Specific conductance of the ILs can also be traced by the abovementioned VFT equation where only (η) will be replaced by specific conductance (κ). A reported trend in conductivity behavior of ILs with varied cations depicted as pyridinium ≤ ammonium < sulfonium ≤ imidazolium [10]. Temperature-dependent variations of conductivity of ILs were studied earlier by several groups, where high electrical conductivity and thermal stability were evidenced [14–16].

Surface tension is another important physicochemical property of ILs and has been under research focus since 2001. Ring (du Noüy) or plate (Wilhelmy) detachment techniques are employed by different groups for getting information on the maximum force required to detach them from the surface of the liquid, which is

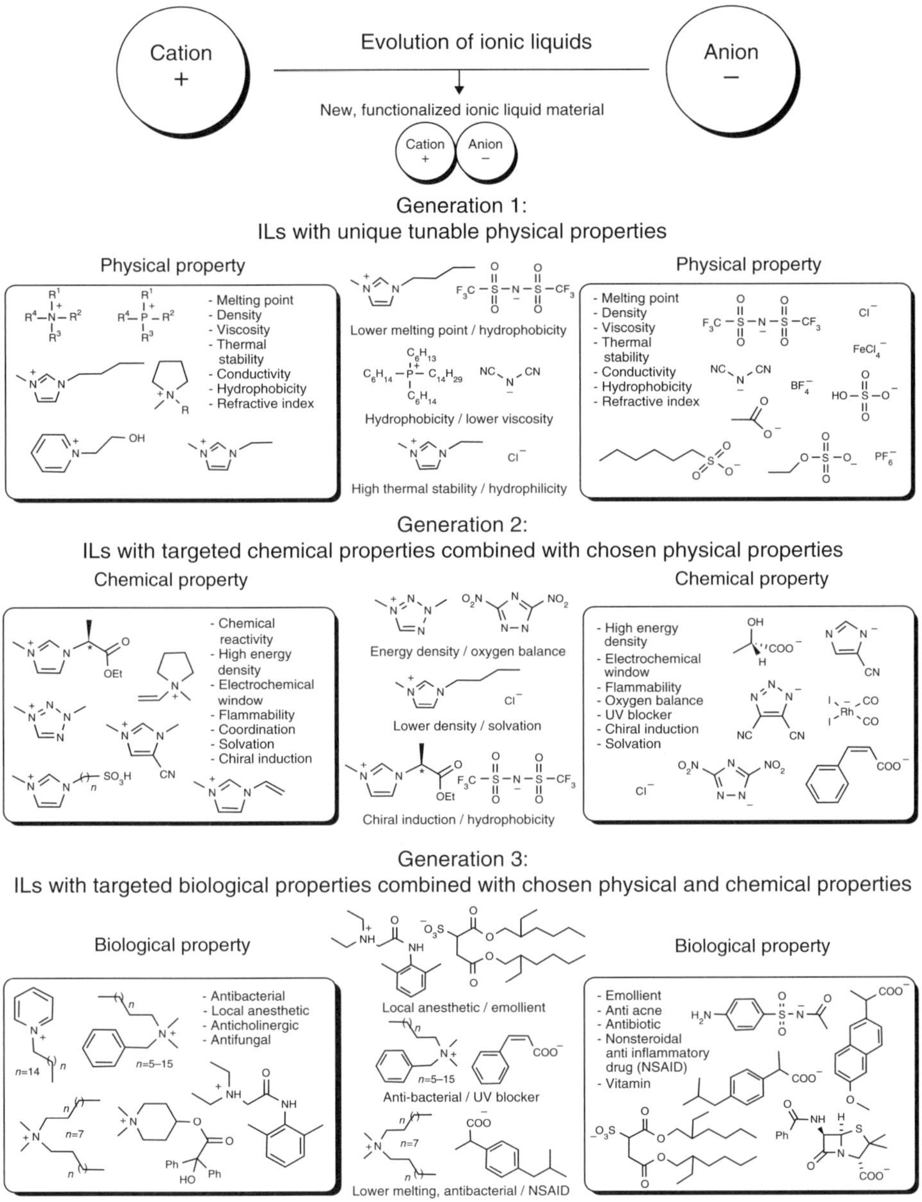

Scheme 10.2 Development of Ionic liquids in three generations. Reproduced from Whitney et al. [7], *New J. Chem.* **2007**, *31*, 1429, with permission from the Royal Society of Chemistry.

proportional to surface tension. In the case of the former technique, the surface tension (σ) is obtained by employing the following equation:

$$\sigma = \frac{F^{\max}}{4\pi R} f\left(\frac{R}{r} \times \frac{R^3}{V}\right)$$

where f is Harkins correction factor [17], which is a function of V (volume of the liquid raised by the ring), R (radius of the ring), and r (radius of the ring's wire). For the Wilhelmy plate, the following equation is valid:

$$W = W_{\text{Plate}} + \sigma P \cos\theta - I$$

where W_{Plate} is the weight of the plate, P is the plate perimeter, θ is the contact angle between the liquid and the plate, and I is the buoyant force.

Surface activity of ILs has generated from their structural and charge anisotropy, whereas lower activity is common for liquid argon and methane for their spherical molecular geometry. This spherical geometry produces negligible surface ordering effect, which strengthens the correlation between surface tension and molecular interaction. On the other side, ILs with alkyl chains either attached with its cationic or anionic moiety have produced dominating surface region ordering effect that leads to the formation of different surface and bulk environments [18–24]. Structural variations in a homologous series of IL, varying temperature, water, and halide contents in them, have proficient effect on their surface activity. It would be worthwhile to present here the variations of surface tension values of a few ILs with 1-alkyl-3-methylimidazolium and phosphonium cations with varied anions at different temperatures (Table 10.1).

Due to their fascinating interfacial activity, their aggregating nature also remained a point of concern for a long period of time. Several research groups focused their researches on exploring the aggregation behavior of ILs and determining their critical micelle concentration (CMC). ILs with varied side chain lengths attached either with anion or cation, and anion–cation combination in their structures can influence their surface activity (Table 10.2) [25–28].

Li et al. [29] have reported aggregation behavior of the long-chain IL, 1-alkyl-3-methylimidazolium bromide ([C_4C_1im][Br]), in another IL, 1-butyl-3-methylimidazolium tetrafluoroborate ([C_4C_1im][BF_4]). [$C_{14}C_1$im][Br] exhibited CMC at 2.5 mM at 298 K in aqueous solution. Another IL, 1-tetradecyl-3-(2,4,6-trimethylphenyl) imidazolium bromide [C_{14}pim][Br], in an aqueous medium has evidenced CMC at 0.61 mM [30]. ILs with same chain length and same anion produced different CMCs due to the differences in their cationic parts. Lower CMC was observed for [C_{14}pim][Br] as compared with 1-tetradecyl-3-methylimidazolium bromide ([C_{14}mim][Br]). Shi et al. [30] interpreted this phenomenon by the fact that enhanced "2,4,6-trimethylphenyl" ring supported π–π interaction in the IL head groups. The phenyl ring compelled [C_{14}pim][Br] to arrange densely at the air–water interface by minimizing

TABLE 10.1 Variation of Surface Tension of Pure ILs with Temperature and Ion Combinations

ILs		γ, mN m^{-1} (T, K)	
[C_4C_1im][BF_4]	46.6 (298) [5]	44.8 (293) [20]	45.4 (284) [21]
[C_4C_1im][Tf_2N]	37.5 (298) [5]	33.6 (293) [20]	33.5 (284) [21]
[C_4C_1im][SCN]	47.5 (298) [18]	46.9 (293) [22]	—
[P_{66614}][Tf_2N]	33.1 (298) [19]	30.8 (293) [23]	—
[P_{66614}]Cl	33.6 (298) [19]	30.7 (293) [24]	—

TABLE 10.2 CMC and γ at CMC (γ_{CMC}) in Aqueous Solution at 298 K of Surfactant-Like Ionic Liquids with Different Alkyl Side Chain Lengths Attached with Their Cations or Anions

Ionic Liquid	CMC, mM	γ_{CMC}, mN m^{-1}	Reference
$[C_{12}C_1im]Br$	10.9	39.4	25
$[C_{14}C_1im]Br$	2.8	39.2	25
$[C_{16}C_1im]Br$	0.55	39.1	25
$[C_{10}C_1py]Cl$	45	44.6	26
$[C_{12}C_1py]Cl$	13	44.6	26
$[C_{14}C_1py]Cl$	3.1	44.5	26
$[C_{16}C_1py]Cl$	0.8	44.6	26
$[C_{18}C_1py]Cl$	0.3	44.7	26
$[pyrr][C_5CO_2]$	9.3	54.2	27
$[pyrr][C_6CO_2]$	4.8	49.0	27
$[pyrr][C_7CO_2]$	2.8	47.8	27
$[pyrr][C_8CO_2]$	2.1	46.4	27
$[im][C_5CO_2]$	11.2	55.5	27
$[im][C_6CO_2]$	5.8	54.0	27
$[im][C_7CO_2]$	4.6	51.6	27
$[im][C_8CO_2]$	2.6	50.6	27
$[C_8C_1im][C_4SO_3]$	140	44.0	28
$[C_8C_1im][C_8SO_3]$	12	28.8	28

the steric hindrance acting among the aromatic rings. A fair degree of delocalized charge of the cationic head was considered to reduce the mutual electrostatic repulsion. In addition, enhanced hydrophobicity due to phenyl ring incorporation favored easier micellization. The phosphonium-based ILs are known to lower the surface tension (γ=30–35 mN m^{-1}) in the temperature range of 298–308 K. They have fair degree of amphiphilicity [31, 32]: the self-assembly of ILs into micelles has been reported [33–41].

IL-1 (triisobutyl(methyl)phosphonium tosylate) is highly soluble in water and has good surface activity but does not form micelle in solution, whereas IL-2 (trihexyl(tetradecyl)phosphoniumbis(2,4,4-trimethylpentyl)phosphinate) is very weakly soluble but micellizes after a critical concentration, that is, CMC. A gradual decrease in γ was observed up to 31.4 mN m^{-1} at [IL-1] = 2.2 M (83% by mass) without showing any CMC (Fig. 10.1a). IL-1 is comparable with tetraalkyl salts, namely, tetramethyl and tetraethylammonium bromides studied by us [42]. Tensiometric measurements (Fig. 10.1b) evidenced a CMC of IL-2 at 0.03 mM. TX-100 produced a CMC of 0.22 mM, sevenfold larger than IL-2. The CMC of Tween-20, Tween-40, and Tween-60 are 0.05, 0.023, and 0.021 mM, respectively, and that of Brij 35 and 700 are 0.06 and 0.02 mM, respectively. The CMC of IL-2 compared well with Tweens and Brijs. Phosphorus-based non-IL surfactants dodecyl-, tetradecyl-, and hexadecyltriphenylphosphonium bromides have shown larger CMC values of 2.53, 0.69, and 0.095 mM, respectively, at 303 K [43]. The differences in the head groups and the counterions are considered the reasons for these variations. The low CMC of IL-2 was caused by large hydrophobicity produced by the attached three hexyl and one tetradecyl chains to the phosphonium cation of the molecule. It is evident from Figure 10.1a and b that at 0.1 mM, IL-1 reduced the surface tension (γ) of water to 68 mN m^{-1}, whereas IL-2

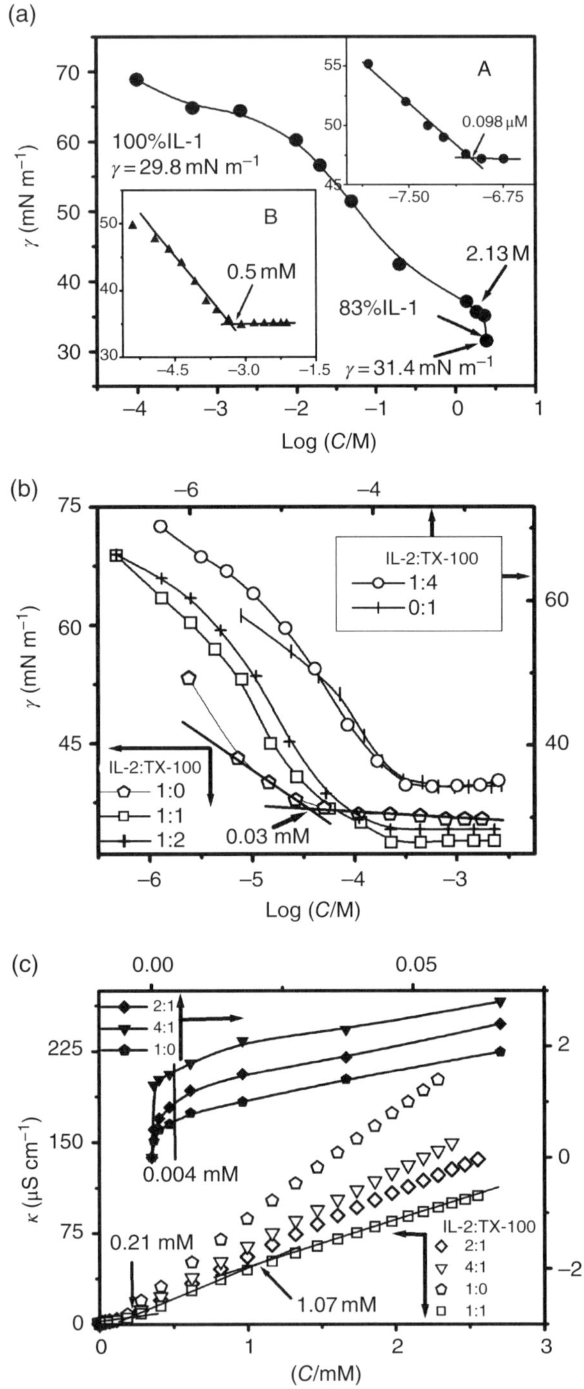

Figure 10.1 (a) γ log C profile of IL-1 in aqueous medium at 298 K. Inset A: γ log C profile of IL-2 in IL-1 at 298 K. Inset B: γ log C profile of aqueous TX-100 with 0.07 M [IL-1] at 298 K. (Coordinates: same as in main plot.) (b) γ log C profiles of IL-2+TX-100 binary mixtures at mole ratios 1:0, 1:1, 1:2, 1:4, and 0:1 at 298 K. (Coordinates are indicated by arrowheads.) (c) Specific conductance (κ) versus concentration (C) plots of binary mixtures of IL-2 and TX-100 at mole ratios 1:0, 4:1, 2:1, and 1:1 at 298 K, with initial stage expanded for mole ratios 1:0, 4:1, and 2:1 in top and right coordinates. (Coordinates are indicated by arrowheads.)

reduced it to 35 mN m^{-1} at 298 K. At this concentration, surface tension of TX-100 (γ_{TX-100}) became 45 mN m^{-1}. In the surface activity scale, IL-2 was well up. In the binary mixtures of IL-2 and TX-100, a systematic increase in CMC of the mixtures with increasing proportion of TX-100 was observed. Pure IL-2 produced three breaks in the conductometric plot, so also its mixtures with TX-100. IL-2 produced breaks at 0.004, 0.16, and 1.07 mM. Of the three breaks, the first at 0.004 mM (Fig. 10.1c) was common to all mixed compositions with TX-100 of mole ratios 1 : 0, 1 : 1, 1 : 2, 1 : 4, 1 : 6, 2 : 1, and 4 : 1; the positions for the second and third breaks were composition dependent (Fig. 10.1c). None of these breaks matched with the CMC points registered in tensiometry (Fig. 10.1b). The conductance behavior revealed variable internal interactions in addition to self-association of the mixed species.

Mutual mixing of two ILs might show fascinating properties. Low concentration (0.3 mM) of IL-1 was found to produce a striking effect on IL-2. In the γ versus log [IL-2] dependence, a CMC-like inflection of IL-2 was observed at a very low concentration of 0.1 μM at 298 K; in aqueous medium, the CMC of IL-2 was found to be 0.03 mM. There were reports of CMC decrease on mutual interaction between two ILs [29]; interaction in the IL-2/IL-1 combination herein found was conspicuous (Fig. 10.1a, inset A). Tensiometric results of CMCs of mixed micelles were tested against Clint's ideal mixed micelle formation model.

The Clint equation is as follows:

$$\frac{1}{\mathrm{CMC}_{\mathrm{mix}}} = \sum_{i=1}^{n} \left(\frac{X_i}{\mathrm{CMC}_i} \right)$$

where CMC_i is the CMC of the ith species with mole fraction X_i in the mixture, n is the number of species present in the mixture, and $\mathrm{CMC}_{\mathrm{mix}}$ is the expected value for the mixed system. Experimentally observed $\mathrm{CMC}_{\mathrm{mix}}$ did not satisfy the Clint equation; mixing was nonideal with positive deviations, suggesting a mutual antagonistic interaction between IL-2 and TX-100 in their mixed forms. Above results were also tested in terms of regular solution theory of Rubingh [42, 43]. This model failed to derive information on the observed nonideality of the mixed system. The mixed system has shown exceptionality in their solution behavior.

In aqueous solution, ionic [44–47], nonionic [48–53], and zwitterionic [54, 55], surfactants have shown significant changes in their CMC and aggregation number (N) of their self-assemblies by the presence of ILs. Behera and Pandey [45] have reported decrease in CMC of aqueous sodium dodecyl sulfate (SDS) with increasing N and average micellar size by the addition of low concentration (≤2 wt.%) of "hydrophilic" IL (1-butyl-3-methylimidazolium tetrafluoroborate [bmim][BF_4]). In addition, they have proposed a concentration-dependent dual CMC influencing effect on SDS using a hydrophilic IL, 1-butyl-3-methylimidazolium tetrafluoroborate. Like a common electrolyte, it decreased the CMC in the concentration range, 0 < [bmim][BF_4] ≤ 2 wt.%. Thereafter, the CMC of SDS increased with increasing concentration of [bmim][BF_4] up to 30 wt.%. Tween-20 has witnessed two different kinds of aggregates in imidazolium-based ILs like [bmim][BF_4] and [bmim][PF_6] [50]. A reported literature [55] also evidenced changes in physicochemical properties of aqueous solution of a zwitterionic surfactant, N-dodecyl-N,N-dimethyl-3-ammonio-1-propanesulfonate (SB-12), on addition of the hydrophilic IL, 1-butyl-3-methylimidazolium

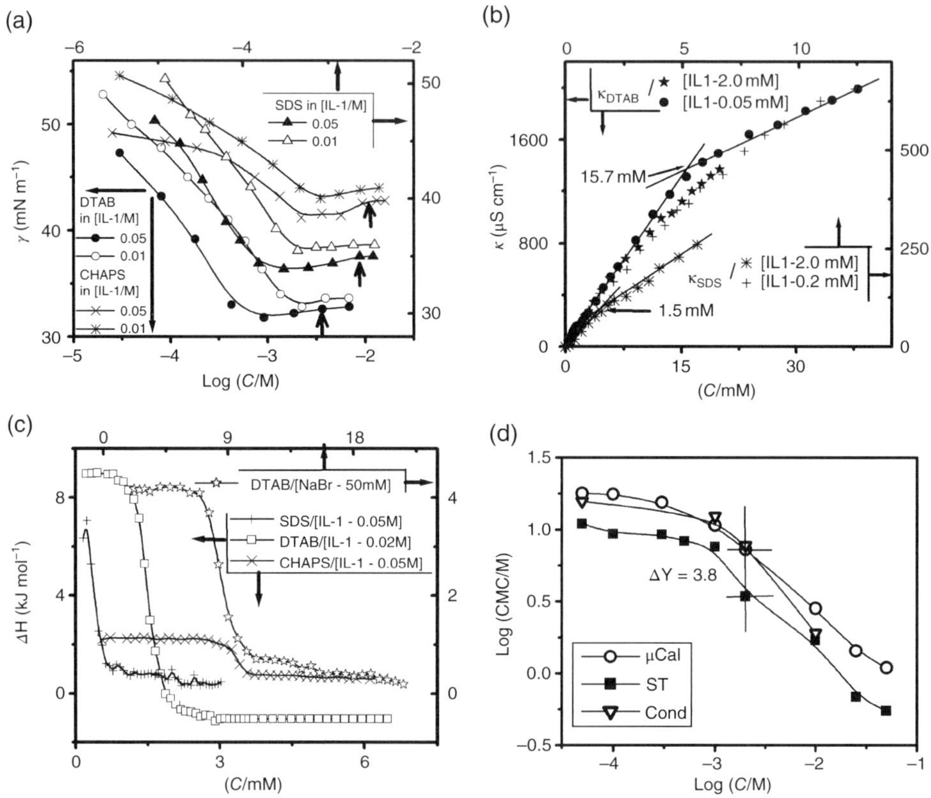

Figure 10.2 (a) IL-1 influenced γ log C profiles of SDS, DTAB, and CHAPS in aqueous medium at 298 K, with broad arrowheads indicating starts of plateau points. [IL-1] used = 10.0 and 50.0 mM. (Coordinates indicated by narrow arrowheads.) (b) IL-1 influenced κ - C profiles of SDS and DTAB in aqueous medium to determine their CMCs at 298 K. [IL-1] used were 0.05 and 0.2 mM for DTAB and 2.0 and 0.2 mM for SDS. (Coordinates are indicated by arrowheads.) (c) Enthalpograms of SDS, DTAB, and CHAPS in aqueous medium with [IL-1] and NaBr at 298 K. [IL-1] used = 20 and 50 mM; [NaBr] = 50 mM. (d) [IL-1] influenced CMC of DTAB obtained by tensiometry (■), conductometry (∇), and microcalorimetry (○) at 298 K.

tetrafluoroborate. We have also studied CMC influencing effect of IL-1 on the surfactants TX-100 (nonionic; Fig. 10.1a, inset B), DTAB (cationic), SDS (anionic), and CHAPS (zwitterionic) (Fig. 10.2a–c) [56]. The CMC of TX-100 in water was increased from 0.22 to 0.5 mM in the presence of 0.07 M IL-1 at 298 K. In the contrary, with increase in [IL-1], the CMCs of all other surfactants (Table 10.3, [56]) decreased [57–59]; the IL-1 facilitated their micelle formation. The reduction of CMC of DTAB was due to the interaction between the anionic tosylate moiety of IL-1 and the cationic DTA^+. The zwitterionic surfactant, CHAPS, balanced the influences of both cations and anions of IL-1 to manifest an overall mild effect. Weak charge interaction between the cationic part of IL-1 and the anionic part of CHAPS and favorable orientation of the triisobutyl groups of the former along the tails of the latter favored the formation of micelle by reducing the CMC. In this connection, the enhancement in CMC of TX-100 by low IL-1 concentration was striking. Magnitudes of CMCs by

TABLE 10.3 Influence of [IL-1] on the CMC, β, ΔH_m^0 of DTAB, SDS, and CHAPS Self-Aggregation Studied by Different Methods (ST, Cond, μCal) at 298 K

	DTAB				SDS				CHAPS		
[IL-1], mM	ST [Plateau]	Cond (β)	μCal	ΔH_m^0	ST [Plateau]	Cond (β)	μCal	$-\Delta H_m^0$	ST [Plateau]	μCal	$-\Delta H_m^0$
0.00	16.2	15.1 (0.71)	16.7	2.95	9.1	8.6 (0.6)	9.9	2.6	5.3	5.1	1.54
0.05	11.0 [30.2]	15.7 (0.67)	17.9	4.61	—	—	—	—	—	—	—
0.10	11.2 [24.6]	—	17.6	3.64	4.7 [12.6]	—	5.5	0.49	—	—	—
0.20	—	—	—	—	—	3.6 (0.32)	4.7	0.66	4.8 [20.0]	4.6	0.31
0.30	11 [19.8]	—	15.4	3.78	—	—	—	—	—	—	—
0.50	8.3 [17.6]	—	—	—	2.2 [4.2]	—	3.8	1.00	—	—	—
1.00	7.4 [19.0]	9.7 (0.35)	10.7	2.24	—	—	—	—	—	—	—
2.00	3.4 [5.6]	6.75 (0.37)	7.31	2.44	1.2 [2.5]	1.5 (0.35)	1.7	2.10	3.6 [14.0]	—	—
10.0	2.4 [4.2]	1.74 (0.42)	2.84	6.28	0.4 [—]	0.8 (0.57)	1.0	2.23	3.5 [11.1]	3.4	0.63
20.0	0.7 [1.4]	—	1.44	9.77	—	—	0.8	2.90	—	—	—
50.0	0.9 [4.0]	—	1.05	9.63	0.3 [2.3]	—	0.3	5.89	2.9 [13.5]	2.8	1.31
200	—	—	—	—	0.1 [0.4]	—	0.1	9.42	—	—	—

Reproduced from Mukherjee et al. [56], *J. Colloid Interface Sci.* **2013**, *395*, 135, with permission from Elsevier Science.
CMCs (by ST, Cond, and μCal) are expressed in mM; ΔH_m^0 are expressed in kJ mol^{-1}.
Average errors in CMC were ±8%; errors in enthalpy ±5%.
CMCs of pure aqueous surfactant solutions were close with earlier reports [57].
$\beta = [1 - (\text{post-CMC slope/pre-CMC slope})]$ in κ versus C plot [58, 59].
ST, surface tension; Cond, conductance; μCal, microcalorimetry.

TABLE 10.4 CMC of Surfactants in ILs and Water Media as Solvents at 298 K

Surfactant	Solvent	CMC/mM	Reference
CTAB	Water	0.96	63
	[emim][Tf_2N]	No aggregation	33
C_{14}PyrB	Water	97.0	64
	EAN	2.7	40
C_5AN	EAN	350	65
C_8AN	EAN	79.0	65
C_{10}AN	EAN	25.0	65
C_{10}AN	EAN	7.9	65
SDS	[bmim][Cl]	48.0	34
	[emim][Tf_2N]	Did not dissolve	33
	Water	8.0	34
Brij-35	[bmim][PF_6]	118	34
	[emim][Tf_2N]	50	33
Brij-700	[bmim][PF_6]	20	34
	[emim][Tf_2N]	10	33
TX-100	[emim][Tf_2N]	>100	33
	Water	0.2	56
Tween-20	[emim][Tf_2N]	50	33
	Water	0.08	66

calorimetry for the ionic surfactants were found to be higher (specifically for the cationic DTAB); the zwitterionic CHAPS produced nearly equivalent CMCs both by ST and calorimetry. CMC values of DTAB by tensiometry were roughly 50% lower than the CMC by conductometry and microcalorimetry. These differences are graphically presented in Figure 10.2d: log CMC_{ST} maintained almost a constant difference (~3.8 unit) with log CMC (calorimetry) and log CMC (conductometry).

CMCs of well-studied surfactants like CTAB, Brijs, TX-100, etc. in IL have shown fascinating behavior compared to their CMCs in aqueous medium. Aggregation of surfactants in IL media in different environmental conditions compared with the association in aqueous medium is driven by weak solvophobic effect generated in ILs along with entropy gain. In ILs, dissolution of amphiphilic hydrocarbons is facilitated that results in higher CMC value, whereas in aqueous medium, solvophobic effect produces easier micellization with lower CMC. Baker and Pandey have reported [60] amphiphilic association in ILs like EAN, [bmim][BF_4], [bmim][PF_6], etc. In another report, Hao and Zemb [61] showed similar solvent behavior of ILs for supporting aggregation of amphiphiles in them as micelles, vesicles, etc. Greaves and Drummond [62] have reported amphiphilic assemblies in ILs, though having better solubility of amphiphiles in them. Few CMC ranges of amphiphiles in IL media are depicted in Table 10.4 [63–66].

10.3 PHYSICOCHEMICAL BEHAVIOR AND BIOLOGICAL APPLICATIONS

Since surface activity is found in a large number of ILs, these classes of compounds claim much prospects in interfacial chemistry and their related applications. More than 10,000 publications and patents exceeding 2000 corroborate the aforesaid facts of ILs in the realm of applications.

Cellulose dissolution, paint additives [67], batteries [68], and solar cells [69] are common domain of IL's applicability with extended prospect in biological fields like functionality in protein stabilization [70], biocatalysis [71], biocidal and antibacterial activities, etc. Initial claim of "greenness" of ILs often failed to satisfy green principles like sustainable synthesis, low toxicity, and limited environmental persistence (green chemistry principles 2, 3, and 10, respectively) [72] due to nebulous concepts. Thus, present researchers started to focus on designing biocompatible ILs with amino acids, carboxylic acids, glucose, etc. [73–75] to get rid of hazards evolved out of normally synthesized ILs. Use of ILs comprising alkyltriphenylphosphonium cation with long alkyl chains as surfactant for cationic exchange of layered silicates has been reported by Livi et al. [76].

Biodegradation, antimicrobial, anticancer, and pharmaceutical activities of ILs are other domains of study proliferating in the current decade. Phosphonium- and ammonium-based ILs evidenced anticancer activity. ILs with longer chain length and specific alkyl substitution on cations were most active and less toxic [77]. Frade et al. have reported on the impact of ILs on human colon carcinoma HT-29 and CaCo-2 cell lines [78]. In this perspective, our findings on the microbial and anticancer activity [11, 12, 56] of the ILs are worth considering. Both IL-1 and IL-2 have produced fair antibacterial activities against *Bacillus subtilis* (Gram-positive) (Fig. 10.3a and b) and *Escherichia coli* (Gram-negative) bacteria (Fig. 10.3c and d) measured by diameter of inhibition zone ("diz") study [11]. Aqueous IL-1 solution of 0.54 g/ml against *B. subtilis* evidenced maximum inhibitory effect ("diz" = 32 mm; Fig. 10.3a), which was reduced against *E. coli* (24 mm; Fig. 10.3c). IL-2 solution (0.54 g/ml) in the isopropyl myristate (IPM) exhibited lower activity against both of the bacterial strains (26 and 21 mm "diz" for *B. subtilis* and *E. coli*, respectively, in Fig. 10.3b and d).

Ungphaiboon et al. [12] have reported a formulation comprising 0.5% w/v turmeric (*Curcuma longa* Linn) extract (a potential disinfectant) in clear liquid soaps, producing 16.7 mm "diz" at ambient temperature against *B. subtilis* after 1 month of storage time. The formulated turmeric liquid soap has been used in skin infection treatment of HIV patients. In comparison, ILs may show better biocidal activities than the clear turmeric liquid soap. Although IL-1 and IL-2 exhibit fair antibacterial activities, their potency is considerably lower than the antibiotic erythromycin, which exhibits "diz" = 32 mm for a 0.003 g/ml aqueous solution. It was observed that 5 mg/ml of IL-1 in 0.9 g% NaCl solution (physical saline) caused 100% killing of Sarcoma-180 (1×10^5 cells/incubation mixture) cell line in 1 h. Half dilution of the aforesaid IL-1 solution causes a 19% reduction in the killing efficacy. Five- and tenfold dilution resulted 47% and 78% reduction of the killing efficacy, respectively. IL-2 had manifested lesser anticancer activity than IL-1. 10 mg/ml IL-2 took almost 2 h to kill 89% of Sarcoma-180 cells and 1 h for 58% of the cells. Cancer is an immunosuppressive disease; hence, cancer patient becomes much susceptible to secondary infections (bacterial infections). Owing to their both anticancer and antibacterial activities, ILs have better prospect in therapeutic applications. Detailed investigations in both *in vitro* and *in vivo* conditions are required.

10.3.1 Phase-Forming Behaviors and Microemulsifications of ILs

In colloidal research, it has been often found that ILs are employed as polar and nonpolar domains; moreover, they can even serve as amphiphilic components in the formation of microheterogeneous assemblies like micelles, microemulsions, etc.

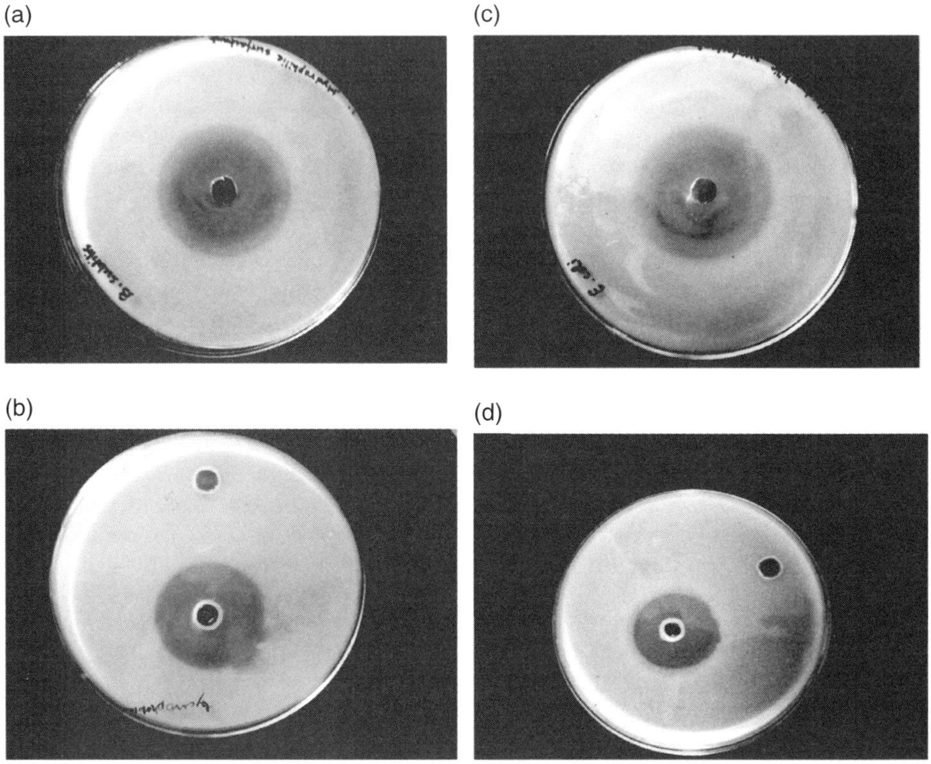

Figure 10.3 Representative images in agar plates showing IL-1 and IL-2 produced inhibition zones for *Bacillus subtilis* (BS) and *Escherichia coli* (EC). (a) Inhibition zone produced by IL-1 against BS. (b) Inhibition zone produced by IL-2 against BS. (c) Inhibition zone produced by IL-1 against EC. (d) Inhibition zone produced by IL-2 against EC. Reproduced from Mukherjee et al. [11], *J. Chem. Eng. Data* **2012**, *57*, 1376, with permission from the American Chemical Society.

Microemulsions are defined as dispersion of either water in oil or oil in water aided by mixed amphiphiles to produce transparent, isotropic, low viscous, thermodynamically stable solution systems. A microheterogeneous system of microemulsion with nanodispersion of one liquid into another must have a very low (about zero) interfacial tension caused by the presence of amphiphiles at the interface. In addition, the interfacial region must be highly flexible, either to permit the curvature required to surround exceedingly small particles or to allow the easy transition from oil-continuous to water-continuous structure, which is an important physicochemical feature of microemulsion. Along with the concept of nanodispersed droplets of either water in oil or oil in water, there is another concept of simultaneous dispersion of both water and oil in a system termed as "bicontinuous" where both the liquids are dispersed nearly in equal proportions. A microemulsion system can be one of the three types depending on the relative ratios of the constituting components. To achieve formation of any such system, phase behavior of the multicomponent (oil, water, amphiphile) systems is required to be studied, which is complex and has been elaborately

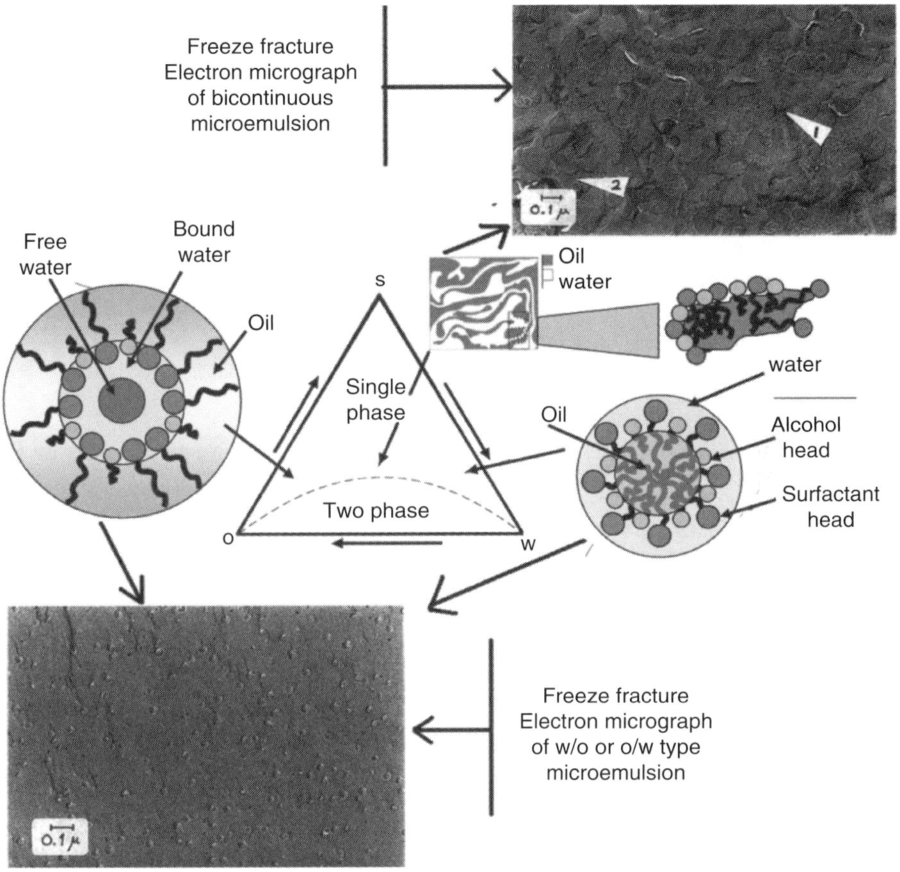

Figure 10.4 Pseudoternary phase diagram showing mono- and biphasic regions of a system comprising surfactant/cosurfactant (s), oil (o), and water (w) and elucidation of w/o-, o/w-, and bicontinuous-type microemulsion zones along with their dispersion level scheme and electron micrographs.

described in literature. A pictorial demonstration of phase-forming behavior of a practical system is depicted in Figure 10.4.

The free energy (driving force) of microemulsion formation can be considered to depend on the extent of lowering of surface tension of the oil–water interface by surfactant and the change in entropy of the system such that

$$\Delta G_{\mathrm{f}} = \gamma \Delta A - T \Delta S$$

where ΔG_{f} is the free energy of formation, γ is the surface tension of the oil–water interface, ΔA is the change in interfacial area on microemulsification, ΔS is the change in entropy of the system that is effectively the dispersion entropy, and T is the absolute temperature. Microemulsification results in the increase in ΔA due to the formation of large number of very small droplets. Basically to form microemulsion, a (transient) negative value of γ is required, but a very small positive value can

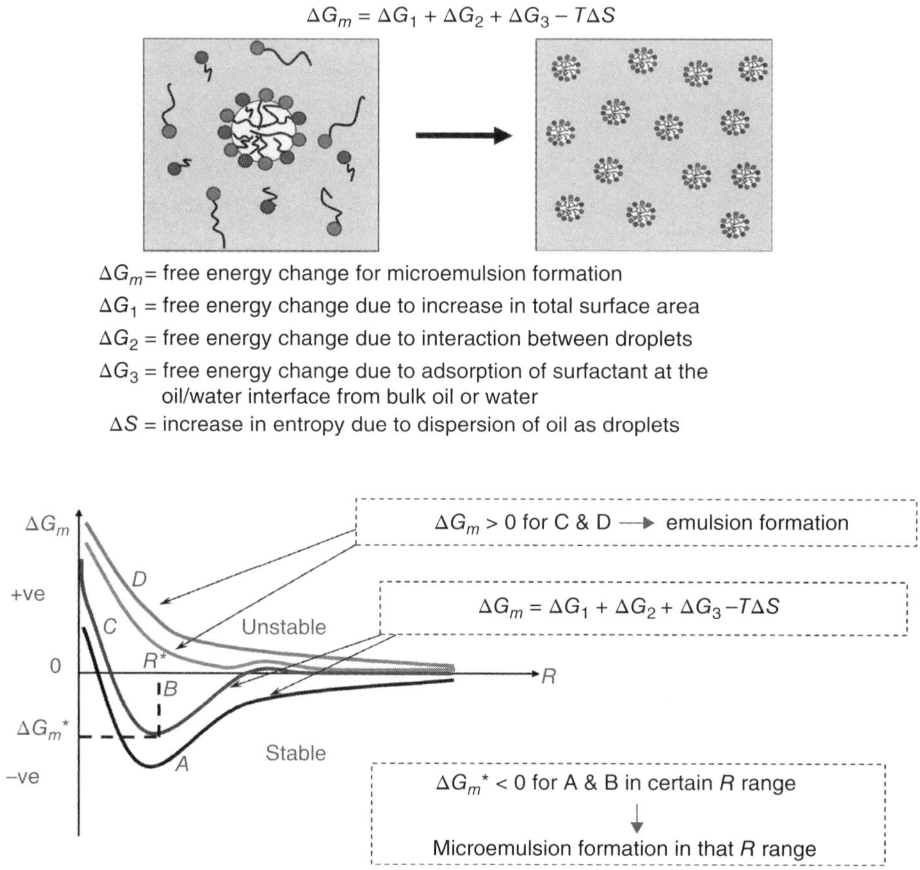

Figure 10.5 Schematic representation of microemulsion droplet formation and energy diagram. Reproduced from Ruckenstein and Chi [79], *Faraday Trans.* **1975**, *II* 71, 1690, with permission from the Royal Society of Chemistry.

serve the purpose since the entropic component can offset the tiny positive free energy to make the system thermodynamically feasible with formation of a stable dispersion [79]. Thermodynamic stability of microemulsion and the related energy diagrams are schematically presented in Figure 10.5.

ILs have shown versatility in the formation of microemulsions. Lü et al. [80] have reported a system with polar domain [bmim][BF_4] (1-butyl-3-methylimidazolium tetrafluoroborate)/AOT/decane. In almost similar fashion, Gao et al. [81] explored a system consisting of 1-butyl-3-methylimidazolium tetrafluoroborate ([bmim][BF_4] polar phase)/TX-100/toluene (oil). They have also studied the changes in phase-forming behavior and physicochemical properties by varying the oil component with cyclohexane and benzene and keeping the polar and the amphiphilic parts the same [82]. Varied hydrophobic environment played the key role in forming oil-in-IL (O/IL) microemulsion, whereas in the case of IL-in-oil (IL/O) microemulsion formation, electrostatic interaction between the positive imidazolium cation and electronegative oxygen of TX-100 compelled [bmim][BF_4] to be entrapped in the core. O/IL droplet got swollen with the incorporation of nonpolar phases (as oils), and after a certain range of addition, it led to the formation of bicontinuous-type microemulsion. Anderson et al. [34] have reported IL/O-type microemulsion droplets of 0.1 μm characterized from light scattering technique. The system [bmim][BF_4]/TX-100/toluene prepared by Li et al. [50] has evidenced microemulsion of IL/O, O/IL, and bicontinuous types, corroborated by DLS and UV–Vis absorbance study. They have concluded that [bmim][BF_4] molecules got fastened with oxyethylene groups of TX-100 in the polar core having different spectral nature as compared with bulk [bmim][BF_4]. IL/O-type droplets have evidenced a solubilization efficacy of biological molecules like riboflavin, metal salts like $Ni(NO_3)_2$, $CoCl_2$, $CuCl_2$, etc. In few reports [53, 83], it was observed that ILs have been used as a nonpolar domain within a system comprising water/Tween-20 or TX-100/[bmim][PF_6] (1-butyl-3-methylimidazolium hexafluorophosphate), where weak electrostatic interaction between $[bmim]^+$ and oxyethylene group of TX-100 played the key role in the formation of microemulsion droplet. Cheng et al. [84] have reported a system comprising EG/TX-100/[bmim][PF_6]. EG in [bmim][PF_6], [bmim][PF_6] in EG, and bicontinuous types were evidenced (Fig. 10.6), and the molar ratio between [bmim][PF_6] and TX-100 influenced the droplet dimensions linearly.

Li et al. [85] have explored a system with [bmim][BF_4]/TX-100/triethylamine. This system has a unique templating capability for synthesizing metal hydroxide or metal oxide in its droplet core, which is rare in using known traditional microemulsions. Very recently, Paul and Panda [86] have reported an interesting system with 1-butyl-3-methylimidazolium methanesulfonate ([bmim][MS] + water)/(Tween-20 + *n*-pentanol)/*n*-heptane. Phase manifestation of it revealed that the area under the clear region depended on IL–water mole ratio. Chai et al. [87] have explored phase behaviors and thermodynamic properties of IL/O microemulsions containing 1-butyl-3-methylimidazolium tetrafluoroborate ([bmim][BF_4])/1-dodecyl-3-methylimidazolium bromide ([C_{12}mim][Br] + pentan-1-ol)/octane and W/O microemulsion by replacing [bmim][BF_4] with water. The area of the single-phase domain in [bmim][BF_4] microemulsion system is larger than that in water-containing system. Cheng et al. reported a system of IL-in-IL-type microemulsion [88] that was comprised of protic IL propylammonium formate/AOT/[bmim][PF_6]. But it exhibited only a tiny single-phase zone. Rojas et al. [89] prepared a system with [emim][$EtOSO_3$], that is, 1-ethyl-3-methylimidazolium ethylsulfate as the polar phase, and [bmim][$OctSO_3$], that is, 1-butyl-3-methylimidazolium octylsulfate, with a surfactant-like counterion as the amphiphile and toluene as oil. Another system was also reported by them [90] as 1-ethyl-3-methylimidazolium *n*-hexyl sulfate (IL)–water (1:1) (polar phase)/SDS/toluene–pentanol (1:1) (oil phase). According to Atkin and Warr [91], microemulsions containing EAN required a higher concentration of surfactant with longer chain length for balancing the reduced solvophobicity of EAN compared with the same system comprising water instead of EAN. Literature reports revealed that TX-100 has been mostly used in the preparation of IL-based microemulsions [53, 92]. Eastoe et al. [93] have reported the presence of IL in cyclohexane droplets in microemulsion that increased with the addition of ILs. For determining droplet dimensions of different IL-based microemulsion systems, SANS study [94] is important. Phosphonium cation-based ILs were developed by us [56] as nonaqueous microemulsion systems comprising triisobutyl(methyl)phosphonium tosylate (IL-1) polar domain/trihexyl(tetradecyl)phosphonium bis(2,4,4-trimethylpentyl)-phosphinate (IL-2) and isopropanol (IP) as surfactant and cosurfactant, respectively, and IPM as oil. A broad clear zone was observed, which remained same when IP was

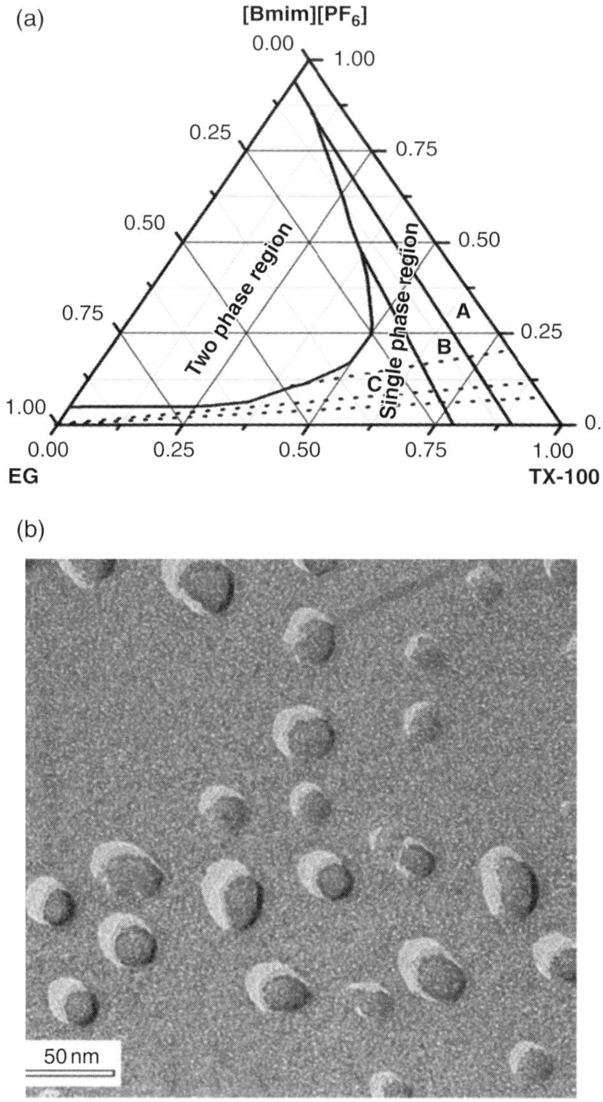

Figure 10.6 (a) Phase diagram of the IL/TX-100/EG ternary system at 30.0°C; the dash lines were chosen for conductivity study. A, B, and C are EG-in-IL, bicontinuous, and IL-in-EG regions, respectively. (b) TEM images of FFEM replicas of IL-in-EG microemulsions stabilized by TX-100 (weight fraction of TX-100 is 0.50 and $R=0.28$). Reproduced from Cheng et al. [84], *Colloids Surf. A* **2007**, *302*, 211, with permission from Elsevier Science.

replaced by butanol (Bu). Our results produced larger clear area compared with that of Cheng et al. [88]. In both the cases, hydrophilic and hydrophobic ILs were used to prepare a nonaqueous-type microemulsion. Another system, where IL-1 was replaced by water and was stabilized by IL-2 and IP, produced a comparatively smaller clear zone [56] than the nonionic formulation mentioned earlier and remained stable up to 323 K (Fig. 10.7).

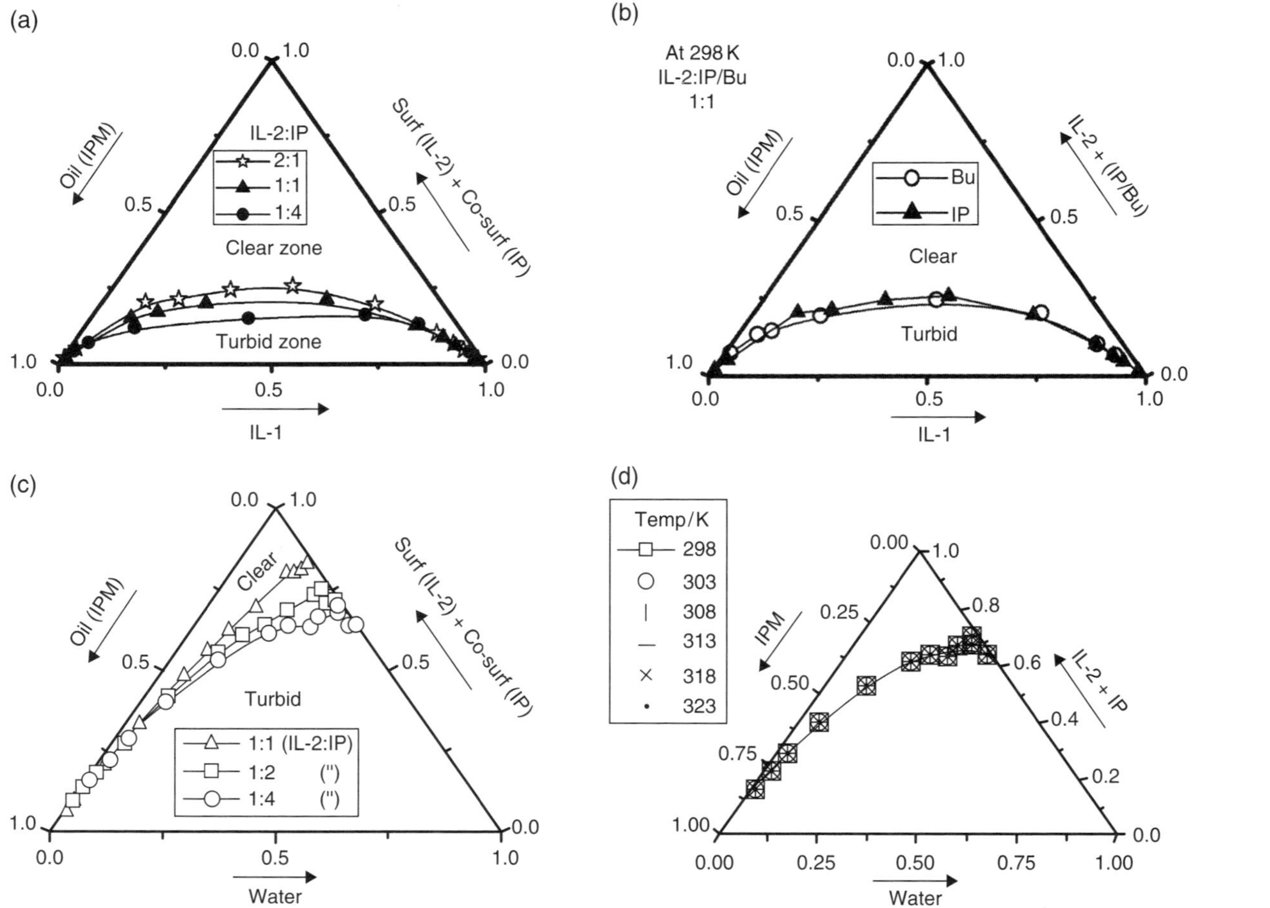

Figure 10.7 Pseudoternary phase diagrams of IL-based microemulsion systems. (a) IL-1/IPM/IL-2 + IP system at IL-2 : IP ratios (wt./wt.) 2 : 1, 1 : 1, and 1 : 4 at 298 K. (b) IL-1/IPM/IL-2 + IP or Bu system at IL-2 : IP or Bu ratios (wt./wt.) 2 : 1 at 298 K. (c) Water/IPM/IL-2 + IP system at IL-2 : IP ratios (wt./wt.) 1 : 1, 1 : 2, and 1 : 4 at 298 K. (d) Water/IPM/IL-2 + IP system at IL-2 : IP ratio (wt./wt.) 1 : 1 within 298–323 K. Reproduced from Mukherjee et al. [56], *J. Colloid Interface Sci.* **2013**, *395*, 135, with permission from Elsevier Science.

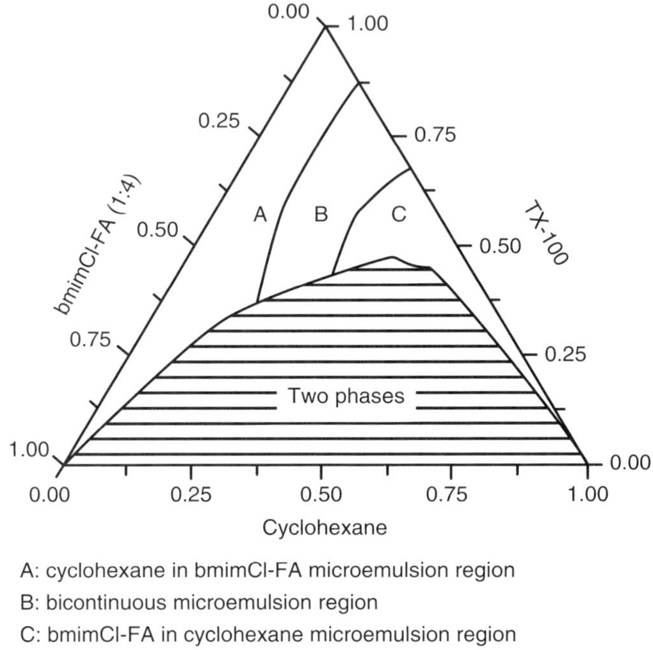

Figure 10.8 Pseudoternary phase diagram of IL-based microemulsion system comprising [bmim][Cl]–formamide/TX-100/cyclohexane at 25 ± 0.1°C. Reproduced from Wei et al. [97], *Colloids Surf. A* **2012**, *414*, 82, with permission from Elsevier Science.

Microemulsions with ILs acted as catalytic activity enhancer for oxidases. Zhou et al. [95] reported a water-in-[bmim][PF_6] microemulsion system stabilized by TX-100 that enhances the catalytic activity of lignin peroxidase (LiP) and laccase. Optimum molar ratios of water to TX-100 were 8 and >20 for LiP and laccase, respectively. Compared to pure or water-saturated [bmim][PF_6], the derived microemulsion evidenced enhanced catalytic activity. Use "O/W" for "oil-in-water" ivation effect of [bmim][PF_6] on LiP and laccase. Xue et al. [96] reported tunable enzyme (laccase) activity in a microemulsion system, water/AOT+TX-100/[bmim][PF_6]. The solution of IL [bmim][Cl] and polar organic solvent formamide (FA) were used to form a nonaqueous microemulsion as [bmim][Cl]–FA/TX-100/cyclohexane (Fig. 10.8) at 25 ± 0.1°C, reported by Wei et al. [97]. By means of electrical conductivity, dynamic light scattering (DLS) and UV–Vis spectroscopy measurements, microstructures, internal phases, and size regime were explored of the aforesaid microemulsion system. UV–Vis studies using $CoCl_2$ indicated metal salt dissolution by microemulsion.

Microemulsion systems with toluene, pentanol, CTAB, and the IL 1-ethyl-3-methylimidazolium ethylsulfate [emim][$EtOSO_3$] or 1-ethyl-3-methylimidazolium hexylsulfate [emim][$HexOSO_3$] were reported by Rabe and Koetz [98]. Increase in pentanol concentration evidenced enlargement of the isotropic region (Fig. 10.9).

As per percent of clear area is concerned (in the phase diagrams of various IL-based microemulsion systems), a comparative study revealed greater extent of clear zone in our system with IL-1/IL-2 + IP/IPM than the other systems presented in Table 10.5 [56]. Replacement of water by IL-1 in the aforesaid

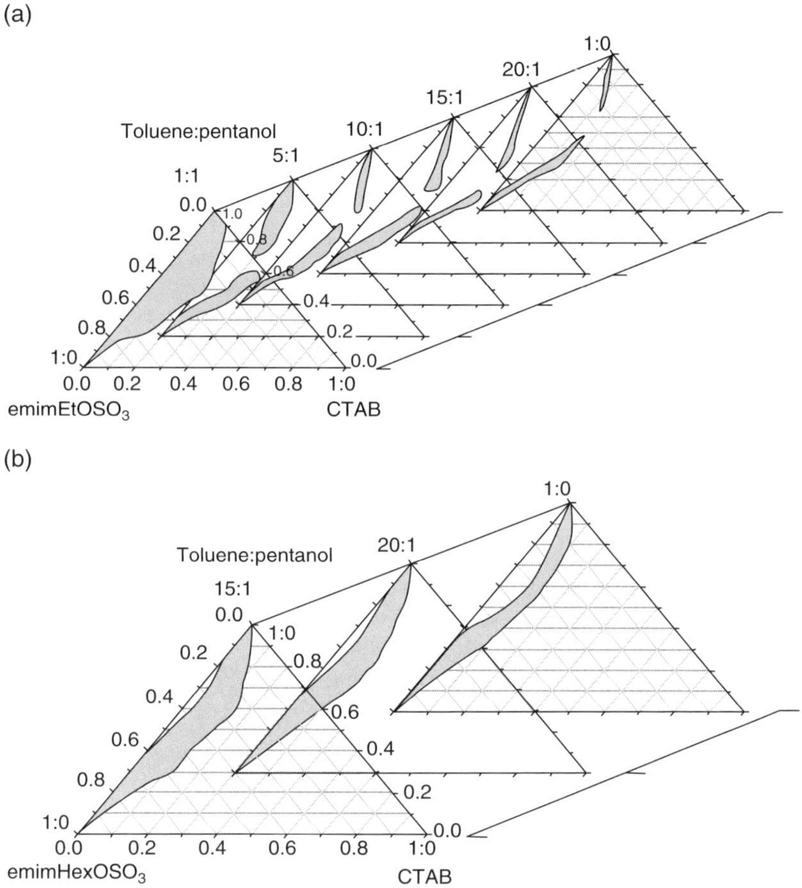

Figure 10.9 At 25°C partial phase diagrams with varying amount of pentanol within the systems comprising (a) CTAB/toluene–pentanol/[emim][EtOSO$_3$] and (b) CTAB/toluene–pentanol/[emim][HexOSO$_3$]. Reproduced from Rabe and Koetz [98], *Colloids Surf. A* **2010**, *354*, 261, with permission from Elsevier Science.

system strikingly increased the area of clear zone from 12 to almost 70%. Favorable presence of (IL-2 + IP) at the IL-1/IPM interface than water/IPM made easier and greater dispersion of IL-1 in IPM than in water. Less cohesive IL-1 than water made its interface more flexible and amenable to easy bending. The observation was unique in nature and brought promise to more trials with different nonmicelle-forming ILs.

Temperature and the counterion hydrophobicity of ILs can also affect microemulsion formation, which in turn produces bicontinuous-type microemulsions. DLS, rheology, and cryo-SEM (Fig. 10.10) elucidated droplet shape and sizes along with the mutual interactions [98].

Safavi et al. [99] have prepared a microemulsion based on IL, 1-octyl-3-methylimidazolium chloride [omim][Cl], as surfactant, hydrophobic IL, 1-butyl-3-methylimidazolium hexafluorophosphate [bmim][PF$_6$], as oil, and water at 25°C.

TABLE 10.5 Percent Area of Clear Zones of Different Microemulsion Systems

System	Area of Clear Zones
(A) Water/IPM/[IL-2: IP (wt./wt.)-1:1, 1:2, 1:4]	1:1 (12.1%), 1:2 (17.1%), 1:4 (21.2%)
(B) IL-1/(IPM)/[IL-2: IP (wt./wt.)-2:1, 1:1, 1:4]	2:1 (63.9%), 1:1 (69.7%), 1:4 (75.9%)
(C) 1-Butyl-3-methylimidazolium tetrafluoroborate ($bmimBF_4$ [polar phase])/TX-100/toluene (oil)[a]	63.2%
(D) 1-Ethyl-3-methylimidazoliumethyl sulfate (polar phase)/1-butyl-3-methylimidazolium octylsulfate (amphiphilic)/toluene[b]	58.4%
(E) 1-Ethyl-3-methylimidazolium *n*-hexyl sulfate (IL):water (1:1)—(polar phase)/sodium dodecyl sulfate (SDS)/toluene–pentanol (1:1)[b]	34.4%
(F) Water/1-octyl-3-methylimidazolium chloride (amphiphilic)/1-butyl-3-methylimidazolium hexafluorophosphate (nonpolar phase)[c]	40.6%

Reproduced from Mukherjee et al. [56], *J. Colloid Interface Sci.* **2013**, *395,* 135, with permission from Elsevier Science.

[a] Reported by Gao et al. [81].

[b] Reported by Rojas et al. [89, 90].

[c] Reported by Safavi et al. [99].

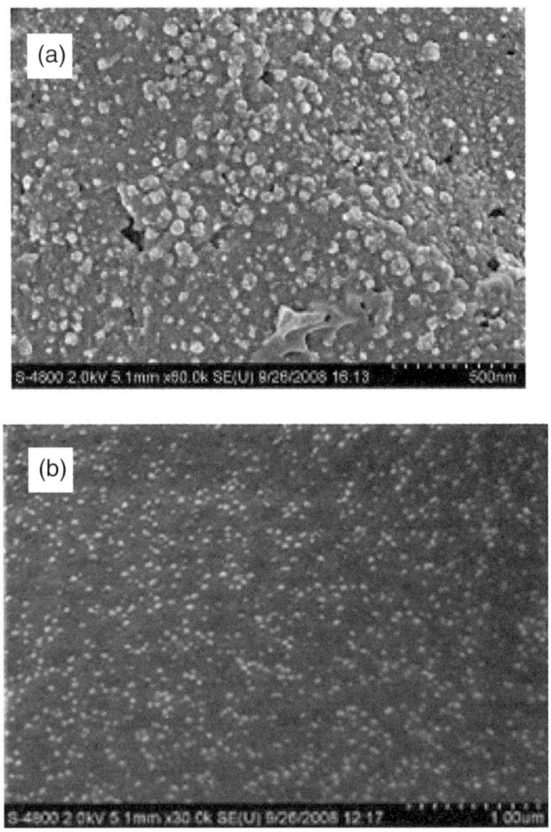

Figure 10.10 Cryo-SEM micrograph of the microemulsion at (a) 15 wt.% of $emimEtOSO_3$, (b) 40 wt.% of $emimEtOSO_3$. Reproduced from Rabe and Koetz [98], *Colloids Surf. A* **2010**, *354*, 261, with permission from Elsevier Science.

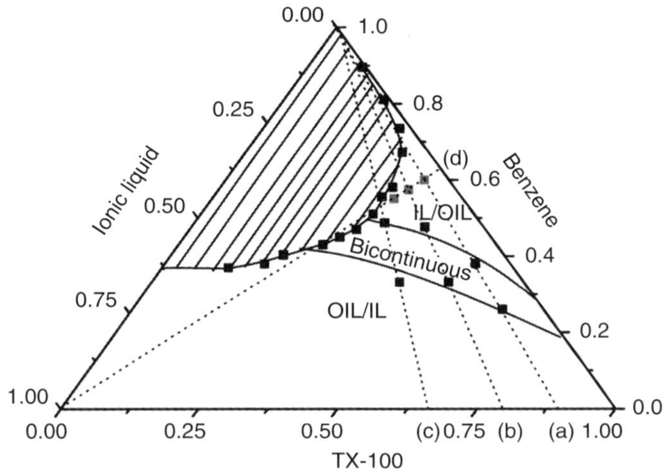

Figure 10.11 Phase diagram of the $[P_{13}][Tf_2N]$/TX-100/benzene three-component system at 25°C. For lines a, b, and c, the initial room-temperature ionic liquid (RTIL) weight fractions are $I = 0.10$, 0.20, and 0.33, respectively. Reproduced from Pramanik et al. [103], *J. Phys. Chem. B* **2011**, *115*, 2322, with permission from the American Chemical Society.

By employing DLS and electrochemical methods (using potassium hexacyanoferrate $K_4Fe(CN)_6$ and ferrocene used as electroactive probes), regions in [bmim] $[PF_6]$-in-water microemulsion were located. On an average, the microemulsion droplet size was approximately 3 nm. The synthesized microemulsion proficiently solubilized a lipophilic dye, 4-(*p*-nitrophenylazo)-pyrocatechol (NAP) and also vitamin K_3 with prospects in biological extractions, and also as solvent for enzymatic reactions.

A typical system composed of EAN/1-hexadecyl-3-methylimidazolium chloride $[C_{16}mim][Cl]$/dodecane at ambient temperature with high thermal stability (stable between 30 and 150°C) was reported by Zech et al. [100]. Inherent properties were characterized by SANS, DLS, generalized indirect Fourier transformation, and Teubner and Strey model. A phase behavioral study with respect to temperature and IL mass fraction of a system comprising water/TX-100/[bmim]$[PF_6]$ was reported by Anjum et al. [101]. SANS and polarized microscopic techniques revealed the existence of discontinuous-type microemulsion droplets in the system.

Chakrabarty et al. [102] have studied solvation dynamics of coumarin 153 in the IL-based microemulsion [bmim]$[BF_4]$/TX-100/cyclohexane. In another work, Pramanik et al. [103] have studied phase behavior along with dynamics of a microemulsion system comprising *N*-methyl-*N*-propylpyrrolidinium bis(trifluoromethanesulfonyl)imide $[P_{13}]$ $[Tf_2N]$/TX-100/benzene (Fig. 10.11).

Sasmal et al. [104] reported deuterium isotope effect on the solvation dynamics and the anisotropy decay of coumarin 480 (C480) in a microemulsion system with 1-pentyl-3-methylimidazolium tetrafluoroborate [pmim]$[BF_4]$ in TX-100/benzene by employing femtosecond upconversion. Retarded solvation dynamics were found with replacement of H_2O by D_2O in the microemulsion identifying water molecules as the main species responsible for solvation.

10.3.2 Applicability of Microemulsions in Colloidal Domain

Since microemulsions are well-known vehicles for drug encapsulation and delivery, recent research has focused on applying IL-based microemulsions as potential drug careers. Moniruzzaman et al. [105] have synthesized IL-based microemulsion with 1-ethyl-3-methylimidazolium tetrafluoroborate (polar phase)/(Tween-80 + Span-20) (stabilizer)/IPM (oil) as a vehicle for sparingly soluble drugs like acyclovir (ACV), methotrexate, dantrolene sodium, etc. (Scheme 10.3). In this study, they concluded that ILs with coordinating anions act as hydrogen-bond acceptor and can efficiently form microemulsion compared with hydrophobic ILs with noncoordinating anions like PF_6^-, $(CF_3SO_2)_2N^-$, etc.

In their other work, Moniruzzaman et al. [106] synthesized microemulsions with surfactants (Tween-80 and Span-20), oil (IPM) and polar phase ($[C_1mim][(CH_3O)_2PO_2]$ for solubilizing the sparingly soluble drug ACV for favorable topical and transdermal delivery. Among various weight ratios of Tween-80 to Span-20 investigated in the microemulsion systems, the ratio 3:2 showed excellent solubility and skin permeation effect of the used model sparingly soluble drug. Study of ACV incorporated in IL droplets as well as other formulations was performed into and across the Yucatan micropig (YMP) porcine skin; the use of IL/O microemulsion has been shown to dramatically increase ACV administration. Cytotoxicity of the aforesaid carrier was evaluated *in vitro* using the reconstructed human epidermal model LabCyteTM EPI-MODEL12. Cell viability of IL/O MEs containing 4 wt.% IL was over 80% compared to Dulbecco's phosphate-buffered salines indicating low cytotoxicity of the carrier. They have concluded that IL-assisted nonaqueous microemulsion could serve as a versatile and efficient nanodelivery system for insoluble or sparingly soluble drug molecules that require solubilizing agents for delivery.

Jiang et al. [107] have reported a technique based on a hydrophilic IL, [bmim] $[BF_4]$, and a hydrophobic IL, [bmim]$[PF_6]$, for extracting penicillin G from its

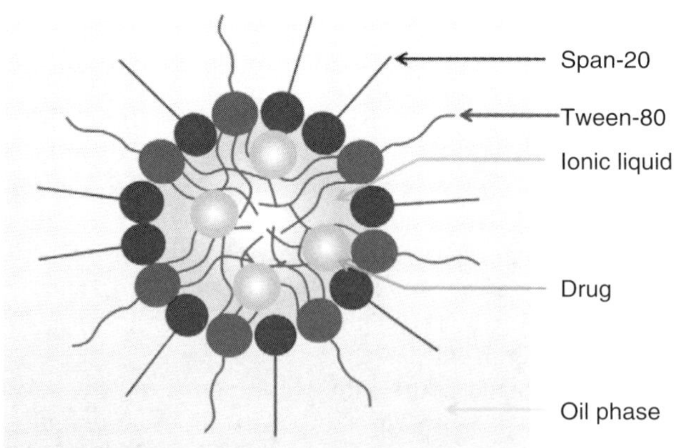

Scheme 10.3 Schematic representation of IL-in-oil microemulsion with drug molecule in the core. Reproduced from Moniruzzaman et al. [105], *J. Colloid Interface Sci.* **2010**, *352*, 136, with permission from Elsevier Science. *See insert for color representation of the figure.*

fermentation broth. Like microemulsion, the pure ILs can also act as novel delivery systems [7]. Lidocaine docusate (LD), a hydrophobic room-temperature IL (RTIL) when compared with lidocaine hydrochloride, exhibits modified solubility, increased thermal stability, and a significant enhancement in the efficacy of topical analgesia in two different models of mouse antinociception. A control release of drugs such as penicillin V potassium, dexamethasone, progesterone, and dehydroepiandrosterone from water-immiscible RTILs like [bmim/hmim/omim][PF_6] has been reported by Jaitely et al. [108], where the ILs acted as reservoirs of the drugs. In a review article, applicability of ILs in pharmaceutical fields has been elaborately discussed by Sekhon [109]. ILs have exhibited several advantages like hindering polymorphisms of crystalline solids, enhancing stability of labile biopharmaceutical products for therapeutic uses, converting active pharmaceutical ingredients to liquids with benign counterion, and being antibiofilm agents. Solubilization of paracetamol and ibuprofen in ILs like [bmim]/[hmim][PF_6] was reported by Smith et al. [110]. Mahajan et al. have reported [111] the binding interactions of drugs like dopamine hydrochloride and acetylcholine chloride with an IL, 1-tetradecyl-3-methylimidazolium bromide [C_{14}mim][Br], and compared it with the structurally similar cationic surfactant tetradecyltrimethylammonium bromide (TTAB).

Moniruzzaman et al. have reported [112] a better enzymatic activity of horseradish peroxidase (HRP) microencapsulated in water-in-ionic liquid (W/IL) microemulsion (water/AOT (sodium bis(2-ethyl-1-hexyl)sulfosuccinate + 1-hexanol)/[C_8mim][Tf_2N](1-octyl-3-methylimidazolium bis(trifluoromethylsulfonyl)amide)) using pyrogallol as the substrate than that in a conventional microemulsion with water/AOT/isooctane. Pavlidis et al. [113] have reported a biocatalysis favorable W/IL-type microemulsion system comprising water/Tween-20 or TX-100/[bmim][PF_6]. This system has manifested better biocatalytic activity for lipases from *C. rugosa*, *C. viscosum*, *and T. lanuginosus* than the commonly reported microemulsion systems due to the presence of IL-based protective environment for the enzymes.

The use of ILs in the preparation of nanomaterials gets extra attention from the researchers. Structure-directing ability as solvents and solvothermal behavior make ILs as potential agents in the preparation of nanostructured materials. ILs with amphiphilic behavior have been employed for synthesizing different microheterogeneous assemblies like micelles, microemulsions, etc. that can be used as templates for synthesizing the desired nanoparticles. Thus, synthesis of nanoparticles in IL-based microemulsions has emerged as a topic of significant interest. Zhao et al. [114] have synthesized silica nanoparticles by employing microemulsion comprising [bmim][BF_4]/TX-100/benzene both in acidic and alkaline media (Fig. 10.12).

Silica microrods with nanosized pores were explored by Li et al. [115] templated in the microemulsion system with water/TX-100/[bmim][PF_6]. Synthesis of silver (Ag) nanoparticles by the photoreduction of silver perchlorate ($AgClO_4$) in water-in-ionic liquid microemulsions of water/Tween-20/([bmim][BF_4]) and 1-octyl-3-methylimidazolium tetrafluoroborate ([omim][BF_4]) was reported by Harada et al. [116]. The average diameters of the metallic Ag particles were 8.9 and 4.9 nm in the water-in-([bmim][BF_4]) and water-in-([omim][BF_4]) microemulsions, respectively. We have [117] synthesized CdS nanoparticles in a bicontinuous-type microemulsion in the system comprising water/IL-2 + IP (1:1-wt./wt.)/IPM as 6.8/86.6/6.6 wt.%. Particle diameter of approximately 3.2 nm had marginal semiconductivity and moderate antibacterial activity against *E. coli.* Li et al. [118] synthesized tetragonal ZrO_2

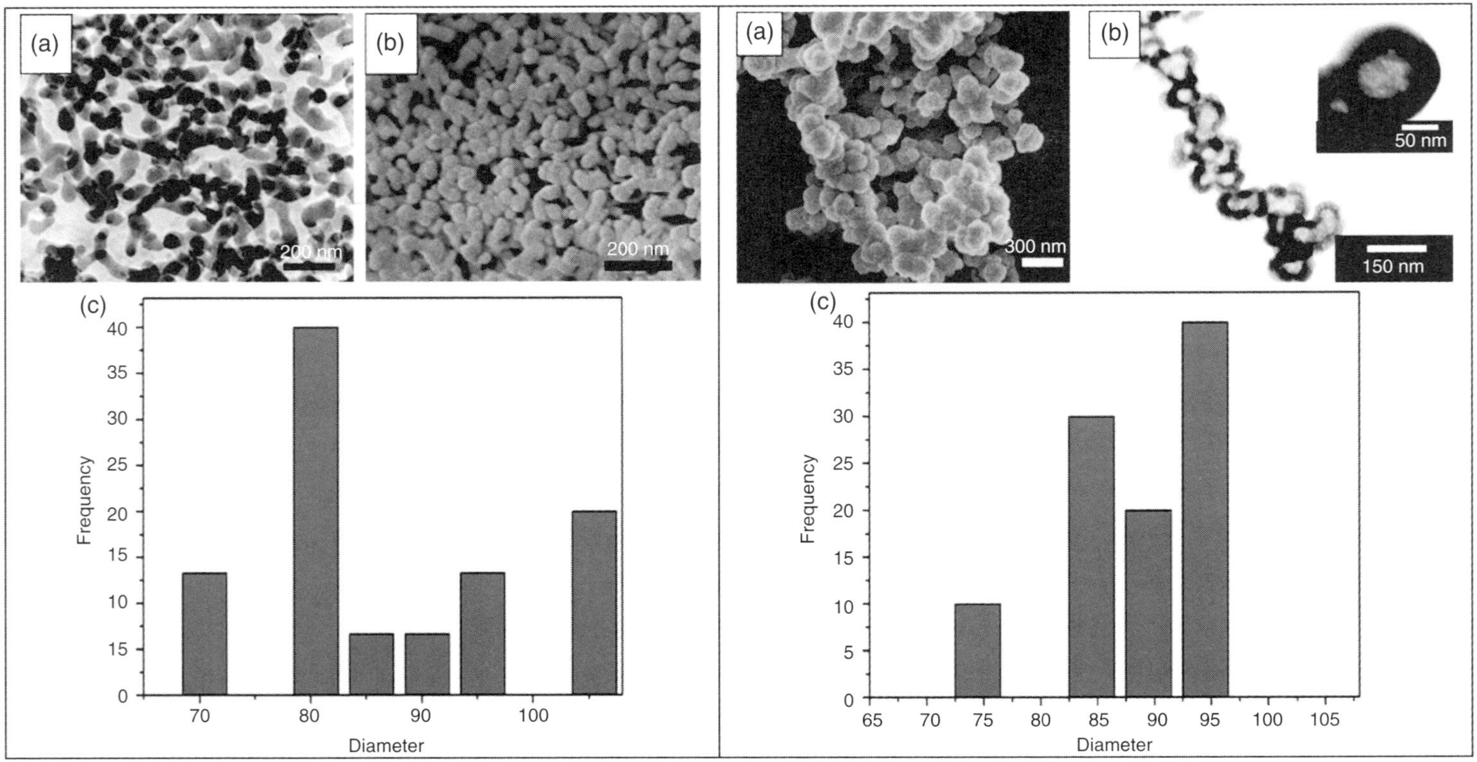

Figure 10.12 TEM and SEM images with particle size distribution of "Si" nanomaterials in acidic and alkaline conditions. Reproduced from Zhao et al. [114], *Colloids Surf. A* **2009**, *346*, 229, with permission from Elsevier Science.

nanoparticles in water-in-IL-type microemulsion system comprising water/TX-100/[bmim][PF_6]. The microemulsion of guanidinium ILs in supercritical CO_2 was employed to prepare spherical gold nanoparticles or gold networks by Liu et al. [119].

10.3.3 Additive Influence on Microemulsion

Effect of additives on the microemulsions often produced interesting observations. Gao et al. have reported [120] that when a small amount of water has been added to the [bmim][BF_4] in cyclohexane-type microemulsion (Fig. 10.13 and Scheme 10.4), droplet size became reduced. Hydrogen bonding between water and peripheral oxyethylene group of TX-100 influenced water molecules to reside at the periphery of the polar core of microemulsion droplets. This resulted in an increase in curvature

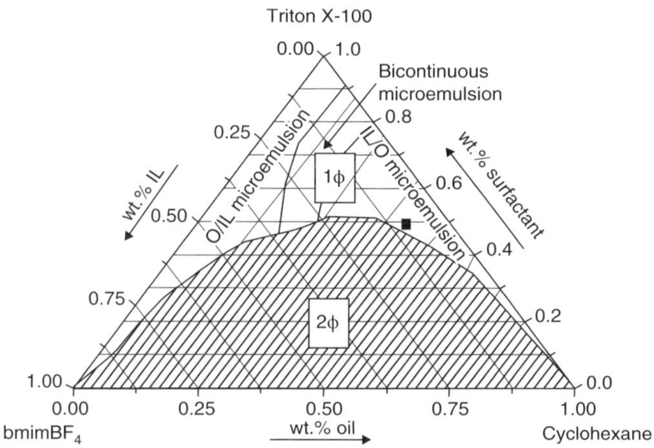

Figure 10.13 Phase diagram of the mixture bmimBF_4/Triton X-100/cyclohexane at 23°C; the studied microemulsion composition marked with ■ is 49.9 wt.% Triton X-100, 40.8 wt.% cyclohexane, 9.3 wt.% bmimBF_4, i.e. [bmimBF_4]/[Triton X-100] molar ratio, R = 0.53. Reproduced from Gao et al. [120], *J. Phys. Chem. B*, with permission from the American Chemical Society.

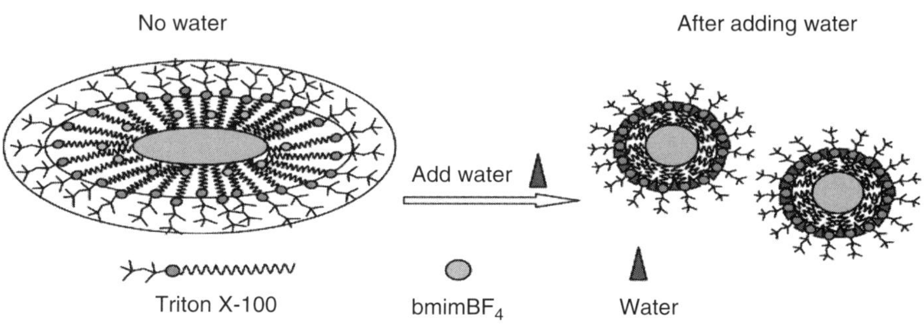

Scheme 10.4 Schematic diagram of the [bmim][BF_4]-in-cyclohexane microemulsion. A large elliptical droplet structure can be assumed in the absence of water, while the addition of water results in small-size spherical microemulsion droplets. Reproduced from Gao et al. [120], *J. Phys. Chem. B* **2008**, *112*, 3711, with permission from the American Chemical Society. *See insert for color representation of the figure.*

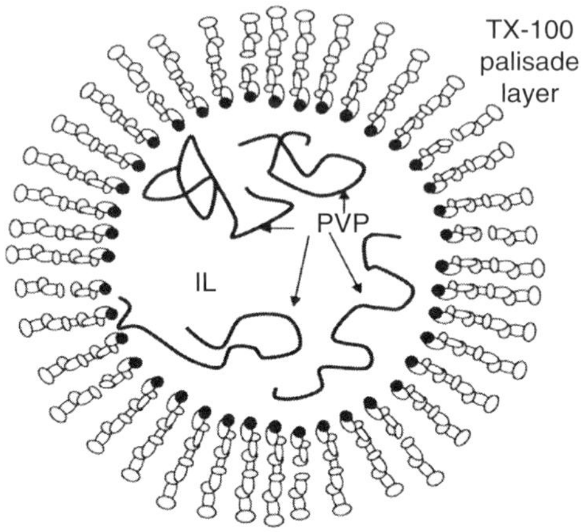

Scheme 10.5 Single droplet of the [bmim][BF_4]-in-cyclohexane microemulsion, illustrating the position of PVP chains in the ionic liquid (IL) core. Reproduced from Gao et al. [121], *Colloids Surf. A* **2008**, *329*, 146, with permission from Elsevier Science.

of the surfactant film of the [bmim][BF_4] at the cyclohexane interface and reduced the droplet diameter. Interfacial ordering of surfactant molecules led to a negative entropy change that was compensated by the increase in enthalpy for the microstructural transition process.

Addition of polymer polyvinylpyrrolidone (PVP) on the microstructure of a microemulsion system consisting of [bmim][BF_4]/TX-100/cyclohexane evidenced a remarkable reduction in the microemulsion zone [121]. An enhancement in droplet diameter and viscosity of the microemulsion was observed with the addition of both [bmim][BF_4] and PVP; the latter was only solubilized inside the polar IL core of the microemulsion droplets (Scheme 10.5).

Stabilizing effect and enlargement of single-phase microemulsion zone by incorporating *n*-butanol (as cosurfactant) composed of [bmim][BF_4]/TX-100/cyclohexane were reported by Cheng et al. [122]. Addition of *n*-butanol increased the [bmim][BF_4] solubilization capacity in the microemulsion. A recent report of Wei et al. [123] described the effect of [bmim][BF_4] on viscosity, conductivity, and droplet diameters of the microemulsion system water/TX-100 + hexanol/cyclohexane.

A fascinating phase behavior of a system comprising 1-ethyl-3-methylimidazolium ethylsulfate ([emim][$EtSO_4$])/polyethylene glycol tert-octylphenyl ether (TX-114 or TX-114)/limonene was reported by Harrar et al. [124]. The system evidenced a firm stability through a large temperature range, down to −35°C. A comparison was made with a previous [125] microemulsion system comprising [emim][$EtSO_4$], limonene, and TX-100 (Fig. 10.14), which was also stable at temperatures well below the freezing point of water. The differences in microstructure, phase behavior, and temperature stability arose from the average repeating number of ethylene oxide units in the surfactant head group, which is smaller for TX-114 compared to TX-100. The reports also depicted a variation in phase behavior of the aforementioned system with varied amphiphilicity.

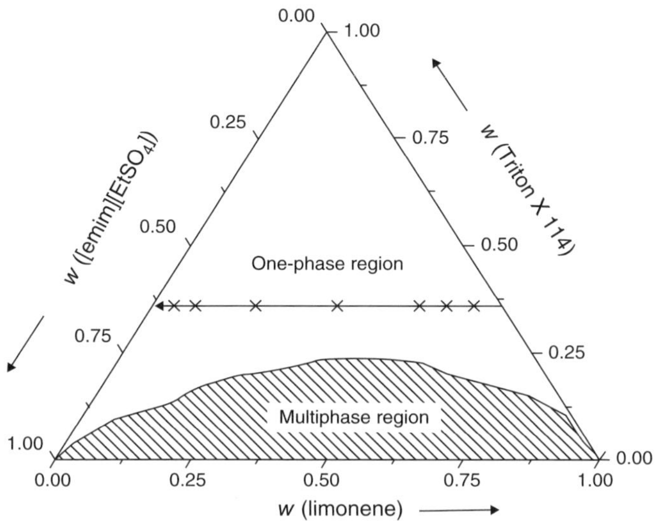

Figure 10.14 Ternary phase diagram at 25°C for the system: [emim][$EtSO_4$]/TX-114/limonene. The black arrow marks the experimental path that was chosen for further experiments and the crosses the samples characterized by SAXS experiments. Reproduced from Harrar et al. [125], *Langmuir* **2011**, *27*, 1635, with permission from the American Chemical Society.

10.4 CONCLUSION

ILs are a fascinating class of compounds with broad range of applicability. Their cation–anion combinations are responsible for their unique physicochemical behaviors in the interfacial and bulk domains. ILs are engineered or designed for fulfilling different aspects of science and technology. Their viscous, conducting, and interfacial behaviors have proved their versatility in the domain of colloids. Their inherent physicochemical properties have enhanced their applicability as polar and nonpolar solvents as well as amphiphilic entities. Assembly formation like micelles, microemulsions, etc. using ILs has been also reported. Utilization of the prepared assemblies in nanomaterial synthesis, drug delivery, oil recovery, enzyme catalysis, etc. has been also of concern. Their self-micellization, mixed micellization, and influence on the micellization of conventional surfactants have been evidenced with specific citations. Phase-forming behaviors of IL-based microemulsions of IL/O, O/IL, and bicontinuous types are discussed along with the effects of additives on them. Formation of large single-phase (clear) zones in the pseudoternary and ternary phase diagrams with ILs has added a special focus in this article. Antibacterial and anticancer activities of ILs and IL-derived systems like micro-emulsions are also presented and discussed in this article for the elucidation of the broad domain of ILs in the field of colloid and interface science.

ACKNOWLEDGMENT

IM thanks the Centre for Surface Science, Department of Chemistry, Jadavpur University for laboratory facilities. SPM thankfully acknowledges both Jadavpur University and Indian National Science Academy for the necessary support.

REFERENCES

[1] Deetlefs, M.; Seddon, K. R. *Chim. Oggi* 2006, **24**, 16.

[2] Plechkova, N. V.; Seddon, K. R., Ionic liquids: Designer solvents for green chemistry, in *Methods and Reagents for Green Chemistry: An Introduction*, ed. P. Tundo, A. Perosa, and F. Zecchini, John Wiley & Sons, Inc., New York 2007, pp. 105–130.

[3] Jin, H.; O'Hare, B.; Dong, J.; Arzhantsev, S.; Baker, G. A.; Wishert, J. F.; Benesi, A. J.; Maroncelli, M. *J. Phys. Chem. B* 2008, **112**, 81.

[4] Forsyth, S. A.; Pringle, J. M.; Mac Farlane, D. R. *Aust. J. Chem.* 2004, **57**, 113.

[5] Huddleston, J. G.; Visser, A. E.; Reichert, W. M.; Willauer, H. D.; Broker, G. A.; Roger, R. D. *Green Chem.* 2001, **3**, 156.

[6] Seddon, K. R.; Stark, A.; Torres, M.-J. *Pure Appl. Chem.* 2000, **72**, 2275.

[7] Whitney, L. H.; Smiglak, M.; Rodríguez, H.; Swatloski, R. P.; Spear, S. K.; Daly, D. T.; Pernak, J.; Grisel, J. E.; Carliss, R. D.; Soutullo, M. D.; Davis, Jr., J. H.; Rogers, R. D. *New J. Chem.* 2007, **31**, 1429.

[8] Welton, T. *Chem. Rev.* 1999, **99**, 2071.

[9] Bonhôte, P.; Dias, A.-P.; Papageorgiou, N.; Kalyanasundaram, K.; Grätzel, M. *Inorg. Chem.* 1996, **35**, 1168.

[10] Wasserscheid, P.; Keim, W. *Angew. Chem. Int. Ed.* 2000, **39**, 3772.

[11] Mukherjee, I.; Manna, K.; Dinda, G.; Ghosh, S.; Moulik, S. P. *J. Chem. Eng. Data* 2012, **57**, 1376.

[12] Ungphaiboon, S.; Supavita, T.; Singchangchai, P.; Sungkarak, S.; Rattanasuwan, P.; Itharat, A. *Songklanakarin J. Sci. Technol.* 2005, **27**, 569.

[13] Angell, C. A. *Science* 1995, **267**, 1924.

[14] Suarez, P. A. Z.; Einloft, S.; Dullius, J. E. L.; de Souza, R. F.; Dupont, J. *J. Chim. Phys.* 1998, **95**, 1626.

[15] Tokuda, H.; Hayamizu, K.; Ishii, K.; Susan, M. A. B. H.; Watanabe, M. *J. Phys. Chem. B* 2004, **108**, 16593.

[16] Yoshida, Y.; Baba, O.; Saito, G. *J. Phys. Chem. B* 2007, **111**, 4724.

[17] Dzyuba, S. V.; Bartsch, R. A. *ChemPhysChem* 2002, **3**, 161.

[18] Domańska, U.; Królikowska, M. *J. Colloid Interface Sci.* 2010, **348**, 661.

[19] Kilaru, P.; Baker G. A.; Scovazzo, P. *J. Chem. Eng. Data* 2007, **52**, 2306.

[20] Freire, M. G.; Carvalho, P. J.; Fernandes, A. M.; Marrucho, I. M.; Queimada, A. J.; Coutinho, J. A. P. *J. Colloid Interface Sci.* 2007, **314**, 621.

[21] Klomfar, J.; Součková, M.; Pátek, J. *J. Chem. Thermodyn.* 2010, **42**, 323.

[22] Sánchez, L. G.; Espel, J. R.; Onink, F.; Meindersma, G. W.; de Haan, A. B. *J. Chem. Eng. Data* 2009, **54**, 2803.

[23] Carvalho, P. J.; Neves, C. M. S. S.; Coutinho, J. A. P. *J. Chem. Eng. Data* 2010, **55**, 3807.

[24] Carrera, G. V. S. M.; Afonso, C. A. M.; Branco, L. C. *J. Chem. Eng. Data* 2010, **55**, 609.

[25] Dong, B.; Zhao, X.; Zheng, L.; Zhang, J.; Li, N.; Inoue, T. *Colloids Surf. A* 2008, **317**, 666.

[26] Blesic, M.; Lopes, A.; Melo, E.; Petrovski, Z.; Plechkova, N. V.; Lopes, J. N. C.; Seddon, K. R.; Rebelo, L. P. N. *J. Phys. Chem. B* 2008, **112**, 8645.

[27] Anouti, M.; Jones, J.; Boisset, A.; Jacquemin, J.; Caillon-Caravanier, M.; Lemordant, D. *J. Colloid Interface Sci.* 2009, **340**, 104.

[28] Blesic, M.; Swadźba-Kwaśny, M.; Holbrey, J. D.; Lopes, J. N. C.; Seddon, K. R.; Rebelo, L. P. N. *Phys. Chem. Chem. Phys.* 2009, **11**, 4260.

[29] Li, N.; Zhang, S.; Zheng, L.; Dong, B.; Li, X.; Yu, L. *Phys. Chem. Chem. Phys.* 2008, **10**, 4375.

[30] Shi, L.; Li, N.; Yan, H.; Gao, Y.; Zheng, L. *Langmuir* 2011, **27**, 1618.

[31] Ramnial, T. Taylor, S. A.; Bender, M. L.; Gorodetsky, B.; Lee, P. T. K.; Dickie, D. A.; McCollum, B. M.; Pye, C. C.; Walsby, C. J.; Clyburne, J. A. C. *J. Org. Chem.* 2008, **73**, 801.

[32] McNulty, J.; Capretta, A.; Cheekoori, S.; Clyburne, J. A. C.; Robertson, A. *Chim. Oggi.* 2004, **22**, 13.

[33] Fletcher, K. A.; Pandey, S. *Langmuir* 2004, **20**, 33.

[34] Anderson, J. L.; Pino, V.; Hagberg, E. C.; Sheares, V. V.; Armstrong, D. W. *Chem. Commun.* 2003, **18**, 2444.

[35] Firestone, M. A.; Dzielawa, J. A.; Zapol, P.; Curtiss, L. A.; Seifert, S.; Dietz, M. L. *Langmuir* 2002, **18**, 7258.

[36] Yoshio, M.; Mukai, T.; Kanie, K.; Yoshizawa, M.; Ohno, H.; Kato, T. *Adv. Mater.* 2002, **14**, 351.

[37] Yoshio, M.; Mukai, T.; Kanie, K.; Yoshizawa, M.; Ohno, H.; Kato, T. *Chem. Lett.* 2002, **31**, 320.

[38] Gordon, C. M.; Holbrey, J. D.; Kennedy, A. R.; Seddon, K. R. *J. Mater. Chem.* 1998, **8**, 2627.

[39] Bowlas, C. J.; Bruce, D. W.; Seddon, K. R. *Chem. Commun.* 1996, **Issue 14**, 1625–26.

[40] Evans, D. F.; Yamauchi, A.; Wei, G. J.; Bloomfield, V. A. *J. Phys. Chem.* 1983, **87**, 3537.

[41] Evans, D. F.; Yamauchi, A.; Roman, R.; Casassa, E. Z. *J. Colloid Interface Sci.* 1982, **88**, 89.

[42] Mitra, D.; Chakraborty, I.; Bhattacharya, S. C.; Moulik, S. P. *Langmuir* 2007, **23**, 3049.

[43] Prasad, M.; MacDonald, A.; Palepu, R.; Moulik, S. P. *J. Phys. Chem. B* 2004, **108**, 355.

[44] Li, N.; Zhang, S.; Zheng, L.; Inoue, T. *Langmuir* 2009, **25**, 10473.

[45] Behera, K.; Pandey, S. *J. Phys. Chem. B* 2007, **111**, 13307.

[46] Behera, K.; Pandey, S. *J. Colloid Interface Sci.* 2007, **316**, 803.

[47] Miskolczy, Z.; Sebok-Nagy, K.; Biczók, L.; Göktürk, S. *Chem. Phys. Lett.* 2004, **400**, 296.

[48] Wu, J.; Li, N.; Zheng, L.; Li, X.; Gao, Y.; Inoue, T. *Langmuir* 2008, **24**, 9314.

[49] Behera, K.; Pandey, S. *J. Colloid Interface Sci.* 2007, **307**, 235.

[50] Li, N.; Gao, Y.; Zheng, L.; Zhang, J.; Yu, L.; Li, X. *Langmuir* 2007, **23**, 1091.

[51] Gao, Y.; Zhang, J.; Xu, H.; Zhao, X.; Zheng, L.; Li, X.; Yu, L. *ChemPhysChem* 2006, **7**, 1554.

[52] Li, J.; Zhang, J.; Gao, H.; Han, B.; Gao, L. *Colloid Polym. Sci.* 2005, **283**, 1371.

[53] Gao, Y.; Han, S.; Han, B.; Li, G.; Shen, D.; Li, Z.; Du, J.; Hou, W.; Zhang, G. *Langmuir* 2005, **21**, 5681.

[54] Behera, K.; Pandey, S. *J. Colloid Interface Sci.* 2009, **331**, 196.

[55] Behera, K.; Pandey, S. *Langmuir* 2008, **24**, 6462.

[56] Mukherjee, I.; Mukherjee, S.; Naskar, B.; Ghosh, S.; Moulik, S. P. *J. Colloid Interface Sci.* 2013, **395**, 135.

[57] Chatterjee, A.; Moulik, S. P.; Sanyal, S. K.; Mishra, B. K.; Puri, P. M.; *J. Phys. Chem. B* 2001, **105**, 12823.

[58] Dan, A.; Chakraborty, I.; Ghosh, S.; Moulik, S. P. *Langmuir* 2007, **23**, 7531.

[59] Hait, S. K.; Moulik, S. P.; Palepu, R. *Langmuir* 2002, **18**, 2471.

[60] Baker, G. A.; Pandey, S. *ACS Symp. Ser.* 2005, **901**, 234.

[61] Hao, J.; Zemb, T. *Curr. Opin. Colloid Interface Sci.* 2007, **12**, 129.

[62] Greaves, T. L.; Drummond, C. *Chem. Soc. Rev.* 2008, **37**, 1709.

[63] Ribeiro, A. C. F.; Lobo, V. M. M.; Valente, A. J. M.; Azevedo, E. F. G.; Miguel, M. D.; Burrows, H. D. *Colloid Polym. Sci.* 2004, **283**, 277.

[64] Škerjanc, J.; Kogel, K.; Cerar, J. *Langmuir* 1999, **15**, 5023.

[65] Velasco, S. B.; Turmine, M.; Di Caprio, D.; Letellier, P. *Colloids Surf. A* 2006, **275**, 50.

[66] Tran, C. D.; Yu, S. F. *J. Colloid Interface Sci.* 2005, **283**, 613.

[67] Plechkova, N. V.; Seddon, K. R. *Chem. Soc. Rev.* 2008, **37**, 123.

[68] Teramoto, D.; Yokoyama, R.; Kagawa, H.; Sada, T.; Ogata, N., A Novel Ionic Liquid-Polymer Electrolyte for the Advanced Lithium Ion Polymer Battery in *Molten Salts and Ionic Liquids: Never the Twain*, ed. M. Gaune-Escard and K. R. Seddon, John Wiley & Sons, Inc., New York, 2010, pp. 367–388.

[69] G24 Innovations, www.g24i.com (accessed March 13, 2015).

[70] Fujita, K.; MacFarlane, D. R.; Forsyth, M. *Chem. Commun.* 2005, **Issue 38**, 4804–06.

[71] Pfruender, H.; Jones, R.; Weuster-Botz, D. *J. Biotechnol.* 2006, **124**, 182.

[72] Anastas, P. T.; Kirchhoff, M. M. *Acc. Chem. Res.* 2002, **35**, 686.

[73] Fukumoto, K.; Yoshizawa, M.; Ohno, H. *J. Am. Chem. Soc.* 2005, **127**, 2398.

[74] Petkovic, M.; Ferguson, J. L.; Gunaratne, H. Q. N.; Ferreira, R.; Leitão, M. C.; Seddon, K. R.; Rebelo, L. P. N.; Silva Pereira, C. *Green Chem.* 2010, **12**, 643.

[75] Poletti, L.; Chiappe, C.; Lay, L.; Pieraccini, D.; Polito, L.; Russo, G.; *Green Chem.* 2007, **9**, 337.

[76] Livi, S.; Duchet-Rumeau, J.; Pham, T. N.; Gérard, J.-F. *J. Colloid Interface Sci.* 2011, **354**, 555.

[77] Kumar, V.; Malhotra, S. V. *Bioorg. Med. Chem. Lett.* 2009, **19**, 4643.

[78] Frade, R. F. M.; Matias, A.; Branco, L. C.; Afonso, C. A. M.; Duarte, C. M. M. *Green Chem.* 2007, **9**, 873.

[79] Ruckenstein, E.; Chi, J. C. *Faraday Trans. II* 1975, **71**, 1690.

[80] Lü, H.; An, X.; Shen, W. *J. Chem. Eng. Data* 2008, **53**, 727.

[81] Gao, Y.; Wang, S.; Zheng, L.; Han, S.; Zhang, X.; Lu, D.; Yu, L.; Ji, Y.; Zhang, G. *J. Colloid Interface Sci.* 2006, **301**, 612.

[82] Gao, Y.; Li, N.; Zhang, S.; Zheng, L.; Li, X.; Dong, B.; Yu, L. *J. Phys. Chem. B* 2009, **113**, 1389.

[83] Gao, Y.; Li, N.; Zheng, L.; Zhao, X.; Zhang, S.; Han, B.; Hou, W.; Li, G. *Green Chem.* 2006, **8**, 43.

[84] Cheng, S.; Fub, X.; Liu, J.; Zhang, J.; Zhang, Z.; Wei, Y.; Han, B. *Colloids Surf. A* 2007, **302**, 211.

[85] Li, N.; Cao, Q.; Gao, Y.; Zhang, J.; Zheng, L.; Bai, X.; Dong, B.; Li, Z.; Zhao, M.; Yu, L. *ChemPhysChem* 2007, **8**, 2211.

[86] Paul, S.; Panda, A. K. *Colloids Surf. A* 2012, **404**, 1.

[87] Chai, J.; Xu, L.; Liu, W.; Zhu, M. *J. Chem. Eng. Data* 2012, **57**, 2394.

[88] Cheng, S.; Zhang, J.; Zhang, Z.; Han, B. *Chem. Commun.* 2007, **Issue 24**, 2497–99.

[89] Rojas, O.; Tiersch, B.; Frasca, S.; Wollenberger, U.; Koetz, J. *Colloids Surf. A* 2010, **369**, 82.

[90] Rojas, O.; Koetz, J.; Kosmella, S.; Tiersch, B.; Wacker, P.; Kramer, M. *J. Colloid Interface Sci.* 2009, **333**, 782.

[91] Atkin, R.; Warr, G. G. *J. Phys. Chem. B* 2007, **111**, 9309.

[92] Gao, H. X.; Li, J. C.; Han, B. X.; Chen, W. N.; Zhang, J. L.; Zhang, R.; Yan, D. D. *Phys. Chem. Chem. Phys.* 2004, **6**, 2914.

[93] Eastoe, J.; Gold, S.; Rogers, S. E.; Paul, A.; Welton, T.; Heenan, R. K.; Grillo, I. *J. Am. Chem. Soc.* 2005, **127**, 7302.

[94] Liu, L.; Baudin, P.; Zemb, T.; Eastoe, J.; Hao, J. *Langmuir* 2009, **25**, 2055.

[95] Zhou, G.-P.; Zhang, Y.; Huang, X.-R.; Shi, C.-H.; Liu, W.-F.; Li, Y.-Z.; Qu, Y.-B.; Gao, P.-J. *Colloids Surf. B* 2008, **66**, 146.

[96] Xue, L.; Qiu, H.; Li, Y.; Lu, L.; Huang, X.; Qu, Y. *Colloids Surf. B* 2011, **82**, 432.

[97] Wei, J.; Su, B.; Liang, R.; Xing, H.; Bao, Z.; Yang, Q.; Yang, Y.; Ren, Q. *Colloids Surf. A* 2012, **414**, 82.

[98] Rabe, C.; Koetz, J. *Colloids Surf. A* 2010, **354**, 261.

[99] Safavi, A.; Maleki, N.; Farjami, F. *Colloids Surf. A* 2010, **355**, 61.

[100] Zech, O.; Thomaier, S.; Kolodziejski, A.; Touraud, D.; Grillo, I.; Kunz, W. *J. Colloid Interface Sci.* 2010, **347**, 227.

[101] Anjum, N.; Guedeau-Boudeville, M.-A.; Stubenrauch, C.; Mourchid, A. *J. Phys. Chem. B* 2009, **113**, 239.

[102] Chakrabarty, D.; Seth, D.; Chakraborty, A.; Sarkar, N. *J. Phys. Chem. B* 2005, **109**, 5753.

[103] Pramanik, R.; Sarkar, S.; Ghatak, C.; Rao, V. G.; Sarkar, N. *J. Phys. Chem. B* 2011, **115**, 2322.

[104] Sasmal, D. K.; Sen Mojumdar, S.; Adhikari, A.; Bhattacharyya, K. *J. Phys. Chem. B* 2010, **114**, 4565.

[105] Moniruzzaman, M.; Kamiya, N.; Goto, M. *J. Colloid Interface Sci.* 2010, **352**, 136.

[106] Moniruzzaman, M.; Tamura, M; Tahara, Y.; Kamiya, N.; Goto, M. *Int. J. Pharm.* 2010, **400**, 243.

[107] Jiang, Y.; Xia, H.; Guo, C.; Mahmood, I.; Liu, H. *Ind. Eng. Chem. Res.* 2007, **46**, 6303.

[108] Jaitely, V.; Karatas, A.; Florence, T. *Int. J. Pharm.* 2008, **354**, 168.

[109] Sekhon, B. S. *Asian J. Pharm. Biol. Res.* 2011, **1**, 395.

[110] Smith, K. B.; Bridson, R. H.; Leeke, G. A. *J. Chem. Eng. Data* 2011, **56**, 2039.

[111] Mahajan, S.; Sharma, R.; Mahajan, R. K. *Langmuir* 2012, **28**, 17238.

[112] Moniruzzaman, M.; Kamiya, N.; Goto, M. *Langmuir* 2009, **25**, 977.

[113] Pavlidis, I. V.; Gournisb, D.; Papadopoulosc, G. K.; Stamatis, H. *J. Mol. Catal. B* 2009, **60**, 50.

[114] Zhao, M.; Zheng, L.; Bai, X.; Li, N.; Yu, L. *Colloids Surf. A* 2009, **346**, 229.

[115] Li, Z. H.; Zhang, J. L.; Du, J. M.; Han, B. X.; Wang, J. Q. *Colloids Surf. A* 2006, **286**, 117.

[116] Harada, M.; Kimura, Y.; Saijo, K.; Ogawa, T., Isodad, S. *J. Colloid Interface Sci.* 2009, **339**, 373.

[117] Mukherjee, I.; Dinda, G.; Ghosh, S.; Moulik, S. P. *J. Nanoparticle Res.* 2012, **14**, 997.

[118] Li, N.; Dong, B.; Yuan, W. L.; Gao, Y.; Zheng, L. Q.; Huang, Y. M.; Wang, S. L. *J. Dispers. Sci. Technol.* 2007, **28**, 1030.

[119] Liu, J. H.; Cheng, S. Q.; Zhang, J. L.; Feng, X. Y.; Fu, X. G.; Han, B. X. *Angew. Chem. Int. Ed.* 2007, **46**, 3313.

[120] Gao, Y.; Hilfert, L.; Voigt, A.; Sundmacher, K. *J. Phys. Chem. B* 2008, **112**, 3711.

[121] Gao, Y.; Voigt, A.; Hilfert, L.; Sundmacher, K. *Colloids Surf. A* 2008, **329**, 146.

[122] Cheng, S.; Han, F.; Wang, Y.; Yan, J. *Colloids Surf. A* 2008, **317**, 457.

[123] Wei, J.; Su, B.; Liang, R.; Xing, H.; Bao, Z.; Yang, Y.; Ren, Q. *J. Chem. Eng. Data* 2012, **57**, 1274.

[124] Harrar, A.; Zech, O.; Klaus, A.; Bauduin, P.; Kunz, W. *J. Colloid Interface Sci.* 2011, **362**, 423.

[125] Harrar, A.; Zech, O.; Hartl, R.; Bauduin, P.; Zemb, T.; Kunz, W. *Langmuir* 2011, **27**, 1635.

CHAPTER 11

Nanostructures of Amphiphiles and Microemulsions in Room-Temperature Ionic Liquids

AHMED MOURCHID

Matière et Systèmes Complexes, UMR 7057 CNRS-Université Paris Diderot, Paris, France

11.1 INTRODUCTION

The growing alarming situation of conventional organic solvents, with regard to their negative environmental impact and potential health hazards, motivates the scientists to explore greener and safer substitutes. The primary objective is to reduce loss to environment that frequently happens by evaporation or associated with the disposal process. One viable option to the organic solvents is room-temperature ionic liquids (RTILs), which are nonvolatile, nonflammable, thermally well stable and environment friendly or green [1, 2]. However, it is because of ease of tunability of the above properties that RTILs possess a broad appeal. The most common RTILs are electrolytes composed of large organic cations whose electrical charge is balanced by a monovalent anion. Either a small difference in the chemical structure of the cation or the substitution of the anion can trigger considerable variation in their intrinsic properties such as their transport properties or compatibility with other chemical compounds [3]. In view of these, RTILs are recognized as good candidates for solubilizing various chemicals and also in chemical processing like organic synthesis, liquid extraction, or separation [4].

Their potential use goes beyond the simple solution formulation and chemical processing. In fact, their tunability puts them at the forefront of advanced strategies in order to produce new and multifunctional materials. Among the new strategies envisioned appear self-assembly and nanostructuring of a broad variety of molecules ranging from small amphiphile surfactants to large multiblock copolymers. However, only a limited number of RTILs were investigated in comparison with the numerous available possibilities. In fact, two classes of RTILs comprising the ammonium or the heterocyclic imidazolium cations constitute a major part of the investigations. The specific goal of such investigations is to take advantage of self-assembly and self-organization of targeted molecules in order to prepare specific

Ionic Liquid-Based Surfactant Science: Formulation, Characterization, and Applications, First Edition. Edited by Bidyut K. Paul and Satya P. Moulik.

nanostructures of different shapes and sizes. Such phenomena are achieved through specific solvent–molecule interactions, which include hydrogen bonding and solvophobic/solvophilic interactions. The potential of RTILs to this field of nanostructuring through self-assembly is widely considered to be vast. In fact, these new solvents are believed to possess more flexibility, which is merely realizable by the subtle selection of the cationic or anionic chemical structure and the combination of the two. The easy implementation of such variability in properties is thus supposed to represent a straightforward means to tune the interactions between a solute and the solvent, prompting or facilitating self-assembly and self-organization of the solute.

On the whole, so far, RTILs are studied as potential solubilizing media or the constituent of dominant or continuous phase. However, one should keep in mind that RTILs could be viewed as excellent candidates to undergo self-assembly and self-organization. Indeed, the inclusion of a sufficiently hydrophobic tail to the cationic moiety confers a strong amphiphilic character to them. Such a characteristic structure yields strong interaction in a polar solvent and favors the self-assembly of the IL molecules [5–8].

In this chapter, we summarize recent work on self-assembly and self-organization of amphiphiles in RTILs employed as selective solvents. A solvent is classified by the IUPAC as selective when it solubilizes at least one component of a mixture but not all components. When the solubilized component constitutes a part of a "surfactant" molecule, that is, a molecule constituted of both solvophilic and solvophobic moieties, a self-assembled structure may result where the solvophilic part is exposed to the solvent.

This chapter focuses only on neutral amphiphilic molecular surfactants in RTILs, although some relevant experimental results from different types of molecular architectures are available. Among them are ionic amphiphiles seen as equivalent to similar ionic surfactants in water under high salinity conditions. The effect of ionic strength consists in screening the electrostatic interactions between charged head groups of the solute, which in turn favors its aggregation. Detailed studies in this field were reported by Greaves et al. [9] and Chen and Bermudez [10]. The scope of this chapter is limited to the discussion on self-assembly in two commonly studied RTILs: the protic ethyl ammonium nitrate or the heterocyclic imidazolium ILs.

11.2 AMPHIPHILES IN RTILs

11.2.1 Micellar Aggregation

Aggregation of amphiphilic molecules into micelles occurs in solvents either polar (water, ethylene glycol, etc.) [11] or nonpolar (benzene, toluene, etc.) [12]. In water, ionic surfactant molecules spontaneously aggregate to form normal micelles by hydrophobic interaction leading to a sharp decrease in the interfacial tension through minimization of unfavorable contacts with water. Nonionic surfactants, bearing oligo(ethylene oxide) polar head groups, form hydrogen bonds with water, which complicates the micellization process to some extent. Physicochemical characteristics, such as the chemical structure or molecular morphology, and the solution parameters, such as concentration or temperature, guide the aggregate shapes and sizes, ranging from spherical, cylindrical, and lamellar. This succession can be rationalized by a geometrical parameter, known as the critical packing parameter [13]. In general, the micelles formed in polar organic solvents are similar in many respects to the micelles in water, although, in organic media, the aggregation is generally not as much efficient as in water. To account for this difference, the term "solvophobic interactions," which are analogous to "hydrophobic interactions" in aqueous media, has been introduced [14, 15].

Self-assembly of amphiphiles in RTILs is interesting and is expected to involve almost all the aspects known to occur in traditional solvents. Per se, their investigation has readily attracted the attention during last two to three decades. One of the earliest studies was carried out by Evans et al. who used ionic alkyltrimethylammonium bromide and alkylpyridinium bromide surfactants in the protic ethylammonium nitrate, EAN [16]. EAN is interesting because of its resemblance to water. This motivated the complete determination of the phase diagram of binary EAN–amphiphile systems and similar protic ILs [16, 17]. The investigation showed the existence of usual phases: micellar isotropic phase, hexagonal phase of infinite cylinders, and lamellar phase of stacked membranes. However, the earlier results did not address the issue of whether the ILs should form a networked hydrogen-bond structure in order to induce self-assembly.

The possibility of self-assembly in aprotic ILs, which constitutes the second important class, was subsequently explored [17]. Several studies have used the IL composed of 1-butyl-3-methylimidazolium as the cation and either chloride, hexafluorophosphate, tetrafluoroborate, or bis(trifluoromethylsulfonyl)imide as the counterion. Nonionic surfactants were chosen, and most of them possess a lipophilic alkyl tail attached to a polar oligo(ethylene oxide) head group, like Triton, or alkyl oligo(ethylene oxide) ethers frequently known as CiEj surfactants. The data show that surfactants in aprotic ILs share many common properties with their equivalent in protic ILs and aqueous systems. However, despite the reported contributions, many fundamental questions are still unanswered. For instance, data on the structure of the formed micelles are still lacking. In the next paragraph, we focus our investigation on the microstructure of the neutral *t*-octylphenoxypolyethoxyethanol (Triton X-100), which is assumed to self-assemble in RTILs. The investigation is carried out in 1-butyl-3-methylimidazolium tetrafluoroborate and 1-butyl-3-methylimidazolium hexafluorophosphate, [bmim][BF_4] and [bmim][PF_6], respectively.

11.2.2 Critical Micelle Concentration

Self-assembly is straightforwardly studied by measuring the critical micelle concentration, CMC, which is the minimum concentration of surfactant molecules needed to induce self-assembly in solution at a given temperature. Below the CMC, the molecules are fully dispersed as monomers. However, if there is an interface such as the air–solvent interface, then the molecules are expected to invade it and have their most solvophobic part pointing toward the air. Surface tension measurements yield accurate determination of the CMC. For TX-100 in [bmim][BF_4], we found that the surface tension decreases as the surfactant concentration increases, which strongly indicates that the amphiphiles have a tendency to locate at the air–[bmim][BF_4] interface at low concentration. Following the initial decrease with increasing concentration, the surface tension reaches a constant value at a specific threshold. The breaking point at which the surface tension levels off corresponds to CMC.

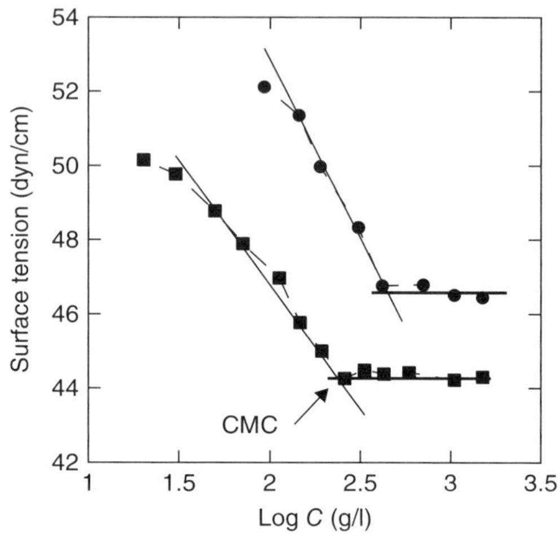

Figure 11.1 Surface tension as a function of concentration for Triton X-100 in [bmim][BF_4] (■) and [bmim][PF_6] (●) at 25°C. The break points indicate the CMCs. The straight lines are linear fits (correlation coefficients: 0.996 and 0.999, respectively).

At this concentration, the molecules self-assemble in order to minimize unfavorable contact with the solvent. Remarkably, this behavior is very similar to the situation observed in water for this family of surfactants. These results are depicted in Figure 11.1 and are also in agreement with those reported for a series of CiEj surfactants in this same IL, with *i* ranging from 10 to 16 and *j* from 4 to 8 [18, 19]. To recall, TX-100 has a more complex chemical structure than alkyl oligo(ethylene oxide) ethers. However, a resemblance could be drawn with surfactants having an alkyl chain of less than 8 carbons, which always display high CMC values in ILs [20]. The effect of exchanging the IL counterion is studied by the substitution of BF_4^- by PF_6^- and is presented in Figure 11.1. We notice that it is necessary to add a huge amount of surfactant molecules in [bmim][PF_6] in order to lower the surface tension to the same level as in [bmim][BF_4]. This observation is associated with a higher CMC in [bmim][PF_6], which is almost two times higher than in [bmim][BF_4].

The quantities determined from surface tension measurements at 25°C are summarized in Table 11.1. The data are compared with those previously reported for TX-100 in water [21]. In addition to CMC and surface tension, we are able to determine the area per molecule, *a*, at the air/solution interface when it is saturated with surfactant molecules. It is calculated from a linear fit to the data below the CMC by using the Langmuir–Szyszkowski adsorption equation. We can notice that both ILs display very high CMC values. Indeed, there is a factor of 1000 between these values and the CMC measured in water.

This behavior is well confirmed since most CiEj surfactants studied always show higher CMC values in RTILs than in water. In order to shed light on the morphology of the micelles, we recall previous data on CiEj surfactants, which showed that the CMC is sensitive to the chemical structure of both the polar head group and the lipophilic alkyl chain [19, 22]. Both CMC and *a* increase when the number *j* of polar head groups increases. In contrast, the CMC decreases with the increase in alkyl chain

TABLE 11.1 Characteristic Parameters of the Surface Tension, Dynamic Light Scattering (DLS), and Small-Angle Neutron Scattering (SANS) Measurements at 25°C

	[bmim][BF_4]	[bmim][PF_6]	H_2O
γ_{CMC} (dyn/cm)	44 ± 2	46 ± 2	30
CMC (mol/l)	0.35 ± 0.05	0.7 ± 0.07	0.00027
CMC (g/l)	219 ± 30	437 ± 40	0.17
a (Å^2)	61 ± 5	43 ± 5	54
R_h (Å)	24 ± 5	18 ± 5	37
R (Å)		16 ± 3	29
N_{agg}		18 ± 5	105

Data in H_2O are compiled from Refs. [17–19].
γ_{CMC} (dyn/cm) is the critical surface tension reached at the CMC (mol/l and g/l); a (Å^2) is the Triton X-100 molecular area at the air/solution interface at saturation; R_h is the hydrodynamic radius from DLS; R and N_{agg} represent the micellar radius and aggregation number, respectively, and are deduced from SANS.

length, *i*. The variation of the CMC with *i* and *j* is in agreement with general behavior found in water [22]. Thus, it can be concluded that the amphiphiles qualitatively act in a similar manner in ILs: they form normal micellar morphologies where the lipophilic alkyl chains are aggregated and protected from being in contact with the solvent by the polar head groups. The high CMC values in ILs strongly suggest that even though self-assembly occurs, the surface-active properties are much weaker than in water because the solvophobicity of the surfactant hydrocarbon chain is very weak in RTILs compared with its hydrophobicity. The alkyl chains are sufficiently soluble in RTILs to compete against self-assembly. Increasing the alkyl chain length is a way to enhance the solvophobicity and facilitate self-assembly.

We mentioned that the nature of the counterion is an important parameter that further influences the self-assembly. It induces a shift in the CMC, which is clearly the case for TX-100 in the hydrophilic [bmim][BF_4] as compared to [bmim][PF_6]. The results in Figure 11.1 suggest that the solvophobic interactions are weaker in [bmim][PF_6]. A similar trend was observed for micellization of CiEj in three imidazolium ILs. The data showed that self-assembly does not occur when the counterion is exchanged to bis(trifluoromethylsulfonyl)imide [19].

To get more insights into the micellar morphologies of TX-100 in our RTILs, we carried out dynamic light scattering (DLS) measurements, which yield the micellar hydrodynamic radius, R_h, along with small-angle neutron scattering (SANS) measurements, which allow the determination of the shape and size of the aggregates. R_h was determined at a surfactant concentration doubled than that of CMC, a relatively dilute regime, where it was independent of the concentration. The measurements yielded satisfactory scattering intensities suggesting the presence of nanometric aggregates. It is noteworthy that the intensity was relatively weak below the CMC. The SANS experiments were performed only in [bmim][PF_6], because [bmim][BF_4] was not transparent to the neutrons due to the presence of absorbing boron. Scattering intensities below the CMC and also on pure [bmim][PF_6] were equally weak.

The data of both scattering measurements are summarized in Table 11.1. The DLS data clearly displays a relatively narrow population of aggregates. We notice that R_h is slightly lower in [bmim][BF_4] and [bmim][PF_6] than in water and always lower in

[bmim][PF_6] than in [bmim][BF_4], in agreement with earliest investigations by Evans et al. who studied ionic surfactants in EAN and previous studies carried out on CiEj surfactants [16, 23]. Despite the difference with aqueous solutions, it is clearly verified that the micellar size is nanometric in both cases. Moreover, data in [bmim][PF_6] is matching well the size determined from the fit to SANS curve by using the polydisperse spherical form factor model. Again, the average micellar radius lies within 1.5 and 2 nm and displays a polydispersity index of 30%. Further, the aggregation number, N_{agg}, has been estimated and presented in Table 11.1. We found that $N_{agg}=18$, which is relatively smaller than the value previously reported in water [23]. The scattering data undoubtedly demonstrate that the micellization process takes place in both of these RTILs. Evans et al. [16] also showed that the average aggregate sizes are smaller than in water, being closer to the values found in formamide solutions. It was evidenced that the solvophobic interactions are very weak in ILs compared to hydrophobic interactions.

11.3 AQUEOUS MICROEMULSIONS WITH ILs

As we have seen, the study of binary solutions reveals that ILs have the ability to dissolve a wide range of chemicals including amphiphiles [24, 25]. Their properties can be easily tuned, for example, [bmim][BF_4] is miscible with water, while [bmim][PF_6] remains immiscible. These distinctive properties automatically lead to their use as solvents to produce microemulsions, replacing either the hydrophilic or the hydrophobic phase [26–30].

11.3.1 Phase Diagram

Two immiscible liquids, such as water and oil, usually do not mix together. The addition of an appropriate amphiphile can yield a three-phase system: an aqueous phase, an organic phase, and an amphiphile-rich middle phase. The last constitutes the microemulsion and forms at well-defined temperature and composition values. What is remarkable is that when the fraction of amphiphiles increases, the middle microemulsion phase spreads to the detriment of the aqueous and organic phases. When the concentration reaches an optimum value, a homogeneous one-phase fluid mixture is obtained. At this concentration, the amphiphiles form a monolayer that separates the two immiscible liquids. The monolayer extension produces a spreading pressure that lowers the interfacial tension to nearly zero. A bicontinuous microemulsion, composed of random interpenetrating domains, forms. The threshold concentration of surfactant at which the bicontinuous phase forms defines its efficiency: this is the minimum surfactant concentration required to solubilize two immiscible solvents. Sottmann and Strey provided a comprehensive review of the general pattern of the phase behavior, properties, and microstructure of microemulsions [31].

Microemulsions are widely used in applications because they are transparent, isotropic, stable low-viscosity mixtures [32]. Kahlweit and Strey performed several investigations drawing a different type of phase diagram (called fish phase diagram) [32–36]. Usually, the polar phase is water, while the nonpolar phase is a *n*-alkane solvent. Recently, the scientific interest has increased many folds for employing RTILs in microemulsion formulations because of their use either as a substitute to water or to oil. Their thermal stability over a wide temperature range is certainly an important driving factor [26–30, 37–39].

Gao et al. used [bmim][BF_4] to prepare nonaqueous [bmim][BF_4]–benzene–TX-100 [26] and [bmim][BF_4]–cyclohexane–TX-100 [30] microemulsions. They reported the phase behavior of IL–oil microemulsion and found physicochemical properties similar to those of water–oil microemulsions [26]. The microstructure was investigated by SANS [28], electron microscopy [30], DLS, UV-Vis, FTIR, and ^{1}H-NMR spectroscopy [26]. An in-depth phase diagram study of EAN–*n*-alkane–CiEj surfactant systems was reported by Atkin and Warr, where the influence of the n-alkane chain length and of the surfactant structure on the efficiency was highlighted [29].

Hydrophobic [bmim][PF_6] was also used to replace typical organic solvents in ternary systems [27, 37, 38]. Gao et al. prepared and characterized water–[bmim][PF_6]–TX-100 [27] as well as water–[bmim][PF_6]–Tween-20 systems [39]. For the former, they identified three different phases by using cyclic voltammetry [27]: water-in-[bmim][PF_6] droplets, bicontinuous structures, and [bmim][PF_6]-in-water droplets. In recent studies, Seth et al. also prepared water–[bmim][PF_6]–TX-100 and studied the interaction of [bmim][PF_6] with water [37, 38].

However, intriguingly all the aforementioned studies, except the one reported by Atkin and Warr [29], were performed at room temperature so that detailed knowledge about the phase diagram and the structure were not explored. To fill this gap, we carried out a thorough study of water–[bmim][PF_6]–TX-100 microemulsions as a function of temperature, T, and surfactant mass fraction at different IL mass fractions. The results are described later where we illustrate the formation of 1, 2, and 3 phases and present the fishtail phase diagram as well as the location of the point of optimal efficiency, called $\tilde{X}$ point. The coordinates of the $\tilde{X}$ point are $\tilde{\gamma}$ and $\tilde{T}$, the fraction of surfactant and temperature, respectively. The reported construction is carried out by inspection of the samples with optical and polarizing microscopy, and the microstructure is studied by SANS. For the sake of completeness, the experimental evidences are compared with data on the phase behavior of water–[bmim][PF_6] ternary systems with two CiEj surfactants.

Figure 11.2 represents the phase diagram of the ternary water–[bmim][PF_6]–TX-100 at 25°C. This diagram clearly shows a two-phase region, which occurs at relatively low surfactant concentration and consists of a water-rich upper phase and a [bmim][PF_6] lower phase. Above the two-phase region, the system is monophasic, transparent, and homogeneous, which hints to the formation of a microemulsion. Thus, we constructed the fish-shaped diagrams by performing temperature scans at different surfactant weight fractions, γ. Each diagram was built for a fixed solvent composition noted α, defined as the ratio of mass of [bmim][PF_6] to the total mass of both solvents. Figure 11.3 shows the fish-shaped diagram at $\alpha = 0.58$, which corresponds to equal [bmim][PF_6] and water volumes. For $\gamma < 0.5$ and $T < 45°C$, a two-phase system noted ($\bar{2}$) is observed composed of a [bmim][PF_6]-in-water droplet microemulsion upper phase coexisting with a [bmim][PF_6] excess lower phase. At an intermediate T, a bicontinuous microemulsion noted (3) coexists with a [bmim][PF_6] excess lower phase and a water excess upper phase. At $T > 75°C$, a water-in-[bmim][PF_6] droplet microemulsion noted ($\underline{2}$) coexists with a water excess phase. At $\gamma > 0.5$, an isotropic, optically transparent one-phase microemulsion noted (1) is observed. A comparison of these phase sequences with those of conventional water–*n*-alkane–CiEj

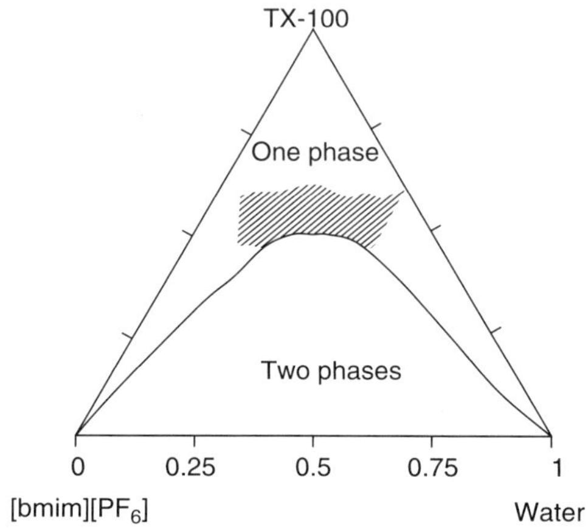

Figure 11.2 Ternary phase diagram of water–[bmim][PF_6]–TX-100 system at 25°C. The hatched area represents the bicontinuous microemulsion region.

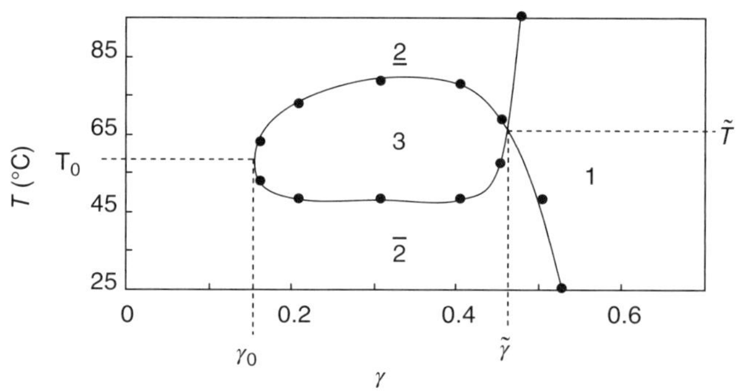

Figure 11.3 Fish-shaped phase diagram (surfactant weight fraction, γ, and temperature, T, plane) of the ternary water–[bmim][PF_6]–TX-100 at equal volumes of water and [bmim][PF_6]. 1, $\underline{2}$, $\bar{2}$, and 3 correspond to one-phase, two-phase (water-in-IL droplet), two-phase (IL-in-water droplet), and three-phase regions, respectively. From Anjum et al. [44].

reveals that they are similar [31]. Indeed, the affinity of the hydrophilic head group of TX-100 toward water decreases with increasing T due to the dehydration of the ethylene oxide units as a consequence of hydrogen-bond breaking. This dehydration simultaneously leads to a decrease in the average head group area and thus to a decrease of the curvature of the interfacial layer from positive in ($\bar{2}$) to negative in ($\underline{2}$) via a zero mean curvature, which triggers the bicontinuous phase (1) in the middle temperature range. The only difference with conventional water–n-alkane–CiEj systems is the larger density of [bmim][PF_6], 1.37 g/cm^3, than that of water. Thus, the [bmim][PF_6]-rich phase is always the lower phase in these samples in contrast to n-alkane systems.

The size and position of the three-phase fish body deserve a close inspection. On the one hand, the coordinates of the fish head, γ_0 and T_0, which represent the

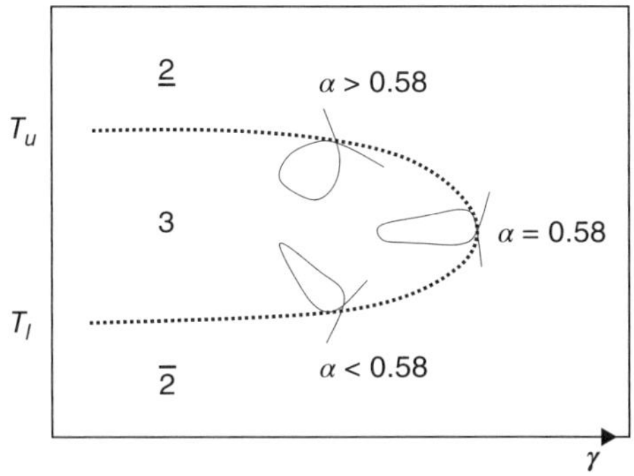

Figure 11.4 Schematic representation of trajectory of the middle phase, dotted line, projected onto the (γ, T) plane. α is the ratio of mass of [bmim][PF_6] to the total mass of solvents.

minimum surfactant concentration required to initiate the microemulsion formation, are shifted toward higher γ and lower T values when α is increased. The reason is the considerably high solvophilicity of TX-100 in [bmim][PF_6]. The more [bmim][PF_6] is added to the solution, the more TX-100 molecules are needed to saturate [bmim][PF_6] with monomers, which, in turn, leads to an increase in γ_0. The decrease in T_0 is certainly due to the fact that addition of [bmim][PF_6] to the solution tends to favor water-in-[bmim][PF_6] droplets. In other words, a zero mean curvature of the interfacial surfactant monolayer occurs when [bmim][PF_6] is added to the solution of [bmim][PF_6]-in-water microemulsion. Comparison of these results with those of water–*n*-dodecane–CiEj system reveals that this inclination to form negatively curved interfacial layers at lower T is equally observed in both systems when [bmim][PF_6] or oil fraction increases [40].

On the other hand, the coordinates of the fishtail are slightly affected by a change of [bmim][PF_6] fraction. The variation of $\tilde{\gamma}$ and $\tilde{T}$ with α allows the determination of the trajectory of the middle phase. Its schematic projection onto the (γ, T) plane is voluntarily amplified and displayed in Figure 11.4. We notice, from this trajectory (dotted line), that the surfactant amount needed to produce one-phase microemulsion is highest when $\alpha = 0.58$, which corresponds to equal water and [bmim][PF_6] volumes. An increase or a decrease in α from 0.58 systematically, although only to a little extent, decreases the optimal amount of surfactant $\tilde{\gamma}$. The small variation in $\tilde{\gamma}$ and $\tilde{T}$ is a significant difference between this system and water–*n*-dodecane systems [40–42]. A careful examination of absolute $\tilde{\gamma}$ values brings the present water–[bmim][PF_6]–TX-100 closer to water–*n*-dodecane–C_4E_1. Indeed, we observe that both are inefficient: nearly 50% of surfactant is needed to solubilize water with the immiscible solvent.

11.3.2 Structure of the Single-Phase Microemulsion

SANS measurements were performed on one-phase samples, prepared in D_2O instead of H_2O, in order to enhance the contrast between microdomains [43]. Two typical SANS spectra of a D_2O–[bmim][PF_6]–TX-100 microemulsion at $\gamma = 0.54$,

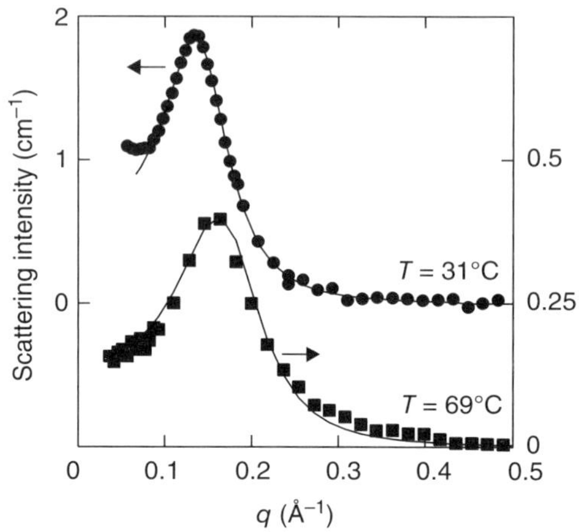

Figure 11.5 Neutron scattering spectra (scattering intensity–q profile) of water–[bmim][PF_6]–TX-100 of equal water and [bmim][PF_6] volumes, at $\gamma=0.54$, $T=31°C$ (●), and $T=69°C$ (■). Solid lines represent the Teubner–Strey model. From Anjum et al. [44].

$\alpha=0.58$, and $T=31$ and 69°C, respectively, are shown in Figure 11.5. The measurements show that SANS intensities always consist of a single broad correlation peak located in the wave vector, q range between 0.13 and 0.16 Å^{-1} [44]. Moreover, above the peak position, the scattering intensity abruptly decreases and follows the Porod law: $I(q) \sim q^{-4}$, demonstrating that the surfactant molecules form a well-defined interface between the two solvents [44]. The evolution of SANS intensities is in agreement with structural investigations usually shown for bicontinuous water–oil microemulsions in the fishtail region.

A meticulous analysis of SANS spectra consists in using the structural model proposed for two random interpenetrating domains, which is the most appropriate for a bicontinuous phase. To model the spectra of Figure 11.5, we used the equation proposed by Teubner and Strey with a limited number of fitting parameters [44, 45]. The parameters include water and IL volume fractions, Φ_1 and Φ_2, respectively, and their corresponding scattering length densities ρ_1 and ρ_2. Thus, $I(q)=\Phi_1\Phi_2(\rho_1-\rho_2)^2(1/(a+bq^2+cq^4))$, where a, b, and c are related to two structural lengths: mean repeat distance d and correlation length ξ. The solid lines in Figure 11.5, which are least squares fits, show that the agreement with experimental data is excellent. The values of d and ξ are 44 Å and 21 Å at $T=31°C$ and 37 Å and 17 Å at $T=69°C$, respectively. Overall, the fits led to different nanometric lengths ($36 \leq d \leq 43$ Å and $16 \leq \xi \leq 23$ Å) depending on sample composition. The small values found for d and ξ are due to the very weak solvophobicity of the surfactant hydrocarbon chain in [bmim][PF_6]. Indeed, a high surfactant concentration is needed to form the water–IL monolayer. As a consequence, the monolayer area is high and the values of d and ξ are small. A comparison of this system and water–oil microemulsions reveals that the present d and ξ are comparable to those usually found in water–n-alkane–CiEj systems with the shortest alkyl lengths.

11.3.3 Comparison of TX-100 with CiEj Surfactants

In order to enhance the microemulsion efficiency, we exchanged the current TX-100 by CiEj surfactants with longer alkyl chains: $C_{12}E_6$ and $C_{14}E_4$, respectively. The investigation clearly confirmed that the use of $C_{12}E_6$ (C_{12} instead of C_8) shifts the position of the fish head to lower γ values. This result, which is classically found in water–oil systems, is also in agreement with data on EAN–*n*-alkane systems [29]. Increasing surfactant alkyl chain to C_{14} shifts the fish head position toward lower γ values. However, in both cases, more surfactant is needed in order to form a one-phase microemulsion in water–[bmim][PF_6]. This behavior conflicts with the results observed in water–*n*-alkane [31] and in EAN–*n*-alkane [29] systems. It is also evident that the increase in the surfactant alkyl chain length has only minor effects on the width of the fish body while it considerably increases its length (extension in T direction vs. γ direction; see Fig. 11.3). In water–oil systems, the fish body shrinks in both directions when longer alkyl chains are used. Again, this effect is most certainly associated with a weaker solvophobicity of CiEj surfactants in [bmim][PF_6] as compared to their hydrophobicity. Moreover, the longest alkyl chains trigger the appearance of a lamellar phase L_α, which was characterized by optical microscopy in water–[bmim][PF_6]–$C_{14}E_4$. We found that L_α dominates the phase diagram up to $\gamma = 0.7$, from between room temperature to 70°C, preventing in this the accurate localization of the $\tilde{X}$ point. A similar trend was observed in EAN–*n*-alkane systems [29]. Such observations are clearly associated with the weak alkyl chains solvophobicity.

11.4 CONCLUSIONS

RTILs are attractive solvents because their properties can be predictably varied by incorporation of appropriate groups. Both hydrophilic and hydrophobic ILs are available. ILs also have the ability to solubilize several species including amphiphilic molecules. However, micellar morphologies of self-assembled surfactants, as well as a range of liquid crystalline phases, are shown to occur in ILs once the CMC is reached. Here, we focused on the micellization process in RTILs, and results showed that the CMC remains, to a large extent, higher than in water. This is a consequence of their ability to solubilize several chemical compounds including hydrocarbons. This limitation does not prevent the preparations of microemulsions. Indeed, both water–IL and IL–oil microemulsions are achievable. The aqueous systems with [bmim][PF_6] presented here confirm the presence of an extended bicontinuous phase. The general pattern of phase behavior and the microstructure of RTIL microemulsions are comparable to conventional systems although the efficiency remains low. This is a consequence of weak solvophobic interactions in ILs. This fact is also reflected in the measured short nanometric repeat distance and correlation length in RTIL microemulsions.

ACKNOWLEDGMENT

The author wishes to acknowledge Drs. Nishat Anjum and Marie-Alice Guedeau-Boudeville, who have helped in carrying out the experimental work. This chapter is dedicated to M. A. Guedeau-Boudeville, on the occasion of her retirement.

REFERENCES

[1] Liu, S., Xiao, J. (2007). *Journal of Molecular Catalysis A: Chemical*, **270**, 1–43.

[2] Sun, P., Armstrong, D. W. (2010). *Analytica Chimica Acta*, **661**, 1–16.

[3] Yoon, H., Lane, G. H., Shekibi, Y., Howlett, P. C., Forsyth, M., Best, A. S., MacFarlane, D. R. (2013). *Energy and Environmental Science*, **6**, 979–986.

[4] Brennecke, J. F., Maginn, E. J. (2001). *AIChE Journal*, **47**, 2384–2389.

[5] Greaves, T. L., Mudie, S. T., Drummond, C. J. (2011). *Physical Chemistry Chemical Physics*, **13**, 20441–20452.

[6] Greaves, T. L., Kennedy, D. F., Mudie, S. T., Drummond, C. J. (2010). *Journal of Physical Chemistry B*, **114**, 10022–10031.

[7] Seguin, C., Eastoe, J., Heenan, R. K., Grillo, I. (2007). *Journal of Colloid and Interface Science*, **315**, 714–720.

[8] El Seoud, O. A., Koschella, A., Fidale, L. C., Dorn, S., Heinze, T. (2007). *Biomacromolecules*, **8**, 2629–2647.

[9] Greaves, T. L., Weerawardena, A., Fong, C., Drummond, C. J. (2007). *Langmuir*, **23**, 402–404.

[10] Chen, L. G., Bermudez, H. (2012). *Langmuir*, **28**, 1157–1162.

[11] Ray, A. (1969). *Journal of the American Chemical Society*, **91**, 6511–6512.

[12] Eicke, H. F., *Surfactants in Nonpolar Solvents* in *Micelles*, Boschke, F. L., ed., Springer,, Berlin, 1980, 85–145.

[13] Hyde, S. T., *Identification of Lyotropic Liquid Crystalline Mesophases in Handbook of Applied Surface and Colloid Chemistry*, Holmberg, K., ed., John Wiley & Sons, Inc., New York, 2001, 299–332.

[14] Kreshec, G. C., *Aqueous Solutions of Amphiphiles and Macromolecules in Water: A Comprehensive Treatise*, Franks, F., ed., Plenum Press, New York, 1975, Vol. **4**, 95–167.

[15] Singh, H. N., Saleem, S. M., Singh, R. P., Birdi, K. S. (1980). *Journal of Physical Chemistry*, **84**, 2194–2199.

[16] Evans, D. F., Yamauchl, A., Jason Wel, G., Bloomfield, V. A. (1983). *Journal of Physical Chemistry*, **87**, 3537–3541.

[17] Anderson, J. L., Pino, V., Hagberg, E. C., Sheares, V. V., Armstrong, D. W. (2003). *Chemical Communications*, 2444–2445.

[18] Misono, T., Sakai, H., Sakai K., Abe, M., Inoue, T. (2011). *Journal of Colloid and Interface Science*, **358**, 527–533.

[19] Patrascu, C., Gauffre, F., Nallet, F., Bordes, R., Oberdisse, J., Lauth-Viguerie, N., Mingotaud, C. (2006). *ChemPhysChem*, **7**, 99–101.

[20] Greaves, T. L., Drummond, C. J. (2013). *Chemical Society Reviews*, **42**, 1096–1120.

[21] Schott, H. (1998). *Journal of Colloid and Interface Science*, **205**, 496–502.

[22] Inoue, T., Yamakawa, H. (2011). *Journal of Colloid and Interface Science*, **356**, 798–802.

[23] Molina-Bolivar, J. A., Aguiar, J., Carnero Ruiz, C. (2001). *Molecular Physics*, **99**, 1729–1741.

[24] Law, G., Watson, P. R. (2001). *Langmuir*, **17**, 6138–6141.

[25] Holbrey, J. D., Seddon, K. R. (1999). *Journal of the Chemical Society, Dalton Transactions*, 2133–2140.

[26] Gao, Y., Li, N., Zheng, L., Zhao, X., Zhang, J., Cao, Q., Zhao, M., Li, Z., Zhang, G. (2007). *Chemistry—A European Journal*, **13**, 2661–2670.

[27] Gao, Y., Han, S., Han, B., Li, G., Shen, D., Li, Z., Du, J., Hou, W., Zhang, G. (2005). *Langmuir*, **21**, 5681–5684.

[28] Eastoe, J., Gold, S., Rogers, S. E., Paul, A., Welton, T., Heenan, R. K., Grillo, I. (2005). *Journal of the American Chemical Society*, **127**, 7302–7303.

[29] Atkin, R., Warr, G. G. (2007). *Journal of Physical Chemistry B*, **111**, 9309–9316.

[30] Gao, H. X., Li, J. C., Han, B. X., Chen, W. N., Zhang, J. L., Zhang, R., Yan, D. D. (2004). *Physical Chemistry Chemical Physics*, **6**, 2914–2916.

[31] Sottmann, T., Strey, R., *Microemulsions* in *Soft Colloids V—Fundamentals in Interface and Colloid Science*, Lyklema, J., ed., Elsevier, Amsterdam, 2005, 5.1–5.96.

[32] Hunter, R. J., *Foundations of colloid science 2nd edition*, Oxford University Press, Oxford, UK, 2001, 1–43.

[33] Kahlweit, M., Strey, R. (1985). *Angewandte Chemie International Edition*, **24**, 654–668.

[34] Kahlweit, M., Strey, R., Firman, P., Haase, D., Jen, J., Schomäcker, R. (1988). *Langmuir*, **4**, 499–511.

[35] Kahlweit, M., Strey, R., Firman, P. (1986). *Journal of Physical Chemistry*, **90**, 671–677.

[36] Kahlweit, M., Strey, R., Busse, G. (1990). *Journal of Physical Chemistry*, **94**, 3881–3894.

[37] Seth, D., Chakraborty, A., Setua, P., Sarkar, N. (2006). *Langmuir*, **22**, 7768–7775.

[38] Seth, D., Chakraborty, A., Setua, P., Sarkar, N. (2007). *Journal of Chemical Physics*, **126**, 224512.

[39] Gao, Y., Li, N., Zheng, L., Zhao, X., Zhang, S., Han, B., Hou, W., Li, G. (2006). *Green Chemistry*, **8**, 43–49.

[40] Burauer, S., Sachert, T., Sottmann, T., Strey, R. (1999). *Physical Chemistry Chemical Physics*, **1**, 4299–4306.

[41] Burauer, S., Belkoura, L., Stubenrauch, C., Strey, R. (2003). *Colloids and Surfaces A: Physicochemical and Engineering Aspects*, **228**, 159–170.

[42] Strey, R. (1994). *Colloid and Polymer Science*, **272**, 1005–1019.

[43] Strey, R., Glatter, O., Schubert, K. V., Kaler, E. W. (1996). *Journal of Chemical Physics*, **105**, 1175–1188.

[44] Anjum, N., Guedeau-Boudeville, M. A., Stubenrauch, C., Mourchid, A. (2009). *Journal of Physical Chemistry B*, **113**, 239–244.

[45] Teubner, M., Strey, R. (1987). *Journal of Chemical Physics*, **87**, 3195–3200.

CHAPTER 12

Microemulsions with Ionic Liquids

JOACHIM KOETZ

Institut für Chemie, Universität Potsdam, Potsdam, Germany

12.1 INTRODUCTION

Microemulsions are thermodynamically stable, isotropic solutions of two nonmiscible liquids in presence of interfacial active substances, first described by Schulman and Stoeckenius in 1959 [1]. The nonmiscible liquids used are normally oil and water. In presence of surfactants and/or cosurfactants, that is, medium-chain alkanols, interfacial tension between both liquids is much reduced, and nanosized droplets are formed. Special features of microemulsion droplets in the size range between 2 and 20 nm are fast percolation process, Newtonian flow, and reversible temperature behavior. Recent advances in analytical techniques, for example, electron microscopy, scattering, and spectroscopic techniques including theoretical models, allow detailed characterization of the morphology of the nanoscalic droplets [2]. Due to the fast exchange processes between the droplets in a microemulsion, they can be used in different technological applications [3], for example, as nanoreactors to form monodisperse nanoparticles [4], as drug delivery systems [5], etc.

Ionic liquids (ILs) with their outstanding solvent properties, that is, high conductivity, tunable polarity, zero volatility, and high thermal stability, are of growing interest in different fields of application including as a solvent component in microemulsions [6, 7]. Thus, the tunable polar properties are of special interest. Therefore, ILs can be used as polar [8–11] as well as nonpolar solvents [12, 13]. Moreover, ILs can show amphiphilic behavior [14] and are called IL surfactants [15]. Their physical properties can be adjusted in a characteristic way by varying the molecular structure of ILs. Since the ILs are sterically mismatched ions, they hinder crystal formation. Different cations, for example, ammonium, phosphonium, sulfonium, imidazolium, pyridinium, or pyrrolidinium, with different anions, for example, chloride, bromide, iodide, fluoroborate, or fluorophosphates, can be selected for a broad range of combinations [6].

The most important cation used is the imidazolium. Noteworthy is that the interfacial properties of the imidazolium cation can be furthermore influenced by varying

Ionic Liquid-Based Surfactant Science: Formulation, Characterization, and Applications, First Edition. Edited by Bidyut K. Paul and Satya P. Moulik.

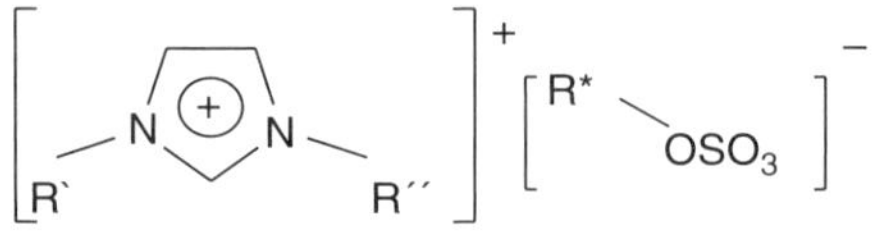

Figure 12.1 Imidazolium salts, with different alkyl moieties, **Emim**$\mathbf{Et}SO_4$ (R′ = **e**thyl; R″ = **m**ethyl; R* = **Et**hyl), **Emim**$\mathbf{Hex}SO_4$ (R′ = **e**thyl; R″ = **m**ethyl; R* = **Hex**yl), **Emim**$\mathbf{Oct}SO_4$ (R′ = **e**thyl; R″ = **m**ethyl; R* = **Oct**yl), **Emim**$\mathbf{Dod}SO_4$ (R′ = **e**thyl; R″ = **m**ethyl; R* = **Dod**ecyl).

the alkyl chains R′ and R″ at both *N*-atoms in the cation (cf. Fig. 12.1). The most used cations by different research groups are 1-ethyl-3-methylimidazolium (Emim) and 1-butyl-3-methylimidazolium (Bmim).

Depending on the utilized anion, the polarity of the IL can also be varied. Thus, in presence of Cl^- and Br^-, water-soluble polar ILs are formed in contrast to the nonpolar ILs containing anions, like BF_4^- or PF_6^-. By using alkylsulfates, for example, ethylsulfate, hexylsulfate, or octylsulfate, the interfacial properties can be controlled (Fig. 12.1). This opens a versatile door for tuning the polarity of the IL.

Based on this knowledge, one can choose the IL in dependence on the requirements of the microemulsion. Taking into account that the formation of microemulsions is a spontaneous self-assembled process, the oil component in the presence or absence of a cosurfactant, as well as the type of surfactant used, that is, nonionic or ionic, is of vital importance.

Gao et al. have already intensively studied the phase diagram in presence of nonionic surfactants. By using Triton X-100, $BmimBF_4$, and cyclohexane, three different types of microemulsions, that is, IL in oil, bicontinuous, and oil in IL, were found to exist [8].

Recently, we have demonstrated that in the presence of ionic surfactants, different types of microemulsions can be formed [16]. Thus, the positive- and negative-charged head groups of the cationic and anionic surfactants were of special relevance. In this chapter, the role of ILs in the presence of ionic surfactants is discussed in detail.

12.2 MICROEMULSIONS CONTAINING IONIC SURFACTANTS AND ILs

12.2.1 ILs Solubilized in Water

BmimCl can be mixed with water in a broad range of concentration. The resulting quasiternary phase diagram with a toluene/pentanol (1:1) mixture and the cationic surfactant CTAB is shown in Figure 12.2. One can see a transparent phase channel between the oil corner and the water corner.

Comparing these results with the phase diagram in the absence of BmimCl (shown in Fig. 12.3), one can conclude that due to the presence of the IL, the isotropic phase range is drastically increased. This means the IL incorporated into the interfacial region can stabilize the surfactant film much better and a so-called palisade layer is formed.

Systematic investigations by varying the surfactant component, that is, by using an amphoteric surfactant, *N*-dodecyl-*N*,*N*-dimethyl-3-ammonio-1-propanesulfonate

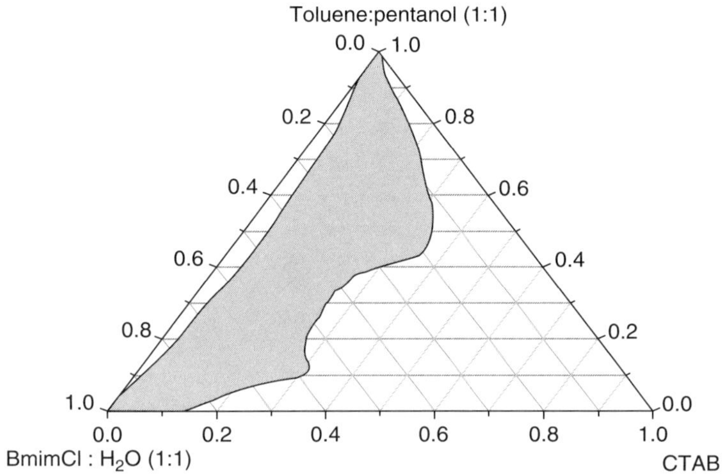

Figure 12.2 Isotropic phase range of the quasiternary system BmimCl/water (1:1)/toluene: pentanol (1:1)/CTAB.

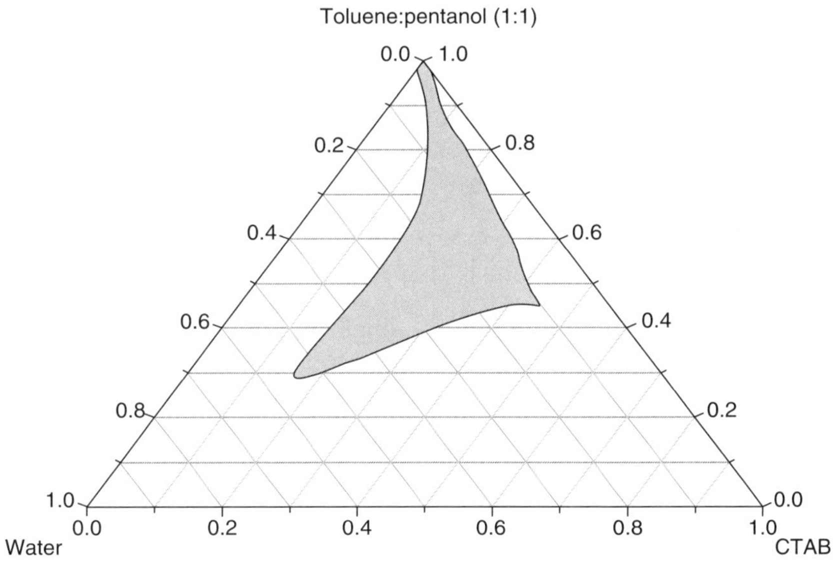

Figure 12.3 Isotropic phase range of the quasiternary system water/toluene/pentanol (1:1)/ CTAB.

(SB-12) with sulfobetaine head groups, show a similar trend. This means that in presence of SB-12, a quite similar isotropic phase channel is observed in comparison with Figure 12.2. These results clearly show that the incorporation of the water-soluble IL in addition to the water-in-oil microemulsion leads to the formation of both oil-in-water and bicontinuous ones.

By using the anionic surfactant, the isotropic phase range is limited to an area in the water corner. This effect can be explained by strong electrostatic interactions

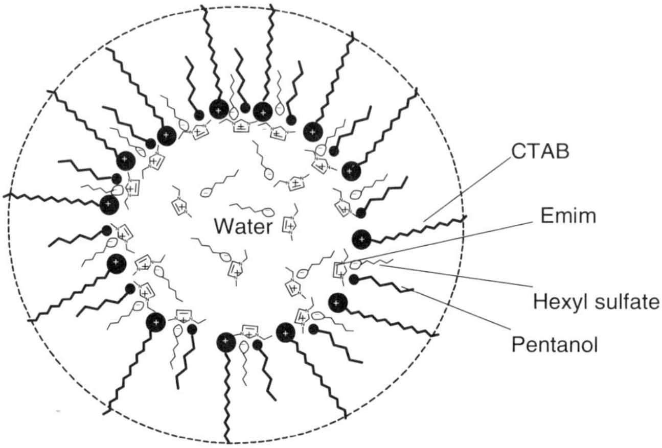

Figure 12.4 Model of an inverse microemulsion droplet in presence of $EmimHexSO_4$.

between the anionic sulfate head groups and the Bmim cation reducing the range of microemulsion.

By making the IL component more surfactant-like, for example, by using $EmimHexSO_4$, in the presence of SDS, a broad isotropic phase channel can be produced [16]. More comprehensive investigations show that all types of microemulsions can be observed in the phase channel. Cryo-SEM micrographs show the transition from droplets to a sponge phase, accompanied by a characteristic change of the reduced self-diffusion coefficient obtained by NMR diffusion experiments.

In general, one can say that the formation of self-assembled microemulsions can be improved by adding water-soluble ILs fulfilling the role of the polar solvent as well as the role of a stabilizer of the interfacial film to form a palisade layer surrounding the microemulsion droplets, as exemplified in Figure 12.4.

12.2.2 ILs as Polar Solvents

$EmimEtSO_4$ as well as $EmimHexSO_4$ can also be used in the absence of water. In that case, the IL fulfills the requirements of the polar phase. Such kinds of microemulsions are of special relevance as noncorrosive solvents [17].

Recently, we have shown the phase diagram of the quasiternary system $EmimEtSO_4$/CTAB/toluene/pentanol (1:1) with an isotropic phase channel between the oil and IL corner, which is somewhat smaller in comparison with the systems in the presence of water, already discussed here and shown in Figure 12.2. It is noteworthy that the phase channel becomes broader by increasing the temperature or by making the IL more surfactant-like, that is, by replacing ethylsulfate with hexylsulfate. The individual $EmimEtSO_4$ droplet structure can be visualized by means of cryo-SEM (Fig. 12.5). Furthermore, our investigations have shown that the role of cosurfactant is of high relevance. This means that to form a bicontinuous microemulsion, a cosurfactant is required in combination with $EmimEtSO_4$, in contrast with the $EmimHexSO_4$, where the hexylsulfate anion can serve this purpose. These results

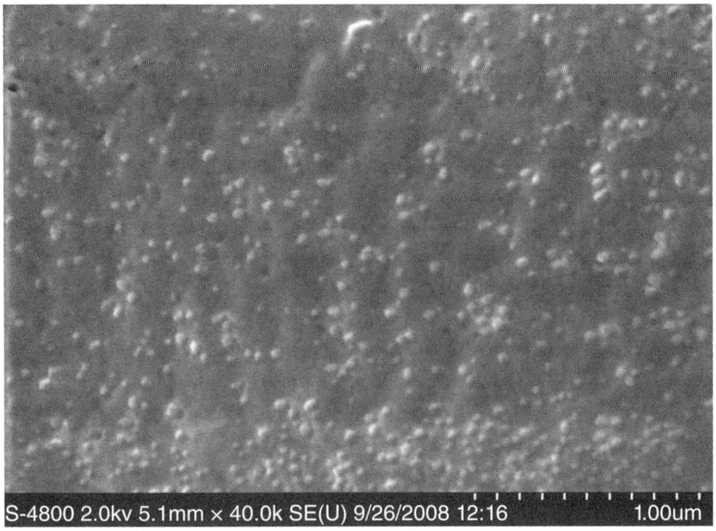

Figure 12.5 Cryo-SEM micrograph of $EmimEtSO_4$ microemulsion droplets.

already show that long-chain anion with amphiphilic properties is of high relevance to form spontaneous microemulsions with ILs.

12.2.3 ILs as Polar Solvents and Surfactants

Recently, we have shown that ILs with long-chain alkylsulfates, that is, Bmim octylsulfate, $BmimOctSO_4$ [18], and Bmim dodecylsulfate, $BmimDodSO_4$ [19], can completely take over the role of the surfactant in the microemulsion formation process. This means nonaqueous, halogen-free microemulsions are formed by mixing two ILs with one oil component. In these systems, $EmimEtSO_4$ is the polar component, and $BmimOctSO_4$ or $BmimDodSO_4$ fulfills the requirements of the interfacial film component. Conductivity measurements established by Gao et al. [8] can be successfully applied to detect the transition between the oil-in-IL and the bicontinuous as well as the bicontinuous and the IL-in-oil microemulsion. Figure 12.6 shows the corresponding phase diagram in the presence of $BmimOctSO_4$.

By using $BmimDodSO_4$, one can increase the isotropic phase range significantly. Dynamic light scattering experiments have shown that individual $EmimEtSO_4$ droplets surrounded by the other more surfactant-like IL, that is, $BmimDodSO_4$, are in the order of few nanometers. These results are in full agreement with our SAXS measurements applying the Teubner–Strey model [19]. It is noteworthy that $BmimDodSO_4$ can form micelles in toluene. A similar phenomenon of appearance of micellar aggregation was observed by Gao et al. by using $BmimBF_4$ or $BmimPF_6$ [20].

In general, one can say that a new type of nonaqueous microemulsions can be formed, consisting of very small IL droplets stabilized by the other more surfactant-like IL. By increasing the amount of IL, a transition to a bicontinuous phase range can be identified by conductivity measurements as well as a transition to oil-in-IL microemulsion droplets.

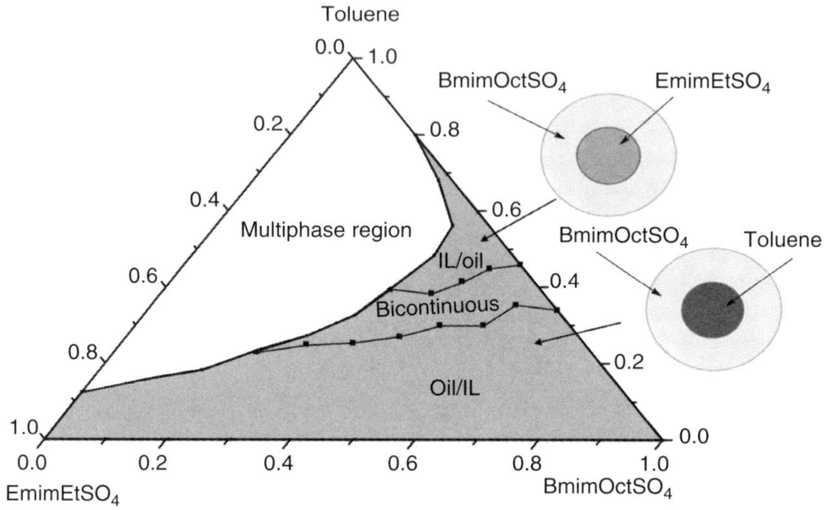

Figure 12.6 Phase diagram of the ternary system $EmimEtSO_4$/toluene/$BmimOctSO_4$.

12.3 CONCLUSIONS

Like water–oil–surfactant, IL–oil–surfactant can also form microemulsions. A special feature of IL-based microemulsions is the flexibility of the IL, which can be the polar or nonpolar as well as the surfactant component. When ILs are incorporated into microemulsions with ionic surfactants, electrostatic interactions between the components can play a crucial role to form a more stable palisade layer surrounding the droplets. By using polar ILs, water-free microemulsions can be formed. The tunable properties of two different ILs may allow spontaneous formation of a new class of IL–oil–IL microemulsions with special features opening new fields of application.

REFERENCES

[1] Schulman, J. H.; Stoeckenius, W. *Journal of Physical Chemistry* 1959, **63**, 1677–1680.

[2] Acharya, D. P.; Hartley, P. G. *Current Opinion in Colloid & Interface Science* 2012, **17**, 274–280.

[3] Mehta, S. K.; Kaur, K. *Indian Journal of Chemistry* 2010, **49A**, 662–684.

[4] Capek, I. *Advances in Colloid and Interface Science* 2004, **110**, 49–74.

[5] Gupta, S.; Moulik, S. P. *Journal of Pharmaceutical Science* 2008, **97**, 22–45.

[6] Kunz, W.; Zemb, T.; Harrar, A. *Current Opinion in Colloid & Interface Science* 2012, **17**, 205–211.

[7] Greaves, T. L.; Drummond, C. J. *Chemical Society Reviews* 2008, **37**, 1709–1726.

[8] Gao, H.; Li, J.; Han, B.; Chen, W.; Zhang, J.; Zhang, R.; Yan, D. *Physical Chemistry Chemical Physics* 2004, **6**, 2914–2916.

[9] Eastoe, J.; Gold, S.; Rogers, S. E.; Paul, A.; Welton, T.; Heenan, R. K.; Grillo, I. *Journal of the American Chemical Society* 2005, **127**, 7302–7303.

[10] Xu, J.; Zhang, L.; Yin, A.; Hou, W.; Yang, Y. *Soft Matter* 2013, **9**, 6497–6504.

[11] Rao, V. G.; Ghosh, S.; Ghatak, C.; Mandal, S.; Brahmachari, U.; Sarkar, N. *Journal of Physical Chemistry B* 2012, **116**, 2850–2855.

[12] Moniruzzaman, M.; Kamiya, N.; Nakashima, K.; Goto, M. *Green Chemistry* 2008, **10**, 497–500.

[13] Behera, K.; Malek, S.; Pandey, S. *ChemPhysChem* 2009, **10**, 3204–3208.

[14] Mukherjee, I.; Mukherjee, S.; Naskar, B.; Ghosh, S.; Moulik, S. P. *Journal of Colloid and Interface Science* 2013, **359**, 135–144.

[15] Asadov, Z. G.; Akhmedova, G. A.; Aga-Zadeh, A. D.; Nasibova, S. M.; Zarbalieva, I. A.; Bagirova, A. M.; Ragimov, R. A. *Russian Journal of General Chemistry* 2012, **82**, 1916–1927.

[16] Rojas, O.; Koetz, J.; Kosmella, S.; Tiersch, B.; Wacker, P.; Kramer, M. *Journal of Colloid and Interface Science* 2009, **333**, 782–790.

[17] Rabe, C.; Koetz, J. *Colloids and Surfaces A: Physicochemical and Engineering Aspects* 2010, **354**, 261–267.

[18] Rojas, O.; Tiersch, B.; Frasca, S.; Wollenberger, U.; Koetz, J. *Colloids and Surfaces A: Physicochemical and Engineering Aspects* 2010, **369**, 82–87.

[19] Rojas, O.; Tiersch, B.; Rabe, C.; Stehle, R.; Hoell, A.; Arlt, B.; Koetz, J. *Langmuir* 2013, **29**, 6833–6839.

[20] Gao, Y.; Li, N.; Li, X.; Zhang S.; Zheng, L.; Bai, X.; Yu, L. *Journal of Physical Chemistry B* 2009, **113**, 123–130.

CHAPTER 13

Properties of Ionic Liquid-Based Microemulsions

MARIA FIGUEIRA-GONZÁLEZ, LUIS GARCÍA-RÍO, MERCEDES PARAJÓ, and PEDRO RODRÍGUEZ-DAFONTE

Departamento de Química Física, Centro Singular de Investigación en Química Biológica y Materiales Moleculares (CiQUS), Universidade de Santiago de Compostela, Santiago, Spain

13.1 INTRODUCTION

Ionic liquids (ILs) are organic salts that are liquids at ambient conditions and are considered an alternative to traditional organic solvents for their properties, such as nonvolatility, nonflammability, and high thermal stability [1]. Some ILs possess the ability to support the self-assembly of amphiphiles. Amphiphiles consist of hydrophobic and hydrophilic moieties and include surfactants, lipids, and block copolymers. The first report of amphiphile self-assembly in a nonaqueous liquid was for the molten salt, pyridinium chloride (melting point 146°C), where micellization of cationic surfactants was demonstrated [2]. This ability was first reported for an IL at ambient temperatures in ethylammonium nitrate (EAN) [3]. There has recently been a very strong renewed interest in this field, with many new ILs reported that can support amphiphile self-assembly. In fact, there are a limited number of molecular solvents possessing this capability. By using molecular solvents, the properties of a self-assembly system are limited to modifying the amphiphile. The use of ILs (many new ILs can be created from different cation and anion combinations) as self-assembly media readily enables the modification of either the solvent or amphiphile. The use of ILs as solvents for amphiphile self-assembly was briefly reviewed [4, 5], highlighting that room-temperature ionic liquids (RTILs) can be used as solvents for amphiphile self-assembly into micelles, vesicles, and liquid crystals, as solvent phases in microemulsions, and in the preparation of nanoparticles (NPs). Significant new work has been reported in this field since these two earlier reviews were published [6].

ILs used in self-assembly systems can be divided in two categories: protic ionic liquids (PILs, solvent properties are similar to water) and aprotic ionic liquids (AILs, largest group most of them derived from disubstituted imidazolium cations). Many surfactant ILs have thermotropic properties, forming liquid crystal phases with no additional solvent present. The solvophobic interaction between PILs and

Ionic Liquid-Based Surfactant Science: Formulation, Characterization, and Applications, First Edition. Edited by Bidyut K. Paul and Satya P. Moulik.

amphiphiles is considered to be similar to water and amphiphiles or other polar self-assembly media. Like water, the PILs are good solvents for the polar regions of the amphiphiles and poor solvents for the apolar regions. For the AILs, which are not capable of forming hydrogen bonds, the solvophobic interaction is based on the AILs being good solvents for polar regions and poor solvents for the apolar regions of the amphiphiles [7–9]. ILs containing long alkyl chains (>5 methylene groups) become amphiphilic in nature and are the main topic to be discussed in this review.

Microemulsions are thermodynamically stable dispersions of two immiscible liquids that, at microscopic level, coexist and form individual domains that are separated by a film of surfactant [10]. Depending on the proportions of the three components, the aggregate structures can be water in oil, bicontinuous, or oil in water. The first nonaqueous microemulsions were reported in 1984 by Rico and Lattes [11] containing formamide instead of water. Since then, numerous studies have been reported on the topic of nonaqueous microemulsions [12–15]. Analogous behavior and phases have been observed for microemulsions containing an IL as one of the solvent phases. All three possible topologies, that is, oil-in-IL (oil is as dispersed phase in ILs), IL-in-oil (IL is as dispersed phase in oil), and bicontinuous regions, were characterized by conductivity measurements. Microemulsions containing ILs are of interest due to the potential to expand the solvating abilities of the neat IL, such as improving the solubility of apolar solutes through having hydrophobic domains for oil-in-IL dispersions. ILs can be used in the microemulsion formulation both in order to promote the self-assembly of amphiphiles and as surfactants.

In this review, we examine recent studies of microemulsions incorporating ILs and some recent applications of such microemulsions. We discuss studies wherein ILs are substituted for the oil component in water and IL microemulsions, for the polar or water component in IL and oil microemulsions, and for the surfactant component, along with new material applications. Taking as starting point the traditional water-in-oil microemulsion, in the new IL microemulsions, the IL can play the role of water, surfactant, or oil (see Scheme 13.1). If the IL substitutes dispersed phase, we have nonaqueous IL microemulsions. If the IL substitutes surfactant and/or continuous phase, we have aqueous IL microemulsions. We establish the following classifications:

(1) Nonaqueous IL microemulsions: oil@surfactant@IL.
(2) Aqueous IL microemulsions: water@surfactant@IL.
(3) IL microemulsions with IL as surfactant. This category includes both aqueous and nonaqueous microemulsions and microemulsions with two ILs.

13.2 NON AQUEOUS MICROEMULSIONS

1-Butyl-3-methylimidazolium tetrafluoroborate, $[C_4mim][BF_4]$, is the IL most frequently used in the preparation of nonaqueous IL microemulsions, but a wide variety of other ILs are suitable for the preparation of IL-based microemulsions. The first microemulsion system with $[C_4mim][BF_4]$ as polar phase was reported by Han et al. in 2004 [16]. SANS experiments carried out by Eastoe et al. [17] on $[C_4mim][BF_4]$/Triton X-100/cyclohexane microemulsions. The effects of confinement

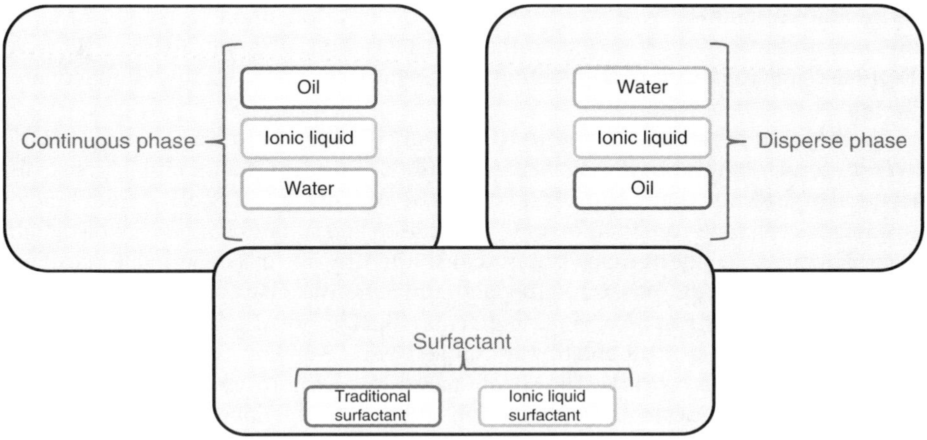

Scheme 13.1 Components of IL microemulsions.

of the IL on solvation dynamics and rotational relaxation of a probe molecule in these microemulsions were explored using steady-state and picosecond time-resolved emission spectroscopy [18].

Dynamic light scattering (DLS) along with freeze-fracture transmission electron microscopy (FF-TEM) measurements revealed that the sizes of single microemulsion droplets vary with temperature [19, 20]. The role of the organic solvent was also investigated for this system changing cyclohexane by benzene, toluene, or p-xylene [21–26].

Addition of a fourth component to [C_4mim][BF_4]/TX-100/cyclohexane microemulsions was also studied. The addition of alcohol as cosurfactant [27] or small amounts of water [28, 29] changes the size and properties of the system. The study of the phase behavior revealed that the existence region of the microemulsion state is remarkably decreased by the addition of PVP to [C_4mim][BF_4]/TX-100/cyclohexane [30]. However, the addition of the polymer PEG-400 increases the size of the microemulsions [31].

Practical applications of [C_4mimBF_4]/TX-100/cyclohexane microemulsions include reaction media for kinetic studies and synthesis of metal NPs. For instance, the rate constant for aminolysis of an ester at the interface of this IL/oil microemulsion is between two and four times higher than in traditional W/oil microemulsions. However, the rate constant is at least five times higher in pure water than in the [C_4mim][BF_4] [32]. An example is a medium to prepare stable silver NPs, with an average diameter of approximately 3 nm, in the absence of any other auxiliary solvent in the whole process [33]. [C_4mim][BF_4]/TX-100/cyclohexane microemulsions were recently used as a new method for the preparation of monodispersed palladium NPs [34].

Combinations of the methyl imidazolium cation with different lengths ([C_4mim]) and anion different from that of [C_4mim][BF_4] can be involved in the preparation of new IL microemulsions. Although [C_4mim][PF_6] is normally taken as the IL in aqueous IL microemulsions, some nonaqueous IL microemulsions were also prepared. The aggregate size of [C_4mim][PF_6]/TX-100/toluene [35] and [C_4mim][PF_6]/TX-100/ethylene glycol [36] microemulsions were characterized further by small-angle

X-ray scattering (SAXS), DLS, freeze-fracture electron microscopy (FFEM), or ultraviolet–visible (UV–Vis) spectroscopy. An example of application of [C_4mim] [PF_6]-based microemulsions is a reaction media synthesis of poly(methyl methacrylate) [37, 38].

The addition of formamide to the IL [C_2mim][Cl] allowed to prepare stable IL/oil and oil/IL microemulsions in cyclohexane with TX-100 as surfactant [39]. The size distribution of these microemulsion droplets was determined by DLS. A different kind of ILs emerges from the combination of $C_n$mim cation and sulfates with a variable length chain. For instance, the ternary system constituted by 1-ethyl-3-methylimidazolium ethylsulfate ([C_2mim][C_2SO_4]), TX-100, and limonene was characterized by means of conductivity, viscosity, and SAXS measurements [40]. Recently, [C_2mim][C_nSO_4]/TX-100/cyclohexane systems (n=4, 6, and 8) were studied by DLS and picosecond time-resolved fluorescence spectroscopy [41]. The sulfate-based ILs can form microemulsions in the presence of other surfactants. For instance, the mixture of two nontoxic nonionic surfactants, polyoxyethylene sorbitan monooleate (Tween-80) and sorbitan laurate (Span-20), in a biological oil phase of isopropyl myristate (IPM), can form microemulsions with the IL [C_2mim][C_4SO_4] [42].

The combination of the anion bis(trifluoromethylsulfonyl)imide ((($CF_3SO_2)_2N^-$), [Tf_2N^-]) with asymmetric cations leads to a new family of ILs that can form aqueous and nonaqueous microemulsions. *N*,*N*,*N*-trimethyl-*N*-propyl ammonium bis(trifluoromethanesulfonyl)imide ([N_{3111}][Tf_2N]; see Scheme 13.2) is one example of this family of ILs. The phase behavior of the ternary system [N_{3111}][Tf_2N]/TX-100/cyclohexane was investigated, and microregions of IL/oil, bicontinuous, and oil/IL were identified by traditional electrical conductivity measurements. DLS revealed the formation of the IL microemulsions [43]. In a similar way, *N*-methyl-*N*-propylpyrrolidinium bis(trifluoromethanesulfonyl)imide ([$P_{13}Tf_2N$]; see Scheme 13.2) was substituted for polar water and formed nonaqueous microemulsions with benzene by the aid of TX-100 [44]. Another example is the IL 1-ethyl-3-methylimidazolium bis(trifluoromethylsulfonyl)imide ([C_2mim][Tf_2N]). The phase behavior and DLS study showed that the [C_2mim][Tf_2N]/TX-100/cyclohexane three-component system can form microemulsion with [C_2mim][Tf_2N] as polar core at suitable condition [45].

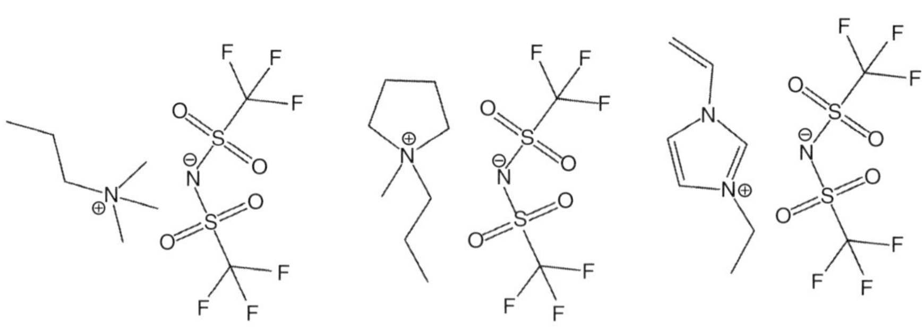

[N_{3111}] [Tf_2N] [P_{13}] [Tf_2N] [veim] [Tf_2N]

Scheme 13.2 Chemical structures of *N*,*N*,*N*-trimethyl-*N*-propyl ammonium bis(trifluoromethanesulfonyl)imide ([N_{3111}][Tf_2N]), *N*-methyl-*N*-propylpyrrolidinium bis(trifluoromethanesulfonyl)imide ([P_{13}][Tf_2N]), and 1-ethyl-3-methylimidazolium bis(trifluoromethanesulfonyl)imide ([C_2mim][Tf_2N]).

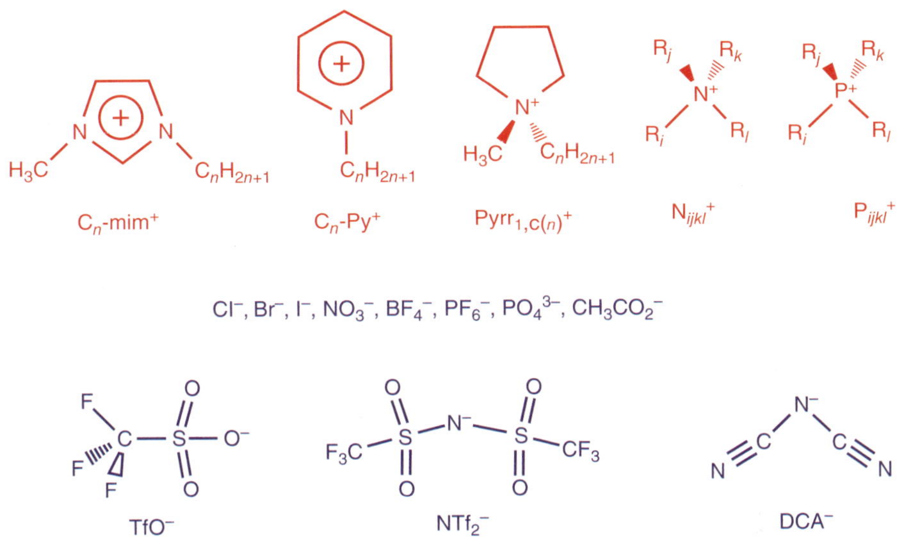

Figure 2.1 Chemical structures for common IL cations (orange) and anions (blue). Source: Castner and Wishart [1] and reproduced with permission from American Institute of Physics.

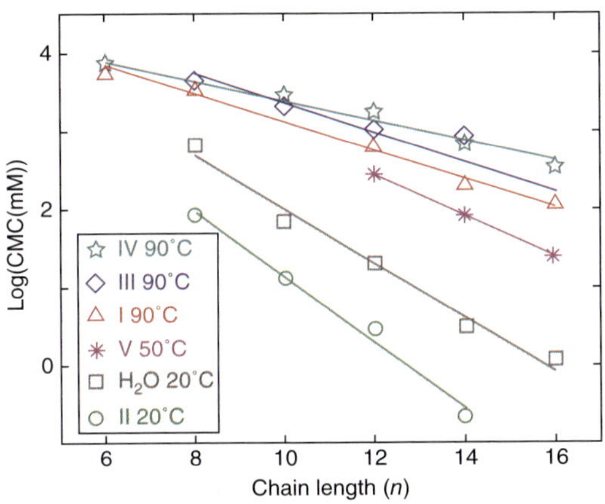

Figure 2.10 Dependence of CMC on C_nTAB alkyl chain length in different ILs: $[C_2C_1IM][EtSO_4]$, I; $[BHEDMA][MeSO_3]$, II; $[C_4C_1IM][BF_4]$, III; $[C_4C_1Pyr][DCA]$, IV; EAN, V. Source: Chen and Bermudez [38] and reproduced with permission from American Chemical Society.

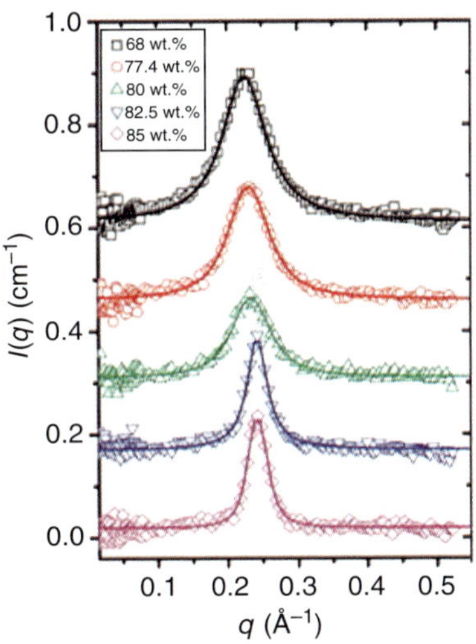

Figure 2.17 SANS profile at 25°C for DDAB/D-EAN solutions at various concentrations. Narrowing and shifting in peak from 80 to 82.5% correspond to the shift from L_3 to $L\alpha$ region. Source: Lopez-Barron et al. [55] and reproduced with permission from American Chemical Society.

Ionic Liquid-Based Surfactant Science: Formulation, Characterization, and Applications, First Edition. Edited by Bidyut K. Paul and Satya P. Moulik.

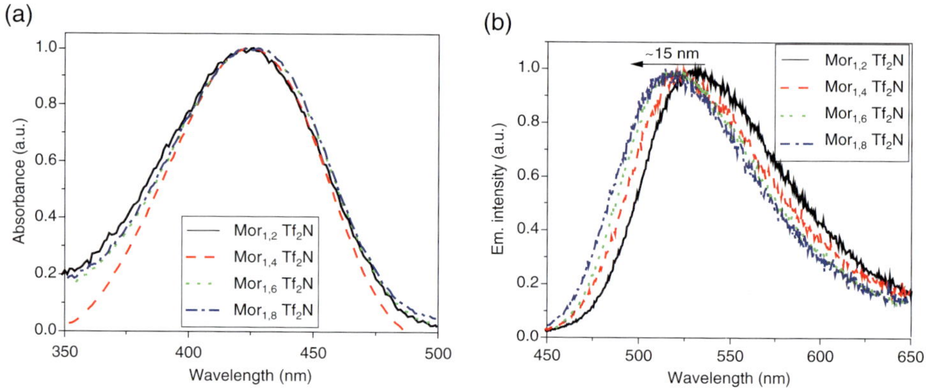

Figure 7.1 See text page 155 for full figure caption.

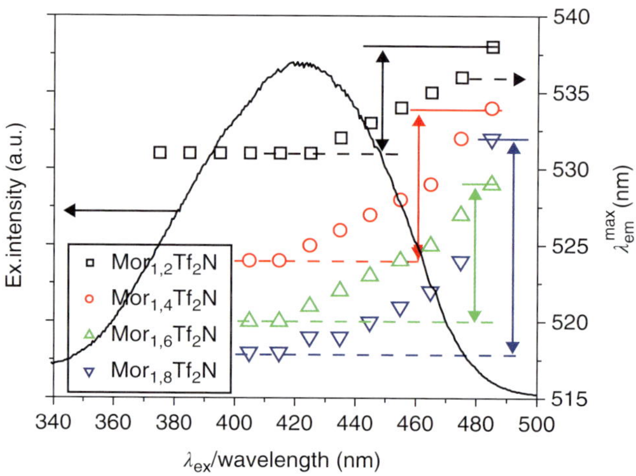

Figure 7.3 See text page 159 for full figure caption.

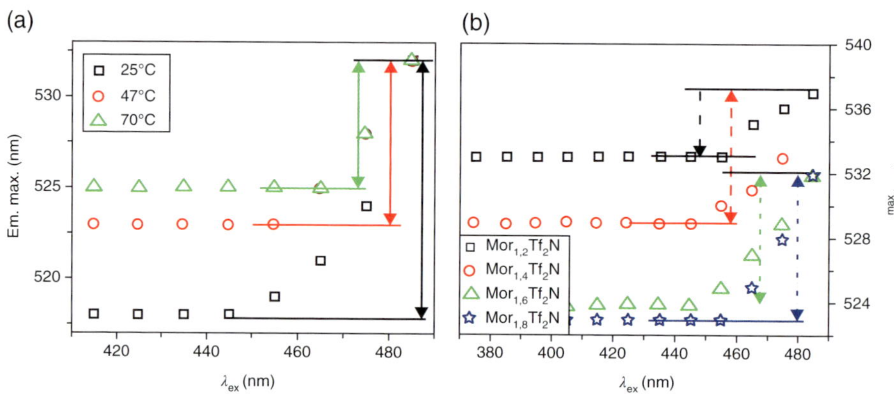

Figure 7.4 See text page 160 for full figure caption.

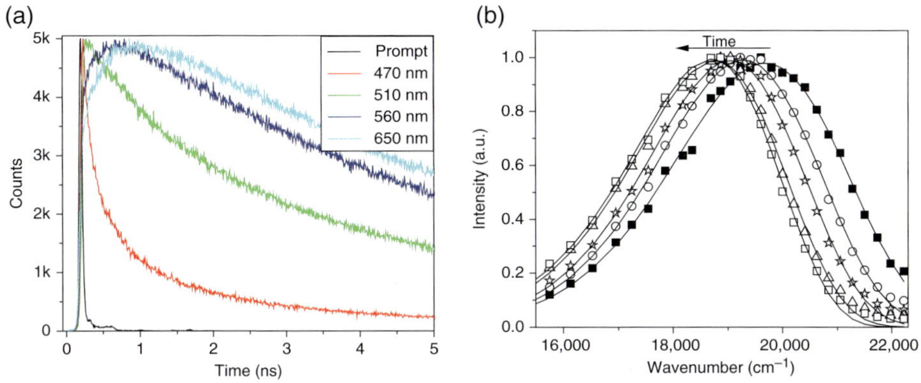

Figure 7.5 See text page 162 for full figure caption.

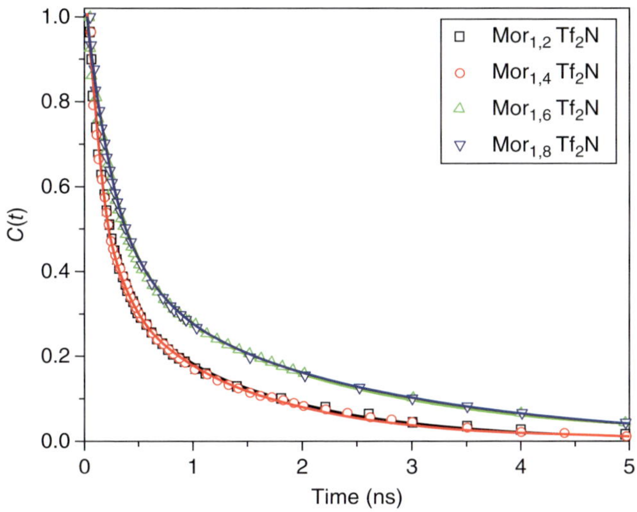

Figure 7.6 See text page 163 for full figure caption.

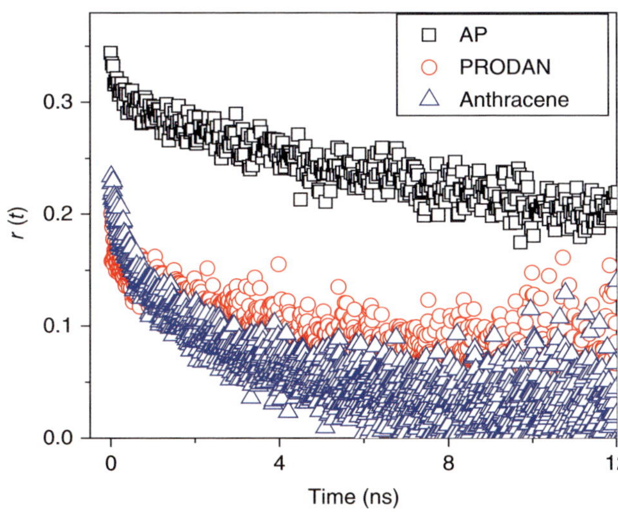

Figure 7.7 See text page 165 for full figure caption.

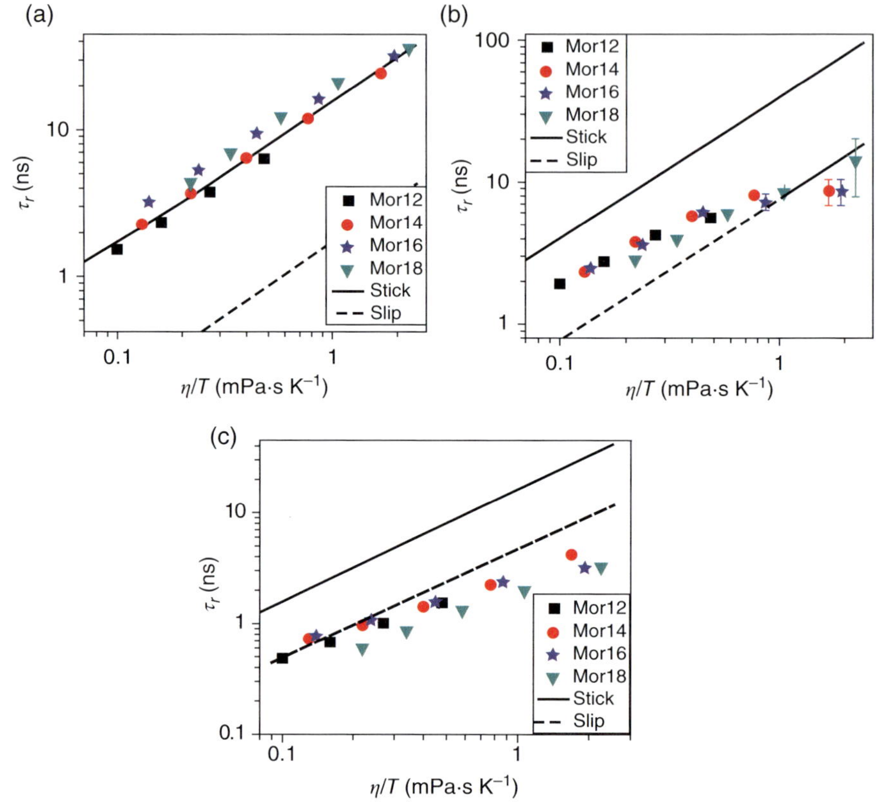

Figure 7.8 Plots of τ_r versus η/T for AP (a), PRODAN (b), and anthracene (c) in four different morpholinium ILs. The points indicate the experimentally measured rotational times. $\lambda_{exc} = 375$nm. Reproduced from Khara et al. [50] with permission from the American Chemical Society.

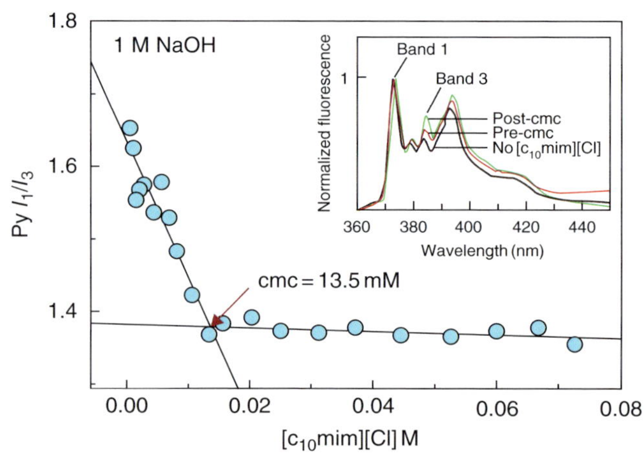

Figure 9.2 Pyrene (1 μM) band 1-to-band 3 emission intensity ratio (Py I_1/I_3) as a function of [C_{10}mim][Cl] concentration in 1 M aqueous NaOH. Inset, normalized fluorescence emission spectra of pyrene (λ_{ex} = 337 nm, slits = 1/1 nm) in 1 M aqueous NaOH in the absence of [C_{10}mim][Cl], in the presence of 1.1 mM [C_{10}mim][Cl] (pre-cmc region), and in the presence of 60 mM [C_{10}mim][Cl] (post-cmc region).

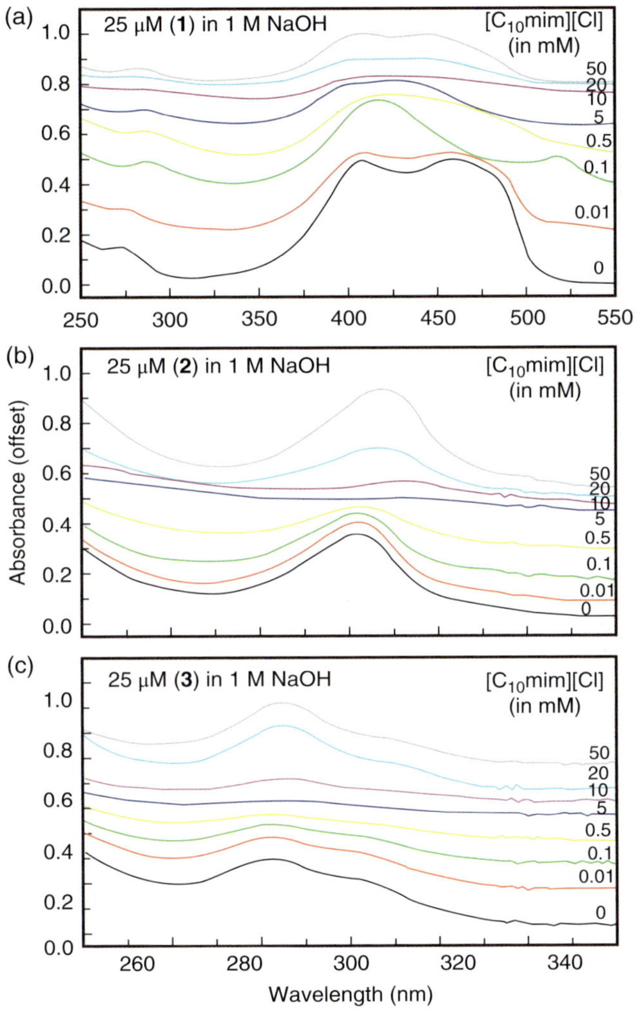

Figure 9.3 (a–c) UV–Vis absorbance spectra of 25 μM each of (**1**), (**2**), and (**3**) dissolved in 1 M aqueous NaOH at varying concentration of [C_{10}mim][Cl].

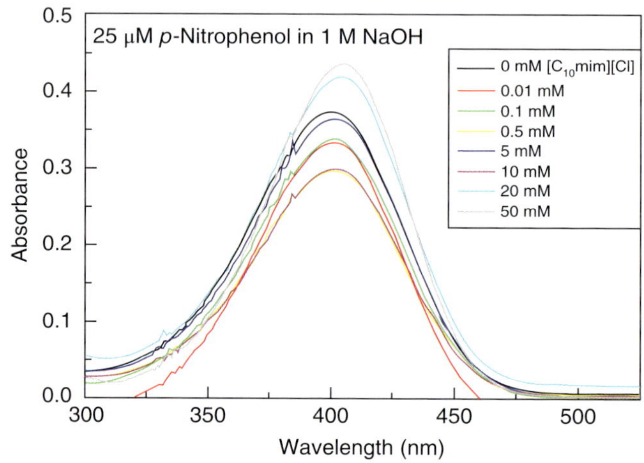

Figure 9.4 UV–Vis absorbance spectra of 25 μM of *p*-nitrophenol dissolved in 1 M aqueous NaOH with varying concentration of [C_{10}mim][Cl].

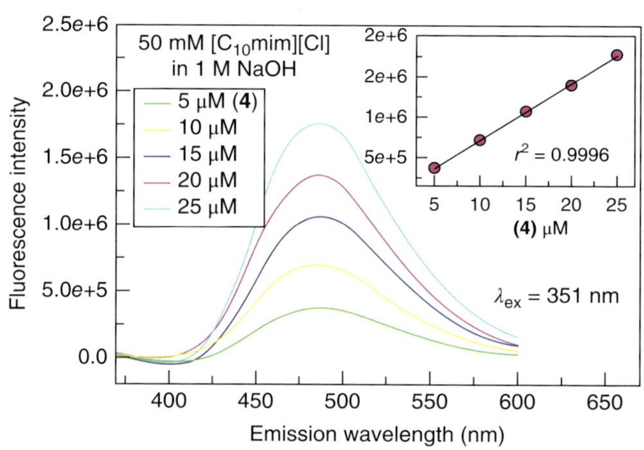

Figure 9.7 Fluorescence emission spectra of (**4**) at different concentrations dissolved in 50 mM [C_{10}mim][Cl] in 1 M aqueous NaOH. λ_{ex} = 351 nm, slits = 3 / 3 nm. Inset, linear regression analysis for fluorescence intensity *versus* concentration of (**4**).

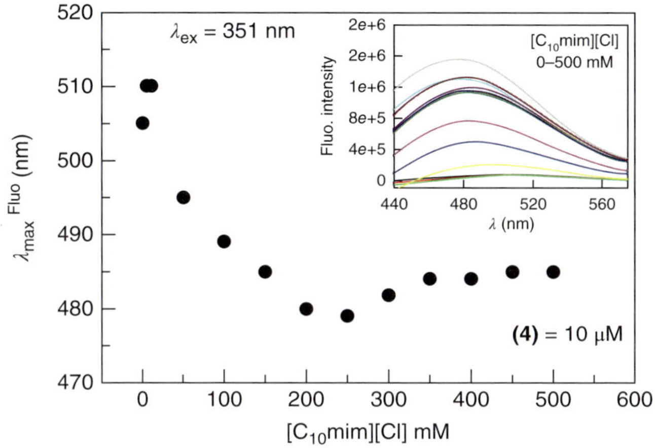

Figure 9.8 Variation in the lowest-energy fluorescence emission maxima (λ_{max}^{Fluo}) of (**4**) (10 μM) with the concentration of [C_{10}mim][Cl] in 1 M aqueous NaOH. λ_{ex} = 351 nm, slits = 3 / 3 nm. Inset, fluorescence emission spectra of (**4**) at different concentrations of [C_{10}mim][Cl] (increasing downward) in 1 M aqueous NaOH.

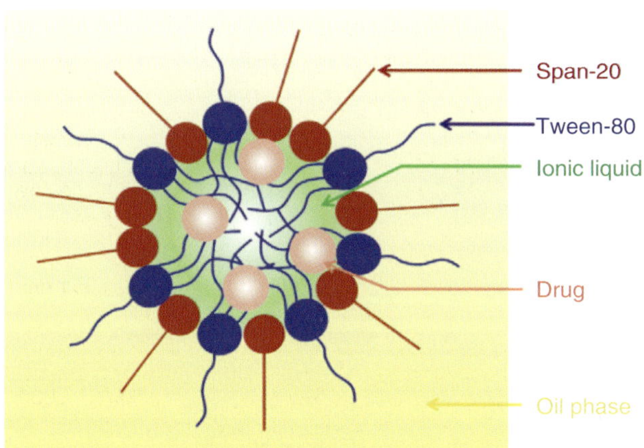

Scheme 10.3 Schematic representation of IL-in-oil microemulsion with drug molecule in the core. Reproduced from Moniruzzaman et al. [105], *J. Colloid Interface Sci.* **2010**, *352*, 136, with permission from Elsevier Science.

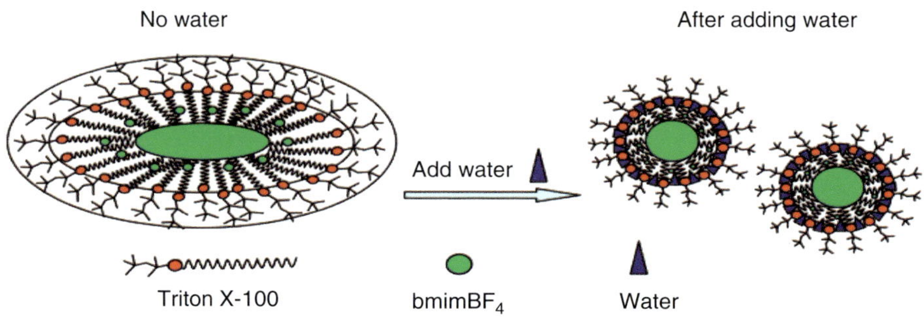

Scheme 10.4 Schematic diagram of the [bmim][BF_4]-in-cyclohexane microemulsion. A large elliptical droplet structure can be assumed in the absence of water, while the addition of water results in small-size spherical microemulsion droplets. Reproduced from Gao et al. [120], *J. Phys. Chem. B* **2008**, *112*, 3711, with permission from the American Chemical Society.

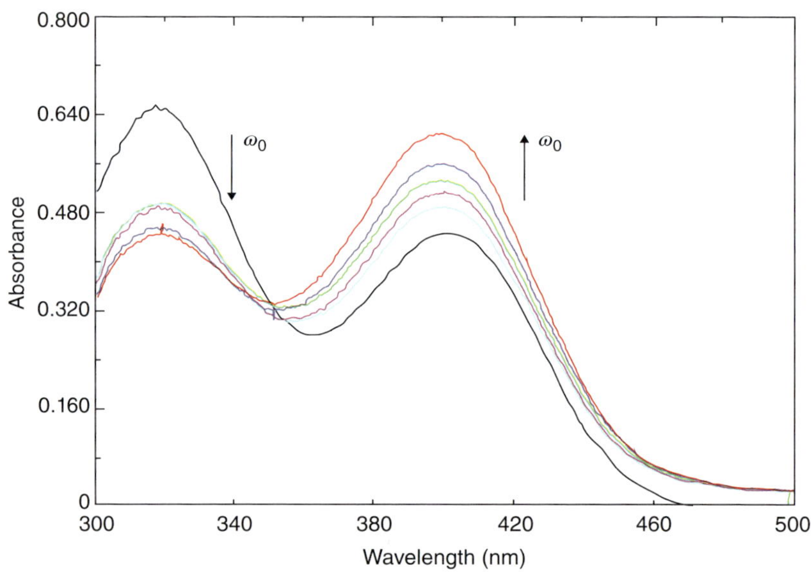

Figure 18.2 Absorption spectra of bioactivator in H_2O/AOT/TX-100/[bmin][PF_6] microemulsion. The water contents (ω_0), indicated by arrows, were 8, 10, 12, 14, and 18, respectively. Reproduced from Xue et al. [37] with permission from Elsevier Science.

There are a wide range of other surfactants that form microemulsions, including nonionic, anionic, and cationic molecules. For instance, nonionic alkyl oligo(ethylene oxide) (C_iE_j) surfactants can be also used in the preparation of microemulsions with the traditional ILs ethylammonium nitrate ($[C_2NH_3][NO_3]$) or propylammonium nitrate ($[C_3NH_3][NO_3]$) and alkanes (octane, decane, dodecane, tetradecane, and hexadecane) [46, 47].

Falcone et al. used the anionic sodium 1,4-bis(2-ethylhexyl) sulfosuccinate (AOT) and the cationic surfactant, benzyl-*n*-hexadecyldimethylammonium chloride (BHDC) to prepare IL microemulsions [48]. The ILs chosen were 1-butyl-3-methylimidazolium trifluoromethanesulfonate ($[C_4mim][CF_3SO_3]$) and 1-butyl-3-methylimidazolium trifluoroacetate ($[C_4mim][CF_3CO_2]$; see Scheme 13.2). DLS experiments reveal the formation of microemulsions. Besides, the FTIR results suggest that the ionic interactions (with the surfactant polar head groups, surfactant counterions, or IL counterions) are substantially modified upon confinement. These interactions produce segregation of IL's ions, altering the composition of the microemulsion interfaces.

Recently, new catanionic surfactants were synthesized and used in the preparation of nonaqueous IL microemulsions. Sarkar et al. synthesized AOT-derived surfactants (AOT, sodium bis(2-ethylhexyl) sulfosuccinate, aerosol-OT), which can be used to produce a large number of IL-in-oil microemulsions. The new surfactants were synthesized from three commercially available traditional surfactants, CTABr, NaAOT, and BHDCl. The combinations of these surfactants with ILs and isopropyl myristate (IPM) not only form an IL/oil microemulsion but also form a unilamellar vesicle [49, 50]. $[P_{13}][Tf_2N]$/CTA-AOT/IPM, $[N_{3111}][Tf_2N]$/CTA-AOT/IPM, $[P_{13}][Tf_2N]$/BHD-AOT/IPM, and $[N_{3111}][Tf_2N]$/BHD-AOT/IPM microemulsions were characterized, and it was found that the polarity of ILs pooled inside the microemulsion is less compared with neat ILs. Later, the fluorescence resonance energy transfer (FRET) study of $[P_{13}][Tf_2N]$/CTA-AOT/IPM and $[N_{3111}][Tf_2N]$/CTA-AOT/IPM ternary systems was carried out [51].

N,*N*-dimethylethanolammonium 1,4-bis(2-ethylhexyl) sulfosuccinate (DAAOT) was also used as surfactant in order to formulate IL-in-oil microemulsions exhibiting stability over a wide range of temperatures. Sarkar et al. characterized $[C_6mim][Tf_2N]$/DAAOT/IPM microemulsions and showed that the amount of IL within the core of aggregates in microemulsions can be easily manipulated to affect the size of the aggregates directly. The use of $[C_6mim][Tf_2N]$ as polar phase allows the tuning of the size of the aggregates in microemulsions from 7.9 to 91.0 nm.

13.3 AQUEOUS IL MICROEMULSIONS

Aqueous microemulsions with the IL as apolar phase are much more interesting than nonaqueous microemulsions. The reason is that both water and the IL are considered green solvents. The IL most commonly used in the preparation of aqueous IL microemulsions is 1-butyl-3-methylimidazolium hexafluorophosphate ($[C_4mim][PF_6]$). The first aqueous IL microemulsion, water/TX-100/$[C_4mim][PF_6]$, was reported for Gao et al. showing water in $[C_4mim][PF_6]$ (w/IL), bicontinuous, and $[C_4mim][PF_6]$ in water (IL/w) subregions (see Fig. 13.1) [52]. These microregions were identified by cyclic voltammetry method using

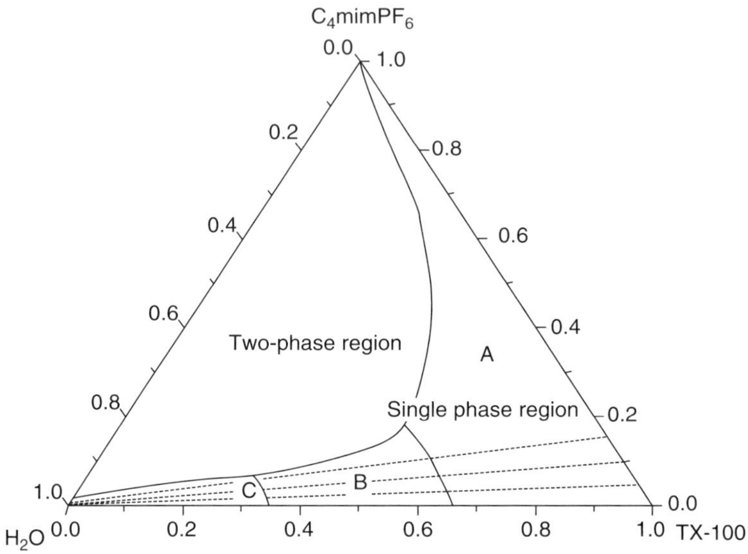

Figure 13.1 Phase diagram of the water/TX-100/[C_4mimPF_6] ternary system at 25°C; the dished lines are chosen for apparent diffusion coefficient study. A, B, and C are w/IL, bicontinuous, and IL/w regions, respectively. Reproduced from Gao et al. [52] with permission from the American Chemical Society. Copyright (2005).

potassium ferrocyanide ($K_4Fe(CN)_6$) as the electroactive probe. DLS study showed that the hydrodynamic diameter of the IL/w microemulsion is nearly independent of water content but increased with increasing IL content due to the swelling of the micelles by the IL. Water domains exist in the w/IL microemulsion, which can dissolve salts such as methyl orange and $K_3Fe(CN)_6$. These salts were used as probe molecules in order to evaluate by UV–Vis the polarity of the water domains of the microemulsions. A similar study was carried out for water/Tween-20/[C_4mimPF_6] microemulsions [53]. Fluorescence techniques can also be employed in the characterization of microemulsions. In this sense, the interaction of IL with water in water/TX-100/[C_4mimPF_6] microemulsions was analyzed by the dynamics of solvent and rotational relaxation of the molecule probes coumarin 153 (C-153) and coumarin 151 (C-151) [54]. Later, with C-153 and coumarin 490 (C-490), similar experiments were carried out [55]. More recently, a broadband dielectric spectroscopy study on the same system was performed [56].

Some remarkable applications for microemulsions constituted by ILs have been recently reported. For instance, aqueous IL microemulsions were used for liquid–liquid extraction. In particular, water/AOT/[C_4mim][PF_6] system has been proved to entail selective extraction of hemoglobin from human whole blood [57] or to develop a synergic microextraction procedure for the preconcentration and determination of glucocorticoid hormones in water samples [58].

In a different line of research, the synthesis of NPs in water/AOT/[C_4mim][PF_6] microemulsions was achieved. Palladium NPs were successfully prepared in situ in this system being the average size of the monodispersed Pd NPs was about 3 nm [59]. The same research group has recently synthesized bimetallic palladium–gold NPs in same ternary system with an average size of 4.5 nm [60]. Metal particles of silver (Ag) have been synthesized by the photoreduction of silver nitrate ($AgNO_3$) in water/Tween-20/[C_8mim][PF_6] and water/TX-100/[C_8mim][PF_6] microemulsions [61]. The formation of microemulsions as well as Ag particles produced by the photoreduction has been investigated by UV–Vis, cryo-TEM, SAXS, and extended X-ray absorption fine structure (EXAFS) measurements.

Water/TX-100/[C_4mim][PF_6] microemulsions were used as reaction media in enzymatic reactions. The catalytic activities of alcohol dehydrogenase in this ternary system were determined, and it was found to be greatly improved as compared with those in pure [Bmim][PF_6] [62]. The same system was used in order to analyze the effect on the catalytic activity of lignin peroxidase and laccase [63]. The catalytic behavior and stability of lipases from *Candida rugosa*, *Chromobacterium viscosum*, and *Thermomyces lanuginosa* in these microemulsions were investigated and compared to other microheterogeneous media used so far for enzyme-catalyzed reactions [64].

An interesting application of aqueous IL microemulsions is to develop IL polymer materials incorporating enzymes that can be used as active, stable, and reusable biocatalysts [65]. The IL 1-vinyl-3-ethylimidazolium bis(trifluoromethylsulfonyl) amide ([veim][Tf_2N]; see Scheme 13.2) was used as the continuous phase. Incorporation of proteins in IL-based polymer frameworks is generally difficult because they are insoluble in most ILs. To overcome this limitation, the authors first employed water/Tween-20/[veim][Tf_2N] microemulsions to solubilize the enzyme in an IL phase and then incorporated the enzyme within these surfactant aggregates into IL polymer frameworks via polymerization in the presence of an IL-soluble cross linker and initiator.

Usually the addition of a fourth component increases the stability of aqueous IL microemulsions. Recently, it was reported the first example of aqueous IL microemulsion generated using a zwitterionic surfactant [66]. The quaternary system was constituted by the surfactant *N*-dodecyl-*N*,*N*-dimethyl-3-ammonio-1-propanesulfonate (dodecyl sulfobetaine, SB-12), water, an IL ([C_4mim][PF_6] or [C_2mim][Tf_2N]), and ethanol. UV–Vis, DLS, NMR, and Fourier transform infrared (FTIR) techniques have demonstrated the formation of w/IL microemulsions in ILs [C_4mim][PF_6] and [C_2mim][Tf_2N] using the zwitterionic surfactant SB-12 in the presence of ethanol. The authors established that the presence of ethanol helps increase uptake of both water and surfactant SB-12 within the ILs initiating the formation of w/IL microemulsions (see Fig. 13.2).

The addition of alcohol, as cosurfactant, to the [C_8mim][Tf_2N]/AOT/water system leads to stable w/IL microemulsions. DLS and protein solubilization experiments confirm the existence of an aqueous nanoenvironment in the IL phase of [C_8mimTf_2N]/AOT/1-hexanol/water microemulsions [67]. The kinetics of the enzymatic reactions were performed in this quaternary system. Specifically, lipase-catalyzed hydrolysis of p-nitrophenyl butyrate (p-NPB) was used as a model reaction [68]. In a similar way, the lipase-catalyzed hydrolysis of p-NPB was investigated to evaluate the catalytic efficiency in water/AOT/Triton X-100/[C_4mim][PF_6] [69]. A large single-phase microemulsion region can be obtained from the combination of two surfactants in IL.

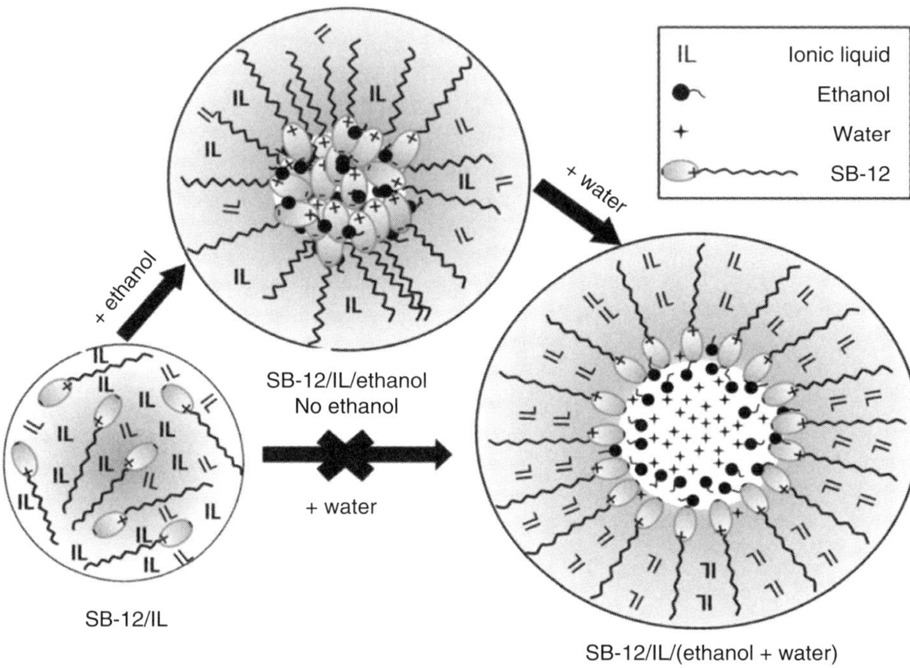

Figure 13.2 Schematic depiction of w/IL microemulsion formation by SB-12 in the presence of ethanol. Reproduced from Rai et al. [66] with permission from John Wiley & Sons. Copyright (2012).

The phase diagram of water/AOT/Triton X-100/[C_4mim][PF_6] was reported previously by Huang et al. [70].

13.4 MICROEMULSIONS WITH IL AS SURFACTANT

Self-assembly of amphiphilic molecules in RTILs has been recently reviewed [71–73]. The ability to form micelles, vesicles, liquid crystal, and microemulsions opens a wide range of applications. The particular advantage of using ILs in the formulation of microemulsions is their versatile use as a polar phase (substituent of water), as an organic solvent (substituent of oil), as a cosurfactant, and/or as an amphiphilic component. In many cases, the microemulsions formulated with ILs show a continuous transition from water-in-oil to oil-in-water types.

More recently, the preparation of ILs with a certain surface activity due to the incorporation of long alkyl chain substituents has expanded the applications range. The so-called IL-like surfactants, ILs based on *N*,*N*-dialkylimidazolium cations, may have considerable advantages in terms of the formation of stable interfaces due to the electrostatic interactions between the anion and the cation. However, reports about the formulation of microemulsions using this kind of surfactants are rather scarce, and only a few have been published so far [74–79]. It has to be mentioned here that two general strategies can be used to come to ILs. On the one hand, long alkyl chains can be incorporated into the imidazolium cation [80], and on the other hand, the counteranion can contain a long alkyl chain.

13.4.1 Long-Chain ILs in Oil-in-Water Microemulsions

Long-chain ILs that are comprised of a hydrophilic imidazolium head group and a hydrophobic tail are good surfactant candidates to form micelles and microemulsions. In comparison with traditional surfactant-based microemulsions, IL systems show several advantages. Firstly, the strong attraction between the imidazolium ring and cosurfactants could facilitate the immobilization of the latter at the fluid/fluid interface, which prevents its leakage into the conjugated water solution. Secondly, the broad imidazolium head shows a higher capacity for solutes than the tert-ammonium cationic system. Thirdly, the imidazolium head is strongly attractive through π–π interaction. Finally, the versatile structures of ILs provide an ample platform to explore the hydrophilic–lipophilic equilibrium in IL reverse micelles, which is a key to optimize the partitioning behaviors in microemulsions. For these reasons, long-chain ILs were used as surfactants in microemulsion systems. ILs based on 1-alkyl-3-methylimidazolium cation [$C_n$mim] with a long alkyl chain possess an inherent amphiphilicity. Thus, as conventional surfactants, these ILs show an interfacial and aggregation behavior. The properties of such ILs in aqueous solution have been examined both experimentally and theoretically by some authors [81–84]. It has been evidenced that they form micelles in dilute aqueous solution; while in concentrated aqueous solution, they produce lamellar lyotropic liquid crystals [85, 86].

Yan and Texter have synthesized amphiphilic ILs (IL-S) that consists of an imidazolium cation polar group and a hydrophobic tail including polymerizable surfactant ILs (Scheme 13.3) [87]. Microemulsions were formulated with IL-S/methyl methacrylate (MMA)/water in highly diluted systems where oil-in-water type aggregates are formed.

Bulk microemulsion polymerization was initiated by 2,2-azobisisobutyronitrile yielding particle aggregation with [C_{12}mim][Br] and polymer latexes without any apparent aggregation in the case of the polymerizable surfactant-based microemulsion. Copolymerization between MMA and the polymerizable surfactant in microemulsions was confirmed by FTIR spectra. Since the polymerizable surfactant is more polar than MMA, it likely forms a polymerized polar shell around the resulting particles, which increases particle stability and allows the particles to be easily redispersed in water and in NaBr solution without any evident aggregation, indicating that the latex surface is hydrophilic. These latexes start to aggregate in $NaBF_4$ solutions

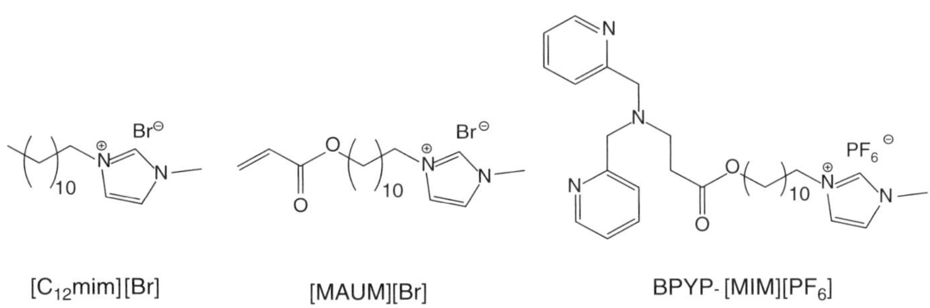

Scheme 13.3 Chemical structures of dodecyl methylimidazolium bromide ([C_{12}mim][Br]), 1-(2-methyl acryloyloxyundecyl)-3-methylimidazolium bromide ([MAUM][Br]), and 3-{11-[(3-(bis(pyridin-2-ylmethyl)amino)propanoyl)oxy]undecyl}-1-methylimidazolium hexafluorophosphorate (BPYP-[MIM][PF_6]).

and dramatically aggregate in KPF_6 solution. This aggregation is due to the surface being transformed from hydrophilic to hydrophobic, by the anion exchange of the Br^- to BF_4^- to PF_6^-. By using this approach, a new class of very stable and stimuli-responsive nanolatexes (based on copolymers of reactive and stimuli-responsive IL surfactants) was synthesized by microemulsion polymerization [88, 89].

Polymerization in these microemulsions can produce polymer particles, gels, and open-cell microporous materials. Surface properties of the resulting copolymers can be modified via anion exchange. Anion exchange from Br^- to PF_6^- changes the copolymer from hydrophilic to hydrophobic at the interfaces between aqueous and polymer domains and thus causes shrinkage at the interface to make pores. This shrinkage results from condensation of IL-S chains on themselves and on the polymeric backbone as the polar imidazolium head groups are transformed from hydrophilic to hydrophobic. The shrunken polymer domains at the interface can be converted back to hydrophilic and expanded in water again after anion exchange from PF_6^- to Br^- and became swollen by imbibing water and partially healing or closing the pores. Polymerization of microemulsions using a surfactant based on a polymerizable IL has been used to obtain cross-linked polymer gels [90]. The resulting polymer gels can be transformed into porous polymers by anion exchange of the IL moiety. These porous polymers can be converted back into polymer gels by immersion in organic solvents, such as DMSO and DMF, and these polymer gels can be reconverted into porous polymers by immersion in water.

Polymerization of microemulsions that contains IL polar cores dispersed in a polymerizable oil has been reported for the system styrene/[C_4mim][BF_4]/[MAUM][Br] where [MAUM][Br] is 1-(2-methyl acryloyloxyundecyl)-3-methylimidazolium bromide and was used to stabilize [C_4mim][BF_4]/styrene microemulsions [91]. In the same way, Texter and coworkers [92] use a ternary microemulsion formed by MMA/[C_4mim][BF_4]/[C_{12}mim][Br] instead of an aqueous microemulsion for atom transfer radical polymerization showing that after the polymerization and isolation of the resultant polymers, the mixture of the catalyst and ILs (continuous phase and surfactant) can be recovered and reused improving the environmental sustainability of the processes. The rate of the polymerization process can be improved by using a new IL, BPYP-[MIM]PF_6, that combines properties of surfactants and catalyst ligands [93] (see Scheme 13.3).

Formation of hexagonal and lamellar phases from a ternary system of [C_{16}mim][Cl], 1-decanol, and water was recently demonstrated [94]. On decreasing the alkyl chain length of the imidazolium cation, [C_8mim][Cl], a lyotropic lamellar phase is formed in water when a suitable amount of alcohol (1-hexanol, 1-octanol, 1-decanol, and 1-dodecanol) is added [95]. The driving force to produce such an ordered assembly is considered as a synergetic result of the hydrophobic force among the hydrocarbon chains and the hydrogen-bonded network comprising an imidazolium ring, Cl^-, water, and alcohol (Fig. 13.3). In these phases, alcohol molecules play important role not only because they could partly penetrate into the palisade layer with their hydroxyl groups extruded to form the network at the aggregate interface but also because they could partly locate themselves in the interior of the hydrophobic bilayers to twist with the alkyl chains of [C_8mim][Cl]. The counterions have much influence on the interactions inside the system and therefore the phase behaviors. With comparison to analogous ternary systems of [C_8mim][BF_4] and [C_8mim][PF_6], it is found that the strong interaction between the counterions and the alcohols favors the appearance of ordered assemblies.

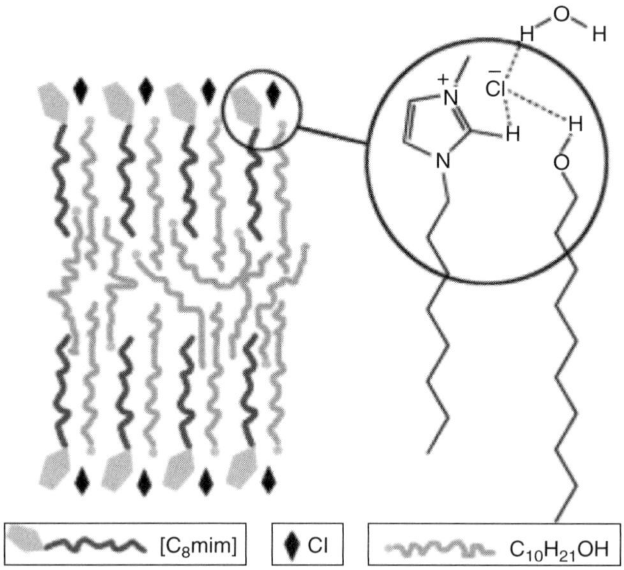

Figure 13.3 Sectional schematic graph of the possible structure formed in the LR phase of the $[C_8mim]Cl-C_{10}H_{21}OH-H_2O$ ternary system. Enlarged inset shows the new H-bonded network comprising the imidazolium cation, Cl^-, H_2O, and $C_{10}H_{21}OH$. Reproduced from Zhang et al. [95] with permission from the American Chemical Society. Copyright (2007).

The use of $[C_n mim]$-based ILs as a surfactant to prepare a ternary microemulsion with water and oil was reported by Zheng and coworkers [96]. These authors have showed that the ternary mixtures of $[C_{16}mim][Br]$/p-xylene/water ($[C_{16}mim]$ can be replaced for $[C_{14}mim]$) form hexagonal and lamellar liquid crystalline phases with small to moderate water content, while at high water contents, the phase diagram shows a microemulsion phase in the water corner. In the absence of oil, p-xylene, $[C_{14}mim][Br]$ forms micelles in water. The addition of xylene to the micelle solution results in significant broadness of the proton signals of the C_{14} alkyl chain, suggesting that the motions of these protons are restricted keeping the signal corresponding to the methyl group or $[C_{14}mim]$ mainly unaffected. These results agree with the p-xylene solubilization into the palisade shells of the micelles forming microemulsions. FF-TEM as well as DLS shows that both micelles and microemulsions (oil-in-water type) are roughly spherical with diameter of 20–30 nm for micelles to 70–80 nm for microemulsions. These O/W microemulsions can incorporate anthracenes with a polar 9-substituent and arrange their molecules at the interface of the microemulsion droplets in such a way that the anthryl moiety is located in the hydrophobic region with the polar substituent toward the water phase. Photoirradiation [97] of such anthracene derivatives in the microemulsions results in great yields of the h–h cyclomers, which is contrasted with the case of the photocyclization in homogeneous solutions.

An IL-in-oil microemulsion [98] can be formulated by using an 1-octyl-3-methylimidazolium chloride, $[C_8mim][Cl]$, as IL-S and $[C_4mim][PF_6]$ as a substitute for traditional organic solvent. DLS was used to confirm the formation of $[C_4mim][PF_6]$-in-water microemulsions with an average size of 3 nm.

13.4.2 Long-Chain ILs in Water-in-Oil Microemulsions

Long-chain ILs ($n > 10$) comprising a hydrophilic imidazolium head group and a hydrophobic tail could serve as good candidates for the phase-formation materials in micelles and water-in-oil type microemulsions [99]. In the same way as with alkyltrimethylammonium surfactants, addition of a cosurfactant (usually a medium-chain alcohol) is necessary in order to stabilize the microemulsion. A quaternary microemulsion water/hexanol/IL-S/oil (cyclohexane, isooctane, or decane) has been tested to solubilize penicillin. [C_{14}mim][Cl], [C_{12}mim][Cl], and [C_{14}dmim][Cl] have been used in comparison with CTAB allowing the authors to conclude that substitution of C_2 hydrogen with a methyl group would decrease the polarity of the Stern layer in the water pool [100]. A microemulsion formed by [C_{14}mim][Br]/cyclohexane/hexanol/water was investigated for Au(III) extraction [101]. In the extraction system, the IL-S bears double functions of surfactant and extractant.

Microemulsions incorporating two ILs (surfactant and polar phase) can be formulated by using long-chain imidazolium IL-S. Kunz and coworkers [102] use 1-hexadecyl-3-methylimidazolium chloride ([C_{16}mim][Cl]) as surfactant, decanol as cosurfactant, dodecane as continuous phase, and RTILs (ethylammonium nitrate (EAN) and 1-butyl-3-methylimidazolium tetrafluoroborate ([C_4mim][BF4]), respectively) as polar microenvironment in the formulation of high thermal stable microemulsions. The hydrodynamic radius increases with the amount of EAN, which indicates a swelling of the formed structures. Both systems exhibit a single broad peak in SAXS, which is typical for microemulsions. The SAXS data confirm the presence of spherical shape IL-in-oil microemulsions in the case of EAN systems, whereas for [C_4mim][BF_4] systems, the data fit a model of repulsive spheres. It was confirmed that changing the nature of the polar IL, EAN and [C_4mim][BF_4], modifies the interfacial rigidity of the microemulsion. The difference between the two systems may be linked to the change in the cohesive forces of the polar ILs or to the cosurfactant property of [C_4mim][BF_4], leading to a reinforcement of the interfacial film.

13.4.3 IL Surfactants with Long-Chain Anions

ILs can be incorporated in the microemulsion formulation as substituents of the polar and nonpolar phases or as surfactants. In recent studies, Koetz and coworkers [103] showed that the role of cosurfactant can also be assumed by an IL in the stabilization of water-in-oil microemulsions. SDS- and CTAB-based water-in-toluene/pentanol microemulsions have been formulated with the aid of ethyl-methylimidazolium hexylsulfate, [C_2mim][C_6SO_4]. Their experimental results showed that replacing water by the IL increases the isotropic phase region of the system. The authors assume the formation of a palisade layer (Scheme 13.4), where the IL plays a similar role like a cosurfactant, changes the spontaneous curvature of the interfacial film, and decreases the droplet size.

The use of NMR self-diffusion coefficients clearly demonstrated that a water-in-oil type microemulsion exists. The transition from water-in-oil to bicontinuous microemulsion and finally to an oil-in-water system by increasing the water–IL content can be detected by considering the change in the diffusion coefficients.

Recent studies [104, 105] have formulated IL-in-oil microemulsions by using IL-S obtained from combination of single- and two-tail anions with imidazolium

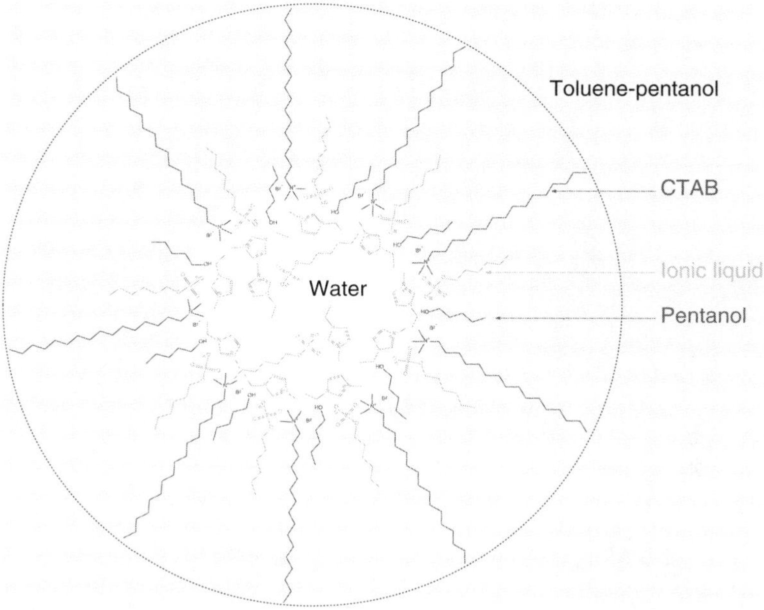

Scheme 13.4 Model of a microemulsion droplet in the water/CTAB-ethyl-methylimidazolium hexylsulfate/toluene–pentanol system.

cation: 1-butyl-3-methylimidazolium octylsulfate ([C_4mim][C_8SO_4]) or 1-butyl-3-methylimidazolium dodecylsulfate ([C_4mim][$C_{12}SO_4$]). Moreover, they formulate an isotropic water-free ternary system consisting of two IL components and one oil component, that is, [C_2mim][C_2SO_4]/toluene/[C_4mim][$C_{12}SO_4$] (Scheme 13.5).

The use of [C_2mim][C_2SO_4] as polar solvent can have technical advantages due to the relatively broad window of fluidity and low toxicity and the low melting temperature and high stability to hydrolysis in comparison with halogen ILs. [C_4mim][$C_{12}SO_4$] fulfills IL-like properties as well as interface-like properties in the ternary system. The results show that the isotropic area can be significantly increased by replacing the anion octylsulfate with dodecylsulfate. Three different regions have been detected in the large isotropic phase: oil-in-IL, bicontinuous, and IL-in-oil microemulsions. DLS and SAXS data show that already in the absence of [C_2mim][C_2SO_4], the surfactant-like IL forms micellar aggregates in the oil component. These micelles containing dodecylsulfate as main surfactant component can solubilize [C_2mim][C_2SO_4], and microemulsion droplets with a core–shell structure are formed. These water- and halogen-free reverse microemulsion droplets, containing two ILs and oil, seem to be of interest as alternative nanoreactors for the preparation of NPs or to perform chemical reactions.

A similar approach for ILs based on an amphiphilic anion and imidazolium cation, [C_4mim][OT], OT being the 1,4-bis(2-ethylhexylsulfosuccinate) anion, has been proposed. The use of sodium 1,4-bis(2-ethylhexyl) sulfosuccinate, AOT, in the formulation of water-in-oil microemulsions is well reported in the literature [106–108]. The incorporation of IL to the formulation of microemulsions has been studied from different points of view: (i) AOT-based microemulsions incorporating

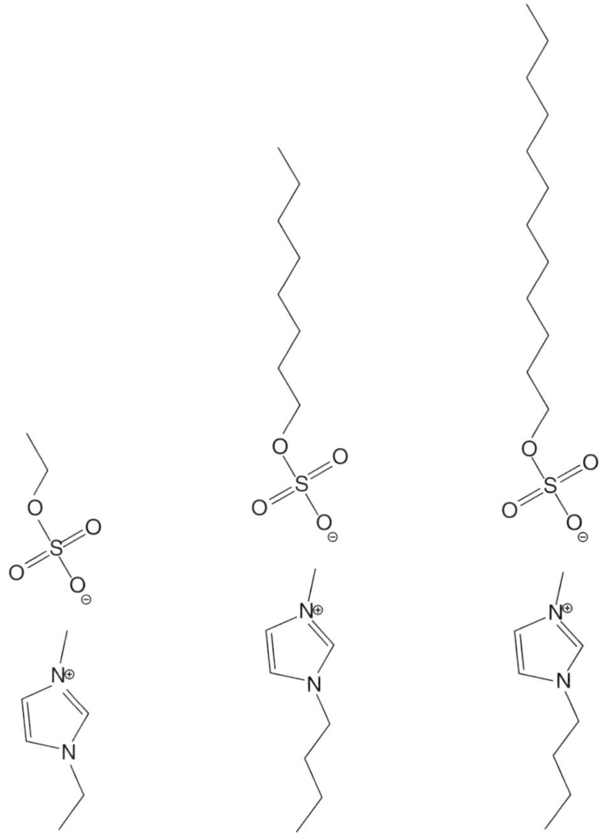

$[C_2mim][C_2SO_4]$ $[C_4mim][C_8SO_4]$ $[C_4mim][C_{12}SO_4]$

Scheme 13.5 Chemical structures of 1-ethyl-3-methylimidazolium ethylsulfate ($[C_2mim][C_2SO_4]$), 1-butyl-3-methylimidazolium octylsulfate ($[C_4mim][C_8SO_4]$), or 1-butyl-3-methylimidazolium dodecylsulfate ($[C_4mim][C_{12}SO_4]$).

ILs and (ii) replacement of the Na^+ counterion by an imidazolium yielding an IL-S. Under conditions for the first approach, AOT can be used to stabilize an IL-in-IL (nonpolar phase dispersed in a polar continuous medium similar to oil-in-water aggregates) microemulsion [109] formulated with two ILs: $[C_4mim][PF_6]$ can be dispersed in hydrophilic IL propylammonium formate (PAF) with the aid of AOT. The size of the aggregates is larger than for conventional microemulsions reaching the values of 30–100 nm in diameter on increasing the molar ratio of $[C_4mim][PF_6]$ to AOT.

The critical temperature of microemulsions, traditionally modified by adding salts, has been decreased by incorporation of $[C_4mim][BF_4]$ in water/AOT/decane microemulsions [110, 111]. This study showed that the amount of $[C_4mim][BF_4]$ needed for curvature adjustment in ionic microemulsions was less than the amount of NaCl normally used. The water solubilization capacity in water/AOT/isooctane microemulsions in the presence of IL has been investigated for $[C_2mim][Cl]$, $[C_4mim][Cl]$, and $[C_8mim][Cl]$ showing an enhanced solubility at low IL concentrations but a decrease at high ones [112, 113]. Moreover, the water solubilization increases on increasing the alkyl chain length of the IL cation. Addition of these ILs also increases the percolation temperature, the effect being greater the larger the alkyl chain of the imidazolium cation. However, in these studies, the surfactant cannot be considered an IL.

The strategy reported by Sarkar and coworkers [114, 115] replaces the Na^+ inorganic cation of AOT with an organic cation, and using different ILs, a polar core allows them to formulate a huge number of microemulsions. Their systematic studies analyze the effect of the alkyl chain of the IL as polar phase of the microemulsion by using [C_2mim][Tf_2N], [C_4mim][TF_2N], and [C_6mim][Tf_2N] as well as the influence of the anions [C_4mim][Tf_2N], [C_4mim][PF_6], and [C_4mim][BF_4] on the properties of the IL/[C_4mim][OT]/benzene microemulsions (Scheme 13.6). In the initial work, the phase diagram of the [C_{4mim}][BF_4]/[C_{4mim}][AOT]/benzene ternary system was obtained at 298 K [116]. The nanodroplet polarity of [C_4mim][BF_4]/[C_4mim][OT]/benzene microemulsions has been studied by using solvatochromic probes. On increasing the [C_4mim][BF_4] to [C_4mim][OT] molar ratio, a shift in the maximum of the absorption spectrum is observed red shift for methyl orange and blue shift for $E_T(30)$, indicating the dye incorporation to the nanodroplet. In both cases, the dye-measured polarity

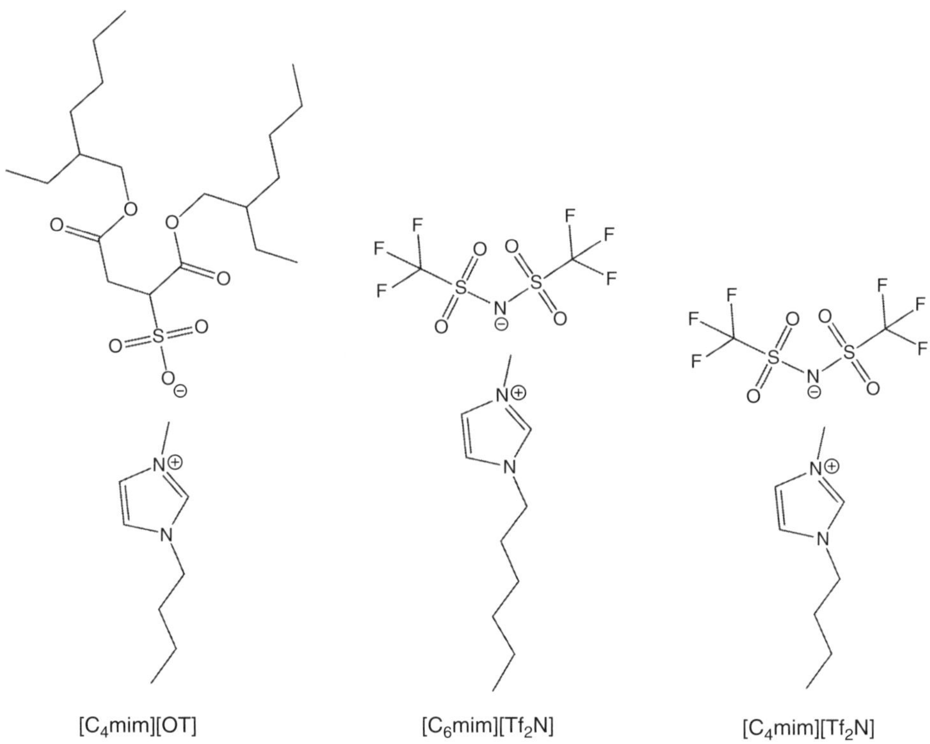

Scheme 13.6 Chemical structures of 1-butyl-3-methylimidazolium 1,4-bis(2-ethylhexylsulfosuccinate) ([C_4mim][OT]), 1-hexyl-3-methylimidazolium bis(trifluoromethylsulfonyl) imide bis(trifluoromethylsulfonyl)imide ([C_6mim][Tf_2N]), and 1-butyl-3-methylimidazolium bis(trifluoromethylsulfonyl)imide ([C_4mim][TF_2N]).

of the nanodroplet was lower than that for pure [C_4mim][BF_4] indicating that dye distribution between the interface and the nanodroplet should be present also for [C_4mim][BF_4] to [C_4mim][OT] molar ratios as high as 1:2.

The area of the single-phase region in the ternary system of IL/[C_4mim][OT]/benzene microemulsions, that is, the IL loading capacity, increases on increasing the alkyl chain length of the IL cation and on decreasing the cation–anion interaction strength of added ILs. The strong interaction leading to pronounced penetration and aligning of the cation of [C_6mim][Tf_2N] with the tail part of [OT^-] anion makes the interfacial layer more rigid; therefore, the IL solubilization capacity as well as the area of single-phase region increases. For ILs with different anions, the area of the single-phase region follows the trend [C_4mim][Tf_2N] > [C_4mim][PF_6] > [C_4mim][BF_4]. Such a difference in behavior arises due to relatively weak interaction between [Tf_2N^-] anion with [C_4mim^+] cation, which causes strong interaction of [C_4mim^+] with [OT^-] anion, making the interfacial layer more rigid. In this way, an accurate selection of the IL used determines the amount of IL within the core of the microemulsion and directly affects the size of the aggregates. Similarly to non-IL AOT-based microemulsions, the size of the aggregate increases with increasing the molar ratio of dispersed phase to surfactant. A different aspect to be remarked is that the size depends on the IL structure following the sequence [C_2mim][Tf_2N] > [C_4mim][Tf_2N] > [C_6mim][Tf_2N], being a consequence of the specific interactions between the IL and the head group of the surfactant and consequently modifying the rigidity of the interfacial layer.

Water addition has important effects on the microstructure of the aggregates. Addition of small water amounts (300 mM) has important effects on the size, being maximum for [C_2mim][Tf_2N] > [C_4mim][Tf_2N] > [C_6mim][Tf_2N] as a consequence of the decrease in the curvature parameter. Water molecules being insoluble in [$C_n$mim][Tf_2N] ILs are inclined to occupy the interfacial layers of the microemulsion and replace some of the [C_nmim^+] cations decreasing the curvature of the surfactant. The maximum perturbation of the surfactant head group was observed with the water addition to [C_2mim][Tf_2N]/[C_4mim][OT]/benzene systems.

The most promising achievement about IL-in-oil microemulsions is their high thermal stability compared to that of aqueous microemulsions. DLS experiments evidence that the aggregates in microemulsions remain monodisperse in nature with increasing temperature, and in all the cases, the size of aggregates decreases with increasing the temperature. The use of benzene as a nonpolar solvent cannot serve the purpose of creating high-temperature stable microemulsions, so that benzene has been replaced with isopropyl myristate (a biologically accepted oil phase with a boiling point of 440 K). Combination of an *N*,*N*-dimethylethanolammonium 1,4-bis(2-ethylhexyl) sulfosuccinate ([DA][OT]) with isopropyl myristate (IPM) and [C_6mim][Tf_2N] IL (Scheme 13.7) allows the formation of thermally stable microemulsions [117] (278 to ≥423 K).

13.5 CONCLUSIONS

ILs can be used in the formulation of microemulsions as disperse medium by replacing water, as continuous medium by replacing alkane solvents, and as a surfactant in those situations where the IL self-aggregates. [C_4mim][BF_4] and [C_4mim][PF_6] are

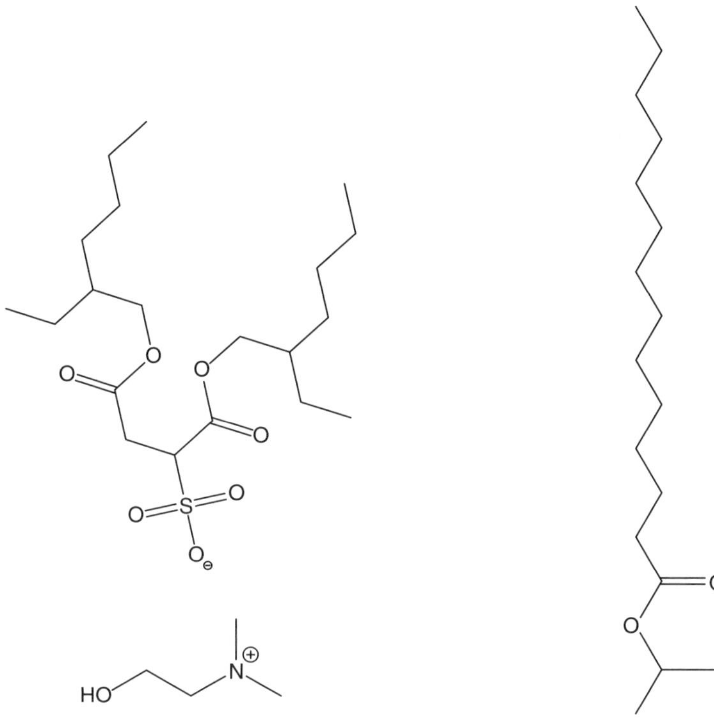

Scheme 13.7 Chemical structures of *N*,*N*-dimethylethanolammonium 1,4-bis(2-ethylhexyl) sulfosuccinate ([DA][OT]) and isopropyl myristate (IPM).

the ILs widely used in the preparation of aqueous and nonaqueous microemulsions, respectively. Recently, new ILs have been designed in order to improve the properties of these aggregates. In addition, ILs with a certain surface activity due to the incorporation of long alkyl chain substituents and that can self-aggregate have expanded the use of ILs in microemulsions. There are two kinds of amphiphilic ILs: (i) ILs with long alkyl chains incorporated into the imidazolium cation and (ii) ILs with the counterions containing a long alkyl chain. Significant applications of microemulsions containing ILs are reported in this chapter as a result of the development of new ILs tuned with the desirable properties.

REFERENCES

[1] Wasserscheid, P. *Nature* 2006, **439**, 797.

[2] Reinsborough, V. C. *Aust. J. Chem.* 1970, **23**, 1473–1475 and references therein.

[3] Evans, D. F.; Yamauchi, A.; Wei, G. J.; Bloomfield, V. A. *J. Phys. Chem.* 1983, **87**, 3537–3541.

[4] Baker, G. A.; Pandey, S. *ACS Symp. Ser.* 2005, **901**, 234–243.

[5] Hao, J.; Zemb, T. *Curr. Opin. Colloid Interface Sci.* 2007, **12**, 129–137.

[6] Greaves, T. L.; Drummond, C. J. *Chem. Soc. Rev.* 2008, **37**, 1709–1726.

[7] Greaves, T. L.; Weerawardena, A.; Fong, C.; Drummond, C. J. *Langmuir* 2007, **23**, 402–404.

[8] Greaves, T. L.; Weerawardena, A.; Fong C.; Drummond, C. J. *J. Phys. Chem. B* 2007, **111**, 4082–4088.

[9] Greaves, T. L.; Weerawardena, A.; Krodkiewska, I.; Drummond, C. J. *J. Phys. Chem. B* 2008, **112**, 896–905.

[10] Luisi, P. L.; Straub, B. E. *Reverse Micelles*, Plenum Press, New York: 1984.

[11] Rico, I.; Lattes, A. *New J. Chem.* 1984, **8**, 429–431.

[12] Falcone, R. D.; Correa, N. M.; Biasutti, M. A.; Silber, J. J. *Langmuir* 2000, **16**, 3070–3076.

[13] Correa, N. M.; Durantini, E. N.; Silber, J. J. *J. Phys. Org. Chem.* 2006, **19**, 805–812.

[14] Falcone, R. D.; Silber, J. J.; Correa, N. M. *Phys. Chem. Chem. Phys.* 2009, **11**, 11096–11100.

[15] Gao, F.; Co, C. C. Non-Aqueous Microemulsions, in C. Stubenrauch, ed., *Microemulsions: Background, New Concepts, Applications, Perspectives*, John Wiley & Sons, Ltd., Chichester: 2009.

[16] Gao, H.; Li, J.; Han, B.; Chen, W.; Zhang, J.; Zhang, R.; Yan, D. *Phys. Chem. Chem. Phys.* 2004, **6**, 2914–2916.

[17] Eastoe, J.; Gold, S.; Rogers, S. E.; Paul, A.; Welton, T.; Heenan, R. K.; Grillo, I. *J. Am. Chem. Soc.* 2005, **127**, 7302–7303.

[18] Chakrabarty, D.; Seth, D.; Chakraborty, A.; Sarkar, N. *J. Phys. Chem. B* 2005, **109**, 5753–5758.

[19] Gao, Y.; Li, N.; Hilfert, L.; Zhang, S.; Zheng, L.; Yu, L. *Langmuir* 2009, **25**, 1360–1365.

[20] Gao, Y. A.; Vogit, A.; Hilfert, L., Sundmacher, K. *ChemPhysChem* 2008, **9**, 1603–1609.

[21] Gao, Y.; Wang, S.; Zheng, L.; Han, S.; Zhang, X.; Lu, D.; Yu, L.; Ji, Y.; Zhang, G. *J. Colloid Interface Sci.* 2006, **301**, 612–616.

[22] Gao, Y.; Zhang, J.; Xu, H.; Zhao, X.; Zheng, L.; Li, X.; Yu, L. *ChemPhysChem* 2006, **7**, 1554–1561.

[23] Gao, Y.; Li, N.; Zheng, L.; Bai, X.; Yu, L.; Zhao, X.; Zhang, J.; Zhao, M.; Li, Z. *J. Phys. Chem. B* 2007, **111**, 2506–2513.

[24] Gao, Y.; Li, N.; Zheng, L.; Zhao, X.; Zhang, J.; Cao, Q.; Zhao, M.; Li, Z.; Zhang, G. *Chem. Eur. J.* 2007, **13**, 2661–2670.

[25] Li, N.; Gao, Y.; Zheng, L.; Zhang, J.; Yu, L.; Li X. *Langmuir* 2007, **23**, 1091–1097.

[26] Gao, Y.; Li, N.; Zhang, S.; Zheng, L.; Li, X.; Dong, B.; Yu, L. *J. Phys. Chem. B* 2009, **113**, 1389–1395.

[27] Cheng, S.; Han, F.; Wang, Y.; Yan, J. *Colloids Surf. A Physicochem. Eng. Asp.* 2008, **317**, 457–461.

[28] Gao, Y.; Hilfert, L.; Voigt, A.; Sundmacher, K. *J. Phys. Chem. B* 2008, **112**, 3711–3719.

[29] Pramanik, R.; Sarkar, S.; Ghatak, C.; Setua, P.; Rao, V. G.; Sarkar, N. *Chem. Phys. Lett.* 2010, **490**, 154–158.

[30] Gao, Y.; Voigt, A.; Hilfert, L.; Sundmacher, K. *Colloids Surf. A Physicochem. Eng. Asp.* 2008, **329**, 146–152.

[31] Pramanik, R.; Sarkar, S.; Ghatak, C.; Setua, P.; Sarkar, N. *Phys. Chem. Chem. Phys.* 2010, **12**, 3878–3886.

[32] Andújar-Matalobos, M.; García-Río, L.; López-García, S.; Rodríguez-Dafonte, P. *J. Colloid Interface Sci.* 2011, **363**, 261–267.

[33] Setua, P.; Pramanik, R.; Sarkar, S.; Ghatak, C.; Rao, V. G.; Sarkar, N.; Das, S. K. *J. Mol. Liq.* 2011, **162**, 33–37.

[34] Banerjee, C.; Mandal, S.; Ghosh, S.; Kuchlyan, J.; Sarkar, N. *Chem. Phys. Lett.* 2013, **580**, 88–93.

[35] Li, J.; Zhang, J.; Gao, H.; Han, B.; Gao, L. *Colloid Polym. Sci.* 2005, **283**, 1371–1375.

[36] Cheng, S. Q.; Fu, X. G.; Liu, J. H.; Zhang, J. L.; Zhang, Z. F.; Wei, Y. L.; Han, B. X. *Colloids Surf. A Physicochem. Eng. Asp.* 2007, **302**, 211–215.

[37] Wang, G.-X.; Lu, M.; Wu, H. *Polymer* 2012, **53**, 1093–1097.

[38] Wang, G.-X.; Lu, M.; Liu, L.-C.; Wu, H.; Zhong, M. *J. Appl. Polym. Sci.* 2013, **128**, 3077–3083.

[39] Wei, J.; Su, B.; Liang, R.; Xing, H.; Bao, Z.; Yang, Q.; Yang, Y.; Ren, Q. *Colloids Surf. A Physicochem. Eng. Asp.* 2012, **414**, 82–87.

[40] Harrar, A.; Zech, O.; Hartl, R.; Bauduin, P.; Zemb, T.; Kunz, W. *Langmuir* 2011, **27**, 1635–1642.

[41] Ghosh, S.; Banerjee, C.; Mandal, S.; Rao, V. G.; Sarkar, N. *J. Phys. Chem. B* 2013, **117**, 5886–5897.

[42] Mandal, S.; Ghosh, S.; Banerjee, C.; Kuchlyan, J.; Banik, D.; Sarkar, N. *J. Phys. Chem. B* 2013, **117**, 3221–3231.

[43] Pramanik, R.; Sarkar, S.; Ghatak, C.; Rao, V. G.; Setua, P.; Sarkar, N. *J. Phys. Chem. B* 2010, **114**, 7579–7586.

[44] Pramanik, R.; Sarkar, S.; Ghatak, C.; Rao, V. G.; Sarkar, N. *J. Phys. Chem. B* 2011, **115**, 2322–2330.

[45] Sarkar, S.; Pramanik, R.; Ghatak, C.; Rao, V. G.; Sarkar, N. *J. Chem. Phys.* 2011, **134**, 074507/1–074507/11.

[46] Atkin, R.; Warr, G. G. *J. Phys. Chem. B* 2007, **111**, 9309–9316.

[47] Atkin, R.; Bobillier, S. M. C.; Warr, G. G. *J. Phys. Chem. B* 2010, **114**, 1350–1360.

[48] Blach, D.; Silber, J. J.; Correa, N. M.; Falcone, R. D. *Phys. Chem. Chem. Phys.* 2013, **15**, 16746–16757.

[49] Banerjee, C.; Mandal, S.; Ghosh, S.; Kuchlyan, J.; Kundu, N.; Sarkar, N. *J. Phys. Chem. B* 2013, **117**, 3927–3934.

[50] Villa, C. C. M.; Moyano, F.; Ceolin, M.; Silber, J. J.; Falcone, R. D.; Correa, N. M. *Chem. Eur. J.* 2012, **18**, 15598–15601.

[51] Banerjee, C.; Kundu, N.; Ghosh, S.; Mandal, S.; Kuchlyan, J.; Sarkar, N. *J. Phys. Chem. B* 2013, **117**, 9508–9517.

[52] Gao, Y.; Han, S.; Han, B.; Li, G.; Shen, D.; Li, Z.; Du, J.; Hou, W.; Zhan, G. *Langmuir* 2005, **21**, 5681–5684.

[53] Gao, Y.; Li, N.; Zheng, L.; Zhao, X.; Zhang, S.; Han, B.; Houa, W.; Li, G. *Green Chem.* 2006, **8**, 43–49.

[54] Seth, D.; Chakraborty, A.; Setua, P.; Sarkar, N. *Langmuir* 2006, **22**, 7768–7775.

[55] Seth, D.; Chakraborty, A.; Setua, P.; Sarkar, N. *J. Chem. Phys.* 2007, **126**, 224512/1–224512/12.

[56] Chen, Z.; Nozaki, R. *J. Chem. Phys.* 2012, **136**, 244505/1–244505/10.

[57] Shu, Y.; Cheng, D.; Chen, X.; Wang, J. *Sep. Purif. Technol.* 2008, **64**, 154–159.

[58] Qin, H.; Li, B.; Liu, M. S.; Yang, Y. L. *J. Sep. Sci.* 2013, **36**, 1463–1469.

[59] Zhang, G.; Zhou, H.; Hu, J.; Liu, M.; Kuang, Y. *Green Chem.* 2009, **11**, 1428–1432.

[60] Zhang, G.; Zhou, H.; An, C.; Liu, D.; Huang, Z.; Kuang, Y. *Colloid Polymer Sci.* 2012, **290**, 1435–1441.

[61] Harada, M.; Yamada, M.; Kimura, Y.; Saijo, K. *J. Colloid Interface Sci.* 2013, **406**, 94–104.

[62] Zhang, Y.; Huang, X. R.; Li, Y. Z. *J. Chem. Technol. Biotechnol.* 2008, **83**, 1230–1235.

[63] Zhou, G. P.; Zhang, Y.; Huang, X. R.; Shi, C. H.; Liu, W. F.; Li, Y. Z.; Qu, Y. B.; Gao, P. J. *Colloid Surf. B Biointerfaces* 2008, **66**, 146–149.

[64] Pavlidis, I. V.; Gouris, D.; Papadopoulos, G. K.; Stamatis, H. *J. Mol. Catal. B* 2009, **60**, 50–56.

[65] Moniruzzaman, M.; Ino, K.; Kamiya, N.; Goto, M. *Org. Biomol. Chem.* 2012, **10**, 7707–7713.

[66] Rai, R.; Pandey, S.; Baker, S.; Vora, S.; Behera, K.; Baker, G. A.; Pandey, S. *Chem. Eur. J.* 2012, **18**, 12213–12217.

[67] Moniruzzaman, M.; Kamiya, N.; Nakashima, K.; Goto, M. *ChemPhysChem* 2008, **9**, 689–692.

[68] Moniruzzaman, M.; Kamiya, N.; Nakashima, K.; Goto, M. *Green Chem.* 2008, **10**, 497–500.

[69] Xue, L.; Li, Y.; Zou, F.; Lu, L.; Zhao, Y.; Huang, X.; Qu, Y. *Colloid Surf. B Biointerfaces* 2012, **92**, 360–366.

[70] Xue, L.; Qiu, H.; Li, Y.; Lu, L.; Huang, X.; Qu, Y. *Colloid Surf. B Biointerfaces* 2011, **82**, 432–437.

[71] Qiu, Z.; Texter, J. *Curr. Opin. Colloid Interface Sci.* 2008, **13**, 252–262.

[72] Rojas, O.; Koetz, J. *J. Surf. Sci. Technol.* 2010, **26**, 173–195.

[73] Kunz, W.; Zemb, T.; Harrar, A. *Curr. Opin. Colloid Interface Sci.* 2012, **17**, 205–211.

[74] Wasserscheid, P.; Hal, R.; van Bösmann, A. *Green Chem.* 2002, **4**, 400–404.

[75] Sarkar, S.; Mandal, S.; Pramanik, R.; Ghatak, C.; Rao, V. G.; Sarkar, N. *J. Phys. Chem. B* 2011, **115**, 6100–6110.

[76] Miskolczy, Z.; Sebok-Nagy, K.; Biczók, L.; Göktürk, S. *Chem. Phys. Lett.* 2004, **400**, 296–300.

[77] Obliosca, J.; Arco, S.; Huang, M. *J. Fluoresc.* 2007, **17**, 613–618.

[78] James, H.; Davis, J.; Fox, P. A. *Chem. Commun.* 2003, 1209–1212.

[79] Himmler, S.; Hörmann, S.; Hal, R.; van Schulz, P. S.; Wasserscheid, P. *Green Chem.* 2006, **8**, 887–894.

[80] Thomaier, S.; Kunz, W. *J. Mol. Liq.* 2007, **130**, 104–107.

[81] Mele, A.; Tran, C. D.; de Paoli Lacerda, S. H. *Angew. Chem. Int. Ed.* 2003, **42**, 4364–4366.

[82] Cammarata, L.; Kazarian, S. G.; Salter, P. A.; Welton, T. *Phys. Chem. Chem. Phys.* 2001, **3**, 5192–5200.

[83] Hanke, C. G.; Lynden-Bell, R. M. *J. Phys. Chem. B* 2003, **107**, 10873–10878.

[84] Dong, B.; Li, N.; Zheng, L.; Yu, L.; Inoue, T. *Langmuir* 2007, **23**, 4178–4182.

[85] Inoue, T.; Dong, B.; Zheng, L. *J. Colloid Interface Sci.* 2007, **307**, 578–581.

[86] Firestone, M. A.; Dzielawa, J. A.; Zapol, P.; Curtiss, L. A.; Seifert, S.; Dietz, M. L. *Langmuir* 2002, **18**, 7258–7260.

[87] Yan, F.; Texter, J. *Chem. Commun.* 2006, 2696–2698.

[88] England, D.; Tambe, N.; Texter, J. *ACS Macro Lett.* 2012, **1**, 310–314.

[89] England, D.; Yan, F.; Texter, J. *Langmuir* 2013, **29**, 12013–12024.

[90] Yan, F.; Texter, J. *Angew. Chem. Int. Ed.* 2007, **46**, 2440–2443.

[91] Yu, S.; Yan, F.; Zhang, X.; You, J.; Wu, P.; Lu, J.; Xu, Q.; Xia, X.; Ma, G. *Macromolecules* 2008, **41**, 3389–3392.

[92] Chen, Z.; Yan, F.; Qiu, L.; Lu, J.; Zhou, Y.; Chen, J.; Tang, Y.; Texter, J. *Langmuir* 2010, **26**, 3803–3806.

[93] Zhou, Y.; Qiu, L.; Deng, Z.; Texter, J.; Yan, F. *Macromolecules* 2011, **44**, 7948–7955.

[94] Zhang, G.; Chen, X.; Xie, Y.; Zhao, Y.; Qiu, H. *J. Colloid Interface Sci.* 2007, **315**, 601–606.

[95] Zhang, G.; Chen, X.; Zhao, Y.; Xie, Y.; Qiu, H. *J. Phys. Chem. B* 2007, **111**, 11708–11713.

[96] Zhang, J.; Dong, B.; Zheng, L.; Li, N.; Li, X. *J. Colloid Interface Sci.* 2008, **321**, 159–165.

[97] Li, X.-W.; Zhang, J.; Zheng, L.-Q.; Chen, B.; Wu, L.-Z.; Lv, F.-F.; Dong, B.; Tung, C.-H. *Langmuir* 2009, **25**, 5484–5490.

[98] Safavi, A.; Maleki, N.; Farjami, F. *Colloids Surf. A Physicochem. Eng. Asp.* 2010, **355**, 61–66.

[99] Glasser, L. *Thermochim. Acta* 2004, **421**, 87–93.

[100] Xia, H.; Yu, J.; Jiang, Y.; Mahmood, I.; Liu, H. *Ind. Eng. Chem. Res.* 2007, **46**, 2112–2116.

[101] Tong, Y.; Han, L.; Yang, Y. *Ind. Eng. Chem. Res.* 2012, **51**, 16438–16443.

[102] Zech, O.; Thomaier, S.; Bauduin, P.; Rück, T.; Touraud, D.; Kunz, W. *J. Phys. Chem. B* 2009, **113**, 465–473.

[103] Rojas, O.; Koetz, J.; Kosmella, S.; Tiersch, B.; Wacker, P.; Kramer, M. *J. Coll. Interface Sci.* 2009, **333**, 782–790.

[104] Rojas, O.; Tiersch, B.; Frasca, S.; Wollenberger, U.; Koetz, J. *Colloids Surf. A* 2010, **369**, 82–87.

[105] Rojas, O.; Tiersch, B.; Rabe, C.; Stehle, R.; Hoell, A.; Arlt, B.; Koetz, J. *Langmuir* 2013, **29**, 6833–6839.

[106] Eastoe, J.; Robinson, B. H.; Steytler, D. C.; Thorn-Leeson, D. *Adv. Colloid Interface Sci.* 1991, **36**, 1–31.

[107] De, T. K.; Maitra, A. *Adv. Colloid Interface Sci.* 1995, **59**, 95–193.

[108] Lopez-Quintela, M. A.; Tojo, C.; Blanco, M. C.; Garcia-Rio, L.; Leis, J. R. *Curr. Opin. Colloid Interface Sci.* 2004, **9**, 264–278.

[109] Cheng, S.; Zhang, J.; Zhang, Z.; Han, B. *Chem. Commun.* 2007, 2497–2499.

[110] Lü, H. H.; An, X. Q.; Shen, W. G. *J. Chem. Eng. Data* 2008, **53**, 727–731.

[111] Liu, L. P.; Bauduin, P.; Zemb, T.; Eastoe, J., Hao, J. C. *Langmuir* 2009, **25**, 2055–2059.

[112] Wei, J.; Su, B.; Yang, J.; Xing, H.; Bao, Z.; Yang, Y.; Ren, Q. *J. Chem. Eng. Data* 2011, **56**, 3698–3702.

[113] Wei, J.; Su, B.; Xing, H.; Bao, Z.; Yang, Y.; Ren, Q. *Colloids Surf. A* 2012, **396**, 213–218.

[114] Rao, V. G.; Mandal, S.; Ghosh, S.; Banerjee, C.; Sarkar, N. *J. Phys. Chem. B* 2012, **116**, 8210–8221

[115] Rao, V. G.; Mandal, S.; Ghosh, S.; Banerjee, C.; Sarkar N. *J. Phys. Chem. B* 2013, **117**, 1480–1493.

[116] Rao, V. G.; Ghosh, S.; Ghatak, C.; Mandal, S.; Brahmachari, U.; Sarkar, N. *J. Phys. Chem. B* 2012, **116**, 2850–2855.

[117] Rao, V. G.; Banerjee, C.; Ghosh, S.; Mandal, S.; Kuchlyan, J.; Sarkar, N. *J. Phys. Chem. B* 2013, **117**, 7472–7480.

CHAPTER 14

Ionic Liquids in Soft Confinement: Effect of Reverse Micelle Interfaces on the Entrapped Ionic Liquid Structure

RUBEN DARIO FALCONE, N. MARIANO CORREA, and JUANA J. SILBER

Departamento de Química, Universidad Nacional de Río Cuarto, Río Cuarto, Argentina

NANCY E. LEVINGER

Department of Chemistry, Colorado State University, Fort Collins, CO, USA

14.1 INTRODUCTION

Microheterogeneous environments, such as those found in reverse micelles (RMs) and microemulsions, have tremendous promise because of the nonstandard environments they present. Often, chemistries that occur in these solutions do not occur in homogeneous liquid solutions [1–4]. Essentially, RMs are spatially ordered macromolecular assemblies of surfactants formed in nonpolar solvents, in which the polar head groups of the surfactants point inward toward a polar core and the hydrocarbon chains point outward toward the nonpolar medium [5, 6] (see schematic representation in Fig. 14.1).

RMs form with a wide variety of surfactants (anionic, cationic, and nonionic), in various nonpolar solvents [1–35]. Examples of classical surfactants include anionic sodium 1,4-bis(2-ethylhexyl) sulfosuccinate (AOT), cationic benzyl-*n*-hexadecyldimethylammonium chloride (BHDC), and nonionic Triton X-100 (TX-100). Molecular structures for AOT, BHDC, and TX-100 are shown in Figure 14.2.

AOT RMs form in aromatic (benzene, toluene, chlorobenzene, xylene) and aliphatic (*n*-heptane, *n*-hexane, isooctane, decane) solvents. BHDC forms RMs in aromatic or mixture of aromatic–aliphatic solvents [27–34]. TX-100, one of the most commonly used nonionic surfactants used to prepare RMs, is able to form this kind of organized system in various hydrocarbon solvents [18–23, 35]. All three surfactants—AOT, BHDC, and TX-100—form RMs without the addition of a cosurfactant. The majority of studies on RMs, solubilize water as the polar component and juxtapose properties such as polarity, viscosity, conductivity and H-bonding for bulk water and confined in RMs [36, 37]. These studies show that the physicochemical

Ionic Liquid-Based Surfactant Science: Formulation, Characterization, and Applications, First Edition. Edited by Bidyut K. Paul and Satya P. Moulik.

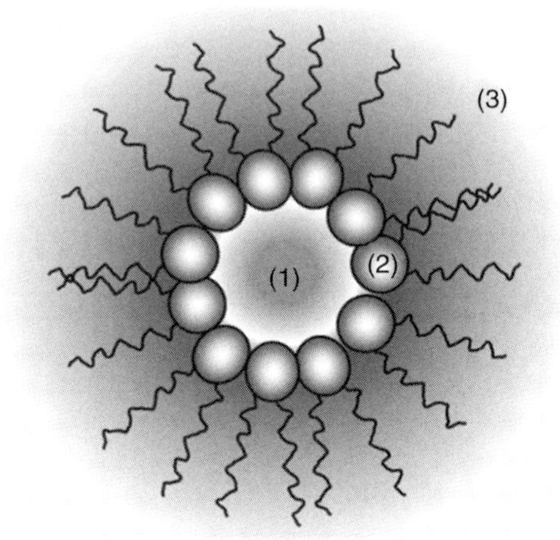

Figure 14.1 Schematic representation of the RMs. (1) Polar pool, (2) interface, and (3) nonpolar pseudophase.

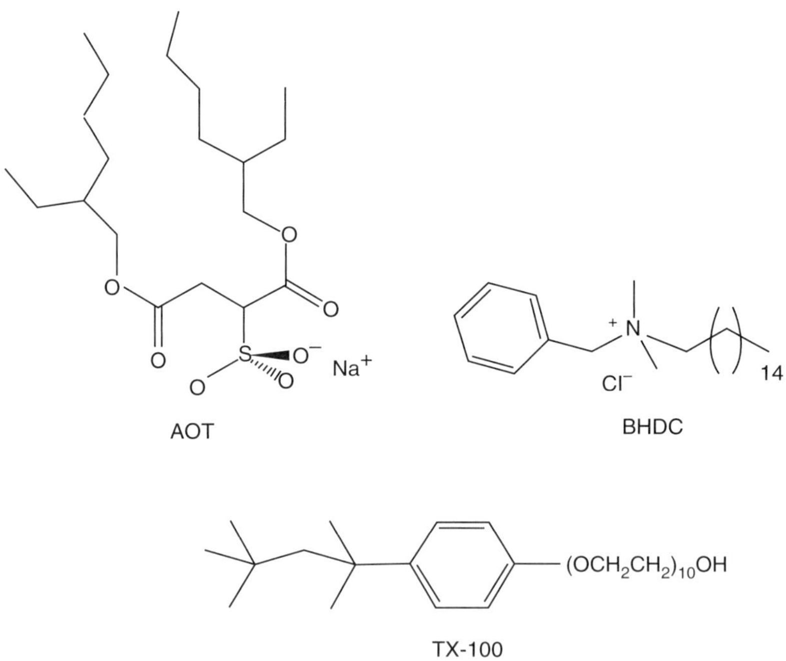

Figure 14.2 Molecular structure of AOT, BHDC, and TX-100 surfactants.

properties of water entrapped inside RMs change dramatically from the bulk, as a result of the specific interactions and confined geometries [38–46]. Moreover, it has been demonstrated that the effect of the kind of surfactant used to create RMs is crucial for the understanding of the water structure [39, 40, 42, 45].

Previous studies performed in our group have shown very peculiar and interesting water properties inside RMs that emerges from the confinement effect and the interaction with the surfactant at the interface [27, 39, 40, 42]. For example, water properties differ between RM systems formed with anionic and cationic surfactants. The water molecules entrapped inside the AOT/benzene RMs show enhanced electron donor ability compared with bulk water, while water entrapped inside the BHDC/benzene RMs appears nonelectron donating due to its interaction with the cationic surfactant polar head group [27, 38, 39, 42]. These results have tremendous impact when RMs are used as nanoreactors [40, 47].

Researchers have also prepared and studied nonaqueous RMs where water has been replaced by polar solvents that possess relatively high dielectric constants and that are immiscible with the continuous nonpolar solvent [6]. These nonaqueous RMs have attracted interest from both fundamental and practical perspectives [48–69]. As reaction media, these RMs are particularly attractive for water-sensitive reactions [70]. In this sense, ionic liquids (ILs) emerge as a powerful and attractive alternative to conventional molecular organic solvents [71–76], and they have received much attention as a class of neoteric solvents [71–83]. The most commonly used ILs are based on *N,N′*-dialkylimidazolium cations, especially 1-butyl-3-methylimidazolium, $[bmim]^+$, with different anions such as tetrafluoroborate $[BF_4]^-$ or bis(trifluoromethylsulfonyl) imide $[Tf_2N]^-$ (see structures in Fig. 14.3). Recent studies on RMs with an IL

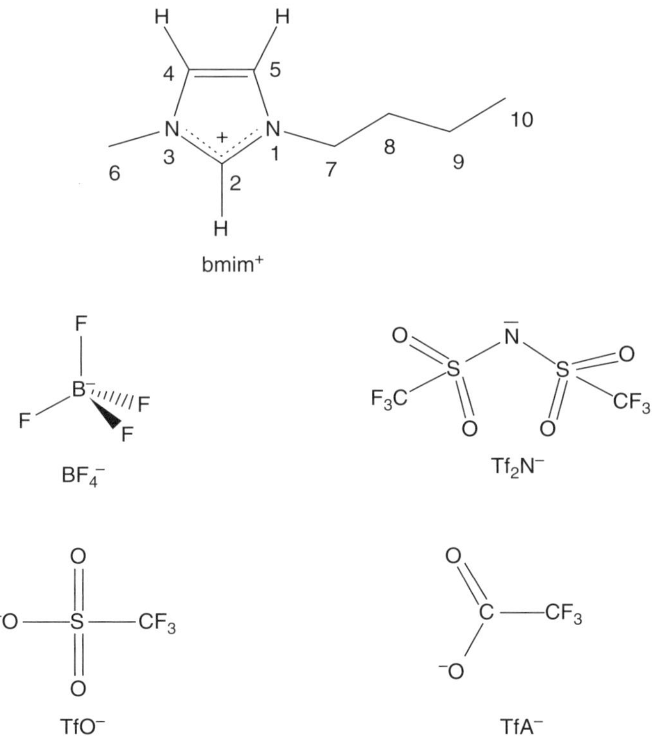

Figure 14.3 Molecular structure of $[bmim]^+$, $[BF_4]^-$, $[Tf_2N]^-$, $[TfO]^-$, and $[TfA]^-$.

component have garnered interest because the ILs can be designed to be used as a surfactant, as the external medium, as a polar phase, or as an additive [84–141]. Systems where surfactants sequester ILs are particularly interesting.

These organized systems generate hydrophobic or hydrophilic nanodomains that expand their potential uses as reaction, separation, and/or extraction media increasing processing versatility while simultaneously reducing costs [111, 139, 142–145]. The possibility of replacing water with ILs to form structures similar to water containing RMs has been demonstrated, and some properties of these complex systems have been characterized [89–139, 142–147].

14.2 EARLY STUDIES OF RM SYSTEMS WITH IL

The most common IL RM system studied entails the nonionic TX-100 surfactant and [bmim][BF_4] mostly in cyclohexane, benzene, or toluene, among others [89, 90, 93, 97, 98, 100, 102, 103, 114, 116, 124, 133]. Dynamic light scattering (DLS) measurements showed that these IL RM samples contained nanosized structures [89, 93, 97, 107, 113, 122–124, 133, 134] and, similar to "classic" RMs that form around water cores, the droplet volume of the IL containing nanosystems increased linearly with added [bmim][BF_4] [89, 93, 97, 124, 134]. In addition, several different studies explored how confinement affects the IL structure. Steady-state and picosecond time-resolved emission spectroscopy of coumarin 153 in [bmim][BF_4]/TX-100/cyclohexane RMs has permitted the investigation of solvation dynamics and rotational relaxation [92]. Gao et al. hypothesized that the interaction between the electronegative oxygen atoms of TX-100 oxyethylene units and the electropositive imidazolium ring may drive [bmim][BF_4] to reside in the core of the TX-100/benzene aggregates [96]. The micropolarities in [bmim][BF_4]/TX-100/toluene RMs investigated with UV–Vis absorption spectroscopy of methyl orange and methylene blue molecular probes indicated increasing polarity of the IL RMs with increasing W_s (W_s = [IL]/[surfactant]) until a real IL pool formed, beyond which the polar microenvironment remains relatively fixed [96]. Although [bmim][BF_4] is probably the IL most studied in RMs, other *N*,*N*′-dialkylimidazolium-derived ILs have been also solubilized inside RMs [99, 119, 120, 122, 123, 137, 138, 142, 145–147]. Li et al. [99] studied the phase behavior of toluene/TX-100/1-butyl-3-methylimidazolium hexafluorophosphate ([bmim][PF_6]) and showed that the RM size depended markedly on the IL content. Gayet et al. [142] studied the phase diagram and microstructure of RMs formed with benzylpyridinium bis(trifluoromethanesulfonyl)imide/nonionic surfactant (octylphenol ethoxylate)/toluene by conductivity, DLS, pulsed-field gradient spin-echo NMR, and SANS techniques. Moniruzzaman et al. [145] explored the effect of various ILs, for example, [mmim][$(MeO)_2PO_2$], on the formation of RMs stabilized by a blend of two nonionic surfactants, polyoxyethylene sorbitan monooleate (Tween-80) and sorbitan laurate (Span-20) in isopropyl myristate. Harrar et al. [119, 120] reported the formation of stable RMs composed by 1-ethyl-3-methylimidazolium ethylsulfate ([emim][$etSO_4$]), limonene, and TX-100 [119] and (1,1,3,3-tetramethylbutyl)phenyl polyethylene glycol (7–8) and TX-114 [120]. Despite the variety of ILs used, the results obtained by different authors suggest that when ILs are entrapped in RMs created by nonionic surfactants, they do not interact strongly with the surfactant at the RMs interface and, in consequence, the IL structure and properties remain quite similar to neat ILs.

Compared to nonionic surfactants like TX-100, studies of IL RMs formed with cationic or anionic surfactants are very scarce. Rojas et al. [146] explored the influence of 1-ethyl-3-methylimidazolium hexylsulfate ([emim][hexSO$_4$]) on RM formation in the optically clear phase region of water/toluene/pentanol mixtures in the presence of the cationic surfactant cetyltrimethylammonium bromide (CTAB). Their results showed that adding the IL significantly expanded the transparent phase region. Rabe and Koetz [147] reported that temperature and counterion hydrophobicity of the IL influenced microemulsion formation in toluene/pentanol/CTAB solutions with the IL [emim][etSO$_4$] or [emim][hexSO$_4$]. Interestingly, both studies required a cosurfactant (pentanol) to guarantee the RM formation.

Only a few reports exist for IL RMs formed with anionic surfactants [85, 86]. Cheng et al. [85] showed that *direct* micelles formed with AOT as the surfactant entrapping [bmim][PF$_6$] as the internal pseudophase and propylammonium formate (PAF) as the external pseudophase. In this case, [bmim][PF$_6$] is dispersed in the nonpolar cores of AOT micelles in a PAF-rich continuous phase, but does not interact with the anionic surfactant polar head group. Consequently, its bulk structure is not modified. Moniruzzaman et al. [86] reported the formation of AOT RMs using 1-octyl-3-methylimidazolium bis(trifluoromethylsulfonyl)imide [omim][Tf$_2$N] as the external continuous phase, 1-hexanol as the cosurfactant, and water as the polar solvent entrapped in the polar core. In this system, [omim][Tf$_2$N] does not interact with the anionic surfactant because it is the external phase and surfactant head groups are located toward the interior of the RMs; thus, the IL properties remain unaffected in the microemulsion.

Very recently, a new class of surfactant-like IL has been developed. Several authors replaced the Na$^+$ counterion in AOT with common IL cations such as [bmim]$^+$ or tetraalkylammonium, leading to very different and attractive properties compared with the precursor. These AOT analogues, synthesized for the first time by Brown et al. [148–150], have been used to create novel RMs prepared in nonpolar solvents entrapping other ILs [129, 130, 135–137]. Rao et al. [129, 130, 135–137] have explored several IL containing RM systems with these AOT analogues. Among others, they characterized the phase diagram of the [bmim][BF$_4$]/[bmim][AOT]/benzene ternary system by DLS and spectroscopies with various molecular probes [130] and investigated RMs formed by *N*,*N*-dimethylethanolammonium 1,4-bis(2-ethylhexyl) sulfosuccinate as a surfactant, isopropyl myristate as the oil phase, and the IL 1-hexyl-3-methylimidazolium bis(trifluoromethanesulfonyl)imide as the polar phase [136]. In all cases, the most promising feature of these systems was their high thermal stability compared to that of aqueous RMs.

14.3 EFFECT OF CONFINEMENT AND INTERFACE ON THE STRUCTURE OF ILs ENTRAPPED IN RMs

One important feature of RMs is their propensity to influence the physicochemical properties of polar solvent entrapped inside as a result of the specific interactions and confined geometries [5, 26, 27, 38–46, 67]. How the polar solvent properties change depends intimately on the nature of the surfactant (anionic, cationic, or nonionic) [38, 39, 42, 45]. We have used a range of methods to characterize the nature of the IL in several different RM systems. We compare and contrast the results with nonionic and ionic surfactants as well as with differing ILs.

14.3.1 Size Characteristics of IL Containing RMs

Our initial studies have focused on [bmim][BF_4] and [bmim][Tf_2N] in RM systems formed using either TX-100 or BHDC in benzene [151]. We measured these systems using DLS. Table 14.1 lists sizes obtained for the RMs explored. The observation that the apparent particle diameter, d_{app}, characterized by W_s, increased with increasing IL content (W_s) confirmed that RMs form in the mixtures (see Fig. 14.4a). RM sizes depend on several variables. To explain the observation (Table 14.1) that [bmim][BF_4] forms larger RMs than [bmim][Tf_2N], we consider parameters governing RM properties, especially the effective packing parameter of the surfactant, $p = v/al_c$, where v and l_c are the volume and length of the hydrocarbon chain, respectively, and a is the surfactant head group area [152]. Smaller p leads to larger RMs [153]. In studies of RMs with nonaqueous polar phases, we have shown that specific interactions of the polar solvent with the surfactants strongly influence p [46, 154]. When a polar solvent encapsulated interacts strongly with the surfactant polar head group, for example, by a hydrogen-bonding interaction, this increases the a value leading p to decrease and therefore an increase in the RM droplet size. Thus, the difference in RM size observed for [bmim][BF_4] compared to [bmim][Tf_2N] (Table 14.1) indicates that [bmim][BF_4] interacts strongly with the surfactant head groups, consequently penetrating the RM interface. This penetration leads to an increase in the effective surfactant head group area, a. Weaker interactions of [bmim][Tf_2N] with the surfactant head groups lead to smaller RMs compared to the [bmim][BF_4].

We have further explored [bmim][Tf_2N] in BHDC/chlorobenzene and AOT/chlorobenzene [155]. These experiments compared the effect of surfactant charge on the properties of the resulting RMs. Data shown in Table 14.1 confirm RM formation for both anionic AOT and cationic BHDC surfactants through the linear increase in the droplet size with increasing W_s. Interestingly, for the same W_s value, RMs formed with anionic AOT appear larger than those with cationic BHDC, but the BHDC RMs showed a stronger dependence of size on W_s. The maximum W_s observed for

TABLE 14.1 Apparent Diameter (d_{app}) Values (nm) of the Different IL RMs from Dynamic Light Scattering Measurements

	d_{app}		
System	($W_s = 1$)	(W_s^{max})	Reference
[bmim][Tf_2N]/Triton X-100/benzene	3.0	7.7 (2)	151
[bmim][BF_4]/Triton X-100/benzene	22.0	58.0 (2)	151
[bmim][Tf_2N]/BHDC/benzene	2.1	6.6 (2)	151
[bmim][BF_4]/BHDC/benzene	11.8	48.0 (2)	151
[bmim][Tf_2N]/BHDC/chlorobenzene	2.5	9.5 (3)	152
[bmim][Tf_2N]/AOT/chlorobenzene	3.9	4.5 (1.5)	152
[bmim][TfA]/AOT/chlorobenzene	2.3	3.5 (3)	153
[bmim][TfO]/AOT/chlorobenzene	4.3	6.5 (1.5)	153
[bmim][TfA]/BHDC/chlorobenzene	2.3	3.6 (3)	153
[bmim][TfO]/BHDC/chlorobenzene	3.5	7.6 (2.5)	153

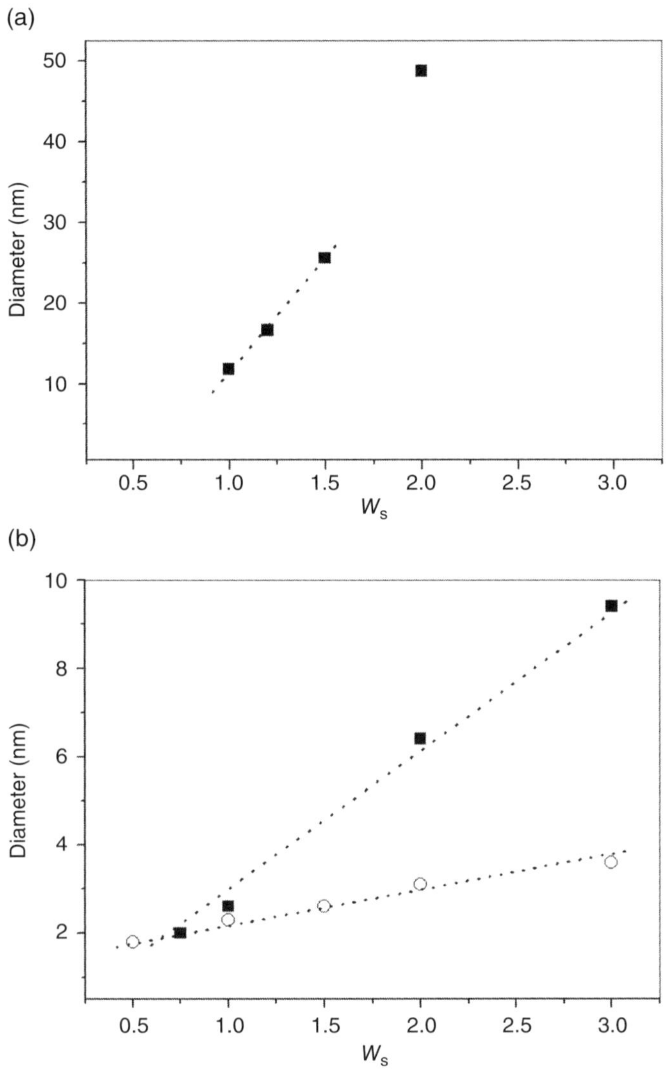

Figure 14.4 Apparent diameter values for different ILs BHDC RMs: (a) benzene/BHDC/[bmim][BF_4] and (b) chlorobenzene/BHDC/[bmim][Tf_2N] (■) and chlorobenzene/BHDC/[bmim][TfA] (○) RMs obtained at 25°C with varying W_s. The straight lines were plotted to guide the eye.

AOT RMs was 1.5, above which phase separation occurs. These investigations mark the first reports of BHDC and AOT RM formation with an aromatic pseudophase with two different ILs.

The most commonly ILs used in RMs, for example, [bmim][BF_4], [bmim][Tf_2N], and [bmim][PF_6], are all polar solvents with low electron donor capacity [156, 157]. To understand the role of electron donor capacity in RM formation and properties, we have investigated RMs formed using ILs whose anions have high electron donor ability, specifically 1-butyl-3-methylimidazolium trifluoromethanesulfonate ([bmim][TfO]) and 1-butyl-3-methylimidazolium trifluoroacetate ([bmim][TfA]) entrapped in anionic (AOT/chlorobenzene) and cationic (BHDC/chlorobenzene) RMs [158]. These ILs have similar polarities, but $[TfA]^-$ has significantly higher electron donor ability compared to $[TfO]^-$ [156, 157]. As noted for other systems, DLS measurements positively identify RM formation based on increasing droplet size with increasing IL content (see examples in Fig. 14.4); moreover, the IL molecules interact with the RM interfaces [6, 151]. We observe larger droplet sizes and a stronger size dependence on W_s for [bmim][TfO] in AOT and BHDC RMs than for [bmim][TfA] and explain this by considering the different IL–surfactant interactions. When [bmim][TfO] is encapsulated in AOT RMs, the anions $[TfO]^-$ interact with the Na^+ counterions; the $[bmim]^+$ cation penetrates the interface toward the SO_3^- group, effectively increasing the interfacial area, decreasing the surfactant packing parameter, and increasing the RM size. In contrast for [bmim][TfA] added to AOT RMs, RM size changes much less suggesting a weak IL–AOT interaction and with $[bmim]^+$ and $[TfA]^-$ essentially located farther of the interface. The main difference between these two ILs (similar polarities) is the higher electron donor ability of $[TfA]^-$ compared to $[TfO]^-$ [159]. The interaction between $[bmim]^+$ and $[TfA]^-$ is stronger upon confinement favoring a strong ion-pair association. Weak interaction with the anionic AOT polar head leads to modest changes in the packing parameter and in the droplet size with varying W_s. We observed similar effects for molecular solvents such as dimethylformamide and dimethylacetamide [154] and IL such as [bmim][Tf_2N][151] encapsulated inside AOT RMs. Likewise, comparing the d_{app} values obtained for both ILs in the BHDC RMs (Table 14.1) suggests that the anion $[TfO]^-$ interacts more strongly with the positive charge of the cationic surfactant polar head group (BHD^+) than that of $[TfA]^-$ that interacts with the surfactant head group. Consequently, $[TfO]^-$ penetrates the RM interface, increasing the effective interfacial area and decreasing the surfactant packing parameter with increase in the RM size. These results also suggest a large ion separation of [bmim][TfO] inside the BHDC RMs, because of a strong IL–BHD interaction. On the other hand, the small changes in the d_{app} values obtained in the chlorobenzene/BHDC/[bmim][TfA] system (Fig. 14.4b) suggest a weaker TfA^-/BHD^+ interaction than the TfO^-/BHD^+, similar to our observations for [bmim][TfA] entrapped in AOT RMs. These results demonstrate how the cation–anion interactions present in the bulk IL structure can be modified upon confinement.

14.3.2 Characteristics of IL in RMs: Response of a Molecular Probe

We have used different methods to characterize and understand the nature of the ILs in the RMs. The measurement of microenvironments in the RMs is one of them. We have used 1-methyl-8-oxyquinolinium betaine (QB) to sense the polar core environment of IL RMs [151]. QB is a useful probe as it is small and very sensitive to different microenvironment properties and it locates exclusively at the RM interface [26, 27]. QB presents two electronic absorption bands [26]: the band in the visible region, B_1, arises from the transition from a predominantly dipolar ground state to an excited state of considerably reduced polarity leading to negative solvatochromism with increasing solvent polarity. Similar to the behavior of $E_{T(30)}$

compound [160], the B_1 spectrum can determine the micropolarity of the RM interface [26, 27, 43, 45, 67, 161]. The other band, B_2 band, observed in the UV region also shifts hypsochromically with the polarity of the solvent although less significantly than B_1, but its intensity is quite sensitive to the environment. The B_2 band primarily senses the hydrogen-bond donating ability of the solvent [26]. As a result, the absorbance ratio of the bands (Abs B_2/Abs B_1) is high for solvents with low H-bond ability and low for solvents with high H-bond ability.

We used QB to probe [bmim][BF_4] and [bmim][Tf_2N] encapsulated in TX-100 and BHDC/benzene [151] and found that the microenvironment of [bmim][BF_4] and [bmim][Tf_2N] encapsulated in TX-100 and BHDC/benzene RMs depends strongly on the surfactant used to create the organized media. In TX-100, both the hypsochromic shift of the B_1 band and trend of the QB B_1 and B_2 absorption band ratio with W_s show almost identical trends for [bmim][BF_4] and [bmim][Tf_2N]. This suggests that the nature of their interactions within the RM structure is similar and that neither of the anions interact appreciably with the TX-100 oxyethylene units [92, 96].

BHDC RMs present a contrasting situation. Here, the QB dye indicates that the properties of the entrapped ILs depend strongly on the IL anion. The ratio of B_1 and B_2 absorbance increases for $[Tf_2N]^-$, similar to the behavior in TX-100 RMs, but decreases slightly for $[BF_4]^-$. This indicates that the two ILs experience differing electrostatic environments. For [bmim][BF_4], its encapsulation disrupts the IL structure due to a strong electrostatic interaction between $[BF_4]^-$ and the cationic head of BHDC. A minimal electrostatic interaction between $[Tf_2N]^-$ and BHDC leads to minor changes in the structure of [bmim][Tf_2N] displaying properties similar to TX-100 encapsulation.

14.3.3 Characteristics of IL in RMs: NMR Spectroscopy

We have employed NMR spectroscopy harnessing the signals from several different nuclei to explore IL RMs. We performed ^1H, ^{11}B, and ^{19}F NMR spectroscopy to study [bmim][BF_4], entrapped in TX-100 and BHDC/benzene RMs, which allows us to observe the effects of the RMs on the IL cation and anion, as well as how the IL affects the surfactant microenvironment [162]. Consistent with our results using the QB molecular probe, the $[BF_4]^-$ ^{11}B and ^{19}F NMR spectra indicate stronger interactions with BHDC than with TX-100 [151]. The ^{11}B and ^{19}F NMR spectra followed how $[BF_4]^-$ interacts with the interfaces in both systems. In TX-100 RMs, we observed minimal differences between the NMR spectra for the IL in RMs and the neat liquid. Indeed, the spectra trend toward the pure liquid with the increase of W_s value. In BHDC, the NMR spectra depart dramatically from those of the pure [bmim][BF_4].

We used ^1H NMR spectroscopy to investigate the IL imidazolium cation microenvironment. Thus, for ILs containing $[bmim]^+$, the $[bmim]^+$ H2 (labeled in Fig. 14.3) provides a signal well separated from other peaks in the RM samples, making it straightforward to follow. For [bmim][BF_4] in TX-100, the $[bmim]^+$ H2 signal is slightly downfield from the signal in pure IL, and it shifts upfield almost to the signal in pure IL with increasing W_s. Although one might expect a significant interaction between the relatively hydrophobic $[bmim]^+$ and the TX-100, these NMR results are consistent with other indicators that the TX-100 environment provides little perturbation to the IL properties.

Signals differ more for IL in RMs formed from ionic surfactants. For example, at the smallest W_s value in BHDC, the H2 [bmim$^+$] signal appears nearly 1.4 ppm downfield compared to its location in TX-100 RMs, and even with substantial upfield shifting with W_s, the H2 signal remains 0.6 ppm downfield of the value for the pure IL. We interpret the results from NMR spectra as indicating strong interactions between the [bmim][BF_4] and the BHDC, which causes microstructures or layering. This phenomenon occurs when [BF_4]$^-$ partitions to the interfacial region, supplanting the regular Cl^- counterion of BHDC. This puts the [BF_4]$^-$ in close proximity to the ammonium surfactant head group and leaves the [bmim]$^+$ further away from the interface and interacting with the Cl^- counterions that migrated into the IL location. This suggests a remarkable difference in the environment sensed by the cation near BHDC.

We have also used ^1H NMR to investigate [bmim][Tf_2N] in BHDC/chlorobenzene and AOT/chlorobenzene [155]; these two systems show different behavior. Strikingly, the [bmim]$^+$ H2 signal in the smallest W_s in AOT RMs differs only slightly from the signal in pure liquid and approaches the value for the pure liquid with increasing IL content. However, similar to our results for [bmim][BF_4] in BHDC, for $W_s = 0.25$, the H2 [bmim]$^+$ signal appears almost 2 ppm downfield compared to pure liquid, and even in the largest BHDC RMs measured, the H2 [bmim]$^+$ signal is more than 0.7 ppm downfield from the value in pure [bmim][Tf_2N] (8.5 ppm).

We also observed modest chemical shift changes for the BHDC surfactant, indicating changes in structural organization. All the signals associated with BHDC protons display upfield shifts with increasing W_s. For example, the benzyl methylene protons shift from 5.68 ppm at $W_s = 0$ to 4.71 ppm at $W_s = 3$. Similar shifts are observed for protons near to the quaternary *N* (*N*-methyl and methylene) in BHDC. On the other hand, all proton signals associated with the AOT surfactant display a smaller overall shift than what we observe for BHDC [155]. Thus, the ILs have relatively stronger interactions with BHDC compared to their interactions with anionic AOT or nonionic TX-100. These results have implications for researchers intending to use these environments for chemistry in confinement and encourage further explorations into these complex and potentially versatile systems [142–145].

14.3.4 Characteristics of IL in RMs: FT-IR Spectroscopy

In addition to DLS, spectroscopy of the QB molecular probe, and NMR spectroscopy, we find that FT-IR spectroscopy provides valuable information about IL in RMs. Our FT-IR studies focus on [bmim][Tf_2N], [155] [bmim][TfO], and [bmim][TfA] [158] as examples of ILs with high and low electron-donating ability. Several different vibrational modes in these molecules reveal details about the IL properties and characteristics in pure liquid and in RMs. Results from our FT-IR studies on these systems are presented in Table 14.2.

We follow spectral signatures of both IL anion and cation to understand how the RM confined environment affects IL properties. For example, following the S–N–S stretching mode, $\nu^{s}_{S\text{-}N\text{-}S}$, in [$Tf_2N$]$^-$ reveals how the [bmim][Tf_2N] interacts with the RM interfaces [155]. Even though the [Tf_2N]$^-$ is generally considered a noncoordinating anion, $\nu^{s}_{S\text{-}N\text{-}S}$ displays an hypsochromic shift in the presence of ions [165, 166]. The $\nu^{s}_{S\text{-}N\text{-}S}$ peak appears at higher frequency for [bmim][Tf_2N] in AOT and BHDC RM environments compared to pure IL (see Table 14.2). It is relatively constant for

TABLE 14.2 Vibrational Frequencies for Ionic Liquids in Reverse Micelles. Uncertainty in Measurements ±0.5 cm^{-1}

System	Mode	Frequency (cm^{-1})	Observation
Pure [bmim][Tf_2N][a]	Tf_2N^- $\nu_{s(S\text{-}N\text{-}S)}$	740	
[bmim][Tf_2N]/AOT[a]	Tf_2N^- $\nu_{s(S\text{-}N\text{-}S)}$	748	Higher frequency than pure IL; constant with W_s
[bmim][Tf_2N]/BHDC[a]	Tf_2N^- $\nu_{s(S\text{-}N\text{-}S)}$	751–748	Higher frequency than pure IL or IL in AOT RMs; shifts to lower frequency with increasing W_s; approaches value in AOT
Pure [bmim][Tf_2N][a]	$bmim^+$ $\nu_{(C(2)\text{-}H)}$	3121	
Pure [bmim][Tf_2N][a]	$bmim^+$ $\nu_{(C(4,5)\text{-}H)}$	3157	
[bmim][Tf_2N]/AOT[a]	$bmim^+$ $\nu_{(C(2)\text{-}H)}$	3117	Does not vary with W_s
[bmim][Tf_2N]/AOT[a]	$bmim^+$ $\nu_{(C(4,5)\text{-}H)}$	3152	Does not vary with W_s
[bmim][Tf_2N]/BHDC[a]	$bmim^+$ $\nu_{(C(2)\text{-}H)}$	3113–3118	Shifts to higher frequency with increasing W_s; approaches value in AOT
[bmim][Tf_2N]/BHDC[a]	$bmim^+$ $\nu_{(C(4,5)\text{-}H)}$	3147–3152	Shifts to higher frequency with increasing W_s; approaches value in AOT
Free TfO^-[b]	$\nu_{asym(SO3)}$	1272	
[bmim][TfO][c]	$\nu_{asym(SO3)}$	1263	Broad peak with no clear splitting
[bmim][TfO]/AOT[c]	$\nu_{asym(SO3)}$	~1270	~33 cm^{-1} band splitting decreases with increasing W_s; larger splitting than in BHDC
[bmim][TfO]/BHDC[c]	$\nu_{asym(SO3)}$	~1270	~19 cm^{-1} band splitting decreases with increasing W_s
Pure [bmim][TfO][c]	$bmim^+$ $\nu_{(C(4,5)\text{-}H)}$	3154	
[bmim][TfO]/AOT[c]	$bmim^+$ $\nu_{(C(4,5)\text{-}H)}$	3149	Does not vary with W_s
[bmim][TfO]/BHDC[c]	$bmim^+$ $\nu_{(C(4,5)\text{-}H)}$	3145–3147	Band shifts slightly to higher frequency with increasing W_s
Pure [bmim][TfA][c]	$\nu_{asym(COO-)}$	1689	
[bmim][TfA]/AOT[c]	$\nu_{asym(COO-)}$	1687	Does not vary with W_s
[bmim][TfA]/BHDC[c]	$\nu_{asym(COO-)}$	1687	Does not vary with W_s
Pure [bmim][TfA][c]	$bmim^+$ $\nu_{(C(4,5)\text{-}H)}$	3149	
[bmim][TfA]/AOT[c]	$bmim^+$ $\nu_{(C(4,5)\text{-}H)}$	3147	Band shifts slightly to lower frequency with increasing W_s
[bmim][TfA]/BHDC[c]	$bmim^+$ $\nu_{(C(4,5)\text{-}H)}$	3146	Band shifts slightly to lower frequency with increasing W_s

[a] Data from Ferreyra et al. [155].
[b] Data from Wendsjö et al. [163] and Bernson and Lindgren [164].
[c] Data from Blach et al. [158].

the AOT/chlorobenzene environment, whereas the peak shifts from higher frequency at low W_s values in BHDC/chlorobenzene, approaching its value in the AOT RMs. Both RM systems alter the IL structure leading to a higher frequency vibration. In the AOT system, the effect is constant, possibly arising from $[Tf_2N]^-$ interactions with the AOT Na^+ counterion. But in the BHDC RMs, the $\nu^{s}_{S\text{-}N\text{-}S}$ peak shifts to lower frequency with decreasing W_s. This suggests that the $[Tf_2N]^-$ interacts preferentially with the BHD^+ head group rather than with the $[bmim]^+$. In the BHDC head group, the positive charge resides localized on the quaternary nitrogen atom compared to $[bmim]^+$ (Fig. 14.3) where the positive charge is delocalized [167]. As a result, the interaction between $[Tf_2N]^-$ and the BHD^+ head group should be stronger than its interaction with $[bmim]^+$, causing the $[Tf_2N]^-$ to locate at the interface and exchange with the regular Cl^- counterion. This result is consistent with other studies showing the propensity of some ILs to form regular structures rather than remaining completely isotropic like standard organic solvents [168]. We also observed the propensity for $[bmim][BF_4]$ to adopt a layered structure when in BHDC RMs [162].

In contrast to $[bmim][Tf_2N]$, the anions of [bmim][TfO] and [bmim][TfA] display high electron-donating character. Comparison of the results from FT-IR spectroscopy of the antisymmetric stretching mode of SO_3^-, $\nu_{asym(SO3)}$, in $[TfO]^-$ and the carboxylate antisymmetric stretching mode in $[TfA]^-$ in AOT/chlorobenzene and BHDC/chlorobenzene RMs provides insight into these self-assembled systems [158]. The $[TfO]^-$ anion is highly sensitive to coordination with cations and exhibits characteristic infrared absorption bands in the vibrational spectra depending on the typical anionic environment [169–172]. Specifically, the local environment in solution can lift the degeneracy of $\nu_{asym(SO3)}$ for free $[TfO]^-$ ion in aqueous solution when the anion interacts with cations in solution [163, 164]. This leads to the broad peak observed for the free anion to split into a spectral doublet. The splitting of the $\nu_{asym(SO3)}$ spectral peak indicates the strength of the ionic interaction. In the pure IL, [bmim][TfO] shows only a broad peak, which indicates that the $[bmim]^+$ ion interaction with the anion is sufficiently weak so that no effect is seen. For [bmim][TfO] in AOT RMs, we observe a significant splitting at low W_s values that decreases as W_s increases (see Table 14.2). The $\nu_{asym(SO3)}$ feature in [bmim][TfO]/BHDC/chlorobenzene also splits but less dramatically than what we observe for AOT RMs. The observed splitting in the RMs and variations with W_s suggest that the $[TfO]^-$ experiences a stronger cation interaction potentially with different ions at the interface, that is, Na^+ in AOT and BHD^+ in BHDC, than observed in pure IL.

We monitored [bmim][TfA] in both RM systems through the $[TfA]^-$ anion antisymmetric carboxylate stretching band ($\nu_{asym(COO-)}$) [158]. Studies reporting the $\nu_{asym(COO-)}$ peak for various metallic salts of $[TfA]^-$ show quite complex behavior. Depending on the nature (size and charge) of the cation investigated, $[TfA]^-$ can act as a monodentate, bidentate, and/or bridging ligand [173]. The $\nu_{asym(COO-)}$ frequency is higher for the purely bridging $[TfA]^-$ compared to those arising from either chelating or both bridging and chelating configurations [174]. Interestingly and in contrast to the behavior for [bmim][TfO], the [bmim][TfA] $\nu_{asym(COO-)}$ mode shows almost no difference when the IL is in either AOT or BHDC RMs compared to pure IL. These results suggest that even though the IL–surfactant interaction is weak, the ion pairing between $bmim^+$ and TfA^- is enhanced upon confinement and interactions with surfactants are not strong enough to disrupt the normal IL structure. Apparently, the electron-donating ability of the ILs is also altered when this class of solvent is entrapped in RMs, and it cannot be predicted considering only the bulk property. In addition to measuring RM effects on IL anions, we have also explored how the RM environment affects the $[bmim]^+$ ions. The signature of the imidazolium C-H stretching modes is evident in the 3100–3200 cm^{-1} spectral window and allows us to evaluate the cation–anion interactions in complex systems such as RMs [162] since $[bmim]^+$ interacting strongly with anions can diminish its positive charge density with concomitant decrease in the ν_{C-H} values. We find the $\nu_{(C(4,5)-H)}$ stretch mode particularly useful to follow. For all three ILs, [bmim][Tf_2N], [bmim][TfO], and [bmim][TfA], the environment in BHDC RMs leads to spectral shifting with changing W_s value (see Table 14.2). This effect is largest for [bmim][Tf_2N]/BHDC/chlorobenzene. Additionally, although both [bmim][Tf_2N] and [bmim][TfO] display the $\nu_{(C(4,5)-H)}$ peak shifting to higher frequency with increasing W_s, the peak shifts to slightly lower frequency for [bmim][TfA] in BHDC/chlorobenzene. Results are more mixed for these ILs in AOT/chlorobenzene RMs. Here, each system displays unique behavior: the $\nu_{(C(4,5)-H)}$ peak does not vary with increasing W_s in [bmim][Tf_2N]/AOT/chlorobenzene and [bmim][TfO]/AOT/chlorobenzene RMs, and it shifts to slightly lower frequency with increasing W_s in [bmim][TfA]/AOT/chlorobenzene. These results indicate that the BHDC provides the most perturbative environment for $[bmim]^+$ in all the ILs investigated, while other environments seem to have a smaller effect.

14.4 SUMMARY

Taken together, the results from our experiments show the versatility of RMs containing ILs. We observe that both the RM environments influence the IL ionic organization and the IL impacts the RM structure. We find that it is not easy to predict what properties will be influenced by confinement. For example, when nonionic surfactants are used to create the RMs, the entrapped IL structure can be expected to be very similar to the bulk ILs. The situation changes when using ionic surfactants. The cationic BHDC RMs entrapping [bmim][BF_4], [bmim][Tf_2N], and [bmim][TfO] display substantial changes upon confinement, because of the interaction with the cationic interface. Consequently, ion layering is found primarily at low IL content. For [bmim][TfA], the confinement shows a more self-associated structure in comparison with the bulk structure. For the anionic AOT, only [bmim][TfO] shows strong interaction with the RM interface.

Valuable information can be obtained investigating the IL RMs in order to improve their ability as nanoreactors because unique microenvironments can be easily created by simply changing the RM components and/or IL content.

ACKNOWLEDGMENTS

We gratefully acknowledge the financial support for this work by the Consejo Nacional de Investigaciones Científicas y Técnicas (CONICET), Agencia Nacional de Promoción Científica y Técnica, Agencia Córdoba Ciencia, Secretaría de Ciencia y Técnica de la Universidad Nacional de Río Cuarto, and National Science Foundation under grant no. 0628260. RDF and NMC also received support by grant CONICET-NSF from CONICET. N.M.C., J.J.S., and R.D.F. hold a research position at CONICET.

REFERENCES

[1] Dwars, T.; Paetzold, E.; Oehme, G. *Angew. Chem. Int. Ed.* 2005, **44**, 7174–7199.

[2] Stamatis, H.; Xenakis, A.; Kolisis, F. N. *Biotechnol. Adv.* 1999, **17**, 293–318.

[3] Holmberg, K. *Adv. Colloid Interface Sci.* 1994, **51**, 137–174.

[4] Crans, D. C.; Trujillo, A. M.; Bonetti, S.; Rithner, C. D.; Baruah, B.; Levinger, N. E. J. *Org. Chem.* 2008, **73**, 9633–9640.

[5] Silber, J. J.; Biasutti, M. A.; Abuin, E.; Lissi, E. *Adv. Colloid Interface Sci.* 1999, **82**, 189–252.

[6] Correa, N. M.; Silber, J. J.; Riter, R. E.; Levinger, N. E. *Chem. Rev.* 2012, **112**, 4569–4602.

[7] De, T. K.; Maitra, A. *Adv. Colloid Interface Sci.* 1995, **59**, 95–193.

[8] Luisi, P. L.; Giomini, M.; Pileni, M. P.; Robinson, B. H. *Biochim. Biophys. Acta* 1988, **947**, 209–246.

[9] Pileni, M. P. J. *Phys. Chem.* 1993, **97**, 6961–6963.

[10] Uskokovic, V.; Drofenik, M. *Adv. Colloid Interface Sci.* 2007, **133**, 23–34.

[11] Levinger, N. E.; Swafford, L. A. *Annu. Rev. Phys. Chem.* 2009, **60**, 385–406.

[12] Jones, M. C.; Leroux, J. C. *Soft Matter* 2010, **6**, 5850–5859.

[13] Fayer, M. D.; Levinger, N. E. *Annu. Rev. Anal. Chem.* 2010, **3**, 89–107.

[14] Moulik, S. P.; Paul, B. K. *Adv. Colloid Interface Sci.* 1998, **78**, 96–195.

[15] Ravey, J. C.; Buzier, M. Influence of Temperature on the Structures of Inverse Nonionic Micelles and Microemulsions, 253–263. *Macro and Microemulsions Theory and Applications*, Shah, D. O., Ed., American Chemical Society, Washington, DC, 1985, p. 253.

[16] Ravey, J. C.; Buzier, M.; Picot, C. J. *Colloid Interface Sci.* 1984, **97**, 9–25.

[17] Friberg, S. E. Stabilization of inverse micelles by nonionic surfactants, 93–114. *Interfacial Phenomena in Apolar Media*, Eicke, H. F.; Parfitt, G. D., Eds., Marcel Dekker, New York, 1987, p. 93.

[18] Zhu, D.; Schelly, Z. A. *Langmuir* 1992, **8**, 48–50.

[19] Gu, J.; Schelly, Z. A. *Langmuir* 1997, **13**, 4256–4266.

[20] Mandal, D.; Datta, A.; Kumar Pal, S.; Bhattacharyya, K. J. *Phys. Chem. B* 1998, **102**, 9070–9073.

[21] Zhu, D. M.; Wu, X.; Schelly, Z. A. *Langmuir* 1992, **8**, 1538–1540.

[22] Zhu, D. M., Feng, K.; Schelly, Z. A. J. *Phys. Chem.* 1992, **96**, 2382–2385.

[23] Zhu, D. M.; Wu, X.; Schelly, Z. A. J. *Phys. Chem.* 1992, **96**, 7121–7126.

[24] Vasilescu, M.; Caragheorgheopol, A.; Almgren, M.; Brown, W.; Alsins, J.; Johannsson, R. *Langmuir* 1995, **11**, 2893–2898.

[25] Sawada, K.; Ueda, M. J. *Chem. Technol. Biotechnol.* 2004, **79**, 369–375.

[26] Correa, N. M.; Biasutti, M. A.; Silber, J. J. J. *Colloid Interface Sci.* 1995, **172**, 71–76.

[27] Correa, N. M.; Biasutti, M. A.; Silber, J. J. J. *Colloid Interface Sci.* 1996, **184**, 570–578.

[28] Novaira, M.; Biasutti, M. A.; Silber, J. J.; Correa, N. M. J. *Phys. Chem. B* 2007, **111**, 748–759.

[29] Costa, S. M. B.; Brookfield, R. L. J. *Chem. Soc. Faraday Trans. 2* 1986, **82**, 991–1002.

[30] Costa, S. M. B.; Aires de Barros, M. R.; Conde, J. P. *J. Photochem.* 1985, **28**, 153–164.

[31] Kikuchi, K.; Thomas, J. K. *Chem. Phys. Lett.* 1988, **148**, 245–248.

[32] Borsarelli, C. D.; Cosa, J. J.; Previtali, C. M. *Langmuir* 1993, **9**, 2895–2901.

[33] Dhami, S.; Cosa, J. J.; Bishop, S. M.; Phillips, D. *Langmuir* 1996, **12**, 293–300.

[34] Borsarelli, C. D.; Previtali, C. M.; Cosa, J. J. J. *Colloid Interface Sci.* 1996, **179**, 34–36.

[35] Pant, D.; Levinger, N. E. *Langmuir* 2000, **16**, 10123–10130.

[36] Hasegawa, M.; Sugimura, T.; Suzaki, Y.; Sindo, Y; Kitahara, A. *J. Phys. Chem.* 1994, **98**, 2120–2124.

[37] Piletic, I. R.; Moilanen, D. E.; Spry, D. B.; Levinger, N. E.; Fayer, M. D. *J. Phys. Chem. A* 2006, **110**, 4985–4999.

[38] Moyano, F.; Quintana, S. S.; Falcone, R. D.; Silber, J. J.; Correa, N. M. *J. Phys. Chem. B* 2009, **113**, 4284–4292.

[39] Quintana, S. S.; Moyano, F.; Falcone, R. D.; Silber, J. J.; Correa, N. M. *J. Phys. Chem. B* 2009, **113**, 6718–6724.

[40] Moyano, F.; Falcone, R. D.; Mejuto, J. C.; Silber, J. J.; Correa, N. M. *Chem. Eur. J.* 2010, **16**, 8887–8893.

[41] Durantini, A. M.; Falcone, R. D.; Silber, J. J.; Correa, N. M. *J. Phys. Chem. B* 2011, **115**, 5894–5902.

[42] Blach, D.; Correa, N. M.; Silber, J. J.; Falcone, R. D. *J. Colloid Interface Sci.* 2011, **355**, 124–130.

[43] Agazzi, F. M.; Falcone, R. D.; Silber, J. J.; Correa, N. M. *J. Phys. Chem. B* 2011, **115**, 12076–12084.

[44] Correa, N. M.; Levinger, N. E. *J. Phys. Chem. B* 2006, **110**, 13050–13061.

[45] Quintana, S. S.; Falcone, R. D.; Silber, J. J.; Correa, N. M. *ChemPhysChem* 2012, **13**, 115–123.

[46] Durantini, A. M.; Falcone, R. D.; Silber, J. J.; Correa, N. M. *J. Phys. Chem. B* 2013, **117**, 3818–3828.

[47] Falcone, R. D.; Biasutti, M. A.; Correa, N. M.; Silber, J. J.; Lissi, E.; Abuin, E. *Langmuir* 2004, **20**, 5732–5737.

[48] Ray, S.; Moulik, S. P. *Langmuir* 1994, **10**, 2511–2515.

[49] Fletcher, P. D. I.; Galal, M. F.; Robinson, B. H. *J. Chem. Soc. Faraday Trans. I* 1985, **81**, 2053–2065.

[50] Fletcher, P. D. I.; Galal, M. F.; Robinson, B. H. *J. Chem. Soc. Faraday Trans. I* 1984, **80**, 3307–3314.

[51] López-Cornejo, P.; Costa, S. M. B. *Langmuir* 1998, **14**, 2042–2049.

[52] Das, K. P.; Ceglie, A.; Lindman, B. *J. Phys. Chem.* 1987, **91**, 2938–2946.

[53] Arcoleo, V.; Aliotta, F.; Goffredi, M.; La Manna, G.; Turco Liveri, V. *Mater. Sci. Eng. C* 1997, **5**, 47–53.

[54] Riter, R. E., Kimmel, J. R., Undiks, E. P.; Levinger, N. E. *J. Phys. Chem. B* 1997, **101**, 8292–8297.

[55] Mathew, C.; Saidi, Z.; Peyrelasse, J.; Boned, C. *Phys. Rev. A* 1991, **43**, 873–882.

[56] Laia, C. A. T.; Lopez-Cornejo, P.; Costa, S. M. B.; d'Oliveira, J.; Martinho, J. M. G. *Langmuir* 1998, **14**, 3531–3537.

[57] Lattes, A.; Rico, I.; de Savignac, A.; Ahamd-Zadeh Samii, A. *Tetrahedron* 1983, **43**, 1725–1735.

[58] Ahamd-Zadeh Samii, A.; de Savignac, A.; Rico, I.; Lattes, A. *Tetrahedron* 1985, **41**, 3683–3688.

[59] Gautier, M.; Rico, I.; Ahamd-Zadeh Samii, A; de Savignac, A.; Lattes, A. *J. Colloid Interface Sci.* 1986, **112**, 484–487.

[60] Rico, I.; Lattes, A.; Das, K. P.; Lindman, B. *J. Am. Chem. Soc.* 1989, **111**, 7266–7267.

[61] Aveyard, R.; Binks, B. P.; Fletcher, P. D. I.; Kirk, A. J.; Swansbury, P. *Langmuir* 1993, **9**, 523–530.

[62] Martino, A.; Kaler, E. W. *Langmuir* 1995, **11**, 779–784.

[63] Tessy, E. I.; Rakshit, A. K. *Bull. Chem. Soc. Jpn.* 1995, **68**, 2137–2141.

[64] Riter, R. E.; Undiks, E. P; Kimmel, J. R.; Levinger, N. E. *J. Phys. Chem. B* 1998, **102**, 7931–7938.

[65] Elles, C.; Levinger, N. *Chem. Phys. Lett.* 2000, **317**, 624–630.

[66] Durantini, A. M.; Falcone, R. D.; Silber, J. J.; Correa, N. M. *ChemPhysChem* 2009, **10**, 2034–2040.

[67] Correa, N. M.; Falcone, R. D.; Biasutti, M. A.; Silber, J. J. *Langmuir* 2000, **16**, 3070–3076.

[68] Silber, J. J.; Falcone, R. D.; Correa, N. M.; Biasutti, M. A.; Abuin, E.; Lissi, E.; Campodonico, P. *Langmuir* 2003, **19**, 2067–2071.

[69] Falcone, R. D.; Correa, N. M.; Biasutti, M. A.; Silber, J. J. *J. Colloid Interface Sci.* 2006, **296**, 356–364.

[70] Correa, N. M.; Durantini, E. N.; Silber, J. J. *J. Phys. Org. Chem.* 2006, **19**, 805–812.

[71] Welton, T. *Green Chem.* 2011, **13**, 225–225.

[72] Hallett, J. P.; Welton, T. *Chem. Rev.* 2011, **111**, 3508–3576.

[73] Wasserscheid, P.; Keim, W. *Angew. Chem. Int. Ed.* 2000, **39**, 3772–3789.

[74] Welton, T. *Chem. Rev.* 1999, **99**, 2071–2084.

[75] Welton, T.; Wasserscheid, P. *Ionic Liquids in Synthesis*, Wiley-VCH Verlag GmbH, Weinheim, 2002.

[76] Welton, T. *Coord. Chem. Rev.* 2004, **248**, 2459–2477.

[77] Zhao, H.; Malhotra, S. V. *Aldrichimica Acta* 2002, **35**, 75–83.

[78] Olivier-Bourbigou, H.; Magna, L. *J. Mol. Catal. A* 2002, **182**, 419–437.

[79] Zhao, D.; Wu, M.; Kou, Y.; Min, E. *Catal. Today* 2002, **74**, 157–189.

[80] Sheldon, R. *Chem. Commun.* 2001, 2399–2407.

[81] Gordon, C. M. *Appl. Catal. Gen.* 2001, **222**, 101–117.

[82] Freemantle, M. *Chem. Eng. News* 1998, **76**, 32–37.

[83] Crowhurst, L.; Falcone, R.; Lancaster, N. L.; Llopis-Mestre, V.; Welton, T. *J. Org. Chem.* 2006, **71**, 8847–8853.

[84] Liu, J.; Cheng, S.; Zhang, J.; Feng, X.; Fu, X.; Han, B. *Angew. Chem. Int. Ed.* 2007, **46**, 3313–3315.

[85] Cheng, S.; Zhang, J.; Zhang, Z.; Han, B. *Chem. Commun.* 2007, 2497–2499.

[86] Moniruzzaman, M.; Kamiya, N.; Nakashima, K.; Goto, M. *ChemPhysChem* 2008, **9**, 689–692.

[87] Moniruzzaman, M.; Kamiya, N.; Nakashima, K.; Goto, M. *Green Chem.* 2008, **10**, 497–500.

[88] Moniruzzaman, M.; Kamiya, N.; Goto, M. *Langmuir* 2009, **25**, 977–982.

[89] Gao, H. X.; Li, J. C.; Han, B. X.; Chen, W. N.; Zhang, J. L.; Zhang, R.; Yan, D. D. *Phys. Chem. Chem. Phys.* 2004, **6**, 2914–2916.

[90] Eastoe, J.; Gold, S.; Rogers, S. E.; Paul, A.; Welton, T.; Heenan, R. K.; Grillo, I. *J. Am. Chem. Soc.* 2005, **127**, 7302–7303.

[91] Gao, Y.; Han, S.; Han, B.; Li, G.; Shen, D.; Li, Z.; Du, J.; Hou, W.; Zhang, G. *Langmuir* 2005, **21**, 5681–5684.

[92] Chakrabarty, D.; Seth, D.; Chakratorty, A.; Sarkar, N. *J. Phys. Chem. B* 2005, **109**, 5753–5758.

[93] Gao, Y. A.; Zhang, J.; Xu, H. Y.; Zhao, X. Y.; Zheng, L. Q.; Li, X. W.; Yu, L. *ChemPhysChem* 2006, **7**, 1554–1561.

[94] Yan, F.; Texter, J. *Chem. Commun.* 2006, 2696–2698.

[95] Gao, Y.; Hilfert, L.; Voigt, A.; Sundmacher, K. *J. Phys. Chem. B* 2008, **112**, 3711–3719.

[96] Gao, Y. A.; Li, N.; Zheng, L. Q.; Zhao, X. Y.; Zhang, J.; Cao, Q.; Zhao, M. W.; Li, Z.; Zhang, G. Y. *Chem. Eur. J.* 2007, **13**, 2661–2670.

[97] Li, N.; Gao, Y.; Zheng, L.; Zhang, J.; Yu, L.; Li, X. *Langmuir* 2007, **23**, 1091–1097.

[98] Li, J. C.; Zhang, J. L.; Han, B. X.; Wang, Y.; Gao, L. *J. Chem. Phys.* 2004, **121**, 7408–7412.

[99] Li, J. C.; Zhang, J. L.; Gao, H. X.; Han, B. X.; Gao, L. *Colloid Polym. Sci.* 2005, **283**, 1371–1375.

[100] Gao, Y.; Wang, S.; Zheng, L.; Han, S.; Zhang, X.; Lu, D.; Yu, L.; Ji, Y.; Zhang, G. *J. Colloid Interface Sci.* 2006, **301**, 612–616.

[101] Adhikari, A.; Sahu, K.; Dey, S.; Ghosh, S.; Mandal, U.; Bhattacharyya, K. *J. Phys. Chem. B* 2007, **111**, 12809–12816.

[102] Gao, Y.; Li, N.; Zheng, L.; Bai, X.; Yu, L.; Zhao, X.; Zhang, J.; Zhao, M.; Li, Z. *J. Phys. Chem. B* 2007, **111**, 2506–2513.

[103] Cheng, S.; Han, F.; Wang, Y.; Yan, H. *Colloids Surf. A* 2008, **317**, 457–461.

[104] Li, N.; Zhang, S.; Zheng, L.; Gao, Y.; Yu, L. *Langmuir* 2008, **24**, 2973–2976.

[105] Lue, H.; An, X.; Shen, W. *J. Chem. Eng. Data* 2008, **53**, 727–731.

[106] Qiu, Z.; Texter, J. *Curr. Opin. Colloid Interface Sci.* 2008, **13**, 252–262.

[107] Seth, D.; Sarkar, S.; Sarkar, N. *Langmuir* 2008, **24**, 7085–7091.

[108] Adhikari, A.; Das, D. K.; Sasmal, D. K.; Bhattacharyya, K. *J. Phys. Chem. A* 2009, **113**, 3737–3743.

[109] Gao, Y.; Li, N.; Hilfert, L.; Zhang, S.; Zheng, L.; Yu, L. *Langmuir* 2009, **25**, 1360–1365.

[110] Gao, Y.; Li, N.; Zhang, S.; Zheng, L.; Li, X.; Dong, B.; Yu, L. *J. Phys. Chem. B* 2009, **113**, 1389–1395.

[111] Zhao, M.; Zheng, L.; Bai, X.; Li, N.; Yu, L. *Colloids Surf. A* 2009, **346**, 229–236.

[112] Zheng, Y. J.; Eli, W. J.; Li, G. *Colloid Polym. Sci.* 2009, **287**, 871–876.

[113] Pramanik, R.; Sarkar, S.; Ghatak, C.; Rao, V. G.; Setua, P.; Sarkar, N. *J. Phys. Chem. B* 2010, **114**, 7579–7586.

[114] Pramanik, R.; Sarkar, S.; Ghatak, C.; Setua, P.; Rao, V. G.; Sarkar, N. *Chem. Phys. Lett.* 2010, **490**, 154–158.

[115] Sasmal, D. K.; Sen Mojumdar, S.; Adhikari, A.; Bhattacharyya, K. *J. Phys. Chem. B* 2010, **114**, 4565–4571.

[116] Sen Mojumdar, S.; Mondal, T.; Das, A. K.; Dey, S.; Bhattacharyya, K. *J. Chem. Phys.* 2010, **132**, 194505-1–194505-8.

[117] Zech, O.; Thomaier, S.; Kolodziejski, A.; Touraud, D.; Grillo, I.; Kunz, W. *Chem. Eur. J.* 2010, **16**, 783–786.

[118] Andujar-Matalobos, M.; Garcia-Rio, L.; Lopez-Garcia, S.; Rodriguez-Dafonte, P. *J. Colloid Interface Sci.* 2011, **363**, 261–267.

[119] Harrar, A.; Zech, O.; Hartl, R.; Bauduin, P.; Zemb, T.; Kunz, W. *Langmuir* 2011, **27**, 1635–1642.

[120] Harrar, A.; Zech, O.; Klaus, A.; Bauduin, P.; Kunz, W. *J. Colloid Interface Sci.* 2011, **362**, 423–429.

[121] Pramanik, R.; Ghatak, C.; Rao, V. G.; Sarkar, S.; Sarkar, N. *J. Phys. Chem. B* 2011, **115**, 5971–5979.

[122] Pramanik, R.; Sarkar, S.; Ghatak, C.; Rao, V. G.; Sarkar, N. *J. Phys. Chem. B* 2011, **115**, 2322–2330.

[123] Sarkar, S.; Pramanik, R.; Ghatak, C.; Rao, V. G.; Sarkar, N. *J. Chem. Phys.* 2011, **134**, 074507-1–074507-11.

[124] Setua, P.; Pramanik, R.; Sarkar, S.; Ghatak, C.; Rao, V. G.; Sarkar, N.; Das, S. K. *J. Mol. Liq.* 2011, **162**, 33–37.

[125] Wang, F.; Zhang, Z.; Li, D.; Yang, J.; Chu, C.; Xu, L. *J. Chem. Eng. Data* 2011, **56**, 3328–3335.

[126] Wei, J.; Su, B.; Yang, J.; Xing, H.; Bao, Z.; Yang, Y.; Ren, Q. *J. Chem. Eng. Data* 2011, **56**, 3698–3702.

[127] Lu, H.; An, X.; Shen, W. *J. Chem. Thermodyn.* 2012, **48**, 48–53.

[128] Lu, H.; An, X.; Yu, J.; Song, X. *J. Phys. Org. Chem.* 2012, **25**, 1210–1216.

[129] Rao, V. G.; Ghosh, S.; Ghatak, C.; Mandal, S.; Brahmachari, U.; Sarkar, N. *J. Phys. Chem. B* 2012, **116**, 2850–2855.

[130] Rao, V. G.; Mandal, S.; Ghosh, S.; Banerjee, C.; Sarkar, N. *J. Phys. Chem. B* 2012, **116**, 8210–8221.

[131] Rao, V. G.; Mandal, S.; Ghosh, S.; Banerjee, C.; Sarkar, N. *J. Phys. Chem. B* 2012, **116**, 13868–13877.

[132] Wei, J.; Su, B.; Liang, R.; Xing, H.; Bao, Z.; Yang, Q.; Yang, Y.; Ren, Q. *Colloids Surf. A* 2012, **414**, 82–87.

[133] Wei, J.; Su, B.; Liang, R.; Xing, H.; Bao, Z.; Yang, Y.; Ren, Q. *J. Chem. Eng. Data* 2012, **57**, 1274–1278.

[134] Wei, J.; Su, B.; Xing, H.; Bao, Z.; Yang, Y.; Ren, Q. *Colloids Surf. A* 2012, **396**, 213–218.

[135] Banerjee, C.; Kundu, N.; Ghosh, S.; Mandal, S.; Kuchlyan, J.; Sarkar, N. *J. Phys. Chem. B* 2013, **117**, 9508–9517.

[136] Rao, V. G.; Banerjee, C.; Ghosh, S.; Mandal, S.; Kuchlyan, J.; Sarkar, N. *J. Phys. Chem. B* 2013, **117**, 7472–7480.

[137] Rao, V. G.; Mandal, S.; Ghosh, S.; Banerjee, C.; Sarkar, N. *J. Phys. Chem. B* 2013, **117**, 1480–1493.

[138] Rojas, O.; Tiersch, B.; Rabe, C.; Stehle, R.; Hoell, A.; Arlt, B.; Koetz, J. *Langmuir* 2013, **29**, 6833–6839.

[139] Rojas, O.; Tiersch, B.; Frasca, S.; Wollenberger, U.; Koetz, J. *Colloids Surf. A* 2010, **369**, 82–87.

[140] Kunz, W.; Zemb, T.; Harrar, A. *Curr. Opin. Colloid Interface Sci.* 2012, **17**, 205–211.

[141] Xue, L.; Zhao, Y.; Yu, L.; Sun, Y.; Yan, K.; Li, Y.; Huang, X.; Qu, Y. *Colloids Surf. B* 2013, **105**, 81–86.

[142] Gayet, F.; El Kalamouni, C.; Lavedan, P.; Marty, J. D.; Brulet, A.; Lauth-de Viguerie, N. *Langmuir* 2009, **25**, 9741–9750.

[143] Yu, S.; Yan, F.; Zhang, X.; You, J.; Wu, P.; Lu, J.; Xu, Q.; Xia, X.; Ma, G. *Macromolecules* 2008, **41**, 3389–3392.

[144] Chen, Z.; Yan, F.; Qiu, L.; Lu, J.; Zhou, Y.; Chen, J.; Tang, Y.; Texter, J. *Langmuir* 2010, **26**, 3803–3806.

[145] Moniruzzaman, M.; Kamiya, N.; Goto, M. *J. Colloid Interface Sci.* 2010, **352**, 136–142.

[146] Rojas, O.; Koetz, J.; Kosmella, S.; Tiersch, B.; Wacker, P.; Kramer, M. *J. Colloid Interface Sci.* 2009, **333**, 782–790.

[147] Rabe, C.; Koetz, J. *Colloids Surf. A* 2010, **354**, 261–267.

[148] Brown, P.; Butts, C.; Dyer, R.; Eastoe, J.; Grillo, I.; Guittard, F.; Rogers, S.; Heenan, R. *Langmuir* 2011, **27**, 4563–4571.

[149] Brown, P.; Butts, C. P.; Eastoe, J.; Fermin, D.; Grillo, I.; Lee, H. C.; Parker, D.; Plana, D.; Richardson, R. M. *Langmuir* 2012, **28**, 2502–2509.

[150] Brown, P.; Butts, C. P.; Eastoe, J.; Grillo, I.; James, C.; Khan, A. *J. Colloid Interface Sci.* 2013, **395**, 185–189.

[151] Falcone, R. D.; Correa, N. M.; Silber, J. J. *Langmuir* 2009, **25**, 10426–10429.

[152] Maitra, A. *J. Phys. Chem.* 1984, **88**, 5122–5125.

[153] Li, Q.; Li, T.; Wu, J. J. *Colloid Interface Sci.* 2001, **239**, 522–527.

[154] Falcone, R. D.; Silber, J. J.; Correa, N. M. *Phys. Chem. Chem. Phys.* 2009, **11**, 11096–11100.

[155] Ferreyra, D. D.; Correa, N. M.; Silber, J. J.; Falcone, R. D. *Phys. Chem. Chem. Phys.* 2012, **14**, 3460–3470.

[156] Lungwitz, R.; Friedrich, M.; Linert, W.; Spange, S. *New J. Chem.* 2008, **32**, 1493–1499.

[157] Lungwitz, R.; Spange, S. *New J. Chem.* 2008, **32**, 392–394.

[158] Blach, D.; Silber, J. J.; Correa, N. M.; Falcone, R. D. *Phys. Chem. Chem. Phys.* 2013, **15**, 16746–16757.

[159] Reichardt, C. *Green Chem.* 2005, **7**, 339–351.

[160] Dimroth, K.; Reichardt, C.; Siepmann, T.; Bohlmann, F. *Liebigs Ann. Chem.* 1963, **661**, 1–37.

[161] Ueda, M.; Schelly, Z. A. *Langmuir* 1989, **5**, 1005–1008.

[162] Falcone, R. D.; Baruah, B.; Gaidamauskas, E.; Rithner, C. D.; Correa, N. M.; Silber, J. J.; Crans, D. C.; Levinger, N. E. *Chem. Eur. J.* 2011, **17**, 6837–6848.

[163] Wendsjö, A.; Lindgren, J.; Thomas, J. O.; Farrington, G. C. *Solid State Ion.* 1992, **53**, 1077–1082.

[164] Bernson, A.; Lindgren, J. *Solid State Ion.* 1993, **60**, 37–41.

[165] Rey, I.; Lassegues, J. C.; Grondin, J.; Servant, L. *Electrochim. Acta* 1998, **43**, 1505–1510.

[166] Abbrent, S.; Lindgren, J.; Tegenfeldt, J.; Wendsjö, A. *Electrochim. Acta* 1998, **43**, 1185–1191.

[167] McFarlane, D. R.; Sun, J.; Golding, J.; Meakin, P.; Forsyth, M. *Electrochim. Acta* 2000, **45**, 1271–1278.

[168] Coleman, S.; Byrne, R.; Minkovska, S.; Diamond, D. J. *Phys. Chem. B* 2009, **113**, 15589–15596.

[169] Suthanthiraraj, S. A.; Kumar, R.; Paul, B. J. *Ionics* 2010, **16**, 145–151.

[170] Johnston, D. H.; Shriver, D. F. *Inorg. Chem.* 1993, **32**, 1045–1047.

[171] Alía, J. M.; Díaz de Mera, Y.; Edwards, H. G. M.; García, F. J.; Lawson, E. E. *J. Mol. Struct.* 1997, **408**, 439–450.

[172] Alía, J. M.; Edwards, H. G. M. *Vib. Spectrosc.* 2000, **24**, 185–200.

[173] Mishra, S.; Ledoux, G.; Jeanneau, E.; Daniele, S.; Joubert, M.-F. *Dalton Trans.* 2012, **41**, 1490–1502.

[174] Mishra, S.; Zhang, J.; Hubert-Pfalzgraf, L. G.; Luneau, D.; Jeanneau, E. *Eur. J. Inorg. Chem.* 2007, **2007**, 602–608.

CHAPTER 15

Designing a New Strategy for the Formation of IL-In-Oil Microemulsions Containing Double Chain Surface-Active Ionic Liquid

VISHAL GOVIND RAO, CHIRANJIB BANERJEE, SURAJIT GHOSH, SARTHAK MANDAL, and NILMONI SARKAR

Department of Chemistry, Indian Institute of Technology, Kharagpur, India

15.1 INTRODUCTION TO IONIC LIQUIDS

Unlike simple inorganic salts like sodium chloride (NaCl), ionic liquids (ILs) are organic salts having low melting points (by convention below 100°C) due to the presence of sterically mismatched ions [1–7] that hinder crystal formation. Their unique properties and the rising necessity of sustainable, "green" chemistry have led to an amazing increase in interest in such salts. Room temperature ionic liquids (RTILs) are an interesting class of tunable, designer solvents with essentially zero volatility. Because the properties of the RTILs are very much dependent on the constituent ions, various RTILs can be designed using different combinations of the cationic and anionic constituents to obtain desired properties for applications. The imidazolium salts have dominated IL research ever since Wilkes et al. [8] had observed that 1,3-dialkylimidazolium electrolytes have lower melting points than their 1-alkylpyridinium counterparts. The properties of the imidazolium salts can be altered easily by varying the alkyl chain length and symmetry of the cation (leading to lowering of melting point) and by changing the anion hydrophobicity. In addition to this, 1-alkyl-3-methylimidazolium salts were the first-studied surface-active ionic liquids (SAILs), which showed aggregation in aqueous solution [9, 10]. SAILs can be defined as functional ILs with combined properties of ILs and surfactants; in other words, ILs bearing long alkyl chains having amphiphilic character are called SAILs [11, 12]. Inoue et al. [13] showed the existence of both lamellar (LR) and hexagonal (H1) lyotropic mesophases in mixtures of the SAIL, 1-dodecyl-3-methylimidazolium

Ionic Liquid-Based Surfactant Science: Formulation, Characterization, and Applications, First Edition. Edited by Bidyut K. Paul and Satya P. Moulik.

bromide with water over a concentration range of 10–50% (v/v), as found in normal surfactants. The imidazolium cations as components of ILs and SAILs are of additional interest due to their unique properties like (i) high stability due to delocalization of the positive charge over the heterocyclic moiety, (ii) presence of a proton substituent at the C-2 position enabling the imidazolium moiety to act as a catalyst by ligating to metals (the planar geometry may also be an important factor for catalyst design), and (iii) ability to provide an acidic medium appropriate for reactions (viz., Diels–Alder and Friedel–Crafts) [14–16].

15.1.1 ILs in Aqueous Microemulsions

Microemulsions are thermodynamically stable, isotropic transparent mixtures of two immiscible liquids (polar and nonpolar) and an amphiphilic component (usually surfactants and cosurfactants). The microheterogeneous environments present in reverse micelles (RMs) and microemulsions hold potential promise for applications in different fields owing to the nonstandard environments they produce. Often, these systems exhibit entirely different chemistry than that observed in homogeneous liquid solutions [17, 18]. Microemulsions are capable of solubilizing both polar and nonpolar substances and have wide applications [19, 20] in various fields such as chemical reactions [21], preparation of nanomaterials [22], and drug delivery [23].

Most studies focus on microemulsions that utilize water as the polar component in combination mostly with the anionic surfactant NaAOT (sodium 1,4-bis(2-ethylhexyl) sulfosuccinate). These studies place side by side the water properties such as polarity, viscosity, conductivity, and hydrogen bonding for bulk water and water confined in RMs [24–39]. However, in recent years, several reports on nonaqueous microemulsions have appeared in the literature [40–46]. The focus has now shifted on RTILs since they play important role in water-in-oil (w/o) microemulsions [47–56]. Lü et al. [47] investigated the critical behavior of water/1-butyl-3-methylimidazolium tetrafluoroborate [C_4mim][BF_4]/NaAOT/decane microemulsions with various concentrations of [C_4mim][BF_4] and found that the critical temperature decreases significantly with the addition of [C_4mim][BF_4]. Liu et al. showed that the amount of IL needed for curvature adjustment in ionic microemulsions was less than the needed amount of NaCl [48]. Using three different ILs, ([C_2mim][Cl], [C_4mim][Cl], and [C_8mim][Cl]), Wei et al. [51] showed that the water solubilization capacity of NaAOT in isooctane microemulsions is enhanced at low IL concentrations but decreases at high concentrations of IL. They further showed that the water solubilization capacity increases with increase of chain length of ILs at low IL concentrations [51]. In a more recent study, they showed (i) increase in percolation temperature (T_p) with the addition of ILs and (ii) increase in percolation temperature with increase in alkyl chain length of the imidazolium cation. Moreover, they also showed that the anion of the ILs has little effect on T_p of microemulsions.

15.1.2 Water-Free IL-Based Microemulsions

While the studies mentioned previously involve water as one of the component, water-free IL-based microemulsions are also studied by many groups [54, 56–68]. In the first report on formation of IL-in-oil microemulsions, Gao et al. [57] prepared [C_4mim][BF_4]/TX-100/cyclohexane microemulsions and characterized them by phase behavior, conductivity measurement, dynamic light scattering (DLS) measurement, freeze-fracturing electron microscopy, and UV–Vis technique. Eastoe et al. [58] further investigated size and shape of the same microemulsions by small-angle neutron scattering (SANS) measurement. They observed regular swelling behavior of microemulsions with the addition of the IL, which indicates that the volume of dispersed nanodomains is proportional to the amount of IL added [58]. This study was followed by several other reports on similar systems. In these investigations, [C_4mim][BF_4], TX-100, and toluene [59] or *p*-xylene [60] or benzene [56] were used as the polar phase, surfactant, and oil phase, respectively.

The water-free IL-based microemulsions with charged surfactants are rather scarce [11, 12, 62, 69–74]. Falcone et al. used cationic surfactant, benzyl-*n*-hexadecyldimethylammonium chloride (BHDC), for the formulations of [C_4mim][BF_4]/BHDC/benzene and [C_4mim][TF_2N]/BHDC/benzene microemulsions and compared their behavior with [C_4mim][BF_4]/TX-100/benzene and [C_4mim][TF_2N]/TX-100/benzene microemulsions [62]. With their recent multinuclear NMR study on the behavior of [C_4mim][BF_4] in [C_4mim][BF_4]/BHDC/benzene and [C_4mim][BF_4]/TX-100/benzene microemulsions [69], they concluded that the structure and behavior of entrapped ILs in the microemulsions depend strongly on the surfactants used [62, 69]. In their recent study, Zech et al. reported that for preparation of microemulsions with high-temperature stability and temperature insensitivity, it is better to use ionic surfactants in combination with ILs [73]. They explored thermal stability of EAN/[C_{16}mim][Cl]/dodecane microemulsions at ambient pressure, which exhibited stability over the temperature range 30–150°C [73]. Due to the interesting features of these kinds of microemulsions, spectroscopic studies are in progress [75, 76]. Water-free IL-based microemulsions also showed their applicability in the synthesis of metal nanoparticles [77, 78].

A new strategy for the formation of IL-in-oil microemulsions with charged surfactants is worth discussing here. We have shown that by replacing the inorganic cation, Na^+ of NaAOT, by an organic cation, and using different ILs as the polar core, a large number of microemulsions can be formulated. We have also used facile methods to adjust the structural parameters of IL-in-oil microemulsions.

15.2 RESULTS AND DISCUSSION

This section has two parts: in the first, we discuss the developed new strategy to prepare and characterize IL-in-oil microemulsions; in the second, we discuss the ways to adjust the structural parameters of microemulsions using different ILs as additives (polar phase).

15.2.1 Novel Strategy for the Preparation of IL-in-Oil Microemulsions

To the best of our knowledge, report of IL-in-oil (oil–benzene, cyclohexane, *n*-heptane, etc.) microemulsions using AOT or AOT-derived surfactants was absent in the literature, while water-in-oil RM and microemulsions using AOT or AOT-derived surfactants are well documented in the literature [28, 79]. It is proposed that due to the existence of favorable molecular interactions between the inorganic cations of AOT (Na^+ NH_4^+, Ca^{2+}, etc.) and water molecules, the formation of water-in-oil microemulsions is quite facile. In the case of ILs, IL-in-oil microemulsion formation is not feasible due to unfavorable interactions of ILs with the cations of AOT (Na^+ NH_4^+, Ca^{2+}, etc.). Consequently, simple inorganic salts are not soluble in ILs [77]. It appears logical that replacement of the inorganic cations of AOT (Na^+ NH_4^+, Ca^{2+}, etc.) by organic cations could lead to the formation of IL-in-oil microemulsions.

Using imidazolium (1-butyl-3-methylimidazolium, $[C_4mim]^+$) moiety as the cationic component, we have synthesized $[C_4mim][AOT]$ by an anion-exchange reaction [80] between NaAOT and 1-butyl-3-methylimidazolium bromide $[C_4mim][Br]$ [81]. We have successfully formed IL-in-oil microemulsions using $[C_4mim][AOT]$ as SAIL, benzene as nonpolar component, and different ILs as the polar component. The results are presented in Table 15.1.

From the series of ILs tabulated in Table 15.1, we have chosen $[C_4mim][BF_4]$ to further characterize the microemulsion. We have characterized the partial phase behavior of the ternary system $[C_4mim][BF_4]$/$[C_4mim][AOT]$/benzene by observing the transition from clear transparent solution to turbid solution in naked eye (Fig. 15.1a). Based on the phase diagram, a series of samples were chosen where we can

TABLE 15.1 Tabulation of Results Obtained by the Addition of ILs to $[C_4mim][AOT]$/ Benzene System

Ionic Liquid Used	Formation of Microemulsion (Yes/No)
1-Butyl-3-methylimidazolium tetrafluoroborate, $[C_4mim][BF_4]$	Yes
1-Hexyl-3-methylimidazolium tetrafluoroborate, $[C_6mim][BF_4]$	Yes
1-Butyl-3-methylimidazolium hexafluorophosphate, $[C_4mim][PF_6]$	Yes
1-Hexyl-3-methylimidazolium hexafluorophosphate, $[C_6mim][PF_6]$	Yes
1-Ethyl-3-methylimidazolium ethylsulfate, $[C_2mim][C_2SO_4]$	Yes
1-Ethyl-3-methylimidazolium *n*-butylsulfate, $[C_2mim][C_4SO_4]$	Yes
1-Ethyl-3-methylimidazolium *n*-hexylsulfate, $[C_2mim][C_6SO_4]$	Yes
1-Ethyl-3-methylimidazolium bis(trifluoromethanesulfonyl)imide, $[C_2mim][TF_2N]$	Yes
1-Butyl-3-methylimidazolium bis(trifluoromethanesulfonyl)imide, $[C_4mim][TF_2N]$	Yes
1-Hexyl-3-methylimidazolium bis(trifluoromethanesulfonyl)imide, $[C_6mim][TF_2N]$	Yes
N-Methyl-*N*-propylpyrrolidinium bis(trifluoromethanesulfonyl)imide, $[Py][TF_2N]$	Yes
N,*N*,*N*-Trimethyl-*N*-propylammonium bis(trifluoromethanesulfonyl)imide, $[N_{3111}][TF2N]$	Yes
N,*N*-Dimethylethanolammonium hexanoate, DAH	Yes
N,*N*-Dimethylethanolammonium formate, DAF	No
Ethylammonium nitrate, EAN	No
Propylammonium formate, PAF	No

Reproduced from Rao et al. [81] with permission from the ACS.

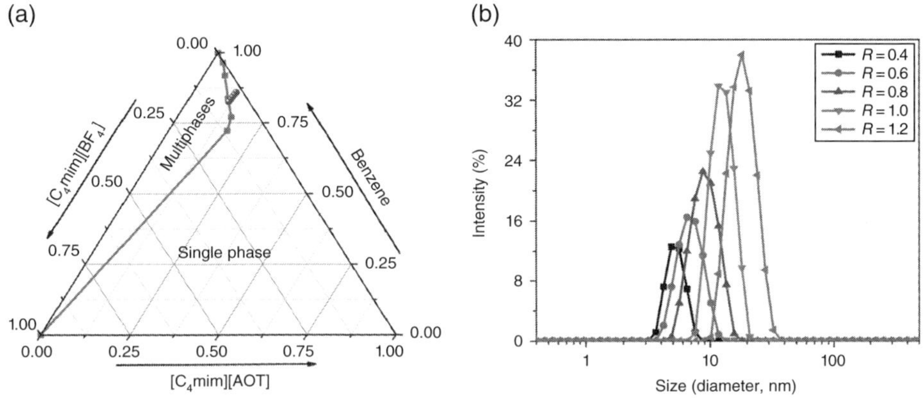

Figure 15.1 (a) Phase diagram of $[C_4mim][BF_4]/[C_4mim][AOT]$/benzene ternary system and (b) size distribution of the droplets of $[C_4mim][BF_4]/[C_4mim][AOT]$/benzene microemulsion at different R values at 298 K. Reproduced from Rao et al. [81] with permission from the ACS.

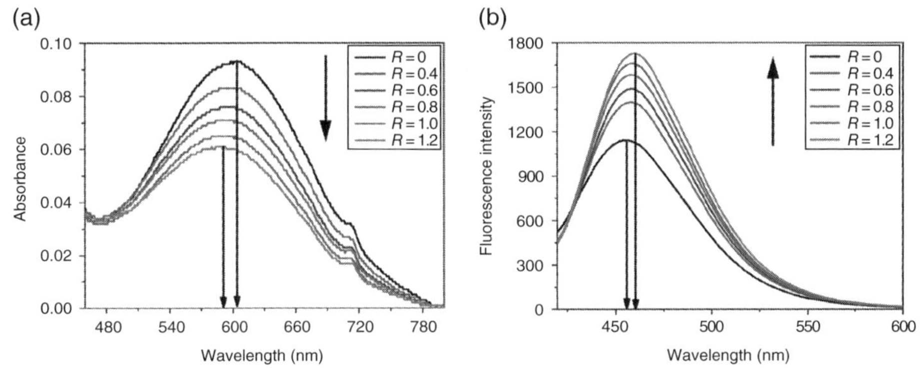

Figure 15.2 (a) Absorption spectra of betaine 30 and (b) fluorescence spectra of C-480 in $[C_4mim][BF_4]/[C_4mim][AOT]$/benzene microemulsion at different R values. Reproduced from Rao et al. [81] with permission from the ACS.

have the possibility of IL/O microemulsions only. The samples were characterized by DLS technique. Owing to the extensive studies on characterization of microemulsion using DLS, we can say that if the IL is really encapsulated to form IL-in-oil microemulsions, the droplets size must increase regularly as the amount of IL increases to a certain level [62, 68, 69, 82, 83]. The IL content of the microemulsions solution, R value, is expressed by the molar ratio of added IL to surfactant; that is, R value = (concentration of IL in molar unit)/(concentration of surfactant in molar unit). The sizes of the aggregates increased from 5.2 to 18.7 nm with increasing R values from 0.4 to 1.2 (Fig. 15.1b, here the R value is defined as the molar ratio of $[C_4mim][BF_4]$ to $[C_4mim][AOT]$).

To further support the formation of IL-in-oil microemulsion, we have utilized the typical polarity probe betaine 30 ($E_T(30)$ probe) [84–86]. For large negative solvatochromism, the $E_T(30)$ probe has been widely used by many groups to determine the polarity of different systems [84–90]. The absorption spectra of $E_T(30)$ probe are shown in Figure 15.2a, which clearly shows that the absorption maximum (λ_{max})

decreases with the addition of $[C_4mim][BF_4]$. With increasing IL content from $R=0$ to 1.2, the absorption maxima blue shifted from 603 to 590 nm. This blue shift in absorption maxima clearly indicates that with increasing R, the probe gets solubilized inside the IL pool in the microemulsion, and the micropolarity around the probe increases. We have also evaluated the $E_T(30)$ value by the use of Equation 15.1 [90]:

$$E_T(30) = hcv_{max}N_A = \frac{28,591\,\text{kcal}}{\lambda_{max}(\text{nm})\,\text{mol}} \tag{15.1}$$

In this equation, h, c, and N_A are Planck's constant, speed of light, and Avogadro's number, respectively. λ_{max} and ν_{max} are the wavelength (in nm) and frequency of the maximum absorption of the $E_T(30)$ probe, respectively. With increasing IL content from $R=0$ to 1.2, the $E_T(30)$ value changes from 47.41 to 48.46 kcal mol^{-1}. Thus, the probe experiences more polar environment. The $E_T(30)$ value of neat $[C_4mim][BF_4]$ was found to be 52.56 kcal mol^{-1}, which is well in agreement with the literature value. The results indicate that $[C_4mim][BF_4]$ in the microemulsion pool is less polar compared to the neat $[C_4mim][BF_4]$.

To further support the formation of IL pools in the IL-in-oil microemulsions, we have used coumarin-480 (C-480) as a probe molecule. It is also sensitive to the local environment, and the emission maximum increases with the polarity of the environment. In recent years, our group has used C-480 for the characterization of different IL-in-oil microemulsions [70, 83, 91]. From these studies, we can say that the red shift in the emission maximum of C-480 with gradual addition of ILs is a clear indication of IL-in-oil microemulsions. In neat benzene, C-480 shows emission maximum at 430 nm, and with the addition of $[C_4mim][AOT]$, the emission maximum gets shifted to 454 nm. This marked red shift of the emission maximum (24 nm) clearly indicates the transfer of the probe molecule from the bulk benzene to the polar interior of the $[C_4mim][AOT]$ RM. Further, addition of IL ($[C_4mim][BF_4]$) to $[C_4mim][AOT]$/benzene RM causes more red shift. With increasing IL content from $R=0$ to 1.0, the emission maximum red shifted from 454 to 459 nm (Fig. 15.2b). The gradual red shift of the emission maximum with the addition of $[C_4mim][BF_4]$ suggests the movement of the probe molecule toward the polar IL pool of the microemulsion. In neat $[C_4mim][BF_4]$, the emission maximum for C-480 at 463 nm indicates higher polarity of neat $[C_4mim][BF_4]$ compared to that of the $[C_4mim][BF_4]$-pool inside the microemulsion.

15.2.2 Methods to Adjust the Structural Parameters of IL-in-Oil Microemulsions

15.2.2.1 Phase Behavior Study For characterization of different microemulsions, we studied their phase behavior. We constructed the partial phase diagram of the ternary systems RTILs/$[C_4mim][AOT]$/benzene at 298 K by visually observing the transition from the clear transparent solution to the turbid solution.

The Influence of Alkyl Side Chain Length of Imidazolium Cation Figure 15.3a, b, and c shows phase diagrams of RTILs/$[C_4mim][AOT]$/benzene ternary systems at 298 K, where RTILs are $[C_2mim][TF_2N]$, $[C_4mim][TF_2N]$, and $[C_6mim][TF_2N]$. Interestingly, the area of the single-phase region was found to be maximum in the

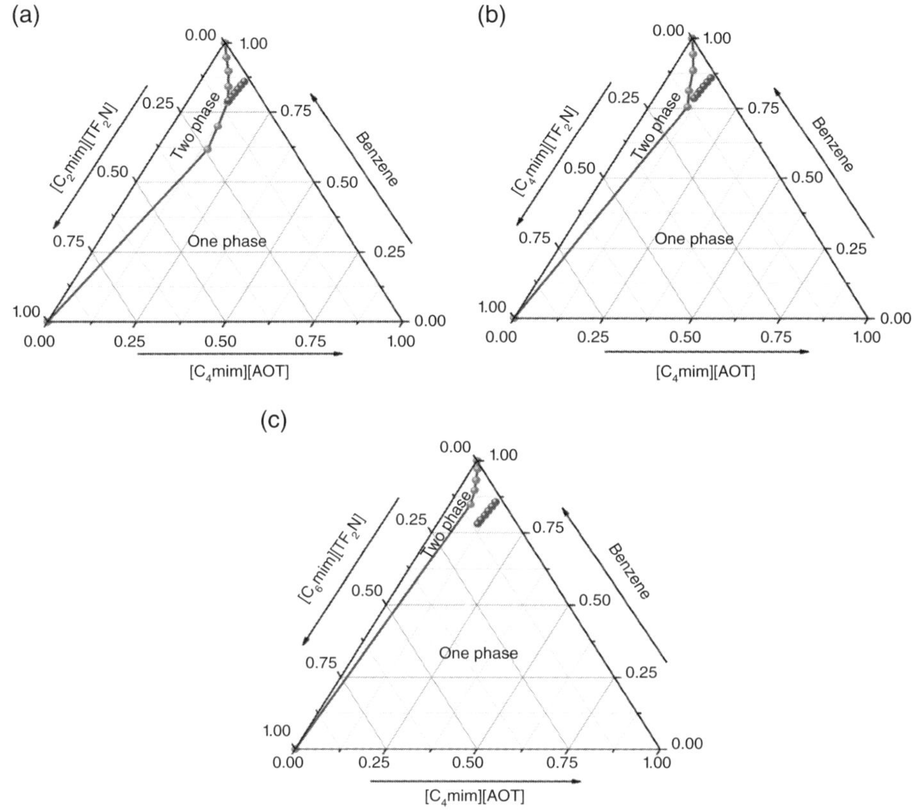

Figure 15.3 Phase diagrams of the RTILs/[C_4mim][AOT]/benzene ternary systems at 298 K, where RTILs are (a) [C_2mim][TF_2N], (b) [C_4mim][TF_2N], and (c) [C_6mim][TF_2N]. Reproduced from Rao et al. [92] with permission from the ACS.

case of [C_6mim][TF_2N]/[C_4mim][AOT]/benzene microemulsions. It is concluded that the area of the single-phase region increases with increasing alkyl chain length of cations of the RTILs.

The observed difference in the properties of the three RTILs arises due to the surface-active properties of their cations since the SAIL is negatively charged. The difference in strength of interaction between their cations and anions arises due to the presence of the hexyl chain on the 1-hexyl-3-methylimidazolium cation responsible for more penetration, because of aligning the cation with tail of the AOT^- anion. However, in the case of 1-ethyl-3-methylimidazolium cation, the ethyl chain is apparently unable to similarly function for the inability of the 1-ethyl-3-methylimidazolium cation to align itself with the tail of AOT^- anion. The difference in strength of interaction between AOT^- anion with the [C_2mim]$^+$ and [C_6mim]$^+$ cation is well supported by ^{1}H-NMR spectra (vide infra). The pronounced penetration and alignment of the cation with the tail of AOT^- anion makes the interfacial layer more rigid; therefore, the RTIL solubilization capacity increases and the area of the single-phase region increases. This increase in rigidity of the interfacial layer with the addition of ILs having long alkyl chain is well supported by Rabe and Koetz [93], and Wei et al. [52].

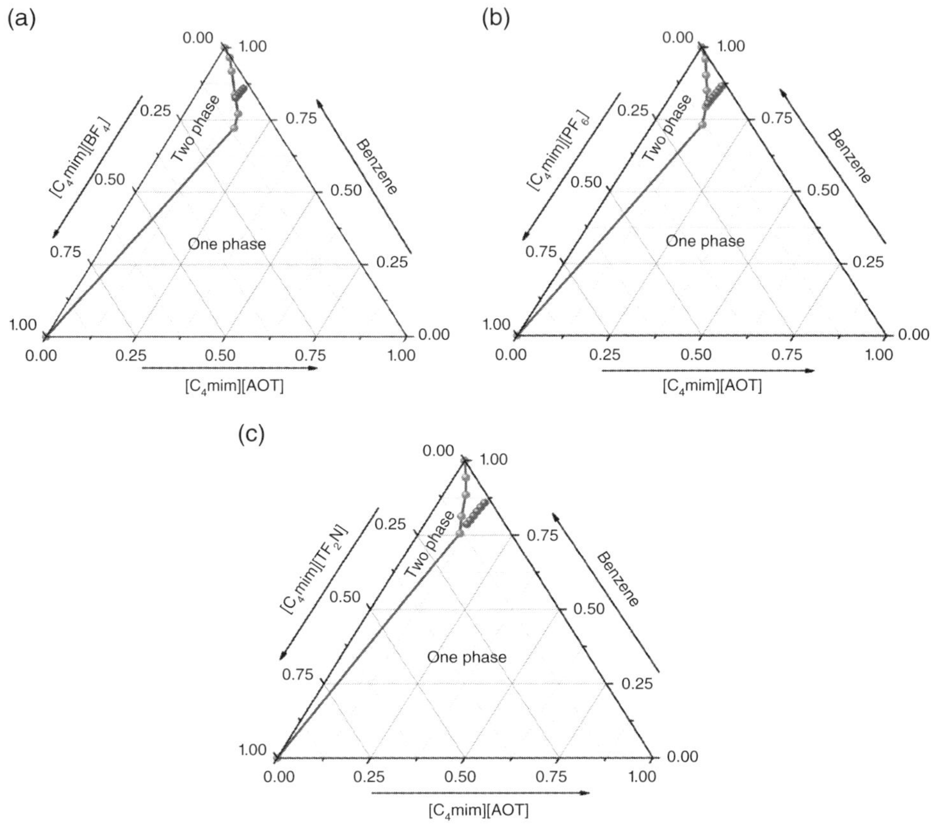

Figure 15.4 Phase diagrams of the RTILs/[C_4mim][AOT]/benzene ternary systems at 298 K, where RTILs are (a) [C_4mim][BF_4], (b) [C_4mim][PF_6], and (c) [C_4mim][TF_2N]. Reproduced from Rao et al. [92] with permission from the ACS.

The Influence of RTILs Anion Figure 15.4a, b, and c shows the phase diagrams of RTILs/[C_4mim][AOT]/benzene ternary systems at 298 K, where RTILs are [C_4mim][TF_2N], [C_4mim][PF_6], and [C_4mim][BF_4]. The area of the single-phase region was found to be maximum in the case of [C_4mim][TF_2N]/[C_4mim][AOT]/benzene microemulsion and follows the trend [C_4mim][TF_2N] > [C_4mim][PF_6] > [C_4mim][BF_4]. Although we observed some difference in the area of the single-phase region, the difference is less pronounced compared to that of IL having different cations. This is due to the absence of attractive interaction between the anion of the RTILs and the negatively charged surfaces of SAIL used for the construction of the IL-in-oil microemulsions.

The observed difference in the area of the single-phase region arises due to different cation–anion interaction strengths of the three ILs [52]. Among the three RTILs, [C_4mim][TF_2N], [C_4mim][PF_6], and [C_4mim][BF_4], TF_2N^- is the least and BF_4^- is the most interacting anion with the C_4mim^+ cation, and they follow the sequence $BF_4^- > PF_6^- > TF_2N^-$ order [94]. TF_2N^- being the least interacting anion with the C_4mim^+ cation causes strong interaction of the C_4mim^+ with the AOT^- anion, which makes the interfacial layer more rigid and allows more RTIL to dissolve inside the polar core of the microemulsions, causing increase in the single-phase region.

15.2.2.2 Size of the Aggregates in Microemulsions: DLS Measurements

Although the phase behavior study indicates formation of microemulsions of RTILs/[C_4mim][AOT]/benzene ternary systems, we need to prove that the IL is effectively dispersed by the surfactants creating IL-in-oil microemulsions. DLS measurements are widely used to determine the RTIL pool size [70,83]. For such measurements, we have used 0.2 M [C_4mim][AOT] in benzene continuum.

The Influence of Alkyl Side Chain Length of Imidazolium Cation Figure 15.5a and b shows the size distributions of RTILs/[C_4mim][AOT]/benzene ternary systems with increasing R at 298 K, where RTILs are [C_2mim][TF_2N], [C_4mim][TF_2N], and [C_6mim][TF_2N]. In all the cases, the aggregate sizes increased from 5.1, 5.4, and 5.0 nm to 44.2, 17.7, and 9.4 nm with increasing R from 0.2 to 1.2 for [C_2mim][TF_2N], [C_4mim][TF_2N], and [C_6mim][TF_2N], respectively. Almost linear variation of size at R values lower than 1.0, 1.2, and 1.6 in the case of [C_2mim][TF_2N], [C_4mim][TF_2N], and [C_6mim][TF_2N], respectively, was observed (Fig. 15.5b (inset)). The observed deviation from linearity at R values higher than 1.0, and 1.2 in the case of [C_2mim][TF_2N] and [C_4mim][TF_2N], respectively, may be attributed to the increased droplet–droplet interaction at higher R value [62]. Interestingly, we did not observe deviation from the linearity in size variation for [C_6mim][TF_2N]/[C_4mim][AOT]/benzene system suggesting that the microemulsion media consist of discrete noninteracting droplets. The difference in the behavior of three ILs can be rationalized by considering the underlying stabilizing interactions that enable the formation of IL-in-oil microemulsions. We mentioned previously that the presence of hexyl chain on 1-hexyl-3-methylimidazolium cation causes more pronounced penetration compared to 1-ethyl-3-methylimidazolium cation, because hexyl chain aids in aligning the cation with the tail part of AOT^- anion (vide supra). The pronounced penetration and alignment of the cation with the tail part of AOT^- anion reduces the effective head group area of the surfactant by screening electrostatic repulsions. This causes increase in the curvature parameter of the surfactant, making the interfacial layer more rigid in the case of [C_6mim][TF_2N]/[C_4mim][AOT]/benzene microemulsions, decreasing attractive interactions between droplets and consequently preventing coalescence of droplets. This is also well supported by observation of Wei et al. [51].

Liu et al. [95] showed that the droplet size of aggregates in microemulsions can directly bring out the effect of additives on the curvature parameter of surfactant. The increase in the droplet size of aggregates in microemulsions corresponds to decrease in the curvature parameter of surfactant.

The Influence of RTILs Anion Figure 15.6a and b shows the size distributions and variation of size of RTILs/[C_4mim][AOT]/benzene ternary systems with increasing R at 298 K. The sizes of the aggregates increased from 5.4, 4.9, and 4.3 nm to 17.7, 21.0, and 18.7 nm with increasing R from 0.2 to 1.2 in the case of [C_4mim][TF_2N], [C_4mim][PF_6], and [C_4mim][BF_4], respectively. We observed almost linear variations of size at R values lower than 1.0 in all the cases (Fig. 15.6b (inset)). Unlike ILs where cations of ILs differ in alkyl chain length, here, the size increase is almost the same with increase in R value. This difference arises due to the absence of attractive interaction

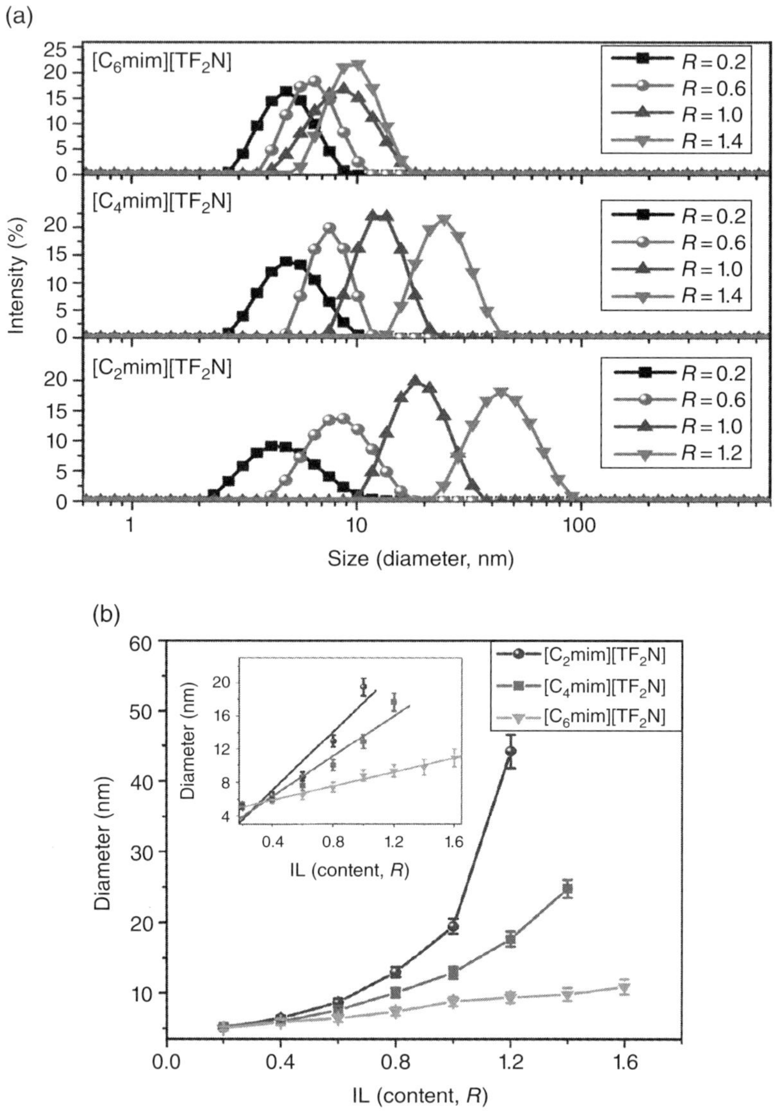

Figure 15.5 (a) Size distribution of the droplets (measured by dynamic light scattering) of RTILs/[C_4mim][AOT]/benzene microemulsions at different *R* values. (b) Diameter of the droplets as a function of RTILs concentration (*R* value). The RTILs are [C_2mim][TF_2N], [C_4mim][TF_2N], and [C_6mim][TF_2N]. Reproduced from Rao et al. [92] with permission from the ACS.

between anion of the RTILs and the negatively charged surfaces of SAIL used for the construction of IL-in-oil microemulsions (vide supra).

15.2.2.3 The Influence of Temperature on the Size of the Aggregates in Microemulsions We investigated the effect of temperature on the size of the aggregates in RTILs/[C_4mim][AOT]/benzene microemulsions at $R = 1.0$ (Fig. 15.7a and b).

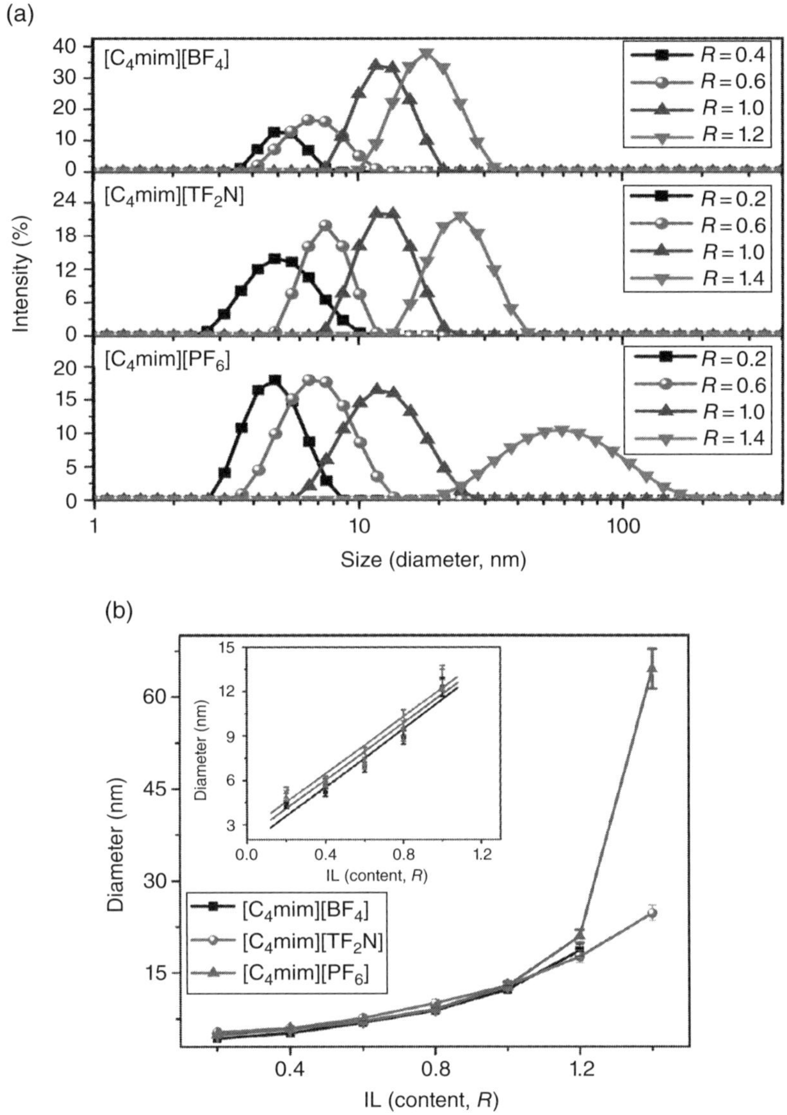

Figure 15.6 (a) Size distribution of the droplets (measured by dynamic light scattering) of RTILs/[C_4mim][AOT]/benzene microemulsions at different *R* values. (b) Diameter of the droplets as a function of RTILs concentration (*R* value). The RTILs are [C_4mim][BF_4], [C_4mim][PF_6], and [C_4mim][TF_2N]. Reproduced from Rao et al. [92] with permission from the ACS.

It is evident from the DLS measurements that the microemulsions are monodisperse and the size of aggregates in all the microemulsions decreases with increasing temperature. This indicates the noninteracting hard sphere nature of the aggregates in RTILs/[C_4mim][AOT]/benzene microemulsions at $R = 1.0$ without droplet coalescing [62, 92, 96]. The microemulsions retain their structural integrity across the temperature range used in the study. The temperature effect on the

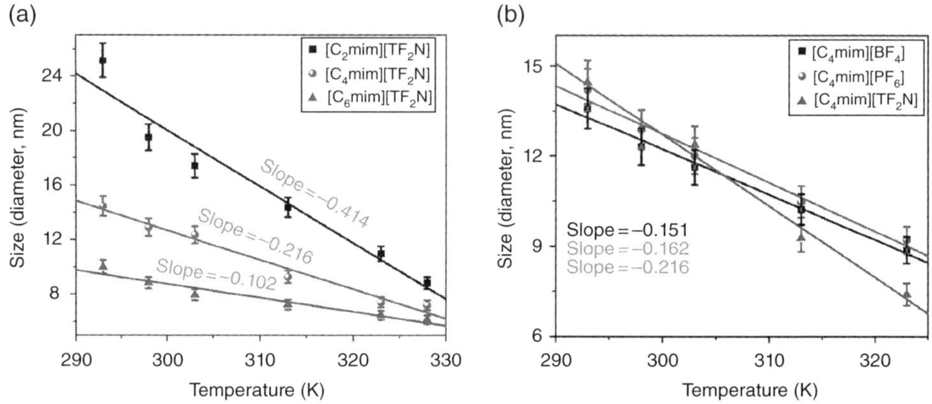

Figure 15.7 Diameter of the droplets of RTILs/[C_4mim][AOT]/benzene microemulsion ($R = 1.0$) as a function of temperature. (a) The RTILs are [C_2mim][TF_2N], [C_4mim][TF_2N], and [C_6mim][TF_2N]. (b) The RTILs are [C_4mim][BF_4], [C_4mim][PF_6], and [C_4mim][TF_2N]. Reproduced from Rao et al. [92] with permission from the ACS.

aggregate size corroborates our earlier reports on the [C_4mim][BF_4]/BHDC/benzene and [Py][TF_2N]/[C_4mim][AOT]/benzene aggregates in microemulsions [68, 92].

At $R = 1.0$, aggregate size decreased from 25.2, 14.5, and 10.0 to 8.8, 7.2, and 6.1 nm with increasing temperature from 293 to 328 K in the case of [C_2mim][TF_2N], [C_4mim][TF_2N], and [C_6mim][TF_2N], respectively. Figure 15.7a clearly indicates that the slopes of the linear fits follow the trend [C_2mim][TF_2N] > [C_4mim][TF_2N] > [C_6mim][TF_2N]. The observed differences arise from the differences in the rigidity of the microemulsions. The [C_6mim][TF_2N]/[C_4mim][AOT]/benzene microemulsion being more rigid in nature (vide supra) shows less temperature dependence compared to the [C_2mim][TF_2N]/[C_4mim][AOT]/benzene microemulsion.

On the other hand, the sizes of the aggregates at $R = 1.0$ decrease from 14.5, 14.2, and 13.6 to 7.4, 9.2, and 8.9 nm with increasing temperature from 293 to 323 K in the case of [C_4mim][TF_2N], [C_4mim][PF_6], and [C_4mim][BF_4], respectively. Almost linear variations of size with temperature but the slope of linear fit almost remains same (Fig. 15.7b). So the observed uniform decrease in size herein observed arises from the nearly same rigidity of microemulsions.

15.2.2.4 The Influence of Water Addition on the Size of the Aggregates in Microemulsions

The Influence of Alkyl Side Chain Length of Imidazolium Cation Considering the extremely limited solubility of [C_2mim][TF_2N], [C_4mim][TF_2N], and [C_6mim][TF_2N] [97], it is interesting to investigate the effect of water addition on the behavior of the microemulsions. It is found from DLS measurements that the microemulsions are monodisperse in nature and the size of the aggregates increases with increasing water content (Fig. 15.8a).

The sizes of the aggregates at $R = 1.0$ (at 298 K) increases from 19.5, 12.9, and 8.8 to 78.3, 21.4, and 12.5 nm with increasing water content from 0 to 300 mM for [C_2mim][TF_2N], [C_4mim][TF_2N], and [C_6mim][TF_2N], respectively. The increase in size with the addition of 300 mM of water is maximum in the case of [C_2mim][TF_2N]/[C_4mim]

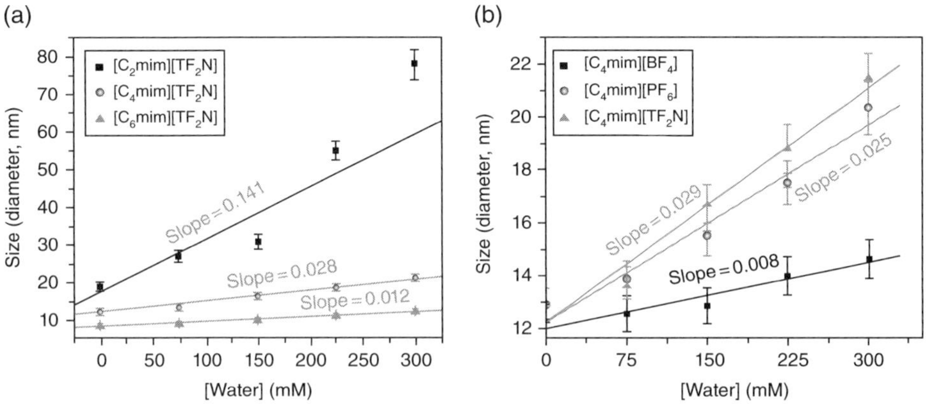

Figure 15.8 Diameter of the droplets of RTILs/$[C_4mim][AOT]$/benzene microemulsion ($R = 1.0$) as a function of water content. (a) The RTILs are $[C_2mim][TF_2N]$, $[C_4mim][TF_2N]$, and $[C_6mim][TF_2N]$. (b) The RTILs are $[C_4mim][BF_4]$, $[C_4mim][PF_6]$, and $[C_4mim][TF_2N]$. Reproduced from Rao et al. [92] with permission from the ACS.

[AOT]/benzene microemulsion and follows the trend $[C_2mim][TF_2N] > [C_4mim][TF_2N] > [C_6mim][TF_2N]$. Consequently, the decrease in the curvature parameter with the addition of water in $[C_nmim][TF_2N]/[C_4mim][AOT]$/benzene microemulsion follows the trend $[C_2mim][TF_2N] > [C_4mim][TF_2N] > [C_6mim][TF_2N]$. The maximum decrease in curvature parameter was observed in the case of $[C_2mim][TF_2N]/[C_4mim][AOT]$/benzene microemulsions, which is also supported by ^{1}H-NMR spectra (vide infra). The observed difference arises from the water insolubility of the $[C_nmim][TF_2N]$ ILs. The added water in $[C_nmim][TF_2N]$ locates itself in the interfacial region of the microemulsion and replaces some of the $[C_nmim]^+$ cation and decreases the curvature of the surfactant. The replacement process of $[C_nmim]^+$ with added water molecules is more efficient in the case of $[C_2mim][TF_2N]/[C_4mim][AOT]$/benzene microemulsion owing to relatively weak interaction between $[C_2mim]^+$ and AOT^- compared to $[C_6mim]^+$ and AOT^-, which causes increase in the effective surface area and decrease in the curvature parameter. This observation is well supported by Wei et al. [51]; they showed that the long-chain IL makes the interfacial layer more rigid compared to that of the small-chain IL (vide supra).

The Influence of RTILs Anion It may be interesting to compare the effect of water addition in different microemulsions containing IL of different hydrophobicity that follows the trend $[C_4mim][TF_2N] > [C_4mim][PF_6] > [C_4mim][BF_4]$ [97]. The sizes of the aggregates at $R = 1.0$ (at 298 K) increase from 12.9, 12.9, and 12.3 to 21.4, 20.3, and 14.6 nm with increasing water content from 0 to 300 mM in the case of $[C_4mim][TF_2N]$, $[C_4mim][PF_6]$, and $[C_4mim][BF_4]$, respectively (Fig. 15.8b). The increase in size with the addition of 300 mM of water was found to be maximum in the case of $[C_4mim][TF_2N]/[C_4mim][AOT]$/benzene microemulsion and follows the trend $[C_4mim][TF_2N] > [C_4mim][PF_6] > [C_4mim][BF_4]$. Thus, the increase in size of aggregates in microemulsions with the addition of water follows exactly the same trend as the hydrophobicity of the IL. The added water molecules being insoluble in $[C_4mim][TF_2N]$ and $[C_4mim][PF_6]$ locate themselves in the interfacial region of the microemulsions and replace some of the $[C_4mim]^+$ cation, which causes increase in the effective surface area and decrease in curvature parameter. The decrease in curvature parameter is manifested by increase in size. In contrast to this, the added water molecules are highly soluble in the polar core ($[C_4mim][BF_4]$) of the $[C_4mim][BF_4]/[C_4mim][AOT]$/benzene microemulsion; thus, the surface area or the curvature parameter remains unchanged and the size increase is minimum.

15.2.3 1H NMR Spectra

1H NMR spectra are taken to provide more detailed information on the intra/intermolecular interactions and thus provide the microstructure characteristics of the IL-in-oil microemulsions including the locations of the RTILs in RTILs/$[C_4mim][AOT]$/benzene microemulsions. For all the measurements, we used 0.2 M $[C_4mim][AOT]$ in benzene. First, we compare the proton signal associated with the imidazolium ring of $[C_4mim][AOT]$ in $[C_4mim][AOT]$/benzene RM and in neat $[C_2mim][TF_2N]$ and $[C_6mim][AOT]$ (Fig. 15.9a and b, Scheme 15.1 shows atom numbering of the respective compounds). The H2 signal in the case of $[C_4mim][AOT]$ appeared at significantly downfield position ($\delta = 9.356$ ppm) compared to that of H2′ of $[C_2mim][TF_2N]$ ($\delta = 8.461$ ppm) and H2″ of $[C_6mim][TF_2N]$ ($\delta = 8.548$ ppm). In contrast to this, the other two protons of the imidazolium ring also shifted to downfield position but the shift is relatively small compared to that of H2 protons. This can be accounted for by the acidity of H2 hydrogen atom, which makes it most sensitive to its environment [69, 96].

Figures 15.9a and b and 15.10a and b show a part of 1H NMR spectra of neat ILs and RTILs/$[C_4mim][AOT]$/benzene microemulsions at different R value. With the addition of IL, $[C_2mim][TF_2N]$ in $[C_4mim][AOT]$/benzene RM, the H2′ proton of $[C_2mim][TF_2N]$ gets significantly shifted downfield compared to the neat $[C_2mim][TF_2N]$ ($\delta = 8.461$ ppm) (Fig. 15.9a). In $[C_2mim][TF_2N]/[C_4mim][AOT]$/benzene microemulsions (at $R = 0.2$), the signal associated with the H2′ proton of $[C_2mim][TF_2N]$ appears at $\delta = 9.356$ ppm, which is 0.895 ppm shifted downfield compared to neat $[C_2mim][TF_2N]$. This huge downfield shift is well supported by many other studies and confirms the incorporation of IL inside the RM [69, 74]. With further addition of IL, $[C_2mim][TF_2N]$, that is, with increasing R value from 0.2 to 1.0, the peaks associated with the H2′ proton of $[C_2mim][TF_2N]$ and H2 proton of $[C_4mim][AOT]$ gets shifted upfield from 9.150 and 9.239 ppm to 8.754 and 8.858 ppm, respectively (Fig. 15.11a). This observation is also supported by Falcone et al. [69] and Ferreyra et al. [74]. Even though the H2′ proton signal of $[C_2mim][TF_2N]$ in microemulsions shifted upfield, toward the neat IL with increasing R value, it remains substantially different from neat $[C_2mim][TF_2N]$. This once again indicates a new environment for $[C_2mim][TF_2N]$. A similar behavior is observed in the case of $[C_4mim][TF_2N]$ and $[C_6mim][TF_2N]$.

We also monitored the signals associated with the surfactant, $[C_4mim][AOT]$ molecule, because the surfactant's protons were used to probe the microenvironment created in the microemulsions [69, 74]. We observed a modest but observable chemical shift change, in the signals associated with the surfactant, $[C_4mim][AOT]$ molecule, which demonstrated significant changes in structural organization [69, 74]. Figures 15.9 and 15.10 show the variation of 1H NMR chemical shift of the proton nearest to the polar head group of RTIL/$[C_4mim][AOT]$/benzene microemulsions studied at different R values. The H1 peak of the anion of the $[C_4mim][AOT]$, which

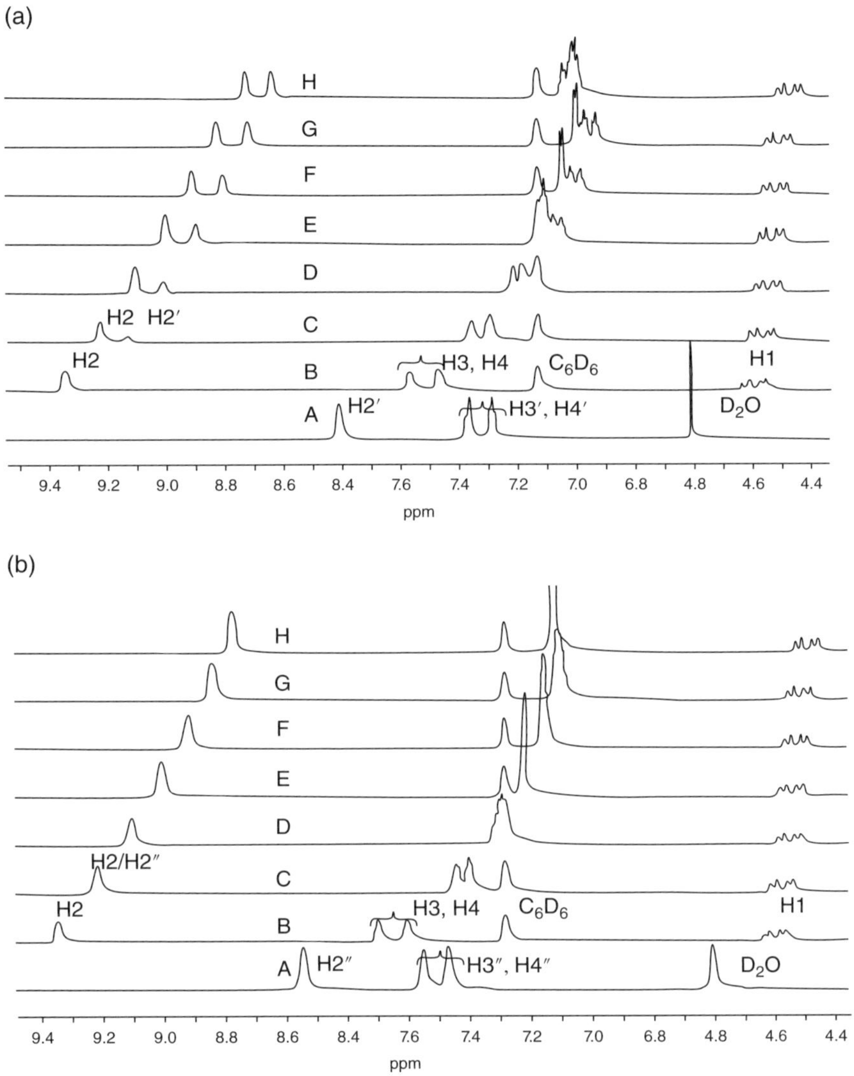

Figure 15.9 Partial 1H NMR spectra of (a) neat [C_2mim][TF_2N] (A) and [C_2mim][TF_2N]/[C_4mim][AOT]/benzene microemulsions at $R=0$ (B), 0.2 (C), 0.4 (D), 0.6 (E), 0.8 (F), and 1.0 (G) and [C_2mim][TF_2N]/[C_4mim][AOT]/benzene microemulsions at $R=1.0$ containing 300 mM of water (H) and (b) neat [C_6mim][TF_2N] (A) and [C_6mim][TF_2N]/[C_4mim][AOT]/benzene microemulsions at $R=0$ (B), 0.2 (C), 0.4 (D), 0.6 (E), 0.8 (F), and 1.0 (G) and [C_6mim][TF_2N]/[C_4mim][AOT]/benzene microemulsions at $R=1.0$ containing 300 mM of water (H). Reproduced from Rao et al. [92] with permission from the ACS.

shows doublet of doublet, displays upfield shifts as R value increases (in the case of all the ILs studied). This upfield shift of the surfactant molecule with the addition of ILs is well supported by Falcone et al. [69] and Ferreyra et al. [74] and suggests the interaction of added ILs with the surfactant molecules. It is noteworthy to mention that with the addition of ILs, we observed significantly larger shifts for the [C_4mim]

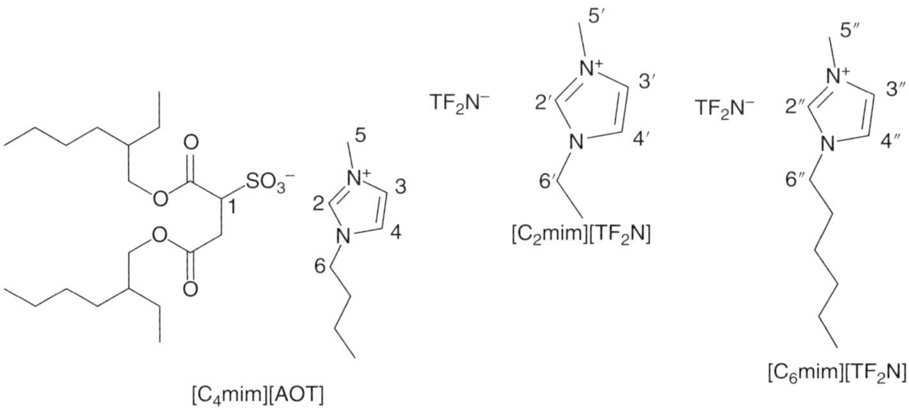

Scheme 15.1 The chemical structure and atom numbering of [C_4mim][AOT], [C_2mim][TF_2N], and [C_6mim][TF_2N].

[AOT] surfactant signals arising from protons, closer to the head group of [C_4mim][AOT] compared to that of the long alkyl chains (negligible change is observed). These observations once again suggest the incorporation of ILs in the polar core of the microemulsions, where the ILs interact specifically with the head group of the surfactant molecules.

We mentioned earlier that the cations of all the ILs interact with the anionic head group of the SAIL, [C_4mim][AOT], due to the attractive cation–anion interaction and the difference in the strength of interaction arises due to the presence of different length of alkyl chains of the added ILs cation ([C_2mim]$^+$, [C_4mim]$^+$, and [C_6mim]$^+$). Thus, depending on the strength of the interaction, we expect difference in the shift of peak positions. In reality, we observed the same. On going from neat [C_2mim][TF_2N] to [C_2mim][TF_2N]/[C_4mim][AOT]/benzene microemulsions at $R=0.2$, the chemical shift value of H5′ proton changes from 3.805 to 3.709 ppm, whereas on going from neat [C_6mim][TF_2N] to [C_6mim][TF_2N]/[C_4mim][AOT]/benzene microemulsions at $R=0.2$, the chemical shift value of H5″ proton changes from 3.848 to 3.695 ppm (Fig. 15.10a and b). So, upfield shifts of 0.096 and 0.153 ppm were observed with the incorporation of [C_2mim][TF_2N] and [C_6mim][TF_2N] in microemulsions (Fig. 15.11a, b, and c). The relatively larger change in the case of [C_6mim][TF_2N] incorporation confirms the stronger interaction with the surfactant.

15.2.3.1 Effect of Water Addition 1H NMR spectroscopic analysis was performed to get detailed information on the interaction between the added water molecules with the surfactant and IL present in the microemulsions and the insight into the solubilization information of the added water [54]. A relatively larger shift in all the NMR peaks were observed with the addition of water to [C_2mim][TF_2N]/[C_4mim][AOT]/benzene microemulsions compared to [C_6mim][TF_2N]/[C_4mim][AOT]/benzene microemulsions (Figs. 15.9 and 15.10). With the addition of water (final [water] = 300 mM) in the [C_2mim][TF_2N]/[C_4mim][AOT]/benzene microemulsions (at $R=1.0$), the chemical shift value of H5 proton of the surfactant changes from 3.534 to 3.590 ppm, whereas in the case of [C_6mim][TF_2N]/[C_4mim][AOT]/benzene microemulsions (at $R=1.0$), it changes from 3.481 to 3.515 ppm. The downfield

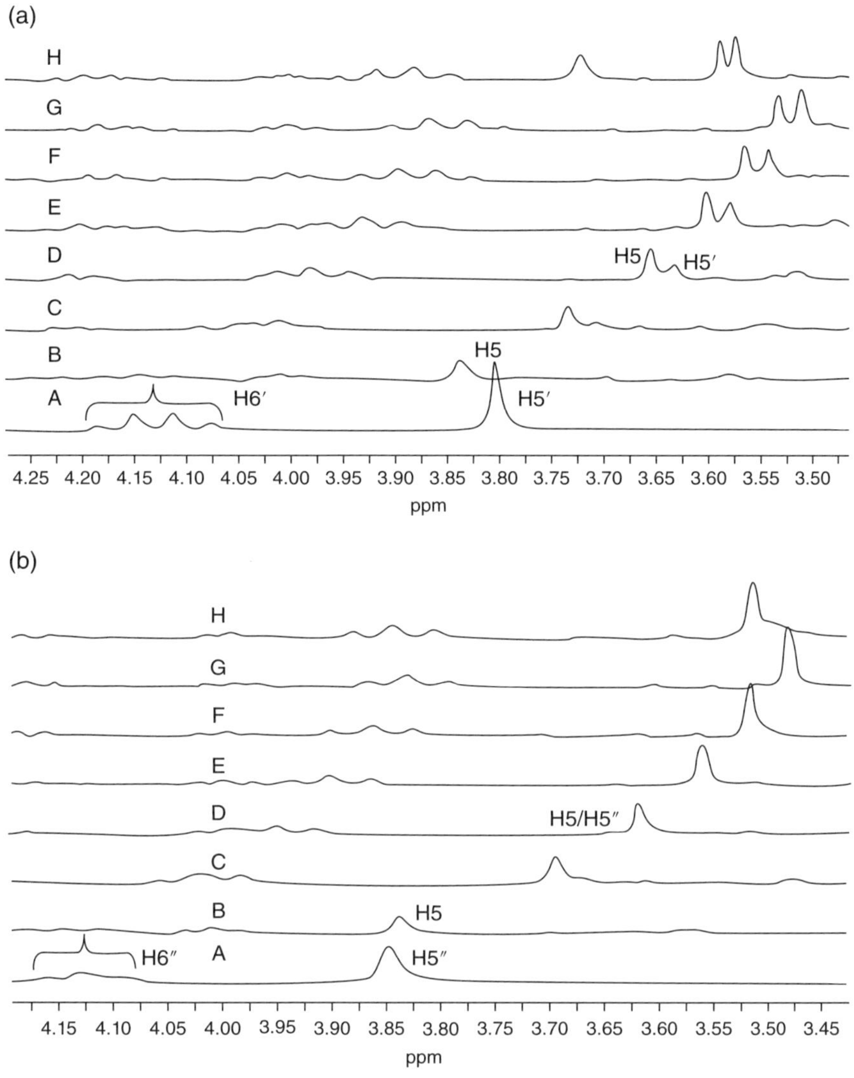

Figure 15.10 Partial 1H NMR spectra showing variation of H5/H5″ and H6′/H6″ (a) neat [C_2mim][TF_2N] (A) and [C_2mim][TF_2N]/[C_4mim][AOT]/benzene microemulsions at $R=0$ (B), 0.2 (C), 0.4 (D), 0.6 (E), 0.8 (F), and 1.0 (G) and [C_2mim][TF_2N]/[C_4mim][AOT]/benzene microemulsions at $R=1.0$ containing 300 mM of water (H) and (b) neat [C_6mim][TF_2N] (A) and [C_6mim][TF_2N]/[C_4mim][AOT]/benzene microemulsions at $R=0$ (B), 0.2 (C), 0.4 (D), 0.6 (E), 0.8 (F), and 1.0 (G) and [C_6mim][TF_2N]/[C_4mim][AOT]/benzene microemulsions at $R=1.0$ containing 300 mM of water (H). Reproduced from Rao et al. [92] with permission from the ACS.

shifts are 0.056 and 0.034 ppm (Fig. 15.10a and b). Similar behavior is observed when we consider the H5′ and H5″ peak of [C_2mim][TF_2N] and [C_6mim][TF_2N], respectively; the downfield shifts of 0.061 and 0.034 ppm, respectively, were observed. In addition to this, the quartet of H6′ proton of [C_2mim][TF_2N] and triplet of H6″ proton of [C_6mim][TF_2N] get downfield shifts of 0.053 and 0.013 ppm, respectively.

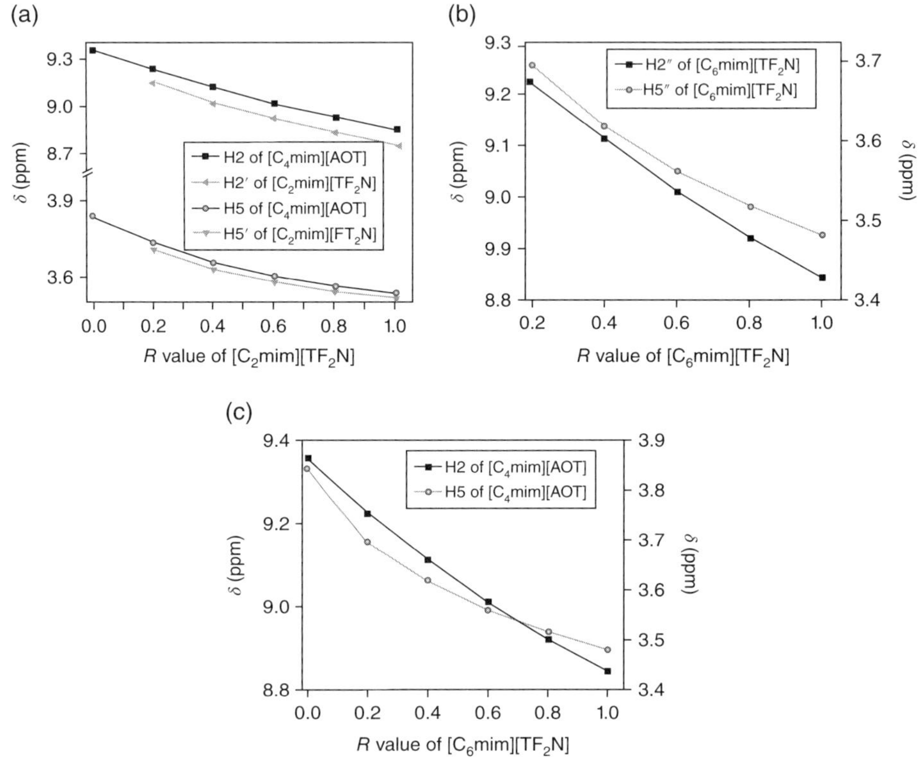

Figure 15.11 Variation of ^{1}H chemical shifts of (a) H2 and H5 of [C_4mim][AOT] and H2′ and H5″ of [C_2mim][TF_2N] in [C_2mim][TF_2N]/[C_4mim][AOT]/benzene microemulsions with increasing *R* value of [C_2mim][TF_2N], (b) H2″ and H5″ of [C_6mim][TF_2N], and (c) H2 and H5 of [C_4mim][AOT] in [C_6mim][TF_2N]/[C_4mim][AOT]/benzene microemulsions with increasing *R* value of [C_6mim][TF_2N]. Reproduced from Rao et al. [92] with permission from the ACS.

So the relatively larger change in the NMR peak position of both surfactant and IL with the addition of water to [C_2mim][TF_2N]/[C_4mim][AOT]/benzene microemulsions compared to the [C_6mim][TF_2N]/[C_4mim][AOT]/benzene microemulsions confirms the larger perturbation in the former. It is noteworthy to mention that with the addition of water, significantly larger shifts for the [C_4mim][AOT] surfactant signals arising from protons closer to the head group of [C_4mim][AOT] compared to that of the long alkyl chains are observed. The results support the location of water molecules in the interfacial region of the microemulsions.

15.3 CONCLUSION

The study presented here describes the formation and characterization of different IL-in-oil microemulsions containing an anionic SAIL, [C_4mim][AOT]. This work opens up the possibility of creating a large number of IL-in-oil microemulsions, just by replacing the inorganic cation, Na^+ of NaAOT, by any organic cation and using different IL as the polar core. It clearly provides different ways to tune the structure of microemulsions, which in turn can provide different routes to alter the size of the prepared nanoparticles/polymers and provide environment for performing organic reactions.

ACKNOWLEDGMENTS

Nilmoni Sarkar is indebted to the Department of Science and Technology (DST) and the Council of Scientific and Industrial Research (CSIR), Government of India, for generous research grants. Vishal Govind Rao, Surajit Ghosh, and Sarthak Mandal are thankful to CSIR, and Chiranjib Banerjee is thankful to UGC for research fellowships. All figures are reprinted with kind permission from the American Chemical Society (ACS).

REFERENCES

[1] Seddon, K. R. *Nature* 2003, **2**, 363–364.
[2] Welton, T. *Chem. Rev.* 1999, **99**, 2071–2084.
[3] Zhao, D.; Wu, M.; Kou, Y.; Min, E. *Catal. Today* 2002, **74**, 157–189.
[4] Parvulescu, V. I.; Hardacre, C. *Chem. Rev.* 2007, **107**, 2615–2665.
[5] Haumann, M.; Riisager, A. *Chem. Rev.* 2008, **108**, 1474–1497.
[6] Plechkova, N. V.; Seddon, K. R. *Chem. Soc. Rev.* 2008, **37**, 123–150.
[7] Rantwijk, F.; van Sheldon, R. A. *Chem. Rev.* 2007, **107**, 2757–2785.
[8] Wilkes, J. S.; Levisky, J. A.; Hussey, C. L.; Druelinger, M. L. *Proc. Int. Symp. Molten Salts* 1980, **81**, 245–255.
[9] Firestone, M. A.; Dzielawa, J. A.; Zapol, P.; Curtiss, L. A.; Seifert, S.; Dietz, M. L. *Langmuir* 2002, **18**, 7258–7260.
[10] Bowers, J.; Butts, C. P.; Martin, P. J.; Vergara-Gutierrez, M. C.; Heenan, R. K. *Langmuir* 2004, **20**, 2191–2198.
[11] El Seoud, O. A.; Pires, P. A. R.; Abdel-Moghny, T.; Bastos, E. L. *J. Colloid Interface Sci.* 2007, **313**, 296–304.
[12] Galgano, P. D.; El Seoud, O. A. *J. Colloid Interface Sci.* 2010, **345**, 1–11.
[13] Inoue, T.; Dong, B.; Zheng, L. Q. *J. Colloid Interface Sci.* 2007, **307**, 578–581.
[14] Yeung, K. S.; Farkas, M. E.; Qiu, Z.; Yang, Z. *Tetrahedron Lett.* 2002, **43**, 5793–5795.
[15] Janus, E.; Goc-Maciejewska, I.; Lozynski, M.; Pernak, J. *Tetrahedron Lett.* 2006, **47**, 4079–4081.
[16] Lee, C. K.; Huang, H. W.; Lin, I. J. B. *Chem. Commun.* 2000, 1911–1912.
[17] Dwars, T.; Paetzold, E.; Oehme, G. *Angew. Chem.* 2005, **117**, 7338–7364.
[18] Crans, D. C.; Trujillo, A. M.; Bonetti, S.; Rithner, C. D.; Baruah, B.; Levinger, N. E. *J. Org. Chem.* 2008, **73**, 9633–9640.
[19] Eastoe, J.; Gold, S.; Rogers, S. E.; Wyatt, P.; Steytler, D. C.; Gurgel, A.; Heenan, R. K.; Fan, X.; Beckman, E. J.; Enick, R. M. *Angew. Chem. Int. Ed.* 2006, **45**, 3675–3677.
[20] Liu, Y.; Jessop, P. G.; Cunningham, M.; Eckert, C. A.; Liotta, C. L. *Science* 2006, **313**, 958–960.
[21] Wagner, G. W.; Procell, L. R.; Yang, Y.-C.; Bunton, C. A. *Langmuir* 2001, **17**, 4809–4811.

[22] Bonini, M.; Bardi, U.; Berti, D.; Neto, C.; Baglioni, P. *J. Phys. Chem. B* 2002, **106**, 6178–6183.

[23] Lv, F. F.; Zheng, L. Q.; Tung, C. *Int. J. Pharm.* 2005, **301**, 237–246.

[24] Altamirano, M. S.; Borsarelli, C. D.; Cosa, J. J.; Previtali, C. M. *J. Colloid Interface Sci.* 1998, **205**, 390–396.

[25] Hasegawa, M.; Sugimura, T.; Suzaki, Y.; Kitahara, A. *J. Phys. Chem.* 1994, **98**, 2120–2124.

[26] Correa, N. M.; Biasutti, M. A.; Silber, J. J. *J. Colloid Interface Sci.* 1996, **184**, 570–578.

[27] Sedgwick, M.; Cole, R. L.; Rithner, C. D.; Crans, D. C.; Levinger, N. E. *J. Am. Chem. Soc.* 2012, **134**, 11904–11907.

[28] Riter, R. E.; Undiks, E. P.; Levinger, N. E. *J. Am. Chem. Soc.* 1998, **120**, 6062–6067.

[29] Fathi, H.; Kelly, J. P.; Vasquez, V. R.; Graeve, O. A. *Langmuir* 2012, **28**, 9267–9274.

[30] Horn, W. D. V.; Ogilvie, M. E.; Flynn, P. F. *J. Am. Chem. Soc.* 2009, **131**, 8030–8039.

[31] Biswas, R.; Chakraborti, T.; Bagchi, B.; Ayappa, K. G. *J. Chem. Phys.* 2012, **137**, 014515–014523.

[32] Bagchi, B. *Chem. Rev.* 2005, **105**, 3197–3219.

[33] Mandal, D.; Datta, A.; Pal, S. K.; Bhattacharyya, K. *J. Phys. Chem. B* 1998, **102**, 9070–9073.

[34] Dutta, P.; Sen, P.; Mukherjee, S.; Halder, A.; Bhattacharyya, K. *J. Phys. Chem. B* 2003, **107**, 10815–10822.

[35] Agazzi, F. M.; Falcone, R. D.; Silber, J. J.; Correa, N. M. *J. Phys. Chem. B* 2011, **115**, 12076–12084.

[36] Nave, S.; Eastoe, J.; Heenan, R. K.; Steytler, D.; Grillo, I. *Langmuir* 2000, **16**, 8741–8748.

[37] Angelico, R.; Ceglie, A.; Colafemmina, G.; Delfine, F.; Olsson, U.; Palazzo, G. *Langmuir* 2004, **20**, 619–631.

[38] Piletic, I. R.; Moilanen, D. E.; Levinger, N. E.; Fayer, M. D. *J. Am. Chem. Soc.* 2006, **128**, 10366–10367.

[39] Baruah, B.; Roden, J. M.; Sedgwick, M.; Correa, N. M.; Crans, D. C.; Levinger, N. E. *J. Am. Chem. Soc.* 2006, **128**, 12758–12765.

[40] Das, K. P.; Ceglie, A.; Lindman, B. *J. Phys. Chem.* 1987, **91**, 2938–2946.

[41] Ray, S.; Moulik, S. P. *Langmuir* 1994, **10**, 2511–2515.

[42] Shirota, H.; Horie, K. *J. Phys. Chem. B* 1999, **103**, 1437–1443.

[43] Shirota, H.; Segawa, H. *Langmuir* 2004, **20**, 329–335.

[44] Riter, R. E.; Kimmel, J. R.; Undiks, E. P.; Levinger, N. E. *J. Phys. Chem. B* 1997, **101**, 8292–8297.

[45] Novaira, M.; Moyano, F.; Biasutti, M. A.; Silber, J. J.; Correa, N. M. *Langmuir* 2008, **24**, 4637–4646.

[46] Silber, J. J.; Falcone, R. D.; Correa, N. M.; Biasutti, M. A.; Abuin, E.; Lissi, E.; Campodonico, P. *Langmuir* 2003, **19**, 2067–2071.

[47] Lü, H. H.; An, X. Q.; Shen, W. G. *J. Chem. Eng. Data* 2008, **53**, 727–731.

[48] Liu, L. P.; Bauduin, P.; Zemb, T.; Eastoe, J.; Hao, J. C. *Langmuir* 2009, **25**, 2055–2059.

[49] Rojas, O.; Koetz, J.; Kosmella, S.; Tiersch, B.; Wacker, P.; Kramer, M. *J. Colloid Interface Sci.* 2009, **333**, 782–790.

[50] Ranke, J.; Stolte, S.; Stormann, R.; Arning, J.; Jastorff, B. *Chem. Rev.* 2007, **107**, 2183–2206.

[51] Wei, J.; Su, B.; Yang, J.; Xing, H.; Bao, Z.; Yang, Y.; Ren, Q. *J. Chem. Eng. Data* 2011, **56**, 3698–3702.

[52] Wei, J.; Su, B.; Xing, H.; Bao, Z.; Yang, Y.; Ren, Q. *Colloids Surf. A* 2012, **396**, 213–218.

[53] Wei, J.; Su, B.; Liang, R.; Xing, H.; Bao, Z.; Yang, Y.; Ren, Q. *J. Chem. Eng. Data* 2012, **57**, 1274–1278.

[54] Gao, Y.; Li, N.; Zheng, L.; Zhao, X.; Zhang, J.; Cao, Q.; Zhao, M.; Li, Z.; Zhang, G. *Chem. Eur. J.* 2007, **13**, 2661–2670.

[55] Li, N.; Cao, Q.; Gao, Y. A.; Zhang, J.; Zheng, L. Q.; Bai, X. T.; Dong, B.; Li, Z.; Zhao, M. W.; Yu, L. *ChemPhysChem* 2007, **8**, 2211–2217.

[56] Gao, Y.; Li, N.; Zheng, L.; Bai, X.; Yu, L.; Zhao, X.; Zhang, J.; Zhao, M.; Li, Z. *J. Phys. Chem. B* 2007, **111**, 2506–2513.

[57] Gao, H. X.; Li, J. C.; Han, B. X.; Chen, W. N.; Zhang, J. L.; Zhang, R.; Yan, D. D. *Phys. Chem. Chem. Phys.* 2004, **6**, 2914–2916.

[58] Eastoe, J.; Gold, S.; Rogers, S. E.; Paul, A.; Welton, T.; Heenan, R. K.; Grillo, I. *J. Am. Chem. Soc.* 2005, **127**, 7302–7303.

[59] Gao, Y. N.; Wang, S. Q.; Zheng, L. Q.; Han, S. B.; Zhang, X.; Lu, D. M.; Yu, L.; Ji, Y. Q.; Zhang, G. Y. *J. Colloid Interface Sci.* 2006, **301**, 612–616.

[60] Gao, Y. A.; Zhang, J.; Xu, H. Y.; Zhao, X. Y.; Zheng, L. Q.; Li, X. W.; Yu, L. *ChemPhysChem* 2006, **7**, 1554–1561.

[61] Li, N.; Gao, Y. A.; Zheng, L. Q.; Zhang, J.; Yu, L.; Li, X. W. *Langmuir* 2007, **23**, 1091–1097.

[62] Falcone, R. D.; Correa, N. M.; Silber, J. J. *Langmuir* 2009, **25**, 10426–10429.

[63] Qiu, Z. M.; Texter, J. *Curr. Opin. Colloid Interface Sci.* 2008, **13**, 252–262.

[64] Sando, G. M.; Dahl, K.; Owrutsky, J. C. *Chem. Phys. Lett.* 2006, **418**, 402–407.

[65] Li, J. C.; Zhang, J. L.; Han, B. X.; Wang, Y.; Gao, L. *J. Chem. Phys.* 2004, **121**, 7408–7412.

[66] Chakrabarty, D.; Seth, D.; Chakraborty, A.; Sarkar, N. *J. Phys. Chem. B* 2005, **109**, 5753–5758.

[67] Liu, J. H.; Cheng, S. Q.; Zhang, J. L.; Feng, X. Y.; Fu, X. G.; Han, B. X. *Angew. Chem. Int. Ed.* 2007, **46**, 3313–3315.

[68] Gao, Y. N.; Li, N.; Hilfert, L.; Zhang, S. H.; Zheng, L. Q.; Yu, L. *Langmuir* 2009, **25**, 1360–1365.

[69] Falcone, R. D.; Baruah, B.; Gaidamauskas, E.; Rithner, C. D.; Correa, N. M.; Silber, J. J.; Crans, D. C.; Levinger, N. E. *Chem. Eur. J.* 2011, **17**, 6837–6846.

[70] Pramanik, R.; Ghatak, C.; Rao, V. G.; Sarkar, S.; Sarkar, N. *J. Phys. Chem. B* 2011, **115**, 5971–5979.

[71] Pramanik, R.; Sarkar, S.; Ghatak, C.; Rao, V. G.; Sarkar, N. *Chem. Phys. Lett.* 2011, **512**, 217–222.

[72] Zech, O.; Thomaier, S.; Bauduin, P.; Rück, T.; Touraud, D.; Kunz, W. *J. Phys. Chem. B* 2009, **113**, 465–473.

[73] Zech, O.; Thomaier, S.; Kolodziejski, A.; Touraud, D.; Grillo, I.; Kunz, W. *Chem. Eur. J.* 2010, **16**, 783–786.

[74] Ferreyra, D. D.; Correa, N. M.; Silber, J. J.; Falcone, R. D. *Phys. Chem. Chem. Phys.* 2012, **14**, 3460–3470.

[75] Adhikari, A.; Sahu, K.; Dey, S.; Ghosh, S.; Mandal, U.; Bhattacharyya, K. *J. Phys. Chem. B* 2007, **111**, 12809–12816.

[76] Adhikari, A.; Das, D. K.; Sasmal, D. K.; Bhattacharyya, K. *J. Phys. Chem. A* 2009, **113**, 3737–3743.

[77] Setua, P.; Pramanik, R.; Sarkar, S.; Ghatak, C.; Rao, V. G.; Sarkar, N.; Das, S. K. *J. Mol. Liq.* 2011, **162**, 33–37.

[78] Banerjee, C.; Mandal, S.; Ghosh, S.; Kuchlyan, J.; Sarkar, N. *Chem. Phys. Lett.* 2013, **580**, 88–93.

[79] Moilanen, D. E.; Levinger, N. E.; Spry, D. B.; Fayer, M. D. *J. Am. Chem. Soc.*2007, **129**, 14311–14318.

[80] Dupont, J.; Consorti, C. S.; Suarez, P. A. Z.; De-Souza, R. F. *Organic Synth.* 2004, **79**, 236–243.

[81] Rao, V. G.; Ghosh, S.; Ghatak, C.; Mandal, S.; Brahmachari, U.; Sarkar, N. *J. Phys. Chem. B* 2012, **116**, 2850–2855.

[82] Gao, Y.; Han, S.; Han, B.; Li, G.; Shen, D.; Li, Z.; Du, J.; Hou, W.; Zhang, G. *Langmuir* 2005, **21**, 5681–5684.

[83] Pramanik, R.; Sarkar, S.; Ghatak, C.; Rao, V. G.; Sarkar, N. *J. Phys. Chem. B* 2011, **115**, 2322–2330.

[84] Martins, C. T.; Sato, B. M.; El Seoud, O. A. *J. Phys. Chem. B* 2008, **112**, 8330–8339.

[85] Sarkar, A.; Pandey, S. *J. Chem. Eng. Data* 2006, **51**, 2051–2055.

[86] Masaki, T.; Nishikawa, K.; Shirota, H. *J. Phys. Chem. B* 2010, **114**, 6323–6331.

[87] Klijn, J. E.; Engberts, J. B. F. N. *J. Am. Chem. Soc.* 2003, **125**, 1825–1833.

[88] Reichardt, C. *Pure Appl. Chem.* 2008, **80**, 1415–1432.

[89] Reichardt, C. *Green Chem.* 2005, **7**, 339–351.

[90] Reichardt, C. *Chem. Rev.* 1994, **94**, 2319–2358.

[91] Pramanik, R.; Sarkar, S.; Ghatak, C.; Rao, V. G.; Setua, P.; Sarkar, N. *J. Phys. Chem. B* 2010, **114**, 7579–7586.

[92] Rao, V. G.; Mandal, S.; Ghosh, S.; Banerjee, C.; Sarkar, N. *J. Phys. Chem. B* 2013, **117**, 1480–1493.

[93] Rabe, C.; Koetz,. J. *Colloids Surf. A* 2010, **354**, 261–267.

[94] Bini, R.; Bortolini, O.; Chiappe, C.; Pieraccini, D.; Siciliano, T. *J. Phys. Chem. B* 2007, **111**, 598–604.

[95] Liu, D.; Ma, J.; Cheng, H.; Zhao, Z. *Colloids Surf. A* 1998, **135**, 157–164.

[96] Headley, A. D.; Jackson, N. M. *J. Phys. Org. Chem.* 2002, **15**, 52–55.

[97] Freire, M. G.; Santos, L. M. N. B. F.; Fernandes, A. M.; Coutinho, J. A. P.; Marrucho, I. M. *Fluid Phase Equilib.* 2007, **261**, 449–454.

CHAPTER 16

Ionic Liquid-Based Microemulsions

JIANLING ZHANG

Institute of Chemistry, Chinese Academy of Sciences, Beijing, China

16.1 INTRODUCTION

The term "microemulsion" was first coined by J. H. Schulman in 1959 [1]. A microemulsion is a thermodynamically stable dispersion of two immiscible fluids stabilized by surfactants. The hydrophilic head groups of the surfactants point to the polar phase, while the hydrophobic tails extend into the nonpolar region. Owing to the capacity to simultaneously host a variety of polar and nonpolar species, microemulsions have been widely applied in drug release [2], nanomaterial synthesis [3], polymerizations and reactions [4], etc. In general, the organic solvent (oil) and water are used as the two immiscible fluids in the formation of microemulsions. They are categorized as water-in-oil (W/O) microemulsions when water is dispersed homogenously in an organic media with the help of the surfactant, oil-in-water (O/W) microemulsions where oil is dispersed in water, and bicontinuous microemulsions where simultaneous dispersions of both water and oil are present.

In recent years, ionic liquids (ILs) have attracted much attention [5–8]. In comparison with the conventional solvents (usually water and organic solvents), ILs have some unique properties. For example, ILs are an interesting class of tunable and designable solvents with essentially zero volatility, wide electrochemical window, nonflammability, high thermal stability, and wide liquid range. These make them unique for uses and applications in the areas of chemistry and chemical engineering.

Microemulsions containing ILs have drawn considerable interest in recent years [9–11] since the first report by Han et al. [12]. The use of IL as dispersed or continuous phase in microemulsions makes them attractive owing to their unusual solvent properties. Up till now, various kinds of IL-based microemulsions have been prepared, and their phase behaviors, microstructures, micropolarities, and thermodynamic and dynamic properties have been investigated. These microemulsions have shown promises in different fields, such as material synthesis, polymerization, biocatalysis, chemical reaction, drug release, protein extraction and capillary electrophoresis, etc.

Ionic Liquid-Based Surfactant Science: Formulation, Characterization, and Applications, First Edition. Edited by Bidyut K. Paul and Satya P. Moulik.

16.2 CHARACTERIZATION OF IL-BASED MICROEMULSIONS

The characterization of the following IL-based microemulsions has been presented as follows: (i) IL-in-oil and oil-in-IL microemulsions, (ii) IL-in-water and water-in-IL microemulsions, (iii) IL-in-ethylene glycol (EG) and EG-in-IL microemulsions, (iv) IL-in-IL microemulsion, and (v) IL-in-supercritical CO_2 and CO_2-in-IL microemulsions.

16.2.1 IL-in-Oil and Oil-in-IL Microemulsions

16.2.1.1 Phase Behavior The investigation of phase behavior of microemulsions is essential to underline the domains of different coexisting phases. Han and coworkers [12] investigated the formation of microemulsions with IL as the nano-sized polar domains. Figure 16.1 shows the ternary phase diagram of 1-butyl-3-methylimidazolium tetrafluoroborate ([bmim][BF_4])/Triton X-100 (TX-100)/cyclohexane system at 35.0 ± 0.1 °C. Above the phase separation boundary curve, the system exists as one phase and the shadow area is the biphasic region. In the IL-in-oil (IL/O) region, the IL droplets are dispersed in cyclohexane, whereas in the oil-in-IL (O/IL) region, the cyclohexane droplets are dispersed in IL. In the bicontinuous region (marked as B in Fig. 16.1), a microemulsion region is in equilibrium with both O/IL and IL/O phases.

Han and coworkers [13] further studied the phase behavior of toluene/TX-100/1-butyl-3-methylimidazolium hexafluorophosphate ([bmim][PF_6]) system. As shown in Figure 16.2, the single-phase microemulsion area covers about 75% of the phase diagram at 25 °C. A transition from IL-in-oil microemulsion (marked as A) via a bicontinuous region (marked as B) to an oil-in-IL microemulsion (marked as C) occurs with the increase of weight fraction of TX-100 and [bmim][PF_6].

Zheng and coworkers [14] investigated the microemulsion formation in nonaqueous [bmim][BF_4]/TX-100/toluene system. As shown in Figure 16.3, three

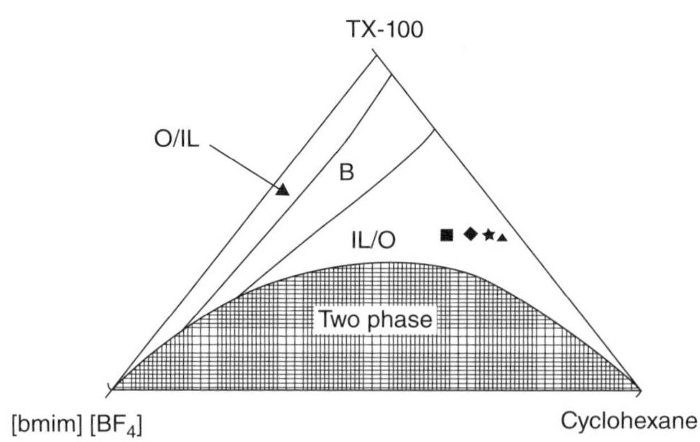

Figure 16.1 Phase diagram of [bmim][BF_4]/TX-100/cyclohexane system (in weight fraction) at 35.0 ± 0.1 °C. Reproduced from Gao et al. [12] with permission from The Royal Society of Chemistry, London.

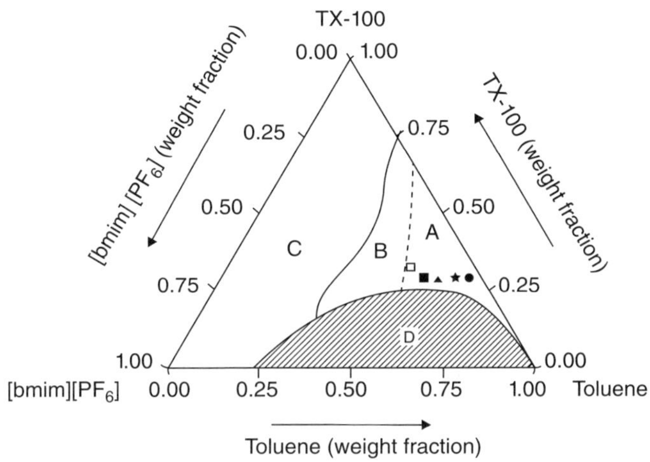

Figure 16.2 Phase diagram of toluene/TX-100/[bmim][PF_6] (in weight fraction) at 25.0 °C. A, IL-in-oil microemulsion; B, bicontinuous region; C, oil-in-IL microemulsion; and D, biphasic region. Reproduced from Li et al. [13] with permission from Elsevier.

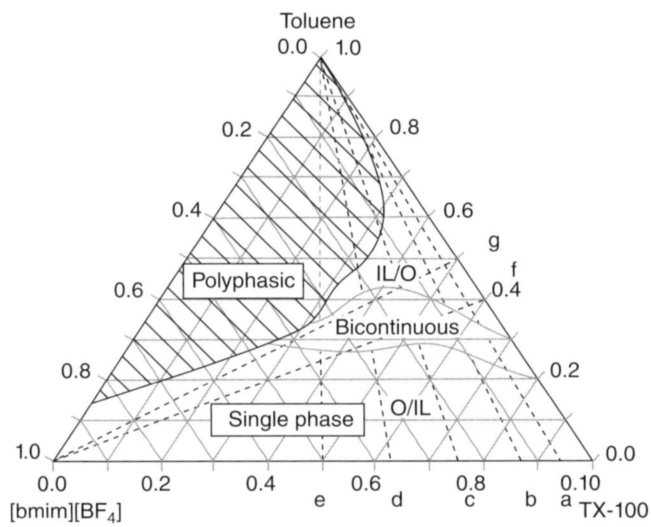

Figure 16.3 Phase diagram of the [bmim][BF_4]/TX-100/toluene system at 25.0 °C. For lines a, b, c, d, and e, the initial TX-100 weight fractions are $I = 0.93$, 0.88, 0.75, 0.63, and 0.50, respectively. For lines f and g, the initial TX-100 weight fractions are 0.40 and 0.50, respectively. Reproduced from Gao et al. [14] with permission from Elsevier.

different microregions, IL-in-oil microemulsions, bicontinuous microemulsions, and oil-in-IL microemulsions, were differentiated, respectively.

Sarkar and coworkers [15] formed nonaqueous microemulsions using *N*,*N*,*N*-trimethyl-*N*-propyl ammonium bis(trifluoromethanesulfonyl) imide ([N_{3111}][Tf_2N]) substituted for water with cyclohexane by the aid of nonionic surfactant TX-100. The ternary phase diagram of [N_{3111}][Tf_2N]/TX-100/cyclohexane system at 25 °C is

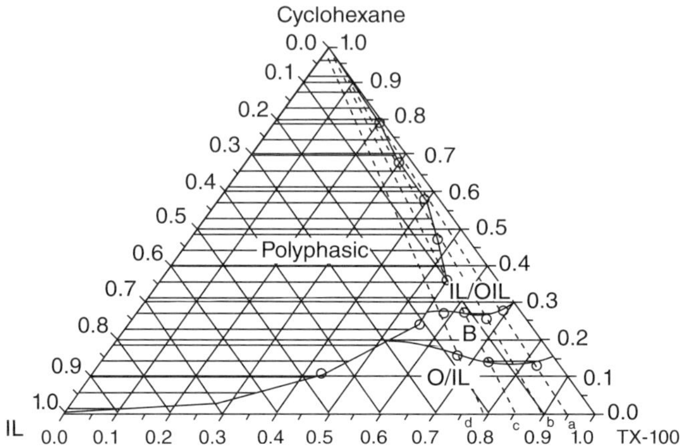

Figure 16.4 Phase diagram of the $[N_{3111}][Tf_2N]$/TX-100/cyclohexane system at 25.0 °C. For lines a, b, c, and d, the initial TX-100 weight fractions are $I = 0.95$, 0.90, 0.85, and 0.80, respectively. Reproduced from Pramanik et al. [15] with permission from the American Chemical Society.

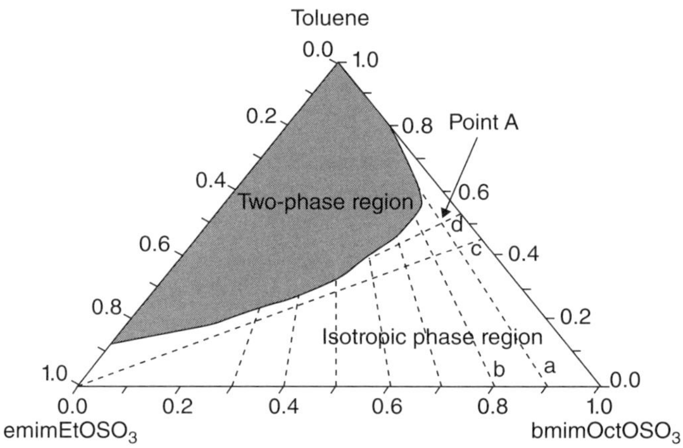

Figure 16.5 Phase diagram of $emimEtOSO_3$/toluene/$bmimOctOSO_3$ system at 25 °C. Reproduced from Rojasa et al. [16] with permission from Elsevier.

illustrated in Figure 16.4. Above the phase separation boundary curve, the system exists as one phase, and the shadow area is the polyphasic region. The monophasic regions are also marked in Figure 16.4. In IL-in-oil area (marked as IL/O), the IL droplets are dispersed in the cyclohexane continuous phase, and in oil-in-IL region (marked as O/IL), the cyclohexane droplets are dispersed in IL. In the bicontinuous region, microemulsion zone coexists with both IL and cyclohexane continuous phases.

Rojasa and coworkers [16] investigated a ternary system consisting of toluene and two ILs, that is, 1-butyl-3-methylimidazolium octylsulfate ($bmimOctOSO_3$) and 1-ethyl-3-methylimidazolium ethylsulfate ($emimEtOSO_3$). The results show that microemulsions can be formed (Fig. 16.5), where $bmimOctOSO_3$ takes over the role of a surfactant in the spontaneous interfacial film formation process.

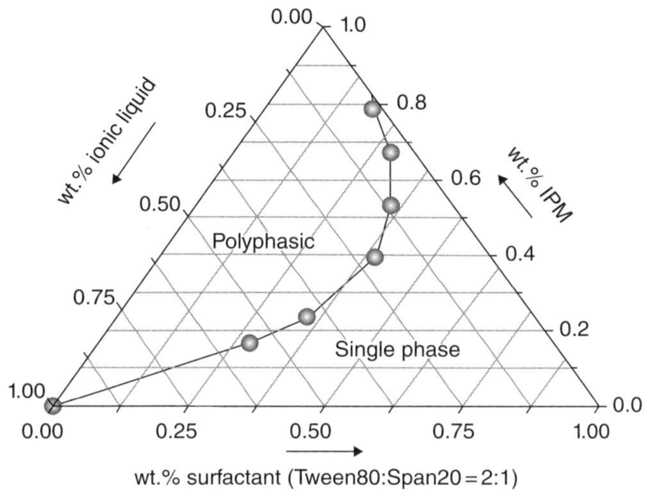

Figure 16.6 Pseudoternary phase diagram of the $[C_2mim][C_4SO_4]$/Tween80/Span20/IPM microemulsion system. Reproduced from Mandal et al. [17] with permission from the American Chemical Society.

The microstructural transitions from IL-in-oil to oil-in-IL microemulsions, passing a bicontinuous phase, can be attributed to the droplet–droplet exchange.

Sarkar and coworkers [17] reported the formulation of a IL-in-oil microemulsion, where the polar core of the IL, 1-ethyl-3-methylimidazolium *n*-butylsulfate ($[C_2mim][C_4SO_4]$), is stabilized by a mixture of two nonionic surfactants, polyoxyethylene sorbitan monooleate (Tween80) and sorbitan laurate (Span20), in a biocompatible oil phase of isopropyl myristate (IPM). The pseudoternary phase diagram of the $[C_2mim][C_4SO_4]$/Tween80/Span20/IPM microemulsion system is shown in Figure 16.6.

16.2.1.2 Microstructure Different techniques have been applied to characterize the microstructures of microemulsion systems, including electrical conductivity, freeze-fracture electron microscopy (FFEM), dynamic light scattering (DLS), small-angle neutron scattering (SANS), and small-angle X-ray scattering (SAXS). The electrical conductivity measurement is efficient in characterizing the subregions of single-phase region of microemulsions. For example, the conductivity of the toluene/TX-100/$[bmim][PF_6]$ system increases with increasing $[bmim][PF_6]$-to-TX-100 molar ratio because the conductivity originates mainly from the IL; therefore, the subareas of single-phase region of microemulsions can be located as IL-in-oil microemulsion, bicontinuous microemulsion, and oil-in-IL microemulsion (Fig. 16.2) [13]. The size and shape of the aggregates in $[bmim][BF_4]$/TX-100/cyclohexane microemulsions were characterized by transmission electron microscopy images of FFEM replicas of microemulsions and DLS [12]. SANS experiments [18] were carried out to demonstrate the formation of dispersed nanodroplets with IL cores in $[bmim][BF_4]$-in-cyclohexane microemulsions. Han and coworkers [13] used SAXS to determine the aggregate size of the $[bmim][PF_6]$-in-oil microemulsions. The apparent gyration radii of the reverse micelles increase with increasing $[bmim][PF_6]$-to-TX-100 molar ratio. Atkin and coworkers [19] characterized microemulsions comprising nonionic alkyl oligoethyleneoxide surfactant, alkane, and IL ethylammonium nitrate using SAXS. The results showed that the microemulsions are highly structured than their aqueous counterparts in general.

16.2.1.3 Micropolarity and Intermolecular Interaction The micropolarities of [bmim][BF_4]-in-toluene microemulsions stabilized by TX-100 [20] and [bmim][PF_6]-in-toluene microemulsions stabilized by Tween80 [21] were investigated by UV–Vis absorbance spectra using methyl orange (MO) and methylene blue (MB) as absorption probes. The results indicated that the polarity of the IL-in-oil microemulsion increased only before the IL pools were formed, whereas a relatively fixed polar microenvironment was obtained in the IL pools of the microemulsions. Zheng and coworkers [22] studied the intermolecular interactions in the toluene-in-[bmim][BF_4] microemulsions by two-dimensional rotating frame nuclear Overhauser effect experiments. The results indicated that there is a strong hydrophobic interaction between the hydrophobic chain of TX-100 and toluene. The hydrophobic interaction between the dispersed organic solvents and the hydrophobic tails of TX-100 might be the driving force for the formation of oil-in-IL microemulsions. Zheng and coworkers [23] measured the heat of dilution of the [bmim][BF_4]/Triton X-100/cyclohexane microemulsions by isothermal titration microcalorimetry, and the second virial coefficient was derived from the heat of dilution and the number density of the IL-based microemulsions. The second virial coefficient was found to be feasible for characterizing the interactions between IL microemulsion droplets. There are much stronger interactions between IL microemulsion droplets than traditional microemulsions, which may be attributed to their relatively larger droplet size compared with that of conventional microemulsions.

16.2.1.4 Thermodynamics of Formation Zhang and coworkers [24] employed the dilution method to investigate the interfacial composition and thermodynamic parameters of IL-in-oil microemulsion formed in [bmim][BF_4]/Brij-35/1-butanol/toluene system. The standard Gibbs free energy change, standard enthalpy change, and standard entropy change for the transfer of 1-butanol from the continuous oil phase (toluene) to the interfacial region of the microemulsion system as a function of temperature were calculated. All of the values of Gibbs free energy change $\left(\Delta G_t^{o}\right)$ were negative, suggesting that IL-in-oil microemulsions formed spontaneously. The absolute values of ΔG_t^{o} increased with increasing IL-to-surfactant molar ratio at a constant temperature. The higher absolute values of ΔG_t^{o} indicated a comparable stronger interaction between surfactant and cosurfactant (1-butanol) at the interface. The IL-in-oil microemulsion systems were mostly endothermic with appreciable positive entropy change, and the microemulsion formation process was entropy driven. Chai and coworkers [25] studied the thermodynamic properties of IL-in-oil microemulsions of 1-dodecyl-3-methylimidazolium bromide [C_{12}mim]Br/1-pentanol/octane/[bmim][BF_4] systems and compared the properties with water-in-oil microemulsions of [C_{12}mim]Br/1-pentanol/octane/water systems. The [bmim][BF_4]-containing systems have smaller absolute values of the standard enthalpy change than the water-containing microemulsions. It indicates that more cosurfactants were required to form an IL-in-oil microemulsion than the conventional water-in-oil microemulsion.

An and coworkers [26] measured the critical concentrations and the critical temperatures for the microemulsion systems consisting of water/[bmim][BF_4]/sodium di(2-ethyl-1-hexyl) sulfosuccinate (AOT)/decane with various compositions of [bmim][BF_4]. By measuring the refractive index at constant pressure in the critical region, the coexistence curves for the quaternary microemulsion were determined. The critical exponent β was deduced from the coexistence curves, which is very close to the 3D Ising exponent in the critical region. The addition of [bmim][BF_4] significantly reduced the lower critical temperature, which was contrary to the water/KCl/AOT/decane system, while the critical composition was almost kept unchanged. They attributed this phenomenon to the unique role of the IL, which is capable of modifying the physicochemical properties of microemulsions. Further, the critical behavior of an IL-in-oil microemulsion of [bmim][BF_4]/cyclohexane/TX-100 system was also reported [27]. This system was considered as a pseudobinary system. The values of critical exponent β of the microemulsion approached the 3D Ising value of 0.327 within the experimental uncertainties, while the temperature was close to the critical point. It was concluded that the refractive index n is a better choice of the concentration variable than the volume fraction for constructing an order parameter.

16.2.1.5 Solvation Dynamics Sarkar and coworkers [28] explored the effects of confinement of IL [bmim][BF_4] on solvation dynamics and rotational relaxation of coumarin 153 (C-153) in TX-100/cyclohexane microemulsions by employing steady-state and picosecond time-resolved emission spectroscopic measurements. The average rotational relaxation time increases with increase in IL-to-TX-100 molar ratio. The solvent relaxation in the core of the microemulsion occurs on two different timescales and is almost insensitive to the increase in IL-to-TX-100 molar ratio. The solvent relaxation is retarded in the pool of the microemulsions compared to the neat solvent. Further, Sarkar and coworkers [29] also studied the solvent and rotational relaxation dynamics in [bmim][BF_4]/TX-100/cyclohexane microemulsions with addition of water by employing the same techniques. With addition of water, the size of the microemulsions decreased, which led to restrict the motion of ions of ILs and consequently slowed down the solvation dynamics of ILs in the microemulsions. Bhattacharyya and coworkers [30] found that the solvation dynamics is slower in a 1-pentyl-3-methylimidazolium tetrafluoroborate ([pmim][BF_4])/water/TX-100/benzene quaternary microemulsion than that of a microemulsion without water. This is ascribed to the smaller size of the water-containing microemulsion. The anisotropy decay in an IL microemulsion is found to be faster in comparison to neat IL. Furthermore, Bhattacharyya and coworkers studied the fluorescence resonance energy transfer from coumarin 480 (C480) to rhodamine 6G [31], deuterium isotope effect on the solvation dynamics as well as the anisotropy decay of C480 [32], and the ultrafast excited state proton transfer of pyranine by femtosecond upconversion inside the [pmim][BF_4]/TX-100/benzene microemulsion [33].

16.2.1.6 Effects of Different Additives Zheng and coworkers [34] studied the effect of water on [bmim][BF_4]-in-benzene microemulsions. The addition of small amount of water to the microemulsion increased the amount of solubilized [bmim][BF_4] in the microemulsion. Sundmacher and coworkers [35] investigated the role of small amounts of water in the [bmim][BF_4]-in-cyclohexane microemulsions. The size of microemulsion droplets decreased remarkably with increasing water content. Cheng and coworkers [36] investigated the effects of n-butanol as a cosurfactant on the phase behavior of cyclohexane/TX-100/[bmim][BF_4] system at 25 °C. The *n*-butanol magnified notably the microemulsion single-phase region and increased the solubilization capacity of [bmim][BF_4] in the microemulsions. *n*-Butanol as the cosurfactant not only affected the micellization process by modifying the solvent but also took part itself as unique component of the micelle aggregates. Zheng and coworkers [37] investigated the effect of polyethylene glycol with molecular weight of 400 (PEG-400) on the microstructure of [bmim][BF_4]/TX-100/cyclohexane microemulsion. The addition of PEG-400 led to the linear increase in size of the microemulsion droplets, showing that PEG-400 was solubilized into the polar interior of the IL microemulsions.

16.2.2 IL-in-Water and Water-in-IL Microemulsions

Han and coworkers [38] determined the phase behavior of the ternary system consisting of [bmim][PF_6], TX-100, and water at 25 °C. By cyclic voltammetry method using potassium ferrocyanide, $K_4Fe(CN)_6$, as the electroactive probe, the water-in-[bmim][PF_6], bicontinuous, and [bmim][PF_6]-in-water microregions of the microemulsions were identified (Fig. 16.7). The hydrodynamic diameter of the [bmim][PF_6]-in-water microemulsions is nearly independent of the water content but increases with increasing [bmim][PF_6] content due to the swelling of the micelles by the IL. Sarkar and coworkers [39–41] reported the solvent and rotational relaxation studies in [bmim][PF_6]-in-water microemulsions and water-in-[bmim][PF_6] microemulsions using different types of probes, coumarin 153 (C-153), coumarin 151 (C-151), and coumarin 490 (C-490). The solvent relaxation time is retarded in the IL-in-water microemulsion compared to that of a neat solvent. The retardation of solvation time of water in the core of the water-in-IL microemulsion is several thousand times compared to pure water. Nozaki and coworkers [42] reported a broadband dielectric spectroscopy study on a microemulsion composed of water,

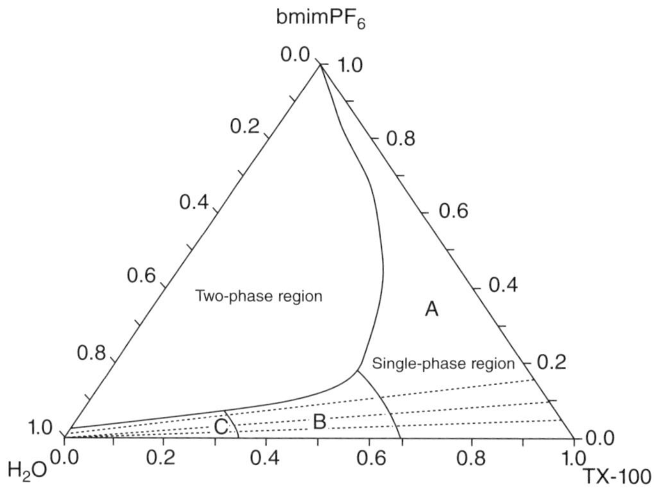

Figure 16.7 Phase diagram of the TX-100/H_2O/[bmim][PF_6] ternary system at 25 °C; A–C are water-in-[bmim][PF_6], bicontinuous, and [bmim][PF_6]-in-water regions, respectively. Reproduced from Gao et al. [38] with permission from the American Chemical Society.

TX-100, and [bmim][PF_6]. It was found that the phase behavior of the microemulsion could be easily identified by its dielectric response. The dielectric behavior of the IL microemulsion in the GHz range is consistent with that of TX-100/water mixtures with comparable water-to-TX-100 weight ratio.

16.2.3 IL-in-EG and EG-in-IL Microemulsions

Han and coworkers [43] determined the phase behavior of the ternary system consisting of EG, [bmim][PF_6], and TX-100. Figure 16.8 shows the phase diagram of EG/TX-100/[bmim][PF_6] ternary system at 30 °C. A mixture in the single-phase region can be considered as microemulsion. As [bmim][PF_6] content increases, the subregion structures transform from [bmim][PF_6]-in-EG (C) to EG-in-[bmim][PF_6] (A) after passing through a bicontinuous region (B), indicative of the existence of three subregions over the entire single-phase microemulsion region.

16.2.4 IL-in-IL Microemulsion

Han and coworkers [44] reported the formation of IL-in-IL microemulsion. The IL [bmim][PF_6] can be dispersed in IL propylammonium formate (PAF) with the aid of surfactant AOT. The ternary phase diagram of the [bmim][PF_6]/AOT/PAF system at 30 °C is illustrated in Figure 16.9. AOT could promote the miscibility of [bmim][PF_6] and PAF in a PAF-rich region, and a single-phase region was observed in the ternary phase diagram, in which the [bmim][PF_6]-in-PAF microemulsions are formed. The droplet size is a linear function of [bmim][PF_6]-to-AOT molar ratio, that is, the microemulsions obey swelling law.

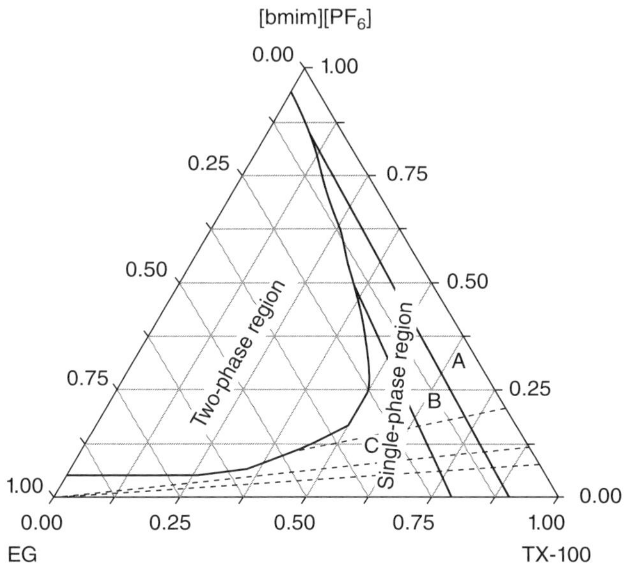

Figure 16.8 Phase diagram of the IL/TX-100/EG ternary system at 30.0 °C; A, B, and C are EG-in-IL, bicontinuous, and IL-in-EG regions, respectively. Reproduced from Cheng et al. [43] with permission from Elsevier.

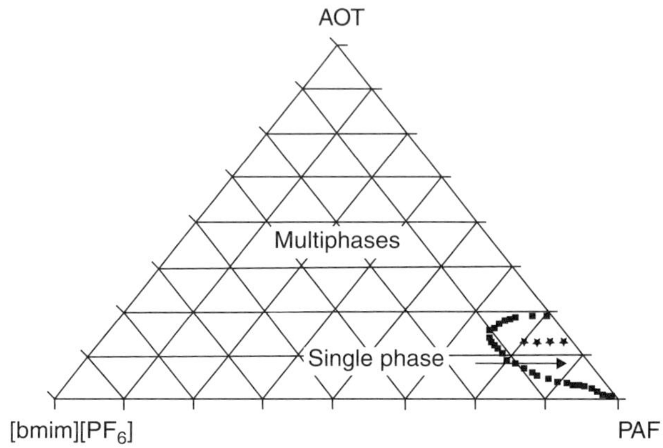

Figure 16.9 Phase diagram of the [bmim][PF_6]/AOT/PAF ternary system at 30 °C. Reproduced from Cheng et al. [44] with permission from The Royal Society of Chemistry, London.

16.2.5 IL-in-Supercritical CO_2 and CO_2-in-IL Microemulsions

Supercritical CO_2 (Sc CO_2) is regarded as a green solvent because it is nontoxic, abundant, tunable, and nonflammable, has moderate critical temperature and pressure, and can be easily recaptured and recycled after use [45–48]. The simultaneous utilization of supercritical (or compressed) CO_2 and IL through the formation of microemulsions can combine the advantages of the two fluids. Han and coworkers [49] constructed the IL-in-Sc CO_2 microemulsions by using surfactant *N*-ethyl perfluorooctylsulfonamide (*N*-EtFOSA) and guanidinium-based ILs. At readily accessible pressures, the IL-to-surfactant molar ratio can reach 0.8 for the CO_2/1,1,3,3-tetramethylguanidinium acetate (TMGA)/*N*-EtFOSA system. Senapati and coworkers [50] investigated the formation of IL-in-Sc CO_2 microemulsions *via* a computer simulation that demonstrates the entire process of self-aggregation at the atomic level. Zhang and coworkers [51] created the CO_2-in-IL microemulsions, with IL as the continuous phase and CO_2 as the dispersed phase. At certain pressure, the micelles have an average size of 47 nm, considerably larger than the CO_2-free micelles. The CO_2-swollen micelles are "tunable," because their size can be easily tuned by the pressure of CO_2.

16.3 APPLICATIONS

IL-based microemulsion as a "green solvent" has been extensively investigated in the fields of material synthesis, polymerization, biocatalysis, organic synthesis, drug release, protein extraction, and capillary electrophoresis.

16.3.1 Material Synthesis

16.3.1.1 Silica Han and coworkers [52] used the water/TX-100/[bmim][PF_6] microemulsions with IL as continuous phase to fabricate silica microrods. Porous silica microrods with nanosized pores were prepared in this microemulsion. Zheng and coworkers [53] synthesized the hollow silica spheres in benzene/TX-100/[bmim]

$[BF_4]$ microemulsion. When TEOS was added to the microemulsion, it dissolved in the benzene core; at the same time, some [bmim]$[BF_4]$ molecules were located in the palisade layers of the microemulsion. Subsequently, TEOS molecules at the interface of the core were hydrolyzed and polymerized, because the [bmim]$[BF_4]$ molecules in the palisade layer could probably be employed as Lewis acid. So, the SiO_2 polymerized and grew thickly around the interface of the benzene cores, forming the hollow silica spheres after calcination. Furthermore, Zheng and coworkers [54] synthesized silica products with two different morphologies using nonaqueous IL microemulsion droplets as templates. By adjusting the reaction conditions, ellipsoidal nanoparticles were formed under acidic condition, while hollow silica spheres were obtained under alkaline condition.

16.3.1.2 ZrO_2 Nanoparticles Zheng and coworkers [55] used the water-in-IL microemulsion to prepare tetragonal ZrO_2 nanoparticles. A number of anomalous spherically dispersed particles (15–40 nm) were obtained. For comparison, the ZrO_2 particles synthesized by water-in-oil microemulsion were congregated after calcinations. It indicates that the IL bmimPF_6 plays an important role in preventing the congregation of the product.

16.3.1.3 Polyaniline/Silver Nanocomposites He and coworkers [56] employed chronopotentiometry to prepare polyaniline/silver (PANI/Ag) nanocomposite films in water-in-IL and IL-in-water microemulsions, by simultaneous oxidative polymerization of aniline to PANI and reduction of silver nitrate to Ag nanoparticles. The PANI/Ag prepared in water-in-IL microemulsion was nanofibrous, and the Ag nanocrystals with 5 nm diameter were dispersed homogeneously, whereas the PANI/Ag prepared in IL-in-water microemulsion exhibited dendritic structure, and the diameter of Ag nanocrystals was 50–100 nm. The special structures of the PANI/Ag nanocomposite resulted in more excellent electrochemical activity than that of the pure PANI.

16.3.1.4 Gold Nanoparticles Han and coworkers [49] proved that the CO_2-continuous microemulsions with IL domains could solubilize $HAuCl_4$ and prepared gold particles by a process of rapid expansion of a supercritical solution into a liquid solvent. Au nanoparticles were formed in the reverse micelles, when the weight ratio of $HAuCl_4$ to 1,1,3,3-tetramethylguanidinium trifluoroacetate (TMGT) (W_{HAuCl_4}/W_{TMGT}) was 0.01. However, Au networks were obtained at a higher concentration of $HAuCl_4$.

16.3.1.5 Bimetallic Palladium/Gold Nanoparticles Zhang and coworkers [57] synthesized bimetallic palladium/gold nanoparticles in the IL-based microemulsion. The palladium and gold precursors were dissolved in dispersive and continuous phase of microemulsion (H_2O/TX-100/[bmim]$[PF_6]$), respectively. $[PdCl_6]^{2-}$ ions were reduced *in situ* by TX-100 in dispersive water phase to prepare Pd nanoparticles and then $[AuCl_4]^-$ crossed through the interface film and reacted with the as-prepared Pd nanoparticles to form Pd_4Au nanoparticles.

16.3.1.6 Mesoporous Metal–Organic Framework Zhang and coworkers [51] explored the application of CO_2-in-IL microemulsions in fabrication of metal–organic frameworks (MOFs), which present great potential in gas storage, separation, and catalysis. The MOFs with bimodal mesoporous structure were obtained. The larger mesopores in the range of 20–50 nm result from the templating effect of micelles, while the smaller mesopores in the range of 2–3 nm can be attributed to the IL–CO_2 solution trapped during the formation of the MOF.

16.3.2 Polymerization

16.3.2.1 Atom Transfer Radical Polymerization Wang and coworkers [58] synthesized poly(methyl methacrylate) by ATRP of MMA in IL-based microemulsion with Tween80 as surfactant. The polymerization was carried out at 25 °C with CCl_4 as initiator, $FeCl_3{\cdot}6H_2O$/*N*,*N*,*N*′,*N*′-tetramethyl-1,2-ethanediamine as catalyst complex in the presence of reducing agent, ascorbic acid. The polymerization obeyed the first-order kinetics, indicating that the polymerization proceeded in a controlled/"living" process. The well-controlled polymers were obtained with narrow polydispersity. Wang and coworkers [59] conducted the copolymerizations of styrene and acrylonitrile in IL microemulsion under reverse atom transfer radical polymerization (ATRP) using 2,2′-azobis(isobutyronitrile) as initiator, $FeCl_3{\cdot}6H_2O$/succinic acid as a complex catalyst, cetyltrimethylammonium bromide (CTAB) as surfactant, and [bmim][PF_6] as IL. The copolymers of styrene and acrylonitrile with predetermined molecular weights and narrow molecular weight distribution were obtained.

16.3.2.2 Electropolymerization Zhang and coworkers [60] studied the electropolymerization behavior of pyrrole in water-in-[bmim][PF_6], bicontinuous, [bmim][PF_6]-in-water microemulsions. It was shown that the water-in-[bmim][PF_6] microemulsion was the optimal medium and its polymerization rate was the fastest. The resultant films electrodeposited in water-in-[bmim][PF_6] microemulsion possessed excellent electrochemical activity and uniform morphology. He and coworkers [61] prepared PANI by cyclic voltammetry in [bmim][PF_6]-in-water microemulsion. The specific capacitance of the PANI film obtained by 50-cycle electropolymerization in [bmim][PF_6]-in-water microemulsion is 334 Fg^{-1} at the charge–discharge current density of 2 $mAcm^{-2}$.

16.3.2.3 Polymerization in the Presence of TiO_2 Nanoparticles Guo and coworkers [62] synthesized the PANI/TiO_2 nanocomposite particles in a microemulsion comprised deionized water as the continuous aqueous phase and an oil solution including aniline and [bmim][PF_6] as the dispersed phase in the presence of anatase TiO_2 nanoparticles. The dispersed TiO_2 nanoparticles adsorbed onto the surface of the microemulsion micelles to form core–shell-type micelles, and subsequently, the polymerization was initiated by ammonium persulfate, during which some chemical bonds generated between TiO_2 and PANI chain. The nanocomposites exhibited spherical morphology with particle sizes about 70 nm, and the nanocrystalline TiO_2 deposited onto the surface of PANI. The PANI/TiO_2 nanocomposite can be used in photocatalysis and electrochemical fields.

16.3.3 Biocatalysis

16.3.3.1 Lipase-Catalyzed Hydrolysis Goto and coworkers [63] proposed an approach for carrying out enzymatic reactions in water-in-IL microemulsion.

The lipase-catalyzed hydrolysis of *p*-nitrophenyl butyrate (*p*-NPB) was used as a model reaction. It was found that the hydrolysis rate was faster in the water-in-IL microemulsions than in the water-in-isooctane microemulsions. The intrinsic activity of lipase in the IL microemulsion was about three times higher than that of water/AOT/isooctane microemulsions of AOT under the given experimental conditions. The enhanced catalytic activity of lipase in water-in-IL microemulsions may be due to (i) aqueous microenvironmental changes, (ii) the partition of the substrate or other molecules involved in the reaction between water and IL phases, and (iii) the existence of 1-hexanol as a cosurfactant.

16.3.3.2 Fungal Oxidase-Catalyzed Oxidation of o-Phenylenediamine Huang and coworkers [64] studied the catalytic activities of two fungal oxidases, lignin peroxidase and laccase for the oxidation of *o*-phenylenediamine, in water-in-[bmim][PF_6] microemulsion stabilized by TX-100. In such a medium, both lignin peroxidase and laccase were catalytically active. Lignin peroxidase and laccase could express their catalytic activity with the optimum molar ratio of water to TX-100 at 8.0 for lignin peroxidase and greater than 20 for laccase and the optimum pH values at 3.2 for lignin peroxidase and 4.2 for laccase, respectively. Comparatively, the two oxidases had negligible catalytic activity in pure or water-saturated [bmim][PF_6], due to the strong inactivating effect of [bmim][PF_6] on both enzymes.

16.3.3.3 Laccase-Catalyzed Oxidation of 2,6-Dimethoxyphenol Huang and coworkers [65] studied the laccase-catalyzed oxidation of 2,6-dimethoxyphenol in the water-in-[bmim][PF_6] microemulsion stabilized by AOT and TX-100. The laccase hosted in the microemulsion exhibited a catalytic activity, and the activity could be regulated by the composition of the interfacial membrane.

16.3.3.4 Lipase-Catalyzed Hydrolysis of 4-Nitrophenyl Butyrate Huang and coworkers [66] investigated the lipase-catalyzed hydrolysis of *p*-NPB in the water-in-[bmim][PF_6] microemulsion stabilized by AOT and TX-100. Under optimum conditions, the catalytic activity of lipase in the microemulsion is much higher than that of water-saturated [bmim][PF_6]. When the partitioning of the substrate in the microemulsion is taken into account, the catalytic efficiency of lipase in this microemulsion is 14.3 times that in water-saturated [bmim][PF_6], which may be due to the significant decrease of the Michaelis constant in the microemulsion.

16.3.4 Chemical Reaction

Zhou and coworkers [67] prepared Pd nanoparticles in situ in the H_2O/TX-100/[bmim][PF_6] microemulsion and used this microemulsion system to catalyze the Heck reaction of butyl acrylate with iodobenzene. The reaction time was decreased compared with conventional solvent system. Other Heck reactions relating to the coupling of iodobenzene and methyl acrylate, ethyl acrylate and styrene were also investigated with high yields. All the results indicated that the H_2O/TX-100/[bmim][PF_6] microemulsion containing Pd nanoparticles was a very efficient catalyst system for the ligand-free Heck reaction.

16.3.5 Drug Release

Goto and coworkers [68] employed IL-in-oil microemulsions to increase the solubility of a sparingly soluble drug to enhance its topical and transdermal delivery. The microemulsion consisted of a blend of nonionic surfactants, Tween80 and Span20, IPM as an oil phase, and IL $[C_1mim][(CH_3O)_2PO_2]$ (dimethylimidazolium dimethylphosphate) as a pseudophase. The IL-in-oil microemulsions showed high solubility of acyclovir, excellent formulation stability, and enhanced permeation to the Yucatan micropig porcine skin. Goto and coworkers [69] also found that some drug molecules, including acyclovir, methotrexate, and 1-[(5-(*p*-nitrophenyl) furfurylidene) amino] hydantoin sodium, that are insoluble or sparingly soluble in most of the conventional solvents have a high degree of solubilization in the IL-in-oil microemulsions.

16.3.6 Extraction of Hemoglobin

Selective extraction of hemoglobin was achieved by employing a water/AOT/[bmim][PF_6] microemulsion system as template by Wang and coworkers [70]. An extraction efficiency of ca. 96% was achieved for a 100 ng·μL^{-1} hemoglobin solution by using an equal volume of the microemulsion, and afterward, a back extraction efficiency of 73% was obtained with a 6 mol·L^{-1} urea solution as stripping reagent. The practical applicability of the microemulsion for sample pretreatment was demonstrated by the successful isolation of hemoglobin from human whole blood.

16.3.7 Capillary Electrophoresis

Cheng and coworkers [71] used the IL-in-water microemulsion as a pseudostationary phase for the detection of a large number of acidic compounds in capillary electrophoresis. A fast and an efficient separation of eight phenolic acids was achieved by using [bmim][PF_6], Tween20, and borate as oil drops, surfactant, and background electrolyte, respectively. The detection efficiency of the analytes by using IL-in-water microemulsion is much higher than that of using oil-in-water microemulsion electrokinetic chromatography. To test the practicality of the proposed IL-in-water microemulsion approach in real sample analysis, the concentrations of phenolic acids in Danshen injection were determined. These acidic compounds were well separated from other components within 6 min.

16.4 CONCLUSIONS AND OUTLOOK

The study on the microemulsions with IL(s) as dispersed or continuous phase is an interesting topic both from the points of view of scientific researches and practical applications. ILs have many advantages, for example, they can dissolve many organic and inorganic substances, and their properties are tunable to satisfy a variety of tasks. Therefore, the IL-based microemulsions will show many unique features and can satisfy many different types of applications. Despite many exciting recent developments, the area of IL-based microemulsions is still in its infancy. Some future researches on this topic are in the following. First, in comparison with the nonaqueous microemulsions consisted of IL, oil, and surfactant, the study on the microemulsions consisted of IL, water, and surfactant is less. The aqueous IL microemulsions involve no volatile organic solvent, more environmentally benign, and should be paid more attention. Second, more IL-based microemulsions with special compositions, structures, and functions should be designed and constructed, for example, IL-in-IL microemulsions, IL-in-Sc CO_2 microemulsions, etc. These novel microemulsions may find many special applications due to their unique properties. Third, the theoretical understanding of IL-based microemulsions is still, by and large, in its infancy stage. Since ILs largely differ from the conventional solvents from physicochemical point of view, hence formation mechanism of these two types of microemulsions should be differed. This could be an interesting topic to be studied in the future, which can pose a significant contribution toward colloid and interface science. Fourth, the applications of the IL-based microemulsions are limited and could be explored in various fields, especially in material synthesis, chemical reaction, and separation science, by combination of both the advantages of IL and microemulsion.

REFERENCES

[1] Schulman, J. H.; Stoeckenius, W.; Prince, L. M. *J. Phys. Chem.*, 1959, **63**, 1677–1680.
[2] Lawrence, M. J.; Rees, G. D. *Adv. Drug Deliv. Rev.*, 2012, **64**, Supplement, 175–193.
[3] Ganguli, A. K.; Ganguly, A.; Vaidya, S. *Chem. Soc. Rev.*, 2010, **39**, 474–485.
[4] Chow, P. Y.; Gan, L. M. *Adv. Polym. Sci.*, 2005, **175**, 257–298.
[5] Smiglak, M.; Metlen, A.; Rogers, R. D. *Acc. Chem. Res.*, 2007, **40**, 1182–1192.
[6] Chaturvedi, D. *Curr. Org. Chem.*, 2011, **8**, 438–471.
[7] Hubbard, C. D.; Illner, P.; Eldik, R. *Chem. Soc. Rev.*, 2011, **40**, 272–290.
[8] Petkovic, M.; Seddon, K. R.; Rebelo, L. P. N.; Pereira, C. S.; Cristina, S. *Chem. Soc. Rev.*, 2011, **40**, 1383–1403.
[9] Greavesa, T. L.; Drummond, C. J. *Chem. Soc. Rev.*, 2008, **37**, 1709–1726.
[10] Kunz, W.; Zemb, T.; Harrar, A. *Curr. Opin. Colloid Interface Sci.*, 2012, **17**, 205–211.
[11] Acharya, D. P.; Hartley, P. G. *Curr. Opin. Colloid Interface Sci.*, 2012, **17**, 274–280.
[12] Gao, H.; Li, J.; Han, B.; Chen, W.; Zhang, J.; Zhang, R.; Yan, D. *Phys. Chem. Chem. Phys.*, 2004, **6**, 2914–2916.
[13] Li, J.; Zhang, J.; Gao, H.; Han, B.; Gao, L. *Colloid Polym. Sci.*, 2005, **283**, 1371–1375.
[14] Gao, Y.; Wang, S.; Zheng, L.; Han, S.; Zhang, X.; Lu, D.; Yu, L.; Ji, Y.; Zhang, G. *J. Colloid Interface Sci.*, 2006, **301**, 612–616.
[15] Pramanik, R.; Sarkar, S.; Ghatak, C.; Rao, V. G.; Setua, P.; Sarkar, N. *J. Phys. Chem. B*, 2010, **114**, 7579–7586.
[16] Rojasa, O.; Tierscha, B.; Frascab, S.; Wollenbergerb, U.; Koetz, J. *Colloids Surf. A Physicochem. Eng. Asp.*, 2010, **369**, 82–87.
[17] Mandal, S.; Ghosh, S.; Banerjee, C.; Kuchlyan, J.; Banik, D.; Sarkar, N. *J. Phys. Chem. B*, 2013, **117**, 3221–3231.
[18] Eastoe, J.; Gold, S.; Rogers, S. E.; Paul, A.; Welton, T.; Heenan, R. K.; Grillo, I. *J. Am. Chem. Soc.*, 2005, **127**, 7302–7303.
[19] Atkin, R.; Warr, G. G. *J. Phys. Chem. B*, 2007, **111**, 9309–9316.
[20] Li, N.; Gao, Y.; Zheng, L.; Zhang, J.; Yu, L.; Li, X. *Langmuir*, 2007, **23**, 1091–1097.
[21] Zheng, Y.; Eli, W. *J. Dispers. Sci. Technol.*, 2009, **30**, 698–703.

[22] Gao, Y.; Li, N.; Zhang, S.; Zheng, L.; Li, X.; Dong, B.; Yu, L. *J. Phys. Chem. B*, 2009, **113**, 1389–1395.

[23] Li, N.; Zhang, S.; Zheng, L.; Gao, Y.; Yu, L. *Langmuir*, 2008, **24**, 2973–2976.

[24] Wang, F.; Zhang, Z.; Li, D.; Yang, J.; Chu, C.; Xu, L. *J. Chem. Eng. Data*, 2011, **56**, 3328–3335.

[25] Chai, J.; Xu, L.; Liu, W.; Zhu, M. *J. Chem. Eng. Data*, 2012, **57**, 2394–2400.

[26] Lü, H.; An, X.; Shen, W. *J. Chem. Eng. Data*, 2008, **53**, 727–731.

[27] Lü, H. H.; An, X. Q.; Shen, W. G. *J. Chem. Eng. Data*, 2011, **56**, 502–507.

[28] Chakrabarty, D.; Seth, D.; Chakraborty, A.; Sarkar, N. *J. Phys. Chem. B*, 2005, **109**, 5753–5758.

[29] Pramanik, R.; Sarkar, S.; Ghatak, C.; Setua, P.; Rao, V. G.; Sarkar, N. *Chem. Phys. Lett.*, 2010, **490**, 154–158.

[30] Adhikari, A.; Sahu, K.; Dey, S.; Ghosh, S.; Mandal, U.; Bhattacharyya, K. *J. Phys. Chem. B*, 2007, **111**, 12809–12816.

[31] Adhikari, A.; Das, D. K.; Sasmal, D. K.; Bhattacharyya, K. *J. Phys. Chem. A*, 2009, **113**, 3737–3743.

[32] Sasmal, D. K.; Mojumdar, S. S.; Adhikari, A.; Bhattacharyya, K. *J. Phys. Chem. B*, 2010, **114**, 4565–4571.

[33 Mojumdar, S. S.; Mondal, T.; Das, A. K.; Dey, S.; Bhattacharyya, K. *J. Chem. Phys.*, 2010, **132**, 194505.

[34] Gao, Y.; Li, N.; Zheng, L.; Bai, X.; Yu, L.; Zhao, X.; Zhang, J.; Zhao, M.; Li, Z. *J. Phys. Chem. B*, 2007, **111**, 2506–2513.

[35] Gao, Y.; Hilfert, L.; Voigt, A.; Sundmacher, K. *J. Phys. Chem. B*, 2008, **112**, 3711–3719.

[36] Cheng, S.; Han, F.; Wang, Y.; Yan, J. *Colloids Surf. A Physicochem. Eng. Asp.*, 2008, **317**, 457–461.

[37] Li, N.; Zhang, S.; Li, X.; Yu, L.; Zheng, L. *Colloid Polym. Sci.*, 2009, **287**, 103–108.

[38] Gao, Y.; Han, S.; Han, B.; Li, G.; Shen, D.; Li, Z.; Du, J.; Hou, W.; Zhang, G. *Langmuir*, 2005, **21**, 5681–5684.

[39] Seth, D.; Chakraborty, A.; Setua, P.; Sarkar, N. *Langmuir*, 2006, **22**, 7768–7775.

[40] Seth, D.; Chakraborty, A.; Setua, P.; Sarkar, N. *J. Chem. Phys.*, 2007, **126**, 224512.

[41] Seth, D.; Setua, P.; Chakraborty, A.; Sarkar, N. *J. Chem. Sci.*, 2007, **119**, 105–111.

[42] Chen, Z.; Nozaki, R. *J. Chem. Phys.*, 2012, **136**, 244505.

[43] Cheng, S.; Fu, X.; Liu, J.; Zhang, J.; Zhang, Z.; Wei, Y.; Han, B. *Colloids Surf. A Physicochem. Eng. Asp.*, 2007, **302**, 211–215.

[44] Cheng, S.; Zhang, J.; Zhang, Z.; Han, B. *Chem. Commun.*, 2007, 2497–2499.

[45] Beckman, E. J. *Ind. Eng. Chem. Res.*, 2003, **42**, 1598–1602.

[46] Gopalan, A. S.; Wai, C. M.; Jacobs, H. K. ed., *Supercritical Carbon Dioxide. Separations and Processes*, ACS Symposium Series, Vol. **860**, American Chemical Society, Washington, DC, 2003.

[47] Jessop, P. G.; Leitner, W. ed., *Chemical Synthesis Using Supercritical Fluids*, Wiley-VCH Verlag GmbH, Weinheim, 2008.

[48] Zhang, J. L.; Han, B. X. *Acc. Chem. Res.*, 2013, **46**, 425–433.

[49] Liu, J. H.; Cheng, S. Q.; Zhang, J. L.; Feng, X. Y.; Fu, X. G.; Han, B. X. *Angew. Chem. Int. Ed.*, 2007, **46**, 3313–3315.

[50] Chandran, A.; Prakash, K.; Senapati, S. *J. Am. Chem. Soc.*, 2010, **132**, 12511–12516.

[51] Zhang, J. L.; Han, B. X.; Li, J. S.; Zhao, Y. J.; Yang, G. Y. *Angew. Chem. Int. Ed.*, 2011, **50**, 9911–9915.

[52] Li, Z.; Zhang, J.; Du, J.; Han, B.; Wang, J. *Colloids Surf. A Physicochem. Eng. Asp.*, 2006, **286**, 117–120.

[53] Zhao, M.; Zheng, L.; Li, N.; Yu, L. *Mater. Lett.*, 2008, **62**, 4591–4593.

[54] Zhao, M., Zheng, L., Bai, X., Li, N., Yu, L., *Colloids Surf. A Physicochem. Eng. Asp.*, 2009, **346**, 229–236.

[55] Li, N.; Dong, B.; Yuan, W.; Gao, Y.; Zheng, L.; Huang, Y.; Wang, S. *J. Dispers. Sci. Technol.*, 2007, **28**, 1030–1033.

[56] Zhou, Z.; He, D.; Guo, Y.; Cui, Z.; Wang, M.; Li, G.; Yang, R. *Thin Solid Films*, 2009, **517**, 6767–6771.

[57] Zhang, G.; Zhou, H.; An, C.; Liu, D.; Huang, Z.; Kuang, Y. *Colloid Polym. Sci.*, 2012, **290**, 1435–1441.

[58] Wang, G.-X.; Lu, M.; Liu, L.-C.; Wu, H.; Zhong, M. *J. Appl. Polym. Sci.*, 2013, **128**, 3077–3083.

[59] Wang, G.-X.; Lu, M.; Hou, Z.-H.; Wu, H. *J. Polym. Res.*, 2013, **20**, 80.

[60] Zhang, G.; Zhou, H.; Zhang, J.; Han, X.; Chen, J.; Kuang, Y. *J. Appl. Polym. Sci.*, 2012, **125**, 2342–2347.

[61] He, D.; Xia, S.; Zhou, Z.; Zhong, J.; Guo, Y.; Yang, R. *J. Exp. Nanosci.*, 2013, **8**, 103–112.

[62] Guo, Y.; He, D.; Xia, S.; Xie, X.; Gao, X.; Zhang, Q. *J. Nanomater.*, 2012, **2012**, 202794.

[63] Moniruzzaman, M.; Kamiya, N.; Nakashima, K.; Goto, M. *Green Chem.*, 2008, **10**, 497–500.

[64] Zhou, G.; Zhang, Y.; Huang, X.; Shi, C.; Liu, W.; Li, Y.; Qu, Y.; Gao, P. *Colloids Surf. B: Biointerfaces*, 2008, **66**, 146–149.

[65] Xue, L.; Qiu, H.; Li, Y.; Lua, L.; Huang, X.; Qu, Y. *Colloids Surf. B: Biointerfaces*, 2011, **82**, 432–437.

[66] Xue, L.; Li, Y.; Zou, F.; Lua, L.; Zhao, Y.; Huang, X.; Qu, Y. *Colloids Surf. B: Biointerfaces*, 2012, **92**, 360–366.

[67] Zhang, G.; Zhou, H.; Hu, J.; Liu, M.; Kuang, Y. *Green Chem.*, 2009, **11**, 1428–1432.

[68] Moniruzzaman, M.; Tamura, M.; Tahara, Y.; Kamiya, N.; Goto, M. *Int. J. Pharm.*, 2010, **400**, 243–250.

[69] Moniruzzaman, M.; Kamiya, N.; Goto, M. *J. Colloid Interface Sci.*, 2010, **352**, 136–142.

[70] Shu, Y.; Cheng, D.; Chen, X.; Wang, J. *Sep. Purif. Technol.*, 2008, **64**, 154–159.

[71] Cao, J.; Qu, H.; Cheng, Y. *Electrophoresis*, 2010, **31**, 3492–3498.

CHAPTER 17

Ionic Liquid-Based Nonaqueous Microemulsion

QILONG REN, QIWEI YANG, BAOGEN SU, ZHIGUO ZHANG, ZONGBI BAO, and HUABIN XING

Key Laboratory of Biomass Chemical Engineering of Ministry of Education, Department of Chemical and Biological Engineering, Zhejiang University, Hangzhou, China

17.1 INTRODUCTION

Microemulsions are clear, isotropic, and thermodynamically stable dispersions of two immiscible solvents and a surfactant, in which nanodroplets of one solvent stabilized by a surfactant are dispersed in another solvent [1]. In view of their unique structure and properties, microemulsions have experienced rapid development in drug dissolution, liquid–liquid extraction, polymerization, preparation of nanoparticles, etc. [2]. Although in most cases water is essential for formation of microemulsions, sometimes, its relatively narrow liquid-state temperature range and the low solubility or instability of many compounds in water limit their applications. In this context, ionic liquid (IL)-based nonaqueous microemulsions where ILs are used as an alternative polar solvent in place of water have emerged with promise. In comparison to other molecular solvents, ILs possess some advantageous physicochemical properties and, as a consequence, have been applied to various chemical processes [3]. With regard to microemulsions, one possible advantage of using ILs for water is that they can be stable over a broader range of temperature [4]. Moreover, IL-based nonaqueous microemulsions may be applied to many more processes, since ILs are better solvents for a large variety of compounds than water. Besides, these systems are also more favorable in water-sensitive systems. ILs can be regarded as "designable solvents," and many task-specific ILs have been devised for reactions and separations. Thus, with designed ILs, the applications of microemulsion technologies can be diversified.

ILs are different from the conventional water solvent, as they consist of cations and anions, and thus, the structure and properties of IL-based nonaqueous microemulsions are unlike those of aqueous microemulsions. In this chapter, we outline recent advances in IL-based nonaqueous microemulsions, with a focus on the structure and properties of these systems.

Ionic Liquid-Based Surfactant Science: Formulation, Characterization, and Applications, First Edition. Edited by Bidyut K. Paul and Satya P. Moulik.

17.2 1-BUTYL-3-METHYLIMIDAZOLIUM TETRAFLUOROBORATE ([bmim]BF_4)-BASED NONAQUEOUS MICROEMULSION SYSTEMS

Compared to the many different types of ILs that have developed so far, a limited variety of those ILs are used in nonaqueous microemulsions, and most of them are imidazolium-based ILs. Among others, 1-butyl-3-methylimidazolium tetrafluoroborate ([bmim]BF_4) has been extensively used in many microemulsion studies since the pioneering work on IL-based microemulsions by Han et al. [5]; probably, it is one of the most classical ILs to date. As a result, studies on [bmim]BF_4-based nonaqueous microemulsions are better characterized than those on other IL-based microemulsions.

As mentioned above, Han and coworkers first reported that ILs can form microemulsions with suitable surfactants and nonpolar solvents [5]. Using Triton X-100 (TX-100) as surfactant, [bmim]BF_4 formed nanosized polar domains in a nonpolar continuous phase of cyclohexane. Electrical conductivity measurements were performed to determine the subregions in the monophasic region as the IL-in-oil microemulsion region, oil-in-IL microemulsion region, or bicontinuous region, respectively. In the IL-in-oil microemulsions, freeze-fracture electron microscopy (FFEM) was used to show directly the images of the microemulsions with different IL-to-TX-100 molar ratios (R) (Fig. 17.1). They showed that the size of polar IL droplets increased with R when the concentration of TX-100 was fixed, which is similar to the behavior of conventional water-in-oil microemulsion systems. This result was further confirmed by dynamic light scattering (DLS) measurements to characterize the size and size distribution of IL droplets. Eastoe et al. [6] investigated the same IL-based microemulsion system by small-angle neutron scattering (SANS). It was reported that the droplets of dispersed IL nanodomains swelled as IL was added and the volume of droplets was directly proportional to the amount of IL, which is in good accordance with common aqueous microemulsions. Using a [bmim]BF_4/TX-100/cyclohexane microemulsion system, Sarkar et al. studied the solvation dynamics and rotational relaxation of coumarin 153 (C-153) [7], which was used as an experimental probe, by steady-state and picosecond time-resolved emission spectroscopy. The average rotational relaxation time of the probe molecules increases with increase in R, implying that the probe molecules were solubilized inside the IL pool. More importantly, the fluorescence spectra showed a red shift of the emission peak of C-153 with the addition of [bmim]BF_4, indicating that the polarity of the microemulsion system increased after the addition of [bmim]BF_4. Furthermore, Li et al. used microcalorimetry to obtain the second virial coefficient of the [bmim]BF_4/TX-100/cyclohexane microemulsion system, through which the interactions between the droplets were described [8]. The results showed that interactions between IL microemulsion droplets were much stronger than those in traditional microemulsions, probably due to the larger size of the microemulsion droplets in the former.

Instead of cyclohexane, Zheng and coworkers developed a new [bmim]BF_4-based nonaqueous microemulsion system using *p*-xylene as the oil phase with the aid of TX-100 [9]. Conductivity measurements were carried out to identify the three microregions of the microemulsions. The DLS characterization showed the same result as that for the [bmim]BF_4/TX-100/cyclohexane microemulsion system. Moreover, the results of FTIR and ^{1}H-NMR revealed the driving force for microemulsion formation. With the addition of [bmim]BF_4, the stretching band of the

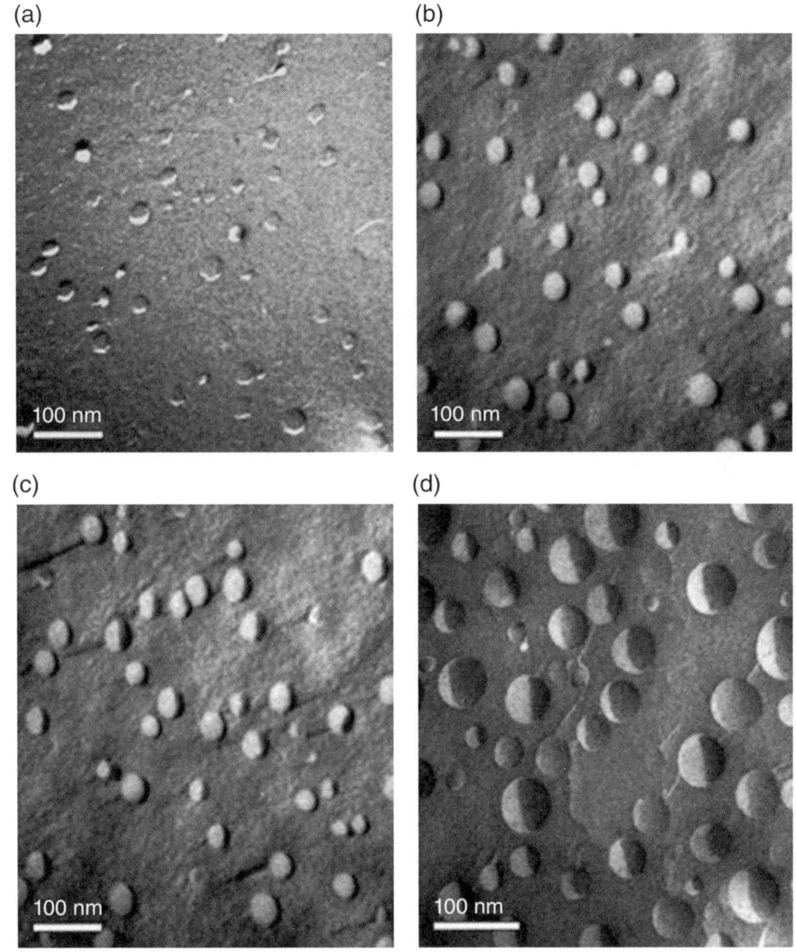

Figure 17.1 TEM images of FFEM replicas of [bmim]BF_4/TX-100/cyclohexane microemulsions (weight fraction of TX-100 is 0.45); (a) $R = 0.2$, (b) $R = 0.5$, (c) $R = 1.0$, and (d) $R = 1.5$. Reproduced from Gao et al. [5] with permission from the Royal Society of Chemistry.

terminal hydroxyl group of TX-100 moved toward the high-frequency region in the FTIR spectrum. They proposed that the electrostatic interaction of [bmim]BF_4 with the oxygen atoms of the EO units hindered hydrogen-bond interactions between the EO units. The proton signals from the imidazolium ring and the OE units of TX-100 in ^{1}H-NMR became increasingly broader with the addition of [bmim]BF_4, which was consistent with the FTIR characterization. From these results, a possible structure of the [bmim]BF_4/TX-100/*p*-xylene microemulsion system was proposed and presented in Figure 17.2.

Zheng et al. reported a [bmim]BF_4/TX-100/toluene microemulsion system that was similar to the IL-based nonaqueous microemulsions mentioned earlier [10]. Cyclic voltammetry experiments were successfully introduced to verify the identification of microregions, in addition to conductivity measurements.

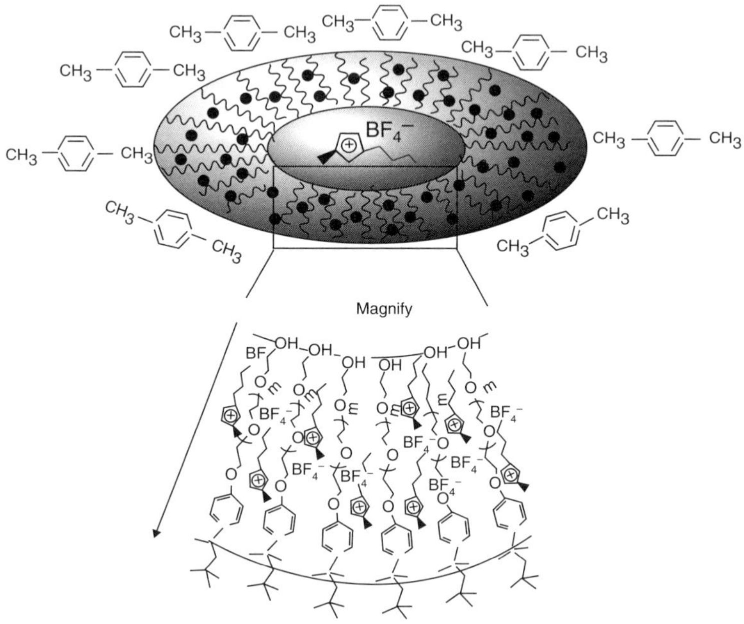

Figure 17.2 Possible structure of the [bmim]BF_4-in-*p*-xylene microemulsion. Reproduced from Gao et al. [9] with permission from Wiley.

Furthermore, using methyl orange (MO) and methylene blue (MB) as absorption probes, the micropolarity of the microemulsion system was investigated by UV–Vis spectroscopy [11]. The results demonstrated that micropolarity increased with the addition of IL only before the IL pools were formed. Once the IL was saturated for the solubilizing ability of the EO units and began to form cores of the microemulsions, the micropolarity became relatively constant.

Zheng et al. investigated how organic solvents (cyclohexane, *p*-xylene, toluene, and benzene) worked in the formation of [bmim]BF_4-based oil-in-IL nonaqueous microemulsion systems [12]. The added molecular solvents provided a nonpolar environment that resulted in the aggregation of the hydrophobic tails of the surfactant TX-100, so that the molecular solvents formed droplets dispersed in the continuous [bmim]BF_4 phase. Results of 2D^{1}H-NMR confirmed that the solvophobic interaction between the molecular solvents and the hydrophobic tails of TX-100 was the driving force in the formation of those oil-in-IL microemulsions.

Nonionic surfactants such as TX-100, polyoxyethylene sorbitan monooleate (Tween-80), or anionic sodium bis(2-ethylhexyl) sulfosuccinate (AOT) are generally regarded as essential components in the formulation of microemulsions. Nevertheless, very recently, Xu et al. made remarkable progress in the development of novel IL-based nonaqueous microemulsions by producing a surfactant-free IL-based nonaqueous microemulsion for the first time [13]. The microemulsion consisted of [bmim]BF_4, ethanol, and toluene, without the involvement of any traditional surfactants. The phase behavior of the [bmim]BF_4/ethanol/toluene system was similar to

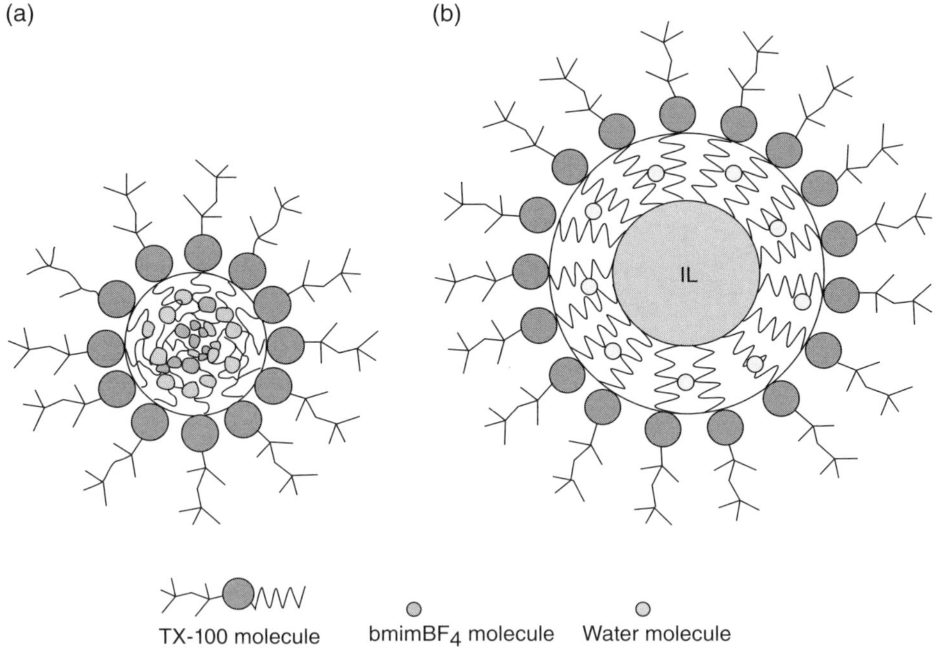

Figure 17.3 A schematic representation of water molecules in two different microenvironments. (a) IL/O microemulsion without IL pools; (b) IL/O microemulsion with an IL pool. Reproduced from Gao et al. [14] with permission from Wiley.

that of common microemulsions stabilized by traditional surfactants. Moreover, the polarity of the [bmim]BF_4 pool in the [bmim]BF_4-in-toluene microemulsion was found to be lower than that of neat [bmim]BF_4, which was also similar to the case of common IL-in-oil microemulsions.

Although [bmim]BF_4 has been used as an important component in formulating IL-based nonaqueous microemulsions, in practice, a small amount of water is likely to be present because of contamination and other reactions in [bmim]BF_4-based microemulsions. Therefore, it is imperative to study the effect of water on the formation and properties of [bmim]BF_4-based microemulsions. Gao et al. [14–16] have investigated the effect of small amounts of water on the structure and properties of [bmim]BF_4/TX-100/toluene microemulsions by UV/Vis spectra, FTIR, and ^{1}H-NMR studies [14, 15]. As a result, a "glue-like" effect of the small amount of dissolved water was demonstrated in [bmim]BF_4/TX-100/benzene microemulsions, implying that water molecules are present between TX-100 and the [bmim]BF_4 pool, that is, in the polar outer shell of the microemulsion, and the interaction with both TX-100 and [bmim]BF_4 gives rise to firmer palisade layers (Fig. 17.3). Consequently, microemulsions containing a small amount of water were more stable than those without water, and the polarity of the [bmim]BF_4 pool in them was the same as that in the water-free samples. Similar results were obtained for [bmim]BF_4/TX-100/benzene microemulsions. However, when the continuous phase was changed from nonpolar benzene to the polar and basic solvent triethylamine, the water molecules were not solubilized in the [bmim]BF_4 pools and remained bound with triethylamine

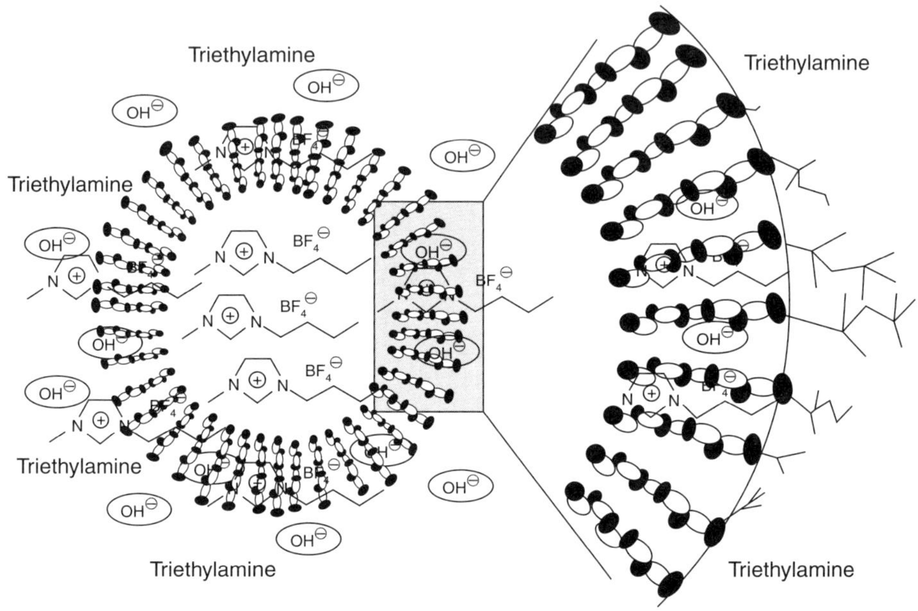

Figure 17.4 Schematic representation of the [bmim]BF_4-in-triethylamine microemulsion in the presence of small amounts of water. Reproduced from Li et al. [16] with permission from Wiley.

to form OH^- ions surrounding the [bmim]BF_4 droplets [16]. As a result, a continuous basic interface was formed in the microemulsions (Fig. 17.4) favorable for performing base-catalyzed reactions or preparing metal hydroxides/oxides in microemulsions.

17.3 OTHER IL-BASED NONAQUEOUS MICROEMULSIONS

1-Butyl-3-methylimidazolium hexafluorophosphate ([bmim]PF_6) is one of the most well-known hydrophobic ILs, and thus, it has been used as the oil phase in formulation of IL-based aqueous microemulsions. [bmim]PF_6 has also been used as the polar phase in the formation of nonaqueous microemulsions because of its immiscibility with nonpolar solvents. Zheng et al. reported nonaqueous microemulsions consisting of [bmim]PF_6, toluene, and a nonionic surfactant Tween-80 [17]. Three different subdomains of the single-phase region have been demonstrated by electrical conductivity measurements; these are [bmim]PF_6-in-toluene, bicontinuous, and toluene-in-[bmim]PF_6, respectively. Comparison of the FTIR spectra of [bmim]PF_6/Tween-80/toluene microemulsion and that of Tween-80/toluene mixture supported at least two interactions between [bmim]PF_6 and Tween-80 in the formulation of the microemulsion. One was the hydrogen-bonding interaction between the hydroxyl hydrogen atom of Tween-80 and the anion of [bmim]PF_6, and the other was the hydrogen-bonding interaction between the acidic hydrogen atom in the imidazolium cation and the oxyethylene units of Tween-80. The polarity of the microemulsion was determined from UV–visible spectroscopy using MO as the absorption probe [18].

The polarity of the [bmim]PF_6-in-toluene microemulsion was found to be higher than that of Tween-80–toluene reverse micelles but became constant after the formation of [bmim]PF_6 pools in the system. Cheng et al. studied formation of microemulsion using [bmim]PF_6 and ethylene glycol as a solvent [19]. Both ethylene glycol and [bmim] PF_6 have a low vapor pressure at room temperature; thus, the microemulsions formed were greener than common microemulsion systems.

Like [bmim]PF_6, ILs comprising a bis(trifluoromethylsulfonyl) anion (NTf_2^-) are also hydrophobic and have been used to construct nonaqueous microemulsions. Falcone et al. constructed microemulsions consisting of 1-butyl-3-methylimidazolium bis(trifluoromethylsulfonyl)imide ([bmim]NTf_2), benzene, and a surfactant [20]. In particular, by performing solvatochromic experiments with 1-methyl-8-oxyquinolinium betaine as a probe, the authors demonstrated that the type of surfactant influences the structure of the sequestrated ILs. When an ionic surfactant, benzyl-*n*-hexadecyldimethylammonium chloride (BHDC), was used, its positively charged head group electrostatically interacted with the anion of [bmim]NTf_2, and it was different from the interaction between a nonionic surfactant TX-100 and [bmim]NTf_2. In benzene/BHDC/[bmim]BF_4 system, the interaction was stronger than benzene/BHDC/[bmim]NTf_2 system because of weaker interaction between BHDC and NTf_2^- than between BHDC and BF_4^-. In subsequent work [21], the authors demonstrated the difference of the [bmim]NTf_2 structure in [bmim]NTf_2-in-chlorobenzene microemulsions when a cationic surfactant BHDC or an anionic surfactant AOT was used. Pramanik et al. [22] investigated the structure and dynamical behavior of *N*,*N*,*N*-trimethyl-*N*-propylammonium bis(trifluoromethanesulfonyl)imide ([N3111]NTf_2)/TX100/cyclohexane microemulsions with different [N3111]NTf_2 content through steady-state and time-resolved fluorescence spectroscopy. It was observed that an increasing amount of [N3111]NTf_2 led to a faster solvation dynamics of IL in the core of microemulsions, as a result of the increased size of droplets. A similar study was also performed by the authors with the *N*-methyl-*N*-propylpyrrolidinium bis(trifluoromethanesulfonyl) imide/TX100/benzene system [23].

Atkin et al. [24, 25] reported microemulsions in which the polar phase was protic ILs, that is, ethylammonium nitrate (EAN) and propylammonium nitrate. The surfactant was nonionic alkyl oligo(ethylene oxide), C_iE_j type, and the nonpolar phase was alkane. In general, the features of these microemulsions were similar to those of corresponding aqueous systems. However, because the interaction between the C_iE_j surfactant and IL was stronger than the C_iE_j–water, the surfactant chain length in EAN system must be longer than that in aqueous system to formulate a strongly structured microemulsion. Moreover, the transition from weakly to strongly structured microemulsions in the EAN system also occurred with a longer surfactant chain length than in an aqueous system. The effect of the chain length of alkanes on the phase behavior of the EAN system was similar to that in aqueous systems. Besides the microemulsion phase, a lamellar phase was also observed in the EAN system with lower temperature and longer alkyl tail in the surfactant.

The effect of alkyl chain length on the properties of neat ILs has been investigated, but few corresponding studies have been performed on IL-based nonaqueous microemulsion systems. Very recently, an important progress in this area was made by Ghosh et al. [26], through the investigation of the phase behavior and properties

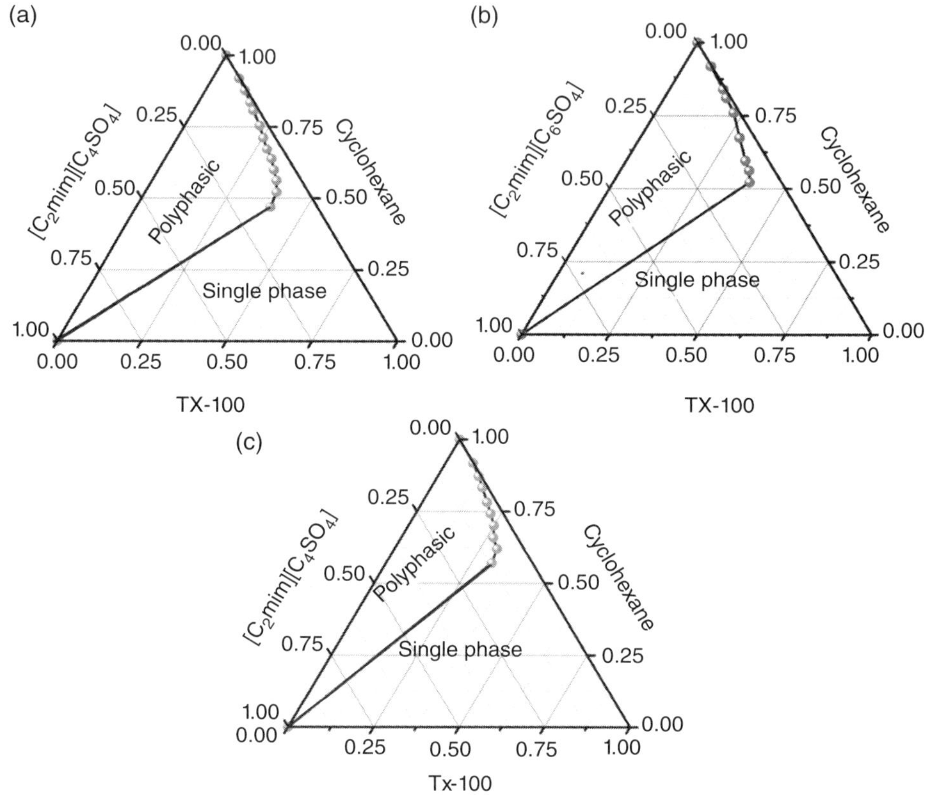

Figure 17.5 Phase diagrams of the (a) [emim][C_4SO_4]/TX-100/cyclohexane, (b) [emim][C_6SO_4]/TX-100/cyclohexane, and (c) [emim][C_8SO_4]/TX-100/cyclohexane three-component system at 298 K. Reproduced from Ghosh et al. [26] with permission from the American Chemical Society.

of 1-ethyl-3-methylimidazolium alkylsulfate ([emim][C_nSO_4])/TX-100/cyclohexane systems with varying n ($n = 2, 4, 6, 8$) in the [C_nSO_4]$^-$ anion. They found that the single-phase region of the [emim][C_nSO_4]/TX-100/cyclohexane system increased with the elongation of the alkyl chain in [C_nSO_4]$^-$ (Fig. 17.5), and [emim][C_2SO_4] did not form microemulsions with TX-100 and cyclohexane. Besides, the average solvation and rotational relaxation time of C-480 in microemulsion also increased with the increase in carbon number similar to that of neat ILs.

17.4 MICROEMULSIONS WITH MIXTURES OF IL AND POLAR SOLVENT AS POLAR PHASE

Among various ILs, chloride (anion)-based ILs have exhibited quite a satisfactory performance in a lot of processes involving hydrogen-bonding interactions because of their relatively high hydrogen-bond basicities compared to other ILs [27–30]. However, the chloride-based ILs are often very viscous or even solid near room temperature, which limits their application in nonaqueous microemulsion system. Therefore, the use of ILs with high viscosity as the polar phase of nonaqueous microemulsions is quite challenging. One solution is to increase the temperature to reduce the viscosity, but the effect is very limited. Moreover, the choice of temperature for microemulsions will be limited by the properties of the oil phase and surfactant.

It has been reported that the addition of a polar solvent to ILs can significantly reduce the viscosity of the IL phase [29, 31]. The physicochemical properties of the IL phase such as polarity and hydrogen-bond basicity can also be adjusted by the addition of miscible solvents [29]. In this context, Ren et al. [32] developed a new type of nonaqueous microemulsion using a mixture of IL and polar solvent as a polar phase to improve the transport properties and diversity of IL-based nonaqueous microemulsions. In this microemulsion system, the mixture of 1-butyl-3-methylimidazolium chloride ([bmim]Cl) and formamide (FA) was used as the polar phase, with TX-100 as the surfactant and cyclohexane as the apolar phase. Detailed characterization of the phase diagram, the particle size, and the microenvironment was done using different techniques. It was found that the addition of [bmim]Cl to the polar solvent increased the single-phase region of the microemulsion, and a higher [bmim]Cl content led to a larger single-phase region. Moreover, the single-phase region of the [bmim]Cl + FA system was larger than that of the classic [bmim]BF_4/TX-100/cyclohexane system. By measuring the conductance of the system, the single-phase region was divided into three subregions. A linear relationship between the droplet size and the molar ratio of the internal phase to surfactant was found by DLS analysis, suggesting a spherical structure of these microemulsions. FTIR experiments showed that there was a relatively strong hydrogen bond between the terminal hydroxyl of the surfactant TX-100 and the mixed internal phase in ([bmim]Cl + FA)/TX-100/cyclohexane microemulsions. By using MO and MB as probes, it was demonstrated that the polarity of the internal phase constantly increased with its content of the polar internal phase, but it was lower than the polarity of the binary mixtures of [bmim]Cl and a polar solvent. The biological molecule riboflavin and the metal salt $CoCl_2$ can be solubilized in these nonaqueous microemulsions comprising IL–polar solvent mixture as the polar phase.

17.5 NONAQUEOUS MICROEMULSIONS WITH IL AS BOTH DISPERSED PHASE AND CONTINUOUS PHASE

As described previously, most IL-based nonaqueous microemulsions consist of a hydrophilic IL, a surfactant, and a nonpolar solvent. Although these microemulsions have been successfully constructed and applied to different fields, the problem of volatility of the environmentally unfriendly organic solvents cannot be avoided. One strategy, to replace both the internal (dispersed) and external (continuous) phases with two immiscible ILs, is theoretically possible, but it may not be easy to find two immiscible ILs because all the ILs are organic salts.

Han and coworkers [33] successfully developed the IL-in-IL microemulsion consisting of two immiscible ILs, [bmim]PF_6 and propylammonium formate (PAF), and AOT. In the presence of surfactant AOT, the hydrophobic IL [bmim]PF_6 (as the internal phase) was dispersed in hydrophilic IL PAF and led to form [bmim]PF_6-in-PAF microemulsions. TEM images of the IL-in-IL microemulsion by means of the FFEM technique showed that the dispersed droplets were spherical. When AOT was maintained at a fixed concentration, the size of the droplets increased with the increase of [bmim]PF_6-to-AOT molar ratio (R), and the average diameter ranged between 30 and ~100 nm for R ranging between 0.2 and 1.0.

Subsequently, Han et al. [34] reported IL-in-IL nanoemulsions for the first time. Assisted with a certain amount of surfactant AOT, the hydrophilic IL PAF and the hydrophobic IL 3-methyl-1-octylimidazolium hexafluorophosphate ([omim]PF_6) formed [omim]PF_6-in-PAF microemulsions when the volume ratio of [omim]PF_6 was relatively low. With the increase in volume ratio of [omim]PF_6, unstable [omim] PF_6-in-PAF nanoemulsions were formed, which were transparent before phase separation. The conductivity of [omim]PF_6-in-PAF nanoemulsions was much lower than that of [omim]PF_6-in-PAF microemulsions. Small-angle X-ray scattering (SAXS) experiment demonstrated that the microdroplets of the nanoemulsion were spherical and the gyration radii of the microdroplets decreased from 19.3 to 15.7 nm when the volume ratio of [omim]PF_6 was reduced from 0.35 to 0.25. Utilizing this IL-in-IL nanoemulsion as the reaction medium, stable and crystallined metal–organic framework (MOF) nanorods were successfully synthesized, indicating potential applications of IL-in-IL nanoemulsions as well as microemulsions in other fields.

17.6 NONAQUEOUS MICROEMULSIONS USING ILS AS BOTH SURFACTANT AND POLAR SOLVENT

ILs with relatively long hydrophobic carbon chains resemble like typical surfactants. The long hydrophobic chain greatly reduces the ionic character of ILs, and as a consequence, they cannot be readily dissolved in common ILs via electrostatic interactions, and aggregation often occurs as formed in their aqueous solutions. Zech et al. [35] reported a microemulsion consisting of 1-hexadecyl-3-methylimidazolium chloride ([C_{16}mim]Cl, as surfactant) and EAN and 1-butyl-3-methylimidazolium tetrafluoroborate ([bmim]BF_4, as polar pseudophase), respectively, and dodecane (as nonpolar phase) and decanol (as cosurfactant). Similar to the 1-hexanol/water/polyoxyethylene(4)lauryl ether (Brij30) system, an irregular boundary between the monophasic region and the multiphasic region was found for the dodecane/([C_{16}mim] Cl + decanol)/EAN system, but the phase diagram of dodecane/([C_{16}mim]Cl + decanol)/[bmim]BF_4 system was regular (Fig. 17.6). A striking difference in the interfacial rigidity was shown for these two systems, showing the critical role of polar IL. In a subsequent study, Zech et al. [36] also demonstrated the high thermal stability of the dodecane/([C_{16}mim]Cl + decanol)/EAN microemulsions by SANS and light scattering measurements. The microemulsion was stable between 30 and 150°C, which exceeded the stable range of aqueous microemulsions (≤100 °C). Nonaqueous microemulsion using ILs as both surfactant and polar solvent was also reported by Rojas et al. [37] without the use of cosurfactant. The surfactant and polar solvent were 1-butyl-3-methylimidazolium dodecylsulfate ([bmim][$DodSO_4$]) and 1-ethyl-3-methylimidazolium ethylsulfate ([emim][$EtSO_4$]), respectively, and the nonpolar solvent was toluene. The size of droplets in the [emim][$EtSO_4$]-in-toluene microemulsions was found by DLS experiments to be less than 10 nm. If the anion of the surfactant-like IL was changed from $[DodSO_4]^-$ to octyl sulfate ($[OctSO_4]^-$), the isotropic area

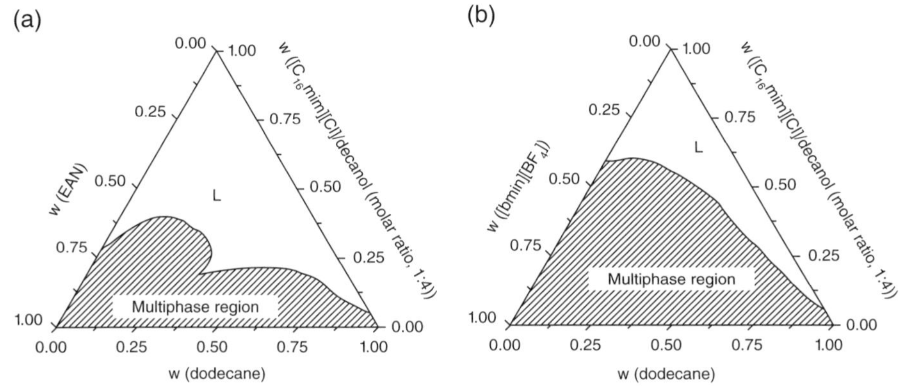

Figure 17.6 Pseudoternary phase diagram of (a) dodecane/([C_{16}mim]Cl + decanol)/EAN and (b) dodecane/([C_{16}mim]Cl + decanol)/[bmim]BF_4 (in weight fraction at 30°C with a [C_{16}mim]Cl/decanol molar ratio of 1:4). Reproduced from Zech et al. [35] with permission from the American Chemical Society.

of the ternary system decreased [38] because the shorter carbon chain in $[OctSO_4]^-$ compared with $[DodSO_4]^-$ caused in decreased surface activity of IL.

Sarkar et al. [39] have made potential contributions in the development of nonaqueous microemulsions using ILs as both surfactant and polar solvent. They synthesized the surface-active IL, [bmim][AOT], and produced a number of IL-in-oil microemulsions consisting of [bmim][AOT] as surfactant and various ILs as a polar pool dispersed in benzene. They compared the polarity of the IL pool in the [bmim]BF_4/[bmim][AOT]/benzene microemulsion to that in neat [bmim]BF_4, by measuring the absorption and fluorescence spectra of three kinds of probes (MO, betaine 30, and coumarin-480), respectively. They suggested that the polarity of neat [bmim]BF_4 was stronger than that of the [bmim]BF_4 pool in the microemulsion. In addition, they investigated the phase diagram of the *N*-methyl-*N*-propylpyrrolidinium bis(trifluoromethylsulfonyl)imide ([Py]NTf_2)/[bmim][AOT]/benzene microemulsion and found that it was stable between 288 and 323 K [40]. The difference between the neat IL and the IL-based microemulsions was investigated by the study of solvent and rotational relaxation of C-153 in both systems. They found that the solvation dynamics of C-153 in neat [Py]NTf_2 was more sensitive to temperature than that in microemulsions containing [Py]NTf_2. Similarly, the average rotational relaxation time of C-153 in [Py]NTf_2-containing microemulsions also exhibited a smaller temperature dependence than that in neat [Py]NTf_2. In order to improve the temperature range for the application of IL-based nonaqueous microemulsions, Sarkar et al. developed a number of microemulsions using a high-boiling-point solvent, isopropyl myristate (IPM), as the oil phase instead of the low-boiling-point benzene. The surfactant in those IPM-containing microemulsions was a surface-active IL *N*,*N*-dimethylethanolammonium 1,4-bis(2-ethylhexyl) sulfosuccinate (DAAOT), and the polar phase was different ILs. It was observed that even heating to 200°C, the 1-hexyl-3-methylimidazolium bis(trifluoromethylsulfonyl) imide ([hmim]NTf_2)/DAAOT/IPM microemulsion was still transparent without phase separation [41]. Besides its good thermal stability, another characteristic of the [hmim]NTf_2/DAAOT/IPM microemulsion was the wide range of droplet size from 7.9 to 91.0 nm as a function of [hmim]NTf_2 concentration. Thus, this type of nonaqueous microemulsions may provide a promising opportunity for the synthesis of nanomaterials, where a relatively high temperature and tunable particle sizes are possible.

17.7 APPLICATION OF IL-BASED NONAQUEOUS MICROEMULSIONS

As we know, conventional microemulsions consisting of surfactants and molecular solvents have been successfully applied to various chemical processes, including dissolution, extraction, catalytic reactions, polymerization, and the synthesis of nanomaterials. ILs have also shown remarkable performances in various fields. IL-based microemulsions have advantages over the conventional microemulsions and have promise for their better effectiveness.

During dissolution/encapsulation is an important application of IL-based nonaqueous microemulsions. The use of many drugs (such as acyclovir, dantrolene sodium, and methotrexate) is restricted owing to their hydrophobicity and oleophobicity, which lead to their low bioavailability. Moniruzzaman and coworkers investigated a series of ILs as the dispersed phase of microemulsions and found that the ILs whose anions have strong hydrogen-bond formation ability can easily form microemulsions [42–44]. In that study, they used Tween-80 and sorbitan laurate (Span-20) as surfactant, and IPM as the oil phase along with an IL-assisted nonaqueous microemulsion system. The IL dimethylimidazolium dimethylphosphate ([C_1mim][$(MeO)_2PO_2$]) replaced water, which clearly improved the solubility of the drugs for transdermal delivery, and the molecules mainly got distributed in the IL core of the microemulsion (Fig. 17.7). More importantly, this microemulsion system had low cytotoxicity [44], which is very important for drug delivery applications.

IL-based nonaqueous microemulsion can also be used for preparation of silicon materials with different morphologies. By optimizing catalytic conditions, Zhao et al. synthesized two different morphologies of silica products using nonaqueous IL microemulsion ([bmim]BF_4+TX-100+benzene system) droplets as templates [45].

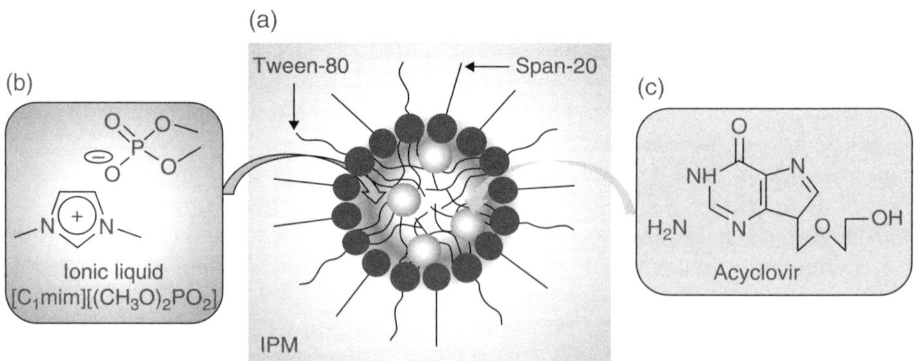

Figure 17.7 (a) Schematic of IL-in-oil microemulsions used as a pharmaceutical carrier. (b) Chemical structure of IL and (c) acyclovir. Reproduced from Moniruzzaman et al. [44] with permission from Elsevier B.V.

The product obtained under acid conditions was an ellipsoid nanoparticle, and that obtained under alkaline condition was hollow silica ensembles. The study revealed that [bmim]BF_4 played a critical role in the process of formation of the mesoporous silica hollow structure. It was thus evident that nonaqueous IL microemulsions can contribute to the synthesis of materials with various structures by providing a reaction environment that is able to control the size and morphology of the materials.

Nonaqueous IL microemulsions were also used as catalysts to improve reaction efficiency. Gayet et al. established an IL-in-oil microemulsion system with benzylpyridinium bis(trifluoromethanesulfonyl)imide ([BnPyr]NTf_2), TX-100, and toluene, in which the Matsuda–Heck reaction between methoxybenzene diazotate and 2,3-dihydrofuran took place [46]. The reaction yield in this IL-in-oil microemulsion was twice as high as that in neat ILs. The results provided a basis for designing a nonaqueous IL microemulsion microreactor and also showed that nonaqueous IL microemulsion might have good prospects of applications in biocatalysis and nanomaterial synthesis.

An important application of IL-based nonaqueous microemulsions is in polymerization. Yu et al. [47] proposed a method for the synthesis of polymer electrolytes with IL-based nonaqueous reverse microemulsions. The microemulsions consisted of styrene as the continuous phase, [bmim]BF_4 as the dispersed phase, and a polymerizable ionic compound, 1-(2-methyl acryloyloxyundecyl)-3-methylimidazolium bromide ([MAUM]Br), as the surfactant. When the radical polymerization of these microemulsions was initiated by 2, 2′-azobis-isobutyronitrile (AIBN) at 160°C, flexible and transparent [bmim]BF_4/polystyrene composite membranes were produced, which showed high conductivity. The efficiency of this approach was justified by those of methyl methacrylate, vinyl acetate, and *N*,*N*-dimethylacrylamide instead of styrene as a polymerizable oil to prepare the [bmim]BF_4-containing microemulsions. The polymerization of all these microemulsions yielded transparent [bmim]BF_4/polymer membranes with high conductivity. The intermolecular interaction and distribution of components in the [bmim]BF_4/polystyrene membranes were investigated by *in situ* FTIR and two-dimensional infrared (2DIR) correlation spectroscopy [48]. It was understood that a polymeric network was formed in the system through the physical aggregation or copolymerization of the alkyl chains in [bmim]BF_4, poly([MAUM]Br), and polystyrene, and the cations of IL were mainly located in the polymeric network, while the anions were stabilized by the anion–cation Coulomb and hydrogen-bonding interactions.

17.8 CONCLUSIONS AND OUTLOOK

In this chapter, we summarized recent progress in the structure, property, and applications of IL-based nonaqueous microemulsions. In most cases, one type of IL was included in microemulsions to act as an alternative to water, but there were also microemulsions containing two different ILs where the second IL acted as either the surfactant or the immiscible solvent. In general, the phase behavior of the IL/surfactant/solvent systems was similar to that of conventional aqueous systems, but differences also existed. The polarity of several IL-in-oil microemulsions was investigated, which are generally lower than that of corresponding neat ILs. Although one expected advantage of IL-based nonaqueous microemulsions over aqueous microemulsions would be in a larger temperature range for applications, the reported IL-based nonaqueous microemulsions that can be used at either high or low temperatures are still few, limited by the choice of appropriate surfactant and oil-phase solvent. Moreover, compared with the diversity of ILs reported and applied in various applications, the varieties of ILs utilized to form nonaqueous microemulsions are still only a few. Most of the reported microemulsions are based on dialkylimidazolium ILs with BF_4^-, PF_6^-, and NTf_2^- as anions, limiting the possibility of obtaining microemulsions with unique physicochemical properties compared with traditional systems. As a result, although there are some studies on the application of IL-based nonaqueous microemulsions in dissolution and reaction, the practical advantages of IL-based nonaqueous microemulsions over aqueous microemulsions are still under investigation. In recent years, ILs are no longer regarded as purely "green" solvents; therefore, the usefulness of such IL-based nonaqueous microemulsions largely depends on whether improved efficiency over conventional systems can be achieved by using them in practical limitations. Thorough future studies are warranted to resolve the issue.

ACKNOWLEDGMENT

The contribution of the graduate students, Mr. Yuanbang Xie, Mr. Wenbin Jin, Mr. Tianxiong Feng, Ms. Jingyi Hu, and Ms. Xiaoqing Xiang, are acknowledged with thanks. The authors are grateful to the National Natural Science Foundation of China (21106127 and 21222601) for financial support.

REFERENCES

[1] Danielsson, I.; Lindman, B. *Colloids and Surfaces* 1981, **3**, 391–392.

[2] Klier, J.; Tucker, C. J.; Kalantar, T. H.; Green, D. P. *Advanced Materials* 2000, **12**, 1751–1757.

[3] Plechkova, N. V.; Seddon, K. R. *Chemical Society Reviews* 2008, **37**, 123–150.

[4] Zech, O.; Kunz, W. *Soft Materials* 2011, **7**, 5507–5513.

[5] Gao, H. X.; Li, J. C.; Han, B. X.; Chen, W. N.; Zhang, J. L.; Zhang, R.; Yan, D. D. *Physical Chemistry Chemical Physics* 2004, **6**, 2914–2916.

[6] Eastoe, J.; Gold, S.; Rogers, S. E.; Paul, A.; Welton, T.; Heenan, R. K.; Grillo, I. *Journal of the American Chemical Society* 2005, **127**, 7302–7303.

[7] Chakrabarty, D.; Seth, D.; Chakraborty, A.; Sarkar, N. *Journal of Physical Chemistry B* 2005, **109**, 5753–5758.

[8] Li, N.; Zhang, S. H.; Zheng, L. Q.; Gao, Y.; Yu, L. *Langmuir* 2008, **24**, 2973–2976.

[9] Gao, Y.; Zhang, J.; Xu, H. Y.; Zhao, X. Y.; Zheng, L. Q.; Li, X. W.; Yu, L. *ChemPhysChem* 2006, **7**, 1554–1561.

[10] Gao, Y. N.; Wang, S. Q.; Zheng, L. Q.; Han, S. B.; Zhang, X.; Lu, D. M.; Yu, L.; Ji, Y. Q.; Zhang, G. Y. *Journal of Colloid and Interface Science* 2006, **301**, 612–616.

[11] Li, N.; Gao, Y. A.; Zheng, L. Q.; Zhang, J.; Yu, L.; Li, X. W. *Langmuir* 2007, **23**, 1091–1097.

[12] Gao, Y. A.; Li, N.; Zhang, S. H.; Zheng, L. Q.; Li, X. W.; Dong, B.; Yu, L. *Journal of Physical Chemistry B* 2009, **113**, 1389–1395.

[13] Xu, J.; Zhang, L.; Yin, A. L.; Hou, W. G.; Yang, Y. *Soft Materials* 2013, **9**, 6497–6504.

[14] Gao, Y.; Li, N.; Zheng, L. Q.; Zhao, X. Y.; Zhang, J.; Cao, Q.; Zhao, M. W.; Li, Z.; Zhang, G. Y. *Chemistry—A European Journal* 2007, **13**, 2661–2670.

[15] Gao, Y.; Li, N.; Zheng, L. Q.; Bai, X. T.; Yu, L.; Zhao, X. Y.; Zhang, J.; Zhao, M. W.; Li, Z. *Journal of Physical Chemistry B* 2007, **111**, 2506–2513.

[16] Li, N.; Cao, Q.; Gao, Y. A.; Zhang, J.; Zheng, L. Q.; Bai, X. T.; Dong, B.; Li, Z.; Zhao, M. W.; Yu, L. *ChemPhysChem* 2007, **8**, 2211–2217.

[17] Zheng, Y. J.; Eli, W. J.; Li, G. *Colloid and Polymer Science* 2009, **287**, 871–876.

[18] Zheng, Y. J.; Eli, W. *Journal of Dispersion Science and Technology* 2009, **30**, 698–703.

[19] Cheng, S. Q.; Fu, X. G.; Liu, J. H.; Zhang, J. L.; Zhang, Z. F.; Wei, Y. L.; Han, B. X. *Colloids and Surfaces A: Physicochemical and Engineering Aspects* 2007, **302**, 211–215.

[20] Falcone, R. D.; Correa, N. M.; Silber, J. J. *Langmuir* 2009, **25**, 10426–10429.

[21] Ferreyra, D. D.; Correa, N. M.; Silber, J. J.; Falcone, R. D. *Physical Chemistry Chemical Physics* 2012, **14**, 3460–3470.

[22] Pramanik, R.; Sarkar, S.; Ghatak, C.; Rao, V. G.; Setua, P.; Sarkar, N. *Journal of Physical Chemistry B* 2010, **114**, 7579–7586.

[23] Pramanik, R.; Sarkar, S.; Ghatak, C.; Rao, V. G.; Sarkar, N. *Journal of Physical Chemistry B* 2011, **115**, 2322–2330.

[24] Atkin, R.; Warr, G. G. *Journal of Physical Chemistry B* 2007, **111**, 9309–9316.

[25] Atkin, R.; Bobillier, S.; Warr, G. G. *Journal of Physical Chemistry B* 2010, **114**, 1350–1360.

[26] Ghosh, S.; Banerjee, C.; Mandal, S.; Rao, V. G.; Sarkar, N. *Journal of Physical Chemistry B* 2013, **117**, 5886–5897.

[27] Ab Rani, M. A.; Brant, A.; Crowhurst, L.; Dolan, A.; Lui, M.; Hassan, N. H.; Hallett, J. P.; Hunt, P. A.; Niedermeyer, H.; Perez-Arlandis, J. M.; Schrems, M.; Welton, T.; Wilding, R. *Physical Chemistry Chemical Physics* 2011, **13**, 16831–16840.

[28] Zhang, H.; Wu, J.; Zhang, J.; He, J. S. *Macromolecules* 2005, **38**, 8272–8277.

[29] Yang, Q. W.; Xing, H. B.; Su, B. G.; Yu, K.; Bao, Z. B.; Yang, Y. W.; Ren, Q. L. *Chemical Engineering Journal* 2012, **181**, 334–342.

[30] Zhang, D. L.; Deng, Y. F.; Li, C. B.; Chen, J. *Industrial & Engineering Chemistry Research* 2008, **47**, 1995–2001.

[31] Gomez, E.; Gonzalez, B.; Dominguez, A.; Tojo, E.; Tojo, J. *Journal of Chemical and Engineering Data* 2006, **51**, 696–701.

[32] Wei, J. J.; Su, B. G.; Liang, R. S.; Xing, H. B.; Bao, Z. B.; Yang, Q. W.; Yang, Y. W.; Ren, Q. L. *Colloids and Surfaces A: Physicochemical and Engineering Aspects* 2012, **414**, 82–87.

[33] Cheng, S. Q.; Zhang, J. L.; Zhang, Z. F.; Han, B. X. *Chemical Communications* 2007, **43**, 2497–2499.

[34] Li, J. S.; Zhang, J. L.; Han, B. X.; Peng, L.; Yang, G. Y. *Chemical Communications* 2012, **48**, 10562–10564.

[35] Zech, O.; Thomaier, S.; Bauduin, P.; Ruck, T.; Touraud, D.; Kunz, W. *Journal of Physical Chemistry B* 2009, **113**, 465–473.

[36] Zech, O.; Thomaier, S.; Kolodziejski, A.; Touraud, D.; Grillo, I.; Kunz, W. *Journal of Colloid and Interface Science* 2010, **347**, 227–232.

[37] Rojas, O.; Tiersch, B.; Rabe, C.; Stehle, R.; Hoell, A.; Arlt, B.; Koetz, J. *Langmuir* 2013, **29**, 6833–6839.

[38] Rojas, O.; Tiersch, B.; Frasca, S.; Wollenberger, U.; Koetz, J. *Colloids and Surfaces A: Physicochemical and Engineering Aspects* 2010, **369**, 82–87.

[39] Rao, V. G.; Ghosh, S.; Ghatak, C.; Mandal, S.; Brahmachari, U.; Sarkar, N. *Journal of Physical Chemistry B* 2012, **116**, 2850–2855.

[40] Rao, V. G.; Mandal, S.; Ghosh, S.; Banerjee, C.; Sarkar, N. *Journal of Physical Chemistry B* 2012, **116**, 8210–8221.

[41] Rao, V. G.; Banerjee, C.; Ghosh, S.; Mandal, S.; Kuchlyan, J.; Sarkar, N. *Journal of Physical Chemistry B* 2013, **117**, 7472–7480.

[42] Moniruzzaman, M.; Kamiya, N.; Goto, M. *Journal of Colloid and Interface Science* 2010, **352**, 136–142.

[43] Moniruzzaman, M.; Tahara, Y.; Tamura, M.; Kamiya, N.; Goto, M. *Chemical Communications* 2010, **46**, 1452–1454.

[44] Moniruzzaman, M.; Tamura, M.; Tahara, Y.; Kamiya, N.; Goto, M. *International Journal of Pharmaceutics* 2010, **400**, 243–250.

[45] Zhao, M. W.; Zheng, L. Q.; Bai, X. T.; Li, N.; Yu, L. *Colloids and Surfaces A: Physicochemical and Engineering Aspects* 2009, **346**, 229–236.

[46] Gayet, F.; El Kalamouni, C.; Lavedan, P.; Marty, J. D.; Brulet, A.; Lauth-De Viguerie, N. *Langmuir* 2009, **25**, 9741–9750.

[47] Yu, S. M.; Yan, F.; Zhang, X. W.; You, J. B.; Wu, P. Y.; Lu, J. M.; Xu, Q. F.; Xia, X. W.; Ma, G. L. *Macromolecules* 2008, **41**, 3389–3392.

[48] Shi, J. Y.; Wu, P. Y.; Yan, F. *Langmuir* 2010, **26**, 11427–11434.

CHAPTER 18

Ionic Liquid Microemulsions and Applications

XUE QIN AN

School of Chemistry and Molecular Engineering, East China University of Science and Technology, Shanghai, China

JUN SHEN

College of Continuing Education, East China University of Science and Technology, Shanghai, China

18.1 INTRODUCTION

Room-temperature ionic liquids (RTILs) have been a topic of great interest since the mid-1990s and have been suggested as potential green alternatives to volatile organic compounds due to negligible vapor pressures and low flammability of ionic liquids (ILs) [1]. RTILs are a special class of salts [2, 3], which usually are 1:1 electrolytes (composed of an anion and a cation) that have melting points below 100°C, and each of these ions can be varied, often by simple ion-exchange procedures [4–7]. RTILs are functional isotropic liquids exhibiting extremely high electrical conductivity, nonflammability, and high thermal stability [8–10]. They continue to attract interest as environmentally benign solvents for green chemistry and electrolytes [11–13]. Their low melting points and low vapor pressures are linked to the intrinsic properties of organic ions having strong electrostatic coupling. ILs have been classified as "green solvents" [14]. They can act as solvents for diverse reactions, and because the ILs are made of ions rather than molecules, such reactions often produce different selectivity and reactivity compared with conventional organic solvents. Moreover, these solvents can be easily modified for particular applications or to obtain certain properties. Hence, the term "designer solvents" has come into common use in describing RTILs. The range of IL uses has been broadened, and there has been a significant increase in the scope of both physical and chemical properties of IL [5, 15]. By taking advantage of their unique properties [4, 16], several IL applications have been described, including reaction media for many organic transformations [4, 17], in separations and extractions [4, 18], as electrolytes in electrochemistry [4, 19], in nanotechnology [4, 20], in biotechnology [4, 21], and in engineering processes [4, 22].

ILs are being increasingly studied in many different chemical applications, particularly in green solvent applications that are extending into microemulsions. The

Ionic Liquid-Based Surfactant Science: Formulation, Characterization, and Applications, First Edition. Edited by Bidyut K. Paul and Satya P. Moulik.

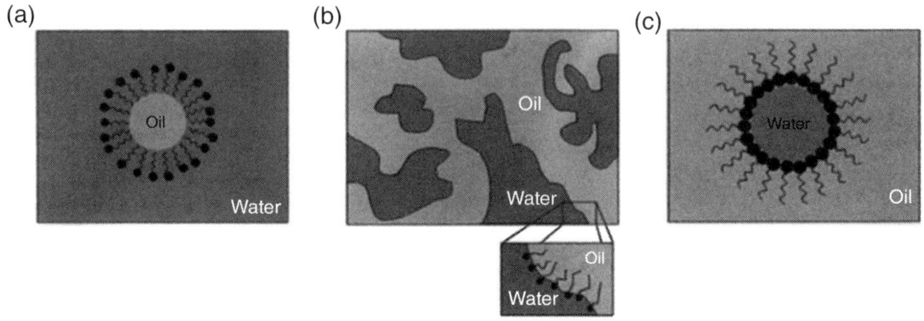

Figure 18.1 Schematic representation of the three most commonly encountered microemulsion microstructures: (a) oil-in-water, (b) bicontinuous, and (c) water-in-oil microemulsions. Reproduced from Lawrence and Rees [23] with permission from Elsevier Science.

microemulsion is a thermodynamically stable isotropic mixture of water, oil, and surfactant, in which either water or oil or both may remain dispersed and the droplet sizes are greater than micelles. The "water-in-oil," "oil-in-water," and "bicontinuous" ensembles are illustrated in Figure 18.1a–c. The term "microemulsion" was proposed by Schulman et al. [24]. Different microemulsions can be prepared by varying the concentrations of the surfactant and the dispersed phases. Microemulsion science heavily relies on characterization techniques, both from the perspective of fundamental elucidation of phase and nanostructural behavior and through identifying the unique properties that make them "fit for function" in the realm of applications [25]. ILs have good prospects in the preparation of useful microemulsions.

18.2 IL-IN-OIL MICROEMULSIONS

ILs' unique solvent properties are comparable with those of highly polar solvents, and they have already found uses in synthesis and catalytic reactions. Despite these features, solubility limitations for apolar solutes remain, which could be overcome by the incorporation of hydrocarbon domains provided by the formation of ionic liquid-in-oil (IL/O) microemulsion. The significance is that nanostructured surfactant assemblies would provide hydrophobic or hydrophilic nanodomains, thereby expanding potential applications of ILs as reaction, separation, and extraction media. In the past few years, the study of such IL-supported nonaqueous microemulsions has attracted much attention due to the favorable physical and chemical properties of ILs. The basic properties of conventional W/O/S microemulsions have led to the development of microemulsion systems in important applications, such as the delivery of food, consumer products, and pharmaceuticals, as well as in oil recovery [26–28]. Recently, IL/O microemulsions formed in mixtures of a nonionic surfactant, octylphenol ethoxylate (Triton X-100, TX-100)/([bmim][BF_4])/cyclohexane, have been reported. The formation of surfactant-stabilized dispersed nanodroplets with IL cores was demonstrated by using light scattering [29] and contrast variation small-angle neutron scattering (SANS) [30]. All three possible topologies, that is, oil-in-IL (oil is as dispersed phase in ILs), IL/O (IL is as dispersed phase in oil), and bicontinuous region, were characterized by conductivity measurements.

18.2.1 Morphology, Size, Property, and Microstructure of IL Microemulsions

The size, shape, formation mechanism, and interior polarity of the different IL microemulsions have been studied extensively by various techniques, such as freeze–fracture electron microscopy (FFEM), dynamic light scattering (DLS), conductivity study, UV–Vis spectroscopy with various probe molecules, and SANS study [30–33].

Several spectroscopic techniques have been used to study different aspects of conventional W/O/S microemulsion structures and properties. The absorption and steady-state emission spectroscopy of probe molecules solubilized in a microemulsion system can find the polarity of the microemulsion at their solvation location [34]. Time-resolved emission spectroscopy also provides information on the dynamics and rotation relaxation of solvent in both classical W/O/S and IL microemulsions [34–36]. Water content, which is defined as the molar ratio of water to total surfactant ([water]/[surfactant]), is one of the key factors in a microemulsion [37]. The ionization degree of bioactivator in IL microemulsion was correlated with the water content (ω_0) using UV–Vis absorption spectra, as shown in Figure 18.2.

The absorption spectra can explore the effect of water content on the micropolarity of IL microemulsion [38]. Formation of water pools of Triton X-100/[bmim][BF_4]/water microemulsion was confirmed by riboflavin using methyl orange (MO) as probe; the absorption spectra of MO in the microemulsion with different water contents are shown in Figure 18.3a [39]. The results suggested that when the water content is in the range of 0–5%, the absorption maximum (λ_{max}) of MO was red-shifted from 430 to 433 nm (Fig. 18.3b) [39]. Therefore, the micropolarity of bulk

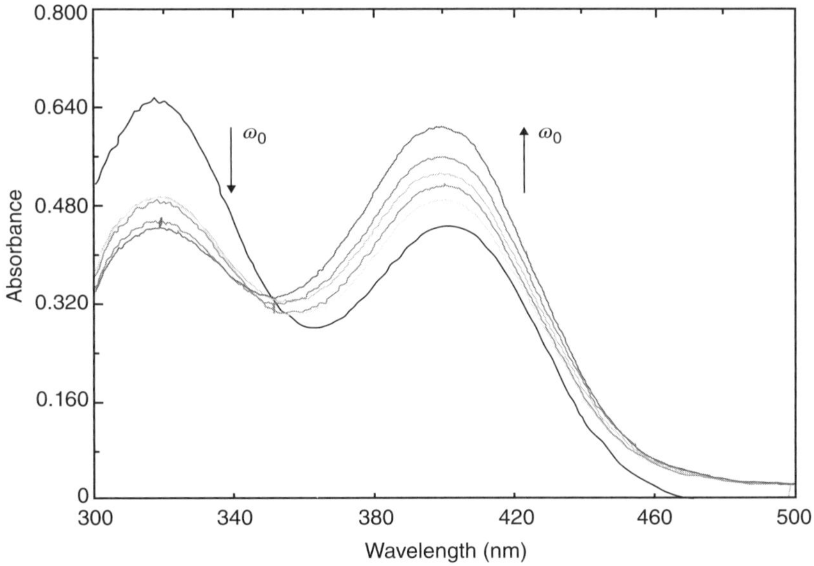

Figure 18.2 Absorption spectra of bioactivator in H_2O/AOT/TX-100/[bmin][PF_6] microemulsion. The water contents (ω_0), indicated by arrows, were 8, 10, 12, 14, and 18, respectively. Reproduced from Xue et al. [37] with permission from Elsevier Science. *See insert for color representation of the figure.*

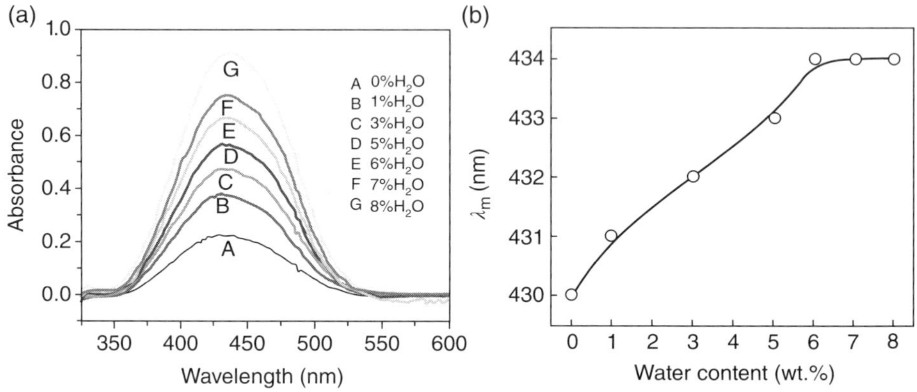

Figure 18.3 (a) Absorption spectra of MO in the Triton X-100/[bmim][BF_4]/water microemulsion as a function of water content. (b) Variation of λ_m of MO in Triton X-100/[bmim][BF_4] micelles as a function of water content. Reproduced from Li et al. [39] with permission from the American Chemical Society.

[bmim][BF_4] increased with the addition of water. The polarity of water was much higher than that of [bmim][BF_4], and the λ_{max} of MO in water was 464 nm. It meant that water was solubilized in bulk [bmim][BF_4].

Microemulsions comprising [bmim][PF_6] as lipophilic component and water stabilized by Tween-20 and by Triton X-100 surfactants were reported [40, 41]. UV–Vis absorption spectra indicated that the polarity of the water domains in water-in-[bmim][PF_6] reverse microemulsions was lower than that of bulk water. The experimental results indicated that the water domains were sufficiently well formed to support the solubility of conventional electrolytes in the water-in-[bmim][PF_6] microemulsions, and spectroscopic data indicated that the ionic species were in the aqueous pseudophase. It was demonstrated that the hydrodynamic diameter of the [bmim][PF_6]-in-water microemulsion droplets was nearly independent of the water content but increased with increasing [bmim][PF_6] content due to the swelling of the micelles by the IL. The size and morphology of IL droplet in the microemulsion depend on the water content (molar ratios of IL/TX-100), and the diameter of droplet increased with the water content as shown in Figure 18.4. The diameters of 40–47, 50–53, 55–57, and 83–90 nm were for the ratios of 0.2, 0.5, 1.0, and 1.5, respectively [29]. The surfactant aggregate sizes were somewhat higher than that of conventional microemulsions.

DLS was used to obtain the hydrodynamic diameter of dispersed droplets in the microemulsion [38]. To investigate the effect of alkyl chain length of IL on the physicochemical properties of microemulsions, ternary microemulsion systems of [C_2mim][C_nSO_4]/TX-100/cyclohexane (n of C_nSO_4:4–6) were investigated by DLS at 298 K. DLS measurements showed that the hydrodynamic diameter of IL/O microemulsion increased linearly with the increasing molar ratio of water to the total surfactant and the hydrodynamic diameter also increased with alkyl chain length of the IL, as shown in Figure 18.5 [42].

In the IL/O microemulsion, a constant polarity of the IL pool was observed after adding small amounts of water to the microemulsion [43], suggesting that the water

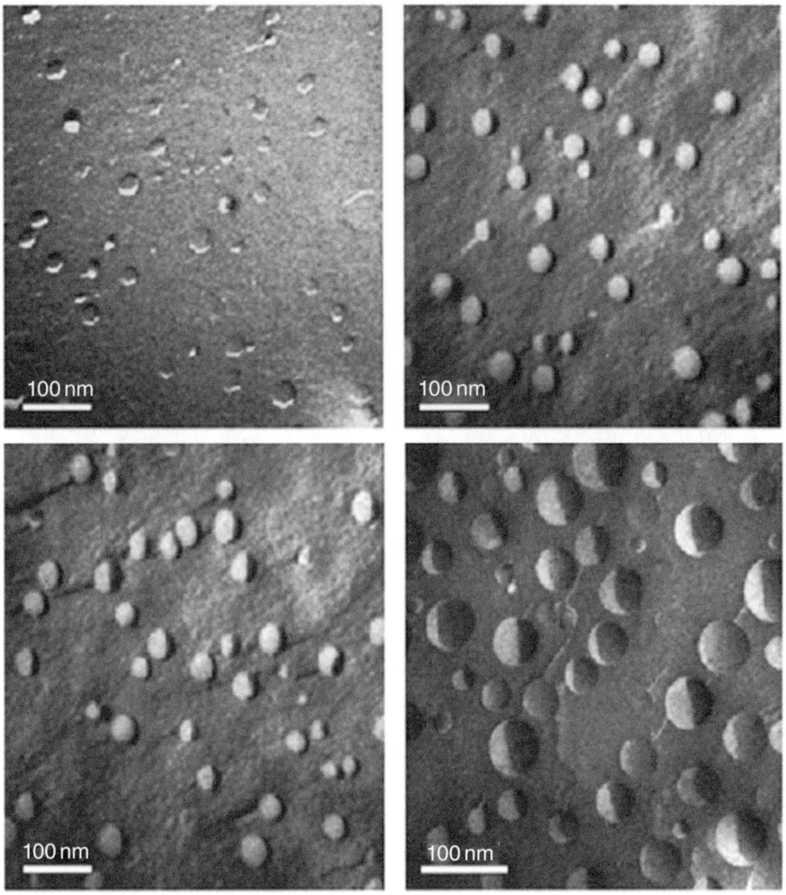

Figure 18.4 FF-TEM images of microemulsion samples of [bmim][BF_4]/TX-100/cyclohexane with the molar ratios of IL/TX-100, 0.2 (top, left), 0.5 (top, right), 1.0 (below, left), and 1.5 (below, right), respectively. The weight fraction of TX-100 is 0.45. Reproduced from Gao et al. [29] with permission from the Royal Society of Chemistry.

molecules were solubilized in the polar outer shell of the microemulsion, as shown in Figure 18.6. This solubilization behavior of added water molecules to the IL-in-benzene microemulsion system may be used as a medium to prepare porous or hollow nanomaterials by hydrolysis reactions of metal alkoxides [43].

The stability of the IL microemulsions is very important for their applications. The stability of IL [C_1mim][$(MeO)_2PO_2$] in the Tween-80/Span-20/IPM system was investigated by monitoring droplet diameter changes as well as visual observations [45]. No turbidity or precipitation was observed up to 45 days. As shown in Figure 18.7, droplet sizes and size distributions did not change significantly with time up to 45 days, confirming the physical stability of the system. This microemulsion may be of potential use for food and drug formulations, especially when water-sensitive compounds are involved.

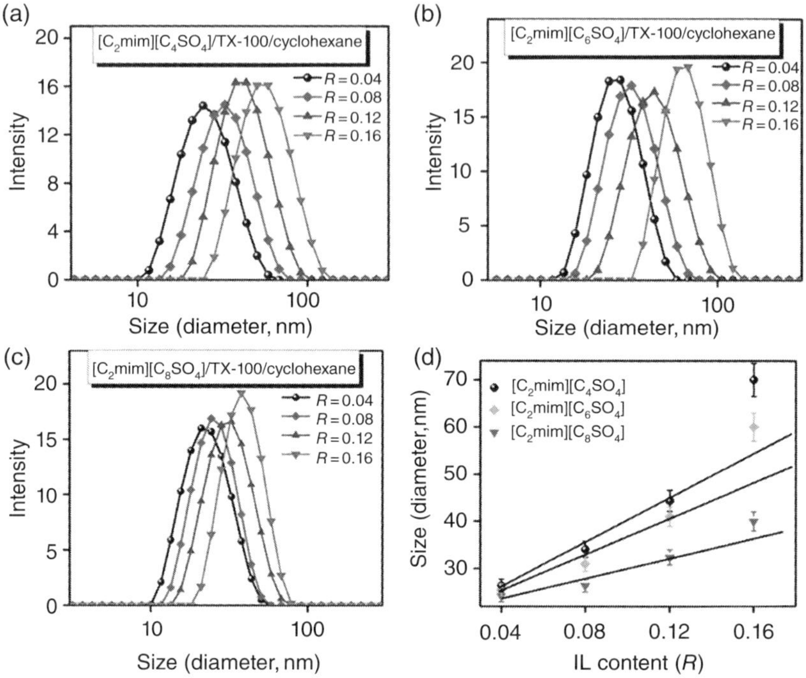

Figure 18.5 Size (diameter) distribution of the droplets (measured by dynamic light scattering) of (a) $[C_2mim][C_4SO_4]$/TX-100/cyclohexane, (b) $[C_2mim][C_6SO_4]$/TX-100/cyclohexane, (c) $[C_2mim][C_8SO_4]$/TX-100/cyclohexane microemulsions at different water content (R), and (d) diameter of the droplets of different microemulsions as a function of water content (R). Reproduced from Ghosh et al. [42] with permission from the American Chemical Society.

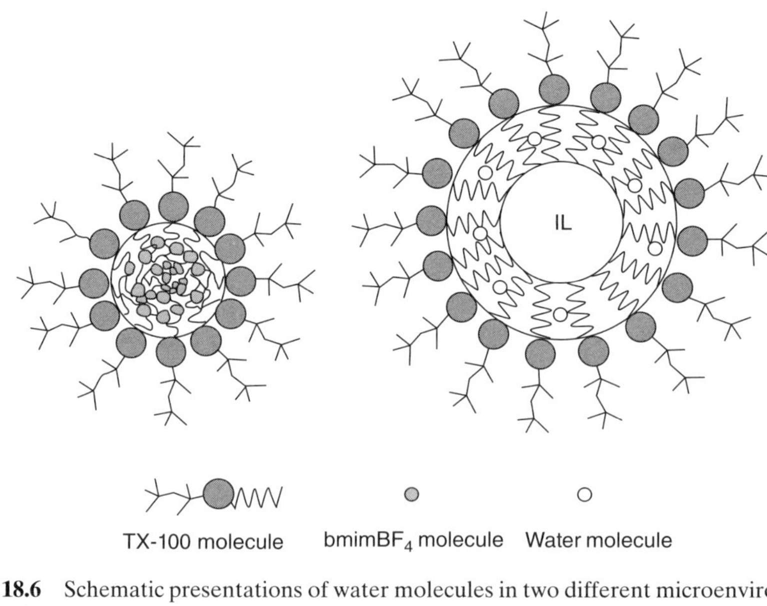

Figure 18.6 Schematic presentations of water molecules in two different microenvironments; (left) IL-in-benzene microemulsion drop lets prior to formation of IL pools; (right) IL-in-benzene microemulsion droplets with bulk-like IL pools. Reproduced from Gao et al. [44] with permission from the Royal Society of Chemistry.

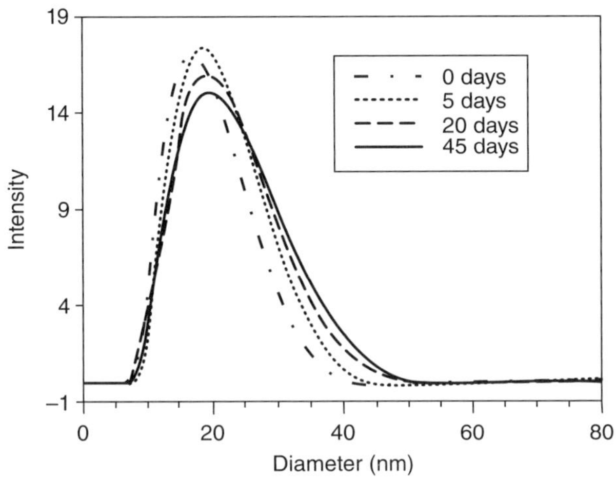

Figure 18.7 The mean hydrodynamic diameter of the Tween-80/Span-20/IL/IPM micelles with fixed water content ($R = 0.5$) was evaluated as a function of time at 25°C to evaluate physical stability. Reproduced from Moniruzzaman et al. [45] with permission from Academic Press Inc. Elsevier Science.

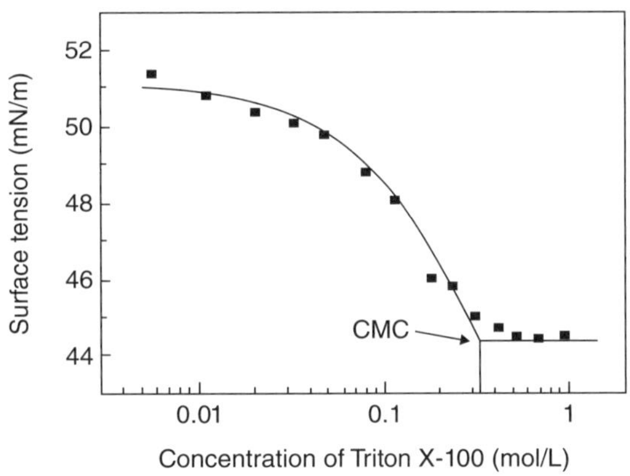

Figure 18.8 Surface tension of bmimBF_4 as a function of Triton X-100 concentration at 25°C. CMC is ca 0.33 mol/L. Reproduced from Gao et al. [46] with permission from the American Chemical Society.

The surface tension measurement is a kind of effective approach for the study of IL aggregation state. The critical micelle (or aggregation) concentration (CMC) can be determined by surface tension measurement. Surface tension measurement revealed that the dissolution of Triton X-100 in IL ([bmim][BF_4]) depressed the surface tension in a manner analogous to aqueous solutions, and a relatively higher CMC was obtained compared to that of water as shown in Figure 18.8 [46].

In order to relate the relationship of critical aggregation concentration (CAC) to thermodynamic parameters, surface tension measurements were carried out for the

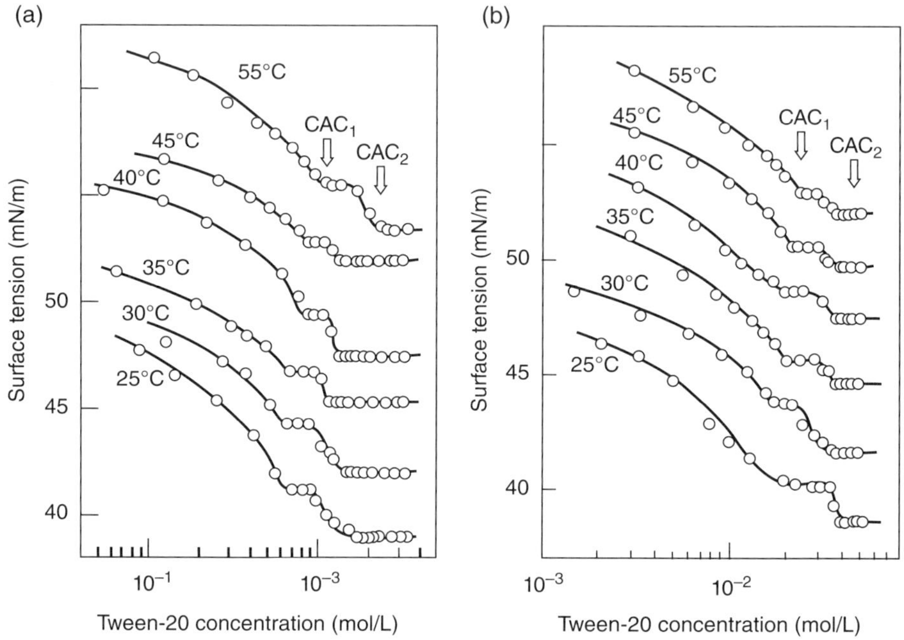

Figure 18.9 Surface tension versus concentration plots obtained for Tween-20 solutions in $bmimBF_4$ (a) and $bmimPF_6$ (b) at various temperatures. Reproduced from Wu et al. [47] with permission from the American Chemical Society.

solutions of polyoxyethylene (20) sorbitan monolaurate (Tween-20) in 1-butyl-3-methylimidazolium tetrafluoroborate ([bmim][BF_4]) and hexafluorophosphate ([bmim][PF_6]) at various temperatures. Two transition points were found in the surface tension versus concentration curves at each temperature (Fig. 18.9). The surfactant concentrations of the two transition points were regarded as the CAC1 and CAC2 (Fig. 18.9). The thermodynamic parameters of the aggregate formation, ΔG^0_{agg}, ΔH^0_{agg}, and ΔS^0_{agg}, were estimated from the CAC values and their temperature dependence. The thermodynamic parameters related to CAC were almost independent of temperature [47].

18.2.2 Phase Equilibrium and Critical Phenomena in the IL-in-Oil Microemulsion

Phase behavior study is an essential step for the characterization of microemulsions and surfactant solutions [42]. IL microemulsions exhibit complex phase behavior in their phase diagrams. Microemulsions comprising a nonionic surfactant, Triton X-100 (TX-100), cyclohexane as nonpolar phase, and RTILs as a polar medium have been recently reported [42]. To investigate the effect of alkyl chain length of IL on the physicochemical properties of microemulsions, the phase behavior of ternary systems was investigated by direct observations of transitions from clear transparent solution to turbid solution by using UV–Vis spectrophotometer at 298 K. The single-phase region was found to increase with the increase in chain length of IL anion.

Phase behavior study showed that with an increase in the alkyl chain of the IL anion, the single-phase region increased. Variation in chain length of IL anions was used for tuning the range of microemulsion area [42].

The formation of IL/O microemulsions in mixtures of [bmim][BF_4] (IL) and cyclohexane, stabilized by the nonionic surfactant, TX-100 has been proved [30]. Three-component mixtures could form IL/O microemulsions of well-defined droplet size determined by fixing the water content (mole ratio of IL to TX-100) [30, 48, 49]. An upper critical point (T_c) was observed in the mixture {([bmim][BF_4]/TX-100) + cyclohexane} with fixed water content (mole ratio of [bmim][BF_4] to TX-100) [50]. The mixture separated into two microemulsion phases of different composition but with the same composition below T_c as occurred in other systems [48]. The microemulsion system, [bmim][BF_4]/TX-100 + cyclohexane, could be regarded as a pseudobinary mixture of [bmim][BF_4]/TX-100 IL droplets dispersed in the cyclohexane continuous phase. Therefore, the phase behavior could be depicted in a two-dimensional diagram with concentration of droplets along the abscissa and temperature along the ordinate. A coexistence curve of temperature (T) against a concentration variable, such as volume fraction (Φ_c), could then be drawn in the same way as it was done for pseudobinary mixtures in AOT/water/decane microemulsions [48].

The coexistence curves (T, n), (T,Φ), and (T, Ψ) (n, Φ, and Ψ are the refractive index, volume fraction, and effective volume fraction, respectively) for the IL microemulsion systems of {polyoxyethylene tert-octylphenyl ether (TX-100) + 1-butyl-3-methylimidazolium tetrafluoroborate ([bmim][BF_4]) + cyclohexane} with water content (various molar ratios of [bmim][BF_4] to TX-100) have been determined by measuring refractive indices at a constant pressure in the critical region (Fig. 18.10) [50]. The critical temperatures (T_c) and critical volume fraction (Φ_c) were obtained

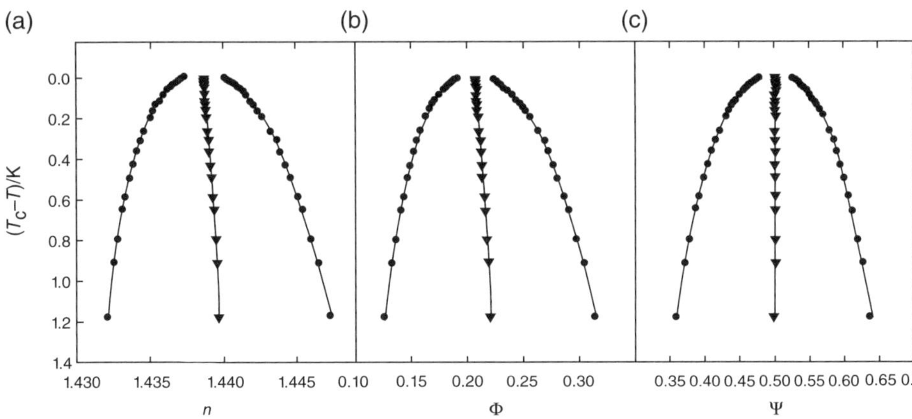

Figure 18.10 Coexistence curves of (T, n), (T, Φ), and (T, Ψ) for the microemulsion of {TX-100 + [bmim][BF_4] + cyclohexane} (a) T versus n; (b) T versus Φ; and (c) T versus Ψ; ● experimental values of concentration variables (ρ) of coexisting phases; ▼, experimental values of diameter (ρ_d) of the coexisting phases; and the lines are concentration variables (ρ_{cal}) and diameter (ρ_{dcal}) calculated from equations. Reproduced from Lü et al. [50] with permission from Academic Press Ltd. Elsevier Science Ltd.

for the IL microemulsions. The critical exponents were deduced precisely from the coexistence curves within about 1 K below T_c, and the values were consistent with the 3D Ising value [49, 50]. It has been found that this system may be taken as a pseudobinary system; the values of critical exponent β of the microemulsion approach the 3D Ising value of 0.327 within the experimental uncertainties when the temperature was close to the critical point.

18.3 IL MICROEMULSION AS REACTION MEDIUM

In recent years, microemulsions have been found to be unique and versatile media for a variety of chemical reactions, such as nanoparticle preparation [41], organic synthesis [51], and bioorganic synthesis [37]. IL microemulsions (a sketch map of the water pool shown in Fig. 18.11) also have prospects in this field. They may be water in oil (IL) or oil (IL) in water as well as bicontinuous. Because of its small size and monodispersion, this special microenvironment is called microreactor or nanoreactor [52]. The size of the microreactor is controlled by the molar ratio of IL/water to surfactant in the IL/W/O microemulsion. Their use as reaction media has the advantage that they constitute efficient solvents for both inorganic salts and nonpolar organic molecules [53–55], they can accelerate a reaction [56–58], alter the course of a reaction in limited quarters [59–61], perturb equilibrium [62, 63], contort conformations [64], and either stabilize reactive reagents [65] or reaction intermediates [66–68]. Intermolecular chemical reactions of two or more substrates encapsulated in the molecular cage can be greatly accelerated and potentially controlled due to the dramatically increased effective molarity and the strictly regulated orientation of

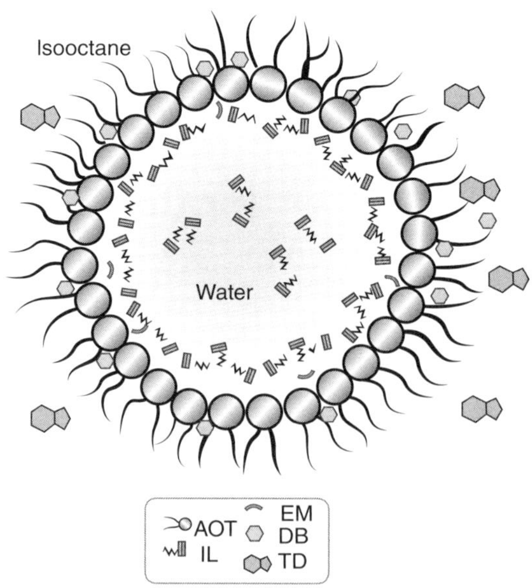

Figure 18.11 Sketch map for water pool of the IL microemulsion system. Reproduced from Lü et al. [52] with permission from Wiley Blackwell.

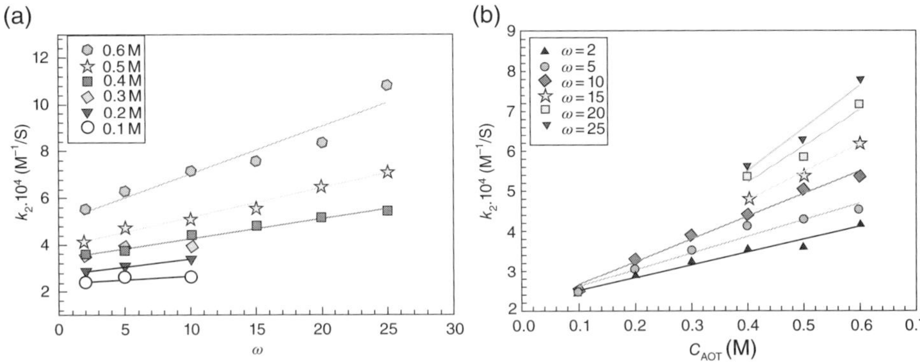

Figure 18.12 The apparent second-order rate (k_2) of Diels–Alder reaction (DAR) between *N*-ethylmaleimide and 2,3-dimethyl-1,3-butadiene at 313.15 K: (a) in the IL microemulsion with various molar ratios of IL to surfactant (ω) and surfactant concentration (C_{AOT}); (b) in the microemulsion with various molar ratios of water to surfactant (ω) and surfactant concentration (C_{AOT}). Reproduced from Lü et al. [52] with permission from Wiley Blackwell.

the substrates inside the cavities [57, 58]. Engberts and coworkers have carried out several Diels–Alder reaction (DAR) kinetic measurements in micelles and microemulsions [69–71], and the reaction rates relatively enhanced compared to that in ordinary organic solvents [69, 72, 73]. IL/water-based microemulsions have special structure and properties and can behave as a new type of template for chemical reactions [74]. Nevertheless, the IL-modified microemulsion used as DAR medium is still very rare.

The DAR between *N*-ethylmaleimide and 2,3-dimethyl-1,3-butadiene was studied in microemulsions with IL [52]. The apparent second-order rate constants were determined by spectrophotometry in the IL microemulsion and AOT microemulsion (Fig. 18.12). They increased with both water content (molar ratio of IL aqueous solution to surfactant) and surfactant concentration (C_{AOT}) in the IL microemulsion systems. The effect of various solvents on the DAR rate was investigated, and results were shown that the rate constants k_2 in both AOT microemulsion and IL microemulsion were roughly four to five times higher than that in the isooctane and the k_2 in pure IL were at least 10 times higher than that in the isooctane as shown in Figure 18.13. The results indicated that high temperature accelerated the reaction and the apparent activation energy, E_a, decreased with the increase of the water content (molar ratio of IL aqueous solution to surfactant).

18.4 IL MICROEMULSION AS DRUG DELIVERY SYSTEMS

The tremendous potentials of microemulsions as modern colloidal carriers for topical and transdermal drug delivery are well recognized [75]. Microemulsions provide a variety of advantages for pharmaceutical use, such as nanometer-sized aggregations, long-term stability, biocompatibility, straightforward preparation, and high solubilization capacity for drug molecules, and enhanced drug delivery [26, 76, 77]. However, drug delivery using such self-assembled systems cannot display its

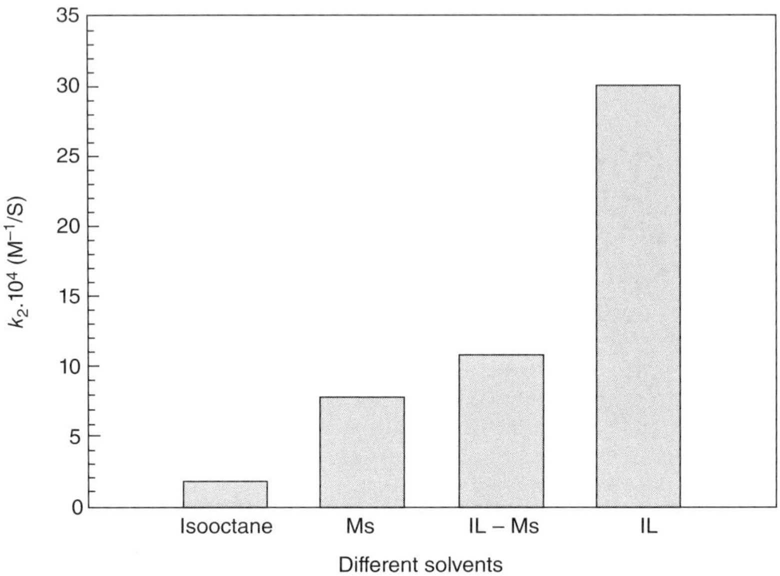

Figure 18.13 Comparison of apparent reaction rate in different solvents at 313.15°C. Reproduced from Lü et al. [52] with permission from Wiley Blackwell.

potential for a large number of drugs, which are insoluble or sparingly soluble in water. These limitations demand the need for alternative solvents that can be used as a polar core in microemulsions for the solubilization of sparingly soluble drug molecules. Considering these points, ILs could be a good candidate because they are capable of dissolving various poorly soluble compounds including drug molecules [78, 79]. IL microemulsion droplets formed in a continuous oil phase using nontoxic surfactants could be efficient carriers for sparingly soluble drugs, which are dissolved in ILs. In fact, IL/O nonaqueous microemulsions would load drug molecules in the core, and the oil phase can be used to obtain the desired features for drug transport behavior. A novel IL-assisted nonaqueous microemulsion prepared with pharmaceutical accepted components can be effectively used in solubilization of a number of drug molecules that are insoluble or poorly soluble in water and most of organic liquids [79]. DLS study clearly demonstrated the formation of spherical micelles with a range of 8–34 nm in diameter. The successful formation of a microemulsion with IL domains may be attributed to the favorable interfacial properties provided by a blend of Tween-80 and Span-20 and the strong tendency of the surfactant head groups to bind IL through hydrogen bonding as well as electrostatic interaction. The effect of IL contents on the solubility of drug in microemulsions was studied. The solubility increased with the increase of IL as shown in Figure 18.14. It meant that IL played the main role in solubilization of drugs in IL-based microemulsions. Since the solubility of all drugs in isopropyl myristate is very low, it could be assumed that the drug molecule existed in the hydrophilic IL core, which was stabilized by the surfactant mixtures creating an interfacial film between IPM and IL (Fig. 18.15).

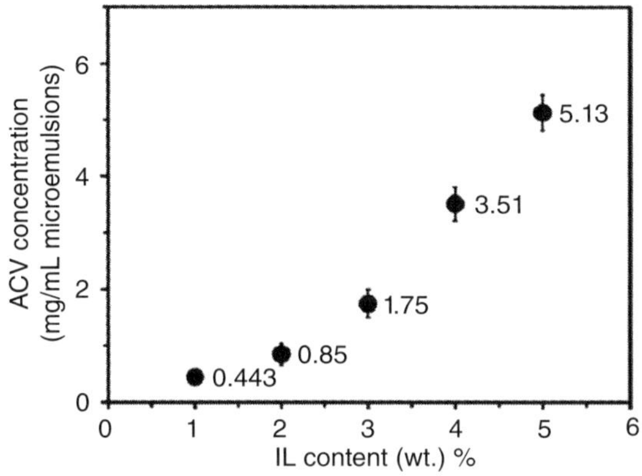

Figure 18.14 Dependence of acyclovir solubility on the ionic liquid content used for formulation of microemulsions. [Surfactant] = 20 wt.%, where the ratio of Tween-80 to Span-20 was 2:3. Reproduced from Moniruzzaman et al. [45] with permission from Academic Press Inc. Elsevier Science.

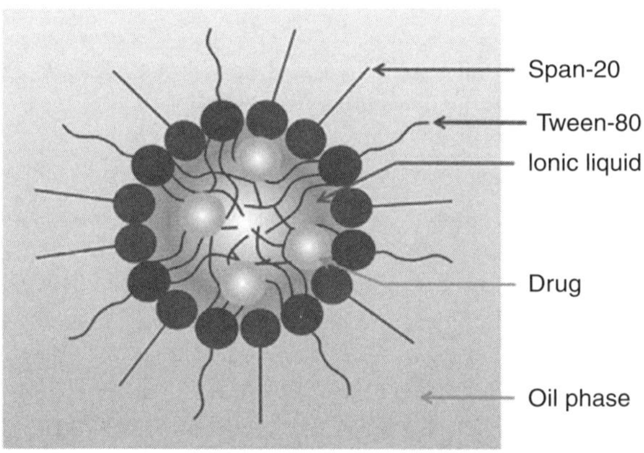

Figure 18.15 Schematic representation of ionic liquid-in-oil (IL/O) microemulsions containing drug molecules. Reproduced from Moniruzzaman et al. [45] with permission from Academic Press Inc. Elsevier Science.

18.5 CONCLUSIONS AND OUTLOOK

IL-derived microemulsions have importance both in basic and applied researches. The research progress of IL microemulsions is reviewed in this article, which includes the morphology, formation, property, and microstructure of IL microemulsions and their applications. The article focuses on phase equilibrium and critical phenomena, functional reaction media, and drug carrier for IL microemulsions. The IL microemulsions may have potential applications in reaction media, extraction processes, drug deliveries, and so on. It is suggested that applying the combination of green properties of ILs with their unique tailor-made physicochemical properties, preparation of potential biocompatible ILs will be available in the near future for the use of

ILs as pharmaceutical solvents, alternative media for reactions, functional solvents for nanoparticle synthesis, extraction, etc.

The issues that should concern us are the following: (i) new method to explore the microstructure of IL microemulsions, (ii) relationship between microstructure and properties, (iii) relationship between microstructure and function, and (iv) newer biocompatible IL microemulsions as drug carriers. In spite of continued researches devoted in the field of IL-derived microheterogeneous systems, the achievement is minor compared to the requirement.

ACKNOWLEDGMENTS

This work was supported by the National High-Tech R&D (863) Program of China (2011AA06A107), the National Natural Science Foundation of China (201473055, 21273073 and 21073063).

REFERENCES

[1] Earle, M. J.; Esperanca, J. M. S. S.; Gilea, M. A.; Lopes, J. N. C.; Rebelo, L. P. N.; Magee, J. W.; Seddon, K. R.; and Widegren, J. A. *Nature* 2006, **439**, 831–834.

[2] Welton, T. *Chem. Rev.* 2005, **9**, 2071–2084.

[3] Handy, S. T. *Curr. Org. Chem.* 2005, **9**, 959–988.

[4] Sowmiah, S.; Srinivasadesikan, V.; Tseng, M.-C.; Chu, Y.-H. *Molecules* 2009, **14**, 3780–3813.

[5] Torimoto, T.; Tsuda, T.; Okazaki, K.; Kuwabata, S. *Adv. Mater.* 2010, **22**, 1196–1221.

[6] MacFarlane, D. R.; Pringle, J. M.; Johansson, K. M.; Forsyth, S. A.; Forsyth, M. *Chem. Commun.* **2006**, 1905–1917.

[7] Rogers, R. D.; Voth, G. A. *Acc. Chem. Res.* 2007, **40**, 1077–1078.

[8] Galinski, M.; Lewandowski, A.; Stepniak, I. *Electrochim. Acta* 2007, **51**, 5567–5580.

[9] Aliaga, C.; Santos, C. S.; Baldelli, S. *Phys. Chem. Chem. Phys.* 2007, **9**, 3683–3700.

[10] Yang, Q. L.; Dionysiou, D. D. *ACS Symp. Ser.* 2005, **902**, 182–198.

[11] Horvath, I. T.; Anastas, P. T. *Chem. Rev.* 2007, **107**, 2169–2173.

[12] Joglekar, H. G.; Rahman, I.; Kulkarni, B. D. *Chem. Eng. Technol.* 2007, **30**, 819–828.

[13] Imperato, G.; Konig, B.; Chiappe, C. *Eur. J. Org. Chem.* 2007, **7**, 1049–1058.

[14] Liu, S. F.; Xiao, J. L. *J. Mol. Catal. A Chem.* 2007, **270**, 1–43.

[15] MacFarlane, D. R.; Seddon, K. R. *Aust. J. Chem.* 2007, **60**, 3–5.

[16] Earle, M. J.; Seddon, K. R. *Pure Appl. Chem.* 2000, **72**, 1391–1398.

[17] (a) Haumann, M.; Riisager, A. *Chem. Rev.* 2008, **108**, 1474–1497; (b) Miao, W. S.; Chan, T. H. *Acc. Chem. Res.* 2006, **39**, 897–908.

[18] (a) Han, X.; Armstrong, D. W. *Acc. Chem. Res.* 2007, **40**, 1079–1086; (b) Huddleston, J. G.; Willauer, H. D.; Swatloski, R. P.; Visser, A. E.; Rogers, R. D. *Chem. Commun.* 1998, **34**, 1765–1766.

[19] (a) Hapiot, P.; Lagrost, C. *Chem. Rev.* 2008, **108**, 2238–2264; (b) Mac-farlane, R.; Forsyth, M. P. C.; Howlett, J. M.; Pringle, J.; Sun, G.; Annat, W.; Neil, E. I.; Izgorodina, E. I. *Acc. Chem. Res.* 2007, **40**, 1165–1173.

[20] (a) Ichikawa, T.; Yoshio, M.; Hamasaki, A.; Mukai, T.; Ohno, H.; Kato, T. *J. Am. Chem. Soc.* 2007, **129**, 10662–10663; (b) Biswas, K.; Rao, C. N. R. *Chem. Eur. J.* 2007, **13**, 6123–6129.

[21] van Rantwijk, F.; Sheldon, R. A. *Chem. Rev.* 2007, **107**, 2757–2785.

[22] (a) Greaves, T. L.; Drummond, C. J. *Chem. Rev.* 2008, **108**, 206–237; (b) Schneider, S.; Hawkins, T.; Rosander, M.; Vaghjiani, G.; Chambreau, S.; Drake, G. *Energy Fuel* 2008, **22**, 2871–2872.

[23] Lawrence, M. J.; Rees, G. D. *Adv. Drug Deliv. Rev.* 2000, **45**, 89–121.

[24] Schulman, J. H.; Stoeckenius, W.; Prince, L. M. *J. Phys. Chem.* 1959, **63**, 1677–1680.

[25] Acharya, D. P.; Hartley, P. G. *Curr. Opin. Colloid Interface Sci.* 2012, **17**, 274–280.

[26] Gupta, S.; Moulik, S. P. *J. Pharm. Sci.* 2008, **97**, 22–45.

[27] Gupta, S. *Curr. Sci.* 2011, **101**, 174–188.

[28] Paul, B. K.; Moulik, S. P. *Curr. Sci.* 2001, **80**, 990–1001.

[29] Gao, H.; Li, J. Han, B.; Chen, W.; Zhang, J.; Zhang, R. et al. *Phys. Chem. Chem. Phys.* 2004, **6**, 2914–2916.

[30] Eastoe, J.; Gold, S.; Rogers, S. E.; Paul, A.; Welton, T.; Heenan, R. K. et al. *J. Am. Chem. Soc.* 2005, **127**, 7302–7303.

[31] Gao, Y. A.; Zhang, J.; Xu, H. Y.; Zhao, X. Y.; Zheng, L. Q.; Li, X. W.; Yu, L. *ChemPhysChem* 2006, **7**, 1554–1559.

[32] Chakrabarty, D.; Seth, D.; Chakraborty, A.; Sarkar, N. *J. Phys. Chem. B* 2005, **109**, 5753–5758.

[33] Zech, O.; Thomaier, S.; Kolodziejski, A.; Touraud, D.; Grillo, I.; Kunz, W. *J. Colloid Interface Sci.* 2010, **347**, 227–232.

[34] Saha, R.; Rakshit, S.; Mitra, R. K., Pal, S. K.; *Langmuir* 2012, **28**, 8309–8317.

[35] Pramanik, R.; Sarkar, S.; Ghatak, C.; Rao, V. G.; Sarkar, N. *Chem. Phys. Lett.* 2011, **512**, 217–222.

[36] Seth, D.; Chakraborty, A.; Setua, P.; Sarkar, N. *J. Phys. Chem. B* 2007, **111**, 4781–4787.

[37] Xue, L. Y.; Li, Y.; Zou, F. X.; Lu, L.; Zhao, Y.; Huang, X. R.; Qu, Y. B. *Colloids Surf. B Biointerfaces* 2012, **92**, 360–366.

[38] Pramanik, R.; Sarkar, S.; Ghatak, C.; Setua, P.; Rao, V. G.; Sarkar, N. *Chem. Phys. Lett.* 2010, **490**, 154–158.

[39] Li, N.; Zhang, S.; Ma, H. Zheng, L. *Langmuir* 2010, **26**, 9315–9320.

[40] Gao, Y.; Li, N.; Zheng, L. Q.; Zhao, X. Y.; Zhang, S. H.; Han, B. X. et al. *Green Chem.* 2006, **8**, 43–49.

[41] Gao, Y.; Han, S. B.; Han, B. X.; Li, G. Z.; Shen, D.; Li, Z. H. et al. *Langmuir* 2005, **21**, 5681–5684.

[42] Ghosh, S.; Banerjee, C.; Mandal, S.; Rao, V. G.; Sarkar, N. *J. Phys. Chem. B* 2013, **117**, 5886–5897.

[43] Qiu, Z.; Texter, J. *Curr. Opin. Colloid Interface Sci.* 2008, **13**, 252–262.

[44] Gao, Y.; Li, N.; Zheng, L. Q.; Zhao, X. Y.; Zhang, J.; Cao, Q. et al. *Chem. Eur. J.* 2007, **13**, 2661–2670.

[45] Moniruzzaman, M.; Kamiya, N.; Goto, M. *J. Colloid Interface Sci.* 2010, **352**, 136–142.

[46] Gao, Y.; Li, N.; Li, X. W. Zhang, S. H.; Zheng, L. Q.; Bai, X. T. and Yu, L. *J. Phys. Chem. B* 2009, **113**, 123–130

[47] Wu, J.; Li, N.; Zheng, L.; Li, X.; Gao, Y.; Inoue, T. *Langmuir* 2008, **24**, 9314–9322.

[48] An, X.; Chen, J.; Huang, Y.; Shen, W. *J. Colloid Interface Sci.* 1998, **203**, 140–145.

[49] Lu, H.; An, X.; Shen, W. *J. Chem. Eng. Data* 2011, **56**, 502–507.

[50] Lü, H.; An, X.; Shen, W. *J. Chem. Thermodyn.* 2012, **48**, 48–53.

[51] Lu, H. L.; An, X. Q.; Chai, H. L. *Colloid J.* 2013, **75**, 181–184.

[52] Lü, H.; An, X.; Yu, J.; Song, X. *J. Phys. Org. Chem.* 2012, **25**, 1210–1216.

[53] Witula, T.; Holmberg, K. *Langmuir* 2005, **21**, 3782–3785.

[54] Häger, M.; Currie, F.; Holmberg, K. *Colloids Surf. A Physicochem. Eng. Asp.* 2004, **250**, 163–170.

[55] Evilevitch, A.; Olsson, U.; Jőnsson, B.; Wennerstrőm, H. *Langmuir* 2000, **16**, 8755–8762.

[56] MacGillivray, L. R.; Atwood, J. L. *Angew. Chem. Int. Ed.* 1999, **38**, 1019–1034.

[57] Yoshizawa, M.; Tamura, M.; Fujita, M. *Science* 2006, **312**, 251–255.

[58] Hastings, C. J.; Fiedler, D.; Bergman, R. G.; Raymond, K. N. *J. Am. Chem. Soc.* 2008, **130**, 10977–10983.

[59] Pluth, M. D.; Bergman, R. G.; Raymond, K. N. *Science* 2007, **316**, 85–89.

[60] Kuil, M.; Soltner, T.; van Leeuwen, P. W. N. M.; Reek, J. N. H. *J. Am. Chem. Soc.* 2006, **128**, 11344–11345.

[61] Yoshizawa, M.; Takeyama, Y.; Okano, T.; Fujita, M. *J. Am. Chem. Soc.* 2003, **125**, 3243–3247.

[62] Dong, V. M.; Fiedler, D.; Carl, B.; Bergman, R. G.; Raymond, K. N. *J. Am. Chem. Soc.* 2006, **128**, 14464–14465.

[63] Trembleau, L.; Rebek, J. J. *Science* 2003, **301**, 1219–1220.

[64] Ajami, D.; Rebek, J. J. *Nat. Chem.* 2009, **1**, 87–90.

[65] Mal, P.; Breiner, B.; Rissanen, K.; Nitschke, J. R. *Science* 2009, **324**, 1697–1701.

[66] Yoshizawa, M.; Kusukawa, T.; Fujita, M.; Yamaguchi, K. *J. Am. Chem. Soc.* 2000, **122**, 6311–6312.

[67] Yoshizawa, M.; Miyagi, S.; Kawano, M.; Ishiguro, K.; Fujita, M. *J. Am. Chem. Soc.* 2004, **126**, 9172–9173.

[68] Garti, N.; Holmberg, K. *Curr. Opin. Colloid Interface Sci.* 2003, **8**, 135–136.

[69] Rispens, T.; Engberts, J. B. F. N. *J. Org. Chem.* 2002, **67**, 7369–7371.

[70] Engberts, J. B. F. N.; Fernández, E.; García-Río, L.; Leis, J. R. *J. Org. Chem.* 2006, **71**, 4111–4117.

[71] Engberts, J. B. F. N.; Fernández, E.; García-Río, L.; Leis, J. R. *J. Org. Chem.* 2006, **71**, 6118–6123.

[72] Gayet, F.; Kalamouni, C. E.; Lavedan, P.; Marty, J. D.; Brûlet, A.; Viguerie, N. L. *Langmuir* 2009, **25**, 9741–9750.

[73] Liu, L.; Bauduin, P.; Zemb, T.; Eastoe, J.; Hao, J. *Langmuir* 2009, **25**, 2055–2059.

[74] Falcone, R. D.; Baruah, B.; Gaidamauskas, E.; Rithner, C. D.; Correa, N. M.; Silber, J. J.; Crans, D. C.; Levinger, N. E. *Chem. Eur. J.* 2011, **17**, 6837–6846.

[75] Rősler, A.; Vandermeulen, G. M. W.; Klok, H. A.; *Adv. Drug Deliv. Rev.* 2001, **53**, 95–108.

[76] Heuschkel, H.; Goebel, A.; Neubert, R. H. H. *J. Pharm. Sci.* 2008, **97**, 603–631.

[77] Kogan, A.; Garti, N. *Adv. Colloid Interface Sci.* 2006, **369**, 123–126.

[78] Jaitely, V.; Karatas, A.; Florence, A. T. *Int. J. Pharm.* 2008, **354**, 168–173.

[79] Moniruzzaman, M.; Tahara, Y.; Tamura, M.; Kamiya, N.; Goto, M. *Chem. Commun.* 2010, **46**, 1452–1454.

CHAPTER 19

Ionic Liquid-In-Oil Microemulsions

DEBOSTUTI GHOSH DASTIDAR and SANJIB SENAPATI

Department of Biotechnology, Indian Institute of Technology Madras, Chennai, India

19.1 INTRODUCTION

Ionic liquids (ILs) are organic salts, which are liquid at temperatures below 100°C. ILs that are liquid at or around room temperature are called room-temperature ionic liquids (RTILs). They are commonly comprised of voluminous organic cations and a wide range of anions and are characterized by low vapor pressure, high chemical and thermal stability, low toxicity, and nonflammability [1–4]. Their thermophysical properties, such as density, viscosity, and hydrophilicity, can be easily tuned through careful selection of cation–anion combinations [5–7]. These "designer" solvents, therefore, continue to find their expanding use in polymer chemistry [2, 8, 9], material chemistry [2, 10], electrochemistry [11, 12], biotechnology [13–18], and various other applications [19–28].

The essentially polar nature of commonly used ILs, however, limits their applications in areas that necessitate the use of certain nonpolar solutes. For example, while ILs have emerged as promising media for performing chemical reactions that require a water-free environment like Diels–Alder reaction, esterification, and polymerization, their applications are sometimes limited due to the poor solvation of nonpolar reactants and products [29]. Likewise, while ILs are known to solubilize several important drugs that are sparingly soluble in water, transdermal delivery of drug–IL formulations is limited by the hydrophobic barriers of the skin against polar ILs [30, 31].

The aforesaid limitations of dissolving apolar solutes in ILs can be overcome by incorporating hydrophobic domains in the IL-containing solution by the formation of microemulsions. A microemulsion is a thermodynamically stable dispersion of two immiscible liquids, a polar and an apolar phase, stabilized by an adsorbed surfactant film at the liquid–liquid interface [32, 33]. Water has been conventionally used as the polar phase in microemulsions, since amphiphilic surfactants spontaneously self-assemble in water [34–46]. In 1982, Evans and coworkers showed, for the first time, the formation of self-assembled micellar structure in the RTIL ethylammonium nitrate (EAN) [47]. Since then, a large number of protic and aprotic ILs have been shown to support the self-assembly of surfactants in IL–surfactant binary

Ionic Liquid-Based Surfactant Science: Formulation, Characterization, and Applications, First Edition. Edited by Bidyut K. Paul and Satya P. Moulik.

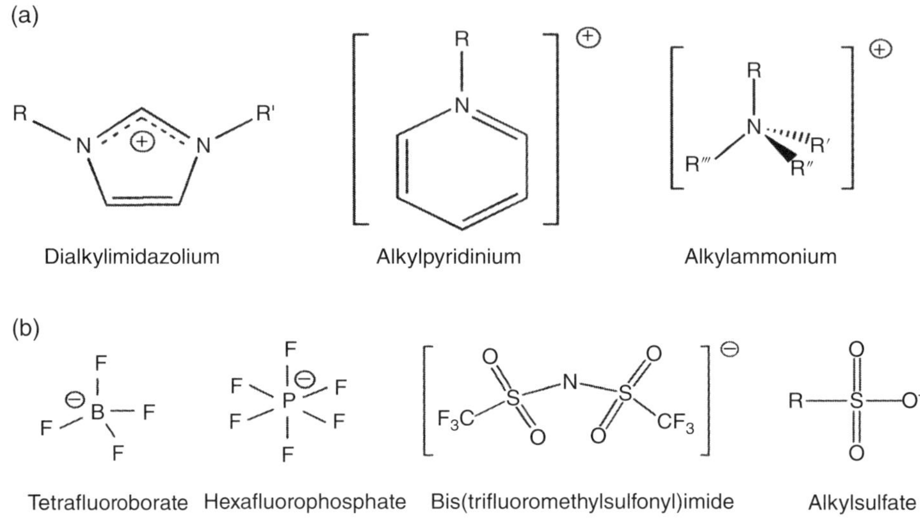

Figure 19.1 (a) Cationic and (b) anionic constituents of ILs that are commonly used in preparation of IL-in-oil microemulsions.

systems [48–57]. Subsequently, a large number of ternary microemulsions with IL as the polar phase have been studied [58–88]. Some ILs commonly used in the preparation of such microemulsions are listed in Figure 19.1. Beginning with the work of Han and coworkers, early IL-based ternary microemulsions predominantly used the aprotic IL [bmim][BF_4] as the polar phase [58–64]. In later years, the possibility of using ILs composed of pyrrolidinium [65], pyridinium [66], and alkyl ammonium [67, 68] cations and hexafluorophosphate (PF_6) [68, 69], bis(trifluoromethanesulfonyl) amide (Tf_2N) [65–67, 70, 71], trifluoromethanesulfonate (TfO), trifluoroacetate (TfA) [72], and alkylsulfate [73–75] anions was explored. The use of alkylsulfate anions made it possible to substitute the widely used, yet comparatively more toxic, halogen-containing ILs to form greener microemulsions [73–75].

The apolar component in an IL-based microemulsion can be an organic or inorganic solvent that is partially miscible or immiscible with the polar IL. The most commonly used solvents for this purpose include cyclohexane [58, 61], benzene [62], toluene [59], *p*-xylene [60], triethylamine [63], and chlorobenzene [70]—generously called as "oil." However, the toxic nature and low boiling point of these oils (e.g., 80°C for benzene and 81°C for cyclohexane) often limit the applications of the resulting microemulsions. This has recently prompted the use of isopropyl myristate (IPM), a biological oil with a boiling point of approximately 167°C, as a suitable substitute for conventional organic solvents in the formation of ionic liquid-in-oil (IL/O) microemulsions [75–79]. Apart from the conventional solvents, newer solvents like supercritical CO_2 [80, 81] and nonpolar ILs [82, 83] are also being used as the apolar phase in recent years. However, the characterization and applications of the corresponding microemulsions are limited and therefore will not be reviewed in this work.

The two immiscible phases in the microemulsion are brought together by means of an appropriate surfactant. Surfactants are amphiphilic molecules that

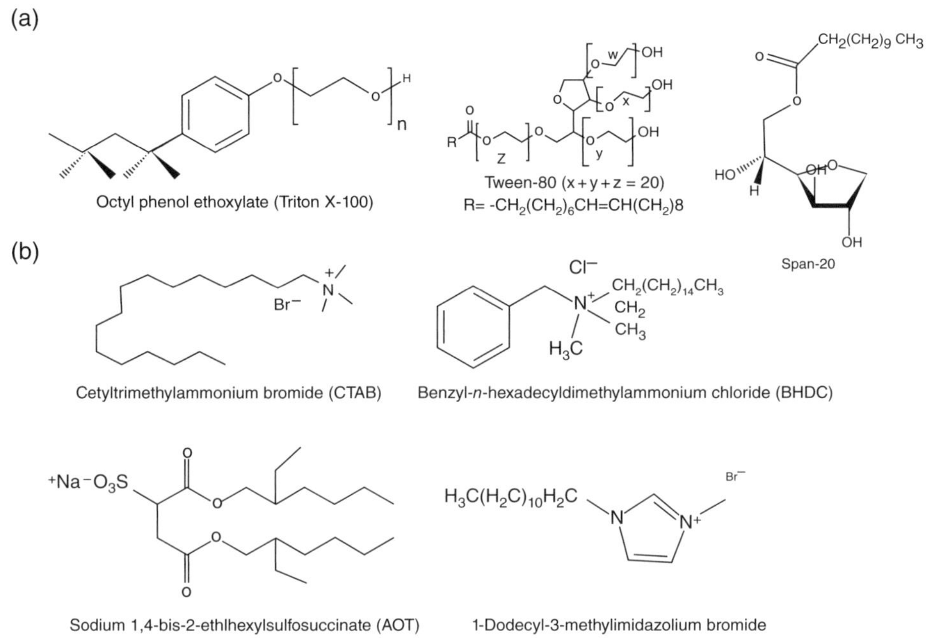

Figure 19.2 Commonly used (a) nonionic and (b) ionic surfactants in the preparation of IL-in-oil microemulsions.

thermodynamically stabilize the microemulsion by reducing interfacial tension between two immiscible liquids. Figure 19.2 lists some of the surfactants commonly used in IL-based microemulsions. These surfactants can be broadly classified into nonionic and ionic surfactants. Most of the first IL/O microemulsions were stabilized using the nonionic octyl phenol ethoxylate (Triton X-100 or TX-100) surfactant [58–63, 69]. More recently, ionic surfactants like cationic cetyltrimethylammonium bromide (CTAB) [84] and benzyl-*n*-hexadecyldimethylammonium chloride (BHDC) [64, 70, 72], anionic sodium 1,4-bis-2-ethylhexyl sulfosuccinate (AOT) [70, 72], and surface-active ionic liquids (SAILs) have been used to generate a series of IL-based microemulsions [73–76, 85–93]. Notably, the study of CTAB-based IL/O ternary mixtures is limited due to the requirement of high temperature and cosurfactants for microemulsion formation and stability [84]. Conversely, a large range of SAIL-based IL/O microemulsions could be prepared at room conditions, simply by fine-tuning the surface-active nature of the constituent surfactants [85, 86, 92–94]. To extend the applications of IL/O systems in pharmaceutics, biocompatible surfactants like polyoxyethylene sorbitan monooleate (Tween-80) and sorbitan laurate (Span-20) have also been used [31, 77–79].

The growing interest on IL/O microemulsions is evident from the publication of many reviews in the last few years [9, 95–100]. Although the structural properties and applications were discussed extensively in these articles, the dynamical behavior of these novel systems was meagerly reviewed. In this chapter, we review some aspects of these microemulsions, with an emphasis on structural characteristics and solvation dynamics of the confined IL pool.

19.2 PHASE BEHAVIOR AND CONDUCTIVITY

In a typical setup for the preparation of IL/O microemulsions, measured quantities of two immiscible liquids are mixed in a glass test tube, and the surfactant is gradually added [58–62, 69, 85–89, 94, 101]. The phase boundaries are determined by observing the transition of the solution from turbidity to transparency. Phase behavior is usually mapped in two dimensions using a Gibbs Triangle, which reflects the composition of the phase across different concentrations of the constituent IL, oil, and surfactant. The phase diagram of a typical IL/O microemulsion composed of [bmim][BF_4]/TX-100/benzene is presented in Figure 19.3a [101]. The broad unshaded area denotes the presence of a single microemulsion phase, which consists of oil-in-IL (O/IL), IL/O, and bicontinuous regions. The shaded area in the diagram represents the existence of a biphasic system, where a microemulsion is present in equilibrium with excess IL or oil.

The area encompassing the single phase is indicative of the extent to which the polar IL can be solubilized by the surfactant/oil mixture to form a stable microemulsion. This stability depends on various factors, including the constituents and the interactions among them. The choice of the apolar phase, however, was found to have minimal effect on the phase behavior, as exemplified from the similar phase diagram of [bmim][BF_4]/TX-100 in cyclohexane [58], toluene [59], *p*-xylene [60], and benzene [62]. The influence of the constituent IL ions on the phase behavior was quite prominent. Although early studies on $[bmim]^+$ cation-based ILs with different anions, $[BF_4]^-$ and $[PF_6]^-$, showed similar phase behavior in TX-100/oil [58–60, 69], the increasing alkyl chain length of [emim][C_nSO_4]-based microemulsions showed a growing and stable single-phase region in TX-100/cyclohexane. The increased single-phase region of these microemulsions was due to favorable hydrophobic interactions between the alkyl tail of the anion and the surfactant [94]. That the IL cations can also influence the phase behavior was evident from the larger single phase formed by $[P_{13}]^+$ compared to $[N_{3111}]^+$ with same $[Tf_2N]^-$ anion in TX-100/oil mixtures [65, 67]. Presumably, the less bulkier $[P_{13}]^+$ could interdigitate into the surfactant

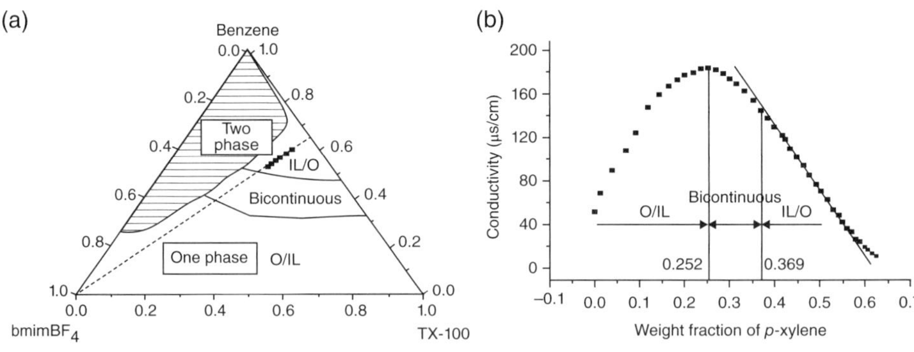

Figure 19.3 (a) Phase diagram of [bmim][BF_4]/TX-100/benzene ternary system at 25°C. Reproduced from Gao et al. [101] with permission from John Wiley & Sons. (b) Electrical conductivity (k) of [bmim][BF_4]/TX-100/*p*-xylene microemulsion with increasing weight fraction of *p*-xylene; TX-100 weight fraction = 0.8. Reproduced from Gao et al. [60] with permission from John Wiley & Sons.

layer, leading to better stability of the IL core. For similar reasons, increasing alkyl chain length of imidazolium-based IL cations showed a widening of the single-phase region in microemulsions stabilized by the AOT-derived SAIL surfactant [bmim][AOT] [76, 85, 86]. The effect of surfactant on the phase behavior could also be understood from the studies on [bmim][X]/oil mixtures (where $[X] = [BF_4]^-$, $[PF_6]^-$) stabilized by [bmim][AOT][86] and TX-100 [69, 101]. Larger single-phase region was obtained in the former, which could be attributed to the enhanced surface-active nature of SAILs. Sarkar and coworkers observed that weaker interionic interactions in the constituent IL can also increase the single-phase region. For example, weakly paired $[PF_6]^-$ allowed the imidazolium cation to interact more strongly with the negatively charged [bmim][AOT] for improved stability [86]. Overall, the range of compositions for which single-phase microemulsions were stable for any IL examined exceeded that of water-in-oil microemulsions [89, 102]. The factors affecting the comparatively higher stability of IL/O microemulsions will be discussed in detail later.

The extended single phase of the IL/O microemulsion, although thermodynamically stable, is not microstructurally homogeneous across the entire composition range. Electrical conductivity (k) is frequently measured to investigate microstructural transitions in microemulsions [58, 69]. As a typical example, Figure 19.3b shows k as a function of added *p*-xylene to [bmim]$[BF_4]$/TX-100 mixture [60], where the conductivity of the solution varies in four successive stages. IL is highly conducting, and the addition of TX-100 greatly depresses its conductivity due to strong interactions between the electronegative EO groups of the surfactant and the positively charged IL cations [60, 101]. Upon addition of *p*-xylene, TX-100 aggregates to form surfactant films, thereby releasing the IL and increasing the conductivity. The initial nonlinear rise in conductivity, in Figure 19.3b, can therefore be attributed to the formation of the O/IL microregion. A decrease in the viscosity of the system upon addition of oil allows easier transport of the conducting ions in the solution and also accounts partially for the rise in conductivity. With increasing oil content, there is a progressive growth and interconnection of the oil microdomains, and the system transits into a bicontinuous phase. The insulating nature of oil causes a nonlinear decrease in conductivity at this stage. With further addition of oil, a linear decrease in k follows, due to the formation of IL/O reverse micelles. Subsequently, the reverse micellar droplets aggregate, and the conductivity continues to decrease nonlinearly with addition of more oil.

The percolation theory is often used to explain how a transition in the microstructure of the microemulsion leads to a change in conductivity [103–105]. Several theoretical and experimental studies have shown that the phase present in lower concentration in the microemulsion forms dispersed globules, which undergo Brownian motion in the continuous phase [106]. According to the percolation theory, interaction between these globules forms the basis of electrical conductivity in these systems. The percolation theory was previously used for water-in-oil microemulsions and was interpreted from the viewpoint of a "static" or a "dynamic" percolation model [107]. According to the dynamic percolation model, attraction between dispersed water globules gives rise to percolation clusters that enable efficient transfer of charge carriers between the globules. This explains changes in conductivity in the IL/O or the O/IL microregions. The static model attributes percolation to the formation of a continuous conducting network of water embedded in an insulating oil phase and explains conductivity in the bicontinuous phase [59, 60].

19.3 POLARITY AND STRUCTURAL CHARACTERISTICS OF THE IL NANODOMAIN

The structural and dynamical properties of ILs confined in the core of IL/O microemulsions differ significantly from that of neat IL in the bulk phase. Steady-state absorption and fluorescence spectroscopic analyses are commonly used to obtain information about the location of IL in the IL/O microemulsion and the polarity of the confined IL nanodomains [63, 85, 87, 108–110]. Some of the probes that are widely used in these studies are shown in Figure 19.4a. These compounds can preferentially localize and probe specific regions of the microemulsion. Electronic transitions of these probes are sensitive to the polarity of the surrounding microenvironment, and their wavelength of maximum absorption (λ_{max}) shifts toward a longer or shorter wavelength as they move from nonpolar to polar environments. Thus, Zheng and coworkers found that the λ_{max} of methyl orange (MO) undergoes considerable red-shift (from 422 to 428 nm) with increasing concentration of [bmim][BF_4] in TX-100/benzene solution, up to a certain IL-to-surfactant molar ratio, R (Fig. 19.4b) [101]. This indicated that the probe experiences increasingly polar environment upon addition of IL. However, upon further increasing R, the λ_{max} remains constant at 428 nm.

The authors explained these observations from differential sensitivities of the azo and the $SO_3^-Na^+$ moieties of MO to the local environments. They have shown that, being insoluble in the apolar oil phase, MO localizes in the interfacial surfactant layer in absence of IL ($R=0$) and its azo group interacts with the electronegative EO head group of TX-100. At low R, the IL molecules also localize in the surfactant layer due to strong interactions of its cations with the EO head groups [60, 101]. This increases the polarity of the microenvironment around the sensitive azo group and causes the initial increase in λ_{max}. Once the EO groups are saturated, excess IL starts moving toward the core of the microemulsion, where the less sensitive $SO_3^-Na^+$ moiety of MO is present. Hence, upon further addition of IL, even though more and

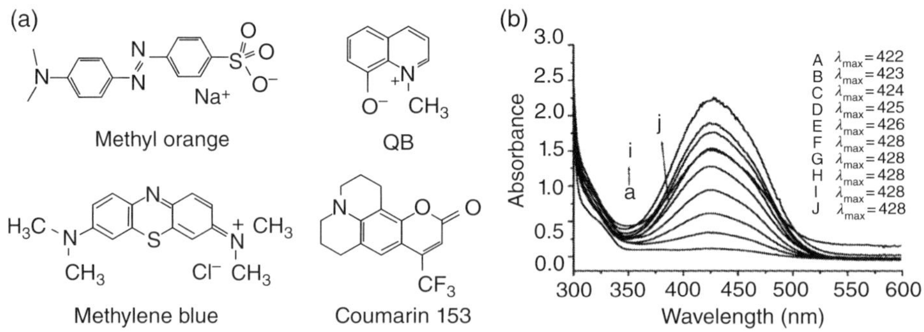

Figure 19.4 (a) Commonly used solvatochromic probes in absorption and emission spectroscopic studies of IL-in-oil microemulsions. (b) Absorption spectra of MO showing long wavelength shift in the [bmim][BF_4]/TX-100/benzene microemulsion with increasing addition of [bmim][BF_4]. Data shown for A, $R=0.17$; B, 0.34; C, 0.52; D, 0.71; E, 0.90; F, 1.11; G, 1.33; H, 1.55; I, 1.66; and J, 1.79. Reproduced from Gao et al. [101] with permission from John Wiley & Sons.

more IL localize in the IL pool, the λ_{max} remains unchanged. Thus, as represented in Figure 19.5, IL is distributed between the surfactant head group (palisade layer) and the core (IL pool) of the microemulsion droplets [60]. Interestingly, at higher IL concentration, the λ_{max} plateaus at 428 nm (Fig. 19.4b), which is close to the λ_{max} of MO in neat [bmim][BF_4] (433 nm). This signifies that the polarity of IL confined in the nanodomain of IL/O microemulsions is not very different from that of the neat IL [101].

Using steady-state absorption studies, several other authors examined the micropolarity of confined IL in microemulsions stabilized by ionic surfactants [64, 85, 87]. For example, Sarkar and coworkers examined [bmim][BF_4]/benzene mixtures stabilized by the anionic SAIL surfactant [bmim][AOT] and observed that, within the studied range, the λ_{max} for solubilized MO continued to undergo redshift with increasing *R* [85, 87]. In another work, Falcone and coworkers compared the micropolarities of [bmim][BF_4]/benzene mixtures stabilized by cationic BHDC and nonionic TX-100 surfactants using 1-methyl-8-oxyquinolinium betaine (QB), a dye that locates mainly at the surfactant interfacial layer [64]. When [bmim][BF_4] was added to both BHDC/benzene and TX-100/benzene systems, a larger hypsochromic shift was sensed by the probe in the former. This implies that the local environments in BHDC/benzene system are more polar. The authors ascribed this phenomenon to the strong electrostatic interactions between the $[BF_4]^-$ anion and the BHD^+ moiety of the cationic surfactant.

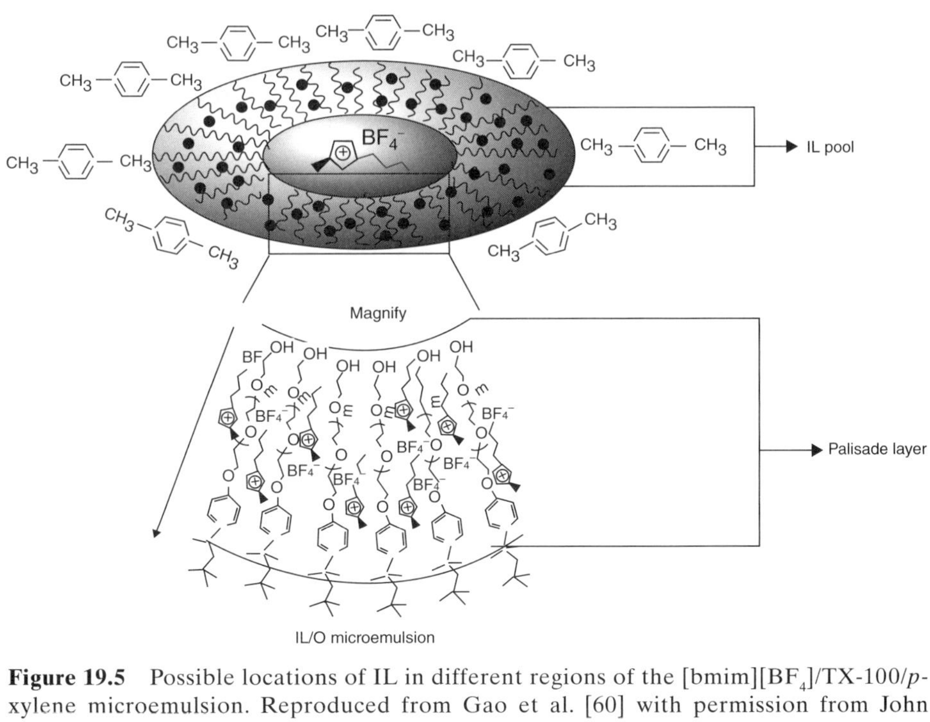

Figure 19.5 Possible locations of IL in different regions of the [bmim][BF_4]/TX-100/*p*-xylene microemulsion. Reproduced from Gao et al. [60] with permission from John Wiley & Sons.

To further probe the micropolarity in the IL nanodomain, Falcone et al. used FTIR and NMR spectroscopic techniques to analyze a series of IL/O microemulsions [70, 72, 111]. The authors observed dramatic changes in the spectroscopic parameters of the constituent IL ions in [bmim][BF_4]/BHDC/benzene microemulsions compared to the neat IL. They proposed an exchange of anion due to favorable interaction of $[BF_4]^-$ of IL with the BHD^+ moiety of the surfactant at the interface. This results in an altered ionic network in the IL pool [111]. Similar changes in the IL nanodomain were observed in [bmim][Tf_2N]/BHDC/chlorobenzene microemulsions [70]. Conversely, with the anionic AOT surfactant, the spectroscopic properties did not show significant perturbations, presumably due to weaker interaction of the $[bmim]^+$ cation with SO_3^- moiety of the surfactant [70]. Similarly, the microstructure of the IL pool in [bmim][BF_4]/benzene mixtures stabilized by the nonionic TX-100 surfactant was largely unaffected [111], in agreement with observations from steady-state spectroscopy. Thus, the interionic interactions in ILs encapsulated by ionic surfactants are considerably perturbed, while ILs confined by nonionic surfactants emulate properties of neat IL. Hence, the micropolarity of the IL pool largely depends on the constituent IL and surfactant.

The size and shape of the droplets in IL/O microemulsions have been characterized through a range of scattering and microscopic techniques, such as dynamic light scattering (DLS), static single-angle scattering, freeze–fracture transmission electron microscopy (FF-TEM), etc. DLS is commonly used for qualitative estimation of hydrodynamic radius and size distribution of microemulsion droplets. Figure 19.6a shows the size distribution of [bmim][BF_4]/TX-100/cyclohexane microemulsion at different *R*, obtained from DLS analysis by Han and coworkers [58]. The size of the microemulsion droplets increased linearly from 10 to 94 nm on increasing *R* from 0.2 to 1.5. This indicated that the system abides by the swelling law of microemulsions

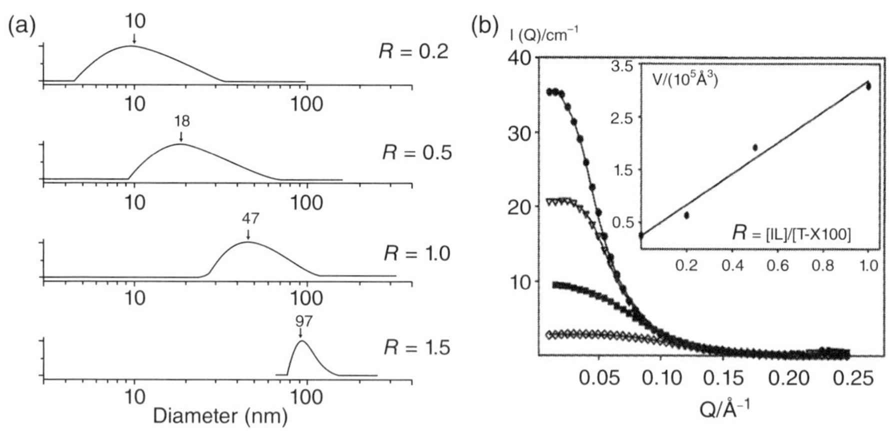

Figure 19.6 (a) Size and size distribution of droplets in [bmim][BF_4]/TX-100/cyclohexane microemulsions at 35°C obtained from DLS. Reproduced from Gao et al. [58] with permission from the Royal Society of Chemistry. (b) SANS from [bmim][BF_4]/TX-100/cyclohexane microemulsions at 55°C. R = 0 (◊), 0.2 (■), 0.5 (∇), and 1.0 (●). The fits shown as lines are to a form factor for homogeneous ellipsoids. Inset shows swelling behavior in terms of ellipsoid volume (*V*) as a function of *R*. Reproduced from Eastoe et al. [61] with permission from the American Chemical Society.

and the droplets are spherical, since according to the swelling law, the size of discrete, noninteracting spherical droplets increases linearly with R [112]. A deviation from linearity, therefore, could imply droplet aggregation or change in the micellar shape. Also, the insensitivity of droplet size to the increasing IL content might indicate that IL is not encapsulated by the surfactants. These principles are often combined with DLS data for inferring the formation of IL nanodroplets in the dispersion phase [58, 64, 70, 72, 76, 85–87].

Small-angle neutron scattering (SANS) and small-angle X-ray scattering (SAXS) techniques are established tools for quantitative measurements of size and shape of IL droplets [61, 69, 73, 113]. Eastoe and coworkers characterized [bmim][BF_4]/TX-100/cyclohexane microemulsions using the SANS technique [61]. As shown in Figure 19.6b, a regular increase in SANS intensity with increasing R (from $R = 0$ to 1) was noted by the authors. The entire range of SANS data could be fitted to a form factor for homogeneous ellipsoids. As shown in the inset to Figure 19.6b, the ellipsoidal volume (V) increased linearly with R, in accordance with the swelling law. However, the semiminor radius of the droplet was found to be 2.4 nm at $R = 1$, which was much smaller than the value of approximately 45 nm reported by Han and coworkers using DLS for the same system [58]. Considering the higher resolution of SANS compared to DLS, Eastoe and coworkers pointed out the greater reliability of the scattering techniques in quantifying the dimensions of microemulsion droplets. However, despite greater accuracy, the use of SANS technique in characterization of IL/O microemulsions is limited due to its sophistication and cost.

Microscopic techniques like FF-TEM [58, 104, 114] and scanning electron microscopy [73] allow direct visualization of the microemulsion droplets and their aggregate morphology. FF-TEM images of [bmim][BF_4]/TX-100/cyclohexane microemulsions generated by Han and coworkers showed that IL/O droplets were of overall spherical shape and the diameter was considerably larger than traditional water-in-oil droplets [58]. From subsequent studies [73, 104, 114], sizes of IL/O microemulsions were seen to range typically from 10 to 150 nm, depending on the specific system and the proportion of IL used. This large size is attributed to the higher amount of surfactants needed to form micelles in IL, the larger size of IL compared to water, and the increased effective area of interaction between ILs and surfactants compared to water and surfactants in water-in-oil microemulsions [60].

19.4 DYNAMICS IN THE IL POOL

The confinement and microviscosity of the IL pool in IL/O microemulsions can have profound effects on its molecular dynamics. Time-resolved anisotropy measurements provide useful information on dynamics by monitoring the rotational relaxation of an excited dye molecule, such as coumarin-based fluorophores. Time-resolved fluorescence anisotropy can be calculated using Equation 19.1:

$$\gamma(t) = \frac{I_{\parallel}(t) - GI_{\perp}(t)}{I_{\parallel}(t) + 2GI_{\perp}(t)} \tag{19.1}$$

where G is the correction factor for detector sensitivity to the polarization direction of emission and $I_{\parallel}(t)$ and $I_{\perp}(t)$ are the fluorescence decays polarized parallel and perpendicular to the polarization of the excitation light, respectively. Thus, Sarkar and coworkers studied the anisotropy decay of coumarin 153 (C-153) in [bmim][BF_4]/TX-100/cyclohexane microemulsions [115]. The anisotropy decay of C-153 in cyclohexane was observed to be quite fast and could be fitted to a single exponential function. The measured average rotational relaxation time was 0.135 ns. In the microemulsion, the average rotational relaxation time increases with the emergence of an additional slow component. For a system with R ([[bmim][BF_4]]/[TX-100]) = 0.2, the average relaxation time was calculated to be 0.582 ns with components 0.348 and 2.15 ns. Having known that the probe specifically localizes in the polar regions of the system, such a biexponential decay function might suggest that the probe experiences two different kinds of polar domains in the microemulsion. In more recent studies, these authors and Gao et al. ascribed the slow component to the IL molecules trapped in the palisade layer, where the positively charged IL cations interact with the electronegative EO head groups of the surfactant [60, 87, 101]. The fast component, on the other hand, arises due to neat IL-like environment in the IL pool. The slow and the fast components of the decay, as well as the average rotational relaxation time of the probe, were shown to increase with increasing IL content.

While anisotropy decay gives information about the rotational dynamics of the fluorophore, solvent dynamics can be determined from the time-resolved fluorescence decay spectra of the dye at different emission wavelengths. As shown by Sarkar and coworkers (Fig. 19.7a), the fluorescence decays of C-153 depend significantly on the emission wavelength [115]. A distinct growth at the red edge of the emission spectra followed by a usual decay and a faster decay at the blue edge were observed by the authors. This wavelength-dependent behavior of temporal decay of the fluorophore is a clear signature of solvation dynamics [87].

The time-resolved emission spectra (TRES) were constructed using a method described elsewhere [116, 117]. Briefly, both the red- and blue-edge decay profiles were best fitted to multiexponents. The TRES at time t, $S(\lambda;t)$, is obtained from the

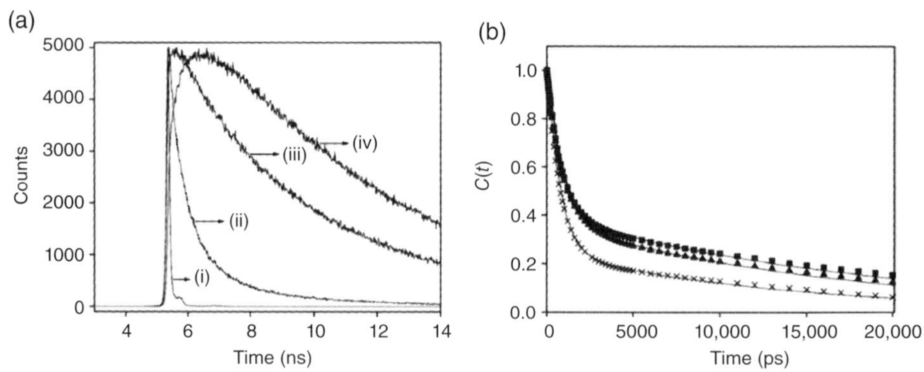

Figure 19.7 (a) Fluorescence decays of C-153 in [bmim][BF_4]/TX-100/cyclohexane microemulsions at R = 1 at (i) instrument response function, (ii) 450, (iii) 510, and (iv) 620 nm. (b) Decay of time correlation function $C(t)$ of C-153 in [bmim][BF_4]/TX-100/cyclohexane at R = 0 (×), 0.2 (■), and 1.5 (▲). Reproduced from Chakrabarty et al. [115] with permission from the American Chemical Society.

fitted decays, $D(t;\lambda)$, by relative normalization to the steady-state spectrum, $S_0(\lambda)$, as given in Equation 19.2:

$$S(\lambda;t) = D(t;\lambda)\frac{S_0(\lambda)}{\int_0^\infty D(t;\lambda)dt} \quad (19.2)$$

Each spectrum (TRES) was then fitted to lognormal line shape function, $g(\nu)$, of the following form to obtain the peak frequency:

$$g(\nu) = g_0 \exp\left[-\ln 2\left(\frac{\ln\left[1+2b(\nu-\nu_p)/\Delta\right]}{b}\right)^2\right] \quad (19.3)$$

where g_0, b, ν_p, and Δ are the peak height, asymmetric parameter, peak frequency, and width parameter, respectively. These frequencies (ν) were used to construct the solvent correlation function $C(t)$ using the relation stated in Equation 19.4:

$$C(t) = \frac{\nu(t)-\nu(\infty)}{\nu(0)-(\infty)} \quad (19.4)$$

where $\nu(0)$, $\nu(t)$, and $\nu(\infty)$ are the peak frequencies at times 0, t, and ∞, respectively. The time dependence of $C(t)$ (Fig. 19.7b) was fitted to a biexponential function of the form

$$C(t) = a_1 e^{-t/\tau_1} + a_2 e^{-t/\tau_2} \quad (19.5)$$

where τ_1 and τ_2 are the two solvation times with amplitudes a_1 and a_2, respectively. The average solvation time (τ_{av}) was calculated using the relation

$$\tau_{av} = a_1\tau_1 + a_2\tau_2 \quad (19.6)$$

The average solvation time of C-153 in neat [bmim][BF_4] is 2.13 ns, with components 0.278 and 3.98 ns [118]. Sarkar and coworkers observed that upon addition of [bmim][BF_4] in the TX-100/cyclohexane solution, the average solvation time of C-153 at $R=0.2$ increased to 8.06 ns, with components 0.815 and 18.94 ns [115]. Several subsequent solvent relaxation studies carried out by these authors on a variety of IL/O microemulsions have shown that the dynamics of IL upon confinement is retarded by a maximum of approximately 10 times [65, 94, 108, 119]. This retardation in IL dynamics is, however, minute compared to the approximately 100–1000-time decrease in the dynamics of water in water-in-oil microemulsions [120–123]. The significant decrease in water-in-oil microemulsions is due to extended hydrogen bonding between water and surfactant head groups and counterions. On the other hand, interactions between IL and surfactant head groups are very similar to (or weaker than) interionic interactions between component ions in neat ILs, and hence, the effect of confinement on solvation dynamics is less prominent. With increasing IL content, the IL pool expands, and the effect of confinement is further reduced, which is reflected in the decrease in the average solvation time with increase in R [67, 87, 115].

19.5 EFFECT OF CONSTITUENTS ON DROPLET STRUCTURE

The versatility of IL/O microemulsions lies in the fact that their physical properties can be tuned by careful selection of the constituents to complement the desired application. For example, the droplet size of a microemulsion is known to be inversely proportional to the interfacial curvature of the surfactant film, which in turn is affected by IL–surfactant interactions [9, 84]. Several studies on a range of IL–surfactant combinations have shown that the hydrophobicity of the IL anion, the nature of interionic interactions in IL, and the interfacial properties of the surfactant are the major factors affecting IL–surfactant interactions [51, 52, 64, 70, 72, 77–79, 86, 94, 98, 124–127]. Falcone and coworkers reported that for microemulsions containing a [bmim]-based IL in TX-100/benzene medium, the change in anion from hydrophilic $[BF_4]^-$ to hydrophobic $[Tf_2N]^-$ reduced the droplet size from 40 to 8 nm [64, 70]. The authors explained that owing to the higher basicity of the $[BF_4]^-$ anion, $[bmim][BF_4]$ interacts strongly with the nonionic TX-100 and therefore penetrates well into the surfactant layer. This causes an increase in the effective head group area of the surfactant, a decrease in the curvature of the interfacial layer, and an increase in the size of the microemulsion. A similar observation was made by changing the surfactant to cationic BHDC.

The role of interionic interactions on microemulsion properties was examined by Falcone and coworkers using [bmim][TfA] and [bmim][TfO] ILs [72]. $[TfO]^-$ with its lower electron-donating ability pairs weakly with $[bmim]^+$ compared to $[TfA]^-$. As a result, $[TfO]^-$ interacts strongly with the surfactant in the cationic BHDC-based microemulsions. This causes an increase in the effective surfactant head group area, leading to an increase in the size of the microemulsions. On the other hand, weak IL–surfactant interactions in [bmim][TfA] give rise to smaller microemulsions. Similarly, in a study using an anionic SAIL surfactant, Sarkar and coworkers found larger-size droplets in microemulsions containing IL $[bmim][Tf_2N]$ than that with $[bmim][BF_4]$, presumably due to weaker interactions of the hydrophobic $[Tf_2N]^-$ with the cation [86]. Hence, the size of microemulsion droplets depends on the IL cation–anion interactions and their interaction with the surfactants.

Interfacial rigidity of the reverse micelles can also influence the droplet size. For example, Sarkar and coworkers observed that IL ions with longer alkyl tails formed microemulsion droplets of smaller size [86, 94]. By penetrating into the interfacial layer, the long alkyl tails enhance surfactant packing by screening off charges, thereby decreasing the effective surfactant head group area. This leads to the formation of rigid microemulsions with smaller size that are less susceptible to changing IL content, compared to microemulsions containing ILs with shorter alkyl chains. The effect of the nature of surfactant on interfacial rigidity was also shown by several authors, who used a surfactant combination of Tween-80/Span-20 to yield IL/O microemulsions of considerably smaller size [77–79]. The interfacial rigidity was due to strong IL–surfactant interactions between IL ions and the multiple hydroxyl groups present on both the surfactants and EO groups on Tween-80. Thus, by careful selection of constituent IL ions and surfactants, the size of droplets in IL/O microemulsions can be tuned. Control over the size of IL/O microemulsion droplets is pivotal in regulating the reaction rate, amount of product formed, or amount of solutes dissolved by the IL domain of the microemulsions.

19.6 EFFECT OF WATER ON DROPLET STRUCTURE AND DYNAMICS

To diversify the possible use of IL/O microemulsions, several groups have studied the effect of added water on these systems. FTIR, ^{1}H-NMR, and UV–visible spectroscopic techniques have been used to probe possible locations of water, their states, and effect on the properties of IL/O microemulsions. For example, Zheng and coworkers used UV–visible absorption spectroscopy to deduce the regions of water localization in [bmim][BF_4]/TX-100/benzene microemulsions [101]. They monitored the change in λ_{max} of MO as a function of water added to the system. With addition of small amounts of water, the λ_{max} remained constant. The authors explained that added water first localize near the head group region of TX-100 and cannot be sensed by MO, which is solubilized in the IL pool. Upon saturation of the palisade layer, water penetrates into the IL pool. The probe senses the increase in polarity of the IL pool due to addition of water, and the λ_{max} undergoes a redshift. This indicates that in presence of water, the IL pool exhibits higher polarity. However, when the authors used triethylamine as the apolar phase, λ_{max} of MO was unchanged [63]. From ^{1}H-NMR studies, they concluded that instead of penetrating the IL pool, water mainly interacted with the triethylamine phase and presumably formed OH^-.

Free and hydrogen-bonded water have molecular vibrations at different IR frequencies, and hence, FTIR spectroscopic analyses provide direct information about the states of solubilized water in the microemulsion [62, 63, 101]. Figure 19.8 depicts changes in the frequencies of O–H stretching vibrations in [bmim][BF_4]/TX-100/benzene system upon gradual addition of water, as seen by Zheng and coworkers [101]. In the absence of water, the O–H stretching band appeared solely due to the terminal hydroxyl group of TX-100 at 3426 cm^{-1}. Upon addition of water up to 1.1 wt.%, this band underwent a significant shift to 3477 cm^{-1}. The authors explained that this large change in frequency was due to the water that was "trapped" in the hydrophobic tail of the surfactant in the palisade layer. Trapped water molecules

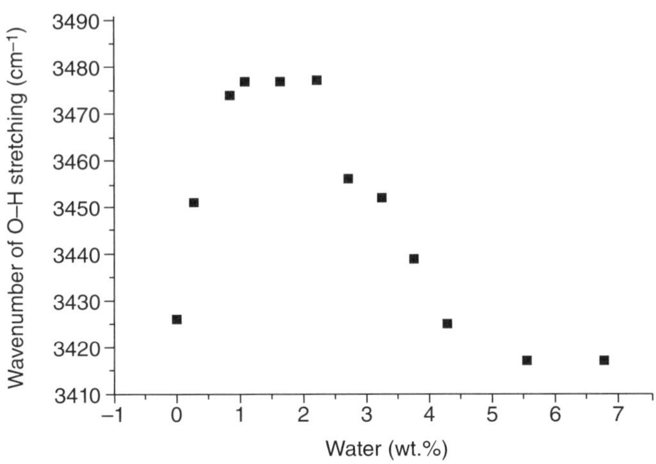

Figure 19.8 Dependence of O–H stretching frequencies of [bmim][BF_4]/TX-100/benzene microemulsion on added water content (wt.%). Reproduced from Gao et al. [101] with permission from John Wiley & Sons.

exist as monomers or dimers, without hydrogen bonding to their surrounding molecules, and thereby absorb in the high-frequency region. With further addition of up to 2.2 wt.% water, the wave number remained unchanged. Finally, a gradual decrease in wave number to nearly 3400 cm^{-1} was observed with increasing water addition. This characteristic change in frequency is due to the increase of "bound" water that is hydrogen bonded to the polar surfactant head groups. Apart from these two types of water in a microemulsion system, there exists "free" water that occupy the core of the surfactant aggregate and are hydrogen bonded among themselves. The O–H stretching band for this water appears at around 3220 cm^{-1} [128].

Several studies showed that water localized in different domains of the IL/O microemulsions affect the size, shape, and stability of these systems [62, 86, 101, 110, 129–131]. In [bmim][BF_4]/TX-100/benzene microemulsion studied by Zheng and coworkers, water residing in the palisade layer was seen to stabilize the microemulsion by mediating hydrogen bonding between the IL cation and the EO groups of the surfactant [62, 101]. Similar observations were made by Zhang and coworkers in their study on [bmim][BF_4]/Brij-35/butanol/toluene pseudoternary IL/O microemulsion, where the water formed extensive hydrogen bonding with the polyoxyethylene chains of the surfactant [129]. Sundmacher and coworkers used ^{1}H-NMR and 2D ROESY techniques to study the localization and effect of water on [bmim][BF_4]/TX-100/cyclohexane microemulsions [110]. Water was found to locate at the periphery of the IL core and behaved like a chock inserted into the palisade layer. This increased the curvature of the interface and decreased the size of the droplets. As a consequence, the hydrophobic tail–tail interactions and the interactions between hydrophilic groups of surfactants were enhanced. Altogether, these effects gave rise to ordered microemulsions, with a consequent change in the shape of the droplets from elliptical to spherical. Bhattacharya and coworkers showed that the decreased droplet size in the presence of water leads to slower solvation dynamics in [pmim][BF_4]/TX-100/benzene microemulsions [108, 130]. In a subsequent study, the authors identified two species of water in these systems, bound and free, with the bound water exhibiting approximately 15-fold slower dynamics than the free water [130]. Sarkar and coworkers observed that the presence of this bound water gives rise to a slow third component in the solvation dynamics of water/[bmim][BF4]/TX-100/cyclohexane system, compared to a biexponential decay seen in the absence of water [115, 131].

Sarkar and coworkers showed that the extent to which water affects the properties of IL/O microemulsions can depend on the nature of the IL [86]. In [$C_n$mim][X]/[bmim][AOT]/benzene microemulsions containing hydrophobic anions, namely, [Tf_2N]$^-$ and [PF_6]$^-$, the droplet size increased with increasing addition of water. From ^{1}H-NMR studies, the authors showed that owing to the hydrophobicity of the IL pool, water localize in the palisade layer of these microemulsions. This causes an increase in the effective surface area of the interfacial layer and a consequent increase in the droplet size. Conversely, in microemulsions containing a hydrophilic IL, water is solubilized in the polar core, causing minimal changes in the droplet size. Thus, addition of small quantities of water to IL/O microemulsions improves the stability of these systems and expands the possibility of regulating their structural and dynamic properties. Moreover, the unique solubilization behavior of water in this class of microemulsions envisages using these systems in preparing porous or hollow nanomaterials by hydrolysis reactions [62, 63, 101, 110].

19.7 EFFECT OF TEMPERATURE ON MICROEMULSION STABILITY

The effect of temperature on IL/O microemulsions is an important area of research, owing to the myriad applications of high- and low-temperature stable microemulsions [76, 103, 114]. Zheng and coworkers studied the effect of temperature on [bmim][BF_4]/cyclohexane mixtures stabilized by the nonionic TX-100 surfactant [114]. The single-phase transparent microemulsion was observed even as the temperature approached the boiling point of cyclohexane at 74°C. The spherical morphology of the droplets, as revealed by FF-TEM studies, was maintained up to 60°C. The size of the droplets increased linearly with temperature, following the regular swelling behavior. Nevertheless, no microstructural transition was observed throughout the temperature range, showing that IL/O microemulsions exhibited significantly higher-temperature stability compared to water-in-oil microemulsions. The hydrogen-bonding interactions between water and TX-100 that stabilize water-in-oil microemulsions are temperature labile, making these systems unstable at higher temperatures. IL/O microemulsions, on the contrary, are stabilized by the temperature-insensitive electrostatic interactions between IL and polar EO head groups of TX-100 [60, 101].

Sarkar and coworkers studied the effect of temperature on a series of IL/O microemulsions stabilized by ionic surfactants [76, 86, 87, 119]. The size of these microemulsions changed only marginally with increase in temperature, unlike the large increase in size observed for IL/O systems stabilized by the nonionic TX-100. They concluded that the electrostatic interactions between an IL and an ionic surfactant was less sensitive to temperature than the ionic-induced dipole interactions with a nonionic surfactant [84]. Apart from the surfactant, the choice of cation was also found to have a significant effect on the extent to which temperature could influence the size of IL droplets [86]. Further, it was shown that microemulsions composed of imidazolium ILs with longer alkyl tails were less sensitive to changes in temperature, owing to the rigidity imposed by the alkyl chains on the interfacial surfactant layer. The choice of anion was found to have no or minimal effect. Effect of temperature on solvation dynamics of the IL pool was also studied by this group [87, 119]. While average solvation time decreased with increasing temperature, the change was not as significant as that in neat IL. This showed that diffusional motion of confined IL in microemulsions is less sensitive to temperature. Similarly, absorption studies showed that the polarity of IL under confinement is less susceptible to temperature changes [87, 119]. Thus, high-temperature stability of IL/O microemulsions and preservation of the static and dynamic properties of the IL core widen the possibility of their potential use as nanoreactors for high-temperature chemical reactions [76].

19.8 APPLICATIONS OF IL/O MICROEMULSIONS AS NANOREACTORS

As described in the earlier sections, the size and shape of IL droplets, interactions among the constituents, thermal stability of the microemulsion, and solvent properties and dynamics in the IL core can be orchestrated to suit the desired application. Hence, IL/O microemulsions can serve as versatile nanoreactors where the kinetics of reactions carried out and the properties of the products formed can be precisely controlled. Lauth-de Viguerie and coworkers carried out the industrially important

Matsuda–Heck arylation reaction between two reactants of opposite polarity, an arenediazonium salt and an olefin, in [BnPyr][Tf_2N]/TX-100/toluene microemulsion [66]. The yield of products was doubled compared to the reaction performed in neat IL or in other conventional solvents. The authors ascribed the higher yield to two factors. Geometric constraints were imposed on the reactants upon confinement in the microemulsion core, and the reactants and catalyst were partitioned between different phases of the microemulsion. Both factors resulted in favorable orientation of reactant and catalyst leading to higher reaction rates. Moreover, the reaction rate could be modulated by varying the size of the polar core upon changing the IL concentration. Garcia-Rio and coworkers compared the rate of aminolysis of esters, a model reaction commonly used to synthesize peptides, in water-in-oil and IL/O microemulsions [102]. The reaction rates were greatly enhanced when carried out in the IL/O microemulsion. The hydrophobic nature of the ester limits its solubility in the aqueous domain of the water/TX-100/cyclohexane microemulsion, and hence, the reaction takes place only in the water/surfactant interfacial layer. In addition, the authors speculated that the strong hydrogen bonding between water and EO groups of TX-100, in the water-in-oil system, increases the electron density on water and decreases its efficiency in stabilizing the negatively charged transition state intermediate, thereby slowing down the reaction kinetics. On the other hand, in the [bmim][BF_4]/TX-100/cyclohexane microemulsion, the reactants are soluble in the IL domain, and the reaction can take place simultaneously at the interface and in the core. Also, the nature of interactions between [bmim][BF_4] and TX-100 is majorly electrostatic, which does not drastically affect the polarity of the IL at the interface, leading to higher reaction rates in this system.

The wide range of thermal stability has further diversified the areas of applications of IL/O microemulsions. For example, the broad liquidus range and high thermal stability of EAN has made it plausible to generate EAN/[C_{16}mim][Cl]/dodecane microemulsions, which were found to be stable across a temperature range of 30–150°C by Kunz and coworkers [132]. The authors proposed the use of such systems in size-controlled synthesis of nanoparticles through thermal decomposition of inorganic precursors. This class of microemulsions also simplifies product recovery, as simple temperature-induced phase separation could enable retrieval of the product-containing phase. In another study, [emim][$EtSO_4$], which exhibits a glass transition temperature of −80°C, and limonene, with a melting point of −96°C, were used to generate limonene/TX-100/[emim][$EtSO_4$] microemulsions, which were stable from ambient to as low as −20°C [103].

19.9 APPLICATIONS OF IL/O MICROEMULSIONS IN SUSTAINABLE POLYMER SYNTHESIS

Sustainable and controlled syntheses of polymers and polyelectrolyte membranes have been carried out in IL/O microemulsions, where the corresponding monomeric organic compounds constitute the oil phase. For example, Yan and coworkers used MMA/[C_{12}mim][Cl]/[bmim][BF_4] microemulsion to carry out atom transfer radical polymerization (ATRP) reaction to generate polymethyl methacrylate (PMMA) [133]. When this reaction was carried out in conventional microemulsions, large quantity of surfactants was needed to stabilize these systems, which rendered the formulation environmentally detrimental and made product recovery tedious. Use of the IL-based microemulsion enabled recovery of the IL, surfactants and catalysts, which could then be recycled for several reaction cycles in a sustainable manner. Additionally, the resultant PMMA polymers were found to have low molecular weight with narrow polydispersity [134]. In another example of polymerization in IL-based microemulsion, IL polar domains were used to generate highly conducting polyelectrolyte membrane by dispersing them in a continuous phase of polymerizable vinyl compounds [135]. The polymerization occurs in the oil phase, with the finely dispersed IL droplets forming highly conducting nanochannels in the membrane.

19.10 APPLICATIONS OF IL/O MICROEMULSIONS AS DRUG DELIVERY MEDIA

The excellent solubilizing capacity of ILs, combined with the high membrane permeability of oil, has led to the successful use of IL/O microemulsions as transdermal drug delivery vehicles [31, 78, 79]. Several drugs against life-threatening diseases, for example, acyclovir, an antiviral drug against herpes simplex virus, and methotrexate, an antimetabolite used in anticancer therapy, are limited in their use due to their low solubility in water. Hydrophilic ILs with strongly coordinating anions, for example, [mmim][$(CH_3O)_2PO_2$], constituting the polar core of IL/O microemulsion overcame this limitation by solubilizing large quantities of the drugs, as studied by Goto and coworkers [31]. The use of biocompatible surfactants, Tween-80 and Span-20, and a biological oil, IPM, as apolar phase makes this microemulsion much less toxic than neat IL or oil [78, 79]. In fact, through cytotoxicity analyses, the authors established that the biocompatibility of these microemulsions was comparable to water-in-oil microemulsions, which are already being used as drug delivery vehicles. The dosage of the drug administered through microemulsions is dependent on the size of the droplet and requires precise control. The unique interfacial properties of the surfactant combination of Tween-80/Span-20 can control the size of the formulations, within an acceptable range of 8–34 nm, and yield droplets of uniform size. The IL content can also be varied to regulate the droplet size and the quantity of drug solubilized by the microemulsion. Moreover, these formulations were found to be stable at 25°C for up to 45 days, a prerequisite for their pharmaceutical application [31, 78, 79].

19.11 SUMMARY AND OUTLOOK

The applications of ILs are broadening with the advent of IL-based microemulsions. The IL-based nonaqueous microemulsions, particularly the IL/O ternary systems, have been considerably studied and subsequently used in industries. This class of microemulsion systems offers several advantages over the corresponding water-in-oil systems, due to the tunable size of the polar droplets, their wide range of temperature stability, and the ease of preparation for specific tasks.

Despite the fact that a large amount of surfactants is required to maintain stability, we foresee a greater attention on these IL-based systems, both in science and technology, in the coming years. There are ample possibilities of designing the "greener" isotypes, consisting of hydrophobic ILs or supercritical CO_2 ($scCO_2$) as the apolar phase. A few studies have already been reported and their applications are being tested [80–83, 136]. With the development of sophisticated experimental techniques, we hope that these novel systems, including the more complex IL-in-CO_2 microemulsions, could be critically examined. Besides, molecular dynamics simulations could play an important role in deciphering atomic-level understanding of these systems to help forming defined structures.

REFERENCES

[1] Rogers, R. D.; Seddon, K. R. *Science* 2003, **302**, 792–793.

[2] Wasserscheid, P.; Welton, T. *Ionic Liquids in Synthesis*, Wiley-VCH Verlag GmbH, Weinheim, 2nd edn, 2008.

[3] Wasserscheid, P.; Keim, W. *Angew. Chem.* 2000, **112**, 3926–3945.

[4] Gorman, J. *Sci. News* 2001, **160**, 156–158.

[5] Davis, J. H. J. *Chem. Lett.* 2004, **33**, 1072–1077.

[6] Fei, Z.; Geldbach, T. J.; Zhao, D.; Dyson, P. J. *Chem. Eur. J.* 2006, **12**, 2122–2130.

[7] Yue, C.; Fang, D.; Liu, L.; Yi, T.-F. *J. Mol. Liq.* 2011, **163**, 99–121.

[8] Kubisa, P. *Prog. Polym. Sci.* 2004, **29**, 3–12.

[9] Greaves, T. L.; Drummond, C. J. *Chem. Soc. Rev.* 2008, **37**, 1709–1726.

[10] Hallett, J. P.; Welton, T. *Chem. Rev.* 2011, **111**, 3508–3576.

[11] Armand, M.; Endres, F.; MacFarlane, D. R.; Ohno, H.; Scrosati, B. *Nat. Mater.* 2009, **8**, 621–629.

[12] Ohno, H. *Electrochemical Aspects of Ionic Liquids*, John Wiley & Sons, Inc., Hoboken, NJ, 2nd edn, 2011.

[13] Park, S.; Kazlauskas, R. J. *Curr. Opin. Biotechnol.* 2003, **14**, 432–437.

[14] van Rantwijk, F.; Sheldon, R. A. *Chem. Rev.* 2007, **107**, 2757–2785.

[15] Moniruzzaman, M.; Nakashima, K.; Kamiya, N.; Goto, M. *Biochem. Eng. J.* 2010, **48**, 295–314.

[16] Mora-Pale, M.; Meli, L.; Doherty, T. V.; Linhardt, R. J.; Dordick, J. S. *Biotechnol. Bioeng.* 2011, **108**, 1229–1245.

[17] Vijayaraghavan, R.; Izgorodin, A.; Ganesh, V.; Surianarayanan, M.; MacFarlane, D. R. *Angew. Chem. Int. Ed.* 2010, **49**, 1631–1633.

[18] Chandran, A.; GhoshDastidar, D.; Senapati, S. *J. Am. Chem. Soc.* 2012, **134**, 20330–20339.

[19] Tang, S.; Baker, G. L.; Zhao, H. *Chem. Soc. Rev.* 2012, **41**, 4030–4066.

[20] Palacio, M.; Bhushan, B. *Tribol. Lett.* 2010, **40**, 247–268.

[21] Werner, S.; Haumann, M.; Wasserscheid, P. *Annu. Rev. Chem. Biomol. Eng.* 2010, **1**, 203–230.

[22] Oppermann, S.; Stein, F.; Kragl, U. *Appl. Microbiol. Biotechnol.* 2011, **89**, 493–499.

[23] Quijano, G.; Couvert, A.; Amrane, A. *Bioresour. Technol.* 2010, **101**, 8923–8930.

[24] Ho, T. D.; Canestraro, A. J.; Anderson, J. L. *Anal. Chim. Acta* 2011, **695**, 18–43.

[25] Giernoth, R. *Angew. Chem. Int. Ed.* 2010, **49**, 2834–28389.

[26] Karadas, F.; Atilhan, M.; Aparicio, S. *Energy Fuel* 2010, **24**, 5817–5828.

[27] de Maria, P. D.; Maugeri, Z. *Curr. Opin. Chem. Biol.* 2011, **15**, 220–225.

[28] Fauzi, A. H. M.; Amin, N. A. S. *Renew. Sustain. Energy Rev.* 2012, **16**, 5770–5786.

[29] Vidis, A.; Ohlin, C. A.; Laurenczy, G.; Kusters, E.; Sedelmeier, G.; Dyson, P. J. *Adv. Synth. Catal.* 2005, **347**, 266–274.

[30] Cevc, G. *Adv. Drug Deliv. Rev.* 2004, **56**, 675–711.

[31] Moniruzzaman, M.; Tahara, Y.; Tamura, M.; Kamiya, N.; Goto, M. *Chem. Commun.* 2010, **46**, 1452–1454.

[32] Bourrel, M.; Schechter, R. S. *Microemulsions and Related Systems: Formulation, Solvency and Physical Properties*, Marcel Dekker, New York, 1988.

[33] Paul, B. K.; Moulik, S. P. *J. Dispers. Sci. Technol.* 1997, **18**, 301–367.

[34] Dutta, P.; Sen, P.; Mukherjee, S.; Halder, A.; Bhattacharya, K. *J. Phys. Chem. B* 2003, **107**, 10815–10822.

[35] Sen, S.; Sukul, D.; Dutta, P.; Bhattacharya, K. *J. Phys. Chem. A* 2002, **106**, 6017–6023.

[36] Nandi, N.; Bhattacharya, K.; Bagchi, B. *Chem. Rev.* 2000, **100**, 2013–2046.

[37] Mitra, R. K.; Paul, B. K. *Colloids Surf. A Physicochem. Eng. Asp.* 2005, **255**, 165–180.

[38] Paul, B. K.; Mitra, R. K. *Colloids Surf. A Physicochem. Eng. Asp.* 2006, **273**, 129–140.

[39] Johnston, K. P.; Harrison, K. L.; Clarke, M. J.; Howdle, S. M.; Heitz, M. P.; Bright, F. V.; Carlier, C.; Randolph, T. W. *Science* 1996, **271**, 624.

[40] DeSimone, J. M.; Maury, E. E.; Menceloglu, Y. Z.; McClain, J. B.; Romack, T. J.; Combes, J. R. *Science* 1994, **265**, 356.

[41] Senapati, S.; Keiper, J. S.; DeSimone, J. M.; Wignall, G. D.; Melnichenko, Y. B.; Frienlinghaus, H.; Berkowitz, M. L. *Langmuir* 2002, **18**, 7371–7376.

[42] Senapati, S.; Berkowitz, M. L. *J. Chem. Phys.* 2003, **118**, 1937–1944.

[43] Senapati, S.; Berkowitz, M. L. *J. Phys. Chem. B* 2003, **107**, 12906–12916.

[44] Senapati, S.; Berkowitz, M. L. *J. Phys. Chem. A* 2004, **108**, 9768–9776.

[45] Senapati, S.. *J. Chem. Phys.* 2007, **126**, 204710–204716.

[46] Chaitanya, V. S. V.; Senapati, S. *J. Am. Chem. Soc.* 2008, **130**, 1866–1870.

[47] Evans, D. F.; Yamauchi, A.; Roman, R.; Casassa, E. Z. *J. Colloid Interface Sci.* 1982, **88**, 89–96.

[48] Anderson, J. L.; Pino, V.; Hagberg, E. C.; Sheares, V. V.; Armstrong, D. W. *Chem. Commun.* 2003, **19**, 2444–2445.

[49] Hao, J.; Song, A.; Wang, J.; Chen, X.; Zhuang, W.; Shi, F.; Zhou, F.; Liu, W. *Chem. Eur. J.* 2005, **11**, 3936–3940.

[50] Greaves, T. L.; Weerawardena, A.; Krodkiewska, I.; Drummond, C. J. *J. Phys. Chem. B* 2008, **112**, 896–905.

[51] Patrascu, C.; Gauffre, F.; Nallet, F.; Bordes, R.; Oberdisse, J.; de Lauth-Viguerie, N.; Mingotaud, C. *ChemPhysChem* 2006, **7**, 99–101.

[52] Greaves, T. L.; Weerawardena, A.; Fong, C.; Drummond, C. J. *Langmuir* 2007, **23**, 402–404.

[53] Greaves, T. L.; Weerawardena, A.; Drummond, C. J. *Phys. Chem. Chem. Phys.* 2011, **13**, 9180–9186.

[54] Fernandez-Castro, B.; Mendez-Morales, T.; Carrete, J.; Fazer, E.; Cabeza, O.; Rodriguez, J. R.; Turmine, M.; Varela, L. M. *J. Phys. Chem. B* 2011, **115**, 8145–8154.

[55] Chen, L. G.; Bermudez, H. *Langmuir* 2012, **28**, 1157–1162.

[56] Yue, X.; Chen, X.; Li, Q. T. *J. Phys. Chem. B* 2012, **116**, 9439–9444.

[57] Wijaya, E. C.; Greaves, T. L.; Drummond, C. J. *Faraday Discuss.* 2013, **167**, 191–215.

[58] Gao, H.; Li, J.; Han, B.; Chen, W.; Zhang, J.; Zhang, R.; Yan, D.; *Phys. Chem. Chem. Phys.* 2004, **6**, 2914–2916.

[59] Gao, Y.; Wang, S.; Zheng, L.; Han, S.; Zhang, X.; Lu, D.; Yu, L.; Ji, Y.; Zhang, G.. *J. Colloid Interface Sci.* 2006, **301**, 612–616.

[60] Gao, Y.; Zhang, J.; Xu, H.; Zhao, X.; Zheng, L.; Li, X.; Yu, L.. *ChemPhysChem* 2006, **7**, 1554–1561.

[61] Eastoe, J.; Gold, S.; Rogers, S. E.; Paul, A.; Welton, T.; Heenan, R. K.; Grillo, I. *J. Am. Chem. Soc.* 2005, **127**, 7302–7303.

[62] Gao, Y.; Li, N.; Zheng, L.; Bai, X.; Yu, L.; Zhao, X.; Zhang, J.; Zhao, M.; Li, Z. *J. Phys. Chem. B* 2007, **111**, 2506–2513.

[63] Li, N.; Cao, Q.; Gao, Y.; Zhang, J.; Zheng, L.; Bai, X.; Dong, B.; Li, Z.; Zhao, M.; Yu, L. *ChemPhysChem* 2007, **8**, 2211–2217.

[64] Falcone, R. D.; Correa, N. M.; Silber, J. J. *Langmuir* 2009, **25**, 10426–10429.

[65] Pramanik, R.; Sarkar, S.; Ghatak, C.; Rao, V. G.; Sarkar, N. *J. Phys. Chem. B* 2011, **115**, 2322–2330.

[66] Gayet, F.; Kalamouni, C. E.; Lavedan, P.; Marty, J.; Brulet, A.; Lauth-de Viguerie, N. *Langmuir* 2009, **25**, 9741–9750.

[67] Pramanik, R.; Sarkar, S.; Ghatak, C.; Rao, V. G.; Setua, P.; Sarkar, N. *J. Phys. Chem. B* 2010, **114**, 7579–7586.

[68] Rao, V. G.; Mandal, S.; Ghosh, S.; Banerjee, C.; Sarkar, N. *J. Phys. Chem. B* 2012, **116**, 13868–13877.

[69] Li, J.; Zhang, J.; Gao, H.; Han, B.; Gao, L. *Colloid Polym. Sci.* 2005, **283**, 1371–1375.

[70] Ferreyra, D. D.; Correa, N. M.; Silber, J. J.; Falcone, R. D. *Phys. Chem. Chem. Phys.* 2012, **14**, 3460–3470.

[71] Sarkar, S.; Pramanik, R.; Ghatak, C.; Rao, V. G.; Sarkar, N. *J. Chem. Phys.* 2011, **134**, 074507–074517.

[72] Blach, D.; Silber, J. J.; Correa, N. M.; Falcone, R. D. *Phys. Chem. Chem. Phys.* 2013, **15**, 16746–16757.

[73] Rojas, O.; Tiersch, B.; Rabe, C.; Stehle, R.; Hoell, A.; Arlt, B.; Koetz, J. *Langmuir* 2013, **29**, 6833–6839.

[74] Rojas, O.; Tiersch, B.; Frasca, S.; Wollenberger, U.; Koetz, J. *Colloids Surf. A Physicochem. Eng. Asp.* 2010, **369**, 82–87.

[75] Banerjee, C.; Kundu, N.; Ghosh, S.; Mandal, S.; Kuchlyan, J.; Sarkar, N. *J. Phys. Chem. B* 2013, **117**, 9508–9517.

[76] Rao, V. G.; Banerjee, C.; Ghosh, S.; Mandal, S.; Kuchlyan, J.; Sarkar, N. *J. Phys. Chem. B* 2013, **117**, 7472–7480.

[77] Mandal, S.; Ghosh, S.; Banerjee, C.; Kuchlyan, J.; Banik, D.; Sarkar, N. *J. Phys. Chem. B* 2013, **117**, 3221–3231.

[78] Moniruzzaman, M.; Kamiya, N.; Goto, M. *J. Colloid Interface Sci.* 2010, **352**, 136–142.

[79] Moniruzzaman, M.; Tamura, M.; Tahara, Y.; Kamiya, N.; Goto, M. *Int. J. Pharm.* 2010, **400**, 243–250.

[80] Liu, J.; Cheng, S.; Zhang, J.; Feng, X.; Fu, X.; Han, B. *Angew. Chem. Int. Ed.* 2007, **46**, 3313–3315.

[81] Zhang, J.; Han, B.; Li, J.; Zhao, Y.; Yang, G. *Angew. Chem. Int. Ed.* 2011, **50**, 9911–9915.

[82] Cheng, S.; Zhang, J.; Zhang, Z.; Han, B. *Chem. Commun.* 2007, 2497–2499.

[83] Li, J.; Zhang, J.; Han, B.; Peng, L.; Yang, G. *Chem. Commun.* 2012, **48**, 10562–10564.

[84] Rabe, C.; Koetz, J. *Colloids Surf. A Physicochem. Eng. Asp.* 2010, **354**, 261–267.

[85] Rao, V. G.; Ghosh, S.; Ghatak, C.; Mandal, S.; Brahmachari, U.; Sarkar, N. *J. Phys. Chem. B* 2012, **116**, 2850–2855.

[86] Rao, V. G.; Mandal, S.; Ghosh, S.; Banerjee, C.; Sarkar, N. *J. Phys. Chem. B* 2013, **117**, 1480–1493.

[87] Rao, V. G.; Mandal, S.; Ghosh, S.; Banerjee, C.; Sarkar, N. *J. Phys. Chem. B* 2012, **116**, 8210–8221.

[88] Banerjee, C.; Mandal, S.; Ghosh, S.; Kuchlyan, J.; Kundu, N.; Sarkar, N. *J. Phys. Chem. B* 2013, **117**, 3927–3934.

[89] Chai, J.; Xu, L.; Liu, W.; Zhu, M. *J. Chem. Eng. Data* 2012, **57**, 2394–2400.

[90] Luczak, J.; Hupka, J.; Thoming, J.; Jungnickel, C. *Colloids Surf. A Physicochem. Eng. Asp.* 2008, **329**, 125–133 (and references therein).

[91] Brown, P.; Butts, C. P.; Dyer, R.; Eastoe, J.; Grillo, I.; Guittard, F.; Rogers, S.; Heenan, R. *Langmuir* 2011, **27**, 4563–4571.

[92] Brown, P.; Butts, C. P.; Eastoe, J.; Fermin, D.; Grillo, I.; Lee, H.-C.; Parker, D.; Plana, D.; Richardson, R. M. *Langmuir* 2012, **28**, 2502–2509.

[93] Brown, P.; Butts, C. P.; Eastoe, J.; Grillo, I.; James, C.; Khan, A. *J. Colloid Interface Sci.* 2013, **395**, 185–189.

[94] Ghosh, S.; Banerjee, C.; Mandal, S.; Rao, V. G.; Sarkar, N. *J. Phys. Chem. B* 2013, **117**, 5886–5897.

[95] Hao, J.; Zemb, T. *Curr. Opin. Colloid Interface Sci.* 2007, **12**, 129–137.

[96] Qiu, Z.; Texter, J. *Curr. Opin. Colloid Interface Sci.* 2008, **13**, 252–262.

[97] Zech, O.; Kunz, W. *Soft Matter* 2011, **7**, 5507–5513.

[98] Greaves, T. L.; Drummond, C. J. *Chem. Soc. Rev.* 2013, **42**, 1096–1120.

[99] Correa, N. M.; Silber, J. J.; Riter, R. E.; Levinger, N. E. *Chem. Rev.* 2012, **112**, 4569–4602.

[100] Kunz, W.; Zemb, T.; Harrar, A. *Curr. Opin. Colloid Interface Sci.* 2012, **17**, 205–211.

[101] Gao, Y.; Li, N.; Zheng, L.; Zhao, X.; Zhang, J.; Cao, Q.; Zhao, M.; Li, Z.; Zhang, G. *Chem. Eur. J.* 2007, **13**, 2661–2670.

[102] Andújar-Matalobos, M.; García-Río, L.; López-García, S.; Rodríguez-Dafonte, P. *J. Colloid Interface Sci.* 2011, **363**, 261–267.

[103] Harrar, A.; Zech, O.; Hartl, R.; Bauduin, P.; Zemb, T.; Kunz, W. *Langmuir* 2011, **27**, 1635–1642.

[104] Gao, Y.; Li, N.; Zhang, S.; Zheng, L.; Li, X.; Dong, B.; Yu, L. *J. Phys. Chem. B* 2009, **113**, 1389–1395.

[105] Lian, Y.; Zhao, K. *J. Phys. Chem. B* 2011, **115**, 11368–11374.

[106] Safran, S. A.; Webman, I.; Grest, G. S. *Phys. Rev. A* 1985, **32**, 506–511.

[107] Grest, G. S.; Webman, I.; Safran, S. A.; Bug, A. L. R. *Phys. Rev. A* 1986, **33**, 2842–2845.

[108] Adhikari, A.; Sahu, K.; Dey, S.; Ghosh, S.; Mandal, U.; Bhattacharyya, K. *J. Phys. Chem. B* 2007, **111**, 12809–12816.

[109] Li, N.; Gao, Y.; Zheng, L.; Zhang, J.; Yu, L.; Li, X.; *Langmuir* 2007, **23**, 1091–1097.

[110] Gao, Y.; Hilfert, L.; Voigt, A.; Sundmacher, K. *J. Phys. Chem. B* 2008, **112**, 3711–3719.

[111] Falcone, R. D.; Baruah, B.; Gaidamauskas, E.; Rithner, C. D.; Correa, N. M.; Silber, J. J.; Crans, D. C.; Levinger, N. E. *Chem. Eur. J.* 2011, **17**, 6837–6846.

[112] Hyde, S. T. *Handbook of Applied Surface and Colloid Chemistry*, John Wiley & Sons, Inc., New York: 2001, p. 299.

[113] Zech, O.; Thomaier, S.; Kolodziejski, A.; Touraud, D.; Grillo, I.; Kunz, W. *Chem. Eur. J.* 2010, **16**, 783–786.

[114] Gao, Y.; Li, N.; Hilfert, L.; Zhang, S.; Zheng, L.; Yu, L. *Langmuir* 2009, **25**, 1360–1365.

[115] Chakrabarty, D.; Seth, D.; Chakraborty, A.; Sarkar, N. *J. Phys. Chem. B* 2005, **109**, 5753–5758.

[116] Maroncelli, M.; Fleming, G. R. *J. Chem. Phys.* 1987, **86**, 6221–6239.

[117] Fee, R. S.; Maroncelli, M. *Chem. Phys.* 1994, **183**, 235–247.

[118] Karmakar, R.; Samanta, A. *J. Phys. Chem. A* 2002, **106**, 4447–4452.

[119] Pramanik, R.; Ghatak, C.; Rao, V. G.; Sarkar, N. *J. Phys. Chem. B* 2011, **115**, 5971–5979.

[120] Bagchi, B. *Chem. Rev.* 2005, **105**, 3197–3219.

[121] Bhattacharyya, K. *Acc. Chem. Res.* 2003, **36**, 95–101.

[122] Baruah, B.; Roden, J. M.; Sedgwick, M.; Correa, N. M.; Crans, D. C.; Levinger, N. E. *J. Am. Chem. Soc.* 2006, **128**, 12758–12765.

[123] Balasubramanian, S.; Pal, S.; Bagchi, B. *Phys. Rev. Lett.* 2002, **89**, 11505.

[124] Inoue, T. *J. Colloid Interface Sci.* 2009, **337**, 240–246.

[125] Inoue, T.; Kawashima, K.; Miyagawa, Y. *J. Colloid Interface Sci.* 2011, **363**, 295–300.

[126] Misono, T.; Sakai, H.; Sakai, K.; Abe, M.; Inoue, T. *J. Colloid Interface Sci.* 2011, **358**, 527–533.

[127] Greaves, T. L.; Mudie, S. T.; Drummond, C. J. *Phys. Chem. Chem. Phys.* 2011, **13**, 20441–20452.

[128] Zheng, H. X.; Li, Z. P.; Wang, H. Q. *Acta Phys. Chim. Sin.* 1999, **15**, 522–526.

[129] Wang, F.; Zhang, Z.; Li, D.; Yang, J.; Chu, C.; Xu, L. *J. Chem. Eng. Data* 2011, **56**, 3328–3335.

[130] Sasmal, D. S.; Mojumdar, S. S.; Adhikari, A.; Bhattacharyya, K. *J. Phys. Chem. B* 2010, **114**, 4565–4571.

[131] Pramanik, R.; Sarkar, S.; Ghatak, C.; Setua, P.; Rao, V. G.; Sarkar, N. *Chem. Phys. Lett.* 2010, **490**, 154–158.

[132] Zech, O.; Thomaier, S.; Kolodziejski, A.; Touraud, D.; Grillo, I.; Kunz, W. *J. Colloid Interface Sci.* 2010, **347**, 227–232.

[133] Chen, Z.; Feng Yan, F.; Qiu, L.; Lu, J.; Zhou, Y.; Chen, J.; Tang, Y.; Texter, J. *Langmuir* 2010, **26**, 3803–3806.

[134] Zhou, Y.; Qiu, L.; Deng, Z.; Texter, J.; Yan, F. *Macromolecules* 2011, **44**, 7948–7955.

[135] Yan, F.; Yu, S.; Zhang, X.; Qiu, L.; Chu, F.; You, J.; Lu, J. *Chem. Mater.* 2009, **21**, 1480–1484.

[136] Chandran, A.; Prakash, K.; Senapati, S. *J. Am. Chem. Soc.* 2010, **132**, 12511–12516.

CHAPTER 20

Recent Advances in Bioionic Liquids and Biocompatible Ionic Liquid-Based Microemulsions

KAUSHIK KUNDU and BIDYUT K. PAUL

Surface and Colloid Science Laboratory, Geological Studies Unit, Indian Statistical Institute, Kolkata, India

SOUMIK BARDHAN and SWAPAN K. SAHA

Department of Chemistry, University of North Bengal, Darjeeling, India

20.1 INTRODUCTION

20.1.1 Background and History of Ionic Liquids

In recent years, the choice of the solvent for performing a reaction has become an increasingly important task; selective reagents are used for chemical transformations, and the choice of the solvent may determine high reaction rates and high selectivity. At the same time, as a response to increasing legislative and social pressure and an increasingly "green"-conscious industrial community, researchers have started to examine more eco-friendly and sustainable chemical processes. Toxicity and recycling considerations are influencing the choice of the solvent to be used for industrial reactions. Ionic liquids (ILs) represent a class of alternative solvents currently receiving serious consideration with the promise of both environmental and technological benefits. Further, the self-assembling phenomena of amphiphilic molecules, (e.g., anionic, nonionic, zwitterionic, and gemini surfactants (GSs) in single or mixed state) in ILs are currently an attractive and useful research area in the field of colloid and interface science, because an understanding of the self-assembling mechanism is of much interest and importance from academic as well as application points of view as well. The field of ILs has been reviewed by several authors, including Welton [1], Holbrey [2], and Seddon [3]. The first room-temperature IL, ethylammonium nitrate, $[EtNH_3][NO_3]$ (m.pt. 12°C), was discovered in 1914 [4], but interest did not develop until the discovery of binary ILs made from mixtures of aluminum (III) chloride and *N*-alkylpyridinium [5] or 1,3-dialkylimidazolium chloride [6].

Ionic Liquid-Based Surfactant Science: Formulation, Characterization, and Applications, First Edition. Edited by Bidyut K. Paul and Satya P. Moulik.

20.1.2 Definition, Types, and Properties of ILs

ILs are a class of novel solvents (or a special class of salts) with low melting points of below 100°C and composed entirely of organic cations (e.g., imidazolium, pyridinium, etc.) and inorganic/organic anions (e.g., chloride (Cl^-), bromide (Br^-) and tetrafluoroborate (BF_4^-), hexafluorophosphate (PF_6^-), bis(trifluoromethylsulfonyl) imide (Tf_2N^-), etc.). The definition of ILs allows distinguishing them from a classical molten salt. A molten salt is mostly a high-melting, highly viscous, and very corrosive substance, while ILs are already liquid at lower temperatures (<100°C) and exhibit in most cases relatively low viscosities [7]. Conventional ILs typically contain bulky organic cations with a low degree of symmetry such as imidazolium, pyrrolidinium, tetraalkyl phosphonium, trialkyl sulfonium, phosphonium, or quaternary ammonium. These cations hinder the regular packing in a crystal lattice. Consequently, the solid crystalline state becomes energetically less favorable, leading to low melting points [8]. This effect enhances further by the implementation of an anion with a delocalized charge, resulting in decreased interionic interactions [9]. Depending on the combination of the cation and the anion, ILs can have either hydrophilic or hydrophobic character. The most frequently investigated ILs are based on imidazolium cations. Anions such as halides, acetate, nitrate, and ethyl sulfate form hydrophilic ILs, while anions such as bis(trifluoromethylsulfonyl)imide (Tf_2N^-) and hexafluorophosphate (PF_6^-) lead to hydrophobic ILs. ILs come in two main categories, namely, simple salts (made of a single anion and cation) and binary ILs (salts where equilibrium is involved). For example, $[EtNH_3][NO_3]$ is a simple salt, whereas mixtures of aluminum(III) chloride and 1,3-dialkylimidazolium chlorides (a binary IL system) contain several different ionic species, and their melting point and properties depend upon the mole fractions of the aluminum(III) chloride and 1,3-dialkylimidazolium chloride present. For the binary systems, the melting point depends upon composition, and this complex behavior has been studied extensively for the archetypal system, [emim]Cl-$AlCl_3$ ($[emim]^+$ = 1-ethyl-3-methylimidazolium) [10]. ILs have been described as designer solvents [11], and this means that their properties can be adjusted to suit the requirements of a particular process. Properties such as melting point, viscosity, density, and hydrophobicity can be varied by simple changes to the structure of the ions. For example, the melting points of 1-alkyl-3-methylimidazolium tetrafluoroborates and hexafluorophosphates [12] are a function of the length of the 1-alkyl group and form liquid crystalline phases for alkyl chain lengths over 12 carbon atoms. Another important property that changes with structure is the miscibility of water in these ILs. For example, 1-alkyl-3-methylimidazolium tetrafluoroborate salts are miscible with water at 25°C where the alkyl chain length is less than six, but at or above six carbon atoms, they form a separate phase when mixed with water. This behavior can be of substantial benefit when carrying out solvent extractions or product separations, as the relative solubilities of the ionic and extraction phase can be adjusted to make the separation as easy as possible. The properties of ILs have been summarized in Table 20.1.

Notably, the inherent synthetic flexibility afforded by pairing different cations with any of a growing number of anions provides the possibility for "fine-tuning" certain IL solvent properties to the particular task at hand. The considerable variation within this general framework is such that conservative estimates of the number

TABLE 20.1 Properties of Ionic Liquids

A salt	Made up of at least two components (the anion and cation)
Freezing point	Preferably below 100°C
Liquidus range	Often >200°C
Thermal stability	Usually high
Viscosity	Normally <100 cP, workable
Dielectric constant	Implied <30
Polarity	Moderate
Specific conductivity	Usually <10 mScm^{-1}, "Good"
Molar conductivity	<10 Scm^2/mol
Electrochemical window	>2 V, even 4.5 V, except for Brønsted acidic systems
Solvent and/or catalyst	Excellent for many organic reactions
Vapor pressure	Usually negligible

of practical ILs that can be easily prepared from relatively inexpensive materials reach the tens of thousands although, in theory, some 10 trillion combinations are possible for these "modular" solvents. However, even though there are 10^9 possible anion–cation combinations that lead to IL formation [13], virtually the majority of studies have focused on imidazolium-based moieties with halide counterions. This has significant practical implications such as relative expense, unknown toxicity, environmentally hazardous starting materials, incompatibility with active metals and strong bases due to the acidity of the C-2 proton of the imidazolium [14–16], and issues with purification [17].

20.1.3 Application of ILs in Various Fields

In 1998, the Rogers group first reported the use of an IL as an alternative to traditional volatile organic solvents for two-phase liquid–liquid separations [18]. The most popular application of ILs in the field of analytical chemistry is in the area of analytical separations. Earlier, Stalcup and Cabovska [19] reviewed some of the unique properties of ILs and their applications in analytical separations. Baker et al. [20] addressed progress made in the application of ILs toward solving of analytical problem, for example, liquid–liquid extraction and liquid-phase microextraction, gas chromatography, liquid chromatography, and capillary electrophoresis. The application of ILs as replacements for the conventional liquid or solid matrixes in matrix-assisted laser desorption/ionization mass spectrometry (MALDI-MS) and also as media for developing optical sensors was recognized [21, 22]. In the last decade, numerous studies focused on the dissolution of natural polymers in ILs demonstrating a great potential of ILs as solvents [23–27]. Cellulose was one of the most studied biopolymers exhibiting a high solubility in a variety of ILs [28]. The solubilities of a wide range of carbohydrates were scrutinized using different ILs and demonstrated that one of the main benefits of using ILs to dissolve carbohydrates is that ILs can be tailored to accomplish dissolution or functionalization of these polymers [28]. Very recently, use of ILs in the development of alternative methods for the extraction and processing of carbohydrates and other compounds from lignocellulosic biomass was reported by da Costa Lopes et al. [29]. The current state of the art on the fractionation of lignocellulosic biomass in ILs as a pretreatment methodology was described. ILs are found to be promising solvents in many of the organic reactions such as Diels–Alder, Bails–Hillman, Heck reaction, esterification, isomerization reactions, and many coupling reactions [30]. Excellent reviews were also published demonstrating the applications of ILs to electrochemical devices such as super capacitors, lithium-ion batteries, polymer electrolyte fuel cells, and dye-sensitized solar cells [31, 32]. Recently, use of ILs with unsaturated fatty acids as anions (oleate and linoleate) for extraction of Li, Na, K, Mn(II), Fe(II), and Zn(II) chlorides from aqueous solution has been reported [33]. Joshi and Anderson highlighted the recent advancements in the use of ILs in separation science including gas chromatography, high-performance liquid chromatography, capillary electrophoresis, extraction and microextraction techniques, as well as mass spectrometry [34].

In the past few years, ILs gained importance for use in several pharmaceutical applications. They are found to be particularly useful for the solubilization of poorly soluble drugs [35–37]. ILs can also be used to synthesize active pharmaceutical ingredients (APIs) with modified solubility, increased thermal stability, and a significant enhancement in the efficacy of topical analgesia compared to their starting materials [38–40]. Furthermore, many ILs exhibit antimicrobial activity [41], which can make them useful as APIs or formulation preservatives and also provide unique opportunities for oral [42] and topical drug delivery [43].

The range of reactions in ILs gives a flavor of what can be achieved in these neoteric solvents. Because of their properties, and associated synthetic flexibility, behavior of the ILs can be adjusted to suit an individual reaction type, so that they can truly be described as "designer solvents" (because it is possible to synthesize an IL with a given required property) [11, 13] and can be served as greener alternatives to conventional toxic organic solvents [44, 45]. Recently, ILs are categorized to be good solvents for dissolution and reconstitution of unmodified biopolymers [46–48]. The significant merits of the procedure to synthesize biopolymer-based materials by using ILs are high solubility of biopolymers that cannot be dissolved in general organic solvents, greener procedures using nonvolatile solvents, and easy production of composites comprising various biopolymers and synthetic polymers. Earlier, Earle and Seddon [13] reported high product yields of various reactions (e.g., Lewis acid-catalyzed Friedel–Crafts reaction, cracking and isomerization reactions, dimerization and oligomerization of olefins, Diels–Alder reaction, Heck reaction, and nucleophilic substitution reaction) by choosing the appropriate IL, and a reduced amount of waste was produced in a given reaction. It was also observed that often an IL is recycled and led to a reduction of the costs of the processes. They also emphasized that reactions in ILs are not difficult to perform, usually required no special apparatus or methodologies. Moreover, the reactions were often quicker and easier to carry out than in conventional organic solvents.

For good reasons, these ILs are often discussed as "green solvents." In contrast to volatile organic solvents and extraction media, ILs have no measurable vapor pressure. Hence, there is no loss of solvent through evaporation. The use of volatile organic solvents can be avoided due to environmental and safety problems, which originate from them. For catalytic applications, where a transition metal catalyst is dissolved in IL, two additional aspects are of interest. Firstly, the special solubility characteristics of an ionic reaction medium enable a biphasic reaction procedure in many cases. The catalyst was allowed to isolate effectively from the product by utilizing the miscibility gap between the ionic catalyst phase and the products that could be reused many times. Secondly, the nonvolatile nature of ILs enables more effective product isolation by distillation. Again, the possibility arises to reuse the isolated ionic catalyst phase. In both cases, the total reactivity of the applied catalysts is increased, and catalyst consumption relative to the generated product is reduced. For example, all these advantages have been convincingly demonstrated for transition metal-catalyzed hydroformylation [49–51].

Besides the aforementioned applications, ILs have attracted interest to use as lubricants [52–54], as solvents for synthesis and catalysis [3, 55], as continuous phases for particle dispersions [56, 57], among many other applications [13, 58–60], and also as solvents and/or (co)solvents and/or reagents in a wide range of applications, including chemical engineering (separation, extraction, and membranes) [61, 62], energy conversion (batteries, fuel cells, and heat storage) [63], and biotechnology (biocatalysis, biomolecule purification, and biofuel production) [64–66].

20.2 TOXICITY OF ILs

The vast majority of applications studies on ILs are available to date have focused mainly on imidazolium cation and highly fluorinated and hydrophobic ones, for example, halides (Cl or F) anion-based ILs. However, the use of the halide-containing ILs may cause serious concern to the environmental impact of the technical processes under certain conditions. This would be explicitly the case for the hydrolysis labile anions $[AlCl_4]^-$ and $[PF_6]^-$, if water is present. In this case, significant amounts of HCl (or HF) would be liberated under reaction conditions. However, this would be the case as well for the potential disposal of technical amounts of spent IL by thermal treatment. In both cases, additional efforts and costs have to be considered to avoid the liberation of toxic and corrosive HF or HCl into the environment. The aspect of IL disposal gets even more important since disposal alternatives, such as the treatment of spent IL in biological wastewater treatment, are equally effected by the presence of halogen atoms in IL. Keeping all these aspects in mind, many research groups decided to look for new, even "greener" ILs for industrial catalytic applications, which should meet the following technically desired combination of properties: (i) melting point or glass point below 40°C, (ii) hydrolysis stable in neutral aqueous solution up to 80°C, (iii) temperature of thermal decomposition above 250°C, (iv) possible disposal by combustion without formation of HF or HCl, (v) possible biodegradation of the used anion in ordinary wastewater treatment, and (vi) synthesis from cheap, technical available raw materials, for example, alkali metal salts [49].

Although a large number of ILs have been synthesized and characterized to date, most of them were based on halogenated counteranions such as hexafluorophosphate (PF_6^-), tetrafluoroborate (BF_4^-), bis(trifluoromethanesulfonyl)amide [$(CF_3SO_2)_2N^-$], and their derivatives. Halogens are effective in reducing negative charge density due to their electron withdrawing effect. Although these halide-containing anions are low melting point ILs, their applications are limited by toxicological, ecological, and economic issues. Most of the ILs studied so far belong to the imidazolium and pyridinium classes, with alkyl or alkyloxy side chains, containing the anions, bromide (Br^-), chloride (Cl^-), hexafluorophosphate (PF_6^-), and tetrafluoroborate (BF_4^-). Preliminary experiments on the effects of ILs on rat cell lines and also human Caco-2 cell lines [67–69] as well as on enzymatic studies were conducted [70]. It was evidenced from these investigations that the toxicity of ILs increases with increasing alkyl chain length. It was explained that increase in lipophilic character of the IL originates from increasing alkyl chain length, which led to disruption of membrane proteins (polar narcosis) by incorporation of IL into biological membranes [71]. The resulting increase in cell membrane permeability might also adversely affect the ability of cells to resist or to repair membrane disruption [72].

Even though the ILs were originally considered to be environmentally benign in nature primarily due to their low vapor pressure, later studies have revealed toxicity of these substances. Many ILs are moderately or highly toxic, particularly to aquatic organisms, and their biodegradability depends on their structural design [73]. Though ILs are often termed as nonvolatile because of their extremely low vapor pressure, it was shown that commonly used imidazolium ILs can be distilled under high vacuum and temperature conditions in pure form [74, 75]. Though the ILs are nonflammable media, highly energetic ILs rich in oxygen and nitrogen ignite under oxygen environment [76]. Even though the ILs appear homogeneous in the naked eye, recent experimental, theoretical, and simulation studies have revealed organized structures of these liquids [77, 78]. As ILs are "designer solvents" [11] and a combination of various cations and anions allows generation of new ILs, it is still possible to think of new ILs with low vapor pressure, low toxicity, and economically viable to serve the purpose of green chemistry. Consequently, the design and development of new ILs and investigation of the physicochemical properties of new ILs continue to remain as an active area of research. Therefore, it is necessary to consolidate all the available data in a single review to locate the groundwork for more in living organisms in a local ecosystem and to meet the health and welfare of our community.

20.3 STRATEGY FOR SYNTHESIS OF BIOIONIC LIQUIDS

ILs have been extensively reviewed in the last 5 years, with particular focus on their synthesis in the area of green solvents from renewable resources [58, 79]. This section presents some recent advances in this area, accompanied by selected works to illustrate the direction of the field moves in.

Earlier, Davis and his coworkers [80] utilized cations derived from the antifungal drug miconazole (a biologically active molecule) in the formulation of new organic ionic liquids (OILs). Overnight reaction of miconazole with neat methyl, ethyl, or *n*-butyliodide at reflux resulted in the quantitative formation of the corresponding *N*(3)-alkylmiconazolium iodides. Similarly, overnight reflux of miconazole with excess of 1,1,1,2,2,3,3,4,4,5,5,6,6-tridecafluoro-8-iodooctane in THF (tetrahydrofuran) produced the expected salt with 52% yield. The miconazole iodides were readily isolated as pale yellow solids by the removal of volatiles *in vacuo* and repeated washing of the residue with ether, followed by hexane. Subsequent anion metathesis with $NaPF_6$ (sodium hexafluorophosphate) in THF or MeOH (methyl alcohol) resulted in the formation of the miconazole-based ILs. These species were isolated as straw-colored liquids by filtration to remove NaI, followed by vacuum evaporation of the solvent. Straightforward verifications of the cation structures and the overall compositions and purities of OILs were accomplished by acquisition of clean ^{1}H- and ^{13}C-NMR spectra as well as by FTIR and fast atom bombardment mass spectrometry (FAB-MS). More than a decade ago, in search of new, even "greener" ILs for industrial catalytic applications, Wasserscheid et al. [81] reported in detail the synthesis and some important physicochemical properties of the halogen-free IL, 1-butyl-3-methylimidazolium ([bmim]) [n-$C_8H_{17}OSO_3$]. First, 1-butyl-3-methylimidazolium chloride ([bmim]Cl) and Na[n-$C_8H_{17}O$-SO_3] were dissolved in hot water. The water was slowly removed under vacuum and a white solid precipitated. Then, CH_2Cl_2 (dichloromethane) was added to extract the product from the mixture and the white solid filtered off. The clear, slightly yellow filtrate was washed several times with water, up to the point where the washing water was chloride-free (test with $AgNO_3$ from acidic solution). The solvent was removed under vacuum. The chloride-free product was obtained in the form of yellowish oil, which corresponds to 73% of the theoretical yield. The structure and water content of the product was confirmed and determined by ^{13}C-NMR and coulometric Karl–Fischer titration. Later, Handy and his coworkers synthesized fructose-derived ILs [82]. Firstly, fructose readily converted into hydroxymethylene imidazole, and from imidazole, double alkylation of this core generates the corresponding imidazolium cation. For larger-scale alkylations, conventional solution-phase alkylations in ethanol with potassium tert-butoxide as the base afforded reproducible results. The second alkylation with methyl iodide in methylene chloride proceeded uneventfully to afford the final iodide salt-based ILs in nearly quantitative yield. Standard anion metathesis using tosylate, trifluoroacetate, acetate, tetrafluoroborate, triflimide, and dicyanimide was employed to afford a range of new room-temperature ionic liquids (RTILs). Pernak et al. [83] also prepared RTILs of choline derivatives by Menschutkin reaction with alkyl chloromethyl ethers and anion changed to bis(trifluoromethylsulfonyl)amide ion. One of the simplest approaches for the preparation of RTILs was reported by Davis and coworkers [84]. Choline chloride, a natural ammonium salt having a high melting point (302°C), has been transformed into ILs by a 1:2 stoichiometric combination with tin(II) chloride or zinc(II) chloride. Both of these salts are liquids at room temperature (but not moisture sensitive) with high viscosity and also show interesting catalytic properties. However, improved results were obtained through the combination of choline salts with urea; the simple 1:2 mixture of choline chloride and urea afforded a material having a freezing point of 12°C (Scheme 20.1).

Amino acids (AA), as one of the most abundant biomaterials in nature, are known to be nontoxic, biodegradable, and biocompatible. They are excellent feedstocks for the synthesis of ILs, because of their reasonable cost and environment-friendly characteristics. In 2005, a group of amino acid ionic liquids (AAILs) was first synthesized by Fukumoto's group from 20 natural AA [85]. They prepared AAILs based on 1-ethyl-3-methylimidazolium cation, [emim], by neutralization method. In this method, [emim]OH was prepared by anion exchange resin, followed by addition of [emim]OH to a slight excess of an equimolar AA aqueous solution to prepare salts and water as a by-product [86]. Further, they synthesized tetrabutylphosphonium-based AAILs ([TBP][AA]) by mixing AA with aqueous solution of tetrabutylphosphonium hydroxide ([TBP][OH]) [87]. Tao et al. [88] synthesized [AAE]NO_3 ILs ([AAE] stands for AA ester as cation and NO_3 stands

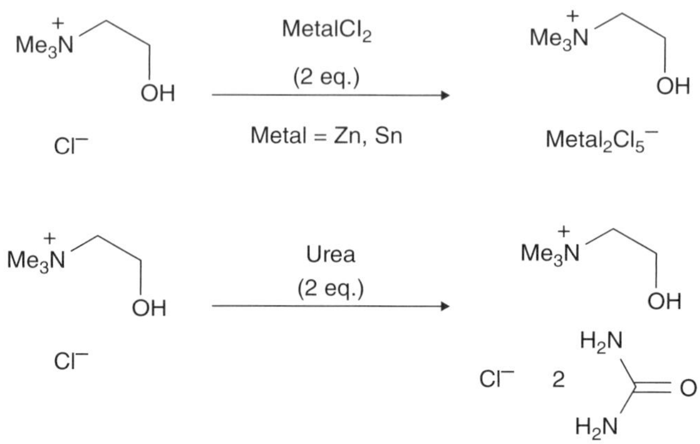

Scheme 20.1 Choline chloride-derived room-temperature ILs. Reproduced from Imperato et al. [79] with permission from Wiley.

for nitrate as anion) by dissolving [AAE]Cl and silver nitrate ($AgNO_3$) in methanol separately. Then, after mixing these two solutions, followed by filtration and evaporation of the resulting liquid *in vacuo*, the product [AAE]NO_3 was obtained. [AAE]Sac ILs (Sac stands for saccharide as anion) were synthesized by the reaction of sodium saccharin and [AAE]Cl in acetone. The resultant precipitate of sodium chloride was filtered off, and acetone was evaporated. Since [AAE]Cl can be obtained commercially, the preparation of both [AAE]NO_3 and [AAE]Sac ILs is a one-step high-yield reaction without any poisonous by-product. Furthermore, it is worthy to mention that the esterification provides a possibility of adjusting the properties of the ILs obtained. Subsequently, they have synthesized L-alanine or valine or leucine or serine or proline - based green ILs, comprising methyl or ethyl or propyl or butyl ester nitrate or saccharinate using these methods. Hannig et al. [89] synthesized chiral ILs (CILs) and imidazol-2-ylidene metal complexes from the proteinogenic AA, L-histidine. Histidinium salt [Bz-His(*n*-propyl)$_2$-OMe$^+$Br$^-$] was prepared by the treatment of the AA derivative Bz–His–OMe (where Bz and His represent *N*-benzoyl and L-histidine, respectively) with excess *n*-propyl bromide. Its subsequent treatment with silver oxide gave the corresponding L-histidine-derived chiral *N*-heterocyclic carbene complex ["(carbene)$_2$Ag . AgBr$_2$"]. Further, chiral-based metal imidazol-2-ylidene complexes ["(carbene)$_2$PdCl$_2$"] and ["(carbene)RhCl(cod)"] were prepared using transmetallation by treatment with $Pd(CH_3CN)_2Cl_2$ or [Rh(cod)Cl]$_2$, respectively. Four diastereomers of the square planar palladium system were observed. Due to the additional chirality center in the L-histidine-derived "Arduengo carbene ligand," two diastereomers of the rhodium carbene complex were formed. Allen et al. [90] synthesized 23 new CILs, by addition of an aqueous solution of tetrabutylammonium hydroxide to an aqueous suspension of the desired AA. The resultant reaction mixture was heated at 60°C for 2h. The water was removed *in vacuo* at 80°C. The resultant residue was dissolved in CH_3CN (acetonitrile) and filtered to remove the unreacted AA. The filtrate was dried over Na_2SO_4 (sodium sulfate) and filtered, and the solvent was removed *in vacuo* to afford the desired product. In this way, they prepared chiral ionic liquids, for example, tetrabutylammonium L-alanate [TBA][L-Ala], tetrabutylammonium L-histidinate [TBA][L-His], tetrabutylammonium L-methioninate

[TBA][L-Met], etc. with high yield and purity. Recently, Morrissey et al. [91] synthesized oxygen-functionalized and amide ILs containing a wide range of ether and polyether esters in two steps followed by anion exchange (Scheme 20.2).

Recently, synthesis, characterization, and toxicological assessment of a new group of environmentally benign ILs were reported by Petkovic et al. [92]. Focusing on the toxic effect of the anion, the ILs were designed by combining the benign cholinium cation, $[NMe_3(CH_2CH_2OH)]^+$, with a range of linear alkanoate anions ($[C_nH_{2n+1}CO_2]^-$, $n = 1–9$), as well as two structural isomers ($n = 3$ or 4). The ILs were prepared by dropwise addition of the corresponding acid (1:1) to aqueous cholinium hydrogen carbonate, stirring at ambient temperature and pressure. Water was then removed under reduced pressure using a rotary evaporator (e.g., 70°C, 30 min) and then stir-heated *in vacuo* (65–70°C, 24 h, *ca.* 0.01 mbar). Chemical route for obtaining the cholinium alkanoate ILs and chemical structures of the anions used are schematically represent in Scheme 20.3. ILs were characterized by ^{1}H- and ^{13}C-NMR spectroscopy (Bruker Avance III spectrometer, 400 MHz) at 25°C. Their purity was confirmed by electrospray ionization mass spectrometry (ESI-MS). The water contents, determined by Karl–Fischer titration, were below 0.1 wt.%.

Recently, Kumar and his coworkers [93] synthesized relatively greener, biodegradable, and nontoxic amino acid ionic liquid surfactant (AAILS) based on natural AA and sodium lauryl sulfate (SLS). L-glycine, L-alanine, L-valine, L-glutamic acid, and L-proline were chosen among natural AA. The AAILs were synthesized by allowing simple variation in the side chain of AA through branching or by the addition of another –COOH group or by cyclization. Because of the incorporation of an ester group into an AA, AAILs showed decrease in melting point as compared to AA, whereas biodegradability significantly increased [94]. The esterification of the AA was carried out by slow addition of thionyl chloride to isopropyl or isobutyl alcohol at 0°C. AA were slowly added to the reaction mixtures, which were then refluxed for 4 h, and resultant mixtures were concentrated in a rotary evaporator. The crude AA ester hydrochlorides were titurated with hexane at 0°C. Pure crystals of amino acid ester hydrochlorides (AAECls) were obtained by recrystallization with methanol/hexane mixture. Equimolar amounts of the AAECls and SLS were then dissolved in hot water. After completion of the reaction, water was removed under vacuum, and AAILSs were extracted by the addition of dichloromethane. All AAILs, except the one obtained from glutamic butyl ester hydrochloride (white

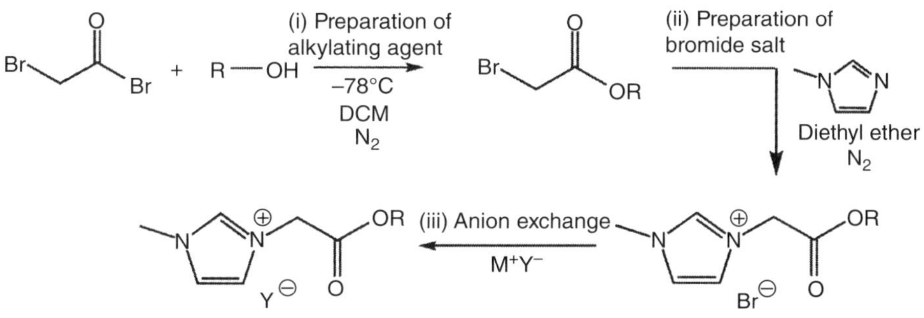

Scheme 20.2 Two-step synthesis of IL followed by anion exchange. Reproduced from Morrissey et al. [91] with permission from the Royal Society of Chemistry.

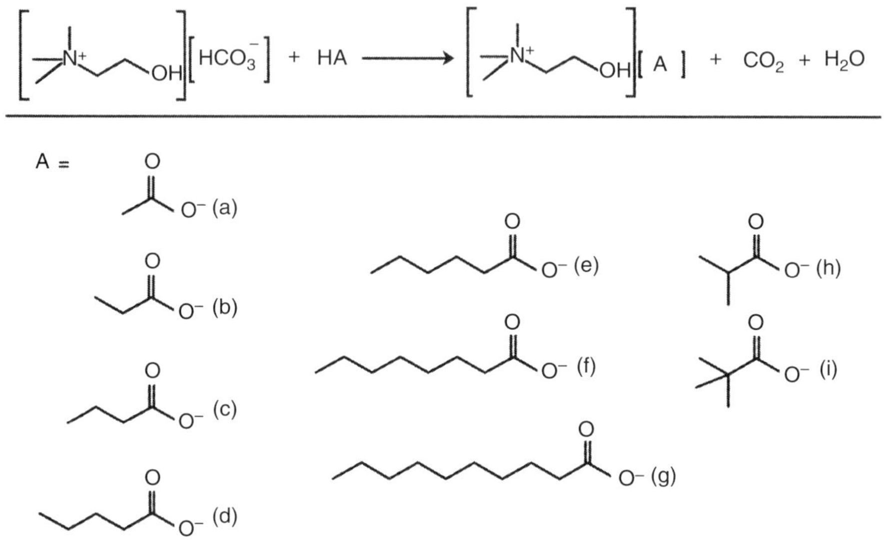

Scheme 20.3 Chemical route for obtaining the cholinium alkanoate ionic liquids, $[NMe_3(CH_2CH_2OH)][C_nH_{2n+1}CO_2]$ ($n = 1–9$), and chemical structures of the anions used: (a) ethanoate, (b) propanoate, (c) butanoate, (d) pentanoate, (e) hexanoate, (f) octanoate, (g) decanoate, (h) 2-methylpropanoate, and (i) 2,2-dimethylpropanoate. Reproduced from Petkovic et al. [92] with permission from the Royal Society of Chemistry.

crystalline solid at room temperature), were clear but slightly viscous liquids at room temperature. The structures of the AAILSs were confirmed by NMR (^1H and ^{13}C), elemental analysis (CHNS), and electrospray ionization mass spectrometry (ESI-MS) [93]. Recently, Muhammad et al. [95] synthesized same AAILs by the procedure as follows. In the first step, [emim][OH] aqueous solution was synthesized by adding 1-ethyl-3-methylimidazolium hydrogen sulfate, $EmimHSO_4$, into an equimolar aqueous solution of $Ba(OH)_2$ (dissolved in boiling water) and stirring for 12 h. In the second step, it was filtered, and [emim][OH]-containing filtrate was neutralized with an equimolar aqueous solution of AA by stirring at room temperature for 12 h. After neutralization, water was evaporated under soft vacuum at 50°C. The excess AA was precipitated by adding ethanol. After filtration, IL was dried using a vacuum rotary and followed by vacuum oven. The estimated purities of the synthesized ILs, namely, EmimGly, EmimAla, EmimSer, and Emimpro, were 97.8%, 97.5%, 97.3%, and 96.9%, respectively The synthesis of these ILs using this method is quite easy as compared to that reported by Fukumoto et al. [85] mentioned earlier. Very recently, Tao et al. [96] synthesized a series of cholinium-based AAILs, [Ch][AA], using the following procedures. In the first step, [Ch]OH aqueous solution was obtained from the metathesis of [Ch]Cl with Ag_2O by using water as the solvent. In the second step, it was centrifuged and filtered; [Ch]OH-containing filtrate was then neutralized with an equimolar aqueous solution of AA by stirring at room temperature for 12 h. After neutralization, water was evaporated under vacuum at 50.15°C. The excess AA was precipitated by adding ethanol. After filtration, the AAIL was kept in a vacuum oven containing P_2O_5 at 90.15°C for 48 h to remove the residual water prior to use. [Ch][AA] ILs were obtained with the yields of more than 80%. Further, the structures of [Ch][AA] ILs were confirmed by ^{1}H-NMR, elemental analysis, and FTIR spectra. The purity of these ILs were estimated from ^{1}H-NMR and elemental analysis. In addition, the water content was reported to be less than 150 ppm, which was analyzed by Karl–Fischer titration.

20.4 PHYSICOCHEMICAL PROPERTIES OF DIFFERENT TYPES OF BILs

20.4.1 AA-Based BILs

Ohno and Fukumoto [86] synthesized for the first time a new class of ILs by coupling of 20 different natural AA (e.g., glycine, leucine, proline, valine, alanine, etc.) as anions, with 1-ethyl-3-methylimidazolium cation. Since the AA contains both an amino group and a carboxylic acid residue in a single molecule, with various side groups/chains and a chiral carbon atom, AA are good candidates to act as a platform for functional ILs [97–99]. Scheme 20.4 represents the use of AA both as cation or anion for synthesis of AA-based ILs.

Further, they observed that the properties (e.g., chiral stability, phase behavior, lower critical separation temperature (LCST) or upper critical separation temperature (UCST) behavior, etc.) of these resulting ionic liquids (AAILs) depend on the side groups/chains of the AA involved. The solubility of organic molecules and thermal properties, such as glass transition temperature, depend on the side chains of the AA anions [86]. While Kamlet–Taft parameters for these ILs were measured, they found that ILs possess a high hydrogen-bond basicity, which is useful for dissolution of biomaterials, such as DNA, cellulose, and other carbohydrates. It was easy to obtain pure AA in large quantities for preparation of AAILs at low cost, which is advantageous from synthesis point of view. Some more advantages of AA include their biodegradability [94] and lower microbial toxicity [100]. The availability of AA as both anions and cations is another advantage (Scheme 20.4). They proposed some applications of these AAILs in the field of biological, medical, and pharmaceutical sciences. AAILs are not only halogen-free but also designer ILs. In another study, Ohno and his coworkers [87] reported further improvement on viscosity and thermal stability in spite of excellent properties of previously synthesized AAILs. AAILs containing 1-ethyl-3-methylimidazolium as a cation ([emim][AA]) were obtained as

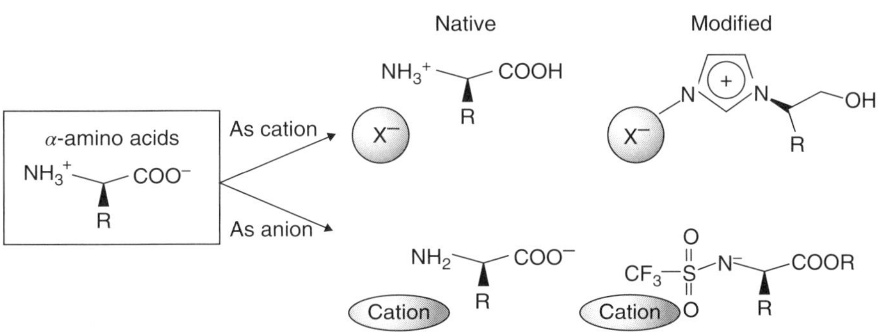

Scheme 20.4 Design of amino acids for functional ionic liquids. Reproduced from Ohno and Fukumoto [86] with permission from the American Chemical Society.

liquids at room temperature, but they were not stable to 250°C and had a narrow electrochemical potential window due to the presence of a proton at the 2-position on the imidazolium ring. The hydrogen bonding at the 2-position of the imidazolium cation ring with its AA anions caused the AAILs to have a relatively high viscosity. In view of this, they reported novel AAILs using a variety of cations to find AAILs with a lower viscosity and better thermal properties. Tetraalkylammonium, pyrrolidinium, and pyridinium cations were used to prepare AAILs. They are generally known as excellent cations to form ILs [101]. They used phosphonium cations, which form salts with better thermal stability [102], and also have a lower toxicity [103] than ammonium salts. As a model experiment, salts were formed by mixing L-alanine (L-Ala) with the butylpyridinium, tetrabutylphosphonium, tetraalkylammonium cations, etc. They reported that the combination of phosphonium cation and AA anion results in an IL with excellent properties, that is, better thermal stability and lower glass transition temperature. In conclusion, they analyzed 20 ILs made by coupling *n*-tetrabutylphosphonium cation with AA anions. Of these, 15 salts were liquid at room temperature. These halogen-free AAILs were proposed to use for various applications because of their high decomposition temperatures [87]. Tao et al. [88] reported two families of a new generation of IL cations, [AA] (AA) and [AAE] (AA ester), directly derived from natural *α*-AA and their ester salts. In principle, combination of these cations with environmentally benign anions might have made "fully green" ILs possible. Moreover, the choice of [AAE] cations will not only decrease the melting points of the products [97] but also increase biodegradability, because of the incorporation of an ester group [94]. Regarding the anions, NO_3^- and Sac^- (saccharide) are best; NO_3^- is nontoxic, pharmaceutically acceptable inorganic anion [104], while Sac^- is nonfluorous and has well-established toxicological profiles [105]. They mainly reported two types of "fully green" ILs, [AAE]NO_3 and [AAE]Sac. Of these, the cation involved directly emerges from natural *α*-AA ester salts, and the anions involved are nontoxic inorganic or organic anions. The synthesized [AAE]NO_3 and [AAE]Sac ILs were found to retain similar physicochemical characteristics as that of conventional imidazolium ILs and also possess same chiralities like natural AA. Thermal stability, phase behavior, viscosity, and miscibility of the representative family members of these ILs were investigated and showed no difference from conventional ILs in general. These ILs (ProNO_3, [AAE]NO_3, and [AAE]Sac) were miscible with high polarity solvents such as water, low molecular weight alcohols, and ketones. These AAILs were used as catalyst and "fully green" solvents in cycloaddition of cyclopentadiene to methyl acrylate, which is a typical Diels–Alder reaction. Hannig et al. [89] developed synthetic pathways of chiral derivatives of *N*-alkylated imidazolium salts based on L-histidine and subsequent deprotonated L-histidine-based chiral imidazol-2-ylidene derivatives or their respective metal complexes. They proposed that the derivatives of such systems might become useful solvents, reagents, or catalysts in reactions proceeding with transfer of their chirality features and also might serve as useful and readily available optically active ILs. Allen et al. [90] demonstrated a straightforward synthesis of a series of ILs with chiral carboxylates. Tetrabutylammonium hydroxide, [TBA]OH, and AA or a chiral carboxylic acid were used for synthesis of CILs. They anticipated that the use of CILs in asymmetric organic syntheses might result in chiral induction without the use of expensive chiral metal catalysts. Almost four decades ago, Seebach and Oei [106] reported that chiral solvents can achieve low enantioselectivities in asymmetric reactions. CILs were liquids at room temperature, and the color of the liquids ranges from colorless to orange. Thermal stability, chemical shift of the α-proton in the starting acid and the subsequent CIL, and the optical rotations, $[\alpha]_D^{20}$, for all of the CILs in aqueous solutions were measured. These CILs were soluble in water, acetonitrile, acetone, dichloromethane, and chloroform. On the other hand, they were insoluble in diethyl ether, hexane, and toluene. In conclusion, they presented a straightforward route toward a series of CILs, which are expected to have a high potential as chiral solvents and also in chiral separations.

Zhao et al. [107] examined the enzyme enantioselectivity in CILs based on D- or L-AA, in the form of $[EMIM^+][$D-$AA^-]$ or $[EMIM^+][$L-$AA^-]$ (EMIM+ is 1-ethyl-3-methylimidazolium). The enzymatic hydrolysis of phenylalanine methyl ester was studied in aqueous solutions of these ILs. These ILs were capable of stabilizing the protease activity and enantioselectivity at low concentrations. Interestingly, higher yields of L-phenylalanine were observed in ILs based on D-AA rather than in those derived from L-isomers. It was argued that D-AA are more kosmotropic than L-isomers. Interestingly, IL-D_2O solution was able to enhance the enzymatic resolution compared to IL–H_2O system. In a recent study, Gardas et al. [108] synthesized a series of ILs based on AA anions and either ammonium or phosphonium-based cations and measured the temperature dependence of density, viscosity, heat capacity, and thermal conductivity of AAILs, namely, tributylmethylammonium serinate ([N4441][Ser]), tributylmethylammonium taurinate ([N4441][Tau]), tributylmethylammonium lysinate ([N4441][Lys]), tributylmethylammonium threonate ([N4441][Thr]), tetrabutylphosphonium serinate ([P4444][Ser]), tetrabutylphosphonium taurinate ([P4444][Tau]), tetrabutylphosphonium lysinate ([P4444][Lys]), tetrabutylphosphonium threonate ([P4444][Thr]), tetrabutylphosphonium prolinate ([P4444][Pro]), tetrabutylphosphonium valinate ([P4444][Val]), and tetrabutylphosphonium cysteinate ([P4444][Cys]). The influence of cations and anions on studied properties was discussed. Subsequently, Muhammad et al. [95] synthesized a series of AAILs, namely, EmimGly, EmimAla, EmimPro, and EmimSer, and subsequently measured their thermophysical properties (such as density, viscosity, and refractive index) at several temperatures under atmospheric pressure. The properties of these AAILs demonstrated applications in the areas of peptide synthesis and chiral catalysis [109] and also employed in reversible CO_2 absorption [110]. Because of dual functionality of AA (i.e., used as cation as well as anion), ILs provided a variety of applications, such as intermediates for peptide synthesis, chiral solvents, functional materials, and biodegradable ILs in the field of industrial and pharmaceutical chemistry [111].

Toxicity and biodegradation remain central issues while dealing with ILs, and the quest for synthesis of greener ILs in terms of these two criteria is increasing day by day [73]. Recently, Kumar and his coworkers [93] selected natural AA, namely, L-glycine, L-alanine, L-valine, L-glutamic acid, and L-proline, and SLS as precursors for synthesis of relatively greener, biodegradable, and nontoxic AAILS architectures. Ionic liquid surfactants based on AA (e.g., $GluC_3LS$, $ValC_3LS$, $ProC_3LS$, $AlaC_3LS$, $GlyC_3LS$, $GluC_4LS$, $ProC_4LS$, $AlaC_4LS$, and $GlyC_4LS$) having a superior surface activity and solvent miscibility were reported for the first time. They used general symbols to represent the AA and use the number of carbon atoms to represent the alkyl groups of the ester (e.g., glycine propyl ester lauryl sulfate is denoted as $GlyC_3LS$). The biodegradability of the synthesized AAILSs was checked by surface tension measurements of AAILSs solutions in natural seawater periodically. The application of representative AAILSs ($AlaC_3LS$ and $AlaC_4LS$) toward the mitigation of harmful algal blooms (HABs) comprising *Amphidinium carterae*, a toxic phytoplankton (dinoflagellate), was studied, which have severe adverse effects on fisheries, aquacultures, drinking water, tourism, and human health. Results showed that very low concentrations of AAILSs are sufficient to provoke an algicidal effect while avoiding detrimental environmental effects. The synthesized AAILs were found to quick disappearance from marine environment after HAB mitigation due to their very high biodegradability. Another potential application of AAILSs was explored. The generation of shape- and size-specific nanostructured CeO_2 materials, which is applied among others automotive three-way catalysis, such as solid oxide fuel cells, metal oxide semiconductor devices, and gas sensors of technological importance [93]. The use of AAILSs as structure-directing agents can be advantageous not only from their green nature, but also they permit distinct morphological control due to the structural and physical properties of AA, such as their ability to coordinate to metal ions, the type and length of the functional groups, or their influence on intrinsic electric fields. They concluded that the synthesized novel highly biodegradable AAILs show better surface activity than conventional surfactants and demonstrated their potential application in the area of the mitigation of HABs from seawater. Their high solubility in water, biodegradability, and nontoxic nature made them attractive for biological and drug delivery applications. In addition, these kinds of AAILs have the ability to bind metallic nanoparticles (NPs), based on which biomolecular devices may be fabricated [93]. In continuation, they also presented the temperature dependence of physical properties such as density, speed of sound, and refractive index of these AAILSs [112]. The refractive indices of ILs were found higher than those of normal organic liquids, but were comparable to long hydrocarbon chain organic solvents. Self-aggregation of AAILSs in aqueous solution was investigated through surface tension, conductivity, steady-state fluorescence, dynamic light scattering (DLS), and transmission electron microscopy (TEM) [113]. The aggregation properties of AAILSs were analyzed as a function of structure of AA and compared with those of analogous ILs and conventional ionic surfactants. Aggregation number and aggregate size of AAILSs were found to depend on the structure of counterion AA. Surface activity of the AAILSs was found superior to that of analogous ILs and conventional ionic surfactants of the same alkyl chain length.

Recently, AAILs based on choline, which are broadly categorized as green natural product, are an essential micronutrient for normal functioning of all cells. AAILs based on choline were synthesized and characterized by different authors. [114–117]. The majority of cholinium-based ILs were reported to be readily biodegradable and low toxic, owing to its cholinium cation containing a hydroxyl group [117]. For example, Elliott and Weaver synthesized five choline phosphate ILs and observed that these cholinium ILs exhibit excellent biocompatibility and biodegradability [116]. The anion size and the presence of moderately long and/or branched alkyl chains were considered to be major factors to affect cell viability. Pereira and Petkovic synthesized several cholinium ILs combined with a range of alkanoate anions. They reported that the cholinium alkanoate ILs showed good biocompatibility and low toxicity to filamentous fungi [92]. In view of these, biocompatible cholinium-based AAILs ([Ch][AA]) synthesized by using choline chloride ([Ch]Cl) as the source for the cation and AA for the different anions. These ILs could provide a variety of applications, such as functional materials, gas absorption, and green catalysis in the fields of industrial and pharmaceutical chemistry. Recently, Calvino-Casilda and coworkers reported on the synthesis of [Ch][AA], which were successfully applied as catalysts for the Knoevenagel condensation reactions between benzaldehyde and different active methylene compounds with high conversion and selectivity [118]. Later on, Liu et al. [119] explored the properties of [Ch][AA] ILs and found that the selective extraction of lignin from lignocellulose was enhanced significantly using the [Ch][Gly] ILs as a solvent.

Very recently, Tao et al. [96] synthesized and characterized a series of cholinium-based AAILs ([Ch][AA]), namely, cholinium glycinate ([Ch][Gly]), cholinium L-alaninate ([Ch][L-Ala]), cholinium β-alaninate ([Ch][β-Ala]), cholinium prolinate ([Ch][Pro]), and cholinium serinate ([Ch][Ser]) by ^{1}H-NMR, Fourier transform infrared (FTIR), elemental analysis, thermogravimetry, and differential scanning calorimetry (DSC) analysis. Physicochemical properties, such as density, viscosity, refractive index, and conductivity of these AAILs, were measured and correlated with the empirical equations in a wide temperature range from the view point of their thermal stability. The thermal expansion coefficient values were also calculated from the acquired experimental density values. From the experimental data, different densities, viscosities, and refractive index values were observed with a fixed cation ($[Ch]^+$) for different AA, because of variation in internal interactions of the constituents (e.g., hydrogen bonding, van der Waals interactions, etc.) in each IL molecule. Thermal gravimetric analysis (TGA) revealed good thermal stability of [Ch][AA]. It can be summarized that all these results supplement to the growing database on AAIL properties and can be applied for the design of many chemical processes such as catalysis, separation, and cellulose dissolution.

20.4.2 Cholinium Cation-Based BILs

One of the strategies to design low toxicity and highly biodegradable ILs is to use cations and anions based on biomolecules, such as those based on choline [120, 121]. Another alternative is to introduce the ether and hydroxyl functionalization on cation moiety, which typically reduces the toxicity [122]. Also, the ester group in the side chain of choline can induce a higher biodegradability [122]. ILs derived from choline have been the subject of many studies. Choline chloride (2-hydroxyethyltrimethylammonium chloride or vitamin B4) is a cheap organic salt, which is used as an animal feed additive. Melting point of choline chloride is very high (298–304°C), impeding its use as an IL. Abbott and coworkers obtained ILs by mixing choline chloride with hydrogen-bond donors, so-called deep eutectic solvent [123]. These are highly conducting solvents capable to solubilize a high quantity of inorganic salts [124]. Eutectic solvents derived from choline acetate or choline chloride glycerol were reported to be highly biodegradable with excellent compatibility with a commercial immobilized *Candida antarctica* lipase B (CALB) [125]. Pernak et al. [126] synthesized a total of 63 choline derivative-based ILs in the forms of chlorides, acesulfamates, and bis(trifluoromethylsulfonyl)imides, and subsequently, their physical properties (viz., density, viscosity, solubility, and thermal stability) were measured. Thirteen of these salts were known chlorides: precursors to the 26 water-soluble acesulfamates, 12 acesulfamates only partially miscible with water, and 12 water-insoluble imides. The crystal structures for two of the chloride salts, such as (2-hydroxyethyl)dimethylundecyloxymethylammonium chloride and cyclododecyloxymethyl(2-hydroxyethyl) dimethylammonium chloride, were determined. Karl–Fisher measurements showed the water contents of these dried $[Ace]^-$ and $[NTf_2]^-$ salts to be less than 600 and 500 ppm, respectively. All the prepared liquids were denser than water and became less dense with increasing size of the alkoxy group. Further, the alkoxymethyl(2-hydroxyethyl)dimethylammonium acesulfamates were shown to be insect feeding deterrents and thus opened up a new generation of synthetic deterrents based on ILs. The alkoxymethyl(2-decanoyloxyethyl) dimethylammonium bis(trifluoromethylsulfonyl) imides were also shown to act as fixatives for soft tissues and reported to be used as substitutes for formalin and also preservatives for blood. Petkovic et al. [92] reported a new group of cholinium ILs from the view points of toxicity and biodegradability of ILs. Focusing on the toxic effect of the anion, the ILs were designed by combining the benign cholinium with a range of linear alkanoate anions ($[C_nH_{2n+1}CO_2]^-$, $n=1$–9), as well as two structural isomers ($n=3$ or 4). The toxicity of these ILs was evaluated using filamentous fungi as model eukaryotic organisms, by determining their minimal inhibitory concentrations (MIC) and minimal fungicidal concentrations (MFC), distinguishing between growth inhibition and death, respectively. *Penicillium brevicompactum* Dierckx (IBETPeA), *P. glandicola* (Oudem) Seifert and Samson (IBETPeB), *P. corylophilum* Dierckx (IBETF6), and *P. diversum* Raper and Fennell (IBETPeE) were used as model eukaryotic organisms. *Penicillium corylophilum* was chosen for studies on IL biodegradability. IL biodegradability in the static fungal cultures (toxicity tests) was analyzed by 1H-NMR spectroscopy, and in some cases, the anion biodegradability was confirmed by liquid chromatography. The anion toxicity was defined by the length of the linear hydrocarbon chain and ranked as follows: ethanoate < propanoate < butanoate < pentanoate < hexanoate < octanoate < decanoate. Branched isomers of butanoate and pentanoate (2-methylpropanoate and 2,2-dimethylpropanoate) exhibited lower inhibitory and fungicidal effects, comparative to the linear isomers. The branched chain anions, 2,2-dimethylpropanoate and 2-methylpropanoate, exhibited less toxicity than their corresponding linear isomers and were more resistant to fungal attack, leading to null or partial degradation. Surprisingly, most of the tested species showed active growth in media containing extremely high IL concentrations, up to molar ranges in some cases. The biodegradability of these ILs was assessed, and new biotechnological applications were proposed, for example, as solvents for biopolymers. Very recently, Isik et al. [127] reported the synthesis of cholinium-based IL methacrylic monomers with halide, lactate, and acetate counteranions and polymerized by using conventional free radical polymerization. The polymer properties were characterized by NMR, TGA, and DSC and compared with eight different cationic polymethacrylic analogs. Polycations with different methacrylic alkyl ammonium backbones having lactate anion displayed comparatively better thermal stability than those with the acetate counter ion. They also exhibited lower glass transition temperatures than their counterparts having acetate and halide counteranions. Further, cholinium lactate methacrylate IL monomer, which shows ability to dissolve cellulose, was used to prepare ion gels by photopolymerization. These results facilitated the preparation of transparent poly(IL)/cellulose composite coatings by photopolymerization. Weaver et al. [116] synthesized choline dihydrogen phosphate (CDP) IL by combining five phosphate-based anion moieties ($H_2PO_4^-$ (DHP), dibutyl phosphate (DBP), bis(2-ethylhexyl) phosphate (BEH), bis(2,4,4-trimethylpentyl) phosphinate (TMP), and *O,O′*-diethyl dithiophosphate (DEP)) with the choline cation (C). Toxicity levels of these ILs were established using a J774 murine macrophage cell line, and further, they reported that these cholinium ILs exhibit excellent biocompatibility and biodegradability. Recently, Calvino-Casilda and coworkers reported on the synthesis of biocompatible cholinium-based AAILs ([Ch][AA]), which were successfully applied as catalysts for the Knoevenagel condensation reaction between benzaldehyde and different active methylene compounds with high conversion and selectivity [118]. A series of ILs were prepared by Fukaya et al. [128] using mainly naturally derived and inexpensive materials. Hydrogen maleate, propionate, tiglate, and hydrogen succinate anions gave RTILs upon coupling with the choline cation. In particular, choline combined with the hydrogen maleate anion gave a RTIL of moderate viscosity. These ILs exhibited higher polarities, especially greater hydrogen-bonding abilities than conventional ILs. Restolho et al. [129] characterized interfacial properties of "bio-ionic liquids" (BILs) using the choline cation and various naturally derived anions. Surface tension and wettability of choline propionate, choline tiglate, and choline H-maleate were presented. Strong hydrogen bonding and low surface tension values of these ILs were obtained in comparison with other ILs with the same characteristic (e.g., [C_2OHmim][BF_4]). In contrast, their polarity was found to be very high when compared with that of water, which confirmed their solvent ability. Recently, Sekar et al. [130] developed several choline-based salts such as choline saccharinate (CS), CDP, choline lactate (CL), and choline tartarate (CT) for use as biocompatible hydrated ILs and successfully employed as the cosubstrate with *S. lentus* in the biodegradation of an azo dye in aqueous solution. Klein et al. [131] studied the thermotropic phase behavior of biocompatible neat choline carboxylates (ChC*m* with alkyl chain lengths, *m* ranging from 12 to 18). The observed phases were characterized by optical polarizing microscopy, NMR spin–spin relaxation, and small-angle X-ray scattering (SAXS) as well as wide-angle X-ray scattering (WAXS) measurements. The thermal stabilities were evaluated by thermogravimetric measurements, while transition temperatures and enthalpies were determined by DSC. Thermal stability of ChC*m* was limited up to about 200°C. The nature of the phases was identified with increasing temperature as crystalline, semicrystalline, and liquid crystalline lamellar. Even long-chain choline carboxylates ($m = 18$) were found to melt into a lamellar liquid crystalline phase below 100°C. With choline as counterion in simple fatty acid soaps, the water solubility of ChC*m* was considerably enhanced, and also the melting points were substantially reduced, which facilitate thermotropic mesomorphism at temperatures between 35 and 95°C. Recently, Klein et al. [132] showed how a new promising green and highly water-soluble surfactant can be designed based on recent progress in the knowledge of counterion–head group binding and crystallization behavior. In view of this, they selected choline (Ch)—a natural human metabolite—as counterion of dodecylsulfate (DS) for synthesis of green surfactant and characterized the resulting surfactant (choline dodecylsulfate (ChDS)) with respect to its Krafft point, cmc, and aqueous self-assembly behavior. Results are compared to values obtained for SDS, potassium dodecylsulfate (KDS), and the alkylcarboxylate homologs ChC12, NaC12, and KC12. ChDS showed an increase in water solubility as compared to its sodium and potassium homologs, in analogy to what was found when choline was employed as counterion in fatty acid soaps. Advantages of choline alkylsulfates over choline carboxylates (ChC*m*) are reported to be due to much less hygroscopic nature of neat ChDS than ChC*m* surfactants and higher thermal stability of ChDS than ChC*m* and even SDS. The phase diagram established for aqueous ChDS corroborates that the degree of counterion–head group binding is significantly higher than ChC*m* soaps. Stronger cation–anion interactions between the "soft" choline and the "soft" alkylsulfate ion suggested a smaller head group area of ChDS as compared to the highly dissociated choline carboxylates. In order to gain information on the biocompatibility of ChDS, its cytotoxicity on cervix carcinoma cells (HeLa) as well as on keratinocytes (SK-Mel-28) was examined. Studies on HeLa cells served as a reasonable alternative to frequently performed *in vivo* eye irritancy tests on rabbits, while analyzed with SK-Mel-28 cells permit conclusions regarding the skin irritancy power. Cytotoxicity studies on two human cell lines revealed that ChDS is harmless as that of SDS. All of these observations highlighted that ChDS is a promising surfactant suitable for widespread applications. Besides its good biocompatibility and water solubility, it was featured by a straightforward synthesis of easily accessible raw materials and an easy handling of the neat surfactant [132] and great potential for the design of green surfactants and ILs [131, 133, 134], encouraging the development of new promising combinations in the future. Very recently, synthesis and characterization of two new choline esters, namely, hydrogen choline chloride succinate ([CholSucc][Cl]) and hydrogen choline chloride glutarate ([CholGlut][Cl]), was reported by Lopes et al. [135]. Their solubilities and solubility of choline chloride ([Chol][Cl]) and choline acetate ([Chol][Ac]) in ethanol, 1-propanol, and 1-butanol at atmospheric pressure were determined. Interestingly, solubility of these choline salts was enhanced with increase in temperature. The relative affinity of the solutes to any of alcohols studied follows the order [Chol][Ac] >> [Chol][Cl] > [CholGlut][Cl] > [CholSucc][Cl]. DSC was employed to measure enthalpies of fusion and melting temperatures for [CholSucc][Cl], [CholGlut][Cl], and [Chol][Ac]. [Chol][Cl] and showed decomposition before melting. Recently Rocha et al. [136] prepared potential electrolytes based on choline carboxylate zwitterionic ILs by two different synthetic routes through esterification of choline salts with a range of anhydrides. Their conjugation with lithium bis(trifluoromethylsulfonyl)imide resulted in the formation of stable RTILs. In view of their use as potential electrolytes, the physicochemical characterization of these attractive ILs was performed. These compounds presented a very low glass transition temperature, excellent conductivity, and good thermal stability. These zwitterionic ILs maintained a liquid state, and no dissociation was observed even after 6 months of storage at room temperature.

20.4.3 Morpholinium Cation-Based BILs

Recently, interest in morpholinium-based ILs has been increased because of their attractive physicochemical and thermal properties and large electrochemical windows. In fact, they are less toxic than the commonly used imidazolium-, pyridinium-, or tetraalkylammonium-based ILs [137]. In particular, ILs based on the morpholinium cation are preferred because of their electrochemical stability [138]. However, little data on the (eco)toxicity of morpholinium-based ILs are available. The functionalized morpholinium compounds tested so far have shown low acute cytotoxic effects up to concentrations of 20 mM in tests with the promyelocytic leukemia rat cell line IPC-81 [68]. Brigouleix and coworkers synthesized new protic ionic liquids (PILs) based on the morpholinium, *N*-methylmorpholinium, and *N*-ethyl morpholinium cations through a simple and atom-economic neutralization reaction between *N*-alkyl morpholine and formic acid [139]. These morpholinium-based PILs were reported to be of low cost and low toxicity. The density, refractive index, thermal property, and electrochemical window of these PILs were measured. The temperature dependence of their dynamic viscosity and ionic conductivity were also determined. The results were allowed to classify these PILs according to a classical Walden diagram and also to evaluate their "fragility." In addition, morpholinium-based PILs exhibited a large electrochemical window as compared to other PILs and possessed relatively high ionic conductivities. They possess wide applicable perspectives as electrolytes for fuel cell devices, thermal transfer fluids, and acid-catalyzed reaction media as replacements of conventional solvents. Recently, Pernak et al. [140] reported on the synthesis of novel 4-benzyl-4-methylmorpholinium [BMmorf] and 4-benzyl-4-ethylmorpholinium [BEmorf] cations (Scheme 20.5) combined with 16 different anions: inorganic (chloride, nitrate, tetrafluoroborate, or hydrogen sulfate), organic aromatic (salicylate or saccharinate) and aliphatic (formate, acetate, hexanoate, octanoate, maleate, crotonate, or lactate) acids, and dodecyl sulfate.

In addition, typical physicochemical properties such as melting points, viscosities, and thermal stabilities, data describing their toxicological properties (acute toxicity toward bacteria, fungi, *in vitro* cell line, and rat), and environmental properties (ready biodegradability) were also presented. The activity of 4-benzyl-4-methylmorpholinium ILs against bacteria and fungi was very low. The cytotoxicity of these compounds was classified as moderate or low, but varied dramatically from 0.25 to 14.13 mM, depending on the anion. The acute toxicity of 4-benzyl-4-ethylmorpholinium acetate was found to be in the range of 300–2000 mg/kg body weight (b.w.) in female rats. This IL was classified as a category 4 (R22 harmful, if swallowed) toxin, according to standard GHS grading (the globally harmonized system of classification and labeling of chemicals). However, 4-benzyl-4-methylmorpholinium cation was not classified as readily biodegradable. Further, 4-benzyl-4-ethylmorpholinium acetate, because of its low viscosity, can effectively be used as a solvent for biomass. Compared to the imidazolium and pyridinium ILs, the low antimicrobial activity of

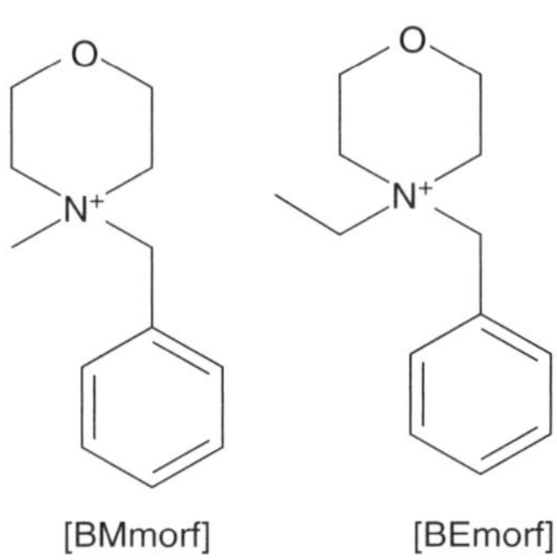

Scheme 20.5 Structure of cations. Reproduced from Pernak et al. [140] with permission from the Royal Society of Chemistry.

morpholinium ILs and the moderate oral acute toxicity toward rats deserve attention, which significantly influences the application of morpholinium-based ILs. Very recently, Khara and Samanta [141] attempted to explore some of the physicochemical properties of four different *N*-alkyl-*N*-methylmorpholinium ILs, abbreviated as $[Mor_{1,2}][Tf_2N]$, $[Mor_{1,4}][Tf_2N]$, $[Mor_{1,6}][Tf_2N]$, and $[Mor_{1,8}][Tf_2N]$, consisting of *N*-alkyl-*N*-methylmorpholinium cation with different alkyl chain lengths and an identical anionic component, $[Tf_2N]^-$, by probing the steady-state and time-resolved fluorescence behavior of a dipolar solute, coumarin-153 (C153), as a function of the excitation wavelength and temperature. They suggested that these *N*-alkyl-*N*-methylmorpholinium ILs are more structured and heterogeneous than the imidazolium ILs. Time-resolved fluorescence anisotropy measurements showed that C153 locates in an environment whose viscosity differs from the bulk viscosity of IL. The solvation reorganization dynamics under isothermal conditions was found to be similar to that in imidazolium ILs. Overall, a more organized domain structure of these *N*-alkyl-*N*-methylmorpholinium ILs, particularly for the long-chain containing ones, was indicated in this report. In a subsequent report, Samanta and his coworkers [142] synthesized *N*-alkyl-*N*-methylmorpholinium (alkyl represents ethyl, butyl, hexyl, and octyl) bis(trifluoromethanesulfonyl)imide (Tf_2N) ILs followed by a two-step procedure. The first step involved preparation of the bromide salt of *N*-alkyl-*N*-methylmorpholinium ion, $[Mor_{1,n}][Br]$, and the second step consisted of exchange of the bromide ion by the bis-(trifluoromethanesulfonyl)imide ion [143]. Rotational dynamics of two dipolar solutes, 4-aminophthalimide (AP) and 6-propionyl-2-dimethylamino naphthalene (PRODAN), and a nonpolar solute, anthracene, were studied in these ILs as a function of temperature and excitation wavelength to probe the microheterogeneous nature of these ILs. The stick to superstick behavior of AP, which was attributed to its H-bonding interaction with the constituent ions of ILs, reflected the most polar environment of the media. The slip to subslip behavior of the nonpolar solute, anthracene, depicted the most nonpolar region of the ILs formed largely by the alkyl group of the cation. On the other hand, the dipolar solute, PRODAN, which exhibited rotational time constant in between the slip and stick behavior, sought to be distributed in both regions. Excitation wavelength dependence of the rotational times of the dipolar probes also supported the heterogeneous environment of these RTILs. Overall, the results provided further insight to the microheterogeneous structure of these ILs, formed by the segregation of the alkyl chains on one hand and the charged components on the other.

20.4.4 Other BILs

In order to obtain "fully green" ILs, the starting materials should be at least nontoxic, while for a perfect solution, they should be renewable [144]. A few successful BILs samples have been reported, for example, lactates [145], artificial sweeteners [146], and sugar substitutes [146, 147] to prepare anions and [82] cations, in which some of them are potentially chiral ions. The details of these ILs are presented in this section.

20.4.4.1 Artificial Sweetener-Based BILs Carter et al. [146] synthesized several ILs by combination of the anions of the sweeteners saccharin (Sac) and acesulfame (Ace) with a variety of organic cations (e.g., C_4MIM, C_6MIM, C_9MIM, $n\text{-}Pr_4N$, etc.). These ILs are relatively new and were characterized by ^1H- and ^{13}C-NMR. The

H-bonding capacity, relative transition metal coordinating ability, and relative basicity of the sweetener anions were compared with the fluorous species, such as fluorous sulfonates and carboxylates, perfluorocarboxylate and perfluorosulfonate, and PF_6^- and $Tf_2N_2^-$. The successful synthesis of ILs by pairing these nontoxic anions with cations were reported to be nontoxic, such as butyrylcholine (2-butanoyloxyethyl-trimethyl-azanium), and further emphasized the possibilities of formation of new ILs that might prove to be wholly benign both environmentally and toxicologically.

20.4.4.2 Fructose-Based BILs In the ongoing quest to improve the environmental compatibility of chemical synthesis of RTILs from biorenewable sources, Handy and his coworkers reported a new family of RTILs based on fructose [82]. Fructose was used as the starting material by pairing with several anions (e.g., tosylate, trifluoroacetate, acetate, tetrafluoroborate, triflimide, and dicyanimide) for the preparation of a fructose-derived ILs. These ILs exhibit tunable solvent properties, like conventional imidazole-based RTILs. Solubility/miscibility properties of different low viscous RTILs with triflimide and dicyanimide anions were investigated. They were employed as recyclable solvents for the Heck reaction of aryl iodides. In view of these, they proposed that a number of other transition metal-catalyzed reactions could be performed equally well in the fructose-derived RTILs, instead of the conventional [bmim] series. At the same time, the protic nature of this new solvent affords the opportunity to explore other features, which were not accessible using conventional RTILs. In order to explore the possible application of fructose-derived ILs as a new class of soluble supports, Handy and Okello [148] focused on the preparation of some simple acrylate-derived Diels–Alder adducts. They concluded that this new type of soluble support (i.e., a hydroxylic IL based on fructose) affords higher potential loadings than other simple homogeneous or heterogeneous supports and can be readily recovered and recycled.

20.4.4.3 Lactate-Based BILs Rogers, Holbrey, and Swatloski [104] suggested that the concepts of nontoxic pharmaceutically acceptable ions and GRAS (generally regarded as safe) materials (taken from the pharmaceutical and food additive industries) should be used to provide guidelines for the development of new ILs for which toxicological and environmental hazards of the individual ions are considered to be low. To that extent, lactates belong to a group of organic ions listed as nontoxic, pharmaceutically acceptable anions. Subsequently, Pernak and Goc [149] synthesized protic ILs in the form of 1-alkyl- as well as 1-alkoxymethylimidazolium lactates and salicylates. Structures of these salts are shown in Scheme 20.6. Some physical properties, for example, melting point, boiling point, density, and their solubility in some common organic solvents and water, were measured. The values of minimum inhibitory concentration (MIC) as well as minimum biocidal concentration (MBC) were determined for 1-alkyl and 1-alkoxymethylimidazolium lactates. Effect of lactate structure on microbiological activity was defined.

In another study, Pernak and his coworkers examined minimum inhibitory concentrations (MIC) and MBC of protic ILs containing asymmetrical monosubstituted imidazolium cation and DL- or L-lactate anion tested against five strains of rods, five strains of cocci, and two strains of fungi [145]. The physicochemical parameters such as density, electrical conductivity, and temperature of onset of decomposition were

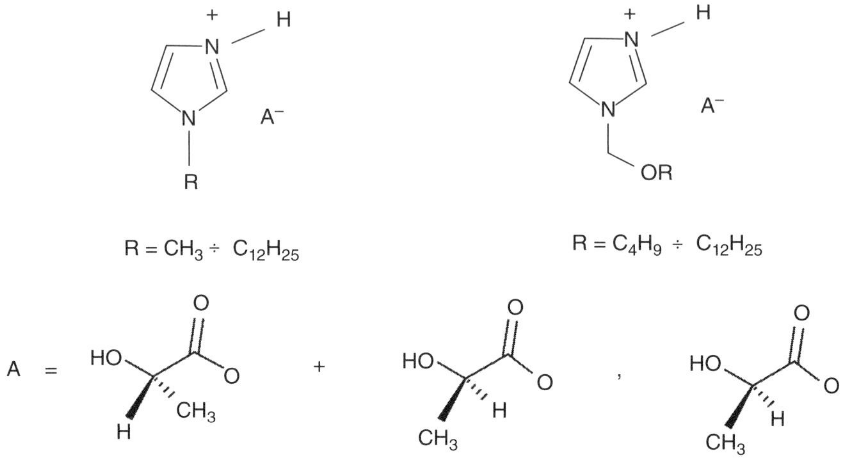

Scheme 20.6 Protic ionic liquids in the form of 1-alkyl- as well as 1-alkoxy methylimidazolium lactates and salicylates. Reproduced from Pernak et al. [145] with permission from the Royal Society of Chemistry.

measured. Testing of selected physical properties demonstrated differences in the values established for DL-lactates and L-lactates. Higher densities were manifested by L-lactates than DL-lactates. A reciprocal relation was noted in case of electrical conductivity. These protic ILs exhibited biological activity against all of the microorganisms. The activity was related mainly to the length of alkyl and alkoxymethyl substituent and thus to molecular weight. The studied lactates manifested a more pronounced bacteriostatic activity, that is, lower MIC values than microbiocidal activity, measured by MBC values. A difference between MIC and MBC values for DL-lactates and L-lactates was observed. 1-Nonyloxymethylimidazolium DL-lactate and 1-decylimidazolium L-lactate were proved to be more effective against five strains of rods. In this case, DL-lactates exhibited much more pronounced growth-inhibiting properties than biocidal properties of microorganism. L-Lactates proved to be most biostatic against two strains of fungi, which included 1-dodecyl- and 1-nonyloxymethylimidazolium, while the most biocidal agents included 1-dodecylimidazolium and 1-undecyloxymethylimidazolium L-lactates. Lactate with methyl, ethyl, propyl, butyl, pentyl, butoxymethyl, and pentyloxymethyl substituents was proved to be inactive. In conclusion, they proposed that the protic ILs, containing an organic anion such as lactate, exhibit similar properties to widely studied aprotic ILs. The choice of an organic anion, such as lactate, was dictated by its nontoxic and pharmaceutically acceptable character, making the resulting ILs environmentally friendly. These ILs are soluble in water and equipped with short substituents, and they do not affect microbiological flora. In addition, their decomposition by ozonation is very effective [145]. In 1999, Seddon and coworkers reported [150] the use of [bmim][lactate] as the solvent in the Diels–Alder reaction between cyclopentadiene and acrylonitrile or ethyl acrylate. The lactate salt provided slightly lower endo/exo selectivity than the corresponding triflate or BF_4^- salts. Further, the reaction rate of those reactions using the lactate salt was in between the other two salts (triflate or BF_4^-). Unfortunately, no enantioselectivity was observed for this cycloaddition process when performed in the chiral lactate salt.

20.4.4.4 Higher Homologues of Alkylsulfate-Based BILs In search of for new, even "greener" ILs for industrial catalytic applications, Wasserscheid et al. [81] reported in detail the synthesis and some important physicochemical properties of the halogen-free IL, 1-butyl-3-methylimidazolium ([bmim]) alkylsulfates, namely, [bmim][n-$C_8H_{17}OSO_3$]. The lower homologues of alkylsulfate anions, namely, methanesulfonate and ethanesulfonate, are known for their sensitivity toward hydrolysis. In the presence of water, these anions formed the corresponding alcohol and hydrogen sulfate at elevated temperatures. Obviously, this process is undesirable for most of the applications, since these IL systems changed dramatically with their properties and an acidic proton was formed that might be a reactive species in many applications. The sensitivity toward hydrolysis was reported to be much lower for alkylsulfates with longer alkyl groups. Consequently, alkylsulfates with 8–18 carbon atoms are used on a multimillion ton scale in detergents, in cosmetics, and in many other applications where water is omnipresent. Hence, a first example for the successful application of this new IL was presented, namely, its use as catalyst solvent in the Rh-catalyzed hydroformylation of 1-octene. The activity of the Rh catalyst was found to be significantly higher with [bmim][n-$C_8H_{17}OSO_3$] when compared with the commonly used hexafluorophosphate and tetrafluoroborate ILs with the same cation. Thus, the report of Wasserscheid et al. [81] demonstrated that [bmim][n-$C_8H_{17}OSO_3$] is a very interesting solvent for Rh-catalyzed hydroformylation. Further, they proposed that hydroformylation research around octylsulfate ILs will be the center of attraction in the future compared to hexafluorophosphate and tetrafluoroborate ILs due to the much lower cost, the better hydrolysis stability, and the wider disposal options related to the octylsulfate ILs. The IL [bmim][n-$C_8H_{17}OSO_3$] was proposed to be applied as an "even greener" IL in the entire range of applications with great promise. These new ILs are clearly characterized by their high technical availability (at relatively low cost) and by their well-documented toxicology and biodegradation (at least for the anion part) due to very cheap anion source Na[n-$C_8H_{17}OSO_3$], which is used in multithousand ton scale as an ingredient in detergents and cosmetics. This makes [bmim][n-$C_8H_{17}OSO_3$] an interesting candidate for industrial and bulk applications of ILs (e.g., as high-performance lubricant, thermofluid, solvent for extraction and extractive distillation, etc.). Moreover, their results demonstrated that [bmim][n-$C_8H_{17}OSO_3$] is a highly promising solvent for transition metal catalysis, if the electrophilicity of the catalytic center is moderate [81].

20.5 ANTIMICROBIAL ACTIVITY, (ECO)TOXICOLOGICAL ASPECT, AND BIODEGRADABILITY OF ILs

In recent years, ILs are found to have potential antimicrobial activities possibly due to their toxicity, which is an important tunable characteristic. In fact, this property allows for appropriate design of ILs that could be used as new and improved antiseptics, disinfectants, and antifouling reagents. A number of methods are reported for the accurate determination of microbial susceptibility to antimicrobial/antibiotic compounds. Such methods yielded vital data regarding fundamental sensitivity or tolerance to a given antimicrobial biocide or antibiotic. These aspects are vital to the successful treatment and management of microbial infections. Furthermore, such tests are useful for determining relative potency of an antimicrobial agent across a range of species and also for identifying antimicrobial synergies. The basic testing procedures, which are used in the assessment of the antimicrobial activity of ILs, are presented in this section. However, the majority of these tests have relied on basic planktonic susceptibility assays (minimum inhibitory concentrations (MIC) or minimum bactericidal/fungicidal concentration (MBC/MFC)) or agar diffusion techniques; the importance of evaluation of antimicrobial activity against microbial biofilms has also been discussed.

To date, a number of publications have emphasized on the antimicrobial activity and biodegradability of ILs as a function of their tunable structural moieties (viz., alkyl chain length, polar head group, counterion, etc.) against various microorganisms. Earlier, Docherty and Kulpa [100] measured microbial toxicity of imidazolium and pyridinium ILs with varying alkyl chain lengths against *Vibrio fischeri* microorganism using the Microtox. They observed that increase in both alkyl group chain length and the number of alkyl groups substituted on the cation ring results in increase in toxicity, whereas variation in the anion identity did not significantly alter toxicity. Further, Pernak and coworkers evaluated the antimicrobial activity of a series of imidazolium ILs with different anions, such as BF_4^-, PF_6^-, bis(trifluoromethanesulfonyl)amide, lactate, etc. against clinically significant pathogens (rods, cocci, and fungi) [41, 83, 151]. Also, Demberelnyamba et al. [152] synthesized three different series of quaternary imidazolium and pyrrolidinonium salts, namely, series A (1-alkyl-3-methylimidazolium chlorides and bromides), series B (1-alkyl-2-methyl-3-hydroxyethylimidazolium chlorides), and series C (*N*-alkyl-*N*-hydroxyethylpyrrolidinonium) for the purpose of developing new disinfectants and antiseptics. These three series were tested to evaluate their bactericidal activity against Gram-positive bacteria, Gram-negative bacteria, and fungi. Seven microbial strains were used in the study: *Escherichia coli* KCTC1924, *Salmonella typhimurium* KCTC1926, *Staphylococcus aureus* 209 KCTC1916, *S. aureus* R209 KCTC1928, *Bacillus subtilis* KCTC1914, *Candida albicans* KCTC1940, and *Chlorella regularis*. The antimicrobial efficiency was measured by bacterial and fungal growth inhibition, expressed as MIC values. Series A and B imidazolium salts showed very good antimicrobial activity against three microbes. However, the pyrrolidinonium salt exhibited low MIC for some of the tested microorganisms. The antibacterial and antifungal active properties of the salts depended upon the structure of functional groups and the alkyl chain length in the imidazolium and pyrrolidinonium rings. Based on published toxicity data, Couling et al. developed the quantitative structure–property relationship models to assess the factors that govern the toxicity of a range of different ILs to two aquatic organisms (*V. fischeri* and *Daphnia magna*) [71]. The toxicity increased slightly with the number of nitrogen atoms in an aromatic cation ring of ILs, in addition to the well-established link between toxicity and alkyl chain length attached with imidazolium-, pyridinium-, and quaternary ammonium-based ILs. The toxicity of ILs depended on cation type and followed the trend ammonium < pyridinium < imidazolium < triazolium < tetrazolium at comparable conditions. In view of these reports, Docherty and his coworkers [153] presented the results of the United States Environmental Protection Agency (US EPA) Ames test for mutagenicity using *S. typhimurium* strains, TA98 and TA100, to 10 ILs with imidazolium, pyridinium, and quaternary ammonium cations. In continuation, they also investigated the biodegradability of three ILs, 1-butyl-3-methylpyridinium bromide, 1-hexyl-3-methylpyridinium bromide, and 1-octyl-3-methylpyridinium bromide, by activated sludge microbial communities. It was reported that these ILs can be fully mineralized, whereas octyl-substituted cation was classified as "readily biodegradable" only [154].

To broaden the restricted knowledge about the hazardous potentials of ILs to human society and the environment, Matzke et al. [155] concentrated on systematic analysis of the anion effect of six different anion moieties (viz., Cl^-, BF_4^-, $(CF_3SO_2)_2N^-$, $(CF_3)_2N^-$, octylsulfate, and bis(1,2-benzenediolato)borate) and the influence of the side chain length associated with the cation of ILs on (eco)toxicity. For these investigations, they used flexible (eco)toxicological test battery containing aquatic and terrestrial compartments as well as different trophic levels, which include enzymes (acetylcholinesterase), mammalian cells (IPC-81), luminescent marine bacteria (*V. fischeri*), limnic unicellular green algae (*Scenedesmus vacuolatus*), wheat (*Triticum aestivum*), cress (*Lepidium sativum*), duckweed (*Lemna minor*), and a soil invertebrate (the spring tail *Folsomia candida*). Petkovic et al. [156] proposed a relationship between the phylogenetic origin of fungal species, their specific response to the presence of ILs, and the modification of the metabolic profile caused by them. The ILs comprised of 1-ethyl or butyl-3-methylimidazolium, 1-ethyl or butylpyridinium, 1-butyl-1-methylpyrrolidinium, and 1-butyl-1-methylpiperidinium as cation and ethylsulfate, chloride, ethanoate, thiocyanate, DL-lactate, and bis(trifluoromethylsulfonyl)amide as anion. The toxicological assessment toward filamentous fungi (*Penicillium* sp.) as model eukaryotic organisms of these ILs (containing an imidazolium, pyridinium, or cholinium cation) was presented. The imidazolium-based ILs exhibited highest toxicity compared to pyridinium- or cholinium-based ILs.

Toxicity to humans and other organisms has obvious significance, while toxicity to microorganisms has the potential to limit biodegradation. The assessment of toxicity of IL is also presented herein, apart from their biodegradability. Earlier, Gathergood et al. [94] reported investigation on the biodegradability of the dialkylimidazolium ILs. They chose dialkylimidazolium ILs as a starting point and incorporated features that improve the biodegradability of other classes of compounds, such as surfactants. They reported that commonly used dialkylimidazolium ILs show negligible biodegradability in the Closed Bottle Test (OECD 301D). The incorporation of an ester in the side chain of the imidazolium cation significantly increased biodegradability. 3-Methyl-1-(pentoxycarbonylmethyl) imidazolium bromide proved to be the most biodegradable compound in this series (32% degradation after 28 days). It was suggested that this improved biodegradation is due to an enzymatic hydrolysis step, which initiates a pathway to further breakdown products. Consistent with this suggestion, the corresponding amide analogues were poorly biodegradable. In continuation, Garcia et al. [157] examined the effect of the anion on a series of ILs containing an ester group in the side chain and compared with ILs, [bmim][Br]. Levels of biodegradation are usually dependent on compound stability and toxicity. The toxicology study of a series of butylmethylimidazolium [bmimX, X = Br, BF_4, PF_6, NTf_2, $N(CN)_2$, and octylOSO_3] and methyl (propoxycarbonyl) imidazolium ILs was performed in order to determine the structural factor of ILs. The effect of both counterion and alkyl chain length on the aquatic toxicity of alkylmethylimidazolium ILs was investigated. The [bmim] series were generally proved to be poorly biodegradable except one member of this series, which showed a little or no degradation after 28 days. However, 3-methyl-1-(propoxymethylcarbonyl)-imidazolium series showed higher levels of biodegradability. Both series of ILs with octylsulfate anion reflected higher levels of biodegradability. In-depth biological studies on methyl- and a few ethylimidazolium ILs in both luminescent bacteria as well as IPC-81 (leukemia cells) and C_6 (glioma cells) rat cell lines were presented by Ranke et al. [158]. In continuation, the influence of the lipophilicity originates from organic cation on the cytotoxicity in IPC-81 leukemia cells from rats for a set of 74 ILs with imidazolium, pyrrolidinium, pyridinium, quinolinium, quaternary phosphonium, and quaternary ammonium cations, and comparatively small anions, Cl^-, Br^-, BF_4^-, or PF_6^-, were also studied [67]. Further, Stolte and his coworkers [159] investigated the influence of different head groups, functionalized side chains, and anions of ILs on marine bacteria (*V. fischeri*), limnic green algae (*S. vacuolatus*), and fresh water plant (*L. minor*) to improve the knowledge base for the molecular design of ILs leading to a reduced (eco)toxicological hazard potential. The test kit comprised of three aromatic head groups (viz., 4-(dimethylamino)pyridinium, pyridinium, and imidazolium), three nonaromatic heterocycles (viz., 4-methylmorpholinium, 1-methylpiperidinium, and 1-methylpyrrolidinium), and one noncyclic quaternary ammonium head group (viz., *N*,*N*-dimethylethylammonium) and substituted with a butyl side chain. In order to draw correlation between lipophilicity and toxicity, imidazolium ILs with different alkyl side chains (-ethyl, -butyl, -hexyl, -octyl, -decyl, -tetradecyl, -hexadecyl, and -octadecyl) were added to the test kit for conducting different biological tests. The anionic component of these ILs were halide (chloride or bromide or iodide) and the $[(CF_3SO_2)_2N]^-$. The nonaromatic compounds exhibited a lower toxicity (concentrations that lead to a 50% reduction of the exposed organisms relative to control, $EC_{50} > 10{,}000\,\mu M$) on *V. fischeri*, whereas the aromatic compounds showed an increased toxicity that lies in a moderate range ($EC_{50} > 1000\,\mu M$), except for 1-butyl-4-(dimethylamino) pyridinium. In contrast, the morpholinium compound exhibited only a moderate effect on *L. minor* ($EC_{50} > 1000\,\mu M$). 1-Butyl-4-(dimethylamino)pyridinium showed drastic toxicity for the algae *S. vacuolatus*, whereas quaternary ammonium and morpholinium compounds exhibited no toxicity. An increased toxicity was found for all tested compounds combined with the $[(CF_3SO_2)_2N]^-$ for *V. fischeri* and *S. vacuolatus*. On the other hand, $[(CF_3SO_2)_2N]^-$ had no influence on the observed effects on *L. minor*. For lipophilic compounds with an alkyl side chain less than C_{10}, a cutoff effect (divergence from correlation between toxicity and lipophilicity) was found for algae and luminescent bacteria. Based on its low lipophilicity/toxicity, the morpholinium head group was proposed to be recommendable, whereas the dimethylaminopyridinium core was proposed to be avoided because of its drastic effects on aquatic organisms. The toxicity and biodegradability of several ILs consisting of an imidazolium cation with two alkyl substituents in positions 3 (R_1) and 1 (R_2) and a counterion (viz., Cl^-, PF_6^-, and XSO_4^-) in aqueous phase were also determined by Romero et al. [160]. Acute toxicity and EC_{50} values of each compound in the aqueous solution were determined by using the Microtox® standard procedure. To analyze the biodegradability of IL, the biochemical oxygen demand for 5 days (BOD_5) of several aqueous samples containing known initial amounts of IL and/or D-glucose was determined.

In addition, antimicrobial activity of new synthesized oxygen-functionalized amide ILs containing a wide range of ether and polyether esters (Scheme 20.7a and b) was reported by Morrissey et al. [91]. Seven strains of bacteria were used to assess the antimicrobial activity of the ILs as follows: four Gram-negative (*Pseudomonas aeruginosa*, *E. coli*, *Klebsiella* sp., and *Salmonella* sp.) and three Gram-positive (*S. aureus*, *Enterococcus* sp., and *B. subtilis*) organisms. These ILs showed significantly lower toxicity than the derivatives without ester and ether or polyether functional groups. HaiBo et al. [161]designed a new family of green ILs based on *N*-methyl-2-pyrrolidone that was modified by the alkyl chains (i.e., *N*-alkyl-*N*-methyl-2-pyrrolidonium bromides (C_nMPnBr, where $n = 10–18$)). They showed benign conductivity, low viscosity, and excellent surface activity. Finally, the acute toxicities of C_nMPnBr ILs were found to be innocuous, which obtained by normative toxicity tests. Further, they reported that these ILs are entirely biodegraded in 28 days by *B. subtilis* using method of GB/T15818-1995 (Surfactant Biodegradability Mensuration issued by National Standard Committee). It was proposed that C_nMPnBr ILs could be used as substitute for molecular solvents because of their good solvency and possibly could reduce environmental pollution with their excellent biocompatibility. Further, a new set of tetrabutylammonium (TBA) IL salts, such as formate, acetate, propionate, butyrate, linoleate, and oleate, were synthesized by Saadeh et al. [162]. The solution of these salts was proposed to be environmentally benign compare to organic solvents, in view of the ever-increasing awareness of

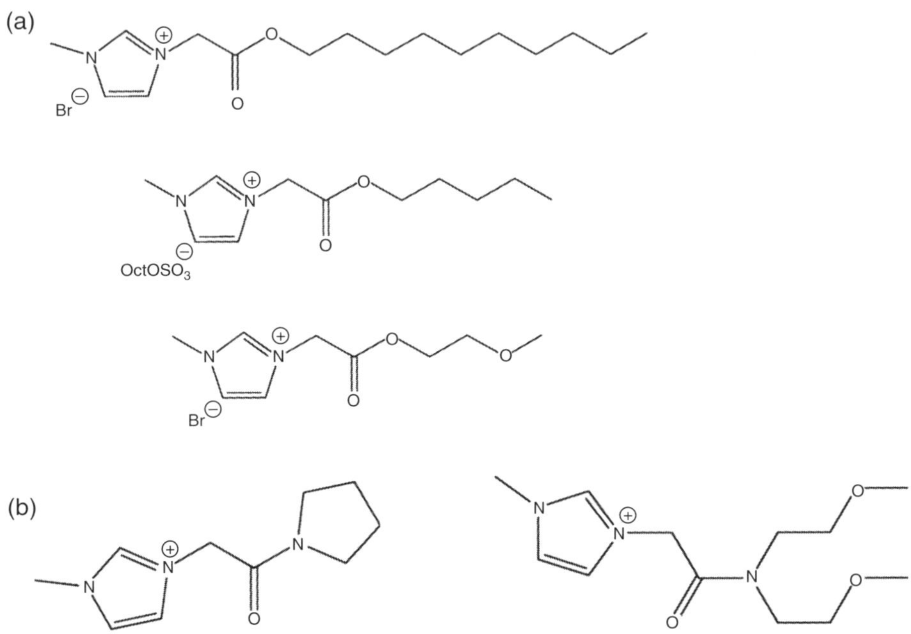

Scheme 20.7 (a) Chemical structure of oxygen-functionalized ILs. (b) Chemical structure of amide ILs. Reproduced from Morrissey et al. [91] with permission from the Royal Society of Chemistry.

the detrimental health and environmental effects of some organic solvents. The antimicrobial activities were tested by the agar diffusion plate method, and the bacteria used for the assay were *Staphylococcus* sp. (Gram-positive bacteria), *Klebsiella* sp., *P. aeruginosa*, *E. coli* (Gram-negative bacteria), and the yeast (*C. albicans*). TBA formate demonstrated an inhibitory effect on the growth of the three types of the Gram-negative bacteria. TBA formate was most effective against *E. coli* (as determined by the clear "growth inhibition" zones surrounding the disks). The Gram-positive *Staphylococcus* sp. was more susceptible to various TBA salt solutions than the Gram-negative *E. coli*. Further, they explored the effect of TBA formate on two proteins, namely, bovine serum albumin (BSA) and catalase (CAT). The change in tryptophan emission spectra as a result of addition of this salt onto both CAT and BSA confirmed the denaturation of both enzymes by the influence of tetrabutylammonium salts.

Recently, Łuczak et al. [163] confirmed the existence of a strong relationship among structure, surface activity, and biological action of imidazolium ILs on Gram-positive and Gram-negative bacteria as well as fungi. The dependence of the antimicrobial activity on chemical structure–chain length and anion type of 30 ILs was determined. The structure–activity comparison showed a strong relationship between antimicrobial efficacy and structure of the imidazolium cation. Shorter substituents on the cation result in low biostatic activities, whereas elongation of the alkyl substituents affects antimicrobial activity and showed a decrease in MIC. The highest activity was shown by compounds with 16 carbon atoms in the alkyl chain and a Cl^- anion. Further, Cornellas and his coworkers [164] synthesized two series of long-chain imidazolium- and pyridinium-based ILs (viz., 1-alkyl-3-methylimidazolium, $C_n$mimBr, and 1-alkylpyridinium bromides, C_nPyrBr) (where $n = 8, 10, 12$, and 14) to investigate the effects of the alkyl chain attached to the methylimidazolium and pyridinium head groups, vis-à-vis structure of cationic head groups on the association phenomena and antimicrobial activity of the amphiphilic ILs in aqueous solution and also to know whether their biological activity is related to the surfactant properties. Seven strains of bacteria, four Gram-positive and three Gram-negative, and one strain of fungi were used to assess the antimicrobial activity of these ILs. Microorganisms were *Micrococcus luteus ATCC9341*, *Staphylococcus epidermidis ATCC155-1*, *S. aureus ATCC25178*, *E. coli ATCC27325*, *Klebsiella pneumonia ATCC9721*, *P. aeruginosa ATCC9721*, *C. albicans ATCC10231*, and *B. subtilis ATCC6633*. The antimicrobial activities were determined *in vitro* on the basis of the MIC values. The ILs with short alkyl substituents were not active against bacteria and fungi, whereas the ILs containing 10, 12, and 14 carbon atoms in the alkyl chain showed significant antimicrobial activity. Further, the report of Rehmann et al. [165] developed a convenient, reproducible, parallel assay for enzyme activity in the presence of water-miscible and water-immiscible ILs using a minimum amount of them to provide a strong incentive to find less toxic biodegradable ILs for use as alternative solvents in laccase-catalyzed reactions. The activity of laccase from *Trametes versicolor* (LTV) was measured in the presence of 63 water-miscible and water-immiscible ILs, including imidazolium, pyridinium, quaternary ammonium and quaternary phosphonium salts. A simple, parallel method was used to determine their activities. Their aim was to obtain semiquantitative data that can be used as a basis to select "enzyme-friendly" ILs for future studies in detailed. Very recently, Bahnous et al. [166] synthesized several new highly functionalized imidazolium derivatives via appropriate synthetic routes, using imidazole, 1-methylimidazole, and 2-phenyl-1-methylimidazole as key intermediates. The antibacterial activity of these compounds was evaluated against *E. coli*, *S. aureus*, *P. aeruginosa*, and *S. typhimurium* using disk diffusion and MIC methods. The microbiological activities of alkyltriphenylphosphonium cation-based ILs with long alkyl chain as surfactants (e.g., triisobutyl(methyl)phosphonium tosylate and trihexyl(tetradecyl)phosphonium bis-2,4,4-(trimethylpentyl)phosphinate) were also studied by Moulik and his coworkers [167] against two bacterial strains, *B. subtilis* and *E. coli*, by measuring the diameter of the inhibition zone ("diz"). They found that both ILs exhibited greater antimicrobial activity than that of conventional nonionic surfactant (TX-100). Very recently, Neumann et al. [168] carried out a systematic investigation on the biodegradability of pyrrolidinium-, morpholinium-, piperidinium-, imidazolium-, and pyridinium-based IL cations substituted with different alkyl or functionalized side chains and halide as counterions. They examined primary degradability of these ILs by specific analysis and/or their ultimate biodegradability using biochemical oxygen demand (BOD) tests according to OECD guideline 301 F [169].

Nowadays, GSs attracted increasing attention from both academic and industrial points of view. In view of this, Li et al. [170] synthesized and subsequently evaluated the antibacterial activities of IL-type GSs based on quaternary ammonium unit. The following surfactants were synthesized: α,ω-bis (dimethylaminomethyl) tetraoxyethylene glycol, α,ω-bis(dimethylaminomethyl) pentaoxyethylene glycol, α,ω-bis(decyldimethylammoniomethyl) tetraoxyethylene glycol dibromide, α,ω-bis(tetra decyldimethylammoniomethyl) pentaoxyethylene glycol dibromide, etc. Further, GSs, n-EO$_m$-n ($m=1, 2$, or 3; $n=8, 10, 12, 14$, or 16) were tested to evaluate their antibacterial property against Gram-negative bacteria strain *E. coli* using diffusion disk method. GSs comprising of long alkyl chains (16-EO$_m$-16) or a short spacer chain (n-EO$_1$-n) did not exhibit any antibacterial activity. GSs with a moderate length of hydrophobic chain (10-EO$_m$-10) demonstrated strongest antibacterial activity. Among these ILs, GS (10-EO$_3$-10) showed the most efficient in antibacterial activity.

Recently, Pham et al. [171] discussed briefly the biodegradability, bioaccumulation, and distribution of ILs in different environmental compartments based on an extensive data sets on (eco)toxicity. The biological effects of ILs evaluated on the basis of toxicological test systems were addressed. The toxicity of ILs toward these test systems comprising of different levels of biological complexity as well as several environmental compartments was also discussed elaborately. The important aspects were dealt as follows: the effect of ILs in an enzyme level, critical inhibitory effects on a variety of bacteria and fungi, the use of algae primary producers to assess the effects of ILs to aquatic environments, cytotoxicity assays of ILs using mainly promyelocytic leukemia rat cell line IPC-81, the toxicity of ILs on human cell lines such as HT-29 and Caco-2 cells, and phytotoxic activity of ILs mostly on the duckweed, *L. minor*, a common aquatic vascular plant. Toxicity of ILs to invertebrates mainly focused on the use of *D. magna*, the snail *Physa acuta*, the spring tail *F. candida* (a soil invertebrate), *Caenorhabditis elegans* (a soil roundworm), and *Dreissena polymorpha* (zebra mussel) as test organisms. Conversely, toxicity of ILs on vertebrates (such as zebrafish (*Danio rerio*)), the frog (*Rana nigromaculata* as an amphibian model), rats, and mice was also dealt. These studies certainly indicated that antimicrobial agents for particular application or specified microorganisms can be prepared using the tunability of ILs.

20.6 ROLE OF ILs IN ENZYMATIC PROCESSES AND BIOTRANSFORMATIONS

It is important to understand the effects of various ILs on protein stability to protein activity and also on biomacromolecules. Polarity and kosmotropicity investigations and various spectroscopic studies are employed to find an explanation for the strong stabilizing ability of ILs on enzymes. All of these methods proved that ILs are able to generate and maintain an excellent nonconventional environment for proteins. That is why they could be used as outstanding solvents in biocatalytic transformations. The stability, activity, selectivity, and enantioselectivity of the enzyme are highly affected by the nature of IL, which can be finely tuned by selecting the appropriate cation and anion. ILs are effective green alternatives for organic solvents, especially those having an ECOENG (Ecology & Engineering) license, which indicates their eco-friendly character. The aim of this section is to summarize the efforts employed on using ILs in pharmaceutical applications.

Earlier, Erbeldinger et al. [172] reported for the first time the enzyme-catalyzed reaction, more specifically thermolysin-catalyzed synthesis of *Z*-aspartame in [bmim][PF_6]. This work presented better enzyme activity in IL compared to that studied in conventional organic solvents. Also, the enzyme was found to maintain a very high stability in [bmim][PF_6]. Since this pioneering work, many reports have been published on enzymatic catalysis in ILs, including the anhydrous transesterification of ethyl butanoate and butanol catalyzed by CALB in 1-butyl-3-methylimidazolium with hexafluorophosphate and tetrafluoroborate anions [173] and the α-chymotrypsin-catalyzed transesterification reaction of *N*-acetyl-L-phenylalanine ethyl ester with 1-propanol in 1-butyl-3-methylimidazolium and 1-octyl-3-methylimidazolium cations in conjunction with the hexafluorophosphate anion [174]. Further, Yang and Pan [64] and Rantwijk and Sheldon [65] also published a detailed report on enzyme catalysis in RTILs. It was found that in pure hydrophilic ILs, enzymes could be dispersed at a single molecule level, but they are usually deactivated. In hydrophobic ILs, enzymes are catalytically active, but they could not be dispersed at a single molecule level. In view of these, it is a challenging problem to reconcile the contradiction between the solubility of an enzyme as well as its retention of activity in RTIL [66, 175, 176].

Biotransformations have been of enormous economic and social importance throughout the history of mankind. The biocatalysis may be the most efficient way of producing fine chemicals. Today, several chemicals like pharmaceuticals, AA, saccharides and polysaccharides, esters, and vitamins are produced by enzymatic biotransformations on industrial scale. For the biocatalysis, five main "green" solvent systems are reported, for example, supercritical fluids (SCFs) (supercritical carbon dioxide ($scCO_2$) and supercritical water (scH_2O)), fluorinated solvents, ILs, water, and solvent-free reactions. Earlier, researchers started to explore RTILs for biocatalysis to identify their unique advantages. Biocatalytic reactions in ILs showed higher selectivity, faster rates, greater enzyme stability, higher yield, and (enantio) selectivity [177, 178]. The stability, activity, selectivity, and (enantio)selectivity of the enzymes were found to be highly affected by the nature of IL, which can be fine-tuned by choosing the appropriate anion and cation. Biphasic (buffer-IL) systems were also proven to provide a higher possibility of more economic processes, by giving the chance for easy recovery of enzymes [179]. Kaar et al. [180] also investigated lipase activity and stability in dialkylimidazolium- and pyrrolidinium-based

ILs with a variety of anions including hexafluorophosphate, acetate, nitrate, methanesulfonate, trifluoroacetate, and trifluoromethylsulfonate by employing lipase-catalyzed transesterification of methyl methacrylate and polytransesterification of divinyl adipate and 1,4-butanediol. Stability studies indicated that lipases exhibit greater stability in ILs than in organic solvents such as hexane, acetonitrile, and tetrahydrofuran. Further, Rantwijk and Sheldon [65] published a well-balanced and comprehensive review that extensively covers the various issues, such as the effect of ILs on the structure and activity of enzymes as well as on their thermal and operational stability and on the (enantio)selectivity of biocatalytic transformations associated with biocatalysis in ILs, in comparison with conventional reaction media. The design of efficient reaction procedures based on the unconventional solvent characteristics of ILs was also reviewed. Many research groups also reported that ILs are able to stabilize enzymes and increase their activity and selectivity [64, 181–185]. Park and Kazlauskas [178] correlated these behaviors to the polarity of ILs characterized by Reichardt's polarity scale and hydrogen-bond basicity. Solvents with higher polarity increased the solubility of polar substrates, led to faster reactions, and changed in selectivity. Several factors of ILs, such as anion nucleophilicity, hydrogen-bond basicity, excipients, impurities, pH, and the overall enzyme–substrate–medium relationship, strongly influenced the stability and activity of enzymes, which made the relationship between ILs and enzymes complicated. Zhao et al. [186] found that enzyme stability follows the Hofmeister series in aqueous solutions of hydrophilic ILs, like inorganic salts. On the other hand, the situation is more complicated in the case of hydrophobic or anhydrous hydrophilic ILs [186]. Machado and Saraiva [184] studied thermal deactivation kinetics of horseradish peroxidase (HRP) from 45 to 90°C in phosphate buffer and 5–25% (v,w/v) 1-butyl-3-methylimidazolium tetrafluoroborate [bmim][BF_4] and 1-butyl-3-methylimidazolium chloride [bmim][Cl]. They found that the presence of 5–10% (v/v) [bmim][BF_4] significantly improved HRP thermal stability with lower activation energies [184]. Alcoholysis between vinyl acetate and 2-phenyl-1-propanol was investigated by Maruyama et al. [187] with lipases from *Pseudomonas cepacia* and *Pseudomonas fluorescens*. Kinetic study revealed that Michaelis constant for 2-phenyl-1-propanol in [bmim]PF_6 was only half of that found in *n*-hexane, suggesting that the IL stabilized the enzyme–substrate complex. Dynamic structure–function relationships in enzyme stabilization were investigated by Lozano et al. [181, 188] by using structural investigating methods, such as intrinsic fluorescence and circular dichroism (CD) spectroscopic techniques. CALB and α-chymotrypsin enzymes were strongly stabilized in [emim]Tf_2N [1-ethyl-3-methylimidazolium bis(trifluoromethanesulfonyl) imide] and [btma]Tf_2N [butyltrimethylammonium bis(trifluoromethanesulfonyl) imide] ILs, where half-life ($t_{1/2}$) increased significantly with respect to those obtained from *n*-propanol and *n*-hexane organic solvents. These spectroscopic studies for both enzymes were correlated to the stabilization phenomena with the maintenance of the native structure of enzymes. These ILs maintained the active structure of an enzyme and prevented it from classical thermal unfolding. For the first time, Du et al. [189] employed a biological sample matrix—human urine—with the aid of an IL/aqueous two-phase system based on the hydrophilic IL, 1-butyl-3-methylimidazolium chloride (bmimCl), and suitable quantities of K_2HPO_4 for direct extraction of proteins from human body fluids. According to their results, this approach offered a clear advantage over traditional liquid extraction procedures with organic solvents, which caused contamination of the obtained proteins. Further, Cheng et al. [190] reported the extraction of heme protein (hemoglobin and myoglobin) into IL, 1-butyl-3-trimethylsilylimidazolium hexafluorophosphate ($BtmsimPF_6$), without using any concomitant reagent or extractant. Hemoglobin at the level of 100 ng/μl was found to be quantitatively extracted into $BtmsimPF_6$ in the absence of any coexisting extractants/additives at pH 7. However, at the same time, other protein species did not get involved in the extraction and remained in the aqueous phase. This system was also applied for selective isolation of hemoglobin from human whole blood without any pretreatment, giving rise to satisfactory results. The extraction of hemoglobin was facilitated by the coordination of the cationic $Btmsim^+$ moiety of IL and the iron atom (Fe) in the heme group. The result also provided a potential procedure for direct isolation of heme proteins from complex biological sample matrix, while the evasion of toxic organic solvents was inclined to offer biocompatibility of the procedure. Recently, Shu et al. [191] reported a simple one-step modification protocol for immobilizing the *N*-methylimidazole (*N*-mim) moiety onto both the surface and bulk of the polyvinyl chloride (PVC) material to generate bound hydrophilic IL for the purpose of protein adsorption (e.g., basic protein; lysozyme (Lys), cytochrome c (cyt-c), and hemoglobin (Hb) and acidic protein; bovine albumin serum (BSA), transferrin (Trf), and immunoglobulin G (IgG)). This bulk immobilization protocol was found to improve significantly the selectivity of basic proteins by effective suppression of nonspecific adsorption of protein species. In addition, the hydrophilic IL–PVC hybrid provided favorable biocompatibility characterized by the evasion of protein denaturation during adsorption process. The impact of Hofmeister effects on enzyme performance in aqueous solution was examined by Yang [192], concerning the specific ion effects on altering the bulk water properties, modifying the protein–water interactions, and interacting with the enzyme molecules. From both kinetic and structural studies, the effects of both cations and anions of ILs on enzyme activity and stability in aqueous IL mixtures were discussed. Finally, a study regarding mushroom tyrosinase in the presence of ILs and their associated anions were performed to illustrate the cooperative functioning of IL cations and anions in affecting enzyme performance, following the Hofmeister series. Design and use of water-mimicking ILs composed of chaotropic cations and kosmotropic anions were proposed to facilitate the applications of ILs in biotransformations.

Recent researches have demonstrated the possibility to carry out integral green biocatalytic processes by combining $scCO_2$ (which is environmentally friendly) and ILs with enzymes. Because of their different miscibilities, $scCO_2$ and ILs produced two-phase systems that showed an exceptional ability to carry out both the biotransformation and the products extraction steps simultaneously [193]. Stabilization of enzymes is a key issue to develop more efficient biocatalysts for industrial, environmental, or biomedical applications. In this respect, encapsulation in sol–gel matrices has been proved as one of the most efficient immobilization methods, considering both activity and biocatalyst stability. Sol–gel encapsulation of lipases is an emerging method to obtain biocatalysts with high activity and enantioselectivity for possible industrial applications. Some of specific ILs could be employed as immobilization additives for enhancing stability, repeated or continuous use, easy separation from the reaction mixture, possible modulation of the catalytic properties, and prevention of protein contamination in the product. These features are very much useful for designing of a more adequate sol–gel matrix to prevent enzyme leakage during operation and preserved the catalytic activity [194]. Further, Rodriguez et al. [195] reported the potential uses of ILs as reaction media of oxidative enzymes, namely, laccase, HRP, oxidase–peroxidase, chloroperoxidase, D-AA oxidase, and soybean peroxidase. The focus was centered in the studies of enzymatic activity and stability, as these are critical properties for biocatalytic processes. It was proposed that enzyme activity might diminish in presence of ILs, but such reduction might be balanced with the increase in solubility of poorly soluble substrates and/or products in ILs and, consequently, led to better performances of the global bioprocess.

In a series of review articles, Goto and his coworkers presented a general overview of the potential approaches to activate enzymes for diverse enzymatic processes and biotransformations in ILs [66, 196, 197]. In these articles, they clearly demonstrated the use of ILs as solvents for the synthesis of drugs, in particular nucleoside-based antiviral drugs, and also discussed advantages of ILs, such as milder reaction conditions, low solvent consumption, easy workup, and less waste production. The high solubility of many sparingly soluble drugs in some ILs (viz., 1-ethyl-3-methylimidazolium acetate ([emim][oAc]),1-vinyl-3-ethylimidazolium bromide ([Veim][Br]), dimethylimidazolium dimethylphosphate [C_1mim][$(MeO)_2PO_2$], etc.) were proposed to open up new functionalization methodology for effective delivery of them, which are not possible in conventional solvents. To improve the enzyme activity as well as stability in ILs, various attempts were made by modifying the form of the enzymes, such as enzyme immobilization onto support materials via adsorption or multipoint attachment, lyophilization in the presence of stabilizing agents, chemical modification with stabilizing agents, formation of cross-linked enzyme aggregates, and pretreatment with polar organic solvents or enzymes combined with suitable surfactants to form microemulsions. Further, one of the most appealing features of ILs for pharmaceutical applications are their use as APIs (with some attractive properties like solubility, purity, thermal stability, and bioavailability) and were discussed briefly by Moniruzzaman and Goto [197].

20.7 APPLICATIONS OF BILs

This section gives a short survey of the relevant literature with respect to historical developments and potential uses of nontoxic ILs and then proceeds to the latest results in colloid chemistry.

Earlier, Dorjnamjin et al. [198] reported a new one-phase method for the synthesis of uniform monodisperse crystalline Ag NPs in aqueous systems that has been developed by using newly synthesized mono and dihydroxylated ILs and cationic surfactants based on 1,3-disubstituted imidazolium cations and halogen anions. The hydroxyl-functionalized ionic liquids (HFILs) and hydroxyl-functionalized cationic surfactants (HFCSs) also simultaneously act as both reductant and protective agent. TEM, electron diffraction, ultraviolet–visible (UV–Vis), and NMR techniques were used for characterization of HFILs, HFCSs, and silver (Ag) NPs. These HFILs, HFCSs, and AgNPs were tested *in vitro* against several Gram-positive and Gram-negative bacteria and fungus. The AgNPs were proved to be environmentally friendly and high antimicrobial activity agents.

Meera et al. [199] reported the preparation and characterization of mesoporous and biocompatible transparent silica aerogel by the sol–gel polymerization of tetraethyl orthosilicate using choline cation-based IL, which allowed the silica framework to form in a noncollapsing environment and controlled the pore size of the gel. Nitrogen sorption measurements of gel prepared with IL exhibited a low surface area of 100.53 m^2/g and high average pore size of 3.74 nm. MTT assay (a colorimetric assay for assessing cell viability) revealed the effectiveness and biocompatibility of the prepared gels. This new nanoporous silica material was proposed to immobilize biological molecules, which may retain their stability over a longer period. A new functional IL (4-acetyl-*N*-butyl pyridinium hexafluorophosphate) was conjugated with nanosphere RBITC@SiO_2 forming a fluorescent core–shell ellipsoidal RBITC@SiO_2-IL NP with anticancer activities through mitochondria involved path [200]. Experimental results provided a new path to design nanomedicines through IL-modified silica NPs to target mitochondria. The effect of ILs on proteins has been gaining immense interest due to easy tunability of cation and anion for generating the desired effect. Very recently, Mehta et al. [201] explored the effect of alkyl imidazolium chloride ILs on collagen at molecular, interfibrillar, and skin matrix level. CD studies revealed that the secondary structure of collagen was not affected by imidazolium ILs and there was no change in thermal stability as well. Further, with increase in concentration of ILs, remarkable decrease in thermal stability of rat tail tendon (RTT) collagen fibers with marginal swelling effect was seen. This kind of exquisite behavior of ILs at different hierarchal order of collagen indicated that ILs are endowed with potential lyotropic action, which can be judiciously employed for biomedical applications.

Very recently, Liang et al. [202] reported the behaviors of nanoscale colloids prepared by long alkyl chain imidazolium-based ionic liquids (LILs), namely, 1-hexadecyl-3-methylimidazolium hexafluorophosphate ($C_{16}MIMPF_6$) and 1-hexadecyl-3-methylimidazolium bromide (C_{16}MIMBr), as well as phosphatidylcholines (PCs), including hydrogenated soybean phosphatidylcholine (HSPC) and egg phosphatidylcholine (EPC) by using morphological observations and measurements of particle size, zeta potential, and fluorescence polarization. According to them, estimation of the cytotoxicity of LILs is an important and basic step before applying them in practice. For better insight into the role of LILs and dispersion concentrations in modifying cell behavior, *in vitro* cytotoxicity of mixed PCs/LILs dispersions was examined using 3-(4,5-di-methylthiazol-2-yl)-5-(3-carboxymethoxyphenyl)-2-(4-sulfophenyl)-2H-tetrazolium (MTS) assay. Human premalignant keratinocytic HaCaT cells, human epidermoid carcinoma A431 cells, and human head and neck squamous carcinoma SCC25 cells were utilized to explore the cytotoxicity of these dispersions. Additionally, the contribution of mixed PC/LIL dispersions to apoptosis in A431 and SCC25 cells was investigated by flow cytometry. Finally, the cell viability and cytotoxicity mechanism were correlated with the physicochemical characteristics of these dispersions. The addition of C_{16}MIMBr into PC vesicles was found to increase the cytotoxicity for HaCaT, A431, and SCC25 cells in a content-dependent manner. Furthermore, mixed PC/C_{16}MIMBr dispersions exhibited a more significant decrease in cell viability than mixed PC/$C_{16}MIMPF_6$ dispersions. This result indicated that former system exhibits more toxicity to these cells than the latter at a comparable molar ratio of PC/LIL. These results inferred that the type of counterion in LIL head groups might play a key role in influencing cytotoxicity. Both of these systems did not specifically inhibit the survival of either an almost normal skin cell (HaCaT) or skin cancer cells (A431 and SCC25). Mixed PC/C_{16}MIMBr dispersions showed smaller diameter, higher positive zeta potential, lower turbidity, smoother particle outline, better storage stability, and higher membrane fluidity than mixed PC/C_{16}MIMPF_6 dispersions, resulting in higher cytotoxicity. Hence, it is cleared that the utilization of HSPC and EPC can reduce the toxicity and increases the safety of future applications of LILs in nanomaterials.

Recently, Liu et al. [203] employed ILs, [bmim][X] (X = Cl, Br, I, OAc, and SCN), as highly efficient reagents for nucleophilic substitution reactions of sulfonate esters (Scheme 20.8) derived from primary and secondary alcohols. The counteranions (X = Cl, Br, I, OAc, and SCN) of these ILs were found to be effectively replaced by the sulfonates affording the corresponding substitution products such as alkyl halides, acetates, and thiocyanides with excellent yields (Table 20.2). The newly developed protocol was proposed to be precisely environmentally attractive, because stoichiometric amounts of ILs were used in these reactions as sole reagents in most cases and did not require additional solvents, any other activating reagents, nonconventional equipment, or special precautions. Moreover, these ILs can readily be recycled without loss of reactivity, making the whole process "greener."

Fabrication of unmodified biopolymer-based materials and composites has been traditionally hampered by the difficulty of dissolving biopolymers due to their highly crystalline nature. Recently, Lee and his coworkers [204] showed that ILs possess a great potential to dissolve biopolymers and developed biopolymer-based materials, because of their synthetic flexibility by changing the combinations of cation and anion and green solvent properties such as nonvolatility, nonflammability, and recyclability. They reported that biopolymers such as cellulose, chitin/chitosan, silk, and gelatin can be easily fabricated into films, membranes, fibers, spheres, and molded shapes by dissolution in ILs and reconstitution in antisolvent. Further, they proposed that biopolymer-based materials with ILs could be useful for the biomedical applications such as tissue engineering scaffolds, wound dressing, drug delivery, implantable devices, and biosensors owing to their inherent biocompatibility and biodegradability. Additionally, it was shown that heparin/biopolymer composites that can be prepared by using ILs mixtures could be prospective to enhance the blood compatibility of biopolymer [204].

Very recently, Isik et al. [127] proposed the preparation of biocompatible ion gels by photopolymerization of the cholinium lactate methacrylic monomer inside the cholinium lactate IL and can be used as drug delivery agents, electrochemical transistors for biosensors, and mediums for biocatalysis. Conversely, the same BIL monomer, 2-cholinium lactate methacrylate, was used to obtain poly(IL)-cellulose composites by a fast UV curing process, which are the major component of plant cell walls, the most abundant biopolymer on earth, and also one of the most exploited natural resources. The use of ILs to replace organic solvents in biocatalytic processes has gained much attention. Cull et al. [205] used 1-butyl-3-methylimidazolium

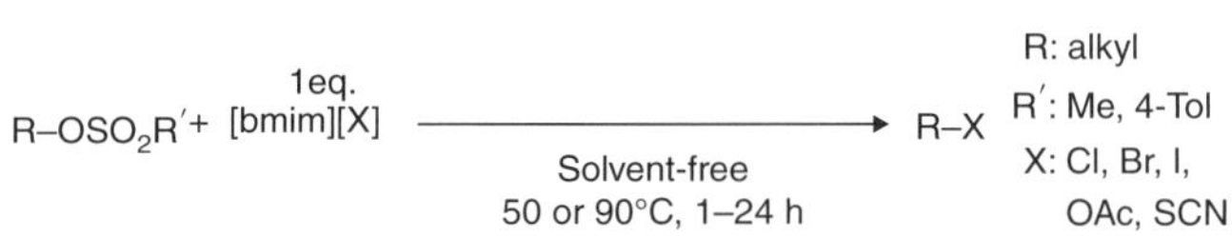

Scheme 20.8 Nucleophilic substitution reactions of sulfonate esters in [bmim][X] (X = Cl, Br, I, OAc, and SCN) media. Reproduced from Liu et al. [203] with permission from the Georg Thieme Verlag.

TABLE 20.2 Product Yield of Nucleophilic Substitution Reactions in 1-Butyl-3-Methylimidazxolium Halide or Acetate or Thiocyanate, [bmim][X] (Where, X = Cl, Br, I, OAc, SCN) Media

Substrate	X	*T* (°C)	*t* (h)	Yield (%, isol.)
	Cl	50	1	99
	Br	50	1	95
	I	50	8	94
	OAc	50	1	96
	SCN	50	5	95
$H_{17}C_8$–OTs	Cl	50	2	98
	Br	50	5	93
	I	50	8	94
	OAc	50	1	99
	SCN	50	5	99
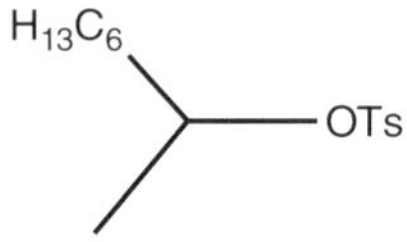	Cl	60	8	90
	Br	60	8	93
	I	60	24	99
	OAc	60	8	98
	SCN	60	24	97
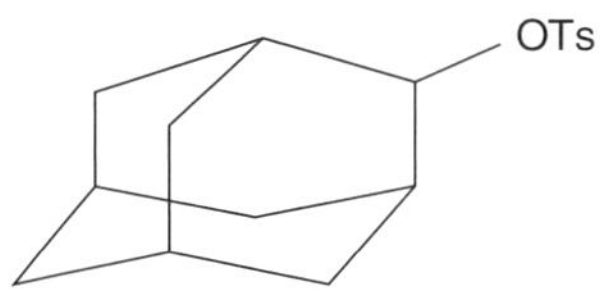	Cl	90	24	92
	Br	90	24	92
	I	90	24	99
	OAc	90	24	98
	SCN	100	24	99

Reproduced from Liu et al. [203] with permission from the Georg Thieme Verlag.

hexafluorophosphate, [bmim][PF_6], for the two-phase biotransformation of 1,3-dicyanobenzene to 3-cyanobenzamide and 3-cyanobenzoic acid using the nitrile hydratase from *Rhodococcus R312*. This work established ILs as potential alternative to organic solvents for multiphase biotransformations.

20.8 BIOCOMPATIBLE IONIC LIQUID-BASED MICROEMULSIONS

In the field of colloid and interface science, the self-assembling phenomena of amphiphilic molecules, (e.g., anionic, nonionic, and zwitterionic surfactants in single or mixed state) in ILs are currently an attractive and useful research area, from basic as well as academic point of view for understanding the mechanism of self-assembly in ILs. However, the tremendous potential of microemulsions as modern colloidal carriers for topical and transdermal drug delivery are well recognized [206]. Microemulsions, a system of water, oil, and amphiphiles, provide a variety of advantages for pharmaceutical use, such as nanometer-sized aggregations, long-term stability, biocompatibility, straightforward preparation and high solubilization capacity for drug molecules, and enhanced drug delivery [207–209]. Oil-soluble drugs can be formulated in oil-in-water (o/w) microemulsions, whereas water-soluble ones are better suited for water-in-oil (w/o) microemulsions. However, self-emulsifying drug delivery systems (e.g., w/o or o/w microemulsion) have not been fully accomplished the drug solubility in water and also in most pharmaceutical grade organic solvents. These limitations demand the need for alternative solvents that can be used as a polar core in microemulsions for the solubilization of sparingly soluble drug molecules. Considering these aspects, ILs could be good candidates because they are capable of dissolution of various poorly soluble compounds including drug molecules [35, 210]. Further, microemulsions in which water is replaced by a polar solvent have attracted great interest, and in this context, the most common solvents used include formamide, dimethylformamide, dimethylacetamide, ethylene glycol, propylene glycol, and glycerol [211, 212]. The reason behind this interest is that there are a large number of organic reactions for which contact with water needs to be avoided. In this regard, nonaqueous IL microemulsions where the ILs are sequestrated by nonionic surfactants represent an interesting area since they provide hydrophobic or hydrophilic microdomains that expand the potential uses of the ILs in microheterogeneous systems [213]. The retention of catalytic activity in IL-based microemulsions is due to the entrapment of enzyme molecules into aqueous microdroplets formed in water-in-ionic liquid (w/IL) microemulsions, indicating that these IL-based reaction systems provide a protective environment for the enzymes. In recent years, great attention has been paid to IL-based microemulsions due to their potential application prospects in biology, pharmaceutical, catalysis, and material synthesis, which have been discussed in subsequent paragraphs.

In an approach to overcome the limitations of solubility of drugs in water, ionic liquid-in-oil (IL/o) microemulsions (mEs) were synthesized and characterized by Moniruzzaman et al. [37] to increase the solubility of a sparingly soluble drug, *acyclovir* (ACV), and to enhance its topical and transdermal delivery. The formulation of microemulsions was composed of a blend of nonionic surfactants, Tween-80 and Span-20, IPM as an oil phase, and IL $[C_1mim][(CH_3O)_2PO_2]$ (dimethylimidazolium dimethylphosphate) as a pseudophase. Among various weight ratios of Tween-80 to Span-20 investigated in the microemulsion systems, the ratio 3:2 showed excellent solubility and skin permeation enhancing effect for ACV, which was used as a model sparingly soluble drug. Taken together these results, they proposed that IL-assisted nonaqueous microemulsions could serve as versatile and efficient nanodelivery systems for insoluble or sparingly soluble drug molecules that require solubilizing agents for delivery. Further, in a series of study, Moniruzzaman and his coworkers formulated a novel mixed Tween-80/Span-20 surfactant-based IL/o microemulsion in IPM to investigate the solubility of drug molecules (e.g., ACV, *methotrexate*, and *1-[(5-(p-nitrophenyl) furfurylidene)amino]hydantoin sodium*) that are insoluble or sparingly soluble in most of the conventional solvents [210, 214]. They concluded that these IL-based microemulsions with unique physical, chemical, and biological properties may offer novel opportunities to develop a potential drug delivery carrier for poorly soluble drug molecules.

In another study, Moniruzzaman et al. [215] explored the use of w/IL microemulsions comprised anionic surfactant, AOT/hydrophobic IL $[C_8mim][Tf_2N]$ (1-octyl-3-methyl imidazolium bis(trifluoromethylsulfonyl)amide)/water/1-hexanol as the reaction medium for the enzymatic oxidation of pyrogallol catalyzed by HRP. The results demonstrated that the rate of HRP-catalyzed reactions in IL microemulsions increases significantly compared with that obtained in conventional oil microemulsions. It was concluded that a w/IL microemulsion may be a very promising system for performing enzymatic reactions with HRP in ILs media. According to them, the findings will be of value for the development of ILs as a medium for the HRP-catalyzed oxidation of a variety of organic compounds such as phenols, biphenols, anilines, benzidines, and related heteroaromatic compounds. Further, they formulated aqueous microemulsion droplets in a hydrophobic IL, 1-octyl-3-methyl imidazolium bis (trifluoromethyl sulfonyl) amide ([C_8mim][Tf_2N]), stabilized by a layer of anionic surfactant (AOT) in the presence of 1-hexanol as a cosurfactant to overcome the insolubility of enzymes in most of the ILs. Further, they reported that the catalytic activity of one of the enzymes (*P. cepacia* lipase or *lipase PS*) became higher in this medium compared to AOT-derived microemulsions in isooctane [216]. The significant enhancement in performance of interfacially active enzymes, for example, *Chromobacterium viscosum* (CV) *lipase* and HRP in CTAB and imidazolium amphiphile-based mixed reverse micelles (RMs) with varying alkyl chains, was reported by Das and his coworkers [217]. Zhou et al. [218] studied the catalytic activities of two fungal oxidases, lignin peroxidase, and laccase for the oxidation of o-phenylenediamine, in the TX-100-stabilized water-in-[bmim][PF_6] microemulsion. Both of these fungal oxidases were catalytically active in this medium. The laccase-catalyzed oxidation of 2,6-dimethoxyphenol in the water-in-[bmim][PF_6] microemulsion exhibited a catalytic activity, which could be regulated by the composition of the interfacial membrane [219]. In a series of studies, Huang and his coworkers [219, 220] formulated and characterized single TX-100 and mixed AOT/TX-100 w/IL microemulsions stabilized in 1-butyl-3-methylimidazolium hexafluorophosphate ([bmim][PF_6]) and subsequently investigated lipase-catalyzed hydrolysis of *lignin peroxidase* (LiP), biomacromolecule *laccase*, and *4-nitrophenyl butyrate* (*p*-NPB) in these systems.

Recently, w/IL microemulsions formulated with nonionic surfactants, Tween-20 and TX-100, in [bmim][PF_6] were successfully used as media for lipase-catalyzed esterification reactions. Lipases from *Candida rugosa*, *C. viscosum*, and *Thermomyces lanuginosa* as biocatalysts were employed. The lipases exhibited higher catalytic activities and operational stability, particularly at higher incubation temperature as compared to that found in nonpolar solvent-based microemulsions. Fourier transform infrared (FTIR) and CD spectroscopic studies suggested that lipases entrapped in IL-based microemulsions retain their native structure or adapt more rigid structure compared to other microemulsions and accounted as the main reason for excellent stability in IL-based microemulsions. Of these, lipase from *T. lanuginosa* retained 90% of its original activity after the reaction repeated for 10 times in Tween-20-based w/IL microemulsions at 30°C [221]. In a subsequent study, Pavlidis et al. [222] investigated the use of w/IL microemulsion-based organogels (w/IL MBGs) as novel supports for the immobilization of lipase B from *C. antarctica* and lipase from *C. viscosum*. Microemulsion-based gels (MBGs) were formulated with w/IL microemulsions (by mixing buffer solution, Tween-20, and [bmim]PF_6) and a biocompatible polymer such as (hydroxypropyl)methyl cellulose (HPMC). These novel lipase-containing w/IL MBGs was found to be more effective as solid-phase biocatalysts in various polar and nonpolar organic solvents or ILs, exhibiting up to 4.4-fold higher esterification activity compared to water-in-oil microemulsion-based organogels. Moreover, FTIR was employed to investigate possible changes in enzyme structure upon immobilization and correlate with the biocatalytic behavior of the immobilized enzymes. FTIR data indicated that the immobilized lipases adopt a more rigid structure, referring to the structure in aqueous solution, and enhanced catalytic behavior was observed.

In recent years, especially with the development of green chemistry, $scCO_2$ and ILs, which are usually regarded as green solvents, have attracted much attention. In comparison with the conventional solvents (usually water and organic solvents), these green solvents have some unique properties. For example, $scCO_2$ is readily available, inexpensive, nontoxic, and nonflammable and has moderate critical temperature and pressure. Most importantly, the physical properties of $scCO_2$ can be adjusted by the pressure and temperature continuously [223–225]. Furthermore, CO_2 can be easily recaptured and recycled after utilization. Such unique properties confer $scCO_2$ and ILs great potential of applications in chemistry and chemical engineering.

Earlier, Liu et al. [226] reported the formation of *N*-ethyl perfluorooctylsulfonamide-based RMs comprising an IL (1,1,3,3-tetramethylguanidinium trifluoroacetate) as the inner component in $scCO_2$. They also prepared gold NPs by solubilization of $HAuCl_4$ in these RMs. The reverse micellar systems found to be advantageous for potential applications of gold NPs, when both $scCO_2$ and ILs were utilized as solvents in these RMs. Further, Harada et al. [227] synthesized silver particles (Ag particles) by the photoreduction of silver perchlorate ($AgClO_4$) in w/IL microemulsions consisting of nonionic surfactant Tween-20, water, and ILs, [1-butyl-3-methylimidazolium tetrafluoroborate ([BMIm][BF_4]) or 1-octyl-3 methylimidazolium tetrafluoroborate ([OMIm][BF_4]), mixed with a high-pressure (25 MPa) CO_2. The time evolution of the Ag particle formation by photoreduction was investigated by UV–Vis, cryo-TEM, extended X-ray absorption fine structure (EXAFS), and SAXS measurements. It was suggested that the water droplet size under high-pressure CO_2 could be more effectively regulated than that under ambient air, thereby prevented Ag particles from aggregation and precipitation. They proposed that the continuous phase might become more hydrophobic during dissolution of CO_2 into IL, which resulted in a higher stability of Ag particles in the water pool of w/IL microemulsion. The average size of Ag particles depended on the weight fraction of Tween-20 ($W_{Tween\text{-}20}$). With increase in $W_{Tween\text{-}20}$, smaller particle sizes were formed in this microemulsion system. Photoreduction of $AgClO_4$ in w/IL microemulsions comprising similar constituents without using CO_2 was reported in a similar study [228].

One of the major targets of modern chemistry is to combine the advantages of both of these green solvents by improving the IL solubility in $scCO_2$. In view of these, Senapati and his coworkers [229] provided the direct evidence of the formation of nanometer-sized, ellipsoidal-shaped IL domains in bulk CO_2 by carrying out detailed molecular dynamics simulations of the mixtures of 1,1,3,3-tetramethylguanidium acetate, [TMG][Ac]/*N*-ethyl perfluorooctylsulfonamide, *N*-EtFOSA/CO_2, and a range of other related IL-in-CO_2 systems [230]. Recently, Zhang and his coworkers [230] explored the application of CO_2-in-IL microemulsions in fabrication of metal–organic frameworks (MOFs), which has great potential in gas storage, separation, and catalysis. Further, they synthesized the bimetallic palladium/gold (Pd_4Au) NPs in TX-100/1-butyl-3-methylimidazolium hexafluorophosphate/water microemulsion by dissolving the palladium and gold precursors. The synthesized Pd_4Au NPs were characterized by TEM, energy-dispersive X-ray spectroscopy, X-ray diffraction, and UV–Vis spectroscopy [231].

20.9 CONCLUDING REMARKS AND FUTURE DIRECTIONS

Ionic liquids (ILs), of which the most often cited attribute is their negligible vapor pressure, have been suggested as a green alternative to traditional organic solvents with the desire to minimize diffusion to the atmosphere. Low volatility, however, does not completely eliminate potential environmental hazards and might pose serious threats to aquatic and terrestrial ecosystems. It is a goal of many researchers to tune the physicochemical properties of ILs via the choice of certain anionic and cationic components when designing a specific IL ideally suited for a specific process. In search for new, even "greener" ILs, many researchers synthesized novel AA, choline, and morpholinium derivative-based ILs with a range of alkanoate, saccharinate (Sac), acesulfamate (Ace), and organic lactate anions. These ILs are expected to develop high impact and more effectiveness toward future biotechnological applications. Also, their properties, namely, density, viscosity, refractive index, conductivity, and thermal stability at atmospheric pressure, are reported in the literature. It is, therefore, believed that these physicochemical properties will be added to the growing database on properties of various BILs and can be applied for the design of many chemical processes, such as catalysis, separation, and cellulose dissolution. The large number of possible structures of ILs and their unique physical, chemical, and biological properties make it feasible to consider them as an opportunity to contribute to the greenness in various fields where they can be applied.

The continual synthesis of new ILs presents an opportunity for the design of compounds that can comply with the most demanding technical requisites while presenting low levels of ecotoxicity and high biodegradability. The following points can be considered separately, to get the most accurate toxicity profile for a particular IL: (i) systematically select test compounds and structural elements according to the "test kit concept," (ii) test the selected substances in a flexible (eco)toxicological test battery at different levels of biological complexity (e.g., enzymes, cells, microorganisms, and organisms), (iii) identify toxicophore substructures in chemicals and use this knowledge in the design of inherently safer chemical products, and (iv) improve the molecular understanding of (eco)toxicological results by relating them to physicochemical properties. Following this approach, researchers investigated the influence of the structural elements side chain, head group, and the anionic moiety on cytotoxicity and (eco)toxicity so far. However, the data collected from various toxicological tests need to be considered together. This data set will cover a wide range of disparate biological systems as well as in a variety of formats. Therefore, visualization, integration, and interpretation of data from different test systems should be considered carefully. Further, the structure–activity comparison showed a strong relationship between antimicrobial efficacy and structure of the imidazolium cation. Shorter substituents on the cation result in low biostatic activities, whereas elongation of the alkyl substituents affects antimicrobial activity and consequently observed as a decrease in MIC. Altering of the anion type has a smaller effect on antimicrobial activity of imidazolium salts, indicating that the biocidal activity of IL is mainly driven by the alkyl chain length in the cation. In commercial applications, ILs may fulfill many functions, such as concurrently exhibiting surface, antimicrobial, and antielectrostatic activity. These create possibilities of developing new disinfectants, antiseptics, and preservatives. In addition, the low antimicrobial activities of short-chained salts indicate that ILs seem to be favorable compounds for bioprocesses, for example, biotransformations, enzymatic reactions, and separation processes as an alternative to organic solvents.

ILs are considered to be excellent host material for immobilizing both native and chemically modified enzymes and will open a new route to fabrication of functional biocatalysts and biochemical sensors. This chapter also brings together a number of techniques that could be used for activation and stabilization of enzymes in ILs, which are receiving increasing attention as "green" solvents for organic synthesis and catalytic processes. Also, it has been shown that ILs can generate and maintain excellent nonconventional environment for proteins. That is why they could be used as outstanding solvents in biocatalytic transformations. The stability, activity, selectivity, and enantioselectivity of the enzymes are highly affected by the nature of IL, which can be fine-tuned by choosing the appropriate cation and anion. It is believed that green and biocompatible ILs will be available in the near future, which will stimulate the use of ILs in industrial biotransformation. Further, it is expected that these findings will be of importance for the development of ILs as a medium for the enzyme-catalyzed oxidation of a variety of organic compounds such as phenols, biphenols, anilines, benzidines, and related heteroaromatic compounds.

Finally, a novel IL-assisted nonaqueous microemulsion prepared by some research groups with pharmaceutical accepted components, which can be effectively used in solubilization of a number of drug molecules that are insoluble or poorly soluble in water and also in most of the organic liquids. This type of microemulsion might have potential application as a carrier in drug delivery. It is believed that applying the combination of green properties of ILs with their unique tailor-made physicochemical properties, biocompatible ILs will be available in near future, which stimulate the use of this kind of ILs as pharmaceutical solvents and/or reagents. Finally, these studies allowed us to conclude that IL/o microemulsions create a new opportunity for the controlled delivery of sparingly soluble drugs that have problems of administration by other routes. A new approach for carrying out enzymatic reactions in ILs has also been reported by formation of aqueous nanometer-sized domains. The IL-based microemulsions possess advantages over conventional microemulsions (prepared in organic solvents) as reaction media for carrying out biotransformations with polar or hydrophilic substrates such as AA and carbohydrates, which are poorly soluble in most organic solvents (e.g., isooctane and hexane). In conclusion, novel IL-based microemulsion is considered to be a fascinating playing field of research. The number of possible systems is close to infinity. However, because of their complexity and several disadvantages, the system with suitable ingredients must be carefully chosen for each application and critically compared to cheaper and easier alternatives.

ACKNOWLEDGMENTS

The financial support in the form of an operating research grant to Professor Bidyut K. Paul and Senior Research Fellowship to Mr. Kaushik Kundu from the authority of Indian Statistical Institute, Kolkata, India, are thankfully acknowledged. Mr. Soumik Bardhan thanks UGC, New Delhi, India, for the financial assistance and UGC-BSR Research Fellowship. Professor Swapan K. Saha also thanks Science and Engineering Research Board (SERB), DST, Govt. of India, for the financial assistance. Our sincere thanks to Professor H. Stamatis, Department of Biological Applications and Technologies, University of Ioannina, Greece, for his valuable suggestions to improve the manuscript.

REFERENCES

[1] Welton, T. *Chem. Rev.* 1999, **99**, 2071–2083.
[2] Holbrey, J.; Seddon, K. R. *Clean Prod. Proc.* 1999, **1**, 223–236.
[3] Seddon, K. R. *J. Chem. Technol. Biotechnol.* 1997, **68**, 351–356.
[4] Walden, P. *Bull. Acad. Imper. Sci. St. Petersburg* 1914, **8**, 405–422.
[5] Chum, H. L.; Koch, V. R.; Miller, L. L.; Osteryoung, R. A. *J. Am. Chem. Soc.* 1975, **97**, 3264–3267.
[6] Wilkes, J. S.; Levisky, J. A.; Wilson, R. A.; Hussey, C. L. *Inorg. Chem.* 1982, **21**, 1236–1264.
[7] Zech, O.; Harrar, A.; Kunz, W. Nonaqueous Microemulsions Containing Ionic Liquids—Properties and Applications. *Ionic Liquids: Theory, Properties, New Approaches*, Kokorin, A., ed., InTech, Rijeka, ch. 11, 2011, p. 245.
[8] Krossing, I.; Slattery, J. M.; Daguenet, C.; Dyson, P. J.; Oleinikova, A.; Weingaertner, H. *J. Am. Chem. Soc.* 2006, **128**, 13427–13434.
[9] Xue, H.; Verma, R.; Shreeve, J. N. M. *J. Fluor. Chem.* 2006, **127**, 159–176.
[10] Hussey, C. L.; Scheffler, T. B.; Wilkes, J. S.; Fannin, Jr., A. A. *J. Electrochem. Soc.* 1986, **133**, 1389–1391.
[11] Freemantle, M. *Chem. Eng. News* 1998, **76**, 32–37.
[12] Gordon, C. M.; Holbrey, J. D.; Kennedy, A. R.; Seddon, K. R. *J. Mater. Chem.* 1998, **8**, 2627–2636.
[13] Earle, M. J.; Seddon, K. R. *Pure Appl. Chem.* 2000, **72**, 1391–1398.
[14] Gorodetsky, B.; Ramnial, T.; Branda, N. R.; Clyburne, J. A. C. *Chem. Commun.* 2004, 1972–1973.
[15] Handy, S. T. *J. Org. Chem.* 2006, **71**, 4659–4662.
[16] Aggarwal, V. K.; Emme, I.; Mereu, A. *Chem. Commun.* 2002, 1612–1613.
[17] Wasserscheid, P.; Welton, T. *Ionic Liquids in Synthesis*, Wiley, Weinheim, 2003.
[18] Huddleston, J. G.; Willauer, H. D.; Swatloski, R. P.; Visser, A. E.; Rogers, R. D. *Chem. Commun.* 1998, 1765–1766.
[19] Stalcup, A. M.; Cabovska, B. *J. Liq. Chromatogr. Relat. Technol.* 2005, **27**, 1443–1459.
[20] Baker, G. A.; Baker, S. N.; Pandey, S.; Bright, F. V. *Analyst* 2005, **130**, 800–808.
[21] Armstrong, D. W.; Zhang, L. K.; He, L.; Gross, M. L. *Anal. Chem.* 2001, **73**, 3679–3686.
[22] Fletcher, K. A.; Pandey, S.; Storey, I. K.; Hendricks, A. E.; Pandey, S. *Anal. Chim. Acta* 2002, **453**, 89–96.
[23] Fukaya, Y.; Sugimoto, A.; Ohno, H. *Biomacromolecules* 2006, **7**, 3295–3297.
[24] Swatloski, R. P.; Spear, S. K.; Holbrey, J. D.; Rogers, R. D. *J. Am. Chem. Soc.* 2002, **124**, 4974–4975.
[25] Kimizuka, N.; Nakashima, T. *Langmuir* 2001, **17**, 6759–6761.
[26] Wu, Y.; Sasaki, T.; Irie, S.; Sakurai, K. *Polymer* 2008, **49**, 2321–2327.
[27] Xie, H.; Zhang, S.; Li, S. *Green Chem.* 2006, **8**, 630–633.
[28] Zakrzewska, M. E.; Bogel-Łukasik, E.; Bogel-Łukasik, R. *Energy Fuel* 2010, **24**, 737–745.
[29] da Costa Lopes, A. M.; João, K. G.; Morais, A. R. C.; Bogel-Łukasik, E.; Bogel-Łukasik, R. *Sustainable Chem. Processes* 2013, **1**, 3–34.

[30] Khupse, N. D.; Kumar, A. *Indian J. Chem.* 2010, **49A**, 635–648 and references therein.

[31] Wang, P.; Dai, Q.; Zakeeruddin, S. M.; Forsyth, M.; MacFarlane, D. R.; Gratzel, M. *J. Am. Chem. Soc.* 2004, **126**, 13590–13591.

[32] Silvester, D.; Compton, R. G. *Z. Phys. Chem.* 2006, **220**, 1247.

[33] Parmentier, D.; Metz, S. J. *Green Chem.* 2013, **15**, 205–209.

[34] Joshi, M. D.; Anderson, J. L. *RSC Adv.* 2012, **2**, 5470–5484.

[35] Jaitely, V.; Karatas, A.; Florence, A. T. *Int. J. Pharm.* 2008, **354**, 168–173.

[36] Mizuuchi, H.; Jaitely, V.; Murdan, S.; Florence, A. T. *Eur. J. Pharm. Sci.* 2008, **33**, 326–331.

[37] Moniruzzaman, M.; Tamura, M.; Tahara, Y.; Kamiya, N.; Goto, M. *Int. J. Pharm.* 2010, **400**, 243–250.

[38] Bica, K.; Rijksen, C.; Nieuwenhuyzen, M.; Rogers, R. D. *Phys. Chem. Chem. Phys.* 2010, **12**, 2011–2017.

[39] Hough, W. L.; Rogers, R. D. *Bull. Chem. Soc. Jpn.* 2007, **80**, 2262–2269.

[40] Hough, W. L.; Smiglak, M.; Rodriguez, H.; Swatloski, R. P.; Spear, S. K.; Daly, D. T.; Pernak, J.; Grisel, J. E.; Carliss, R. D.; Soutullo, M. D.; Davis, J. J. H.; Rogers, R. D. *New J. Chem.* 2007, **31**, 1429–1436.

[41] Pernak, J.; Sobaszkiewicz, K.; Mirska, I. *Green Chem.* 2003, **5**, 52–56.

[42] Williams, H. D.; Sahbaz, Y.; Ford, L.; Nguyen, T. H.; Scammells, P. J.; Porter, C. J. H. *Chem. Commun.* 2014, **50**, 1688–1690.

[43] Dobler, D.; Schmidts, T.; Klingenhöfer, I.; Runkel, F. *Int. J. Pharm.* 2013, **441**, 620–627.

[44] Sowmiah, S.; Srinivasadesikan, V.; Tseng, M. S.; Chu, Y. H. *Molecules* 2009, **14**, 3780–3813.

[45] Zhao, D. B.; Wu, M.; Kou, Y.; Min, E. *Catal. Today* 2002, **74**, 157–189.

[46] Novoselov, N. P.; Sashina, E. S.; Kuźmina, O. G.; Troshenkova, S. V. *Russ. J. Gen. Chem.* 2007, **77**, 1395–1405.

[47] El Seoud, O. A.; Koschella, A.; Fidale, L. C.; Dorn, S.; Heinze, T. *Biomacromolecules* 2007, **8**, 2629–2647.

[48] Mantz, R. A.; Fox, D. M.; Green, J. M.; Fylstra, P. A.; De Long, H. C.; Trulove, P. C. *Z. Naturforsch., A: Phys. Sci.* 2007, **62**, 275–280.

[49] Wasserscheid, P.; Waffenschmidt, H.; Machnitzki, P.; Kottsieper, K. W.; Stelzer, O. *Chem. Commun.* 2001, 451–452.

[50] Favre, F.; Bourbigou, H. O.; Commereuc, D.; Saussine, L. *Chem. Commun.* 2001, 1360–1361.

[51] Brauer, D. J.; Kottsieper, K. W.; Liek, C.; Stelzer, O.; Waffenschmidt, H.; Wasserscheid, P. *J. Organomet. Chem.* 2001, **630**, 177–184.

[52] Bhushan, B.; Palacio, M.; Kinzig, B. *J. Colloid Interface Sci.* 2008, **317**, 275–287.

[53] Perkin, S.; Albrecht, T.; Klein, J. *Phys. Chem. Chem. Phys.* 2010, **12**, 1243–1247.

[54] Zhou, F.; Liang, Y.; Liu, W. *Chem. Soc. Rev.* 2009, **38**, 2590–2599.

[55] Sheldon, R. *Chem. Commun.* 2001, 2399–2407.

[56] Ueno, K.; Imaizumi, S.; Hata, K.; Watanabe, M. *Langmuir* 2009, **25**, 825–831.

[57] Wang, B.; Wang, X.; Lou, W.; Hao, J. *Nanoscale Res. Lett.* 2011, **6**, 259–268.

[58] Dupont, J.; De, S. R. F.; Suarez, P. Z. *Chem. Rev.* 2002, **102**, 3667–3692.

[59] Jacquemin, J.; Anouti, M.; Lemordant, D. *J. Chem. Eng. Data* 2011, **56**, 556–564.

[60] Kokorin, A., ed. *Ionic Liquids: Theory, Properties, New Approaches*, InTech, Rijeka, Croatia, 2011.

[61] Anderson, J. L.; Armstrong, D. W. *Anal. Chem.* 2005, **77**, 6453–6462.

[62] Visser, A. E.; Swatloski, R. P.; Griffin, S. T.; Hartman, D. H.; Rogers, R. D. *Sep. Sci. Technol.* 2001, **36**, 785–804.

[63] Belieres, J. P.; Gervasio, D.; Angell, C. A. *Chem. Commun.* 2006, 4799–4801.

[64] Yang, Z.; Pan, W. *Enzyme Microb. Technol.* 2005, **37**, 19–28.

[65] van Rantwijk, F.; Sheldon, R. A. *Chem. Rev.* 2007, **107**, 2757–2785.

[66] Moniruzzaman, M.; Kamiya, N.; Goto, M. *Org. Biomol. Chem.* 2010, **8**, 2887–2899.

[67] Ranke, J.; Muller, A.; Weber, U. B.; Stock, F.; Stolte, S.; Arning, J.; Stormann, R.; Jastorff, B. *Ecotoxicol. Environ. Saf.* 2007, **67**, 430–438.

[68] Stolte, S.; Arnig, J.; Weber, U. B.; Muller, A.; Pitner, W.; Biermann, U. W.; Jastorff, B.; Ranke, J. *Green Chem.* 2007, **9**, 760–767.

[69] Lorenzo, A. G.; Tozo, E.; Tojo, J.; Teijeira, M.; Rodríguez-Berrocal, F. J.; González, M. P.; Martínez-Zorzano, V. S. *Green Chem.* 2008, **10**, 508–516

[70] Stock, F.; Hoffmann, J.; Ranke, J.; Stormann, R.; Ondruschka, B.; Jastorff, B. *Green Chem.* 2004, **6**, 286–290

[71] Couling, D. J.; Bernot, R. J.; Docherty, K. M.; Dixon, J. K.; Maginn, E. J. *Green Chem.* 2006, **8**, 82–90

[72] Lamberti, G.; Kulacki, K. *Green Chem.* 2008, **10**, 104–110.

[73] Coleman, D.; Gathergood, N. *Chem. Soc. Rev.* 2010, **39**, 600–637

[74] Chambreau, S. D.; Vaghjiani, G. L.; Koh, C. J.; Golan, A.; Leone, S. R. *J. Phys. Chem. Lett.* 2012, **3**, 2910–2914.

[75] Taylor, A. W.; Lovelock, K. R. J.; Deyko, A.; Licence, P.; Jones, R. G. *Phys. Chem. Chem. Phys.* 2010, **12**, 1772–1783.

[76] Fox, D. M.; Gilman, J. W.; Morgan, A. B.; Shields, J. R.; Maupin, P. H.; Lyon, R. E.; Long, H. C. D.; Trulove, P. C. *Ind. Eng. Chem. Res.* 2008, **47**, 6327–6332.

[77] Paul, A.; Samanta, A. *J. Phys. Chem. B* 2008, **112**, 16626–16632.

[78] Samanta, A. *J. Phys. Chem. B* 2006, **110**, 13704–13716.

[79] Imperato, G.; König, B.; Chiappe, C. *Eur. J. Org. Chem.* 2007, **7**, 1049–1058.

[80] Davis, Jr., J. H.; Forrester, K. J.; Merrigan, T. *Tetrahedron Lett.* 1998, **39**, 8955–8958.

[81] Wasserscheid, P.; van Hal, R.; Bösmann, A. *Green Chem.* 2002, **4**, 400–404.

[82] Handy, S. T.; Okello, M.; Dickenson, G. *Org. Lett.* 2003, **5**, 2513–2515.

[83] Pernak, J.; Chwala, P.; Syguda, A. *Pol. J. Chem.* 2004, **78**, 539–546.

[84] Abbott, A. P.; Capper, G.; Davis, D. L.; Rasheed, R. K.; Tambyrajah, V. *Green Chem.* 2002, **4**, 24–26.

[85] Fukumoto, K.; Yoshizawa, M.; Ohno, H. *J. Am. Chem. Soc.* 2005, **127**, 2398–2399.

[86] Ohno, H.; Fukumoto, K. *Acc. Chem. Res.* 2007, **40**, 1122–1129.

[87] Kagimoto, J.; Fukumoto, K.; Ohno, H. *Chem. Commun.* 2006, 2254–2256.

[88] Tao, G. H.; He, L.; Liu, W. S.; Xu, L.; Xiong, W.; Wang, T.; Kou, Y. *Green Chem.* 2006, **8**, 639–646.

[89] Hannig, F.; Kehr, G.; Frohlich, R.; Erker, G. *J. Org. Chem.* 2005, **690**, 5959–5972.

[90] Allen, C. R.; Richard, P. L.; Ward, A. J.; van de Water, L. G. A.; Masters, A. F.; Maschmeyer, T. *Tetrahedron Lett.* 2006, **47**, 7367–7370.

[91] Morrissey, S.; Pegot, B.; Coleman, D.; Garcia, M. T.; Ferguson, D.; Quilty, B.; Gathergood, N. *Green Chem.* 2009, **11**, 475–483.

[92] Petkovic, M.; Ferguson, J. L.; Gunaratne, H. Q. N.; Ferreira, R.; Leitão, M. C.; Seddon, K. R.; Rebelo, L. P. N.; Pereira, C. S. *Green Chem.* 2010, **12**, 643–649.

[93] Trivedi, T. J.; Rao, K. S.; Singh, T.; Mandal, S. K.; Sutradhar, N.; Panda, A. B.; Kumar, A. *ChemSusChem* 2011, **4**, 604–608.

[94] Gathergood, N.; Garcia, M. T.; Scammells, P. J. *Green Chem.* 2004, **6**, 166–175.

[95] Muhammad, N.; Man, Z. B.; Bustam, M. A.; Mutalib, M. I. A.; Wilfred, C. D.; Rafiq, S. *J. Chem. Eng. Data* 2011, **56**, 3157–3162.

[96] Tao, D. J.; Cheng, Z.; Chen, F. F.; Li, Z. M.; Hu, N.; Chen, X. S. *J. Chem. Eng. Data* 2013, **58**, 1542–1548.

[97] Tao, G.; He, L.; Sun, N.; Kou, Y. *Chem. Commun.* 2005, 3562–3564.

[98] Luo, S.; Zu, D.; Yue, H.; Wang, L.; Yang, W.; Xu, Z. *Tetrahedron Asymmetry* 2006, **17**, 2028–2033.

[99] Guillen, F.; Brégeon, D.; Plaquevent, J.-C. *Tetrahedron Lett.* 2006, **47**, 1245–1248.

[100] Docherty, K. M.; Kulpa, Jr., C. F. *Green Chem.* 2005, **7**, 185–189.

[101] Noda, A.; Hayamizu, K.; Watanabe, M. *J. Phys. Chem. B* 2001, **105**, 4603–4610.

[102] Del Sesto, R. E.; Corley, C.; Robertson, A.; Wilkes, J. S. *J. Organomet. Chem.* 2005, **690**, 2536–2542.

[103] Rosonkiewicz, A. C.; Pernak, J.; Feder, J. K.; Ramani, A.; Robertson, A. J.; Seddon, K. R. *Green Chem.* 2005, **7**, 855–862.

[104] Swatloski, R. P.; Holbrey, J. D.; Rogers, R. D. *Green Chem.* 2003, **5**, 361–363.

[105] von Rymon Lipinski, G. R., ed., *Ullman's Encyclopedia of Industrial Chemistry*, Wiley-VCH Verlag GmbH, Weinheim, 2003, vol. **35**, p. 407.

[106] Seebach, D.; Oei, H. A. *Angew. Chem. Int. Ed.* 1975, **14**, 634–636.

[107] Zhao, H.; Jackson, L.; Song, Z.; Olubajo, O. *Tetrahedron Asymmetry* 2006, **17**, 1549–1553.

[108] Gardas, R. L.; Ge, R.; Goodrich, P.; Hardacre, C.; Hussain, A.; Rooney, D. W. *J. Chem. Eng. Data* 2010, **55**, 1505–1515.

[109] Plaquevent, J. C.; Levillain, J.; Guillen, F.; Malhiac, C.; Gaumont, A.-C. *Chem. Rev.* 2008, **108**, 5035–5060.

[110] Zhang, J. M.; Zhang, S. J.; Dong, K.; Zhang, Y. Q.; Shen, Y. Q.; Lv, X. M. *Chem. Eur. J.* 2006, **12**, 4021–4026.

[111] Yang, J. Z.; Zhang, Q. G.; Wang, B.; Tong, J. *J. Phys. Chem. B* 2006, **110**, 22521–22524.

[112] Trivedi, T. J.; Bharmoria, P.; Singh, T.; Kumar, A. *J. Chem. Eng. Data* 2012, **57**, 317–323.

[113] Rao, K. S.; Singh, T.; Trivedi, T. J.; Kumar, A. *J. Phys. Chem. B* 2011, **115**, 13847–13853.

[114] Hu, S.; Jiang, T.; Zhang, Z.; Zhu, A.; Han, B.; Song, J.; Xie, Y.; Li, W. *Tetrahedron Lett.* 2007, **48**, 5613–5617.

[115] Yu, Y. H.; Lu, X. M.; Zhou, Q.; Dong, K.; Yao, H. W.; Zhang, S. J. *Chem. Eur. J.* 2008, **14**, 11174–11182.

[116] Weaver, K. D.; Kim, H. J.; Sun, J.; MacFarlane, D. R.; Elliott, G. D. *Green Chem.* 2010, **12**, 507–513.

[117] Petkovic, M.; Seddon, K. R.; Rebeloa, L. P. N.; Pereira, C. S. *Chem. Soc. Rev.* 2011, **40**, 1383–1403.

[118] Moriel, P.; García-Suárez, E. J.; Martínez, M.; García, A. B.; Montes-Morán, M. A.; Calvino-Casilda, V.; Bañares, M. A. *Tetrahedron Lett.* 2010, **51**, 4877–4881.

[119] Liu, Q.; Hou, X.; Li, N.; Zong, M. *Green Chem.* 2012, **14**, 304–307.

[120] Abbott, A. P.; Boothby, D.; Carrea, G.; Pantwijk, F.; Secundo, F.; Sheldon, R. *J. Am. Chem. Soc.* 2004, **126**, 9142–9147.

[121] Lau, R.; Sorgedrager, M.; Carrea, G.; Pantwijk, F.; Secundo, F.; Sheldon, R. *Green Chem.* 2004, **6**, 483–487.

[122] Tang, S.; Baker, G. A.; Zhao, H. *Chem. Soc. Rev.* 2012, **41**, 4030–4066.

[123] Abbott, A. P.; Frisch, G.; Hartley, J.; Ryder, K. S. *Green Chem.* 2011, **13**, 471–481.

[124] Abbott, A. P.; Capper, G.; Davies, D. L.; Rasheed, R. K.; Tambyrajah, V. *Chem. Commun.* 2003, 70–71.

[125] Zhao, H.; Baker, G. A.; Holmes, S. *Org. Biomol. Chem.* 2011, **9**, 1908–1916.

[126] Pernak, J.; Syguda, A.; Mirska, I.; Pernak, A.; Nawrot, J.; Pradzynska, A.; Griffin, S. T.; Rogers, R. D. *Chem. Eur. J.* 2007, **13**, 6817–6827.

[127] Isik, M.; Gracia, R.; Kollnus, L. C.; Tomé, L. C.; Marrucho, I. M.; Mecerreyes, D. *ACS Macro Lett.* 2013, **2**, 975–979 and references therein.

[128] Fukaya, Y.; Iizuka, Y.; Sekikawa, K.; Ohno, H. *Green Chem.* 2007, **9**, 1155–1157.

[129] Restolho, J.; Mata, J. L.; Saramago, B. *Fluid Phase Equilib.* 2012, **322–323**, 142–147.

[130] Sekar, S.; Surianarayanana, M.; Ranganathan, V.; MacFarlane, D. R.; Mandal, A. B. *Environ. Sci. Technol.* 2012, **46**, 4902–4908.

[131] Klein, R.; Dutton, H.; Diat, O.; Tiddy, G. J. T.; Kunz, W. *J. Phys. Chem. B* 2011, **115**, 3838–3847.

[132] Klein, R.; Kellermeier, M.; Touraud, D.; Müller, E.; Kunz, W. *J. Colloid Interface Sci.* 2013, **392**, 274–280.

[133] Klein, R.; Touraud, D.; Kunz, W. *Green Chem.* 2008, **10**, 433–435.

[134] Klein, R.; Tiddy, G. J. T.; Maurer, E.; Touraud, D.; Esquena, J.; Tache, O.; Kunz, W. *Soft Matter* 2011, **7**, 6973–6983.

[135] Lopes, J. M.; Paninho, A. B.; Molho, M. F.; Nunes, A. V. M.; Rocha, A.; Lourenco, N. M. T.; Visak, V. N. *J. Chem. Thermodyn.* 2013, **67**, 99–105.

[136] Rocha, A.; Carvalho, T.; Vidinha, P.; Lourenco, N. M. T. *ChemPlusChem* 2012, **77**, 1106–1111.

[137] Pretti, C.; Chiappe, C.; Baldetti, I.; Brunini, S.; Monni, G.; Intorre, L. *Ecotoxicol. Environ. Saf.* 2009, **72**, 1170–1176.

[138] Wu, T.-Y.; Su, S.-G.; Gung, S.-T.; Lin, M.-W.; Lin, Y.-C.; Lai, C.-A.; Sun, I.-W. *Electrochim. Acta* 2010, **55**, 4475–4482.

[139] Brigouleix, C.; Anouti, M.; Jacquemin, J.; Caillon-Caravanier, M.; Galiano, H.; Lemordant, D. *J. Phys. Chem. B* 2010, **114**, 1757–1766.

[140] Pernak, J.; Borucka, N.; Walkiewicz, F.; Markiewicz, B.; Fochtman, P.; Stolte, S.; Steudte, S.; Stepnowski, P. *Green Chem.* 2011, **13**, 2901–2910.

[141] Khara, D. C.; Samanta, A. *J. Phys. Chem. B* 2012, **116**, 13430–13438.

[142] Khara, D. C.; Kumar, J. P.; Mondal, N.; Samanta, A. *J. Phys. Chem. B* 2013, **117**, 5156–5164.

[143] Kim, K. S.; Choi, S.; Demberelnyamba, D.; Lee, H.; Oh, J.; Lee, B. B.; Mun, S. J. *Chem. Commun.* 2004, 828–829.

[144] Handy, S. T. *Chem. Eur. J.* 2003, **9**, 2938–2944.

[145] Pernak, J.; Goc, I.; Mirska, I. *Green Chem.* 2004, **6**, 323–329.

[146] Carter, E. B.; Culver, S. L.; Fox, P. A.; Goode, R. D.; Ntai, I.; Tickell, M. D.; Traylor, R. K.; Hoffman, N. W.; Davis, J. H. *Chem. Commun.* 2004, 630–631.

[147] Pernak, J.; Stefaniak, F.; Weglewski, J. *Eur. J. Org. Chem.* 2005, **2005**, 650–652.

[148] Handy, S. T.; Okello, M. *Tetrahedron Lett.* 2003, **44**, 8399–8402.

[149] Pernak, J.; Goc, I. *Pol. J. Chem.* 2003, **77**, 975–984.

[150] Earle, M. J.; McCormac, P. B.; Seddon, K. R. *Green Chem.* 1999, **1**, 23–25.

[151] Pernak, J.; Czepukowicz, A.; Poźniak, R. *Ind. Eng. Chem. Res.* 2001, **40**, 2379–2383.

[152] Demberelnyamba, D.; Kim, K. S.; Choi, S.; Park, S. Y.; Lee, H.; Kim, C. J.; Yoo, I. D. *Bioorg. Med. Chem.* 2004, **12**, 853–857.

[153] Docherty, K. M.; Hebbeler, S. Z.; Kulpa, Jr., C. F. *Green Chem.* 2006, **8**, 560–567.

[154] Docherty, K. M.; Joyce, M. V.; Kulacki, K. J.; Kulpa, C. F. *Green Chem.* 2010, **12**, 701–712.

[155] Matzke, M.; Stolte, S.; Thiele, K.; Juffernholz, T.; Arning, J.; Ranke, J.; Welz-Biermann, U.; Jastorff, B. *Green Chem.* 2007, **9**, 1198–1207.

[156] Petkovic, M.; Ferguson, J.; Bohn, A.; Trindade, J.; Martins, I.; Carvalho, M. B.; Leitao, M. C.; Rodrigues, C.; Garcia, H.; Ferreira, R.; Seddon, K. R.; Rebelo, L. P. N.; Pereira, C. S. *Green Chem.* 2009, **11**, 889–894.

[157] Garcia, M. T.; Gathergood, N.; Scammells, P. J. *Green Chem.* 2005, **7**, 9–14.

[158] Ranke, J.; Molter, K.; Stock, F.; Bottin-Weber, U.; Poczobutt, J.; Hoffmann, J.; Ondruschka, B.; Filser, J.; Jastorff, B. *Ecotoxicol. Environ. Saf.* 2004, **58**, 396–404.

[159] Stolte, S.; Matzke, M.; Arning, J.; Boschen, A.; Pitner, W.-R.; Biermann, U. W.; Jastorff, B.; Ranke, J. *Green Chem.* 2007, **9**, 1170–1179.

[160] Romero, A.; Santos, A.; Tojo, J.; Rodriguez, A. *J. Hazard. Mater.* 2008, **151**, 268–273.

[161] HaiBo, Z.; XiaoHai, Z.; JinFeng, D.; GaoYong, Z.; CunXin, W. *Sci. China, Ser. B: Chem.* 2007, **50**, 238–242.

[162] Saadeh, S. M.; Yasseen, Z.; Sharif, F. A.; Abu Shawish, H. M. *Ecotoxicol. Environ. Saf.* 2009, **72**, 1805–1809.

[163] Łuczak, J.; Jungnickel, C.; Łacka, I.; Stolte, S.; Hupka, J. *Green Chem.* 2010, **12**, 593–601.

[164] Cornellas, A.; Perez, L.; Comelles, F.; Ribosaa, I.; Manresa, A.; Garcia, M. T. *J. Colloid Interface Sci.* 2011, **355**, 164–171.

[165] Rehmann, L.; Ivanova, E.; Ferguson, J. L.; Gunaratne, H. Q. N.; Seddon, K. R.; Stephens, G. M. *Green Chem.* 2012, **14**, 725–733.

[166] Bahnous, M.; Bouraioua, A.; Chelghouma, M.; Bouacida, S.; Roisnel, T.; Smati, F.; Bentchouala, C.; Gros, P. C.; Belfaitah, A. *Bioorg. Med. Chem. Lett.* 2013, **23**, 1274–1278.

[167] Mukherjee, I.; Manna, K.; Dinda, G.; Ghosh, S.; Moulik, S. P. *J. Chem. Eng. Data* 2012, **57**, 1376–1386.

[168] Neumann, J.; Steudte, S.; Cho, C. W.; Thöming, J.; Stolte, S. *Green Chem.* 2014, **16**, 2174–2184.

[169] Organization for Economic Co-operation and Development (OECD). OECD Guideline for Testing of Chemicals: 301 - Ready Biodegradability, OECD, Paris Cedex, 1992.

[170] Li, H.; Yu, C.; Chen, R.; Li, J.; Li, J. *Colloids Surf. A Physicochem. Eng. Asp.* 2012, **395**, 116–124.

[171] Pham, T. P. T.; Cho, C. W.; Yun, Y. S. *Water Res.* 2010, **44**, 352–372.

[172] Erbeldinger, M.; Mesiano, A. J.; Russell, A. J. *Biotechnol. Prog.* 2000, **16**, 1129–1131.

[173] Lau, R. M.; van Rantwijk, F.; Seddon, K. R.; Sheldon, R. A. *Org. Lett.* 2000, **2**, 4189–4191.

[174] Laszlo, J. A.; Compton, D. L. *Biotechnol. Bioeng.* 2001, **75**, 181–186.

[175] Zhao, H. *J. Chem. Technol. Biotechnol.* 2010, **85**, 891–907.

[176] Yu, X.; Sun, Y.; Xue, L.; Huang, X.; Qu, Y. *Top. Catal.* 2014, **57**, 923–934.

[177] Kragl, U.; Eckstein, M.; Kaftzik, N. *Curr. Opin. Biotechnol.* 2002, **13**, 565–571.

[178] Park, S.; Kazlauskas, R. J. *Curr. Opin. Biotechnol.* 2003, **14**, 432–437.

[179] Feher, E.; Major, B.; Belafi-Bako, K.; Gubicza, L. *Biochem. Soc. Trans.* 2007, **35**, 1624–1627.

[180] Kaar, J. L.; Jesionowski, A. M.; Berberich, J. A.; Moulton, R.; Russell, A. J. *J. Am. Chem. Soc.* 2003, **125**, 4125–4131.

[181] Lozano, P.; de Diego, T.; Gmouh, S.; Vaultier, M.; Iborra, J. L. *Biocatal. Biotransform.* 2005, **23**, 169–176.

[182] Okochi, M.; Nakagawa, I.; Kobayashi, T.; Hayashi, S.; Furasaki, S.; Honda, H. *J. Biotechnol.* 2007, **128**, 376–382.

[183] Ulbert, O.; Bé lafi-Bakó, K.; Tonova, K.; Gubicza, L. *Biocatal. Biotransform.* 2005, **23**, 177–183.

[184] Machado, M. F.; Saraiva, J. M. *Biotechnol. Lett.* 2005, **27**, 1233–1239.

[185] Jain, N.; Kumar, A.; Chauhan, S.; Chauhan, S. M. S. *Tetrahedron* 2005, **61**, 1015–1060.

[186] Zhao, H. *J. Mol. Catal. B: Enzym.* 2005, **37**, 16–25.

[187] Maruyama, T.; Yamamura, H.; Kotani, T.; Kamiya, N.; Goto, M. *Org. Biomol. Chem.* 2004, **2**, 1239–1244.

[188] de Diego, T.; Lozano, P.; Gmouh, S.; Vaultier, M.; Iborra, J. L. *Biomacromolecules* 2005, **6**, 1457–1464.

[189] Du, Z.; Yu, Y.-L.; Wang, J.-H. *Chem. Eur. J.* 2007, **13**, 2130–2137.

[190] Cheng, D. H.; Chen, X. W.; Shu, Y.; Wang, J. H. *Talanta* 2008, **75**, 1270–1278.

[191] Shu, Y.; Chen, X. W.; Wang, J. H. *Talanta* 2010, **81**, 637–642.

[192] Yang, Z. *J. Biotechnol.* 2009, **144**, 12–22.

[193] Habulin, M.; Primožič, M.; Knez, Z. Application of Ionic Liquids in Biocatalysis. *Ionic Liquids: Applications and Perspectives*, Kokorin, A., ed., InTech, Rijeka, ch. 19, 2011, p. 461.

[194] Péter, F.; Paul, C.; Ursoiu, A. Applications of Ionic Liquids to Increase the Efficiency of Lipase Biocatalysis. *Ionic Liquids: Applications and Perspectives*, Kokorin, A., ed., InTech, Rijeka, ch. 20, 2011, p. 481.

[195] Rodriguez, O.; Tavares, A. P. M.; Cristóvão, R.; Macedo, E. A. Ionic Liquids: Alternative Reactive Media for Oxidative Enzymes. *Ionic Liquids: Applications and Perspectives*, Kokorin, A., ed., InTech, Rijeka, ch. 21, 2011, p. 499.

[196] Moniruzzaman, M.; Nakashima, K.; Kamiya, N.; Goto, M. *Biochem. Eng. J.* 2010, **48**, 295–314.

[197] Moniruzzaman, M.; Goto, M. *J. Chem. Eng. Jpn.* 2011, **44**, 370–381.

[198] Dorjnamjin, D.; Ariunaa, M.; Shim, Y. K. *Int. J. Mol. Sci.* 2008, **9**, 807–820.

[199] Meera, K. M. S.; Sankar, R. M.; Jaisankar, S. N.; Mandal, A. B. *Colloids Surf. B Biointerfaces* 2011, **86**, 292–297.

[200] Yang, X.; Chen, Q. Y.; Li, X.; Gao, J. *Colloids Surf. B Biointerfaces* 2012, **98**, 91–96.

[201] Mehta, A.; Rao, J. R.; Fathima, N. N. *Colloids Surf. B Biointerfaces* 2014, **117**, 376–382.

[202] Liang, C.-H.; Ho, W.-Y.; Yeh, L.-H.; Cheng, Y.-S.; Chou, T.-H. *Colloids Surf. A Physicochem. Eng. Asp.* 2013, **436**, 1083–1091.

[203] Liu, Y.; Xu, Y.; Jung, S. H.; Chae, J. *Synlett* 2012, **23**, 2692–2698.

[204] Lee, S. H.; Miyauchi, M.; Dordick, J. S.; Linhardt, R. J. Preparation of Biopolymer-Based Materials Using Ionic Liquids for the Biomedical Application. *Ionic Liquid Applications: Pharmaceuticals, Therapeutics, and Biotechnology*, Malhotra, S., ed., ACS Symposium Series, American Chemical Society, Washington, DC, Vol. **1038**, ch. 10, 2010, p. 115.

[205] Cull, S. G.; Holbrey, J. D.; Vargas-Mora, V.; Seddon, K. R.; Lye, G. J. *Biotechnol. Bioeng.* 2000, **69**, 227–233.

[206] Rosler, A.; Vandermeulen, G. M. W.; Klok, H. A. *Adv. Drug Deliv. Rev.* 2001, **53**, 95–108.

[207] Gupta, S.; Moulik, S. P. *J. Pharm. Sci.* 2008, **97**, 22–45.

[208] Heuschkel, H.; Goebel, A.; Neubert, R. H. H. *J. Pharm. Sci.* 2008, **97**, 603–631.

[209] Kogan, A.; Garti, N. *Adv. Colloid Interface Sci.* 2006, **123–126**, 369–385.

[210] Moniruzzaman, M.; Tahara, Y.; Tamura, M.; Kamiya, N.; Goto, M. *Chem. Commun.* 2010, **46**, 1452–1454.

[211] Friberg, E.; Podzimek, M. *Colloid Polym. Sci.* 1984, **262**, 252–253.

[212] Correa, N. M.; Silber, J. J.; Riter, R. E.; Levinger, N. E. *Chem. Rev.* 2012, **112**, 4569–4602.

[213] Yan, F.; Texter, J. *Chem. Commun.* 2006, 2696–2698.

[214] Moniruzzaman, M.; Kamiya, N.; Goto, M. *J. Colloid Interface Sci.* 2010, **352**, 136–142.

[215] Moniruzzaman, M.; Kamiya, N.; Goto, M. *Langmuir* 2009, **25**, 977–982.

[216] Moniruzzaman, M.; Kamiya, N.; Nakashima, K.; Goto, M. *Green Chem.* 2008, **10**, 497–500.

[217] Das, D.; Das, D.; Das, P. K. *Biochimie* 2008, **90**, 820–829.

[218] Zhou, G.; Zhang, Y.; Huang, X.; Shi, C.; Liu, W.; Li, Y.; Qu, Y.; Gao, P. *Colloids Surf. B Biointerfaces* 2008, **66**, 146–149.

[219] Xue, L.; Qiu, H.; Li, Y.; Lua, L.; Huang, X.; Qu, Y. *Colloids Surf. B Biointerfaces* 2011, **82**, 432–437.

[220] Xue, L.; Li, Y.; Zou, F.; Lu, L.; Zhao, Y.; Huang, X.; Qu, Y. *Colloids Surf. B Biointerfaces* 2012, **92**, 360–366.

[221] Pavlidis, I. V.; Gouris, D.; Papadopoulos, G. K.; Stamatis, H. *J. Mol. Catal. B: Enzym.* 2009, **60**, 50–56.

[222] Pavlidis, I. V.; Tzafestas, K.; Stamatis, H. *Biotechnol. J.* 2010, **5**, 805–812.

[223] Ford, J. W.; Janakat, M. E.; Lu, J.; Liotta, C. L.; Eckert, C. A. *J. Org. Chem.* 2008, **73**, 3364–3368.

[224] Chen, Q. L.; Beckman, E. J. *Green Chem.* 2008, **10**, 934–938.

[225] Jessop, P. G.; Subramaniam, B. *Chem. Rev.* 2007, **107**, 2666–2694.

[226] Liu, J.; Cheng, S.; Zhang, J.; Feng, X.; Fu, X.; Han, B. *Angew. Chem. Int. Ed.* 2007, **46**, 3313–3315.

[227] Harada, M.; Kawasaki, C.; Saijo, K.; Demizu, M.; Kimura, Y. *J. Colloid Interface Sci.* 2010, **343**, 537–545.

[228] Harada, M.; Kimura, Y.; Saijo, K.; Ogawa, T.; Isoda, S. *J. Colloid Interface Sci.* 2009, **339**, 373–381.

[229] Chandran, A.; Prakash, K.; Senapati, S. *J. Am. Chem. Soc.* 2010, **132**, 12511–12516.

[230] Zhang, J. L.; Han, B.; Li, J.; Zhao, Y.; Yang, G. *Angew. Chem. Int. Ed.* 2011, **50**, 9911–9915.

[231] Zhang, G.; Zhou, H.; An, C.; Liu, D.; Huang, Z.; Kuang, Y. *Colloid Polym. Sci.* 2012, **290**, 1435–1441.

CHAPTER 21

Density Prediction of Ternary Mixtures of Ethanol + Water + Ionic Liquid Using Backpropagation Artificial Neural Networks

J. MORALES, O. A. MOLDES, M. A. IGLESIAS-OTERO, and J. C. MEJUTO

Department of Physical Chemistry, Faculty of Science, University of Vigo at Ourense, Ourense, Spain

G. ASTRAY

Department of Physical Chemistry, Faculty of Science, University of Vigo at Ourense, Ourense, Spain
Department of Geological Sciences, College of Arts and Sciences, Ohio University, Athens, OH, USA
Faculty of Law, International University of La Rioja, Logroño, Spain

A. CID

Chemistry Department, REQUIMTE-CQFB, Faculty of Science and Technology, New University of Lisbon, Caparica, Portugal

21.1 INTRODUCTION

Ionic liquids are composed of ions and they are typically fused salts or molten oxides whose properties as a solvent for chemical processes differ from those of aqueous and organic solvents, such as the electrochemical reduction of aluminum from alumina that is impossible in aqueous solutions [1]. Therefore, ionic liquids have great advantages because of their unique combination properties [1], among them small vapor pressure, which minimizes loss by volatilization and its environmental damage. Unfortunately, experimental data availability for these ionic liquids is still limited [2, 3] because of their relatively recent study for the huge amount of ion combination possibilities.

Although there are some literature about physical properties of pure ionic liquids and binary mixtures [1, 4–6], data are insufficient [2, 3] and so, it prevents its use in industrial applications [7] such as in hydraulic, fluid transport, and mass and energy

Ionic Liquid-Based Surfactant Science: Formulation, Characterization, and Applications, First Edition. Edited by Bidyut K. Paul and Satya P. Moulik.

transfers, among many other processes [8]. However, ionic liquids may be sources of environmental problems, not by vapor pressure, but also ionic liquids have significant solubility in water [8–11].

To sum up, the correlation and prediction of the multicomponent mixtures properties are a developing area of present-day research [2]. Taking into account the experience of our research group predicting chemical variables, we proposed the possibility to determine different physical and chemical properties using artificial neural networks (ANNs) in order to avoid cumbersome and expensive experimental work [12, 13].

ANNs are data mining techniques designed to extract information from a nontrivial large database in order to identify relevant information for attempting to solve the problem. These techniques can be defined as a system for the treatment of information whose processing units are inspired in the biological neuron. Overall, ANNs form a network with a very high throughput, so ANNs are able to solve problems that are otherwise difficult to solve.

The use of ANN is highly developed due their great advantage compared with traditional computing systems. ANNs have a flexible structure, capable to make a nonlinear mapping between input and output data sets. In fact, multilayer perceptrons, one of the more extended neural network architectures, are universal approximators for complex problems [12]. The application of this is reflected in the literature devoted to prediction of many physical and chemical parameters, such as nanofluids density [14], density of binary mixtures of ionic liquids [15], electrical percolation temperature [16], molecular diffusivity of nonelectrolytes [17], vegetable oils viscosity [18], esters flash point prediction [12], polarity parameter in binary mixed solvents systems [19], etc.

Nowadays, a lot of references have been developed to predict properties of binary mixtures of ionic liquids by ANNs such as heat capacity [20, 21], activity coefficients at infinite dilution [22], or melting point [23] and less for ternary solutions to determine electrical conductivity [24] or viscosity [25].

The aim of this chapter is to demonstrate the good predictive power of ANNs for determination of the density of ternary mixtures of ethanol + water + ionic liquid.

21.2 MATERIALS AND METHODS

21.2.1 Dataset

Density of ternary mixtures used in the present chapter corresponds with experimental values carried out at 298.15 K and at atmospheric pressure. These data have been compiled from the literature [2, 3, 7, 8, 26], being a total of 327 cases from the ternary systems comprising ethanol + water + ionic liquid. The available dataset has been divided into two groups: the first one with 283 cases used for training and learning process, 1-octyl-3-methylimidazolium chloride ([C_8mim][Cl]), 1-ethyl-3-methylimidazolium ethylsulfate (EMISE), 1,3-dimethylimidazolium methylsulfate ([MMIM][$MeSO_4$]), 1-butyl-3-methylimidazolium methylsulfate ([Bmim][$MeSO_4$]), and 1-butyl-3-methylimidazolium chloride ([C_4mim][Cl]), and a second group with 44 cases (validation cases, cases not previously used) designed to validate obtained neural network models and test their predictive power for 1-hexyl-3-methylimidazolium chloride ([C_6mim][Cl]). The testing set represents 13.4% of full data.

21.2.2 Software Package

To develop multiple linear regression (MLR), we used the IBM SPSS Statistics v.19 software, tool for managing statistics, researching, and spotting trends in data. For the implementation of ANN, we have used the commercial packages provided by Neural Planner Software and Alyuda Research, LLC. Neural networks were implemented with different activation functions (sigmoidal, linear, or hyperbolic tangent) and different training algorithms such as backpropagation, Levenberg–Marquardt, or quasi-Newton algorithm.

21.2.3 MLR

MLR is a common primary tool in physical chemistry studies about properties of chemical compounds and their mixtures (Eq. 21.1) [27]. IBM SPSS Statistics has been used in order to obtain the density values of mixtures ($\rho_{\text{predicted}}$) from the other variables involved in this way:

$$\rho_{\text{predicted}} = cte + a_0 b_1 + a_1 b_2 + a_2 b_3 + a_3 b_4 \tag{21.1}$$

The MLR method was developed used the training data. The aim of this method consists in obtaining the constant value and the a_x coefficients for each input variable (b_x). After that, the parameter ρ for other mixtures that have not entered the training phase can be calculated easily.

21.2.4 ANN

Basic unit of a neural network is the artificial neuron, shown in Figure 21.1. An artificial neuron has inputs, denoted as x, and outputs, denoted as y. Information inputs are weighted by a value (w) corresponding to the specificity of each neuron input [12].

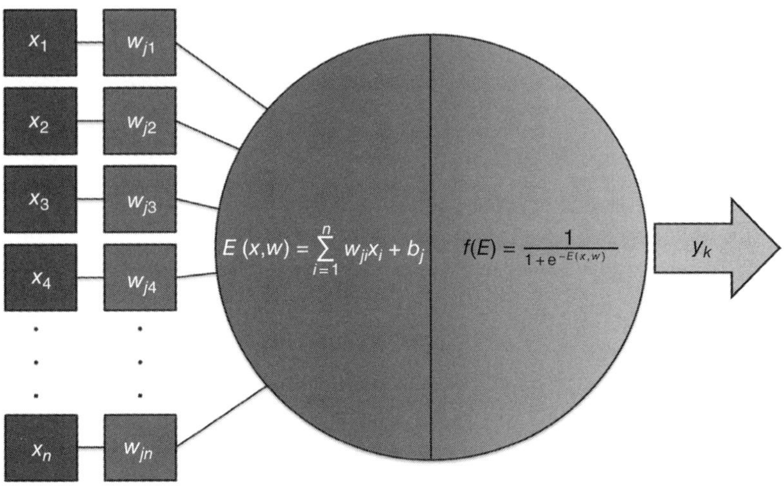

Figure 21.1 Basic structure of an artificial neuron.

In the neuron, the propagation function, E, defines the internal state of the neuron (Eq. 21.2). N is the number of neurons in the input layer, w_{ji} is the weight between the input neuron i and the intermediate neuron j, and b_j is the value of the bias associated to intermediate neuron j [12]. This equation unifies all inputs in a single value that will then be treated by the activation function (Eq. 21.3) that generate an output value of the neuron:

$$E(x,w) = \sum_{i=1}^{N} w_{ji} x_n + b_j \tag{21.2}$$

$$f(E) = \frac{1}{1+e^{-E(x,w)}} \tag{21.3}$$

The last neuron of the neural network generates a value ($y_{\text{predicted}}$, output value) that is checked with the experimental value (y_{exp}) to establish the error value (Eq. 21.4). Then, it is used to determine the end of the training phase [27]. In this equation, p is the number of neurons in the output layer:

$$\text{Error} = \frac{1}{2}\sum_{k=1}^{p} (y_{\text{exp}} - y_{\text{predicted}})^2 \tag{21.4}$$

An example of ANN is the following ANN that corresponds with a network with five neurons in the input layer; corresponds with the input variables (x_1, x_2, x_3, M_w, and ρ_0), nine neurons in the intermediate layers, and finally one neuron in the output layer; and corresponds with the density of ternary mixtures ($\rho_{\text{predicted}}$) (Fig. 21.2).

21.3 RESULTS AND DISCUSSIONS

Once the ternary mixture data were collected [2, 3, 7, 8, 26] and the molecular weights were compiled [28–30], we decided which of them would be the variables that will define the different models implemented. In this sense, five input variables have been used: mole fraction of ethanol (x_1), water (x_2), and ionic liquid (x_3), molecular weight of ionic liquid (M_W), and the density of pure ionic liquid (ρ_0).

21.3.1 Multiple Linear Regressions (MLRs)

The first step to develop an MLR in order to predict the density of ternary mixtures of ionic liquid consists on performing a Pearson correlation analysis to determine which variables can be better fitted to the linear model (Table 21.1).

In view of the data provided by the Pearson analysis, it precedes the development of an MLR model. We used 283 mixtures cases for the development of the MLR model. Each ionic liquid mixture is characterized by the following variables: (i) mole fraction of ethanol (x_1) and (ii) ionic liquid (x_3) and density of pure ionic liquid (ρ_0). We dispose all these variables for training set, except the density of pure ionic liquid for 1-butyl-3-methylimidazolium chloride ([C_4mim][Cl]). The ignorance of this variable will serve us to compare the predictive power of an MLR model against an ANN model when one of the input variables is unknown.

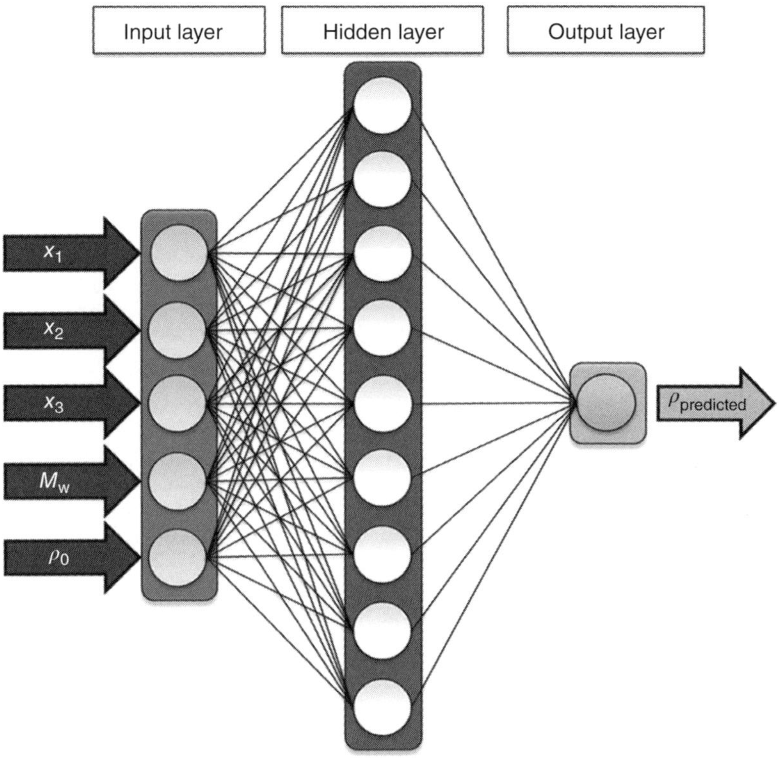

Figure 21.2 Artificial neural network with an input layer with five neurons, one hidden layer with nine neurons and one input layer with one neuron to determinate the density of ternary mixtures.

TABLE 21.1 Pearson Correlation Analysis Results for Ternary Mixtures of Ethanol +Water + Ionic Liquid for 283 Cases

	x_1	x_2	x_3	M_W	ρ_0	ρ
x_1	1	−0.781**	−0.200**	0.056	−0.058	−0.597**
x_2		1	−0.455**	−0.034	6.50×10^{-5}	0.090
x_3			1	−0.026	0.076	0.710**
M_W				1	−0.298**	0.053
ρ_0					1	0.446**
ρ						1

**$p < 0.05$.

Once the MLR model has been developed for these 283 mixtures cases, we obtain the equation which can be seen in Table 21.2. This table shows the R^2 coefficient and the root mean square error (RMSE) and the average percentage deviation (APD) calculated by Equations 21.5 and 21.6, respectively, where N corresponds with number of data cases:

$$\mathrm{RMSE} = \sqrt{\frac{\sum_{i=1}^{N}\left(\rho_{\mathrm{exp}} - \rho_{\mathrm{predicted}}\right)^2}{N}} \quad (21.5)$$

TABLE 21.2 MLR Model Implemented

	Parameter	Standard Error
Constant	0.626	0.028
x_1	−0.204	0.013
x_3	0.379	0.017
ρ_0	0.327	0.023
	Adjustments (283 cases)	
R^2	0.328	
RMSE	0.180	
APD	11.660	
	Adjustments (212 cases)	
R^2	0.853	
RMSE	0.040	
APD	3.170	

Adjustments for 283 cases (all cases) and for 212 cases (without $[C_4mim][Cl]$). R^2 is the square correlation coefficient, RMSE is the root mean square error, and APD is the average percentage deviation for MLR model developed.

$$\mathrm{APD} = \frac{\sum_{i=1}^{N}\left(\left(\left|\frac{\rho_{\exp} - \rho_{\text{predicted}}}{\rho_{\exp}}\right|\right) \times 100\right)}{N} \tag{21.6}$$

As we can see in Table 21.2, the different fit parameters for the MLR model developed are poor, presenting an R^2 of 0.328 with a RMSE of $0.180\,\mathrm{g \cdot cm^{-3}}$ and an average percentage deviation of 11.66%. These MLR settings are so low for the inclusion of $[C_4mim][Cl]$ mixtures in data cases to develop the MLR model. For this ionic liquid ($[C_4mim][Cl]$), the viscosity of the pure ionic liquid is not reported. For this reason, in Figure 21.3, we can see a number of points (71 data cases), corresponding to $[C_4mim][Cl]$, which are beyond the predictive model MLR. The behavior of these points gives us to understand that when an ionic liquid has some unknown variable, the model makes big errors in prediction.

If we only take into account the ionic liquid mixtures with complete data (212 cases), we can see that the fit parameters are better than when we do not consider the incomplete ionic liquid data $[C_4mim][Cl]$. In this case, the MLR model has good prediction power, R^2 of 0.853 with an RMSE of $0.040\,\mathrm{g \cdot cm^{-3}}$ and an APD of 3.17%.

Once the MLR model is developed, we proceed to see its behavior in ternary solution ionic liquid previously reserved (validation cases). For this case, we used the second ternary mixtures group (44 cases) to validate the model, and consequently, we obtain the results shown in Figure 21.4.

The ionic liquid used in this phase, 1-hexyl-3-methylimidazolium chloride ($[C_6mim][Cl]$), has not been used in the training phase of MLR model. For these mixtures, we know all input variables to apply the MLR model. As we can see, the fits for these ternary solutions show good correlation index R^2 (0.912), with a low RMSE ($0.035\,\mathrm{g \cdot cm^{-3}}$) and an average percentage deviation of 2.79% (Table 21.3). It can be seen that in case of a ternary mixture of ionic liquid not previously seen, the MLR model is able to provide an acceptable fit.

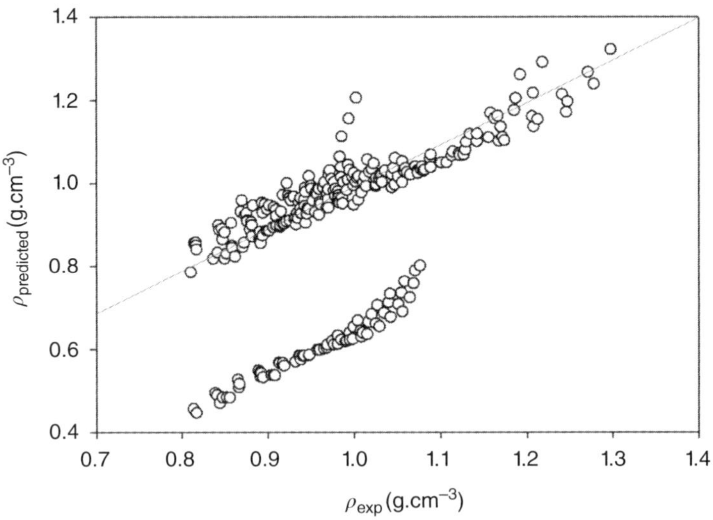

Figure 21.3 Experimental values of density (ρ_{exp}) versus predicted values ($\rho_{predicted}$) for the training phase of MLR model. The values farthest from the trend line one (continuous line) correspond to ionic liquid [C_4mim][Cl], which the initial density is unknown.

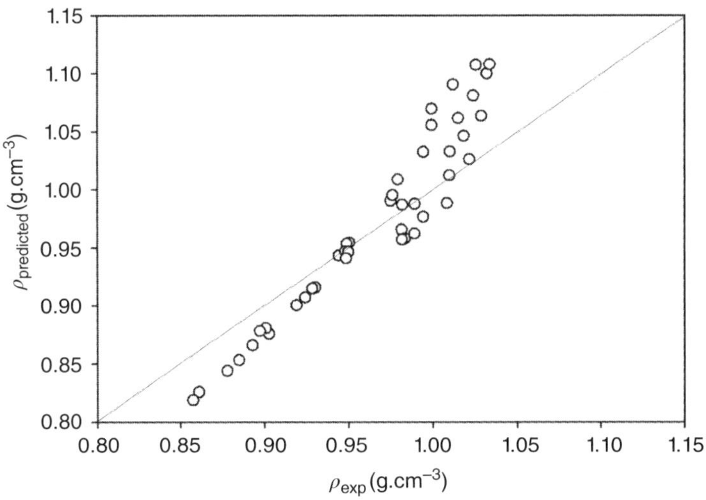

Figure 21.4 Experimental values of density (ρ_{exp}) versus predicted values ($\rho_{predicted}$) for the validation phase of MLR model. Continuous line represents the trend line one.

21.3.2 Artificial Neural Network (ANN)

We have implemented several ANNs with different architectures to predict the density of ternary mixtures of ionic liquid and to overcome the MLR model fits. The tests consisted of carrying out different ANNs with different number of training cycles, modifying the level of error and also modifying the number of layers and the neurons used in each one [27].

TABLE 21.3 Fits for Validation Cases (44 Cases) Using MLR Implemented

	Adjustments
R^2	0.912
RMSE	0.035
APD	2.79

R^2 is the square correlation coefficient, RMSE is the root mean square error, and APD is the average percentage deviation for MLR model developed.

The ANN model was developed with the five variables available in the input layer, corresponding to the (i) mole fraction of ethanol (x_1), (ii) mole fraction of water (x_2), (iii) mole fraction of ionic liquid (x_3), (iv) molecular weight of pure ionic liquid (M_w), and (v) density of pure ionic liquid (ρ). As mentioned earlier, these variables were compiled from the literature [2, 3, 7, 8, 26], and the molecular weights were compiled from ChemExper, ChemicalBook, and chemBlink websites [28–30].

ANNs were developed with different neurons in the intermediate layer. As the number of neurons is more difficult to determinate, we have used the following expression (Eq. 21.7) to determinate by trial and error, which is the best topology to predict the density of ternary mixtures of ionic liquids:

$$1 < n < 2N + 1 \tag{21.7}$$

In Equation 21.7, N corresponds with the number of input neurons (five variables) and n denotes the number of neurons in the intermediate layers. All neural networks are defined using the following nomenclature (Eq. 21.8):

$$N_{\text{in}} - (N_{h1} - N_{h2} - N_{h3}) - N_{\text{out}} \tag{21.8}$$

In Equation 21.8, N_{in} and N_{out} correspond with the numbers of neurons in the input and output layer, respectively. N_{h1}, N_{h2}, and N_{h3} correspond with the number of neurons in the intermediate layers and g corresponds with the number of intermediate layers.

Table 21.4 shows the topology and the results for the five best neural networks implemented. To evaluate the accuracy of the ANN models developed, we have used the RMSEs in validation (RMSE_V). As it can be observed in Table 21.4, the neural networks with lower RMSEv is, in this case, the neural network with topology 5-$(4)_1$-1. We choose this ANN because it presents a lower RMSEv and lower APD_V. The best topology developed, 5-$(4)_1$-1, consists in five input neurons, one middle layer with four neurons and one output neuron in the output layer. To train the $\text{ANN}_{5\text{-}(4)\text{-}1}$ model, a maximum number of 750 training cycles was established; the learning rate was set at 0.60 and the momentum value at 0.80.

According to the results shown in Table 21.4, $\text{ANN}_{5\text{-}(4)\text{-}1}$ gives a high level of correlation in terms of R^2 for training (0.982), with a low value of RMSE ($0.024\,\text{g}\cdot\text{cm}^{-3}$) and an average percentage deviation in training phase of 2.17%. We can see the good correlation for predicted data, even in the case of ionic liquid [C_4mim][Cl] (Fig. 21.5). If we compare the figures to the development phase of the MLR model (Fig. 21.3) and neural network model (Fig. 21.5), we can see the great difference in the prediction values of the [C4mim][Cl] density mixtures. We must remember that in the case of mixtures of

TABLE 21.4 Topology of the Five Best Neural Networks Implemented for This Chapter

Cycles	Architectures	R_T^2	$RMSE_T$	APD_T	R_V^2	$RMSE_V$	APD_V
750	$5\text{-}(4)_1\text{-}1$	0.982	0.024	2.17	0.977	0.008	0.64
1000	$5\text{-}(3)_1\text{-}1$	0.974	0.036	3.35	0.967	0.009	0.87
750	$5\text{-}(3)_1\text{-}1$	0.977	0.034	3.17	0.966	0.009	0.82
1250	$5\text{-}(3)_1\text{-}1$	0.972	0.038	3.47	0.968	0.010	0.94
1000	$5\text{-}(4)_1\text{-}1$	0.974	0.029	2.62	0.978	0.010	0.89

R^2 is the square correlation coefficient for training data (R_T^2) and validation data (R_V^2), root mean square errors (RMSE) of training ($RMSE_T$) and testing data ($RMSE_V$), and average percentage deviation for training phase (APD_T) and validation phase (APD_V) for different artificial neural network models developed.

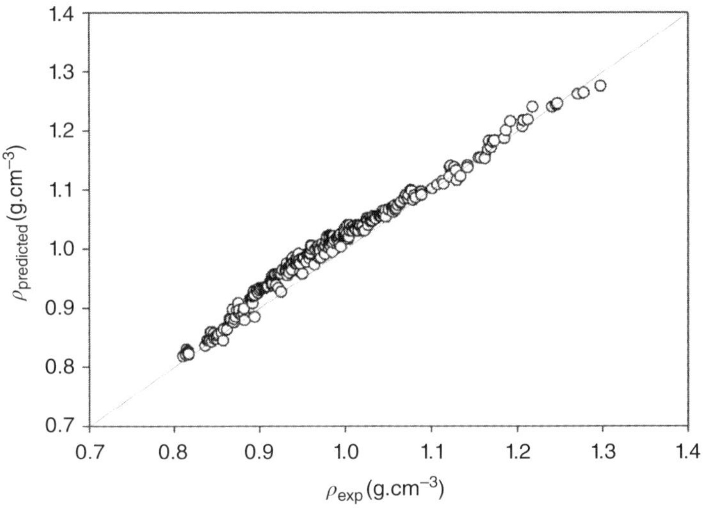

Figure 21.5 Experimental values of density (ρ_{exp}) versus predicted values ($\rho_{predicted}$) for the training phase of the ANN model. Continuous line represents the trend line one.

[C4mim][Cl], we do not know the density value of pure ionic liquid, and as expected, the MLR model makes big mistakes for these ionic liquid mixture (Fig. 21.3). However, the large capacity of the ANN to adapt to the absence of some input variable allows the values predicted by our ANN are close to the expected values (Fig. 21.5).

For validation phase, we used the same ionic liquid mixture cases that were used in the MLR model validation phase, 71 cases corresponding with the mixtures of ionic liquid, 1-hexyl-3-methylimidazolium chloride ([C_6mim][Cl]), that has not been used in the training phase of ANN model. In Table 21.4, we can see the good results for validation phase with a R^2 equal to 0.977, $RMSE_V$ of 0.008 g·cm^{-3}, and an APD_V of 0.64%. It can be seen that in this case (Fig. 21.6), the ANN model is able to provide a better fit for new ionic liquid mixture that is not previously known and compared with the MLR model (Fig. 21.4).

These results represent an improvement, in terms of average percentage deviation, of 86.66% in training phase and 77.80% in validation phase by using $ANN_{5\text{-}(4)\text{-}1}$ compared to using implemented MRL model (Table 21.5). In this sense, we understand that the $ANN_{5\text{-}(4)\text{-}1}$ is a good model to determine the density of ternary

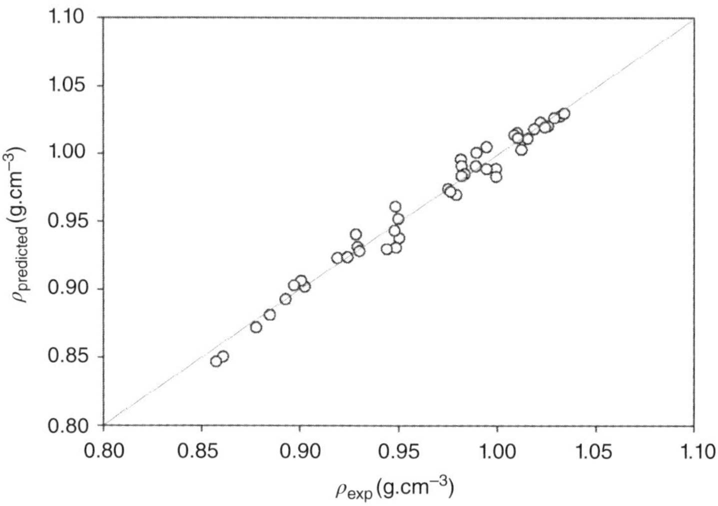

Figure 21.6 Experimental values of density (ρ_{exp}) versus predicted values ($\rho_{predicted}$) for the validation phase of the ANN model. Continuous line represents the trend line one.

TABLE 21.5 Comparison between MLR Model and ANN$_{5\text{-}(4)\text{-}1}$ Model Implemented in This Chapter

Model	R^2_T	RMSE$_T$	APD$_T$	R^2_V	RMSE$_V$	APD$_V$
MLR (283 cases)	0.328	0.180	11.66	0.912	0.035	2.79
ANN$_{5\text{-}(4)\text{-}1}$	0.982	0.024	2.17	0.977	0.008	0.64
Improvement (%)		86.66	81.43		77.80	77.06

R^2 is the square correlation coefficient for training data (R^2_T) and validation data (R^2_V), root mean square errors (RMSE) of training (RMSE$_T$) and testing data (RMSE$_V$), and average percentage deviation for training phase (APD$_T$) and validation phase (APD$_V$).

TABLE 21.6 Weight Matrix of the Best ANN Model Developed (ANN$_{5\text{-}(4)\text{-}1}$)

x_1	x_2	x_3	M_W	ρ_0
1.586	−0.006	−6.550	0.024	−0.038
−1.295	−0.763	−3.143	0.510	2.481
0.818	−0.315	−0.236	−0.759	−1.354
2.467	0.957	0.050	−0.340	−2.328
−3.870	−2.521	−1.362	−2.055	

systems of ionic liquid and we can say that the variables used in this chapter for the determination of density ternary mixtures of ionic liquids have a good power system characterization.

The importance of every connection between all neurons was denoted by a non-dimensional value so-called weight, and it is presented in Table 21.6 as a matrix. These values represent the synaptic strength of every connection that can take positive or negative values.

TABLE 21.7 Variable Importance for the Best ANN Developed ($ANN_{5\text{-}(4)\text{-}1}$)

Variable	Importance (%)
x_3	38.35
$\rho 0$	23.83
x_1	23.70
x_2	7.84
M_W	6.28

The sum of the absolute value of all weights for each input neuron gives us an idea of the importance of input neuron in the neural system. Taking into account the weight matrix (Table 21.6), we found that the most important variables to predict the density of ternary mixtures of ionic liquid were the mole fraction of ionic liquid (38.35), density of pure ionic liquid (23.83), and molar fraction of ethanol (23.70) (Table 21.7). Molar fraction of water and molecular weight of ionic liquid present an importance of 7.84 and 6.28, respectively, and they had not shown significant to determine the density of ternary mixtures of ionic liquid (Table 21.7).

21.4 CONCLUSIONS

As quoted previously, densities of ternary mixtures of ethanol + water + ionic liquid were obtained using ANNs. The model that achieves better predictions consists in five input neurons, one middle layer with four neurons and one output neuron. This model presents RMSEs of $0.024\,g\cdot cm^{-3}$ ($R^2=0.982$) for the training set and $0.008\,g\cdot cm^{-3}$ ($R^2=0.977$) for the validation set. The average percentage deviation for training and validation phase is 1.96%.

It has been demonstrated that ANN models can predict chemical values and physical properties even when some data is unknown compared with the MLR developed. This predictive procedure shows an improvement of prediction results around 86.66% in comparison to classical MLR model, that is, the neural network will be able to adjust more precisely those ternary solutions of ionic liquids in which one can ignore any value.

With these results, we can recommend ANN as a suitable predicting method for densities of ternary mixtures of ethanol + water + ionic liquid against MLR procedures. In addition, ANNs can predict accurate values for physical properties of these ternary mixtures.

ACKNOWLEDGMENTS

Astray G. thanks Xunta de Galicia, Consellería de Cultura, Educación e Ordenación Universitaria, for the postdoctoral grant (Plan I2C), P.P.0000 421S 14008 and CIA Project P.P. K224 131H 6450211 for financial support. J. Morales, M.A. Iglesias-Otero, and O.A. Moldes thank the University of Vigo for their predoctoral fellowship, which supported this research. Antonio Cid acknowledges the postdoctoral grant SFRH/BPD/78849/2011, given by the Portuguese Foundation for Science and Technology (Fundação para a ciência e tecnologia (FCT)) from Portugal.

REFERENCES

[1] Marsh, K.N.; Boxall, J.A.; Lichtenthaler, R. *Fluid Phase Equilib.*, 2008, **219**, 93–98.

[2] Calvar, N.; Gómez, E.; González, B.; Domínguez, A. *J. Chem. Eng. Data*, 2007, **52**, 2529–2535.

[3] Calvar, N.; González, B.; Domínguez, A.; Tojo, J. *J. Solut. Chem.*, 2006, **35**, 1217–1225.

[4] Sirieix-Plénet, J.; Gaillon, L.; Letellier, P. *Talanta*, 2004, **63**, 979–986.

[5] Heintz, A.. *J. Chem. Thermodyn.*, 2005, **37**, 525–529.

[6] Gómez, E.; González, B.; Domínguez, A.; Tojo, E.; Tojo, J. *J. Chem. Eng. Data*, 2006, **51**, 696–701.

[7] Gómez, E.; Calvar, N.; Domínguez, I.; Domínguez, A. *Phys. Chem. Liq.*, 2006, **44**, 409–417.

[8] González, B.; Calvar, N.; Gómez, E.; Domínguez, A. *J. Chem. Thermodyn.*, 2008, **40**, 1274–1281.

[9] Anthony, J.L.; Maginn, E.J.; Brennecke, J.F. *J. Phys. Chem. B*, 2001, **105**, 10942–10949.

[10] McFarlane, J.; Ridenour, W.B.; Luo, H.; Hunt, R.D.; Depaoli, D.W.; Ren, R.X. *Sep. Sci. Technol.*, 2005, **40**, 1245–1265.

[11] Wong, D.S.H.; Chen, J.P.; Chang, J.M.; Chou, C.H. *Fluid Phase Equilib.*, 2002, **194–197**, 1089–1095.

[12] Astray, G.; Gálvez, J.F.; Mejuto, J.C.; Moldes, O.A.; Montoya, I. *J. Comput. Chem.*, 2013, **34**, 355–359.

[13] Astray, G.; Caderno, P.V.; Ferreiro-Lage, J.A.; Galvez, J.F.; Mejuto, J.C. *J. Chem. Eng. Data*, 2010, **55**, 3542–3547.

[14] Karimi, H.; Yousefi, F. *Fluid Phase Equilib.*, 2012, **336**, 79–83.

[15] Lashkarbolooki, M.; Hezave, A.Z.; Babapoor, A. *Korean J. Chem. Eng.*, 2013, **30**(1) 213–220.

[16] Cid, A.; Astray, G.; Manso, J.A.; Mejuto, J.C.; Moldes, O.A. *Tenside. Surfactants. Deterg.*, 2011, **48**, 477–483.

[17] Gharagheizi, F.; Eslamimanesh, A.; Mohammadi, A.H.; Richon, D. *J. Chem. Eng. Data*, 2011, **56**, 1741–1750.

[18] Aksoya, F.; Yabanovab, I.; Bayrakçekena, H. *Indian J. Chem. Technol.*, 2011, **18**, 227–233.

[19] Habibi-Yangjeh, A.; Nooshyar, M. *Phys. Chem. Liq.*, 2005, **43**, 239–247.

[20] Lashkarbolooki, M.; Hezave, A.Z.; Ayatollahi, S. *Fluid Phase Equilib.*, 2012, **324**, 102–107.

[21] Valderrama, J.O.; Martinez, G.; Faúndez, C.A. *Int. J. Thermophys.*, 2011, **32**, 942–956.

[22] Nami, F.; Deyhimi, F. *J. Chem. Thermodyn.*, 2011, **43**, 22–27.

[23] Torrecilla, J.S.; Rodríguez, F.; Bravo, J.L.; Rothenberg, G.; Seddon, K.R.; López-Martin, I. *Phys. Chem. Chem. Phys.*, 2008, **10**, 5826–5831.

[24] Hezave, A.Z.; Lashkarbolooki, M.; Raeissi, S. *Fluid Phase Equilib.*, 2012, **314**, 128–133.

[25] Lashkarbolooki, M.; Hezave, A.Z.; Al-Ajmi, A.M.; Ayatollahi, S. *Fluid Phase Equilib.*, 2012, **326**, 15–20.

[26] Gómez, E.; González, B.; Calvar, N.; Domínguez, A. *J. Chem. Thermodyn.*, 2008, **40**, 1208–1216.

[27] Astray, G.; Iglesias-Otero, M.A.; Morales, J.; Mejuto, J.C. *Mediterr. J. Chem.*, 2012, **2**, 388–400.

[28] ChemExper, http://www.chemexper.com, Belgium (accessed on March 16, 2015).

[29] ChemicalBook Inc., http://www.chemicalbook.com, People's Republic of China (accessed on March 16, 2015).

[30] chemBlink Inc., http://www.chemblink.com, United States of America (accessed on March 16, 2015).

CHAPTER 22

Effect of Ionic Liquids on Catalytic Properties and Structure of Biocatalysts

MARIA H. KATSOURA, ATHENA A. PAPADOPOULOU, ANGELIKI C. POLYDERA, and HARALAMPOS STAMATIS

Laboratory of Biotechnology, Department of Biological Applications and Technologies, School of Health Sciences, University of Ioannina, University Campus, Ioannina, Greece

22.1 INTRODUCTION

Ionic liquids (ILs) are salts in a liquid state below 100°C. ILs present numerous unique physicochemical properties that distinguish them from molecular solvents. Their negligible vapor pressure, nonflammability, and high thermal stability have rendered ILs as a "green" alternative to harmful volatile organic solvents. On account of their environmental friendliness together with the possibility of fine-tuning of their properties, based on the appropriate choice of ion composition, ILs have required much interest over the last decade.

The research efforts of Wilkes and Zaworotko [1] generated a class of promising ILs with halides or weakly coordinating anions (BF_4^- and PF_6^-). In the years to follow, a wide range of cationic moieties in combination with various anions have been reported. These so-called second-generation ILs have been extensively studied, and a significant amount of information regarding their properties, such as viscosity, density, melting point, polarity, miscibility with solvents and gases, thermodynamic data, toxicity data, etc., is available today. The knowledge of their properties has made second-generation ILs attractive for several applications (e.g., electrolytes, thermal fluids, lubricants, solvents for separation, and solvents for organic synthesis) [2].

One particular field in which ILs have been widely applied is that of biocatalysis (Fig. 22.1). The reasons for the use of ILs as reaction media for biocatalytic reactions lie behind the need to replace polar organic solvents with media of similar polarity that are able to dissolve various substrates without inactivating enzymes. Works employing various enzymes have demonstrated the advantages of these media, in

Ionic Liquid-Based Surfactant Science: Formulation, Characterization, and Applications, First Edition. Edited by Bidyut K. Paul and Satya P. Moulik.

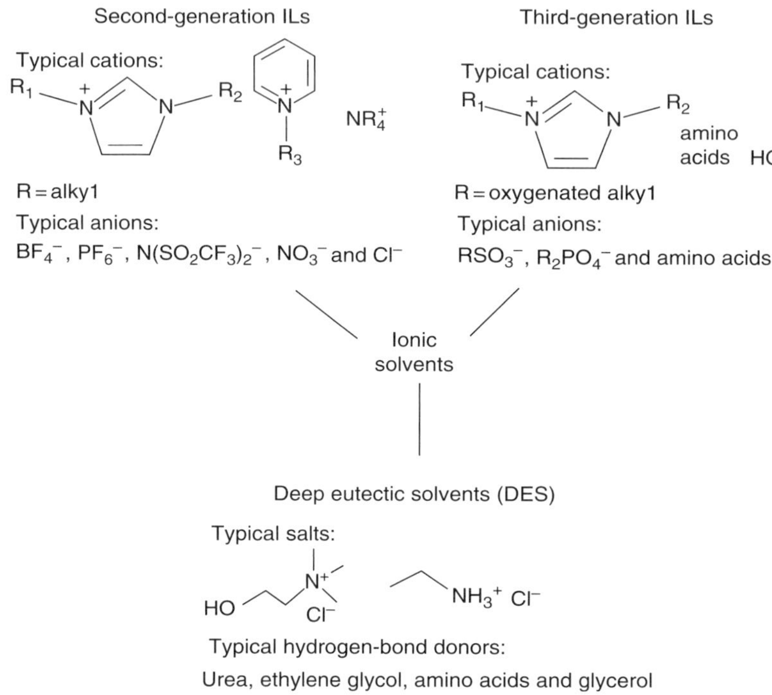

Figure 22.1 Typical structures of ionic solvents employed as reaction media for biocatalytic reactions.

terms of enzymatic activity, selectivity, and stability [3–6]. However, despite the remarkable results obtained with second-generation ILs, several reports have recently pointed out their disadvantages such as high cost, purity, and toxicity. The ILs are at least 10-fold costlier than many organic solvents. Their purification is of vital importance, and even trace amounts of impurities can affect the physical properties of ILs [7]. Moreover, the synthesis of second-generation ILs cannot be considered environmentally friendly, and at the same time, there is growing evidence that these media are relatively toxic and poorly biodegradable [8, 9].

In order to overcome these drawbacks, the third-generation ILs have recently emerged. They comprise readily available, biodegradable, and lower toxic components (sugars, amino acids, organic acids, alkylsulfates, alkylphosphates, hydroxyl ammoniums, etc.) [7, 10]. Another class of third-generation ionic solvent systems is the deep eutectic solvents (DES). DES contain quaternary ammonium salts and uncharged hydrogen donors, and they are not entirely ionic [11]. Examples of enzymatic reactions performed in third-generation ionic solvents are still scarce, but interest in these promising novel media is steadily growing.

It is expected that ILs can have actual applications in the years to come. In the following sections, the role of selected ILs on the catalytic *behavior* (activity, selectivity, and stability) and the structure of several enzymes will be presented and discussed.

22.2 EFFECT OF ILs ON CATALYTIC PROPERTIES OF ENZYMES

ILs present a variety of structures comprising a diverse range of anions and cations. Depending on their ion composition, their physicochemical properties vary significantly, thus exerting different effects on the catalytic properties (activity, selectivity, and stability) of enzymes. The majority of data refer to the well-studied second-generation ILs that will be reviewed in the following sections.

22.2.1 Factors Affecting Enzyme Activity and Selectivity

Several biocatalytic reactions have been investigated in ILs or IL-based systems. From these studies, a significant amount of information has been accumulated regarding the physicochemical properties of ILs (viscosity, polarity, hydrophobicity, nucleophilicity, H-bond basicity, and kosmotropicity/chaotropicity) that affect the activity and selectivity of enzymes. However, one should keep in mind that physical and chemical characteristics of ILs can be affected by several factors, for instance, by the presence of impurities (halide ions, acids, residual solvents, etc.) [12], which can further affect the enzyme activity and ultimately the performance of biocatalytic transformations [13–15].

22.2.1.1 Viscosity ILs are much more viscous than conventional organic solvents. Their viscosity, which is governed by H-bonding structural and van der Waals interactions, ranges from 10 to more than 1000 cP at room temperature [3, 16]. For enzyme-catalyzed reactions in ILs, viscosity is considered to be a key parameter, since it can lead to handling difficulties and most importantly affect enzyme activity through mass transfer limitations. In many cases, a decrease of initial rates of several biocatalytic reactions was observed in pure ILs or IL-containing mixtures compared to organic solvents, and it was attributed to the increased viscosity of the medium [14, 17–23].

The viscosity of ILs depends on their ion composition. As a general trend, the viscosity increases with the increase of alkyl chain length of the substituents on the cation and depends on the symmetry and size of the anion [3, 24, 25]. The increase of viscosity seems to be directly correlated to the decrease of enzyme activity observed by different groups. For instance, as far as the cation is concerned, Lozano et al. [26] reported that a higher α-chymotrypsin activity was observed in the less viscous [emim]NTf_2 than in the more viscous [mtoa]NTf_2. Similarly, Chen et al. [27] obtained a higher degree of substitution of konjac glucomannan in the less viscous, water-miscible [emim] and [bmim] tetrafluoroborates, as compared to the more viscous [omim]BF_4. Moreover, Kim and Koo [28] observed a decrease in the reaction rate of CaLB-catalyzed butyl acetate synthesis in the order of increasing viscosity [bmim]TfO < [hmim]TfO < [omim]TfO of the medium. With regard to the anion, it has been reported that the increase of viscosity from [bmim]NTf_2 to [bmim]PF_6 was responsible for the decrease of the activity of various lipases [20] and penicillin G acylase [29]. Moreover, Novozym 435 activity was lower in the more viscous PF_6^--compared to the BF_4^--containing imidazolium ILs of lower viscosity [27].

The usual methods employed in order to lower the viscosity of ILs are the increase of reaction temperature [27, 30–33] and the addition of an organic solvent [34, 35], both of which positively affect enzyme activity. Often, the increase of agitation speed has the same results as it limits the mass transfer phenomena [36–38]. Finally, the presence of water seems to affect IL viscosity, since dried ILs usually are more viscous [18, 29, 39].

22.2.1.2 *Polarity and Hydrophobicity* The polarity of several ILs has been studied, using solvatochromic empirical scales, and ILs have been described as solvents with polarities comparable to those of lower alcohols (methanol, ethanol, etc.) [24, 40]. In general, the polarity of ILs depends on their ion composition and increases with the decrease of both the chain length of the substituents of the cation and the size of the anion [40]. Medium polarity can affect the active conformation of an enzyme and therefore, indirectly, its activity. Although polar organic solvents inactivate enzymes, ILs, surprisingly, do not. This fact enables enzymatic reactions with various substrates to be performed in a polarity range previously inaccessible, and in many cases, enhanced reactivity and selectivity have been observed, explained by the increased solubility of substrates and/or a more favorable structure of enzymes in polar ILs [3, 41–44]. Comparing different ILs, there have been a few reports that correlate the increase of polarity with an improvement of enzyme activity [31, 41, 45–48]. However, many other studies have failed to establish any correlation between enzyme activity or selectivity and IL polarity [15, 17, 27, 38, 41, 49, 50].

The hydrophobicity of ILs is another parameter often discussed in order to explain the effect of these media on enzyme activity. Although hydrophobicity is considered a narrower concept of polarity, the terms must be distinguished. Hydrophobicity often refers to the miscibility of the IL with water, which can vary widely depending on the ion composition of the IL [25]. Several groups have reported that various hydrolytic enzymes present higher activity and selectivity or, in some cases, they are active only in hydrophobic PF_6^-- or NTf_2^--containing ILs, as compared to water-miscible ILs (comprising BF_4^-, dca^-, NO_3^-, CH_3COO^-, and CF_3COO^- anions) [29, 49, 51–57]. In some cases, a general trend with respect to the cation was also established, toward higher enzyme activity and selectivity as the cation size, and thus the hydrophobicity of ILs, increased [50, 53, 58, 59]. However, other works involving the use of imidazolium and pyridinium ILs with varying cation chain lengths and different anions suggest a bell-shaped pattern, where lipase activity increases up to a maximum and then decreases with further increase of the hydrophobicity of ILs [15, 60]. The positive effect of the increase of hydrophobicity is, in all cases, explained by a reduced tendency of hydrophobic ILs to remove water from the vicinity of enzymes, thus leading to sufficient hydration and enhancement in enzyme activity [61]. On the other hand, a high hydrophobicity of the IL could also lead to specific interactions with the substrates (such as a thermodynamic ground-state stabilization) or to interactions with the enzyme through electrostatic forces, resulting in poor enzymatic performance [10, 15, 62].

Taking into consideration a number of other studies reporting higher enzyme activities in hydrophilic ILs [41, 63–65] or even no correlation between IL hydrophobicity and enzyme activity [66], it seems elusive to extract general trends with respect to hydrophobicity alone, and multiple other factors must be considered in order to explain enzyme activity and selectivity in these media.

22.2.1.3 *Ion-Specific Effects* Several studies have pointed out that enzymatic performance in ILs often depends strongly on the nature of anion, rather than the cation. Sheldon's group initially, and other groups later, showed that ILs with strongly coordinating anions, such as nitrate, trifluoromethanesulfonate, acetate, and lactate, dissolve enzymes leading to a (usually reversible upon dissolution in water) loss of activity, probably as a result of conformational changes in the enzyme induced by the coordination of these anions to the enzyme surface [48, 49, 62, 67]. Kim and Koo [28] studying the CaLB-catalyzed synthesis of butyl acetate in TfO^--containing imidazolium ILs offered further insight suggesting that the nature of cation also plays a significant role, since strongly coordinating anions probably cannot be offset by small weakly coordinating cations (such as $[emim]^+$), while bigger cations (such as $[bmim]^+$ or $[omim]^+$) can efficiently counteract them, thus retaining enzyme activity.

Other concepts often considered to explain the different enzyme activities in ILs are nucleophilicity and H-bond basicity. Strongly coordinating anions are also more nucleophilic than weakly coordinating ones [49, 67]. The nucleophilicity of ILs is strictly anion dependent [68], and its decrease can be correlated with an increase of enzyme activity in ILs, since less nucleophilic anions would show lower tendency to change the enzyme's conformation by interacting with the positively charged sites in the enzyme structure [38, 47, 69–71]. Even though this reverse correlation can explain the low enzyme activity in ILs with nucleophilic halide, nitrate, acetate, and other anions, in many other cases, when widely employed BF_4^--, PF_6^--, and NTf_2^--containing ILs were used, no correlations between nucleophilicity and activity could be established. For instance, CaLB was shown by different authors to have higher activity in ILs based on NTf_2^- compared to ILs based on PF_6^- or BF_4^-, although the nucleophilicity order is $PF_6^- < BF_4^- < NTf_2^-$ (considering the polarity of B–F, P–F, and C–F bonds) [15, 19, 72]. For ILs with these common anions, the order of decreasing H-bond basicity $BF_4^- > PF_6^- > NTf_2^-$ (considering the spreading of negative charge over fluorine atoms) is more consistent with the positive effect on enzyme activities [15, 19, 71, 73]. This order represents the strength of an anion to interact with solvated cations through ionic attraction or could represent the strength of interactions between the anion and the changed surface of enzymes [15]. The H-bond basicity, like anion nucleophilicity, can also explain the loss of activity in enzyme-incompatible ILs, since they present increased interference with the internal hydrogen bonds of an enzyme leading to its inactivation [14, 47, 74]. However, the exact mechanism of this ion–protein interaction is not fully understood. For instance, even though TfO^--based ILs present higher H-bond basicity than that of PF_6^--, BF_4^--, and NTf_2^--based ILs, they lead to high lipase activity [56, 74–77], pointing to the fact that multiple other factors can affect enzyme activity. For example, it has been reported that different anions can interact with substrates and/or products through the formation of H bonds, thus changing their solubility or their interaction with the active site of enzymes, which in turn, indirectly, can affect enzyme activities and selectivities [22, 23, 27, 33, 34, 78, 79].

Ion-specific effects on enzyme activity could be explained through their kosmotropicity (Hofmeister series) [80]. The Hofmeister series describes the strength of interaction of ions with water. Kosmotropic ions interact strongly with water and tend to produce ordered water structures, while chaotropic ions interact less with water than water itself. The ability of ions to modify the water structure influences the protein hydration environment, and combined with direct interactions between ions and proteins, the protein structure can be affected. The suggested hypothesis is that kosmotropic anions and chaotropic cations stabilize enzymes, while chaotropic anions and kosmotropic cations destabilize them [80]. In a number of reports, enzyme behavior in aqueous IL solutions corresponds well to the Hofmeister series [58, 80–84]. However, depending on the enzyme and mainly on the ion composition of the IL, the Hofmeister series has been reversed or certain irregularities were observed. For instance, only when the number of carbon atoms of the alkyl chain of cation was less than 10 and the IL concentrations were less than the optimum values did the influence of ILs on enzyme activity show a Hofmeister effect [71, 79, 85]. Moreover, kosmotropicity effect has only been observed in aqueous ILs or some concentrated ILs, but enzyme activity does not seem to follow the Hofmeister series when studies are conducted in hydrophobic, anhydrous, or highly concentrated hydrophilic ILs [58, 71, 86–88].

22.2.2 Factors Affecting Enzyme Stability

The stability of enzymes in ILs is a crucial factor to be considered in view of the practical applications of ILs as reaction media for biotransformations. Several groups have studied enzyme stability in ILs. In some ILs, higher stability or even activation of enzymes has been reported, compared to that reported in water or organic solvents [17, 26, 27, 31, 45, 49, 50, 55, 84, 89–95], while in others a substantial loss of activity has been observed [49, 69, 84, 96–98].

The stabilizing or destabilizing effects of ILs on enzymes have been shown to depend on several parameters. Initial studies explained the high enzyme stability observed in ILs through the existence of electrostatic interactions between the IL and the protein, leading to a more rigid protein, which results in a high kinetic barrier for the unfolding of the enzyme [17, 27, 55, 89, 93, 95, 99]. More recent studies point out that the strength of electrostatic interactions is the crucial factor for the stabilization or destabilization of enzymes in ILs. Klähn et al. [62] have shown, through molecular dynamics (MD) studies, that enzyme–IL interactions are primarily determined by the Coulomb interactions of the anions. Their study reveals that among eight different ILs, those containing NO_3^- denature CaLB, due to strong electrostatic interactions of nitrate anion with the protein. The denaturing effect becomes more intense when the ILs contain smaller cations, since, in this case, the number of anions interacting with the enzyme surface increases. Nordwald and Kaar [100] offered further insight, succeeding the stabilization of enzymes in ILs through modifications that reduce the ratio of positive-to-negative surface charges of the enzymes, thus inhibiting the anion binding with enzymes.

Hydrogen bonding is another concept often employed to explain stabilization effects. It has been pointed out that IL–protein interaction should not be too strong because it can lead to the dissociation of the internal hydrogen bonds, which maintain the secondary structure integrity, and cause the protein to unfold [69]. The role of anion appears again to be an important factor in these effects. According to MD studies, ion–ion and ion–enzyme hydrogen bonds in ILs are dominated by electrostatic interactions [62]. Spreading the negative charge of the anion over multiple atoms weakens its ability to hydrogen bond with the protein, making it less likely that the IL will denature the enzyme [101]. For this reason, high enzyme stability has been reported in NTf_2^--, PF_6^--, and dhp^--containing ILs [26, 27, 31, 45, 49, 50, 53, 55, 70, 89–94, 97, 102–104], while NO_3^--, $MeSO_4^-$-, and $EtSO_4^-$-containing ILs, lactate ILs [47, 62, 79, 83, 94, 105], and, in some cases, BF_4^--containing ILs destabilize enzymes [49, 50, 53, 55, 69, 89, 97]. In some cases, both the H-bond accepting and the

H-bond donating properties of the anion and cation, respectively, seem to affect enzyme stability in ILs. In the case of CaLB, for example, the H-bond activity of the Et_3MeN^+ cation was shown to exert a stabilizing effect, even in the presence of the mildly H-bond accepting $MeSO_4^-$ anion [69]. According to Lau and coworkers, a balance of mild hydrogen-bond accepting and donating properties is required to maintain enzyme activity in ILs [69].

ILs with NTf_2^- or PF_6^- anions, besides their mild hydrogen-bond accepting properties, are also hydrophobic in nature. Hydrophobic ILs are superior to hydrophilic ones, in terms of stability, since they tend to preserve the essential water layer around the protein, thus maintaining the catalytically active conformation of enzymes [26, 27, 31, 45, 49, 53, 55, 89, 90, 102] or reducing direct protein–ion interactions [53, 55, 105]. It has been specifically pointed out, in the case of water/[bmim]PF_6 microemulsions, that a small amount of the IL could be soluble in the aqueous microphase, thus exerting a protective effect on the enzyme microenvironment [106]. On the other hand, MD studies reveal a significant reduction of the size of water clusters at the protein surface and their replacement by the IL in media (such as [bmim]Cl or [bmim]NO_3) may result a destabilization effects [104, 105].

However, the anion effect on enzyme stability presents several variations, depending strongly on the water content of the medium. At high water contents, conformational changes seem to occur [105] and the flexibility of proteins has been reported to increase [104], justifying the lower stability even in hydrophobic ILs. The stabilizing effects of NTf_2^--, PF_6^--, or dhp-containing ILs are, in most cases, observed to make nearly anhydrous conditions at low water contents (up to 20–25%) or low water activities [26, 45, 70, 90, 91, 102–105]. The beneficial effect of low water contents is also evident in microheterogeneous systems. Pavlidis et al. [106] observed higher lipase stability in water/[bmim]PF_6 microemulsions with low water contents, related to the higher rigidity of protein molecules.

In the presence of water, stabilizing effects can also be explained in view of ion kosmotropicity. ILs dissociate into individual ions in aqueous systems. In these systems, the combination of a chaotropic cation and a strong kosmotropic anion has been reported to have a positive effect on enzyme stability. PF_6^-, BF_4^-, and NO_3^- are all chaotropic anions, Cl^- and Br^- are borderline ions, while $MeSO_3$, $MeSO_4^-$, $EtSO_4^-$, CH_3COO^-, and CF_3COO^- are kosmotropic. Following the Hofmeister series, the stability of protease P6 was higher in CH_3COO^-- and CF_3COO^-- compared to BF_4^--containing ILs [107], glucose dehydrogenase half times were higher in $EtSO_4^-$- than in PF_6^-- and BF_4^--containing ILs [108], and feruloyl esterases were also rapidly inactivated in BF_4^--based systems [103]. However, there are several cases where the Hofmeister series could not explain the observed results or that it has even been reversed. For example, horseradish peroxidase presented good thermal stability in water-containing [bmim]BF_4 systems [99, 109]; papain stability was improved in [$C_n$mim]BF_4 (n=2–6) compared to ILs containing more kosmotropic anions (like Cl^- or CH_3COO^-) [47]; tyrosinase stability in aqueous solutions varied in the order [bmim]PF_6 > [bmim]BF_4 > [bmim]$MeSO_4$, even though $MeSO_4^-$ is more kosmotropic than PF_6^- or BF_4^- [79]; and protease MS-7 was stable in the presence of 50% of [hmim]BF_4 but completely lost its activity in the presence of 50% of [bmim]Br [97]. Several reasons have been proposed to explain these exceptions. For instance, the influence of cations has been pointed out, in the sense that the stabilizing kosmotropic behavior of anions could be lessened in the presence of kosmotropic cations, like $[bmim]^+$ or other imidazolium cations with large alkyl chains, which tend to ion pair with the kosmotropic counterions, thus reducing their abundance in the aqueous bulk solution [79, 82]. Another important factor to be considered is whether ILs can totally dissociate into ions in aqueous solutions. It is believed that hydrophobic ILs, which have low solubility in water, present limited ion dissociation [58, 82]. This fact can explain why various enzymes present high stability in PF_6^--containing ILs, despite its highly chaotropic nature.

From a different point of view, "wet" hydrophobic ILs have been described as nanostructured materials, where cations and anions are connected together with hydrogen bonds, forming an extended network of polar and nonpolar regions [110]. In concentrated aqueous solutions of ILs, these organized structures can be maintained, and under these conditions, Hofmeister series concept could not be properly applied [82]. Moreover, enzyme molecules have been considered to be "included" into these IL networks, thus maintaining their native structures and the essential water molecules. In a sense, these ILs could be considered both as solvents and liquid immobilization supports in which high enzyme stability can be achieved [45, 70, 90, 110, 111].

22.3 EFFECT OF ILs ON ENZYME STRUCTURE

Conformational changes of the structural integrity of the active site and binding pocket(s) of enzymes could significantly affect their stability, activity, and selectivity. Therefore, before the design and the development of effective biocatalytic processes in ILs, it is important to identify what happens to the structure of the enzyme in these media by analyzing its internal structural changes and the dynamics of the substrate binding pockets and active sites. Detailed understanding of IL effect on the structure and dynamics of enzymes was a subject of interest of several research groups.

Structural information on proteins can be obtained with X-ray crystallography and nuclear magnetic resonance (NMR) spectroscopy; however, due to various limitations intrinsic to these high-resolution structural methods, only alternate spectroscopic techniques (especially fluorescence, circular dichroism (CD), Raman, and infrared spectroscopic techniques) can be employed for conformational characterization of proteins as well as stabilization and denaturation of enzymes.

CD is one of the most widely used techniques for determining the structure of proteins. The far-UV CD spectra mainly reflect the secondary structure of a protein. The near-UV CD spectra, derived from the aromatic residues of protein and its disulfide bonds in the asymmetric environment, are often used to reveal the tertiary structure.

On the other hand, Fourier transform infrared (FTIR) spectroscopy is a well-established technique for analysis of the secondary structure of proteins in water, as well as in organic and IL media. Two regions of the IR spectrum, called amide I ($1600–1700\,cm^{-1}$) and amide III ($1215–1335\,cm^{-1}$), have been used to study the individual elements of secondary structure and their changes. The amide I mode of the peptide bond is particularly relevant for protein analysis since it is conformationally sensitive. Dynamic structure–function relationships in enzyme stabilization were investigated by several research groups as summarized in Table 22.1.

TABLE 22.1 Studies of Dynamic Structure–Function Relationships in Enzyme Stabilization

Enzyme	Ionic Liquid	Technique	Observation	Reference
Cyt c	ILs with dicyanamide, saccharinate, and dihydrogen phosphate anions	FTIR	Retention of secondary structure to extreme temperatures. The nature of the anion appears to be the important factor	[112]
CaLB	ILs with alkylsulfate, nitrate, and lactate anions	FTIR	Changes of the secondary structure depend on the H-bond accepting and donating properties of anions as well as on ions size	[69, 113]
Papain	ILs with $[C_n mim]^+$ and various anions	FTIR and fluorescence	The nature of the cation and especially the anion affects the structure of the protein	[47]
CrL	Various ILs	FTIR	Changes in secondary structure of enzyme	[114]
Trc	[bmim]Cl	Fluorescence	IL leads to unfolded and inactive enzyme	[115]
TsLac	$[bmim]^+/[bmpyr]^+/[tma]^+$ with TfO^-	Fluorescence and CD	Effect of ILs on enzyme structure was associated with the chaotropicity of their cations	[84]
Cbd	$[mmim]Me_2PO_4$, $[bmim]CH_3COO$, $[mmim]CH_3HPO_2(OCH_3)$	Fluorescence and UV–Vis	ILs induce strong structural changes	[116]
β-Gal Tt	$[bmim]/[omim]PF_6$, [cpma]MS, $[mtoa]NTf_2$	Fluorescence and CD	Changes on the enzyme secondary and tertiary structure	[117]

Several groups have employed scattering techniques, such as small-angle X-ray scattering (SAXS) and small-angle neutron scattering (SANS), as well as dynamic light scattering (DLS), to investigate whether the enzyme is undergoing aggregation or large-scale denaturation in ionic solvents [118]. For instance, the structure of *Candida antarctica* lipase B in various imidazolium-based ILs ($[emim]EtOSO_3$, $[emim]NO_3$, and $[emim]N(CN)_2$) was investigated using SANS and DLS [119]. This study indicated that some ILs caused formation of large aggregates. The average size of the CaLB aggregates increased with hydrogen-bonding potential of the ILs' anion, and this was associated with loss of enzymatic activity.

Using computer simulations, several groups investigated the effect of ionic solvents on catalytic, structural, and dynamic properties of enzymes in these media. MD simulations, in particular, are a valuable tool to gain molecular scale insights of the interactions between various solvents and biomolecules [120, 121].

Micaêlo and Soares [105] reported for the first time a molecular simulation study of the serine protease cutinase from *Fusarium solani* in two ILs, [bmim]PF_6 and [bmim]NO_3. It was found that the enzyme structure is highly dependent on the amount of water present in the IL media and that the protein seems to display structural properties following a bell-shape-like behavior relative to the amount of water present in the system. ILs can interact with protein molecules, and this interaction is governed by the affinity of anions toward the protein main chain.

In a similar manner, Latif et al. [104] utilized MD simulations in order to gain insight into the structural properties and dynamics of α-chymotrypsin in imidazolium-based ILs with different types of anions ([bmim]PF_6, [bmim]BF_4, [bmim]Cl, [bmim]TfO, and [bmim]NTf_2). At a low water content, the conformation of enzyme was closer to its native structure in the presence of ILs, presenting also a bell-shaped dependency on water content. However, no major conformational changes at the active site of the enzyme were observed. The solvation of the enzyme in water-immiscible ILs led to a higher enzyme flexibility at increased water content.

Klähn et al. [122], using MD simulations, monitored the folding of CaLB in various ILs at high temperatures. They found that the interaction between the protein and ILs is dominated by electrostatic interactions with anions. The same group found that strong electrostatic and van der Waals interactions between ILs and the protein core disrupt its structure [62].

The effect of two ILs, [amim]Cl and [omim]Cl, on the structure of adenosine deaminase (ADA) was investigated by spectroscopic techniques, and the results were compared with docking and MD simulation studies [123]. The docking results have manifested that the presence of ILs can trigger enzyme deactivation acting as competitive or noncompetitive inhibitors. MD studies indicated that ILs reduce intermolecular hydrogen bonds and unfold enzyme structure depending on their hydrophobic tail.

Recently, Burney and Pfaendtner [120] reported results from MD simulations of *Candida rugosa* lipase in two ILs ([bmim]PF_6 and [bmim]NO_3). The analysis showed a strong interaction between the protein surface and IL anions due to a broader spatial distribution of positively charged protein residues and reduced mobility of the cation due to the size of the imidazolium ring.

MD simulations of xylanase II from *Trichoderma longibrachiatum* in mixtures of ILs [emim]OAc or [emim]$EtSO_4$ and water showed that the enzyme solvated in higher concentrations of ILs generally remains more stable than when solvated in water on the timescales studied. The study indicated that [emim] cation binds strongly in the binding pocket of the enzyme, acting as a competitive inhibitor [118].

22.4 BIOCATALYSIS IN NOVEL IONIC SOLVENTS

Over the last decade, due to economic and ecological requirements, a considerable amount of attention has been focused on the development of a novel class of ILs composed of natural compounds, known as third-generation ILs. These media, also called advanced ILs, are based on biodegradable, readily available, and behavior ions, such as sugars, amino or organic acids, natural bases, sugars, and naturally occurring carboxylic acids [7,10,11]. Only a few examples concerning the catalytic behavior of enzymes in these media have been reported.

In particular, Fujita and coworkers [92, 112, 124, 125] evaluated the effect of dhp-containing ILs on the activity and stability of metalloproteins. They showed that cytochrome c, horseradish peroxidase, and ascorbate oxidase retain their activity in hydrated choline dhp ILs. Moreover, structural studies (ATR-FTIR, Raman, and CD) have shown that the surroundings of the active site and the secondary structure of proteins dissolved in these media are similar to those in buffer solution. The positive effect of third-generation ILs on enzyme activity and stability was also confirmed for hydrolases, such as lipase, protease, lysozyme, and α-chymotrypsin [78, 126–129]. Recently, SAXS studies of lysozyme in choline dihydrogen phosphate IL have indicated that charge effects on the surface of the protein play a role in protein stability in this medium [130].

In addition to third-generation ILs, a new class of ionic solvent systems, called DES, has been developed. DES are typically formed by mixing a quaternary ammonium salt, such as choline chloride, and an uncharged hydrogen-bond donor, such as urea, carboxylic acids (e.g., oxalic or amino acids), or polyols (e.g., glycerol) [11, 131]. Although DES are not entirely composed of ionic species, their physicochemical properties are very close to those of common ILs [132].

In 2008, Kazlauskas and coworkers were the first to report that several hydrolases and epoxide hydrolases presented high activities in various DES based on choline chloride or ethylammonium chloride paired with hydrogen-bond donors such as acetamide, ethylene glycol, glycerol, urea, and malonic acid [133]. Since then, several enzyme-catalyzed processes using DES as solvents have been reported. In 2010, Lindberg et al. [134] used DES as cosolvents in aqueous solutions for epoxide hydrolase. It was pointed out that the appropriate selection of the cosolvent can improve the reaction yield and regioselectivity. Later, in 2011, Zhao and coworkers [135] studied lipase-catalyzed transesterification reactions in choline acetate- and glycerol-based DES. It was shown that these media have advantageous properties, such as low viscosity and excellent biocompatibility with lipases, compared to choline chloride–urea media. The same group demonstrated that cross-linked proteases can be successfully used in choline acetate and choline chloride paired with glycerol DES, showing high activity and stability as compared to *t*-butanol. Although acetate and chloride anions have been shown to denature enzymes, their detrimental effects can be lowered in DES, due to the strong hydrogen bonds that are formed between those anions and glycerol [136].

Recently, Durand et al. [137] evaluated the advantages and limitations of various DES in lipase-catalyzed reactions. Immobilized lipase from *C. antarctica* was highly stable in choline chloride-based DES containing urea or glycerol. However, this study showed that some DES can compete with the substrates and react, leading to by-product formation. This side reaction can alter the physicochemical properties of the medium, such as viscosity, and thus the performance of the reaction.

More recently, choline-based DES have been successfully assessed as reaction media for the chymotrypsin-catalyzed peptide synthesis, leading to high productivity and selectivity through the reduction of the hydrolysis pathway observed in aqueous media [138].

In an attempt to demonstrate the feasibility of biodiesel preparation in DES, Zhao and coworkers in 2013 [139] studied the enzymatic transesterification of soybean oil in choline-based DES by Novozym 435. It was found that under optimal conditions, high triglyceride conversion was achieved in choline chloride–glycerol DES.

22.5 CONCLUSIONS

ILs have obvious potential as reaction media for many enzymatic reactions because of their unusual characteristics, as reported in this chapter. These solvents can be beneficial regarding to activity, selectivity, and stability of enzymes. Several parameters such as polarity, hydrogen-bonding capacity, viscosity, kosmotropicity/chaotropicity, and hydrophobicity were investigated, and various spectroscopic and scattering studies were used in order to explore the structural and conformational dynamics of enzymes in these media and understand how ILs affect the stability and activity of enzymes. The study of individual IL properties, although useful, cannot lead to a unified theory, as in many cases contradictory results have been reported, depending on several parameters such as the presence of water, the enzyme studied, and even in some cases the substrates used. Further information regarding structural and conformational dynamics of proteins could be helpful before choosing or developing an IL to serve as solvent media for enzymatic reaction.

In most cases, ILs that have been used are based on imidazolium, pyridinium, and ammonium cations, associated with several polyatomic anions. Although these salts have beneficial effects on the properties of the enzymes, many reports pointed out the hazardous toxicity and the very poor biodegradability of these ILs. It is expected that the development of green and biodegradable ILs formulated with compounds derived from renewable resources may further stimulate their uses in industrial biocatalytic processes taking into account of both ecological and economic requirements. The researches on the development and application of the third-generation ILs and DES, as media for enzymatic reactions, aim to a future step in this area, to generate greener chemical and biochemical industries.

REFERENCES

[1] Wilkes, J.; Zaworotko, M. *J Chem Soc Chem Commun* 1992, **13**, 965–967.
[2] Patel, D. D.; Lee, J. M. *Chem Rec* 2012, **12**, 329–355.
[3] Park, S.; Kazlauskas, R. J. *Curr Opin Biotechnol* 2003, **14**, 432–437.
[4] Yang, Z.; Pan, W. *Enzyme Microb Technol* 2005, **37**, 19–28.
[5] Roosen, C.; Müller, P.; Greiner, L. *Appl Microbiol Biotechnol* 2008, **81**, 607–614.
[6] Ha, S. H.; Koo, Y. M. *Korean J Chem Eng* 2011, **28**, 2095–2101.
[7] Gorke, J.; Srienc, F.; Kazlauskas, R. *Biotechnol Bioprocess Eng* 2010, **15**, 40–53.
[8] Romero, A.; Santos, A.; Tojo, J.; Rodríguez, A. *J Hazard Mater* 2008, **151**, 268–273.
[9] Pham, T. P. T.; Cho, C. W.; Yun, Y. S. *Water Res* 2010, **44**, 352–372.
[10] Tavares, A. P. M.; Rodríguez, O.; Macedo, E. A. *Ionic Liquids—New Aspects for the Future*; INTECH, Rijeka, Croatia, 2013, p. 537.
[11] Domínguez De María, P.; Maugeri, Z. *Curr Opin Chem Biol* 2011, **15**, 220–225.
[12] Freemantle, M., Properties of Ionic Liquids, ed., *An Introduction to Ionic Liquids*; The Royal Society of Chemistry, London, UK, 2010, p 31.
[13] Seddon, K. R.; Stark, A.; Torres, M. J. *Pure Appl Chem* 2000, **72**, 2275–2287.
[14] Lee, S. H.; Ha, S. H.; Lee, S. B.; Koo, Y. M. *Biotechnol Lett* 2006, **28**, 1335–1339.
[15] Zhao, H.; Baker, G.; Song, Z.; Olubazo, O.; Zanders, L.; Campbell, S. M. *J Mol Catal B: Enzym* 2009, **57**, 149–157.

[16] Wilkes, J. *J Mol Catal A Chem* 2004, **214**, 11–17.

[17] Lou, W.; Zong, M.; Wu, H. *Biocatal Biotransformation* 2004, **22**, 171–176.

[18] Izák, P.; Mateus, N. M. M.; Afonso, C. A. M.; Crespo, J. G. *Sep Purif Technol* 2005, **41**, 141–145.

[19] Lou, W. Y.; Zong, M. H.; Wu, H.; Xu, R.; Wang, J. F. *Green Chem* 2005, **7**, 500–506.

[20] Mantarosie, L.; Coman, S.; Parvulescu, V. I. *J Mol Catal A Chem* 2008, **279**, 223–229.

[21] Vafiadi, C.; Topakas, E.; Nahmias, V. R.; Faulds, C. B.; Christakopoulos, P. *J Biotechnol* 2009, **139**, 124–129.

[22] Galonde, N.; Nott, K.; Debuigne, A.; Deleu, M.; Jerôme, C.; Paquot, M.; Wathelet, J. P. *J Chem Technol Biotechnol* 2012, **87**, 451–471.

[23] Wang, J.; Sun, G. X.; Yu, L.; Wu, F. A.; Guo, X. J. *Bioresour Technol* 2013, **128**, 156–163.

[24] Bonhôte, P.; Dias, A. P.; Papageorgiou, N.; Kalyanasundaram, K.; Grätzel, M. *Inorg Chem* 1996, **35**, 1168–1178.

[25] Van Rantwijk, F.; Madeira Lau, R.; Sheldon, R. A. *Trends Biotechnol* 2003, **21**, 131–138.

[26] Lozano, P.; De Diego, T.; Guegan, J. P.; Vaultier, M.; Iborra, J. L. *Biotechnol Bioeng* 2001, **75**, 563–569.

[27] Chen, Z. G.; Zong, M. H.; Li, G. J. *J Chem Technol Biotechnol* 2006, **81**, 1225–1231.

[28] Kim, H. S.; Koo, Y. M. *Korean J Chem Eng* 2012, **29**, 1610–1614.

[29] Pereira, S. C.; Bussamara, R.; Marin, G.; Giordano, R. L. C.; Dupont, J.; Giordano, R. C. *Green Chem* 2012, **14**, 3146–3156.

[30] Lozano, P.; De Diego, T.; Carrié, D.; Vaultier, M.; Iborra, J. L. *J Mol Catal A Chem* 2004, **214**, 113–119.

[31] Shan, H.; Li, Z.; Li, M.; Ren, G.; Fang, Y. *J Chem Technol Biotechnol* 2008, **83**, 886–891.

[32] Tian, J.; Wang, Q.; Zhang, Z. *Eur Food Res Technol* 2009, **229**, 357–363.

[33] Yang, Z.; Schultz, L.; Guo, Z.; Jacobsen, C.; Xu, X. *Open Biotechnol J* 2012, **6**, 5–12.

[34] Chen, B.; Guo, Z.; Let, M. B.; Lue, B. M.; Xu, X. *Org Biomol Chem* 2008, **6**, 3196–3201.

[35] Yang, Z.; Huang, Z. L. *Catal Sci Technol* 2012, **2**, 1767–1775.

[36] Gubicza, L.; Nemestóthy, N.; Fráter, T.; Bélafi-Bakó, K. *Green Chem* 2003, **5**, 236–239.

[37] Guo, Z.; Xu, X. *Green Chem* 2006, **8**, 54–62.

[38] Lou, W. Y.; Zong, M. H.; Liu, Y. Y.; Wang, J. F. *J Biotechnol* 2006, **125**, 64–74.

[39] Huddleston, J. G.; Visser, A. E.; Reichert, W. M.; Willauer, H. D.; Broker, G. A.; Rogers, R. D. *Green Chem* 2001, **3**, 156–164.

[40] Carmichael, A. J.; Seddon, K. R. *J Phys Org Chem* 2000, **13**, 591–595.

[41] Park, S.; Kazlauskas, R. J. *J Org Chem* 2001, **66**, 8395–8401.

[42] Kim, M. J.; Choi, M. Y.; Lee, J. K.; Ahn, Y. *J Mol Catal B: Enzym* 2003, **26**, 115–118.

[43] Katsoura, M. H.; Polydera, A. C.; Tsironis, L.; Tselepis, A. D.; Stamatis, H. *J Biotechnol* 2006, **123**, 491–503.

[44] Gervaise, C.; Daniellou, R.; Nugier-Chauvin, C.; Ferrières, V. *Tetrahedron Lett* 2009, **50**, 2083–2085.

[45] Lozano, P.; De Diego, T.; Carrié, D.; Vaultier, M.; Iborra, J. L. *Biotechnol Lett* 2001, **23**, 1529–1533.

[46] Lozano, P.; De Diego, T.; Carrié, D.; Vaultier, M.; Iborra, J. L. *J Mol Catal B: Enzym* 2003, **21**, 9–13.

[47] Lou, W. Y.; Zong, M. H.; Smith, T. J.; Wua, H.; Wang, J. F. *Green Chem* 2006, **8**, 509–512.

[48] Toral, A. R.; de los Ríos, A. P.; Hernández, F. J.; Janssen, M. H. A.; Schoevaart, R.; van Rantwijk, F.; Sheldon, R. A. *Enzyme Microb Technol* 2007, **40**, 1095–1099.

[49] Kaar, J. L.; Jesionowski, A. M.; Berberich, J. A.; Moulton, R.; Russel, A. J. *J Am Chem Soc* 2003, **125**, 4125–4131.

[50] Katsoura, M. H.; Polydera, A. C.; Tsironis, L. D.; Petraki, M. P.; Rajačić, S. K.; Tselepis, A. D.; Stamatis, H. *N Biotechnol* 2009, **26**, 83–91.

[51] Nara, S. J.; Harjani, J. R.; Salunkhe, M. M. *Tetrahedron Lett* 2002, **43**, 2979–2982.

[52] Nakashima, K.; Maruyama, T.; Kamiya, N.; Goto, M. *Org Biomol Chem* 2006, **4**, 3462–3467.

[53] de los Ríos, A. P.; Hernández-Fernández, F. J.; Rubio, M.; Gómez, D.; Víllora, G. *J Chem Technol Biotechnol* 2007, **82**, 190–195.

[54] Shen, Z. L.; Zhou, W. J.; Liu, Y. T.; Ji, S. J.; Loh, T. P. *Green Chem* 2008, **10**, 283–286.

[55] Vidya, P.; Chadha, A. *J Mol Catal B: Enzym* 2009, **57**, 145–148.

[56] Pang, N.; Gu, S. S.; Wang, J.; Cui, H. S.; Wang, F. Q.; Liu, X.; Zhao, X. Y.; Wu, F. A. *Bioresour Technol* 2013, **139**, 337–342.

[57] Papadopoulou, A. A.; Katsoura, M. H.; Chatzikonstantinou, A.; Kyriakou, E.; Polydera, A. C.; Tzakos, A. G.; Stamatis, H. *Bioresour Technol* 2013, **136**, 41–48.

[58] Zhao, H. *J Mol Catal B: Enzym* 2005, **37**, 16–25.

[59] Habulin, M.; Knez, Ž. *J Mol Catal B: Enzym* 2009, **58**, 24–28.

[60] Lou, W. Y.; Zong, M. H. *Chirality* 2006, **18**, 814–821.

[61] Halling, P. J. *Philos Trans R Soc Lond B Biol Sci* 2004, **359**, 1287–1296.

[62] Klähn, M.; Lim, G. S.; Seduraman, A.; Wu, P. *Phys Chem Chem Phys* 2011, **13**, 1649–1662.

[63] Jain, N.; Kumar, A.; Chauhan, S.; Chauhan, S. M. S. *Tetrahedron* 2005, **61**, 1015–1060.

[64] Tavares, A. P. M.; Rodriguez, O.; Macedo, E. A. *Biotechnol Bioeng* 2008, **101**, 201–207.

[65] Rodríguez, O.; Cristóvão, R. O.; Tavares, A. P. M.; Macedo, E. A. *Appl Biochem Biotechnol* 2011, **164**, 524–533.

[66] Ulbert, O.; Fráter, T.; Bélafi-Bakó, K.; Gubicza, L. *J Mol Catal B: Enzym* 2004, **31**, 39–45.

[67] Sheldon, R. A.; Lau, R. M.; Sorgedrager, M. J.; van Rantwijk, F.; Sheddon, K. R. *Green Chem* 2002, **4**, 147–151.

[68] Muldoon, M. J.; Gordon, C. M.; Dunkin, I. R. *J Chem Soc Perkin Trans* 2001, **2**, 433–435.

[69] Lau, R. M.; Sorgedrager, M. J.; Carrea, G.; van Rantwijk, F.; Secundo, F.; Sheldon, R. A. *Green Chem* 2004, **6**, 483–487.

[70] De Diego, T.; Lozano, P.; Gmouh, S.; Vaultier, M.; Iborra, J. L. *Biomacromolecules* 2005, **6**, 1457–1464.

[71] Li, N.; Du, W.; Huang, Z.; Zhao, W.; Wang, S. *Chinese J Catal* 2013, **34**, 769–780.

[72] Hernández-Fernández, F. J.; de los Ríos, A. P.; Rubio, M.; Gómez, D.; Víllora, G. *J Chem Technol Biotechnol* 2007, **82**, 882–887.

[73] Katsoura, M. H.; Polydera, A. C.; Katapodis, P.; Kolisis, F. N.; Stamatis, H. *Process Biochem* 2007, **42**, 1326–1334.

[74] Ventura, S. P. M.; Santos, L. D. F.; Saraiva, J. A.; Coutinho, J. A. P. *World J Microbiol Biotechnol* 2012, **28**, 2303–2310.

[75] Irimescu, R.; Kato, K. *J Mol Catal B: Enzym* 2004, **30**, 189–194.

[76] Ha, S. H.; Lan, M. N.; Lee, S. H.; Hwang, S. M.; Koo, Y. *Enzyme Microb Technol* 2007, **41**, 480–483.

[77] Lee, S. H.; Ha, S. H.; Hiep, N. M.; Woo-Jin, C.; Yoon-Mo, K. *J Biotechnol* 2008, **133**, 486–489.

[78] Guo, Z.; Chen, B.; López Murillo, R.; Tan, T.; Xu, X. *Org Biomol Chem* 2006, **4**, 2772–2776.

[79] Yang, Z.; Yue, Y. J.; Huang, W. C.; Zhuang, X. M.; Chen, Z. T.; Xing, M. *J Biochem* 2009, **145**, 355–364.

[80] Constantinescu, D.; Weingärtner, H.; Herrmann, C. *Angew Chem Int Ed Engl* 2007, **46**, 8887–8889.

[81] Zhao, H.; Song, Z. *J Chem Technol Biotechnol* 2007, **82**, 304–312.

[82] Yang, Z. *J Biotechnol* 2009, **144**, 12–22.

[83] Lai, J. Q.; Li, Z.; Lü, Y. H.; Yang, Z. *Green Chem* 2011, **13**, 1860–1868.

[84] Yu, X.; Zou, F.; Li, Y.; Lu, L.; Huang, X.; Qu, Y. *Int J Biol Macromol* 2013, **56**, 62–68.

[85] Bi, Y.; Wang, Z.; Mao, Y.; Zheng, S.; Zhang, H.; Shi, H. *Chin J Catal* 2012, **33**, 1161–1165.

[86] Shipovskov, S.; Gunaratne, H. Q. N.; Seddon, K. R.; Stephens, G. *Green Chem* 2008, **10**, 806–810.

[87] Yang, Z. *Biotransformations and Organocatalysis*; John Wiley & Sons, Inc., Hoboken, NJ, 2012, p. 15.

[88] Goldfeder, M.; Egozy, M.; Shuster Ben-Yosef, V.; Adir, N.; Fishman, A. *Appl Microbiol Biotechnol* 2013, **97**, 1953–1961.

[89] Persson, M.; Bornscheuer, U. T. *J Mol Catal B: Enzym* 2003, **22**, 21–27.

[90] De Diego, T.; Lozano, P.; Gmouh, S.; Vaultier, M.; Iborra, J. L. *Biotechnol Bioeng* 2004, **88**, 916–924.

[91] Ulbert, O.; Bélafi-Bakó, K.; Tonova, K.; Gubicza, L. *Biocatal Biotransformation* 2005, **23**, 177–183.

[92] Fujita, K.; Forsyth, M.; Macfarlane, D. R.; Reid, R. W.; Elliot, G. D. *Biotechnol Bioeng* 2006, **94**, 1209–1213.

[93] Yuan, Y.; Bai, S.; Sun, Y. *Food Chem* 2006, **97**, 324–330.

[94] Hernández-Fernández, F. J.; de los Ríos, A. P.; Tomás-Alonso, F.; Gómez, D.; Víllora, G. *Can J Chem Eng* 2009, **87**, 910–914.

[95] Housaindokht, M. R.; Monhemi, H.; Hosseini, H. E.; Googheri, M. S. S.; Najafabadi, R. I.; Ashraf, N.; Gholizadeh, M. *J Mol Liq* 2013, **187**, 30–42.

[96] Basso, A.; Cantone, S.; Linda, P.; Ebert, C. *Green Chem* 2005, **7**, 671–676.

[97] Karbalaei-Heidari, H. R.; Shahbazi, M.; Absalan, G. *Appl Biochem Biotechnol* 2013, **170**, 573–586.

[98] Machado, M. F.; Queirós, R. P.; Santos, M. D.; Fidalgo, L. G.; Delgadillo, I.; Saraiva, J. A. *World J Microbiol Biotechnol* 2014, **30**, 487–494.

[99] Wang, S. F.; Chen, T.; Zhang, Z. L.; Pang, D. W. *Electrochem Commun* 2007, **9**, 1337–1342.

[100] Nordwald, E. M.; Kaar, J. L. *Biotechnol Bioeng* 2013, **110**, 2352–2360.

[101] van Rantwijk, F.; Sheldon, R. A. *Chem Rev* 2007, **107**, 2757–2785.

[102] Lozano, P.; De Diego, T.; Gmouh, S.; Vaultier, M.; Iborra, J. L. *Biocatal Biotransformation* 2005, **23**, 169–176.

[103] Zeuner, B.; Ståhlberg, T.; van Buu, O. N.; Kunov-Kruse, A. J.; Riisager, A.; Meyer, A. S. *Green Chem* 2011, **13**, 1550–1557.

[104] Latif, M. A. M.; Tejo, B. A.; Abedikargiban, R.; Rahman, M. B. A.; Micaêlo, N. M. *J Biomol Struct Dyn* 2014, **32**, 1263–1273.

[105] Micaêlo, N. M.; Soares, C. M. *J Phys Chem B* 2008, **112**, 2566–2572.

[106] Pavlidis, I. V.; Gournis, D.; Papadopoulos, G. K.; Stamatis, H. *J Mol Catal B: Enzym* 2009, **60**, 50–56.

[107] Zhao, H.; Olubajo, O.; Song, Z.; Sims, A. L.; Person, T. E.; Lawal, R. A.; Holley, L. A. *Bioorg Chem* 2006, **34**, 15–25.

[108] Hussain, W.; Pollard, D. J.; Truppo, M.; Lye, G. J. *J Mol Catal B: Enzym* 2008, **55**, 19–29.

[109] Machado, M. F.; Saraiva, J. M. *Biotechnol Lett* 2005, **27**, 1233–1239.

[110] Dupont, J. *J Braz Chem Soc* 2004, **15**, 341–350.

[111] Chang, H. C.; Jiang, J. C.; Liou, Y. C.; Hung, C. H.; Lai, T. Y.; Lin, S. H. *J Chem Phys* 2008, **129**, 0445061–0445066.

[112] Fujita, K.; MacFarlane, D. R.; Forsyth, M. *Chem Commun* 2005, **38**, 4804–4806.

[113] van Rantwijk, F.; Secundo, F.; Sheldon, R. *Green Chem* 2006, **8**, 282–286.

[114] Liu, Y.; Chen, D.; Yan, Y. *J Mol Catal B: Enzym* 2013, **90**, 123–127.

[115] Turner, M. B.; Spear, S. K.; Huddleston, J. G.; Holbrey, J. D.; Rogers, R. D. *Green Chem* 2003, **5**, 443–447.

[116] Bekhouche, M.; Blum, L. J.; Doumèche, B. *J Phys Chem B* 2012, **116**, 413–423.

[117] Sandoval, M.; Cortés, Á.; Civera, C.; Treviño, J.; Ferreras, E.; Vaultier, M.; Berenguer, J.; Lozano, P.; Hernáiz, M. *RSC Adv* 2012, **2**, 6306–6314.

[118] Jaeger, V. W.; Pfaendtner, J. *ACS Chem Biol* 2013, **8**, 1179–1186.

[119] Sate, D.; Janssen, M. H. A.; Stephens, G.; Sheldon, R. A.; Seddon, K. R.; Lu, J. R. *Green Chem* 2007, **9**, 859–867.

[120] Burney, P. R.; Pfaendtner, J. *J Phys Chem B* 2013, **117**, 2662–2670.

[121] Lousa, D.; Baptista, A. M.; Soares, C. M. *Phys Chem Chem Phys* 2013, **15**, 13723–13736.

[122] Klähn, M.; Lim, G. S.; Wu, P. *Phys Chem Chem Phys* 2011, **13**, 18647–18660.

[123] Ajloo, D.; Sangian, M.; Ghadamgahi, M.; Evini, M.; Saboury, A. A. *Int J Biol Macromol* 2013, **55**, 47–61.

[124] Fujita, K.; MacFarlane, D. R.; Forsyth, M.; Yoshizawa-Fujita, M.; Murata, K.; Nakamura, N.; Ohno, H. *Biomacromolecules* 2007, **8**, 2080–2086.

[125] Fujita, K.; Ohno, H. *Biopolymers* 2010, **93**, 1093–1099.

[126] Guo, Z.; Xu, X. *Org Biomol Chem* 2005, **3**, 2615–2619.

[127] Zhao, H.; Jackson, L.; Song, Z.; Olubajo, O. *Tetrahedron: Asymmetry* 2006, **17**, 1549–1553.

[128] Mann, J. P.; McCluskey, A.; Atkin, R. *Green Chem* 2009, **11**, 785–792.

[129] Attri, P.; Venkatesu, P.; Kumar, A. *Phys Chem Chem Phys* 2011, **13**, 2788–2796.

[130] Weaver, K. D.; Vrikkis, R. M.; Van Vorst, M. P.; Trullinger, J.; Vijayaraghavan, R.; Foureau, D. M.; McKillop, I. H.; MacFarlane, D. R.; Krueger, J. K.; Elliott, G. D. *Phys Chem Chem Phys* 2012, **14**, 790–801.

[131] Abbott, A. P.; Boothby, D.; Capper, G.; Davies, D. L.; Rasheed, R. K. *J Am Chem Soc* 2004, **126**, 9142–9147.

[132] Zhang, Q.; De Oliveira Vigier, K.; Royer, S.; Jérôme, F. *Chem Soc Rev* 2012, **41**, 7108–7146.

[133] Gorke, J. T.; Srienc, F.; Kazlauskas, R. J. *Chem Commun* 2008, **10**, 1235–1237.

[134] Lindberg, D.; de la Fuente Revenga, M.; Widersten, M. *J Biotechnol* 2010, **147**, 169–171.

[135] Zhao, H.; Baker, G. A.; Holmes, S. *Org Biomol Chem* 2011, **9**, 1908–1916.

[136] Zhao, H.; Baker, G. A.; Holmes, S. *J Mol Catal B: Enzym* 2011, **72**, 163–167.

[137] Durand, E.; Lecomte, J.; Baréa, B.; Piombo, G.; Dubreucq, E.; Villeneuve, P. *Process Biochem* 2012, **47**, 2081–2089.

[138] Maugeri, Z.; Leitner, W.; Domínguez de María, P. *Eur J Org Chem* 2013, **20**, 4223–4228.

[139] Zhao, H.; Zhang, C.; Crittle, T. D. *J Mol Catal B: Enzym* 2013, **85–86**, 243–247.

CHAPTER 23

Analytical Applications of Ionic Liquid-Based Surfactants in Separation Science

MARÍA J. TRUJILLO-RODRÍGUEZ, PROVIDENCIA GONZÁLEZ-HERNÁNDEZ, and VERÓNICA PINO

Departamento de Química Analítica, Nutrición y Bromatología, Universidad de La Laguna (ULL), La Laguna, Tenerife, Spain

23.1 CONVENTIONAL CATIONIC SURFACTANTS IN ANALYTICAL CHEMISTRY: AN OVERVIEW

From an analytical chemistry point of view, one important property of micelles is their ability to dissolve different kinds of compounds. Compounds with low solubility in water, or even insoluble, can be dissolved and dispersed by the micelles in aqueous solutions. In analytical chemistry, main applications of cationic surfactants are linked to their solubilization power [1]. In fact, the first applications were mainly focused on the use of micelles to solubilize proteins. The solubilization power of surfactants is obviously observed when working with surfactant concentrations higher than the critical micelle concentration (CMC) value. Micellar media make possible not only to change the solubility of several analytes but also to change their microenvironment and thus to control several physicochemical phenomena.

Cationic surfactants are found to be good solubilizing media for a wide variety of analytes. If this ability is accompanied by a low CMC value, the situation becomes ideal in order to work with dilute analyte solutions. For analytical purposes, nonfoaming surfactants are more convenient in practice.

Micellar solutions of cationic surfactants have been widely employed in separation science: in spectroscopic methods [2], in extraction–preconcentration sample preparation schemes [3], in chromatographic techniques, etc., mainly as convenient substitutes of toxic conventional organic solvents.

With respect to chromatographic techniques, cationic surfactant micelles have been used as mobile phase additives in the well-known micellar liquid chromatography (MLC) [4]. They have also been employed as pseudostationary phases in micellar electrokinetic capillary chromatography (MEKC) [5].

Ionic Liquid-Based Surfactant Science: Formulation, Characterization, and Applications, First Edition. Edited by Bidyut K. Paul and Satya P. Moulik.

Extraction and concentration schemes based on analyte solubilization in cationic surfactant micelles, such as micellar-assisted extraction, coacervative extraction (CAE) [6], micellar-enhanced ultrafiltration (MEUF), surfactant-assisted transport of solutes across liquid membranes (LSM), surfactant-mediated solid-phase extractions (MSPEs), and micellar solid-phase microextractions (MSPMEs), can be cited [3].

23.2 IONIC LIQUID-BASED SURFACTANTS: TYPE AND PROPERTIES

Ionic liquids (ILs) are a class of low melting point ionic compounds, which have a variety of properties, qualifying many of them to be sustainable green solvents. These nonmolecular solvents possess high thermal stabilities and negligible vapor pressures at room temperature, making them attractive alternatives to environment unfriendly solvents that generate volatile organic compounds (VOCs). They possess unique solvation properties and can be structurally tailored for specific applications in a relatively simple mode. In turn, this has resulted in increasing studies of ILs in many areas of fundamental and applied chemistry.

A number of ILs containing long alkyl chain substituents have been reported to be surface active and are currently termed as IL-based surfactants [7]. In fact, the term IL-based surfactants is more precise than IL-surfactants, especially considering that the majority of the IL-based surfactants have melting points above 100°C. IL-based surfactants, which possess characteristics of cationic surfactants, undoubtedly constitute a new class of surfactant system. It must be considered that there is only a limited number of conventional cationic surfactants that have wide applicability in separation science.

Obviously, once an IL is dissolved in water, the anion/cation pair no longer constitutes a neat IL. Indeed, there will be only cations and anions in solution. However, the ability to modify the micellar properties of IL-based surfactant solutions by changing easily the structural makeup of the IL-based surfactant is clearly correlated with the tunability of neat ILs. In other words, simple modifications in the structure of the anion/cation belonging to the IL-based surfactant are accompanied by significant changes in the micellar properties. This is hardly observed with classical surfactant solutions. It is well known that simple modifications in ILs structures are accompanied by huge changes in their properties.

Table 23.1 [8–40] includes a list of the most successful IL-based surfactants employed in analytical separations and applications. Their structures and abbreviations are accompanied with their reported CMCs and available aggregation number (N_{agg}).

23.3 ANALYTICAL APPLICATIONS OF IONIC LIQUID-BASED SURFACTANTS

At present, a number of analytical applications have emerged in separation science regarding the use of IL-based surfactant, not only as solubilization media for proteins [41], but in analytical techniques (such as chromatographic or electrophoretic methods), and also in sample preparation schemes. Given their different performances, it is important to study each application field in an independent way.

TABLE 23.1 IL-Based Surfactants Most Widely Used in Analytical Separation Science Applications

IL-Based Surfactant Cation	IL-Based Surfactant Anion	IL-Based Surfactant Abbreviation	CMC Values ($mmol \cdot L^{-1}$)	N_{agg} Values
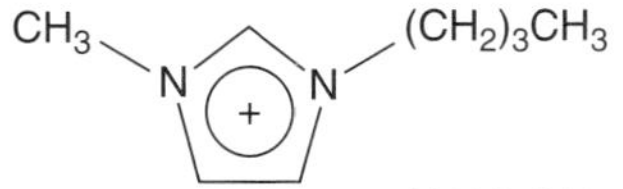	BF_4^-	$C_4MIm\text{-}BF_4$	729 [8], 800–820 [9]	10–27 [9]
	MSO_4^-	$C_4MIm\text{-}MSO_4$	80 [10]	—
	Br^-	$C_6MIm\text{-}Br$	400 [11], 470 [12], 600 [11], 797 [13]	—
	BF_4^-	$C_6MIm\text{-}BF_4$	1.99 $mmol \cdot mol^{-1}$ [14]	—
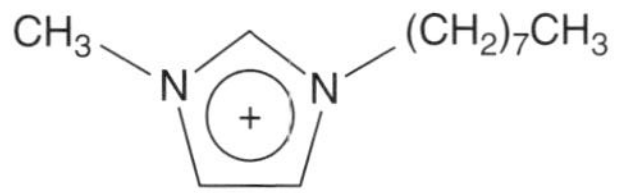	Cl^-	$C_8MIm\text{-}Cl$	90 [9], 100 [8, 9, 15], 102 [16], 210 [17], 220 [18], 234 [18, 19]	23 [11]
	Br^-	$C_8MIm\text{-}Br$	120 [16], 121 [12, 20], 141 [13], 150 [11], 170 [21], 210 [22], 205 [23]	25 [13], 21–27 [11]
	BF_4^-	$C_8MIm\text{-}BF_4$	28 [8]	—
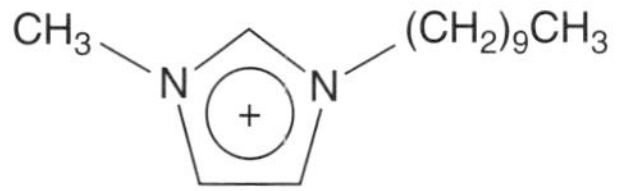	Br^-	$C_{10}MIm\text{-}Br$	20 [12], 29.3 [24], 30 [11], 31 [13], 32.9 [24], 39.7 [22], 40 [11, 25], 41 [26], 47 [23]	27 [13], 42 [26], 37–47 [11]
	BF_4^-	$C_{10}MIm\text{-}BF_4$	0.11 $mmol \cdot mol^{-1}$ [14]	
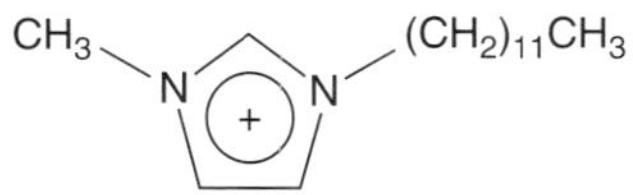	Cl^-	$C_{12}MIm\text{-}Cl$	12 [17], 12.5 [27], 13.2 [28], 14.14 [29], 14.80 [30], 15 [12], 15.5 [19], 16.1 [19], 19.96 [30]	51 [28], 58 [28], 67 [28]
	Br^-	$C_{12}MIm\text{-}Br$	4.3 [12], 8.5 [24], 9.10 [22], 9.53 [31], 9.8 [26, 27], 10.9 [24, 32], 11 [13, 23], 11.2 [32], 11.21 [33]	37 [32], 43 [31], 44 [26], 46 [13]

(Continued)

TABLE 23.1 (*Continued*)

IL-Based Surfactant Cation	IL-Based Surfactant Anion	IL-Based Surfactant Abbreviation	CMC Values (mmol·L^{-1})	N_{agg} Values
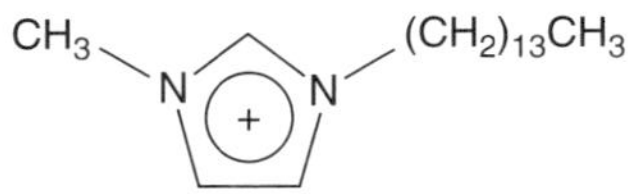	Cl^-	C_{14}MIm-Cl	3.0 [28], 3.15 [18], 3.2 [19], 3.32 [29], 3.4 [18], 3.5 [27], 4 [17]	60 [28], 79 [28], 89 [28]
	Br^-	C_{14}MIm-Br	1.9 [21], 2.3 [23], 2.5 [26, 27], 2.61 [31], 2.78 [22], 2.8 [32], 2.9 [32]	48 [32], 59 [26], 61 [31]
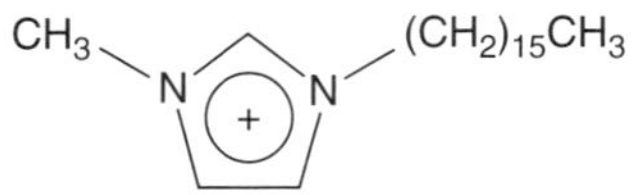	Br^-	C_{16}MIm-Br	0.51 [34], 0.55 [32], 0.55–0.84 [35], 0.61 [22, 26], 0.653 [31], 0.76 [36], 0.8 [12], 0.84 [32], 2.66 [37]	64 [32], 66 [26], 76 [31]
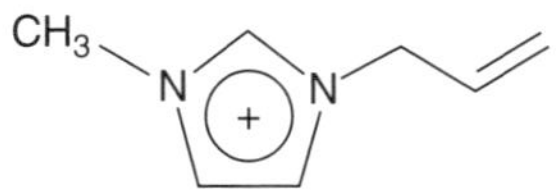	BF_4^-	AllMIm-BF_4	—	—
	Br^-	$C_{12}C_{12}$Im-Br	0.1 [12]	—
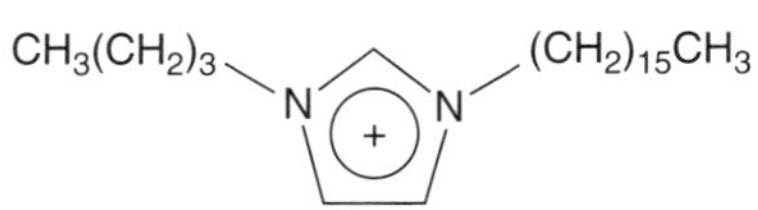	Br^-	$C_{16}C_4$Im-Br	0.08 [12], 0.1 [12]	—

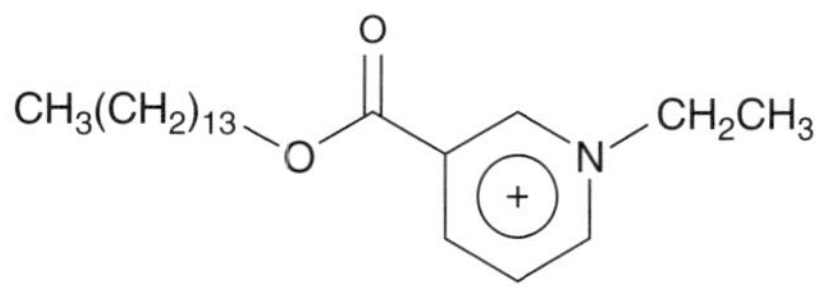	Br^-	$C_2COOC_{14}Py$-Br	—	—
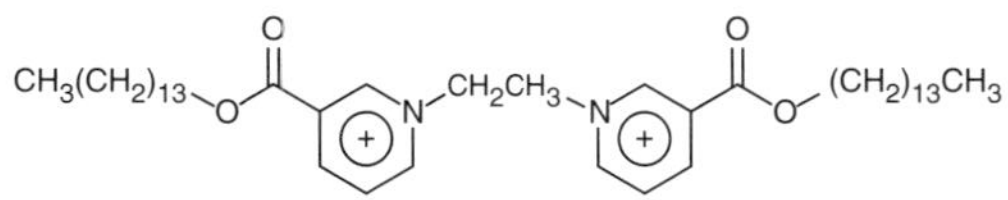	Br^- Br^-	$(COOC_{14}Py)_2C_4$-2Br	—	—
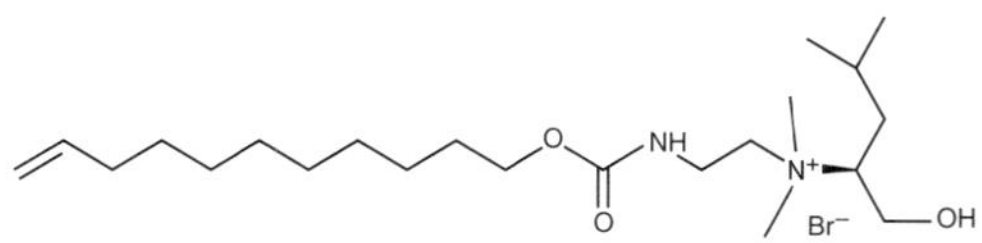	Br^-	L-UCLB	0.84 [38]	97 [38]
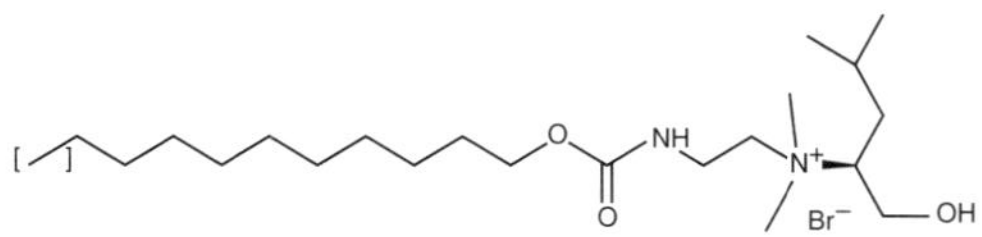	Br^-	poly-L-UCLB	—	25 [38]
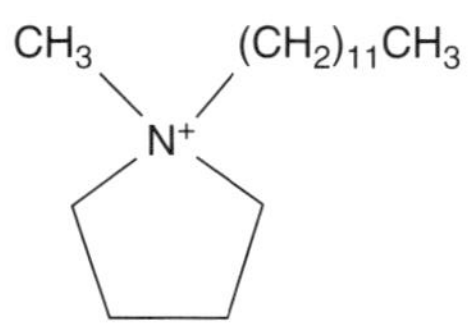	Br^-	$C_{12}MPyrr$-Br	13.6 [39]	—

(*Continued*)

TABLE 23.1 (*Continued*)

IL-Based Surfactant Cation	IL-Based Surfactant Anion	IL-Based Surfactant Abbreviation	CMC Values (mmol·L^{-1})	N_{agg} Values
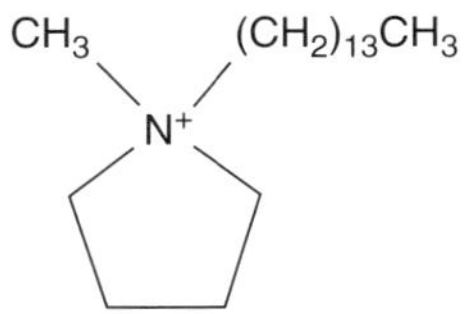	Br^-	C_{14}MPyrr-Br	3.30 [39]	—
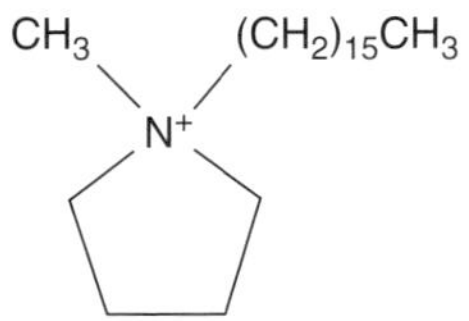	Br^-	C_{16}MPyrr-Br	0.83 [39]	—
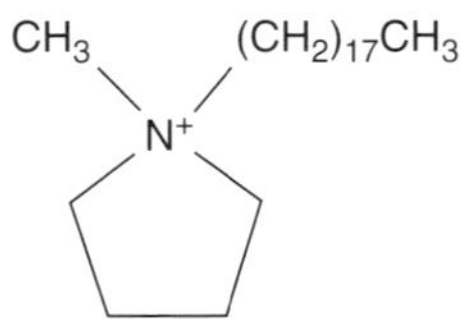	Br^-	C_{18}MPyrr-Br	0.25 [39]	—
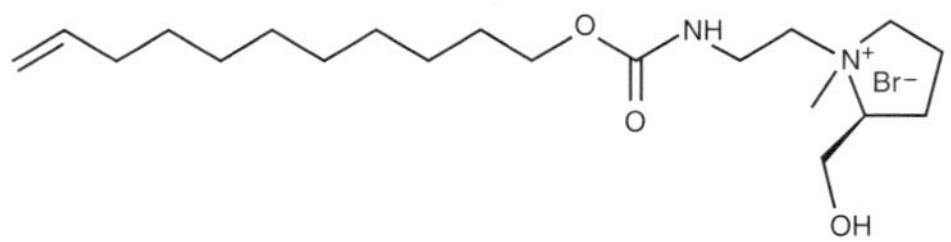	Br^-	L-UCPB	1.15 [38]	95 [38]

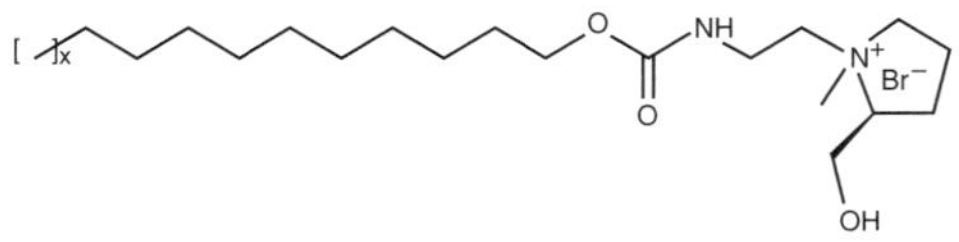	Br^-	poly L-UCPB	—	34 [38]
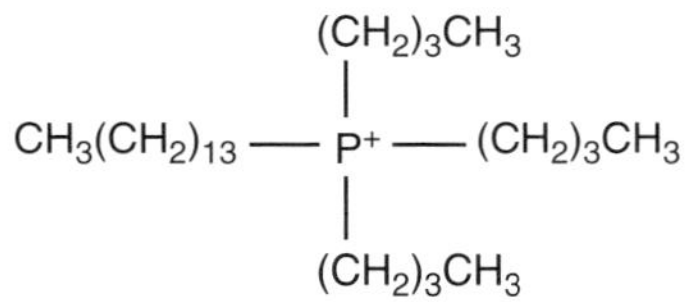	Cl^-	$P_{14,4,4,4}Cl$	0.6 [40]	—
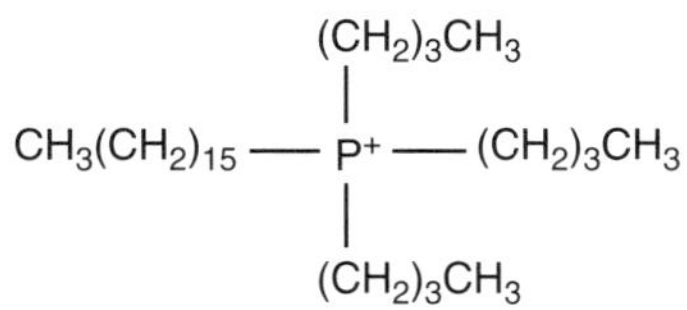	Cl^-	$P_{16,4,4,4}Cl$	—	—

23.3.1 Analytical Applications of IL-Based Surfactants in Analytical Techniques

Lately, IL-based surfactants have been used in separation techniques of wide applicability, such as high-performance liquid chromatography (HPLC) and capillary electrophoresis (CE). This section will only focus on those applications that take advantages of the micellar properties of IL-based surfactants, that is, works that use micellar solutions of IL-based surfactants in HPLC or in CE. In this consideration, applications that employ IL-based surfactant as mere background electrolytes have not been taken into account.

23.3.1.1 Utilization of IL-Based Surfactants in HPLC MLC is a mode of HPLC in which micelles are used as pseudophases in the mobile phase. Surfactants valid for MLC applications should have a low CMC value, which is a common trend for many IL-based surfactants (Table 23.1).

First application of IL-based surfactants as pseudophases in MLC was reported in 2009 [42]. It was a preliminary study regarding the retention behavior of benzene using three different LC stationary phases (Eclipse XDB C8, Zorbax C8, and Gemini C18) and $C_{16}C_4$Im-Br. A comparison was also made with the conventional cationic surfactant cetyltrimethylammonium bromide (CTAB) showing similar performance.

More recently, Flieger et al. [43] used C_{12}MIm-Cl as a pseudophase in MLC for the separation of eight derivatives of 1,4-thiosemicarbazides, employing two different LC stationary phases (Kromasil C18 and Zorbax SB-CN). In this case, the comparison was established with the most common surfactant in MLC, the anionic surfactant sodium dodecyl sulfate (SDS). SDS also showed adequate behavior.

These pioneer studies point out to the potential of IL-based surfactants in MLC, since they are able to compete with both cationic and anionic conventional surfactants.

Apart from their uses as pseudophases in MLC, Qui et al. have also reported the uses of C_{12}MIm-Br and C_{14}MIm-Br as modifiers of the LC stationary phase ODS [44]. The application was the ion chromatography (IC) separation of eight anions. The authors also stated that the performance of the C_{12}MIm-Br was much better than that of the conventional cationic surfactant dodecyltrimethylammonium bromide (DTAB).

23.3.1.2 Utilization of IL-Based Surfactants in CE IL-based surfactants have been widely employed as background electrolytes in capillary zone electrophoresis (CZE). However, these applications do not take advantage of the micellar properties of IL-based surfactants, and so their performance is not related to any micellar property.

Micellar solutions have been employed in CE as pseudostationary phases in MEKC and also in online preconcentration techniques based on the use of micelles.

The application of the linear solvation energy relationship (LSER) model to MEKC [45] indicated that polar analytes experience stronger interactions when using cationic surfactants with large head groups. Clearly, many IL-based surfactants are good candidates for MEKC due to their large cationic head group and also due to their low CMC values. Main applications of IL-based surfactants in MEKC are summarized in Table 23.2 [27, 35, 38–40, 46]. It is worthy of mentioning that the

TABLE 23.2 Analytical Applications of IL-Based Surfactants in MEKC

IL-Based Surfactant (Concentration, mmol·L^{-1})	Analytes (Number)	Separation Time (min)	Detection	Comments	Comparison with Conventional Surfactants	Reference
L-UCPB (25)[a] L-UCLB (25)[b] poly-L-UCPB (25) poly-L-UCLB (25)	Acidic analytes: ±-phenoxypropionic Acid herbicide and ±-bromophenylacetic acid	~27	DAD	Chiral resolution of compounds	—	[38]
C_{12}MPyrr-Br (50) C_{14}MPyrr-Br (25) C_{16}MPyrr-Br (15) C_{18}MPyrr-Br (10)	Probe organic compounds (35)	—	DAD	Study of the linear solvation energy relationships (LSERs)	CTAB[c]	[39]
C_{12}MIm-Cl (15–20)	Resorcinol, methylresorcinol isomers (3) and benzene derivates (3)	~13 (for resorcinol and methylresorcinol isomers) and ~6 (for benzene derivates)	UV	—	DTAC[d] and TTAC[e]	[27]
C_{16}MIm-Br (10) C_{14}MIm-Br (10)	Phenols (7)	~17 (with C_{16}MIm-Br) and ~13 (with C_{14}MIm-Br)	DAD	—	CTAB[c]	[35]
$P_{14,4,4,4}$Cl (10) $P_{16,4,4,4}$Br (10)	Benzene derivates (4)	~7	UV	—	—	[40]
C_{14}MIm-Br (20)	Urinary nucleosides (7)	~12	UV	—	TTAB[f]	[46]

[a] Undecenoxycarbonyl-L-leucinol bromide.
[b] Undecenoxycarbonyl-L-pyrrolidinol bromide.
[c] Cationic surfactant cetyltrimethylammonium bromide.
[d] Cationic surfactant dodecyltrimethylammonium chloride.
[e] Cationic surfactant tetradecyltrimethylammonium chloride.
[f] Cationic surfactant tetradecyltrimethylammonium bromide.

majority of works include a comparison with conventional cationic surfactant analogues, showing improved performance over them [27, 35, 39, 46].

It is interesting to note that the first application involved the use of chiral IL-based surfactants and their polymeric analogues to separate chiral acidic analytes [38] with successful results in this difficult task.

IL-based surfactants have been also used in online preconcentration techniques of CE that are based on micelles. For example, they have been used by Su et al. in sweeping MEKC [47]. C_{16}MIm-Br and C_{16}MPyrr-Br were tested for the separation and preconcentration of seven benzodiazepines. The authors observed that the performance of C_{16}MPyrr-Br was highly superior to that of CTAB and that the use of sweeping MEKC led to high sensitivity with detection limits ranging from 4.68 to 9.75 $ng \cdot mL^{-1}$. The optimum conditions of the separation, and the superior performance of sweeping MEKC *versus* MEKC, can be observed in Figure 23.1a.

Another successful mode of online preconcentration technique of CE based on the use of micelles in which IL-based surfactants have found an interesting application is micelle to solvent stacking (MSS). Quirino et al. [48] utilized the IL-based surfactant C_{12}MIm-BF_4 for the preconcentration and separation of a group of profens and herbicides. The authors obtained up to 10 times better performance with this IL-based surfactant compared to the conventional cationic surfactant CTAB as MSS carrier due to (i) the higher affinity between the analytes and the IL-based surfactants (high partition coefficient value) and (ii) the good solubility of the IL-based surfactant in high percentages of organic solvent, which facilitated a more effective reversal of mobility in CE.

Figure 23.1b also includes the electrophoretic separation obtained for profens and herbicides using C_{12}MIm-BF_4 under optimum conditions in MSS. More recently, Wang et al. [49] have proposed two-step stacking by sweeping and MSS using C_{16}MPyrr-Br in CZE for the separation and preconcentration of benzoic acid, 4-nitrophenol, and 4-chlorophenol getting limits of detection between 0.025 and 0.25 $mg \cdot L^{-1}$. All these novel applications are opening a new field for IL-based surfactants in CE.

23.3.2 Analytical Applications of IL-Based Surfactants in Sample Extraction and Preconcentration

IL-based surfactants have also found a large application field in sample preparation, apart from their intense uses in the previously mentioned analytical techniques. In sample preparation, IL-based surfactants are mainly used as extracting solvents, acting as adequate substitutes for toxic organic solvents.

23.3.2.1 Partitioning Behavior of Analytes to IL-Based Surfactants The successful utilization of IL-based surfactants as extracting solvents of analytes lies in the strength of the analyte–micelle interaction of the IL-based surfactants, as pointed out in Figure 23.2. The higher the partitioning of an analyte for a specific micelle of an IL-based surfactant, the higher is the efficiency of an extraction scheme.

Thus, obtaining the partition coefficient value ($K_{M,m}$) that accounts for such interaction (analyte–micelle of IL-based surfactants) is of vital importance for developing analytical extraction methods for IL-based surfactants. The partition

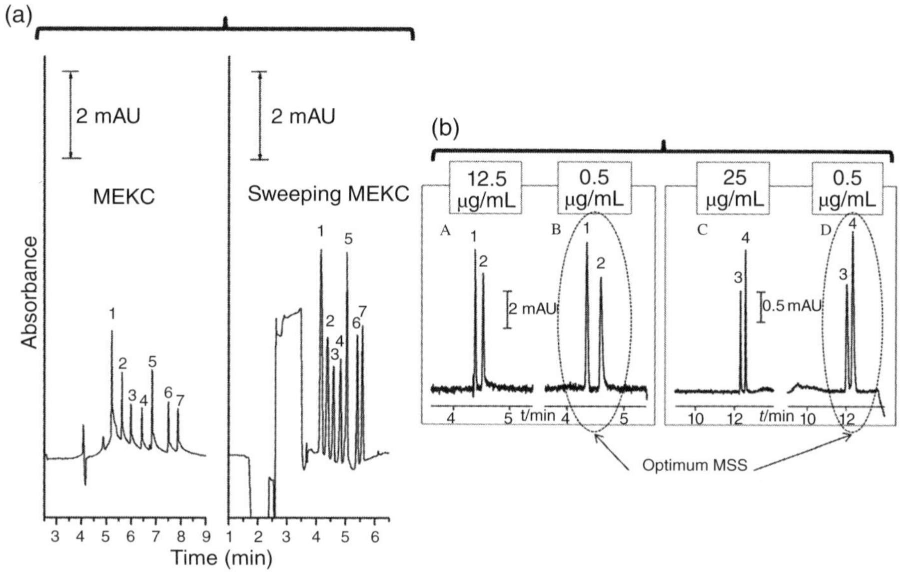

Figure 23.1 Examples of applications of IL-based surfactants in CE preconcentration strategies based on micelles: (a) Regular MEKC *versus* sweeping MEKC for the determination of seven benzodiazepines (labeled from 1 to 7) using C_{16}MPyrr-Br. MEKC conditions: BGE, 15 mM borate (pH 9.0) containing 30% MeOH and 20 mM C_{16}MPyrr-Br; sample injection at 2.07 kPa for 3 s; sample concentration, 50 g·mL^{-1}; injection length, 0.9 mm. Sweeping MEKC conditions: BGE, 15 mM borate (pH 9.0) containing 30% MeOH and 20 mM C_{16}MPyrr-B; sample injection at 3.45 kPa for 300 s; sample concentration, 1 µg·mL^{-1} in BGE (without C_{16}MPyrr-Br); injection length, 150 mm. Reproduced from Su et al. [47] with permission from Elsevier. (b) Micelle to solvent stacking (MSS) for the determination of two profens (analytes 1 and 2) and two herbicides (analytes 3 and 4) using as stacking matrix 7.5 mM of C_{12}MIm-BF_4 and 7.5 mM of ammonium acetate. Injection: 3 (A), 90 (B), and 5 (C) s at 50 mbar and 35 s at approximately 1 bar. Capillary length: 50 cm total and 41.5 cm effective (A and B) and 75 cm total and 66.5 cm effective (C and D). Reproduced from Quirino et al. [48] with permission from Elsevier.

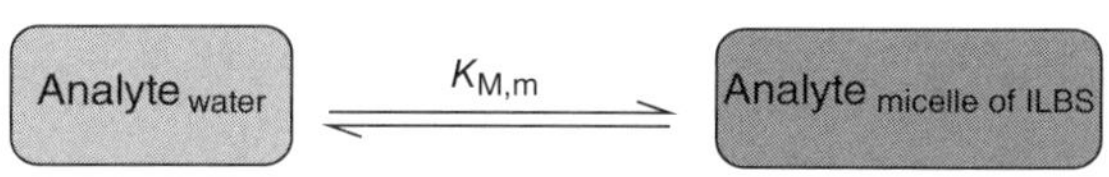

Figure 23.2 Partition of an analyte from water to an IL-based surfactant micelle.

coefficient of an analyte from an aqueous sample to an IL-based surfactant micelle is defined as

$$K_{M,m} = \frac{C_m \text{ in the micellar phase M}}{C_m \text{ in water}} \tag{23.1}$$

where C_m is the concentration of the specific analyte (m) in the micellar phase (M) or in water.

Common methods to determine partition coefficients for conventional micelles are dialysis, ultradialysis, and centrifugation [50, 51]. Solid-phase microextraction

(SPME) has been also applied successfully to the measurement of partition coefficients for conventional surfactants [52]; the amount of analyte extracted by SPME is so small that in most cases it will not disturb the equilibrium of the system. SPME offers the advantage of being a simpler and more efficient technique compared to conventional methods.

When introducing a SPME fiber in an aqueous solution containing analytes and IL-based surfactant micelles, the following equation is applicable:

$$\frac{1}{C_{\mathrm{W,m}}} = \frac{1}{C_{\mathrm{total,m}}} + \frac{K_{\mathrm{M,m}}}{C_{\mathrm{total,m}}} C_{\mathrm{M}} \tag{23.2}$$

where $C_{\mathrm{W,m}}$ is the free concentration of analyte in water (measured by SPME); C_{M} corresponds to the overall IL-based surfactant micellar concentration in solution, which is calculated by subtracting the CMC value from the total IL-based surfactant concentration used in the experiments ($C_{\mathrm{M}} = C_{\mathrm{total,ILBS}} - \mathrm{CMC}$); and $C_{\mathrm{total,m}}$ is the total analyte concentration in the experiments.

If an analyte does not suffer partitioning to an IL-based surfactant micelle, the plot $1/C_{\mathrm{W,m}}$ *versus* C_{M} (Eq. 23.2) is a constant, as it can be observed in Figure 23.3a. However, if an analyte suffers partitioning, the plot reflects such interaction (Fig. 23.3b), and calculations can be performed to obtain $K_{\mathrm{M,m}}$ (Fig. 23.3c). Therefore, $K_{\mathrm{M,m}}$ can be determined from the slope obtained by plotting $1/C_{\mathrm{W,m}}$ *versus* C_{M}. In addition, the obtained intercepts should be close to the theoretical value, which is the inverse of the total concentration of the analyte, which is kept constant in the experiments while increasing the IL-based surfactant concentration.

The SPME method has been applied to the determination of $K_{\mathrm{M,m}}$ values for a variety of analytes (mainly organic contaminants such as hydrocarbons, phenols, and esters) and different IL-based surfactant micelles (C_{16}MIm-Br, $C_{16}C_4$Im-Br, $C_{12}C_4$Im-Br, $C_{12}C_{12}$Im-Br, $(C_{10}Im)_2C_4$-2Br, and $(C_{10}Im)_2C_{12}$-2Br) using gas chromatography (GC) [53] and HPLC [54]. Using this approach, $K_{\mathrm{M,m}}$ values varied from 30 for methyl caproate and dimethyl phthalate with the dicationic IL-based surfactant $(C_{10}Im)_2C_4$-2Br, and being also of 30 for *o*-cresol with the monocationic IL-based surfactant $C_{12}C_4$Im-Br [53], to 6140 for acenaphthene with C_{16}MIm-Br [54]. Clearly, hydrophobic analytes presented higher partition coefficient values over polar

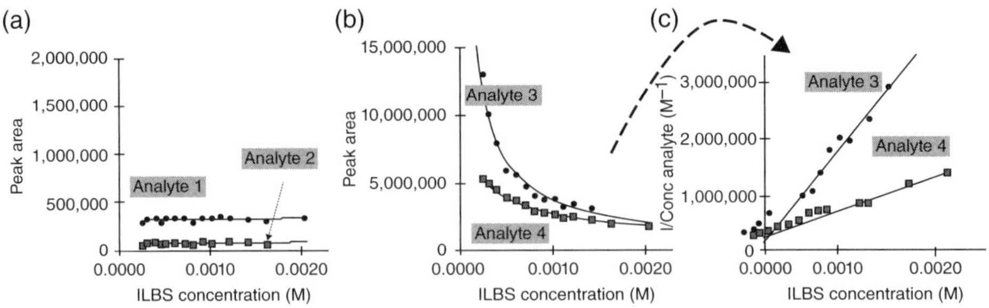

Figure 23.3 Peak area obtained in solid-phase microextraction at increasing IL-based surfactant concentrations keeping constant the analyte concentration in solution for (a) analytes that do not suffer partition to the IL-based surfactant micelles and (b) analytes that suffer partitioning to the IL-based surfactant micelles. (c) Plots to calculate partition coefficients (from the slope) for analytes that suffer partitioning.

analytes to the IL-based surfactants studied, which could eventually lead to the development of selective extraction methods (hydrophobic *versus* polar compounds). It is interesting to note that for the group of IL-based surfactants evaluated, higher partition coefficients were obtained for monocationics (C_{16}MIm-Br, $C_{16}C_4$Im-Br, $C_{12}C_4$Im-Br, and $C_{12}C_{12}$Im-Br) than dicationics ((C_{10}Im)$_2C_4$-2Br and (C_{10}Im)$_2C_{12}$-2Br).

Furthermore, differences in the obtained $K_{M,m}$ values were also attributable to the micellar structure of IL-based surfactants. These results aimed to employ IL-based surfactant in efficient and selective extraction methods by adequate selection of the IL-based surfactant nature and/or the group of compounds to be determined.

It is worthy of mentioning that Yao et al. [54] also proved that the partition coefficient values obtained for a group of analytes and IL-based surfactants were higher than those obtained for the conventional cationic surfactant CTAB, which points out to the superior performance in terms of extraction abilities for these IL-based surfactants.

23.3.2.2 Food and Environmental Applications Recent trends in analytical chemistry are aimed to develop processes that eliminate, or at least minimize, the consumption of organic solvents during the extraction step, due to toxicity and environmental issues associated with the conventional organic solvents and the VOCs generated during their use. In this sense, current approaches are based on microextraction methods, in which the organic solvent consumption is eliminated or, at least, reduced to a few microliters. Another important trend within green analytical chemistry is the substitution of organic solvents by less harmful reagents [55]. In this sense, the utilization of IL-based surfactants as adequate substitutes of toxic conventional organic solvents in extraction methods, in combination with microextraction approaches, seems to be a convenient and exciting strategy in sample preparation, which is the common bottleneck in an analytical method [7]. Thus, IL-based surfactants have proved to be particularly helpful as substitutes of organic solvents when dealing with the extraction of analytes from complex solid samples, which highlight the usefulness of these novel reagents in sample preparation.

Organic compounds present in a complex solid sample can be extracted by utilizing aqueous solutions of IL-based surfactants (at a concentration above the CMC) as extractant solvents, rather than organic solvents, allowing a positive interaction of analyte–micelle. The strength of this interaction is evidently dictated by the corresponding affinity, which can be estimated through the partition coefficient. At this point, the application of microwaves or ultrasounds is very helpful to improve the kinetics of the transference of analyte from the solid sample to the IL-based surfactant micelle. Afterward, the analytes extracted are present in the aqueous solution of the IL-based surfactant, which can be separated from the solid sample simply by centrifugation and/or filtration, and subjected to the determination using HPLC, or GC if solid-phase microextraction is used. Figure 23.4a summarizes the overall performance of the extraction methods based on IL-based surfactants. It must be observed that the method avoids the use of chlorinated organic solvents during the overall extraction step, despite the complexity of the original solid sample.

More recently, an *in situ* preconcentration strategy to improve the sensitivity of the overall extraction method for solid samples using IL-based surfactants as extracting solvents has been proposed [56]. The method is based on transforming a water-soluble IL-based surfactant into a water-insoluble IL-based surfactant by means of

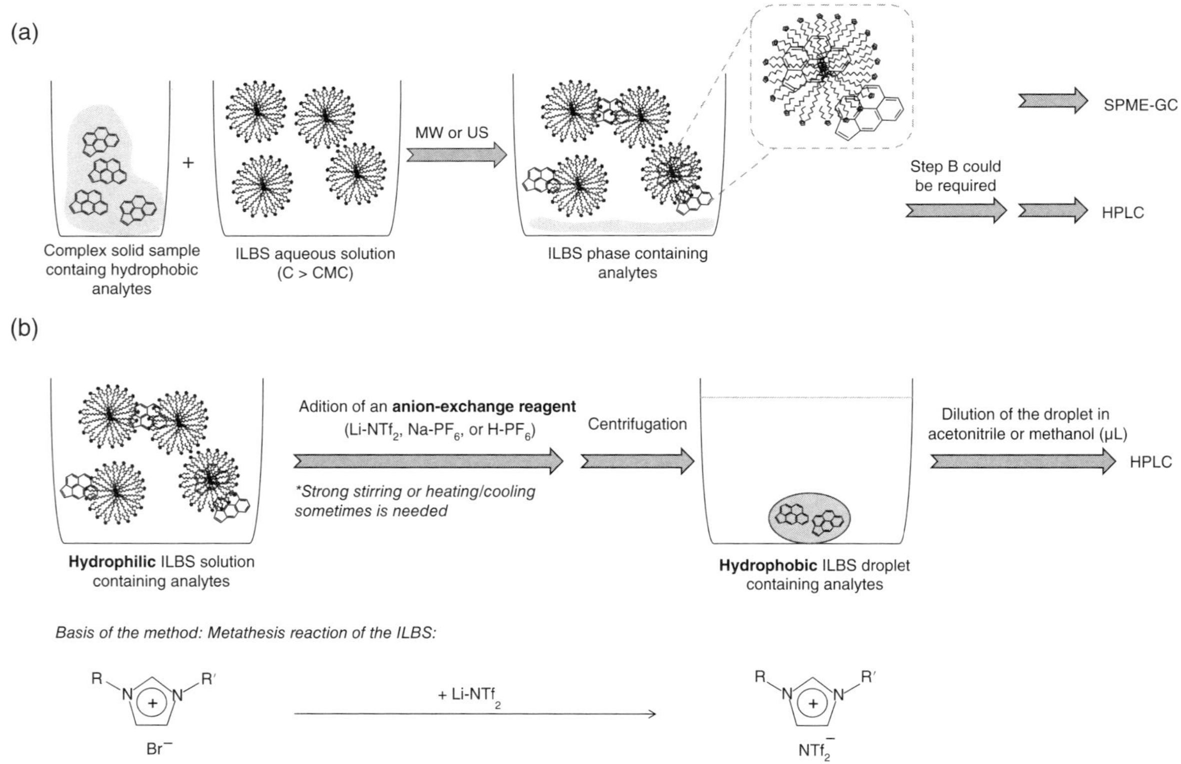

Figure 23.4 (a) Overall extraction scheme for analytes contained in solid samples when using aqueous solutions of IL-based surfactants (at concentrations above the CMC) as extracting solvents. No chlorinated organic solvent are needed in the extraction step and the matrix can be finally removed by filtration or centrifugation. (b) *In situ* preconcentration step with ILBSs. R and R′ are long alkyl chains.

a metathesis reaction using an anion-exchange reagent. In fact, the ability to transform an IL-based surfactant from water soluble to water insoluble by modifying simply the nature of the anion totally resembles the behavior of the neat ILs. Figure 23.4b summarizes the performance of the *in situ* preconcentration method, which presents characteristics of an IL-dispersive liquid–liquid microextraction (DLLME) procedure [57]. Thus, once analytes have been extracted from a solid sample using an IL-based surfactant as extraction solvent, the adequate addition of an anion-exchange reagent such as lithium bis(trifluoromethanesulfonyl)imide (Li-NTf_2), normally in a 1:1 ratio with the IL-based surfactant, generates a water-insoluble IL-based surfactant that precipitates in the form of a microdroplet, containing the extracted analytes. This droplet can be later be subjected to HPLC injection after dilution with acetonitrile or methanol. It is important to note that the overall extraction–preconcentration method with IL-based surfactants for complex solid samples is carried out avoiding the use of toxic conventional organic solvents.

The applications of IL-based surfactants have not been exclusively associated to their utilization as extraction solvents in extraction and/or preconcentration schemes. In fact, another interesting approach consists of utilizing IL-based surfactants as the modifiers of extraction of sorbents when linked to the solid supports forming hemimicelles or admicelles. The modified sorbents perform better in the solid-phase extraction (SPE) approaches than the conventional sorbents. Gangula et al. termed this methodology as admicellar or hemimicellar SPE in a review published in 2010, covering mainly applications with conventional surfactants [58].

In recent years, a number of methods that use IL-based surfactants in sample preparation schemes have been described for both environmental and food analysis (Tables 23.3 and 23.4).

The IL-based surfactants employed for these analyses are commonly composed of methyl- or butylimidazolium cations containing long alkyl chains in their structures: between 4 and 16 carbon atoms [20, 23, 36, 56, 59–73]. The most common anion in these applications is bromide [20, 23, 36, 56, 60, 62, 64, 67, 68, 71, 73] followed by the tetrafluoroborate anion [65, 66, 69]. There is only one application utilizing a pyridinium-based IL-based surfactants [63], in which a comparison of the performance of a monocationic and a dicationic pyridinium-based IL-based surfactant was made.

In any case, it is important to highlight that low amounts of IL-based surfactants are required to achieve successful extractions (between 200 μL and 20 mL of aqueous solutions of IL-based surfactants), which is within the microextraction trend of green chemistry.

The applications involving IL-based surfactants with environmental samples are summarized in Table 23.3 [23, 36, 59–64]. The first application was published in 2008 by Pino et al. regarding the determination of polycyclic aromatic hydrocarbons (PAHs) from marine sediments using focused microwave-assisted extraction [36], wherein the IL-based surfactant $C_{16}C_4$Im-Br was used as the extracting solvent. Since then, a number of methods have been described focusing not only on the determination of PAHs [60, 62] but other organic compounds such as phthalates [59], alkylphenols [62], parabens [62], glucocorticoids [64], or methyl orange [63]. Besides, IL-based surfactants have been employed in the determination of metals such as Zn(II) [23]. In this particular case, it is necessary to use a chelating reagent for the metal before carrying out the spectrophotometric determination.

TABLE 23.3 Environmental Applications of IL-Based Surfactants

IL-Based Surfactant (Amount)	Analytes (Number)	Sample (Amount)/ Sample Requirement	Extraction Technique	Clean-Up Steps	Analysis	Comments	RR[a] (%)/RSD[b] in % (spiked level)/LOD[c]	Reference
$C_{16}C_4$Im-Br (9 mL, 45 mmol·L^{-1})	PAHs[d] (7)	Marine sediments (0.1 g)/-	Focused MW (180 W, 90°C, ~7 min)	Filtration	HPLC-FD	• Validation with a CRM[e] • Comparison with CTAB[f]	70.1–108.3 (being of 43.7 for indeno (1,2,3-cd)pyrene)/ RSD < 10.4 (–)/0.8–53 ng·g^{-1}	[36]
C_{12}MIm-Br[g] (25 mg)	Phthalates (5)	Water (25 mL)/ pH 6.5	IL-based surfactant hemimicelles in SPE[h] (IL-based surfactant coated on silica)	Elution using vacuum and MeOH (3 mL, pH 2), drying with a stream of N_2 (60°C), and redissolution in MeOH (0.5 mL) applying US (5 min)	HPLC-DAD	Study of adsorption mechanism	85–107/RSD < 6 (1 μg·L^{-1}) /0.12–0.17 μg·L^{-1}	[59]
C_{16}MIm-Br[g] (9 mL, 45 mmol·L^{-1}) or $C_{16}C_4$Im-Br[g] (9 mL, 45 mmol·L^{-1})	PAHs[d] (7)	Certified sediment (0.45 g)/-	Focused MW (180 W, 90°C, ~7 min) and SPME[i] using a 100 μm PDMS[j] fiber	After MW: filtration and dilution to 10 mmol·L^{-1} for C_{16}MIm-Br and to 1.6 mmol·L^{-1} for $C_{16}C_4$Im-Br	GC-MS	• Validation with a CRM[e] • Comparison with CTAB[f]	75.8–117/RSD < 19 (–)/0.05–0.45 μg·g^{-1}	[60]

C_{16}MIm-Br[g] (50 mg)	PAHs[d] (3)	Water (300 mL) or soil (0.5 g)/ NaOH (1 mol·L^{-1}, pH 10)	• Soil extraction: 10 mL acetone/ dichloromethane (1:1, v/v) and US (10 min) (×3 times) • SPE[h] procedure: IL-based surfactant mixed hemimicelles in SPE[h] (IL-based surfactant coated on 80 mg Fe_3O_4 MNPs*k*)	• After soil extraction: combination of the extracts, evaporation of the solvent under N_2, and redissolution in water (300 mL) • After SPE[h] procedure: US (5 min), equilibration (10 min), separation of the IL-coated MNPs[j] with a magnet (10 min), elution using ACN with 1% HAc[l], drying with a stream of N_2 (55°C), and redissolution with ACN (0.5 mL)	HPLC-FD	• Comparison with CTAB[f] • Study of adsorption mechanism	73–105/RSD < 6.9 (–)/0.33–8.33 ng·L^{-1}	[61]

(*Continued*)

TABLE 23.3 (*Continued*)

IL-Based Surfactant (Amount)	Analytes (Number)	Sample (Amount)/ Sample Requirement	Extraction Technique	Clean-Up Steps	Analysis	Comments	RR[a] (%)/RSD[b] in % (spiked level)/LOD[c]	Reference
C_{16}MIm-Br[g] (5 mL, 40 mmol·L^{-1})	PAHs[d] (3), alkylphenols (5) and parabens (1)	Marine sediments (0.1 g)/-	MW (140 W, 90°C, 6 min) followed by *in situ* ILs-DLLME[m] using 4.0 mL of the filtrate, Li-NTf_2 (92 μL, 0.5 g·mL^{-1}), and ACN (800 μL)	• After MW: centrifugation (3600 rpm, 5 min) and filtration • After *in situ* IL-DLLME[m]: heating (65°C, 5 min), vortex (3 min), centrifugation (3400 rpm, 4 min), dilution of the droplet up to 200 μL with ACN, and vortex (1 min)	HPLC-DAD	• Li-NTf_2 as anion-exchange reagent • Validation with CR M[e]	91.1–127/RSD < 3.4 (0.8–2.3 mg·L^{-1})/ 0.04–1 mg·kg^{-1}	[62]
C_{14}MIm-Br[g] (1.00 mg·mL^{-1})	Zn(II)	Calcium–zinc gluconate oral solution (–)/dilution up to 100 mL with HCl, 1:1 v/v, and NH_3-NH_4Cl buffer (pH 7.5)	—	Quelating reaction (10 min) before determination	Spectrophotometry	• 5-Br-PADAP[n] (1 mL, $5.0 \cdot 10^{-4}$ mol·L^{-1}) as quelating reagent • The IL-based surfactant enhances the absorbance signal	98.7 – 103.0/RSD < 4.65 (–) /2.7 ng·mL^{-1}	[23]

$(COOC_{14}Py)_2C_4$-2Br[o] (3.6 mmol·L^{-1}) and C_2COOC_{14}Py-Br[p] (8.0 mmol·L^{-1})	Methyl orange	Wastewater (–)/pH 6.3	IL-based surfactants as modifiers of Na-Bt[q] surfaces, followed by extraction/removal of methyl orange	Heating (30°C, 30 min) before determination	Spectro-photometry	Study of adsorption mechanism	9.68 for $(COOC_{14}Py)_2C_4$-2Br[o] surface and 99.88 for C_2COOC_{14}Py-Br[p] surface[r]/-/-	[63]
C_4MIm-BF_4 (200 µL)	Glucocorti-coids (3)	Water (5.0 mL)/Triton X-100[s] (500 µL, 0.05%, v/v) and Na_2HPO_4 (1% w/v)	Vortex (3 min)	Cooling in an ice bath (5 min) and dilution of the IL-based surfactant phase with ACN (500 µL)	HPLC-UV	Triton X-100[s] as synergic reagent	97.2–102.2/RSD < 1.81 (1.0–100.0 ng·mL^{-1})/4.11–9.19 ng·mL^{-1}	[64]

[a] Relative recovery.
[b] Relative standard deviation.
[c] Limit of detection.
[d] Polycyclic aromatic hydrocarbons.
[e] Certified reference material.
[f] Cationic surfactant cetyltrimethylammonium bromide.
[g] IL-based surfactants selected (others were also tried in the study).
[h] Solid-phase extraction.
[i] Solid-phase microextraction.
[j] Poly(dimethyl)siloxane.
[k] Magnetic nanoparticles.
[l] Acetic acid.
[m] Ionic liquid dispersive liquid–liquid microextraction.
[n] 2-[(5-bromo-2-pyridyl)-azo]-5-(diethyl-amino)phenol.
[o] Dicationic IL-based surfactant 1,10-(butane-1,4-diyl)-bis(3-(tetradecyloxycarbonyl)pyridinium) dibromide.
[p] Monocationic 1-ethyl-3-(tetradecyloxycarbonyl)pyridinium bromide.
[q] Sodium bentonite.
[r] Methyl orange removal efficiency at a dosage of 0.06 g.
[s] Commercial nonionic surfactant octyl phenol ethoxylate.

TABLE 23.4 Food Applications of IL-Based Surfactants

IL-based surfactant (Amount)	Analytes (Number)	Sample (Amount)/ Sample Requirement	Extraction Technique	Clean-Up Steps	Analysis	Comments	RR[a] (%)/RSD[b] in % (spiked level)/LOD[c]	Reference
$C_4MIm\text{-}BF_4$[d] (10 mL, 1.5 mol·L^{-1}) and $C_6MIm\text{-}BF_4$[d] (10 mL, 1.0 mol·L^{-1})	Phenolic alkaloids (3)	Medicinal plants (1 g)/-	MW (280 W, 1.5 min)	Dilution up to 50 mL with water and filtration	HPLC-UV	Solid:liquid ratio 1:10 for $C_4MIm\text{-}BF_4$ and 1:15 for $C_6MIm\text{-}BF_4$	98–105/RSD < 5.4 (–)/0.11–0.13[e] μg·g^{-1}	[65]
$C_4MIm\text{-}BF_4$[d] (10 mL, 2 mol·L^{-1})	Piperidine	White pepper (1.0 g)/-	US (500 W, 30 min)	Filtration, dilution up to 50 mL with water and refiltration	UHPLC-UV	Solid:liquid ratio 1:15	93.5–98.7/RSD < 0.87 (35.7 mg·g^{-1})/2[e] μg·g^{-1}	[66]
$C_{16}MIm\text{-}Br$[d] (4 mL, 50 mmol·L^{-1})	Tanshinones (3)	Medicinal plant (0.05 g)/-	US (30 min) and centrifugation (5000 rpm, 5 min) (x 3 times)	Combination of the 3 extracts, dilution (15 mL, 5 mmol·L^{-1} of IL-based surfactant) and filtration	HPLC-DAD	• Solid:liquid ratio of 80 mL·g^{-1} • Comparison with CTAB[f]	87.5–107.6/RSD < 6 (–)/3–15[e] μg·g^{-1}	[20]
$C_6MIm\text{-}Br$[d] (20 mL, 1.5 mol·L^{-1})	Dehydrocavidine	Medicinal plant (1.0 g)/-	MW (70°C, ~10 min), filtration and dilution up to 50 mL with water	—	HPLC-UV	• Solid:liquid ratio 20:1 • Comparison with other techniques	93.8–109.3/RSD[g] < 6.5 (10 mg·L^{-1}) /1.75[e] μg·g^{-1}	[67]

$C_{12}MIm\text{-}Br$[d] (5 mL, 0.8 mol·L^{-1})	Biphenyl cyclooctene lignans (4)	Medicinal plant (0.5 g)/-	US (50 kHz, 200 W, 30 min ×3)	• Before US: soaking (4 h) of medicinal plant in the IL-based surfactant • Before HPLC injection: filtration	HPLC-UV	—	96.5–104.2/RSD <2.8 (–)/-	[68]
$C_4MIm\text{-}BF_4$[d] (10 mL, 0.6 g·mL^{-1}), $C_{10}MIm\text{-}BF_4$[d] (10 mL, 0.8 g·mL^{-1}) or $AllMIm\text{-}BF_4$[d] (10 mL, 0.4 g·mL^{-1})	Podophyllotoxin	3 types of medicinal plants (0.1 g)/-	MW (50–60°C, 10–15 min)	Centrifugation (10,000 rpm, 30 min) and dilution up to 25 mL with water	HPLC-UV-DAD	Comparison with other techniques	97.1–102/RSD[g] < 5.8 (–)/0.51–0.56[e] mg·g^{-1}	[69]
$C_8MIm\text{-}Cl$[d] (0.50 mol·L^{-1})	Tanshinones (3)	Medicinal plant (0.05 g)/-	US (105 W, 80 min) followed by *in situ* IL-DLLME[h] using 1.0 mL of the filtrate, 20 μL $H\text{-}PF_6$, and NaCl (15%, w/v), also assisted by US (135 W, 6 min)	• After US: filtration • After *in situ* IL-DLLME[h]: centrifugation (10,000 rpm, 5 min)	HPLC-UV	• Solid: liquid ratio of 1:40 (g·mL^{-1}) • $H\text{-}PF_6$ as anion-exchange reagent • Comparison with other techniques	84.3 – 96.2/RSD[g] < 4.7 (–)/3.20–3.88[e] μg·g^{-1}	[70]

(*Continued*)

TABLE 23.4 (*Continued*)

IL-based surfactant (Amount)	Analytes (Number)	Sample (Amount)/ Sample Requirement	Extraction Technique	Clean-Up Steps	Analysis	Comments	RR[a] (%)/RSD[b] in % (spiked level)/LOD[c]	Reference
$C_{16}C_4$Im-Br (4.5 mL, 40 mmol·L^{-1})	PAHs[i] (16)	Toasted cereals (0.1 g)/-	MW (40 W, 80°C, ~14 min)	Centrifugation (4000 rpm, 5 min)	HPLC-UV-FD	—	70.1–109/RSD < 12.6 (0.04 mg·kg^{-1})/ 0.001–1.037 μg·g^{-1}	[71]
C_4MIm-MSO_4[d] (10 mL, 1 mol·L^{-1})	Flavonoids (2)	Celery (0.5 g)/ pH 1.0	US (20 kHz, 200 W, 90 min)	—	HPLC-UV	Solid : liquid ratio 1:10	72.7–89.5/RSD < 3.1 (10 μg·mL^{-1})/ 1.4–2.2[e] μg·g^{-1}	[72]
$C_{16}C_4$Im-Br (4.5 mL, 40 mmol·L^{-1})	PAHs[i] (16)	Toasted cereals (0.1 g)/-	MW (50 W, 80°C, ~14 min) followed by *in situ* IL-DLLME[h] using 1.5 mL of the filtrate and Li-NTf_2 (34 μL, 0.5 g·mL^{-1})	• After MW: centrifugation (4000 rpm, 5 min) and filtration of the IL-based surfactant phase • After *in situ* IL-DLLME[h]: vortex (2.5 min), centrifugation (3400 rpm, 4 min), cooling in the freezer (1 h), and dilution of the droplet with ACN (up to 95 μL)	HPLC-UV-FD	• Li-NTf_2 as anion-exchange reagent • Comparison with CTAB[f]	87.9–128/RSD < 22 (9–4500 μg·kg^{-1})/ 0.03–83 μg kg^{-1}	[56]

$C_8MIm\text{-}Br$[d] ($0.75\,mol \cdot L^{-1}$)	Phloro-glucinols (4)	Medicinal plant(1.0g)/-	MW (600 W, 50°C, 7 min)	Cooling to room temperature, dilution up to 20 mL, and filtration	HPLC-DAD	• Solid:liquid ratio of 1:12 ($g \cdot mL^{-1}$) • Comparison with other techniques	96.9–103.5/RSD < 4.7 (–)/5.52–9.36[e] $\mu g \cdot g^{-1}$	[73]
$C_{14}MIm\text{-}Br$[d] ($1.00\,mg \cdot mL^{-1}$)	Zn(II)	Milk powder (1.0 g)/dry, heated (600°C, 6 h), dilution up to 100 mL with HCl, 1:1 v/v, and NH_3-NH_4Cl buffer (pH 7.5)	—	Quelating reaction (10 min) before determina-tion	Spectro-photometry	• 5-Br-PADAP[j] (1 mL, $5.0 \cdot 10^{-4}$ $mol \cdot L^{-1}$) as quelating reagent • The IL-based surfactant enhances the absorbance signal	98.7–103.0/RSD < 4.65 (–)/0.27[e] $\mu g \cdot g^{-1}$	[23]

[a] Relative recovery.
[b] Relative standard deviation.
[c] Limit of detection.
[d] IL-based surfactants selected (others were also tried in the study).
[e] Calculated as LOD of analyte in the sample solution *x* dilution factor (the ratio volume of sample solution to weight of solid sample dissolved in it).
[f] Cationic surfactant cetyltrimethylammonium bromide.
[g] Interday precision.
[h] Ionic liquid dispersive liquid–liquid microextraction.
[i] Polycyclic aromatic hydrocarbons.
[j] 2-[(5-bromo-2-pyridyl)-azo]-5-(diethyl-amino)phenol.

Moreover, the methods have been described for a variety of environmental solid samples, sediments [36, 60, 62] and soils [61], and for different environmental liquid samples, water [59, 61, 64], wastewater [63], and personal care products [23]. Low amounts of samples are always required, which support the usefulness of the methods.

In some cases, a preconcentration step has been required to obtain the adequate analytical sensitivity. Thus, Delgado et al. used the *in situ* IL-DLLME method with Li-NTf_2 as anion-exchange reagent [62] to determine organic contaminants in marine sediments.

There is also an interesting application that uses $(COOC_{14}Py)_2C_4$-2Br and C_2COOC_{14}Py-Br as modifier of sodium bentonite (Na-Bt) surfaces to extract and remove quantitatively methyl orange from wastewater [63]. Besides, IL-based surfactant hemimicelles have been used as modifiers of sorbents in SPE, coated on solid supports like silica [59] or Fe_3O_4 magnetic nanoparticles (MNPs) [61]. It is interesting to note that when MNPs are used as the solid support, the elution step is facilitated because the adsorbent material is easily removed from the matrix with the help of a magnet. In any case, all works involving IL-based surfactants as modifiers of sorbents require a deep study of the adsorption isotherms to ensure proper performance.

The majority of these environmental methodologies involving IL-based surfactants have been carried out in combination with HPLC determination [36, 59, 61, 62, 64] or spectrophotometry [23, 63]. Nevertheless, Guerra-Abreu et al. employed SPME-GC-MS [60]. Thus, IL-based surfactants can be used in combination with any determination technique, which enormously expands the applications of these reagents.

With regard to the use of IL-based surfactant in the extraction schemes for food analysis, Table 23.4 summarizes the main applications found in the literature [20, 23, 56, 65–73].

The majority of the works are focused in the determination of organic compounds with biogenic activity [20, 23, 65–70, 72, 73] or organic contaminants [56, 71]. There is only one method that determines a metal, specifically Zn(II) in milk [23].

Regarding food applications, medicinal plants [20, 65, 67–70, 73] and cereals [56, 71, 72] have been extensively studied. Low amounts of food samples between 0.05 and 1 g are also required, as it happened with environmental applications.

In the use of IL-based surfactants as extraction solvents in food analysis, the extraction of the analytes is normally undertaken with the aid of microwaves [56, 65, 67, 69, 71, 73] or ultrasounds [20, 66–68, 70, 72].

The utilization of the *in situ* preconcentration via IL-DLLME has been described using H-PF_6 [70] or Li-NTf_2 [56] as the anion-exchange reagent. The separation technique of choice when employing IL-based surfactants in food analysis is HPLC with ultraviolet (UV) detection [56, 65, 67, 68, 70–72], diode array detection (DAD) [20, 69, 73], or fluorescence (FD) detection [56, 71].

It is important to note that several works, in both environmental and food analysis, compare the extraction performance of the IL-based surfactant with the conventional cationic surfactant CTAB [20, 36, 56, 60, 61]. For example, Pino et al. have reported much better extraction efficiencies of PAHs from marine sediments when using $C_{16}C_4$Im-Br (being quantitative) than when using CTAB under the same extraction conditions [36]. Moreover, this application was validated using a certified reference material (CRM).

Methods based on IL-based surfactants (Tables 23.3 and 23.4) are generally characterized for presenting short extraction times (mainly when compared to conventional extraction methods), relative recoveries around 100%, high precision values (with low relative standard deviation values, <22%), and low limits of detections (with LOD values down to low $\mu g \cdot g^{-1}$ for solid samples). Methods have been also compared with conventional extraction methods, with methods using conventional surfactants, and even validated with CRMs, thus proving their successful performance in separation science.

23.4 CONCLUDING REMARKS

IL-based surfactants are mainly characterized for presenting low CMC values and also higher interaction affinities for a variety of compounds when compared to conventional cationic surfactant analogues. These two key factors, together with their relatively simple synthesis, justify the current number of diverse analytical applications based on IL-based surfactants, which we believe would continue to flourish.

REFERENCES

[1] Sosa-Ferrera, Z.; Padrón-Sanz, C.; Mahugo-Santana, C.; Santana-Rodríguez, J. J. *Trends Anal. Chem.* 2004, **23**, 469–479.

[2] Ruan, Q.; Sai, X.-Z.; Ruan, Z. *Yunnan Huagong* 2011, **38**, 23–29.

[3] Yazdi, A. S. *Trends Anal. Chem.* 2011, **30**, 918–929.

[4] Ruiz-Ángel, M. J.; Carda-Broch, S.; García-Álvarez-Coque, M. C. *Sep. Purif. Rev.* 2013, **42**, 1–27.

[5] Rizvi, S. A. A.; Do, D. P.; Saleh, A. M. *Eur. J. Chem.* 2011, **2**, 276–281.

[6] Paleologos, E. K.; Giokas, D. L.; Karayannis, M. I. *Trends Anal. Chem.* 2005, **24**, 426–436.

[7] Pino, V.; Germán-Hernández, M.; Martín-Pérez, A.; Anderson, J. L. *Sep. Sci. Technol.* 2012, **47**, 264–276.

[8] Singh, T.; Kumar, A. *Colloids Surf. A Physicochem. Eng. Asp.* 2008, **318**, 263–268.

[9] Miskolczy, Z.; Sebök-Nagy, K.; Biczók, L.; Göktürk, S. *Chem. Phys. Lett.* 2004, **400**, 296–300.

[10] Modaressi, A.; Sifaoui, H.; Mielcarz, M.; Domanska, U.; Rogalski, M. *Colloids Surf. A Physicochem. Eng. Asp.* 2007, **302**, 181–185.

[11] Goodchild, I.; Collier, L.; Millar, S. L.; Prokes, I.; Lord, J. C. D.; Butts, C. P.; Bowers, J.; Webster, J. R. P.; Heenan, R. K. *J. Colloid Interface Sci.* 2007, **307**, 455–468.

[12] Baltazar, Q. Q.; Chandawalla, J.; Sawyer, K.; Anderson, J. L. *Colloids Surf. A Physicochem. Eng. Asp.* 2007, **302**, 150–156.

[13] Zhao, Y.; Gao, S.; Wang, J.; Tang, J. *J. Phys. Chem. B* 2008, **112**, 2031–2039.

[14] Maia, F. M.; Rodríguez, O.; Macedo, E. A. *J. Chem. Thermodyn.* 2012, **48**, 221–228.

[15] Clavero, E.; Rodriguez, J. *J. Mol. Liq.* 2011, **163**, 64–69.

[16] Vaghela, N. M.; Sastry, N. V.; Aswal, V. K. *Colloids Surf. A Physicochem. Eng. Asp.* 2011, **373**, 101–109.

[17] Blesic, M.; Marques, M. H.; Plechkova, N. V.; Seddon, K. R.; Rebelo, L. P. N.; Lopes, A. *Green Chem.* 2007, **9**, 481–490.

[18] Jungnickel, C.; Łuczak, J.; Ranke, J.; Fernández, J. F.; Müller, A.; Thöming, J. *Colloids Surf. A Physicochem. Eng. Asp.* 2008, **316**, 278–284.

[19] Łuczak, J.; Jungnickel, C.; Markiewicz, M.; Hupka, J. *J. Phys. Chem. B* 2013, **117**, 5653–5658.

[20] Wu, K.; Zhang, Q.; Liu, Q.; Tang, F.; Long, Y.; Yao, S. *J. Sep. Sci.* 2009, **32**, 4220–4226.

[21] Cornellas, A.; Perez, L.; Comelles, F.; Ribosa, I.; Manresa, A.; Garcia, M. T. *J. Colloid Interface Sci.* 2011, **355**, 164–171.

[22] Liu, J.; Zhao, M.; Zhang, Q.; Sun, D.; Wei, X.; Zheng, L. *Colloid Polym. Sci.* 2011, **289**, 1711–1718.

[23] Hou, T.; Zhu, X. *J. Mol. Liq.* 2012, **166**, 17–21.

[24] Dong, B.; Li, N.; Zheng, L. Q.; Yu, L.; Inoue, T. *Langmuir* 2007, **23**, 4178–4182.

[25] Sirieix-Plénet, J.; Gaillon, L.; Letellier, P. *Talanta* 2004, **63**, 979–986.

[26] Vanyúr, R.; Biczók, L.; Miskolczy, Z. *Colloids Surf. A Physicochem. Eng. Asp.* 2007, **299**, 256–261.

[27] Borissova, M.; Palk, K.; Koel, M. *J. Chromatogr. A* 2008, **1183**, 192–195.

[28] El Seoud, O. A.; Pires, P. A. R.; Abdel-Moghny, T.; Bastos, E. L. *J. Colloid Interface Sci.* 2007, **313**, 296–304.

[29] Galgano, P. D.; El Seoud, O. A. *J. Colloid Interface Sci.* 2011, **361**, 186–194.

[30] Sastry, N. V.; Vaghela, N. M.; Aswal, V. K. *Fluid Phase Equilib.* 2012, **327**, 22–29.

[31] Inoue, T.; Ebina, H.; Dong, B.; Zheng, L. *J. Colloid Interface Sci.* 2007, **314**, 236–241.

[32] Dong, B.; Zhao, X.; Zheng, L.; Zhang, J.; Li, N.; Inoue, T. *Colloids Surf. A Physicochem. Eng. Asp.* 2008, **317**, 666–672.

[33] Gu, Y.; Shi, L.; Cheng, X.; Lu, F.; Zheng, L. *Langmuir* 2013, **29**, 6213–6220.

[34] Geng, F.; Liu, J.; Zheng, L.; Yu, L.; Li, Z.; Li, G.; Tung, C. *J. Chem. Eng. Data* 2010, **55**, 147–151.

[35] Niu, J.; Qiu, H.; Li, J.; Liu, X.; Jiang, S. *Chromatographia* 2009, **69**, 1093–1096.

[36] Pino, V.; Anderson, J. L.; Ayala, J. H.; González, V.; Afonso, A. M. *J. Chromatogr. A* 2008, **1182**, 145–152.

[37] Mahajan, S.; Sharma, R.; Mahajan, R. K. *Langmuir* 2012, **28**, 17238–17246.

[38] Rizvi, S. A. A.; Shamsi, S. A. *Anal. Chem.* 2006, **78**, 7061–7069.

[39] Schnee, V. P.; Baker, G. A.; Rauk, E.; Palmer, C. P. *Electrophoresis* 2006, **27**, 4141–4148.

[40] Wiedmera, S. K.; King, A. W. T.; Riekkola, M.-L. *J. Chromatogr. A* 2012, **1253**, 171–176.

[41] Mester, P.; Wagner, M.; Rossmanith, P. *Sep. Purif. Technol.* 2012, **97**, 211–215.

[42] Pino, V.; Yao, C.; Anderson, J. L. *J. Colloid Interface Sci.* 2009, **333**, 548–556.

[43] Flieger, J.; Siwek, A.; Pizoń, M.; Czajkowska-Zelazko, A. *J. Sep. Sci.* 2013, **36**, 1530–1536.

[44] Qui, H.; Zhang, Q.; Chen, L.; Liu, X.; Jiang, S. *J. Sep. Sci.* 2008, **31**, 2791–2796.

[45] Schnee, V. P.; Palmer, C. P. *Electrophoresis* 2008, **29**, 767–776.

[46] Rageh, A. H.; Pyell, U. *J. Chromatogr. A* 2013, **1316**, 135–146.

[47] Su, H.-L.; Lan, M.-T.; Hsieh, Y.-Z. *J. Chromatogr. A* 2009, **1216**, 5313–5319.

[48] Quirino, J. P.; Anres, P.; Sirieix-Plénet, J.; Delaunay, N.; Gareil, P. *J. Chromatogr. A* 2011, **1218**, 5718–5724.

[49] Wang, Q.; Qiu, H.; Han, H.; Liu, X.; Jiang, S. *J. Sep. Sci.* 2012, **35**, 589–595.

[50] Pacifici, G. M.; Viani, A. *Clin. Pharmacokinet.* 1992, **23**, 449–468.

[51] Jepson, G. W.; Hoover, D. K.; Black, R. K.; McCafferty, J. K.; Mahle, D. A.; Gearhart, J. M. *Fundam. Appl. Toxicol.* 1994, **22**, 519–524.

[52] Pino, V.; Ayala, J. H.; González, V.; Afonso, A. M. *Anal. Chem.* 2004, **76**, 4572–4578.

[53] Pino, V.; Baltazar, Q. Q.; Anderson, J. L. *J. Chromatogr. A* 2007, **1148**, 92–99.

[54] Yao, C.; Pino, V.; Anderson, J. L. *J. Chromatogr. A* 2009, **1216**, 948–955.

[55] Armenta, S.; Garrigues, S.; De la Guardia, M. *Trends Anal. Chem.* 2008, **27**, 497–511.

[56] Germán-Hernández, M.; Pino, V.; Anderson, J. L.; Afonso, A. M. *J. Chromatogr. A* 2012, **1227**, 29–37.

[57] Trujillo-Rodríguez, M. J.; Rocío-Bautista, P.; Pino, V.; Afonso, A. M. *Trends Anal. Chem.* 2013, **51**, 87–106.

[58] Gangula, S.; Suen, S.-Y.; Conte, E. D. *Microchem. J.* 2010, **95**, 2–4.

[59] Li, J.; Cai, Y.; Shi, Y.; Mou, S.; Jiang, J. *Talanta* 2008, **74**, 498–504.

[60] Guerra-Abreu, L.; Pino, V.; Anderson, J. L.; Afonso, A. M. *J. Chromatogr. A* 2008, **1214**, 23–29.

[61] Zhang, Q.; Yang, F.; Tang, F.; Zeng, K.; Wu, K.; Cai, K.; Yao, S. *Analyst* 2010, **135**, 2426–2433.

[62] Delgado, B.; Pino, V.; Anderson, J. L.; Ayala, J. H.; Afonso, A. M.; González, V. *Talanta* 2012, **99**, 972–983.

[63] Wang, C.; Jiang, X.; Zhou, L.; Xia, G.; Chen, Z.; Duan, M.; Jiang, X. *Chem. Eng. J.* 2013, **219**, 469–477.

[64] Qin, H.; Li, B.; Liu, M. S.; Yang, Y. L. *J. Sep. Sci.* 2013, **36**, 1463–1469.

[65] Lu, Y.; Ma, W.; Hu, R.; Dai, X.; Pan, Y. *J. Chromatogr. A* 2008, **1208**, 42–46.

[66] Cao, X.; Ye, X.; Lu, Y.; Yu, Y.; Mo, W. *Anal. Chim. Acta* 2009, **640**, 47–51.

[67] Du, F.-Y.; Xiao, X.-H.; Xu, P.-P.; Li, G.-K. *Acta Chromatogr.* 2010, **22**, 459–471.

[68] Ma, C.-H.; Liu, T.-T.; Yang, L.; Zu, Y.-G.; Wang, S.-Y.; Zhang, R.-R. *Anal. Chim. Acta* 2011, **689**, 110–116.

[69] Yuan, Y.; Wang, Y.; Xu, R.; Huang, M.; Zeng, H. *Analyst* 2011, **136**, 2294–2305.

[70] Bi, W.; Tian, M.; Row, K. H. *Talanta* 2011, **85**, 701–706.

[71] Germán-Hernández, M.; Pino, V.; Anderson, J. L.; Afonso, A. M. *Talanta* 2011, **85**, 1199–1206.

[72] Han, D.; Row, K. H. *J. Sci. Food Agric.* 2011, **91**, 2888–2892.

[73] Li, X.-J.; Yu, H.-M.; Gao, C.; Zu, Y.-G.; Zu, Y.-G.; Wang, W.; Luo, M.; Gu, C.-B.; Zhao, C.-J.; Fu, Y.-J. *J. Sep. Sci.* 2012, **35**, 3600–3608.

CHAPTER 24

Ionic Liquids: Surfactant Agents for Layered Silicates

SÉBASTIEN LIVI, JEAN-FRANÇOIS GÉRARD, and JANNICK DUCHET-RUMEAU

Université de Lyon, Lyon, France
INSA Lyon, Villeurbanne, France
Laboratoire d'Ingénierie des Matériaux Polymères, CNRS, UMR 5223, Villeurbanne, France

24.1 INTRODUCTION

There is a challenge in academic and industrial research to create and design new polymer materials with significant improvements in their physical properties. One of the innovative pathways is by way of addition of ionic liquids (ILs) within the polymer matrix. ILs having melting points below 100°C have unique physicochemical properties suitable for numerous applications wherein conventional organic solvents and additives are mostly ineffective. Excellent thermal and chemical stability, inflammability, low vapor pressure, and high ionic conductivity of ILs make them prospective processing aids for advanced materials. Many (not all) ILs are "green solvents" because they have negligible volatility. Further, they are nonexplosive and nontoxic and can be recycled. The great advantage of ILs is to design them for specific purposes by changing a profound number of cation–anion combinations. ILs offer media (new and efficient) for synthesis and modification of polymers [1–4]. Studies in the literature report the uses of ILs as multifunctional agents of polymers for processing advanced materials [5–7]. ILs based on ammonium cations have been commonly used as surfactants for dispersion of layered silicates during the last 20 years since the discovery of the nanocomposites by Okada et al. [8]. This surface treatment generates organically modified clays (or organoclays) having improved compatibility with most of the polymers, which is a prerequisite for a successful formation of polymer–clay nanocomposites including a control over the distribution in size and dispersion of the nanofillers. Nevertheless, the number of publications combining polymers and ILs still is only limited. The large number of possible cation–anion combinations of ILs offers a unique opportunity for the tunability of ILs toward polymers. This chapter aims to discuss the state-of-the-art approaches of polymers combined with ILs generating materials of diverse properties and

Ionic Liquid-Based Surfactant Science: Formulation, Characterization, and Applications, First Edition. Edited by Bidyut K. Paul and Satya P. Moulik.

applications of such advanced materials. In this chapter, the use of thermostable ILs mainly based on pyridinium, imidazolium, and phosphonium cations as surface-active agents in the formation of polymer nanocomposites based on layered silicates have been focused.

24.2 ILs: INTERCALANTS FOR LAYERED SILICATES

24.2.1 Structure and Properties of Layered Silicates

For the preparation of polymer–layered silicate (PLS) nanocomposites, the most commonly used are layered silicates (montmorillonite (MMT), hectorite, and saponite), which belong to phyllosilicates 2:1. Their crystal lattice consists in layers made up of an octahedral sheet of either aluminum or magnesium hydroxide confined between two tetrahedrally coordinated silicon atoms layers. The layer thickness is about few nanometers, and the other dimensions of these platelets vary from tens of nanometers to few micrometers, depending on the nature of the layered silicate. Their nanoscale dimensions are responsible for a *high specific surface* (from 400 up to 700 $m^2.g^{-1}$), and their particular morphology confers them a *high aspect ratio* (from 100 up to 1000). The consequence of the stack of the sheets leads to a van der Waals gap between the layers denoted interlayer space, *d*-spacing, or clay gallery. In fact, isomorphic substitutions (Al^{3+} by Mg^{2+}) create locally negative charges on the layer surface that are compensated for positive ions (alkali or alkaline cations) localized inside the clay galleries. These exchangeable cations are easily replaced by other organic cations that separate the platelets, and the whole dimension (basal spacing d_{001}) depends on the nature of the cations and the degree of the clay hydration. In conclusion, the layered silicates, for example, the MMT with chemical formula $M_x(Al_{4-x}Mg_x)Si_8O_{20}(OH)_4$, have the ability to exchange cations, which are key parameter for the preparation of PLS nanocomposites. In fact, the type of layered silicates is characterized by a specific property known as the cationic exchange capacity (CEC) and generally expressed in milliequivalent per gram (meq/g). The CEC corresponds to the substitution of the sodium or calcium compensating for cations initially present at the platelet surface by monovalent cations in 100 g of clay. Generally, the typical value of the CEC is included between 60 and 200 meq/100 g.

24.2.2 Organic Modification of Layered Silicates

The low cost of layered silicates, especially MMT, is the driving force for their use in the field of nanocomposites. However, it is well known that this nanofiller is only compatible with hydrophilic polymers such as poly(ethylene oxide) (PEO) [9] and poly(vinylalcohol) (PVA) [10]. Regarding hydrophobic polymers, a nonmiscibility, analogous to polymer blends, is observed, which results from the poor interfacial interactions between organic and inorganic compounds. Such a low affinity leads to poor thermal and mechanical properties of the nanocomposites, which display a microcomposite-like morphology. It is therefore necessary to modify the surface of clays in order to improve interfacial interactions. One of the most commonly used surface treatments involves cationic exchange.

24.2.2.1 Cationic Exchange of Layered Silicates The intercalation of inorganic exchangeable cations of layered silicates by organic cations in aqueous medium has been proposed for the first time by Gieseking [11] and Hendricks [12]. The mineral host structure and the nature of the intercalating agents are two important parameters determining the success of the cationic exchange. Generally, the cationic exchange takes place in water or water/alcohol mixture. In fact, the use of ultrapure water is preferred in order to facilitate the clay swelling and to get rid of undesired cation. Sometimes, it is necessary to use an organic solvent as a cosolvent to get a complete dissolution of the intercalating agents. In the first step, the layered silicates are mixed and vigorously stirred in the aqueous medium. A required amount of the modifier agent based on the CEC of the lamellar silicates usually equivalent to two times of CEC previously dissolved in the organic solvent is added to the clay suspension. Recently, a more environment-friendly cationic exchange process has been developed with supercritical carbon dioxide (sc-CO_2) [13]. Several parameters such as temperature, organic modifier concentration, and reaction time ensure the success of the cationic exchange. Indeed, the cationic exchange involves several types of interactions: electrostatic and also weak intermolecular interactions between the clay surface and the surfactant. The nature of the counter ions and the steric hindrance of the intercalating agents play a crucial role in the distribution of the organic species as the physically adsorbed species on the clay surface and the intercalated species in the clay intergalleries [14, 15].

Identification of the Organic Species Thermogravimetric analysis (TGA) may be employed for analysis of clay by investigating the degradation mechanisms and the identification of the interactions between the clay mineral and the intercalating agents. According to the literature [16], two types of interactions take place that can be evidenced from thermal degradation kinetics: (i) degradations from 150 to 350°C that correspond to the organic species physically adsorbed on the clay surface via van der Waals interactions and (ii) degradations from 350 to 550°C associated with the intercalated species via ionic interactions in the clay galleries. For example, Figure 24.1 illustrates the degradation mechanisms of MMT modified with an imidazolium IL [17].

Selection of the IL For several years, many of organic salts have been used as modifier agents of layered silicates and for the preparation of polymer–clay nanocomposites. The most commonly used surfactants are alkylammonium salts [18–26]. Recently, alkylpyridinium, alkylimidazolium, and alkylphosphonium compounds have emerged. They reduce the surface energy of nanoclay and can improve their compatibility with the polymer matrix [27]. Besides, the presence of long chains in them induces a larger interlayer spacing, which allows the diffusion of polymer chains between the clay layers making them to dissociate [28, 29].

Ammonium ILs Currently, an intensive research is done on the modification of layered silicates by several varieties of modifiers associated with chloride and bromide anions [18, 19]. The alkylammonium salts derived from the complete alkylation of amine are the most used ILs to prepare organoclays. The main role of quaternary alkylammonium is to lower the surface energy of the modified clays and to increase the basal spacing, that is, to allow the diffusion of the polymer between

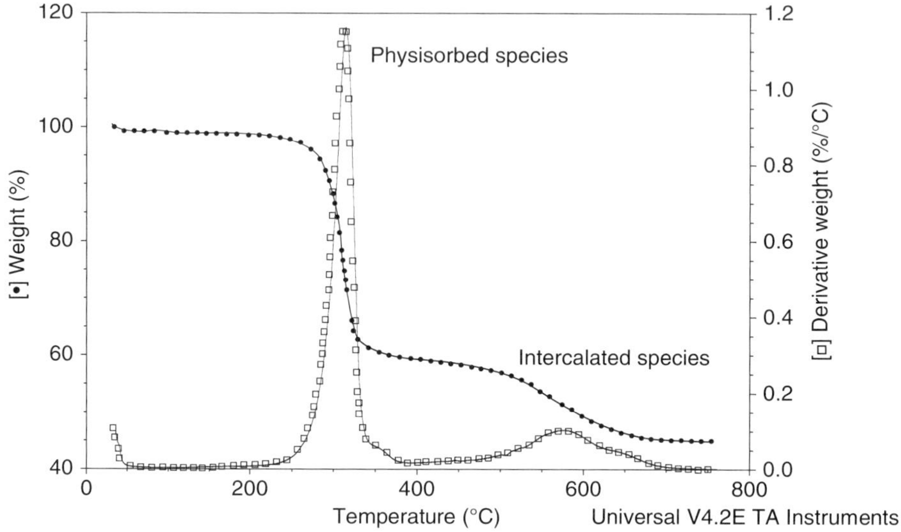

Figure 24.1 Weight loss and weight loss derivative as a function of temperature (TGA and DTG) of the imidazolium-treated montmorillonite.

the nanoplatelets. The chemical nature of cation, length of carbon chains, and size and shape of the polar head have major effects on the cationic exchange efficiency.

In the case of polymer/clay nanocomposites, alkylammonium exchange species influence the affinity between the polymer and the clay surface. For example, it was reported that clays treated with dialkyl dimethylammonium halides, in particular with two chains of about 18 carbon atoms, have a surface energy similar to poly(olefins) such as polypropylene (PP) and polyethylene (PE) [27]. Polar polymers as polyamides (PA) have been recommended to get better interactions as reported by Toyota [8]. The alkyl chain length is related with increase in interlayer space required for the intercalation of polymer chains. Because of the nonpolar nature of their chains, they reduce the electrostatic interactions between the silicate layers and lower the surface energy of the layered silicates. As a consequence, an optimal diffusion of the polymer to dissociate the stacked clay layers, that is, an exfoliation process, can be obtained. Despite the compatibility of MMT modified by long alkyl chain quaternary ammonium with hydrophobic polymers (PE and PP), conventional alkylammonium ions show low thermal stability, that is, an onset decomposition temperature is close to 180°C. This poor thermal stability could limit their use in the preparation of PLS with matrices processed at high temperatures such as PA, poly(ethylene terephthalate) (PET), and poly(ether ether ketone) (PEEK) [30].

Thermostable ILs Polymer nanocomposites are often prepared from thermoplastics processed at high temperature or from thermosets with high polymerization temperatures [31, 32]. In order to get thermal stability, the commonly used surfactants combine organic cations containing nitrogen such as pyridinium and imidazolium ILs [33]. Other chemical compounds, such as pyrrolidinium [34] and phosphonium [35, 36], were also considered due to their excellent thermal stability compared to ammonium-based ones [37].

Slade et al. [38] used hexadecylpyridinium IL to modify vermiculite for the adsorption of nonpolar molecules. The adsorption of pollutants such as chlorobenzene or phenanthrene by dodecylpyridinium-modified bentonite was also studied [39, 40]. But thermal stability of modified nanoclays and their uses in the preparation of PLS nanocomposites have been scarcely studied. Nevertheless, the importance of the chemical nature of the anion on the thermal stability of the modified clays has been shown. Lee and Kim [41] have demonstrated that the combination of pyridinium cation with fluorinated anion such as BF_4^- leads to an increase in the thermal stability of the exchanged clays.

The imidazolium ILs, opposite to the pyridinium salts, have been the subject of several studies for the preparation of organically modified clays, compared to the conventional quaternary ammonium-treated ones [31, 42, 43]. The research focused on the thermal decomposition of the imidazolium salts and the role of their chemical structure in it. According to the reports [44, 45], at high temperature, the imidazole cation is resistant to ring fission during thermal rearrangements of 1-alkyl and 1-arylimidazoles. The results support a better thermal stability of the imidazolium cation compared to the ammonium salts. Recently, Livi et al. have demonstrated that imidazolium ILs have a better thermal stability compared to ammonium salts [17]. In fact, the imidazolium-treated MMT and the MMT bearing a dimethyl ditallow quaternary ammonium as cation were compared. Increases in the degradation temperatures were clearly observed (+50°C for physisorbed species as well as for intercalated ones).

The influence of the anion type on the thermal stability of the imidazolium ILs has been studied by Awad et al. [32]. They have demonstrated that for the hexafluorophosphate PF_6^-, tetrafluoroborate BF_4^-, and bis(trifluoromethylsulfonyl)amide $[(CF_3SO_2)_2]N^-$, increase of 100°C in the degradation temperature was obtained compared to the use of the halide salts (bromide and chloride). For nanoclays modified with imidazolium salts as surfactants, the works reported in the literature demonstrate a large increase of degradation temperature under nitrogen and oxidative atmosphere compared to ammonium-treated MMT. However, despite the intrinsic stability of imidazolium ILs induced by the substitution of halide anions by fluorinated ones, no significant improvement in the thermal stability of the MMT treated by ILs having tetrafluoroborate or hexafluorophosphate anions has been reported.

Because of the well-known properties of phosphorous compounds, such as flame retardancy and heat stabilization, the use of phosphonium IL as intercalating agents for layered silicates has been widely studied [46–50]. According to the literature, the large increase (about 50–70°C) of the degradation temperature for phosphonium-modified clays compared to ammonium-treated ones is relevant for the preparation of PLS in the melt blending, which involves high temperatures [13, 51]. In 2007, Patel et al. [51] prepared a series of MMT modified with different bromide quaternary phosphonium salts, denoted methyl, ethyl, propyltriphenylphosphonium, tetraphenylphosphonium, and tributyltetradecylphosphonium, displaying a higher thermal stability than the ammonium-treated MMT. In particular, a large increase up to 350–450°C of the thermal stability of the tetrabutylphosphonium- and tetraphenylphosphonium-modified MMT has been shown. Recently, other authors have highlighted that the combination of the anion associated with the organic cation based on phosphonium salt has a significant role on the thermal stability of IL. In fact, the use of hexafluorophosphate (PF_6^-) combined with the phosphonium cation causes a significant increase of temperature about 140°C compared to the halide salts (Cl^- and Br^-) [43].

In conclusion, it was demonstrated that the use of ILs as intercalating agents of layered silicates was developed by using quaternary alkylammonium. However, the use of these ILs is limited by their low thermal stability, which implies to focus on more thermostable ILs based on pyridinium, imidazolium, and phosphonium cations. Nevertheless, does the addition of such highly thermostable IL surfactants improve the dispersion of nanofillers in polymer matrices as well as the final properties of the resulting nanocomposites?

24.3 ILs-TREATED LAYERED SILICATES FOR POLYMER NANOCOMPOSITES

24.3.1 PLS Nanocomposites

The organic–inorganic hybrid materials have shown significant increases in properties compared to the conventional composites or neat polymers. The degree of dispersion of nanofillers in a polymer matrix and the processing method play a key role on the final properties of the materials. Key objectives of researches are to design nanocomposites with enhanced properties at low filler contents. Different modified clays have been used in view of these objectives [31, 52]. There are reports on the use of ammonium-treated layered silicates [52, 53], whereas the use of thermostable ILs such as pyridinium, imidazolium, or phosphonium is poorly reported. However, their combinations with poly(styrene) (PS) [54], PE [55], PP [56], poly(vinylidene fluoride) (PVDF) [57], and PET matrices [58] have been reported in the literature.

24.3.2 Processing Routes to Design PLS Nanocomposites

The preparation of PLS nanocomposites is strongly dependent on the nature of the polymer matrix such as its surface energy, molar mass, the presence of reactive groups, characteristics of the layered silicates, and the chemical nature of the intercalants. Three main routes were reported to combine layered silicates with a polymer matrix (Fig. 24.2).

24.3.2.1 Solution Intercalation Solution intercalation requires the use of a solvent in which the polymer is firstly dissolved, while modified clays or pristine clays are dispersed in the same solvent suitable to allow the swelling of the clay and to facilitate the insertion of polymer chains between the clay layers. The solvent is then evaporated to form the nanocomposites. This procedure was considered for water soluble polymers such as PEO, poly(acrylic acid) (PAA), and PVA [9, 10, 59, 60] and for polymers soluble in organic solvents such as PS.

24.3.2.2 In Situ Intercalative Polymerization *In situ* polymerization is the processing route used by Toyota [8, 61]. The monomer introduced between clay layers undergoes the polymerization in the presence of an initiator. A variety of

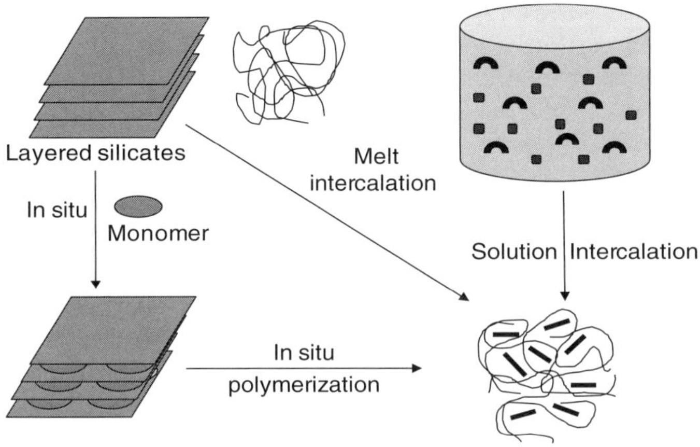

Figure 24.2 Different routes for the preparation of the nanocomposites.

polymer nanocomposites has been prepared using this method, that is, PA [62], PS [63], or poly(olefins) (PP or PE) in layered silicates [56].

24.3.2.3 Melt Intercalation The great advantage of the melt intercalation technique does not require a solvent. In this process, layered silicates or modified layered silicates can be mixed with the polymer matrix in the molten state using methods like extrusion and injection. In the molten state, polymer chains are intercalated between nanoplatelets and lead to exfoliated nanocomposites. A wide range of PLS nanocomposites such as poly(olefins) [56] and PVDF [57] have been generated from such route.

24.4 APPLICATION OF ILs IN POLYMER NANOCOMPOSITES

24.4.1 PS/IL-Modified Clay Nanocomposites

Nanocomposites based on acrylonitrile–butadiene–styrene (ABS) matrix and ammonium- or imidazolium-modified layered silicates (5–15 wt.%) have been prepared in boiling acetone by ultrasonic processing to improve the efficiency of mixing [64, 65]. Nevertheless, the XRD spectra of the resulting nanocomposites of both ABS/ammonium- and imidazolium-treated MMT are similar with a higher degree of 3D distribution. Dynamical mechanical spectroscopy (DMA) measurements reveal an increase of the storage modulus by 40% at 25°C for the ABS/imidazolium-treated MMT. These have been attributed to the good dispersion level of clay layers in the polymer matrix and also to the excellent thermal stability of the imidazolium cation, which is not affected during processing at 200°C for 10 min, compared to ammonium-modified clays.

Rheology of nanoparticle suspension in molten state is also used to characterize the dispersion level, that is, to differentiate the degree of exfoliation such as for PS/modified clay nanocomposites [65]. A better dispersion of nanoclays is evidenced when they are treated with imidazolium surfactant.

In the literature, other methods for preparation of nanocomposites based on PS by melt compounding [66–68] or *in situ* polymerization [69, 70] along with various modifiers such as pyridinium, quinolinium, or phosphonium ILs in order to study their effects on the thermal, physical, and mechanical properties of the nanocomposites have been reported. Bottino et al. [69] have prepared PS/MMT nanocomposites at 3 wt.% via *in situ* polymerization. The influence of the surfactant (ammonium or imidazolium) and the effects of alkyl chain length of the imidazolium salt (C_{12}, C_{16}, and C_{18}) on the morphology and the photooxidation properties have been studied. TEM micrographs of PS/MMT treated with an imidazolium salt having a short alkyl chain (C_{12}) show an intercalated morphology with very few exfoliated layers in the polymer matrix compared to the imidazolium functionalized with long alkyl chains (C_{16} and C_{18}) where partially or fully exfoliated morphology is obtained. The effects of imidazolium-treated MMT on the thermal stability of nanocomposites are identical with the ammonium-treated MMT. The photooxidation properties of the PS/modified layered silicates are lower than the neat PS. More recently, Pucci et al. have used three different polymerizable alkylvinylbenzylimidazolium chlorides as cation exchange monomers for MMT to prepare PS/clay nanocomposites [71]. Thus, a better thermal stability is obtained with an increased degradation temperature of 66°C (440°C instead of 374°C for neat PS). These authors have also demonstrated that ILs are good candidates for fire barrier properties. Similar improvement with imidazolium functionalized with polyhedral oligomeric silsesquioxane (POSS) for fire retardancy properties of nanocomposites [72] has been reported. Chu et al. prepared poly(styrene-co-acrylonitrile) (SAN)/ammonium-, imidazolium-, and phosphonium-modified clays by melt compounding procedure nanocomposites [66]. They have concluded that ammonium-bearing polar functions (OH) or imidazolium- and phosphonium-bearing aromatic functions are the most relevant for SAN nanocomposite processing leading to better clay dispersion. Also, the ammonium functionalized hydroxyl groups improve the thermal and the fire resistance properties of the nanocomposites. These results highlighted the importance of the nature of the organic cation, in particular its functionalization (alkyl chains vs. hydroxyl groups vs. epoxy vs. vinyl) as a function of the chemical nature of the matrix considered.

24.4.2 PVDF/IL-Modified Clay Nanocomposites

PVDF nanocomposites based on ammonium-, pyridinium-, imidazolium-, or phosphonium-modified MMT have been prepared by melt intercalation [57, 64–73]. The influence of the modified clays on the morphologies and their consequence on the physical properties have been investigated. XRD and TEM analyses evidence different morphologies as a function of the nature of the surfactant. An exfoliated structure is obtained with ammonium-, pyridinium-, and imidazolium-treated MMT, while phosphonium IL-modified MMT leads to a partially exfoliated morphology. The mechanical properties of nanocomposites are higher than the neat PVDF. However, the influence of thermostable ILs on mechanical properties is limited. In fact, the storage modulus of the neat PVDF and the PVDF nanocomposites remains the same whatever be the nature of the IL intercalant. Nevertheless, the elongation at break increases (+175%) with 5 wt.% of pyridinium-treated MMT compared to +200% of strain at break obtained for ammonium-modified MMT. Recently,

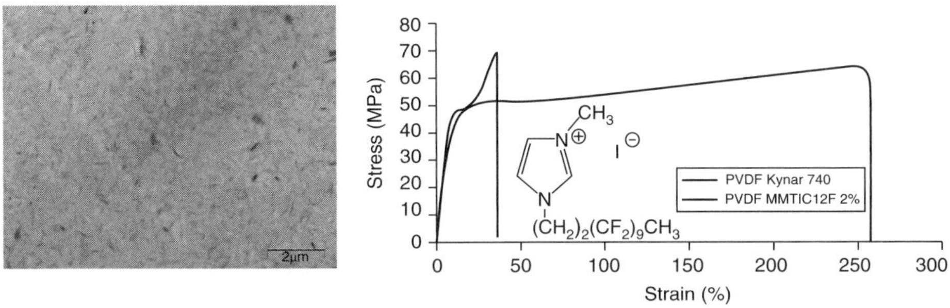

Figure 24.3 Morphology and mechanical behavior of PVDF filled with imidazolium-treated MMT.

Livi et al. demonstrated that the use of perfluoroalkylimidazolium as surfactant agent of MMT (1%) allows a dramatic increase of mechanical properties with a strain at break of 250% compared to 30% for neat PVDF [73]. The impact of MMT modification by the imidazolium IL on the structuration as well as on the mechanical properties of PVDF is shown in Figure 24.3.

In both cases, the authors are also interested on the effects of modified clays on the polymorphic structure of the PVDF mainly composed of the α and β phases. They have noticed that the use of fluorinated imidazolium and phosphonium clays increase the melting and crystallization temperatures of PVDF. Moreover, phosphonium- and imidazolium-treated MMT appear to be the most efficient nucleating agents and to generate β-phase, which is the crystalline phase required for dielectric applications. In fact, we have demonstrated that the addition of only 1 wt.% of imidazolium functionalized with a perfluorinated chain-treated MMT leads to the formation of β phase, which opens new applications in the field of fuel cells.

24.4.3 Poly(olefins)/IL-Modified Clay Nanocomposites

PP/imidazolium-modified clay nanocomposites have been prepared using melt compounding method [56]. Morphological analyses of the nanocomposites containing 3 vol.% of organically treated montmorillonite (OMMT) by TEM showed a mixed intercalated/exfoliated morphology with few tactoïds and the presence of individual clay layers. Young's modulus increases by 35% for PP/OMMT containing 4 vol.% of fillers, while a decrease of strain at break is observed, and an improvement in the relative oxygen permeation is noticed. TGA analysis shows that the addition of only 1 vol.% of imidazolium-modified MMT improves the thermal behavior of PP, but this improvement is optimized for PP/OMMT with 4 vol.% of fillers (+90°C). Mittal [56] has shown that the use of imidazolium-treated clay allowed an increase of the thermal barrier and mechanical properties. Other studies on PP nanocomposites based on imidazolium-modified MMT have been reported [74, 75]. Ding et al. [74] have used imidazolium salt as intercalating agents of MMT for the preparation of nanocomposites by *in situ* dissolution of isotactic PP (xylene used as solvent). The influence of solvent (water vs. xylene) during the MMT modification was also investigated. XRD and TEM analyses show well-exfoliated MMT platelets, which are disorderly dispersed in the PP matrix. The addition of OMMT leads to a significant increase of 166°C for the degradation temperature compared to virgin PP when the treatment of the surface of MMT is carried out in xylene compared to +62°C when water is used.

24.4.4 Polyester/IL-Modified Clay Nanocomposites

Nanocomposites from poly(ε-caprolactone) (PCL) have been prepared by *in situ* polymerization with dibutyltin dimethoxide as an initiator [76]. The materials showed an excellent dispersion of clay layers with highly exfoliated nanocomposite morphology. The authors have demonstrated that the use of imidazolium functionalized with hydroxyl function as intercalating agent of MMT bearing a long flexible alkyl chain allowed the grafting of polymer chains on the MMT surface. This good affinity between the polymer and the modified clay leads to a significant increase of 150% of storage modulus at 30°C.

24.4.5 PVC/IL-Modified Clay Nanocomposites

The influence of modified layered silicates on the thermal stability, especially the thermal behavior of the poly(vinyl chloride) (PVC) nanocomposites containing 2wt.% of clays treated with ionic modifiers as imidazolium, ammonium, or phosphonium salts and nonionic species such as poly(propylene glycol) or glycerol monostearate, was investigated [77]. The nonionic compounds did not affect the thermal stability, whereas the use of cationic surfactants accelerated the dehydrochlorination of PVC in the following order: phosphonium ion > imidazolium ion > alkyl ammonium ion > ethoxy/alkyl ammonium ion. These results show the importance of the choice of the intercalant with respect to the matrix. In this example, for the PVC plasticized with phthalates, an improved thermal stability has been reported [78, 79]. Instead of using ILs as intercalating agents for lamellar silicates to improve the thermal and mechanical properties of the PVC nanocomposites, it may be more relevant to consider only the use of ILs as reinforcing and plasticizing agents [80, 81].

24.5 CONCLUSION

In summary, layered silicates are one-dimensional nanofillers smaller than 100nm with a significant aspect ratio and a high specific surface commonly used to overcome the limitations of conventional polymer composites. However, in most of the cases, pristine layered silicates are not compatible with organic polymers, especially with hydrophobic polymers. This poor affinity between the clay mineral and the polymer leads to composites of poor properties. In order to have better properties, the dispersion of layered silicates in polymer matrices and the interactions between the surface of layered silicates and the polymer matrices require to be improved from organic treatment of clays. The surface modification of layered silicates using modifying agents based on organic cations, especially ammonium salts, has been widely reported.

Nevertheless, the low thermal stability of the quaternary alkylammonium salts due to the Hofmann's elimination has a limiting effect on their use in the processing of the thermoplastic and thermosetting nanocomposites requiring higher temperatures of processing or curing. Recently, ILs based on pyridinium, imidazolium, or phosphonium cations are emerging as new alternatives for the design of thermally stable organically modified clays. The preparation, characterization, and properties of different nanocomposites using these new surfactants to modify the fillers according to the nature of the polymer matrices have been herein reported. We have demonstrated that the PLS nanocomposites with low IL-modified clay amounts (<5 wt.%) have displayed properties mostly superior to the neat polymer matrices. In conclusion, the true potential of ILs based on the infinite cation–anion combinations is yet to be fully exploited. The reduced choice and the high cost of the ILs functionalized with different groups (epoxy, hydroxy, vinyl, and fluorinated chains) are the main reasons for this limitation. To optimize PLS nanocomposites, more planned studies are needed.

REFERENCES

[1] Stanzione, J.F.; Jensen, R.E.; Costanzo, P.J.; Palmese, G.R., *ACS Appl. Mater. Interfaces*, 2012, **4**, 6142–6150.

[2] Mecerreyes, D., *Prog. Polym. Sci.*, 2011, **36**, 1629–1648.

[3] Tonova, K., *Sep. Purif. Technol.*, 2012, **89**, 57–65.

[4] Wilpiszewska, K.; Spychaj, T., *Carbohydr. Polym.*, 2011, **86**, 424–428.

[5] Jimenez, A.E.; Bermudez, M.D.; Iglesias, P.; Carrion, F.J.; Martinez-Nicolas, G., *Wear*, 2006, **260**, 766–782.

[6] Scott, M.P.; Rahman, M.; Brazel, C.S., *Eur. Polym. J.*, 2003, **39**, 1947–1953.

[7] Livi, S.; Duchet-Rumeau, J.; Gérard, J.F., *Chem. Commun.*, 2011, **47**, 3589–3591.

[8] Okada, A.; Fukushina, Y.; Kawasumi, M.; Inagaki, S.; Usuki, A.; Sugiyama, S.; Kurauchi, T.; Kamigaito, O. Composite material and process for manufacturing same US patent no. 4,729,007, 1988.

[9] Aranda, P.; Ruiz-Hitzky, E., *Chem. Mater.*, 1992, **4**, 1395–1403.

[10] Greenland, D.J., *J. Colloid Sci.*, 1963, **18**, 647–664.

[11] Gieseking, J.E., *Soil Sci.*, 1939, **47**, 1–14.

[12] Hendricks, S.B., *J. Phys. Chem.*, 1941, **45**, 65–81.

[13] Livi, S.; Duchet-Rumeau, J.; Gérard, J.F., *J. Colloid Interface Sci.*, 2011, **353**, 225–230.

[14] Le Pluart, L.; Duchet, J.; Sautereau, H.; Gérard, J.F. *J. Adhes.*, 2002, **78**, 645.

[15] Le Pluart, L.; Duchet, J.; Sautereau, H.; Gérard, J.F., *Macromol. Symp.*, 2003, **194**, 155–160.

[16] Davis, R.D.; Gilman, J.W.; Sutto, T.E., *Clay Clay Miner.*, 2004, **52**, 171.

[17] Livi, S.; Pham, T.N.; Duchet-Rumeau, J.; Gérard, J.F., *J. Colloid Interface Sci.*, 2010, **349**, 424–433.

[18] He, H.; Frost, R.L.; Bostrom, T.; Yuan, P.; Duong, L.; Yang, D.; Xi, Y.; Kloprogge, J.T., *Applied Clay Science*, 2006, **31**, 262–271.

[19] Lee, S.Y.; Cho, W.J.; Kim, K.J.; Ahn, J.H.; Lee, M., *J. Colloid Interface Sci.*, 2005, **284**, 667.

[20] Elban, W.L.; Howarter, J.A.; Richardson, M.C.; Stutzman, P.E.; Forster, A.M.; Nolte, A.J.; Holmes, G.A., *Appl. Clay Sci.*, 2012, **61**, 29–36.

[21] Xidas, P.I.; Triantafyllidis, K.S., *Eur. Polym. J.*, 2010, **46**, 404–417.

[22] Kim, M.; Kim, D.-K.; Bineesh, K.V.; Kim, D.-W.; Selvaraj, M.; Park, D.-W., *Catal. Today*, 2013, **1**, 24–29.

[23] Sun, H.; Zhang, J.; Li, L.; Xu, J.; Sun, D., *Colloids Surf. A Physicochem. Eng. Asp.*, 2013, **426**, 26–32.

[24] Pálková, H.; Hronský, V.; Jankovič, L.; Madejová, J., *J. Colloid Interface Sci.*, 2013, **395**, 166–175.

[25] Hoshino, J.; Limpanart, S.; Khunthon, S.; Osotchan, T.; Traiphol, R.; Srikhirin, T., *Mater. Chem. Phys.*, 2010, **123**, 706–713.

[26] Hu, Z.; He, G.; Liu, Y.; Dong, C.; Wu, X.; Zhao, W., *Appl. Clay Sci.*, 2013, **75**, 134–140.

[27] Boucard, S.; Duchet, J.; Gérard, J.F.; Prele, P.; Gonzalez, S., *Macromol. Symp.*, 2002, **194**(1), 241.

[28] Vermogen, A.; Masenelli-Varlot, K.; Séguéla, R.; Duchet-Rumeau, J.; Boucard, S.; Prele, P., *Macromolecules*, 2005, **38**, 9661–9669.

[29] Vermogen, A.; Masenelli-Varlot, K.; Vigier, G.; Sixou, B.; Thollet, G.; Duchet-Rumeau, J., *J. Nanosci. Nanotechnol.*, 2007, **7**, 3160–3171.

[30] Xie, W., Gao, Z.; Pan, W.P.; Hunter, D.; Singh, A.; Vaia, R., *Chem. Mater.*, 2001, **13**, 2979.

[31] Pavlidou, S.; Papaspyrides, C.D., *Prog. Polym. Sci.*, 2008, **33**, 1119–1198.

[32] Awad, W.H.; Gilman, J.W.; Nyden, M.; Harris, R.H.; Sutto, T.E.; Callahan, J.; Trulove, P.C.; DeLong, H.C.; Fox, D.M., *Thermochim. Acta*, 2004, **409**, 3–11.

[33] Chigwada, G.; Wang, D.; Wilkie, C.A., *Polym. Degrad. Stab.*, 2006, **91**, 848–855.

[34] Avalos, F.; Ortiz, J.C.; Zitzumbo, R.; López-Manchado, M.A.; Verdejo, R.; Arroyo, M., *Appl. Clay Sci.*, 2009, **43**, 27–32.

[35] Stoeffler, K.; Lafleur, P.G.; Denault, J., Proceedings of 3rd International Symposium on Polymer Nanocomposites Sci. and Technol, Quebec, Canada, 2005.

[36] Mittal, V., *Appl. Clay Sci.*, 2012, **56**, 103–109.

[37] Plechkova, N.V.; Seddon, K.R., *Chem. Soc. Rev.*, 2008, **37**, 123.

[38] Slade, P.G.; Raupach, M.; Emerson, W.W., *Clay Clay Miner.*, 1978, **26**, 125–134.

[39] Wagner, J.; Chen, H.; Brownawell, B.J.; Westall, J.C., *Environ. Sci. Technol.*, 1994, **28**, 231–237.

[40] Changchaivong, S.; Khaodhiar, S., *Appl. Clay Sci.*, 2009, **49**, 317–321.

[41] Lee, S.Y.; Kim, S.J., *Clay Clay Miner.*, 2002, **50**, 435–445.

[42] Gilman, J.W.; Awad, W.H.; Davis, R.D.; Shields, J.; Harris Jr., R.H.; Davis, C.; Morgan, A.B.; Sutto, T.E.; Callahan, J.; Trulove, P.C.; Delong, H.C., *Chem. Mater.*, 2002, **14**, 3776.

[43] Livi, S.; Pham, T.N.; Duchet-Rumeau, J.; Gérard, J.F., *J. Colloid Interface Sci.*, 2011, **354**, 555–562.

[44] Ngo, H.; Le Compte, K.; Hargens, L.; Mc Ewen, A.B., *Thermochim. Acta*, 2000, **97**, 357.

[45] Begg, C.G.; Grimmett, M.R.; Wethey, P.D., *Aust. J. Chem.*, 1973, **26**, 2435.

[46] Naveau, E.; Calberg, C.; Detrembleur, C.; Bourbigot, S.; Jérôme, C.; Alexandre, M., *Polymer*, 2009, **50**, 1438–1446.

[47] Hedley, C.B.; Yuan, G.; Theng, B.K.G., *Appl. Clay Sci.*, 2007, **35**, 180–188.

[48] Abdallah, W.; Yilmazer, U., *Thermochim. Acta*, 2011, **525**, 129–140.

[49] Suin, S.; Shrivastava, N.K.; Maiti, S.; Khatua, B.B., *Eur. Polym. J.*, 2013, **49**, 49–60.

[50] Jeschke, F.; Meleshyn, A., *Geoderma*, 2011, **169**, 33–40.

[51] Patel, H.A.; Somani, R.S.; Bajaj, H.C.; Jasra, R.V., *Appl. Clay Sci.*, 2007, **35**, 194–200.

[52] Awad, W.H.; Beyer, G.; Benderly, D.; Ijdo, W.L.; Songtipya, P.; Jimenez-Gasco, M.d.M.; Manias, E.; Wilkie, C.A., *Polymer*, 2009, **50**, 1857–1867.

[53] Gârea, S.-A.; Iovu, H.; Bulearca, A., *Polym. Test.*, 2008, **27**, 100–113.

[54] He, H.; Duchet, J.; Galy, J.; Gerard, J.-F., *J. Colloid Interface Sci.*, 2005, **288**, 171–176.

[55] Stoeffler, K.; Lafleur, P.G.; Denault, J., *Polym. Eng. Sci.*, 2008, **48**, 1449–1466.

[56] Mittal, V., *Eur. Polym. J.*, 2007, **43**, 3727–3736.
[57] Patro, T.U.; Mhalgi, M.V.; Khakhar, D.V.; Misra, A., *Polymer*, 2008, **49**, 3486–3499.
[58] Stoeffler, K.; Lafleur, P.G.; Denault, J., *Polym. Degrad. Stab.*, 2008, **93**, 1332–1350.
[59] Francis, C.W., *Soil Sci.*, 1973, **115**, 40–54.
[60] Ding, Y.; Zhang, X.; Xiong, R.; Wu, S.; Zha, M.; Tang, H., *Eur. Polym. J.*, 2008, **44**, 24–31.
[61] Usuki, A.; Kawasumi, M.; Okada, A.; Fukushima, Y.; Kurauchi, T.; Kamigaito, O., *J. Mater. Res.*, 1993, **8**, 1179–1184.
[62] Cui, L.; Bara, J.E.; Brun, Y.; Yoo, Y.; Yoon, P.J.; Paul, D.R., *Polymer*, 2009, **50**, 2492–2502.
[63] Morgan, A.B.; Harris, J.D., *Polymer*, 2004, **45**, 8695–8703.
[64] Modesti, M.; Besco, S.; Lorenzetti, A.; Causin, V.; Marega, C.; Gilman, J.W.; Fox, D.M.; Trulove, P.C.; De Long, H.C.; Zammarano, M., *Polym. Degrad. Stab.*, 2007, **92**, 2206–2213.
[65] Modesti, M.; Besco, S.; Lorenzetti, A.; Zammarano, M.; Causin, V.; Marega, C.; Gilman, J.W.; Fox, D.M.; Trulove, P.C.; De Long, H.C.; Maupin, P.H., *Polym. Adv. Technol.*, 2008, **19**, 1576–1583.
[66] Chu, L.-L.; Anderson, S.K.; Harris, J.D.; Beach, M.W.; Morgan, A.B., *Polymer*, 2004, **45**, 4051–4061.
[67] Wang, Z.M.; Chung, T.C.; Gilman, J.W.; Manias, E., *J. Polym. Sci. B*, 2003, **41**, 3173–3187.
[68] Costache, M.C.; Heidecker, M.J.; Manias, E.; Gupta, R.K.; Wilkie, C.A., *Polym. Degrad. Stab.*, 2007, **92**, 1753–1762.
[69] Bottino, F.A.; Di Pasquale, G.; Fabbri, E.; Orestano, A.; Pollicino, A., *Polym. Degrad. Stab.*, 2009, **94**, 369–374.
[70] Bottino, F.A.; Fabbri, E.; Fragala, I.L.; Malandrino, G.; Orestano, A.; Pilati, F.; Pollicino, A., *Macromol. Rapid Commun.*, 2003, **24**, 1079–1084.
[71] Pucci, A.; Liuzzo, V.; Melai, B.; Pomelli, C.S.; Chiappe, C., *Polym. Int.*, 2012, **61**, 426–433.
[72] Fox, D.M.; Maupin, P.H.; Harris, R.H.; Gilman, J.W.; Eldred, D.V.; Katsoulis, D.; Trulove, P.C.; De Long, H.C., *Langmuir*, 2002, **23**, 7707–7714.
[73] Livi, S.; Duchet-Rumeau, J.; Gérard, J.F., *Eur. Polym. J.*, 2011, **47**, 1361–1369.
[74] Ding, Y.; Guo, C.; Dong, J.Y.; Wang, Z., *J. Appl. Polym. Sci.*, 2006, **102**, 4314–4320.
[75] He, A.; Hu, H.; Huang, Y.; Dong, J.Y.; Han, C.C., *Macromol. Rapid Commun.*, 2004, **25**, 2008–2013.
[76] Tarkin-Tas, E.; Goswami, S.K.; Nayak, B.R.; Mathias, L.J., *J. Appl. Polym. Sci.*, 2008, **107**, 976–984.
[77] Sterky, K.; Hjertberg, T.; Jacobsen, H., *Polym. Degrad. Stab.*, 2009, **94**, 1564–1570.
[78] Benderly, D., *J. Vinyl Addit. Technol.*, 2008, **14**, 155.
[79] Kovarova, L.; Kalendova, A.; Gerard, J.F.; Malac, J.; Simonik, J.; Weiss, Z., *Macromol. Symp.*, 2005, **221**, 105.
[80] Dias, A.M.A.; Marceneiro, S.; Braga, M.E.M.; Coelho, J.F.J.; Ferreira, A.G.M.; Simões, P.N.; Veiga, H.I.M.; Tomé, L.C.; Marrucho, I.M.; Esperança, J.M.S.S.; Matias, A.A.; Duarte, C.M.M.; Rebelo, L.P.N.; De Sousa, H.C., *Acta Biomater.*, 2012, **8**, 1366–1379.
[81] Chiellini, F.; Ferri, M.; Morelli, A.; Dipaola, L.; Latini, G., *Prog. Polym. Sci.*, 2013, **38**, 1067–1088.

CHAPTER 25

Deep Eutectic Solvents as a New Reaction Medium for Biotransformations

ZHEN YANG AND QING WEN

College of Life Sciences, Shenzhen University, Shenzhen, Guangdong, China

25.1 INTRODUCTION

Searching for inexpensive, sustainable, and ecofriendly solvents for chemical and biochemical processes has always been an important issue. Ionic liquids (ILs) have been regarded as potential green alternatives to conventional organic solvents due to their negligible vapor pressures and excellent chemical/thermal inertness and, therefore, have gained extensive investigations in a variety of applications including biotransformations [1–4]. Deep eutectic solvents (DESs) make up a new class of nonaqueous solvents derived from ILs.

The term of DES was first described in 2003 by Abbott's group [5]. It is actually a eutectic mixture of an ammonium salt (such as choline chloride) and a hydrogen-bond donor (HBD, such as urea and glycerol), the melting point of which is usually much lower than that of any of its individual components, thus allowing its use as a solvent at ambient temperature. A typical example is given by the mixture of choline chloride ($T_m = 302°C$) and urea ($T_m = 133°C$) at a molar ratio of 1:2, resulting in a rather low freezing point of 12°C [5]. The main driving force of this phenomenon involves the interaction between the HBD and the salt anion through hydrogen bonding, thus forming an extensive H-bond network throughout the solvent [6]. Figure 25.1 shows the typical salts and HBDs used for DES formation.

Although holding a molecular component, DESs share some attractive IL-like solvent properties such as low melting point, low volatility, high thermal stability, high solubility for various substances, and the "designer solvent" property (i.e., the physicochemical properties of the solvent can be fine-tuned by altering the DES components). Besides, as the two major components (ammonium salts and HBDs) used for preparing a DES are rich in nature and environmentally benign and a DES can be simply prepared by thermal mixing or freeze-drying a mixture of the two without the requirement of further separation and purification, using DESs

Ionic Liquid-Based Surfactant Science: Formulation, Characterization, and Applications, First Edition. Edited by Bidyut K. Paul and Satya P. Moulik.

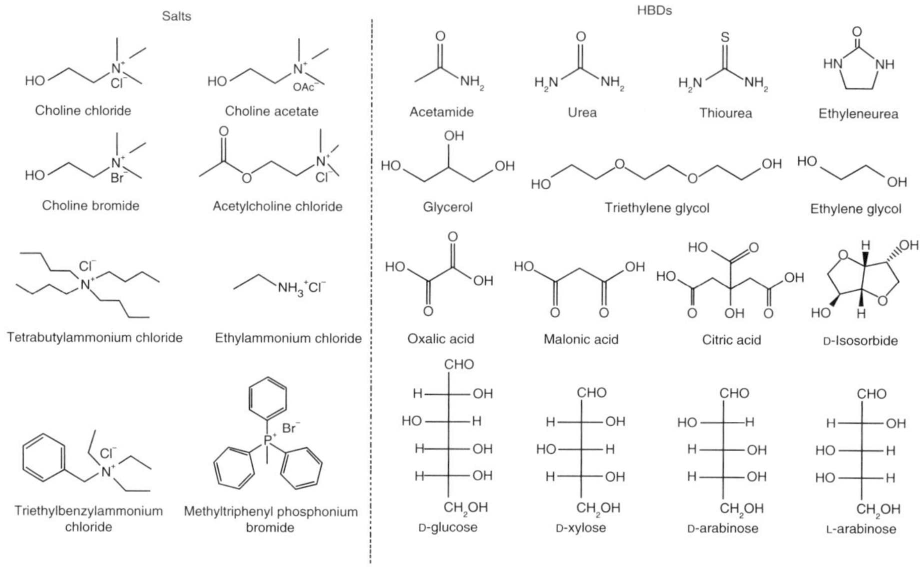

Figure 25.1 Typical salts and hydrogen-bond donors used for DES formation.

possesses some additional advantages over the use of conventional ILs such as (i) lower price, (ii) easier preparation with higher purity, (iii) higher biodegradability and lower toxicity, and (iv) outstanding designability with a broader selection (with choices of cations, anions and HBDs at different salt/HBD molar ratios). It is all these unique features that have made DESs promising and attractive as a new type of nonaqueous solvent/cosolvent for a broad range of applications such as catalysis, organic synthesis, dissolution and extraction processes, electrochemistry, material chemistry, and, more recently, biocatalysis and biotransformation [4, 7].

So far, there are only a few publications that have reported the use of DESs, alone or as cosolvent, as media for biocatalytic reactions [8–14]. Although only at a very early stage of development, these results have suggested that DESs deserve consideration as a potential green solvent for enzymatic processes. In this chapter, we discuss the physicochemical properties of DESs; we review what has been known so far about the use of DESs as solvents or cosolvents for biocatalytic reactions; and we introduce a new type of DESs, natural deep eutectic solvents (NADESs), that have an enormous potential for applications due to their nontoxicity, sustainability, and environmental friendliness.

25.2 PROPERTIES OF DESs

Like conventional ILs, DESs are chemically tailorable solvents, that is, their physical/chemical properties can be fine-tuned by altering their compositions such as the cation and anion of the salt, the HBD, and the salt/HBD molar ratio. Therefore, it is important to have a good understanding about how the DES properties can be manipulated, so that we can design a specific DES with properties required for a specific chemical or biochemical reaction, such as to increase the substrate solubility, to tailor the reaction rate, or to modify the enzymatic performance including the activity, stability, and selectivity. Although many efforts have been devoted to it, the physicochemical characterization of DESs is still far from complete. Here, we report some of the preliminary results so far obtained on this.

25.2.1 Freezing Point

The freezing point (T_f) of a DES determines directly whether this DES is feasible or not to be used as a real solvent. All the DESs reported to date have a freezing point lower than 150°C, but those that remain liquid under ambient temperature are still very limited [7].

In a recent review article by Zhang et al. [7], the freezing points of various DESs have been listed; these DESs are composed of ammonium and phosphonium salts with HBDs such as amides, amines, alcohols, and carboxylic acids. The T_f values of all these DESs varied in a very broad range (from −66 to +149°C) depending very much on the anion and cation of the salt, the HBD, and the salt/HBD molar ratio as presented in Table 25.1. In terms of the salt anion, the T_f depression order $F^- > NO_3^- > Cl^- > BF_4^-$ suggests that a correlation may exist between the freezing point of a DES and the hydrogen-bond strength of the salt anion [5].

For the first DES that was formed between choline chloride (ChCl) and urea (U), there is a V-shaped relation between the freezing point and the ChCl/U ratio, producing a eutectic with the lowest T_f of −12°C at a molar ratio of 1:2 [5]. This indicates that for a molecule of choline chloride, two molecules of urea are required to form the eutectic with the most depressed freezing point. When carboxylic acids were used as the HBD in place of urea [19], one can find that a eutectic was formed at a salt/HBD

TABLE 25.1 Impact of the Cation and Anion of the Salt, the Hydrogen-Bond Donor (HBD), and the Salt/HBD Molar Ratio on Freezing Points (T_f) of DESs

Salt Cation	Salt Anion	HBD	Salt/HBD Molar Ratio	T_f (°C)	Reference
Choline	Cl^-	Glycerol	1:2	−8	[15]
N,N-Diethylenethanol ammonium				31	[15]
Methyltriphenylphosphonium				−4	[15]
Choline	F^-	Urea	1:2	1	[5]
	NO_3^-			4	[5]
	Cl^-			12	[5]
	BF_4^-			67	[5]
Choline	Cl^-	Urea	1:2	12	[5]
		Acetamide		51	[5]
		Ethylene glycol		−66	[16]
		Glycerol		−40	[17]
Choline	Cl^-	Glycerol	1:1	−8	[15]
			1:2	36	[15]
			1:3	33	[15]

Data were obtained from Dai et al. [18].

molar ratio of 1:2 for a monocarboxylic acid such as phenylacetic acid (similar as in the case of urea), 1:1 for a dicarboxylic acid such as malonic acid, and 2:1 for a tricarboxylic acid such as citric acid. In general, the freezing point of the DES is more depressed when an acid is used as the HBD with a lower molecular weight or in a lower mass fraction.

Taking into consideration that a DES is formed by self-association mainly through H-bonding between the salt anion and the HBD, it is not so difficult to understand the above phenomena. Abbott et al. [19] have proposed that the freezing point of a DES should depend on (i) the lattice energies of the salt and the HBD, (ii) the interaction between the salt anion and the HBD, and (iii) the entropy changes arising from forming a DES liquid.

25.2.2 Density

Density is an important physical property for a solvent. Generally, almost all of the DESs reported so far present higher densities than water [7]. But the results may vary according to a change in the temperature, the pressure, or the DES compositions. Leron et al. [20] have demonstrated that the density of the choline chloride/glycol DES decreased with increasing temperature and decreasing pressure. But a DES may have its density increased or decreased by an increase in the salt/HBD molar ratio, depending on whether the density of the HBD is respectively lower or higher than that of the corresponding DES [15].

25.2.3 Viscosity

Most of the DESs exhibit relatively high viscosities at room temperature (>100 cP, in the same range as conventional ILs, as compared to 0.3 and 0.9 cP for acetone and water at 25°C, respectively) [7], and the major factor responsible for this is always believed to be the extensive hydrogen-bonding network throughout the solvent. This is one of the largest obstacles for the application of DESs as a reaction medium, simply because the reaction would be seriously hindered by the mass transfer limitation. Therefore, development of DESs with low viscosities is highly desirable.

The viscosity of DES is affected by the temperature, the water content, and the chemical nature of the DES components (the salt, the HBD, and the salt/HBD molar ratio). An increase in temperature [19] or water content [18] is accompanied by a significantly lower viscosity. On the other hand, HBD also plays an important role in manipulating the viscosity of the DES. Use of glycerol and ethylene glycol, respectively, as the HBD for forming DES with choline chloride (ChCl) at a salt/HBD molar ratio of 1:3 results in very distinct viscosities, 450 and 19 cP, at 20°C, and addition of the salt (ChCl) to the two DES systems also leads to a change in their viscosities in opposite directions [18, 21] (see Fig. 25.2). In comparison, use of derived sugars (e.g., sorbitol) or carboxylic acids (e.g., malonic acid) as HBDs led to DESs exhibiting high viscosities (12,730 cP for ChCl/sorbitol (1:1) at 30°C [22] and 1124 cP for ChCl/malonic acid (1:2) at 25°C [23]), presumably due to the robust 3D intermolecular hydrogen-bond network. On the other hand, Zhao et al. [11] have discovered that use of acetate to replace chloride as the anion, coupled with glycerol as the HBD, yielded DESs with relatively low viscosities (e.g., 80 cP

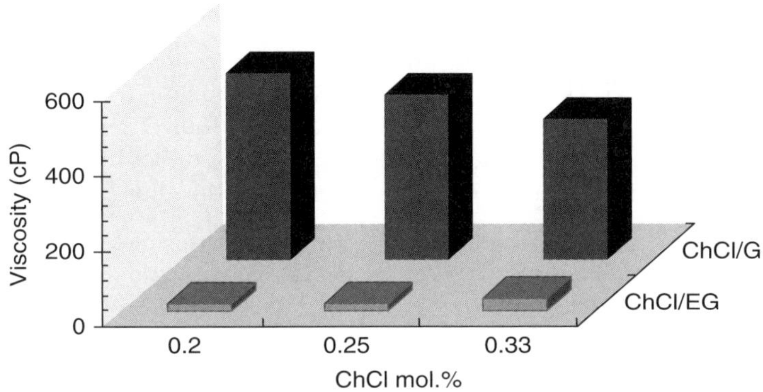

Figure 25.2 Variation of the viscosity of the ChCl-based DES formed with different HBDs upon an increase in the ChCl content in the DES. This figure is drawn by using the data taken from Abbott et al. [17, 21].

for ChAc/G (1:2) at 50°C). The use of glycerol as the HBD in forming DESs also offers advantages such as low toxicity, high biodegradability, and rich availability. Moreover, the lowest viscosity of a DES is often found in the vicinity of the eutectic composition [18].

25.2.4 Surface Tension

So far, studies regarding the surface tension of DES have been limited. All the reported data are greater than the results for normal organic solvents but are comparable to those of conventional ILs [7]. In fact, a high surface tension for a DES is not surprising because like viscosity, it is also strictly dependent on the intermolecular interactions that are responsible for the DES formation.

25.2.5 Polarity

The polarity of a solvent is generally characterized by its polarity scale, E_T (30), which is the electronic transition energy of a probe dye (e.g., Reichardt's dye 30) in a solvent. By examining the polarity data for the ChCl/glycerol DESs, Abbott et al. [17] have found that these DESs are similarly polar as the ammonium ILs, such as $RNH_3^+X^-$ and $R_2NH_2^+X^-$, and that addition of the salt results in an increase in the polarity of the DES. Very recently, Pandey et al. [24] have also verified that DESs formed from choline chloride combined with 1,2-ethanediol, glycerol, urea, and malonic acid, respectively, in 1:2 molar ratios are fairly dipolar in nature. The high polarity allows DES to dissolve a wide range of different substances such as inorganic salts and metal oxides, aromatic acids, amino acids, proteins, DNA and RNA, drugs, surfactants, and sugars [5, 25–27]. In fact, these are the compounds that are capable of accepting or donating protons to form hydrogen bonds. Meanwhile, the high polarity may also be part of the reason why DESs are miscible with those solvents (e.g., methanol and ethanol) that are capable of forming strong hydrogen bonds with their anions such as Cl^- [5].

25.2.6 Hygroscopicity

DESs are so hydrophilic that they are always very hygroscopic, very often much more than conventional ILs [11]. For instance, after exposure to humid air at 50°C for 24h, a high water content of 3.5–6wt.% was achieved by the four DESs based on choline chloride and acetate, while two Tf_2N^--based ILs yielded a rather lower water content (≤2wt.%). This hygroscopic nature may be affected by the choice of DES compositions such as the salt cation and anion, the HBD, and the salt/HBD molar ratio.

25.2.7 Toxicity

Toxicity is an important issue that has to be considered when introducing DESs as a new solvent type into either chemical or biochemical reactions. Because the components used for preparing DESs are considered to be safe (for instance, choline chloride, so-called vitamin B_4, is an essential nutrient for both human and animals), people always take it for granted that DESs are nontoxic, eco-friendly, and biocompatible. Hayyan's group was the first to question whether DESs are benign or toxic. They investigated the toxicity (on four bacteria) and cytotoxicity (on *Artemia salina* leach) of choline chloride-based DESs combined with four HBDs including glycerol, ethylene glycol, triethylene glycol, and urea [28]. All the tested DESs did not exhibit toxic effect on the bacteria but showed higher potent cytotoxicity than their individual components. More recently, this group also carried out a similar investigation on phosphonium-based DESs [29]. The DESs were prepared by using methyltriphenylphosphonium bromide (MTPB) combined with three HBDs including glycerol, ethylene glycol, and triethylene glycol. While confirming the much higher cytotoxicity of DESs than that of their individual components, this study also revealed that the phosphonium DESs exhibited toxic effect on the tested bacteria.

Recently, we have explored the toxicity of various choline-based DESs by testing their effect on the growth of *Hydra sinensis*, a sensitive test animal for toxicity screening [30]. Our preliminary results suggest that a DES might be less toxic than its components [31] (see also Fig. 25.3). An extensive investigation regarding the toxicity, cytotoxicity, and biodegradability of DESs are being undertaken in our laboratory. Nevertheless, the results obtained previously have given us a warning that

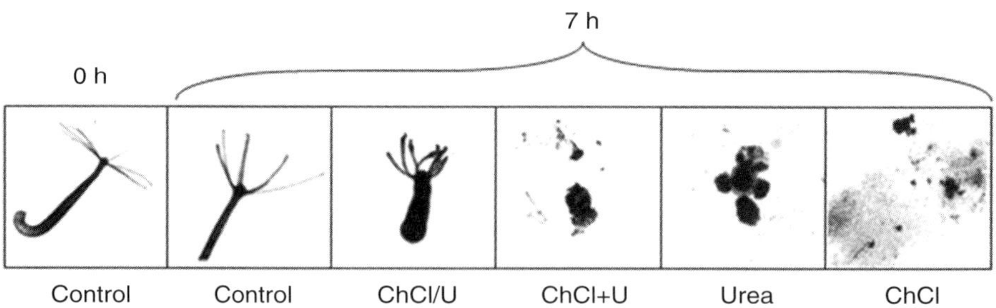

Figure 25.3 Impact of DES and its components on the growth of *Hydra sinensis*. The hydra survived after incubation for 7h in aqueous solution containing the DES ChCl/U (1:1) but was disintegrated in the presence of choline chloride (ChCl), urea (U), and a mixture of ChCl and U at a molar ratio of 1:1 (ChCl+U).

attention has to be paid to the potential toxicity of both DESs and their components (such as ChCl). This has to be seriously considered when designing DESs for (bio) chemical reactions and processes.

25.3 BIOCATALYSIS IN DES SYSTEMS

Although DESs have emerged as a promising new type of green solvents for a decade already, their application in biocatalysis is only in its infancy. Since the first report about biotransformations in DESs given in 2008 [8], up till now, there are only seven research papers published on this aspect [8–14], which however have successfully demonstrated that DESs can work as either a solvent or a cosolvent for biocatalytic transformations. Reactions studied involve several hydrolase-catalyzed processes including (trans)esterification, aminolysis, epoxidation, and ring-opening lactone polymerizations. However, because the experimental data regarding these processes have been so limited, all being focused on the use of hydrolases, in particular lipases, it is still impossible to work out a general guide about how to select a proper DES for a particular enzymatic reaction. The following is a brief review of the use of DESs as solvent/cosolvent for enzymatic reactions.

25.3.1 Use of DESs as Solvent for Biocatalytic Reactions

As the first proof of concept, Gorke et al. [8] have successfully carried out lipase-catalyzed transesterification and aminolysis in DESs. Three lipases have been screened for catalyzing the transesterification of ethyl valerate with 1-butanol in eight DESs with chloride as the salt anion (see Fig. 25.4), and the results have shown that

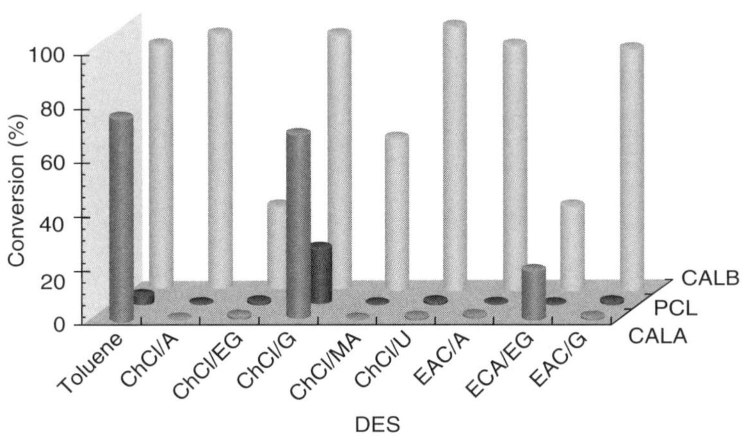

Figure 25.4 Conversion obtained in the transesterification reaction between ethyl valerate and 1-butanol in eight DESs as compared to that obtained in toluene, catalyzed by three lipases: *Candida antarctica* lipase B (CALB), *C. antarctica* lipase A (CALA), and *Pseudomonas cepacia* lipase (PCL). ChCl, choline chloride; EAC, ethylammonium chloride; A, acetamide; G, glycerol; EG, ethylene glycol; U, urea; MA, malonic acid. This figure is drawn by using the data taken from Gorke et al. [8].

(i) among the three lipases tested, *Candida antarctica* lipase B (CALB) was the most active enzyme, which was much more stable in the DES (ChCl/U) than in the aqueous solution containing either of the DES components (10 M urea or 5 M choline chloride); (ii) among the eight DESs tested, the ChCl/glycerol DES was the most efficient one for all the three lipases, yielding the highest conversions that are greater than, or comparable to, the results obtained in either organic solvents such as toluene or conventional ILs such as [BMIm][Tf_2N] and [BMIm][BF_4]; and (iii) as a competitor for the transesterification reaction, glycerol and ethylene glycol were much less reactive when incorporated in the DES as the HBD. ChCl/glycerol was also the DES in which the highest conversion was achieved in the CALB-catalyzed aminolysis reaction.

Recently, Zhao's group [11] has synthesized new choline-based DESs with acetate as the anion in place of the commonly used chloride, coupled with glycerol as the HBD. What they found was that these new DESs provide a reaction medium that is not only less viscous (80 cP for ChAc/glycerol (1:2) vs. 120 cP for ChCl/urea (1:2) at 50°C) but also more efficient in promoting the lipase-catalyzed transesterification reactions: when catalyzing a transesterification reaction between ethyl sorbate and 1-propanol, Novozym 435 obtained a high initial activity (1.02 μmol/min/g) with a high selectivity (>99%) in the DES ChAc/glycerol (1:1.5), superior to the results obtained in conventional ILs such as [BMIm][Tf_2N] and [Me$(OEt)_3$-Me-Et-Im] [Tf_2N], the latter two having been shown to be good solvents for this reaction; when applied to the lipase-catalyzed biodiesel production from Miglyol oil, ChAc/glycerol (1:1.5) was also the DES that promoted the highest conversion. A more recent study by Zhao's group has further confirmed the use of choline-based DESs as reaction medium for biodiesel production [12]: Novozym 435 was proven to be the most efficient enzyme among the 10 different lipases tested, and ChCl/G (1:2) was the best DES in which a high biodiesel production yield of 88% was achieved from soybean oil under optimal conditions.

The glycerol-based DESs derived from choline chloride or acetate have also been found to be effective solvents for proteases such as subtilisin and α-chymotrypsin, but subtilisin seems to be more compatible than α-chymotrypsin in these DESs [10]. Another finding from this work is that water is critical in affecting both activity (for catalyzing a transesterification reaction of *N*-acetyl-L-phenylalanine ethyl ester with 1-propanol) and selectivity (the aforementioned transesterification reaction in competition with the hydrolysis reaction) of subtilisin in the DES system: an increase in water content in both ChCl/glycerol (1:2) and ChAc/glycerol (1:2) caused the enzyme to present an enhancing activity but a reducing selectivity.

The recent study by Durand et al. [13] has further verified that the ChCl/urea and ChCl/glycerol DESs are suitable solvents for both activity and stability of lipase. More recently, the same group has furthered their study by determining the role of different DES components (e.g., ChCl, urea, and water) in lipase-catalyzed alcoholysis reactions in DESs [14].

Regarding the role of water, it has been well known that a certain amount of water is required for biocatalysis in nonaqueous environment, but different enzymes require different amounts of water for their optimal catalytic performance. It is the amount of water associated with the enzyme, rather than the total water content in the reaction system, that is the key determinant of the enzyme properties (e.g., activity, stability, and specificity) [32], and thermodynamic water activity (a_w) has been generally considered as a parameter to quantify the hydration level on the enzyme [33].

Taking the high hygroscopicity of DESs into consideration, it is reasonable to assume that the water added to the DES is structurally tied up with the solvent, thus limiting its free availability in the DES system. In fact, by determining the thermodynamic water activity of choline chloride (ChCl) in aqueous solution and examining the ChCl–water association by using the FT-IR spectroscopy, Durand et al. [14] have found out that water associates with ChCl in two stages: when its level is low (molar ratio of ChCl/H_2O < 1:1.75), water is strongly associated with the salt by hydrogen bonding; when the water content gets higher, the intensity of ChCl–water association becomes weaker, and more free water molecules become available in the solvent. Therefore, in the presence of the same water content, the actual thermodynamic water activity (a_w) of DESs is anticipated to be lower than that of conventional organic solvents and ILs, which may facilitate the enzymatic reaction that require a low water activity; and the availability of water in the DES to participate in the competing hydrolysis reaction should be lower as well, thus alleviating this side reaction. These explanations can be used to account for the experimental results regarding the impact of water on the enzyme in the DES system.

25.3.2 Use of DESs as Cosolvent for Biocatalytic Reactions

DESs have also proven to be suitable as cosolvent for enzymatic reactions in aqueous solution. Gorke et al. [8] have reported that upon addition of 10% of the ChCl/glycerol DES, the hydrolysis of *p*-nitrophenyl acetate was accelerated moderately when catalyzed by several esterases: a threefold increase for pig liver esterase (PLE) and *Rhizopus oryzae* esterase (ROE) and a 25% increase for *Pseudomonas fluorescens* esterase (PFE) and CALB. The presence of the same DES in aqueous solution also triggered an up to 20-fold enhancement in the rate of the epoxide hydrolase (from *Agrobacterium radiobacter*)-catalyzed hydrolysis of styrene oxide. In fact, there was a bell-shaped relationship between the reaction rate and the DES content, as shown in Figure 25.5.

In a later study by Lindberg et al. [9], however, both the activity and stability for the same enzyme, epoxide hydrolase from potato (stEH1), decreased with an increase in the concentrations of all the three DESs they tested, in the order of ChCl/glycerol > ChCl/ethylene glycol > ChCl/urea, all in 1:2 stoichiometric ratios. These DES effects seem to be not related to the polarity of the DES but to the chemical nature of the HBD in the DES. The DES deactivated the enzyme more by elevating the K_m and less by altering the k_{cat}, thus leading to a serious decrease in the catalytic efficiency, k_{cat}/K_m. Therefore, the loss in the enzyme activity at higher DES concentrations should not be attributed to protein denaturation, but rather to destabilization of enzyme–substrate or reaction intermediate complexes. The regioselectivity in the stEH1-catalyzed hydrolysis of (1R,2R)-2-MeSO can be improved by addition of ChCl/glycerol and ChCl/ethylene glycol rather than ChCl/urea, and the underlying reason for this solvent-dependent effect on regioselectivity is still unknown.

Very recently, our group has investigated the effects of DESs and their components on the activity and stability of both *Penicillium expansum* lipase (PEL) and horseradish peroxidase (HRP) by preparing 24 DESs with two cholinium salts (cholinium chloride (ChCl) and cholinium acetate (ChAc)) combined with four HBDs (urea, glycerol, acetamide, and ethylene glycol) at three molar ratios (1:2, 1:1, and 2:1) [31, 34]. These studies have demonstrated that DESs can be viable

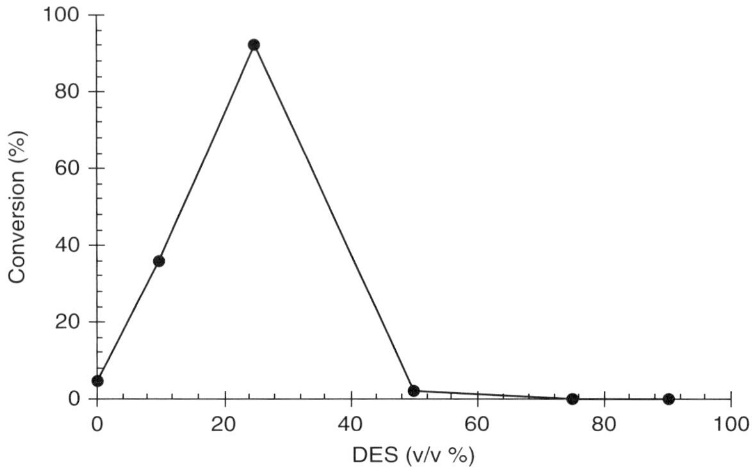

Figure 25.5 Conversion obtained in the hydrolysis of styrene oxide catalyzed by epoxide hydrolase AD1 from *Agrobacterium radiobacter* as a function of the volume fraction of the ChCl/glycerol DES in aqueous solution. This figure is drawn by using the data taken from Gorke et al. [8].

enzyme activators and stabilizers, depending on their compositions: the cation and anion of the salt, the HBD, and the molar ratio of the salt to HBD. For lipase catalysts [31], ChAc/glycerol (1:2) turns out to be a good choice, when being used either alone as the reaction medium for transesterification reaction (e.g., biodiesel production) or as an additive/cosolvent for hydrolytic reaction in aqueous solution; and the superiority of ChAc over ChCl and that of glycerol and ethylene glycol over urea and acetamide has been (at least partially) ascribed to their influence on the enzymatic mechanism. In terms of HRP [34], however, the enzyme activity is favored in the presence of the ChCl-based DESs rather than the ChAc-based ones, while for a DES composed of the same cholinium salt and HBD, a higher salt/HBD molar ratio is more beneficial to the enzyme activity. Therefore, designing/selecting an appropriate DES for a specific biocatalytic application may be a complicated situation, and a careful examination concerning various factors (such as structure and mechanism of the enzyme, choice of the substrates and products, toxicity and biodegradability of the DES, etc.) has to be seriously considered. Our study has also demonstrated that the aforementioned DES effects arise from the formation of DES rather than from the synergetic action contributed by each of its components, thereby supporting the idea that the extensive H-bonding network throughout the DES is sufficiently strong to prevent the DES from dissociation in aqueous solution.

25.4 NADES

The concept of NADES was introduced very recently in 2011, when Choi and coworkers [35] noticed that a few simple molecules like choline, amino acids, sugars, and some organic acids such as malate, citrate, and succinate, which are potential ingredients of normal ILs and DESs, are always present in considerable amounts in all microbial, mammalian, and plant cells. They hypothesized that these metabolites may form a new liquid phase (in addition to water and lipids), referred to as an NADES, for solubilizing, storing, and transporting nonwater-soluble metabolites in living cells and organisms. Indeed, over 100 different NADESs can be formed by different combinations of these common metabolites that are abundantly available in all types of cells and organisms [18]; they change their states from solid to liquid when mixed in proper ratios *in vitro*. Circumstantial evidence has been collected to really demonstrate that NADES does play an important role in cellular metabolism. Its occurrence can explain many biological phenomena such as the biosynthesis and storage of poorly water-soluble metabolites and macromolecules in the aqueous environments of cells and the survival of organisms in extreme drought and/or cold conditions.

25.4.1 Preparation and Structure of NADES

Like normal DES, NADES can also be prepared by using two methods: heating and freeze-drying [18]. Both methods show the same chemical profile for the liquid obtained as demonstrated by the ^{1}H nuclear magnetic resonance (NMR) spectra [18], but the first one is comparatively both cheaper and safer.

NMR spectroscopy has been applied in search for molecular interactions in NADES. The existence of the hydrogen bond between the fluoride ion from choline fluoride and protons from urea was observed by using the heteronuclear Overhauser enhancement spectroscopy (HOESY) [5]. The 2D NMR spectra of 1,2-propanediol–choline chloride–water (1:1:1, molar ratio) have also revealed the strong interactions among the three components involved [35]. The formation and stability of NADES can be seriously affected by the type, the number, and the spatial orientation of the HBD/acceptor groups [18].

25.4.2 Physicochemical Properties of NADES

In their very recent study, Dai et al. [18] have determined a series of physicochemical properties of more than 100 NADESs they prepared. All these eutectic mixtures have a glass transition point (T_g) of lower than −50°C but without a melting point, reassuring that they can be used as a solvent under ambient temperature (at least 0–100°C). The densities of these NADESs are all higher than that of water. They all present a rather high viscosity (33–581 mm^2/s at 40°C, relative to ~1 mm^2/s for water), which is undoubtedly one of the largest obstacles for the application of NADES as a solvent, but the viscosity can be effectively reduced by adding water and increasing temperature. For instance, the viscosity of glucose–choline chloride–water decreased to 1/10 and 1/3 of the original value by addition of 10% water and by increasing the temperature from 20 to 40°C, respectively. NADESs are also highly polar, which accounts for their high solubilizing capacity. The polarity data showed that the organic acid-based NADESs are the most polar ones (44.81 kcal/mol), followed by amino acids and pure sugar-based NADES with a polarity similar to water (48.21 kcal/mol), while both sugar- and polyalcohol-based NADES are the least polar, with a polarity close to that of methanol (51.89 kcal/mol). The polarity of an NADES can be tailored by

TABLE 25.2 Comparison of the Solubilities of Some Compounds in NADES Relative to That in Water

Solute	Solubility in Water	Solubility in NADES	NADES Used for Solubilization
Rutina	0.01 mM	120.6 mM	Glc–ChCl–H_2O
Quercetin	0.0004 mM	166.95 mM	Xylitol–ChCl–H_2O
Cinnamic acid	0.13 mM	128.47 mM	1,2-Propanediol–ChCl–H_2O
Carthamin	0.02 mM	6.77 mM	Xylitol–ChCl–H_2O
1,8-Dihydroxyanthraquinone	0.00 mM	0.41 mM	Lactate–Glc–H_2O
Taxol	0.0002 mM	3.45 mM	Xylitol–ChCl–H_2O
Ginkgolide B	0.01 mM	38.34 mM	1,2-Propanediol–ChCl–H_2O
Glutin	0.03 $g/mol_{solvent}$	2.64 $g/mol_{solvent}$	Lactate–Glc–H_2O
DNA	4.56 $g/mol_{solvent}$	157.01 $g/mol_{solvent}$	Lactate–Glc–H_2O
Starch	Not detected	7.55 $g/mol_{solvent}$	Glc–ChCl–H_2O

water addition, very likely due to the rupture of the existing hydrogen bonds between the two components.

The excellent solubilization ability is undoubtedly one of the most attractive features of NADES. They are capable of dissolving a wide range of poorly water-soluble metabolites or natural products with small (e.g., flavonoids) or big (e.g., cellulose) molecular weights (see Table 25.2), and the solubilities of these compounds can be increased for up to 460,000-fold as compared to that in water, which can be further increased by adding a certain amount of water or increasing the temperature [18]. In fact, there is a bell-shaped relationship between the solubility of a solute and the water content in the NADES, and the optimal water content varies from solute to solute. An NMR study has suggested that the hydrogen bonding between the NADES and the solute should be responsible for the enhanced solubility of these compounds [18].

25.4.3 Effect of Water on Both Structure and Properties of NADES

It is noteworthy that water plays an important role in maintaining the structure of NADES and in altering their properties. It has been found that in some NADES, water is present as part of the solvent (e.g., Glc–ChCl–H_2O, 1:1:1), which cannot be simply removed by evaporation. The tight incorporation of water in the NADES has been confirmed by the ^1H–^1H nuclear Overhauser enhancement spectroscopy (NOESY) of 1,2-propanediol–choline chloride–H_2O, which indicates a strong interaction (very possibly through hydrogen bonding) between the protons on the hydroxyl groups from the three components involved [18]. This is possibly why the thermodynamic water activities of most NADES are much lower than that of pure water [18]. As expected, an NADES shows a higher water activity when more water is incorporated in it (Fig. 25.6). However, extensive dilution of NADES with water will lead to the loss of the existing hydrogen bonds and hence the disruption of its original networking structure [36].

Addition of a proper amount of water to the NADES can also influence its physicochemical properties such as a decrease in the viscosity, a change in the polarity, and an increase in the solubilizing ability as mentioned in Section 25.4.2.

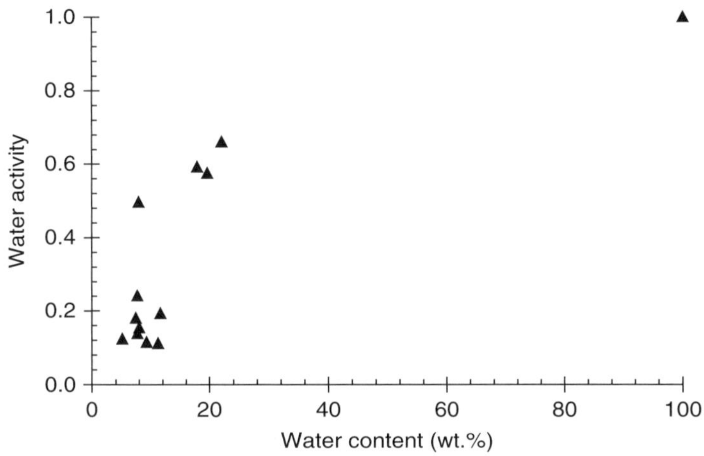

Figure 25.6 Thermodynamic water activities of NADESs as a function of the water contents (wt.%) in the eutectic mixtures. This figure is drawn by using the data taken from Dai et al. [18].

25.4.4 Applications of NADES

As has been noted previously, NADES composed of natural compounds are sustainable and biodegradable and exhibit an acceptable pharmaceutical toxicity profile and a high solubilization power of substances of diverse polarity. It is all these excellent characteristic features that have made NADES very interesting alternatives to the currently used toxic organic solvents and synthetic ILs. NADESs hold a tremendous potential of applications: they can be utilized as truly green solvents for extractions, synthetic organic chemistry, and enzymatic reactions.

The research on the applications of NADES was again pioneered by the Choi's group in using it for extraction [37]. A series of NADES were investigated for the extraction of phenolic compounds of varying polarity from safflowers. The extraction yield in the NADES has been found to be comparable to, or higher than, that obtained in other conventional solvents such as water and ethanol. The high extractability of phenolic compounds with NADES may be attributed to the H-bonding interactions between the molecules of NADES and the phenolic compounds, and the extraction efficiency is clearly governed by the physical properties of the NADES, including polarity and viscosity. For instance, among all the tested NADES, 1,2-propanediol–choline chloride has the lowest polarity, and yet it showed a low efficiency for extracting polar compounds, while a high one for nonpolar phenolics. The high viscosity of NADES is indeed a hindrance to using it as an extraction solvent, but this can be effectively alleviated by dilution with water or by an increase in temperature.

25.5 CONCLUDING REMARKS

Earlier, we have described what is currently known about DES (and NADES), a new type of IL-related and potentially green solvent, its physicochemical properties, and its applications in biocatalysis and biotransformation. Although only in its infancy, the research on the use of DES as a reaction medium for biochemical reactions has shown very encouraging results.

However, in order to take full advantage of the perceived benefits of DESs such as their greenness and cheapness, their excellent designability, and their remarkable power of solubilizing substances of varying polarities, it is necessary to understand in more details the fundamentals of biocatalysis in such a unique solvent system, including (i) the correlation between the structure and composition of a DES and its physicochemical properties, (ii) the correlation between the structure of a DES and its interactions with an enzyme, and (iii) the correlation between the DES structure and enzyme function. Meanwhile, practical issues such as the rather high viscosity and the toxicity and ecological impacts have to be tackled. Nevertheless, DES and NADES hold a promising bright future for being developed as a truly green designer solvent for biocatalytic processes.

REFERENCES

[1] Park, S., Kazlauskas, R. *Current Opinion in Biotechnology*, 2003, **14**, 432–437.

[2] Yang, Z., Pan, W. *Enzyme and Microbial Technology*, 2005, **37**, 19–28.

[3] van Rantwijk, F., Sheldon, R.A. *Chemical Reviews*, 2007, **107**, 2757–2785.

[4] Domínguez de María, P., *Ionic liquids in biotransformations and organocatalysis: solvents and beyond*, Hoboken, NJ: John Wiley & Sons, Inc., 2011.

[5] Abbott, A.P., Capper, G., Davies, D.L., Rasheed, R.K., Tambyrajah, V. Chemical Communications, 2003, 70–71.

[6] Francisco, M., van den Bruinhorst, A., Kroon, M.C. *Angewandte Chemie International Edition*, 2013, **52**, 3074–3085.

[7] Zhang, Q., Vigier, K.D.O., Royer, S., Jérôme, F. *Chemical Society Reviews*, 2012, **41**, 7108–7146

[8] Gorke, J.T., Srienc, F., Kazlauskas, R.J. Chemical Communications, 2008, 1235–1237.

[9] Lindberg, D., de la Fuente Revenga, M., Widersten, M. *Journal of Biotechnology*, 2010, **147**, 169–171.

[10] Zhao, H., Baker, G.A., Holmes, S. *Journal of Molecular Catalysis B: Enzymatic*, 2011, **72**, 163–167.

[11] Zhao, H., Baker, G.A., Holmes, S. *Organic & Biomolecular Chemistry*, 2011, **9**, 1908–1916.

[12] Zhao, H., Zhang, C., Crittle, T.D. *Journal of Molecular Catalysis B: Enzymatic*, 2013, **85–86**, 243–247.

[13] Durand, E., Lecomte, J., Baréa, B., Piombo, G., Dubreucq, E., Villeneuve, P. *Process Biochemistry*, 2012, **47**, 2081–2089.

[14] Durand, E., Lecomte, J., Villeneuve, P. *European Journal of Lipid Science and Technology*, 2013, **115**, 379–385.

[15] Shahbaz, K., Baroutian, S., Mjalli, F.S., Hashim, M.A., AlNashef, I.M. *Thermochimica Acta*, 2012, **527**, 59–66.

[16] Shahbaz, K., Mjalli, F.S., Hashim, M.A., AlNashef, I.M. *Thermochimica Acta*, 2011, **515**, 67–72.

[17] Abbott, A.P., Harris, R.C., Ryder, K.S., D'Agostino, C., Gladden, L.F., Mantle, M.D. *Green Chemistry*, 2011, **13**, 82–90.

[18] Dai, Y., van Spronsen, J., Witkamp, G.J., Verpoorte, R., Choi, Y.H. *Analytica Chimica Acta*, 2013, **766**, 61–68.

[19] Abbott, A.P., Boothby, D., Capper, G., Davies, D.L., Rasheed, R.K. *Journal of the American Chemical Society*, 2004, **126**, 9142–9147.

[20] Leron, R.B., Wong, D.S.H., Li, M.H. *Fluid Phase Equilibria*, 2012, **335**, 32–38.

[21] Abbott, A.P., Harris, R.C., Ryder, K.S. *Journal of Physical Chemistry B*, 2007, **111**, 4910–4913.

[22] Maugeri, Z., Domínguez de María, P., *RSC Advances*, 2012, **2**, 421–425.

[23] D'Agostino, C., Harris, R.C., Abbott, A.P., Gladden, L.F., Mantle, M.D. *Physical Chemistry Chemical Physics*, 2011, **13**, 21383–21391.

[24] Pandey, A., Rai, R., Pal, M., Pandey, S. *Physical Chemistry Chemical Physics*, 2014, **16**, 1559–1568.

[25] Mamajanov, I., Engelhart, A.E., Bean, H.D., Hud, N.V. *Angewandte Chemie International Edition*, 2010, **49**, 6310–6314.

[26] Guo, W., Hou, Y., Wu, W., Ren, S., Marsh, K.N. *Green Chemistry*, 2013, **15**, 226–229.

[27] Morrison, H.G., Sun, C.C., Neervannan, S. *International Journal of Pharmaceutics*, 2009, **378**, 136–139.

[28] Hayyan, M., Hashim, M.A., Hayyan, A., Al-Saadi, M.A., AlNashef, I.M., Mirghani, M.E.S., Saheed, O.K. *Chemosphere*, 2013, **90**, 2193–2195.

[29] Hayyan, M., Hashim, M.A., Hayyan, A., Al-Saadi, M.A., AlNashef, I.M., Mirghani, M.E.S. *Chemosphere*, 2013, **93**, 455–459.

[30] Beach, M.J., Pascoe, D. *Water Research*, 1998, **32**, 101–106.

[31] Huang, Z.L., Wu, B.P., Wen, Q., Yang, T.X., Yang, Z. *Journal of Chemical Technology and Biotechnology*, 2014, **89**, 1975–1981.

[32] Zaks, A., Klibanov, A.M. *Journal of Biological Chemistry*, 1988, **263**, 8017–8021.

[33] Halling, P.J. *Trends in Biotechnology*, 1989, **7**, 50–51.

[34] Wu, B.-P., Wen, Q., Xu, H., Yang, Z. *Journal of Molecular Catalysis B: Enzymatic*, 2014, **101**, 101–107.

[35] Choi, Y.H., van Spronsen, J., Dai, Y., Verberne, M., Hollmann, F., Arends, I.W.C.E., Witkamp, G.-J., Verpoorte, R. *Plant Physiology*, 2011, **156**, 1701–1705.

[36] Gutiérrez, M.C., Ferrer, M.L., Mateo, C.R., del Monte, F. *Langmuir*, 2009, **25**, 5509–5515.

[37] Dai, Y., Witkamp, G.J., Verpoorte, R., Choi, Y.H. *Analytical Chemistry*, 2013, **85**, 6272–6278.

INDEX

Note: Page numbers in *italics* refer to Figures; those in **bold** to Tables.

AA-based BILs
  biodegradability, 410
  cholinium-based, 410, 411
  enzymatic hydrolysis, phenylalanine methyl ester, 409
  hydrogen-bond basicity, 407, 408
  properties, 407
  self-aggregation, 410
  as structure-directing agents, 410
  synthesis, 407
AAILs *see* amino acid ionic liquids (AAILs)
AAILSs *see* amino acid-derived counterions (AAILSs)
aggregation, amphiphilic molecules
  nonionic surfactants, 240
  "solvophobic interactions", 241
aggregation behavior, IL-based gemini surfactants
  aggregate structure, 137
  counterion binding degree, 136
  ionic liquid crystals (ILCs), 137–8
  Krafft temperatures and melting points, 138
  micellization
    ammonium-based gemini surfactants, 133
    imidazolium gemini surfactants, 133
    N-(2-hydroxyalkyl)-*N'*-(2-hydroxyethyl)imidazolium surfactants, 133
    surface activity parameters, 131, **132**
  surface activity parameters
    quaternary ammonium gemini surfactants, 134
    surface tension measurement, 133
    surfactant efficiency, 134
  thermodynamic parameters, **135**
    CMC value dependence, 134
    enthalpy-entropy compensation, 136
    molecule conformational change, 136
air-liquid interfaces, ILs
  Clausius–Clapeyron relation, 119, *120*
  Eotvos equation, 119
  Guggenheim equation, 119
  Langmuir's principle of independent surface action, 117
  nanosegregated nature, 117
  spherical molecules, 117
  ST values, 117
  surface tension, *118*
alkylsulfate-based BILs, 419
amino acid-derived counterions (AAILSs), 176, 185, 189, 403, 405, 406, 407, 408, 409, 410–411, 413

Ionic Liquid-Based Surfactant Science: Formulation, Characterization, and Applications, First Edition. Edited by Bidyut K. Paul and Satya P. Moulik.

amino acid ionic liquids (AAILs), 176, 185, 189, 403, 405, 406, 407, 408, 409, 410, 411, 413
ammonium ILs, 34, 521
  alkylammonium salts, 505
  polyamides (PA), 506
  thermal stability, 506
amphiphiles, 261
  critical micelle concentration (CMC)
    CiEj surfactants, 242–3
    scattering measurements data, **243**, 243–4
    self-assembly, 241
    surface tension, Triton X-100, 242, *242*
  micellar aggregation
    ethylammonium nitrate (EAN), 241
    ionic surfactant molecules, 240
    nonionic surfactants, 240, 241
    self-assembly, 241
    "solvophobic interactions", 241
anionic surfactant-like ILs (ASAILs)
  self-assembly
    amino acid-derived counterions (AAILSs), 185
    CMC and surface parameters, 182, 183, **185**
    [C_4mim]DS and [ProC_3]DS, 186
    double-chain, 184
    long-chain [C_4mim]DS, 184
    lower limiting surface tension values, 184
    medium-chain, 184
    melting point, 182
    molecular structures, cations and anions, 182, *183*
    shape-controlled CeO_2 nanomaterials preparation, 185
    single-chain anion DS, 185
    surface excess concentration and minimum area per molecule, 184, **185**
ANNs *see* artificial neural networks (ANNs)
aqueous microemulsions, 304
  [C_4mim][PF_6], 265
  fluorescence techniques, 266
  IL polymer materials development, 267
  liquid-liquid extraction, 266
  palladium NPs, 266–7
  phase diagram, 244–7
    bicontinuous microemulsion, 244
    [bmim][PF_6], 247, *247*
    EAN-*n*-alkane-CiEj surfactant systems, 245
    fish phase diagram, 244, *246*
    hydrophobic [bmim][PF_6], 245
    water-[bmim][PF_6]-TX-100 microemulsions, 245, *246*
  single-phase microemulsion, 247–8
  TX-100, CiEj surfactants, 249
  water/TX-100/[C_4mimPF_6] ternary system, *266,* 267
  w/IL microemulsions formation, 267, *268*
artificial neural networks (ANNs), 448, 449
  see also density of ternary mixtures
artificial sweetener-based BILs, 416–17
atomic force microscopy (AFM), 18, 24, 30, *30, 31,* 56, 121, 122, *123,* 139, *141*

BASIL *see* Biphasic Acid Scavenging by Ionic Liquids (BASIL)
BILs *see* bioionic liquids (BILs)
biocatalysis
  fungal oxidase-catalyzed oxidation, *o*-phenylenediamine, 337
  lipase-catalyzed hydrolysis
    4-nitrophenyl butyrate, 337
    p-nitrophenyl butyrate (*p*-NPB), 336–7
biocatalysis, DESs
  choline-based DES, 469
  cosolvent
    p-nitrophenyl acetate hydrolysis, 525
    stEH1-catalyzed hydrolysis, regioselectivity, 525
    styrene oxide hydrolysis, 525, 526
  hydrolase-catalyzed processes, 523–6
  solvents
    choline-based DESs, 524
    glycerol-based DESs, 524
    lipase activity and stability, 524
    transesterification reaction, 523–4
    water content, role of, 524–5
biochemical oxygen demand (BOD) tests, 425
biocompatible ionic liquid-based microemulsions
  bimetallic palladium/gold (Pd_4Au) NPs, 435
  fungal oxidases catalytic activities, 434
  ionic liquid-in-oil (IL/o) microemulsions, 433
  lipase-catalyzed esterification reactions, 434
  N-ethyl perfluorooctylsulfonamide based RMs, 435
  oil-soluble drugs, 432
  $scCO_2$, 435

self-assembling phenomena, amphiphilic molecules, 432
self-emulsifying drug delivery systems, 432–3
single TX-100 and mixed AOT/TX-100 w/IL microemulsions, 434
Tween-80/Span-20 surfactant-based IL/o microemulsion, 433
w/IL microemulsions, 433, 434
bioionic liquids (BILs)
AA-based, 407–11
alkylsulfate-based, 419
antimicrobial activity, (eco)toxicological aspect, and biodegradability
alkyltriphenylphosphonium cation-based ILs, 425
anion effect, anion moieties, 421
antibacterial activity, 425
biochemical oxygen demand (BOD) tests, 425
C_nMPnBr ILs, 423
dialkylimidazolium ILs, 421
disinfectants and antiseptics development, 420
"enzyme-friendly" ILs, 424
green ILs, 423
imidazolium ILs, 424
long-chain imidazolium- and pyridinium-based ILs, 424
3-methyl-1-(propoxymethylcarbonyl)-imidazolium series, 422
nitrogen atom number, aromatic cation ring, 420
nonaromatic compounds, 422
octyl-substituted cation, 421
oxygen-functionalized amide ILs, 423
quaternary imidazolium and pyrrolidinonium salts, 420
tetrabutylammonium (TBA) IL salts, 423
applications
fabrication, unmodified biopolymer-based materials and composites, 431
fluorescent core-shell ellipsoidal RBITC@SiO_2 -IL NP, 430
heparin/biopolymer composites, 431
hydroxyl-functionalized cationic surfactants (HFCSs), 429
hydroxyl-functionalized ionic liquids (HFILs), 429
long alkyl chain imidazolium-based ionic liquids (LILs), 430–431
lyotropic action, 430
nucleophilic substitution reactions, 431, **432**
silica aerogel, 429–30
uniform monodisperse crystalline Ag NPs, 429
artificial sweetener-based, 416–17
biocompatible ionic liquid-based microemulsions, 432–5
cholinium cation-based, 411–14
enzymatic processes and biotransformations, 426–9
fructose-based, 417
lactate-based, 417–19
morpholinium-based, 414–16
synthesis
[AAE]Sac ILs, 404
amino acid ionic liquids (AAILs), 403, 405, 406
chiral ILs (CILs), 404–5
choline chloride-derived room-temperature ILs, 403, 404
cholinium alkanoate ILs, 405, 406
cholinium-based AAILs, 406, 407
environmentally benign ILs, 405
fructose-derived, 403
halogen-free IL, 403
miconazole-based ILs, 402
oxygen-functionalized and amide ILs, 405
RTILs, choline derivatives, 403
bio-related applications, ILBSs
antimicrobial activity, 86, *86*
microbial biofilms, 85
series A and B imidazolium salts, 86
structure/activity relationships (SAR), 85, 87–8
Biphasic Acid Scavenging by Ionic Liquids (BASIL), 151–2

calixarenes
dansylated, [C_{10}mim][Cl]
covalent attachment, fluorophoric moiety, 201
fluorescence emission spectra, 201, *202*
hydrophobicity, 203
lowest-energy fluorescence emission maxima, 202, *202*
twisted intramolecular charge transfer (TICT) excited state, 203
fluorescence, probe pyrene
pyrene polarity scale, 199–200
quenching, 200–201, *201*

calixarenes (*cont'd*)
molecular absorbance, $[C_{10}mim][Cl]$ presence
electron-withdrawing $-NO_2$ functionality, 198
p-nitrophenol, UV-vis absorbance spectra, 199, *199*
UV-vis absorbance spectra, 197, *198*
synthesis, 196
capillary electrophoresis (CE), 338
capillary zone electrophoresis (CZE), 482
micellar electrokinetic capillary chromatography (MEKC), 482, 483
micellar solutions, 482
micelle to solvent stacking (MSS), 484
online preconcentration techniques, 484
capillary zone electrophoresis (CZE), 482, 484
CASAILs *see* catanionic surfactant-like ILs (CASAILs)
catalytic properties, enzymes
ion-specific effects, 462–4
polarity and hydrophobicity, 462
stability, 464–6
viscosity, 461–2
catanionic surfactant-like ILs (CASAILs)
colloidal solutions, 189
imidazolium-based, 189
long-chain counterions, 189
synthesis, 188
cationic exchange capacity (CEC), 504, 505
cationic exchange, layered silicates
ammonium ILs, 505–6
electrostatic and weak intermolecular interactions, 505
ILs selection, 505
mineral host structure and nature, intercalating agents, 505
organic species, identification, 505, 506
thermostable ILs, 506–8
cationic surfactant-like ILs (CSAILs)
with anionic surfactants/ASAILs, 186
CMC and surface parameters, **179–81,** 182
dicationic surface-active ILs (DCSAILs), 177
head groups, 177
long-chain imidazolium and pyridinium, 182
molecular structures, 177, *178*
pyridinium head groups, 189
surface excess concentration and minimum area per molecule, 178
cationic surfactants
micellar electrokinetic capillary chromatography (MEKC), 475
micellar solutions, 475
mobile phase additives, 475
solubilizing media, analytes, 475, 476
CEC *see* cationic exchange capacity (CEC)
cholinium cation-based BILs, 411–14
Clint equation, 212
cryo-transmission electron microscopy (cryo-TEM), 42–3, *43,* 137, 267, 435
CSAILs *see* cationic surfactant-like ILs (CSAILs)
CZE *see* capillary zone electrophoresis (CZE)

DAR *see* Diels–Alder reaction (DAR)
DCSAILs *see* dicationic surface-active ILs (DCSAILs)
Debye-Hückel-Onsager conductivity theory, 70
deep eutectic solvents (DESs), 460, *460*
advantages, conventional ILs, 517–18
ammonium salts and HBDs, 517, 518
biocatalysis, 523–6
enzymatic processes, 518
NADES, 526–9
nonaqueous solvents, ILs, 517
properties
density, 520
freezing point, 519–20
hygroscopicity, 522
polarity, 521
surface tension, 521
toxicity, 522–3
viscosity, 520–521
density of ternary mixtures
artificial neural network (ANN)
accuracy evaluation, 454
density *vs.* predicted values, *455, 456*
information inputs, 449
input, hidden and output layer, 450, *451*
output value of neuron, 450
structure, 449, *449*
topology, **455**
validation phase, 455
variable importance, 457, **457**
variables, model, 454
weight matrix, **456,** 456–7
dataset, 448
multiple linear regression (MLR)
vs. $ANN_{5\text{-}(4)\text{-}1}$ model, **456**

average percentage deviation (APD), 451, 452
density *vs.* predicted values, training and validation phase, 452, *453*
IBM SPSS Statistics, 449
implemented model, 451–2, **452**
Pearson correlation analysis, 450, **451**
root mean square error (RMSE), 451
variables, ionic liquid mixture, 450
software package, 449
DESs *see* deep eutectic solvents (DESs)
dicationic surface-active ILs (DCSAILs), 175–7, *178*
didodecyldimethylammonium bromide (DDAB), 39, *39*
Diels–Alder reaction (DAR), 304, 369, *369,* 375, 400, 408, 417, 418
dynamic light scattering (DLS), 35–6, *37,* 137, 224, 225, 227, 243, **243,** 263–7, 271, 273, 276, 286, 287, 288, 289, 292, 305, 307, 311–12, 329, 344, 351, 352, 361, 362, 370, 382, 383, 410, 467

edge excitation shift (EES), 157, *159*
effective cell model (ECM), 13, *14,* 15, 16
electrospray ionization mass spectrometry (ESI-MS), 405
Eotvos equation, 119, *120*
ESI-MS *see* electrospray ionization mass spectrometry (ESI-MS)

fluorescence resonance energy transfer (FRET) study, 265
fluorescence spectroscopy, 33–4, *34,* 264, 349
Fourier transform infrared (FTIR) spectroscopy, 245, 265, 267, 269, 344, 345, 347, 348, 351, 355, 382, 387, 403, 407, 411, 434, 466
freeze-fracture TEM (FF-TEM), 42, 263, 271, *363,* 382, 383, 389
freezing point, DESs, 519–20
fructose-based BILs, 417
FT-IR spectroscopy, RMs
electron-donating character, [bmim][TfO] and [bmim][TfA], 293
imidazolium C-H stretching modes, 295
S-N-S stretching mode, 292–3
vibrational frequencies, **294**

gemini surfactants *see also* IL-based gemini surfactants
cmc values, 82
definition, 127
symmetrical and unsymmetrical, *128*
Guggenheim equation, 119, *120*

HBDs *see* hydrogen-bond donor (HBDs)
high-performance liquid chromatography (HPLC)
cationic and anionic conventional surfactants, 482
MLC mode, 482
pseudophase C_{12}MIm-Cl, 482
Hofmeister series, 20, 427, 428, 463–4, 465, 466
HPLC *see* high-performance liquid chromatography (HPLC)
hydrogen-bond donor (HBDs)
methyltriphenylphosphonium bromide (MTPB), 522
salt/HBD molar ratio, 519
typical salts and, 517, 518
viscosity, DES, 520, 521
hydroxyl-functionalized ionic liquids (HFILs), 429
hygroscopicity, DESs, 522, 525

IL-based gemini surfactants
advantages, 127–8
aggregation behavior
aggregate structure, 137
counterion binding degree, 136
ILCs, 137–8
Krafft temperatures and melting points, 138
micellization, 131–3
surface activity parameters, 133–4
thermodynamic parameters, 134, **135,** 136
behavior, interface
air/solution interface, 139
solid/liquid interface, 139–40, *141*
and biomacromolecules interactions
DNA complexes, 140, 142–3
protein complexes, 143–4, *145*
gene delivery vectors, 128
synthesis
cationic, 129
hydrophilic head groups with spacer, 129–30
hydrophobic alkyl chains to hydrophilic head group, 129, *130*
hydrophobic chains by spacer, 130–131

IL-based nonaqueous microemulsions
application
acyclovir, *354*
as catalysts, 355
dissolution/encapsulation, 354
IL-in-oil microemulsions, pharmaceutical carrier, *354*
polymerization, 355
silicon materials preparation, 354
benzene/BHDC/[bmim]BF_4 system, 349
bis(trifluoromethylsulfonyl) anion (NTf_2^-), 349
[bmim]PF_6-in-toluene microemulsion, 348, 349
1-butyl-3-methylimidazolium hexafluorophosphate ([bmim] PF_6), 348
1-butyl-3-methylimidazolium tetrafluoroborate ([bmim]BF_4), 344–8
[emim][C_nSO_4]/TX-100/cyclohexane, 350, *350*
ethylammonium nitrate (EAN), 349
IL and polar solvent as polar phase, 350–351
IL as dispersed and continuous phase, 351–2
IL as surfactant and polar solvent, 352–4
N-methyl-*N*-propylpyrrolidinium bis(trifluoromethanesulfonyl) imide/ TX100/benzene system, 349
N,N,N-trimethyl-*N*-propylammonium bis(trifluoromethanesulfonyl) imide, 349
IL-based surfactants
analytical separation science applications, 476–81
analytical techniques
in capillary electrophoresis (CE), 482–4
in high-performance liquid chromatography (HPLC), 482
description, 476
low melting point ionic compounds, 476
micellar properties, 476
ILBSs *see* ionic liquid-based surfactants (ILBSs)
IL-in-oil (IL/O) and oil-in-IL (O/IL) microemulsions
additive effects, 331–2
micropolarity and intermolecular interaction, 330
microstructures, 329–30
phase behavior, 326–9
solvation dynamics, 331
thermodynamics of formation, 330–331
IL-in-water and water-in-IL microemulsions, 332–3
IL microemulsions
as drug delivery systems
acyclovir solubility dependence, 370, *371*
advantages, 369
DLS study, 370
IL/O microemulsions, 370, *371*
nonaqueous microemulsions, 370
as reaction medium
Diels-Alder reaction (DAR) kinetic measurements, 369, *369*
microreactor/nanoreactor, 368
reaction rate, 369, *370*
water pool, 368, *368*
IL/O microemulsions *see* ionic liquid-in-oil (IL/O) microemulsions
ILs *see* ionic liquids (ILs)
in situ intercalative polymerization, 508–9
ionic liquid-based surfactants (ILBSs)
antitumor activity, 90–91, *92*
bio-related applications, 85–9
cationic, 65
evolution of publications, 63, *64*
gemini, 66
interactions, biomacromolecules, 92
micellar properties, 71
micellized *see* micellized ILBS
molecular structure, *65*
nanotechnology, 92–5
structures, 66
synthesis
absence of solvents, 66
ether-containing gemini ILBS, 68
functionalized side chain, 66
gemini-type, 66, 67
routes, 65–6
with side ester/amide group side chain, 69
thioether-containing gemini ILBS, 68
zwitterionic, 66, 69
zwitterionic, 66, 69
ionic liquid-in-oil (IL/O) microemulsions
aggregates size
alkyl side chain length, imidazolium cation, 311, *312*
RTILs anion, 311–12, *313*
constituents effect, droplet structure, 386

as drug delivery media, 391
1H NMR spectra
[C_4mim][AOT] molecule, 316, *317,* 318
[C_2mim][TF_2N], 317, 318
[C_6mim][TF_2N], 318
water addition, 318–20
IL pool dynamics
average solvation time, 385
fluorescence decays, C-153, 384, *384*
time-resolved emission spectra (TRES), 384–5
time-resolved fluorescence anisotropy, 383
morphology, size, property, and microstructure
absorption spectra, 361, *361, 362*
methyl orange (MO) absorption spectra, 361–2, *362*
time-resolved emission spectroscopy, 361
water content, 361, *361*
as nanoreactors, 389–90
phase behavior and conductivity, 378–9
phase behavior study
alkyl side chain length, imidazolium cation, 308–9
RTILs anion, *310,* 310–311
phase equilibrium and critical phenomena
polarity and structural characteristics
anionic AOT surfactant, 382
[bmim][BF_4]/benzene mixtures, 381
[bmim][BF_4]/TX-100/cyclohexane microemulsions, 382, *382*
[bmim][BF_4]/TX-100/*p*-xylene microemulsion, 381, *381*
dynamic light scattering (DLS), 382
FF-TEM, 383
FTIR and NMR spectroscopic techniques, 382
MO absorption spectra, [bmim][BF_4]/TX-100/benzene microemulsion, *380*
small-angle neutron scattering (SANS), 383
solvatochromic probes, absorption and emission spectroscopic studies, 380, *380*
steady-state absorption and fluorescence spectroscopic analyses, 380, 381
preparation
betaine 30, absorption spectra, *307*
[C_4mim][AOT], 306, **306**
[C_4mim][BF_4]/[C_4mim][AOT]/benzene ternary system, *307*
coumarin-480 (C-480), *307,* 308
E_T(30) probe, 307–8
imidazolium (1-butyl-3-methylimidazolium, [C_4mim]$^+$) moiety, 306
nonionic and ionic surfactants, 377, *377*
in sustainable polymer synthesis, 390–391
temperature, 312–14, *314*
temperature, microemulsion stability, 389
water addition
alkyl side chain length, imidazolium cation, 314–15, *315*
RTILs anion, 315–16
water, droplet structure and dynamics, 387–8
ionic liquids (ILs) *see also* long-chain ILs
aggregate formation
aqueous solutions, ILs, 110
[C_8C_1im]Cl, [$C_{10}C_1$im]Cl, and [$C_{12}C_1$im]Cl, 110
[C_1C_{12}Pyrr]Br and [C_4C_{12}Pyrr]Br CMC values, 104, 105
[C_4C_8Pyrr]Br, enthalpy of micellization, 105
critical micelle concentration (CMC), 103, **103,** 105–6, *106*
demicellization, 104
diluted aqueous solutions, IL-based surfactants, 105
double-chained surfactants, 109
enthalpy change, *104*
excess volume, solutions, 116, *116*
free energies and enthalpies, *108*
global volumetric behavior, 110, *111*
isothermal titration calorimetry (ITC), 103
longer-chain surfactants, 108
molar volumes, 111, 112, *112*
mole fraction, bound counterions, *107*
partial molar volumes, *113, 114,* 115
"polar" head group nature, 106
tetraalkylammonium- and pyrrolidinium-based surfactants, 108
uni-univalent ionic surfactants, 107
air-liquid interfaces, 116–20
amphiphiles *see* self-assembled amphiphiles
amphiphilic behavior, 253
applications, 1

ionic liquids (ILs) (*cont'd*)
active pharmaceutical ingredients (APIs) synthesis, 400
analytical separations, 399
catalytic applications, 400–401
cellulose, 399
as "designer solvents", 400
fractionation, lignocellulosic biomass, 400
as "green solvents", 400
as lubricants, 401
pharmaceutical applications, 400
separation science, 400
aqueous microemulsions
phase diagram, 244–7
single-phase microemulsion, 247–8
TX-100, CiEj surfactants, 249
in aqueous microemulsions, 304
bulk solvation properties, 23
categories, 398
cationic and anionic constituents, *376*
commonly used constituent ions, *152*
confinement *see* reverse micelles (RMs)
definition, 398
description, 101
dicationic *(gemini)* ionic liquids, 67
enzyme structure
circular dichroism (CD), 466
dynamic structure-function relationships, 466, **467**
Fourier transform infrared (FTIR) spectroscopy, 466
MD simulations, 468
scattering techniques, 467
imidazolium
interfacial properties, 253–4, *254*
polarity, 254
imidazolium-based, 2
imidazolium-based room-temperature, 4
inherent fluorescence properties, 33
mesoscopic order, 4
miscibility of water, 398
polar nature, 375
as polar solvents
$EmimEtSO_4$/CTAB/toluene/pentanol quasiternary system, 256
$EmimEtSO_4$ droplet structure, 256, *257*
and surfactants, 257, *258*
properties, 398, **399**
protic and aprotic, 375
as pure bulk phases, 124
purity distribution, 63, *64*
second-generation, 66, 459, *460*
self-assembly systems, 261
solid surfaces
adsorbed layers, 120–121
$[C_{12}C_1im]$Br ILs, 122
$[C_{10}C_1im]$Cl, $[C_{12}C_1im]$Cl and $[C_{14}C_1im]$Cl, 123, *123*
frequency change, 121, *122*
hemimicelle formation, 121
quartz crystal microbalance (QCM), 121
solubilized in water
BmimCl, isotropic phase range, 254, *255*
inverse microemulsion droplet, $EmimHexSO_4$, 256, *256*
self-assembled microemulsions, 256
water/toluene/pentanol, isotropic phase range, 254, *255*
solvent properties, 253
surfactant properties, 32
synthetic routes, 65–6
volatility, 23
ionic liquids in colloidal regime
biological applications
additive influence, 231–3
microemulsions, colloidal domain, 228–31
phase-forming behaviors and microemulsifications, 217–27
phosphonium- and ammonium-based ILs, 217
categories, generations, 207, 209
cations and anions, 207, 208
physicochemical properties (bulk and interfacial)
aggregation behavior, 210
amphiphiles CMC ranges, IL media, 216, **216**
Clint equation, 212
IL-1 (triisobutyl(methyl)phosphonium tosylate), 211
influence of [IL-1], 214, **215**
ionic, nonionic and zwitterionic surfactants, 212
g log *C* profile, IL-1 in aqueous medium, *213, 214*
phosphonium-based ILs, 211
phosphorus-based non-IL surfactants, 211
specific conductance, 208
surface activity, 210, 211

surface tension, 208–9, **210,** 212
tensiometric measurements, 211
Vogel–Fulcher–Tammann equation, 208
isothermal titration calorimetry (ITC), 71, 82, 92, 103

Knudsen effusion method, 120

lactate-based BILs, 417–19
Langmuir's principle of independent surface action, 117
layered silicates
organic modification
cationic exchange of, 505–8
hydrophilic and hydrophobic polymers, 504
structure and properties, 504
linear solvation energy relationship (LSER) model, 482
long-chain ILs
oil-in-water microemulsions
amphiphilic ILs, 269
[C_{12}mim][Br], [MAUM][Br] and BPYP[MIM][PF_6], 269
[C_8mim]Cl-$C_{10}H_{21}OH$-H_2O ternary system, 270, *271*
IL-in-oil microemulsion, 271
ordered assemblies, 270
polymerization, 270
water-in-oil microemulsions, 272
lower critical solution temperature (LCST), 36, 407
LSER *see* linear solvation energy relationship (LSER) model

matrix-assisted laser desorption/ionization mass spectrometry (MALDI-MS), 399
MEKC *see* micellar electrokinetic capillary chromatography (MEKC)
melt intercalation, 509
Menschutkin reaction, 403
micellar electrokinetic capillary chromatography (MEKC), 194, 475, 482, **483,** 484, *485*
micellar properties, ILBSs, 71
cmc values, gemini surfactants, 82
ΔG^0_{mic} dependence, number of methylene groups, 84, *84*
ln(cmc) dependence, number of carbon atoms, 82, *83*
surfactants, functionalized side chains, 85
micelle to solvent stacking (MSS), 484, *485*
micellized ILBS
aqueous solution, micellar properties, 71, **72–8**
critical micelle concentration, 70
enthalpies of micellization, 71
Evans equation, 70, 71
Frahm equation, 70, 71
isothermal titration calorimetry (ITC), 71
thermodynamic parameters, 71, **79–81**
microemulsion(s) *see also* ionic liquid-in-oil microemulsions; nonaqueous microemulsions; phase-forming behaviors, IL-based microemulsions
additive influence
[bmim][BF_4]-in-cyclohexane microemulsion, *231,* 231–2
[emim][$EtSO_4$]/TX-114/limonene, 232, *233*
aqueous *see* aqueous microemulsions
biocatalysis
fungal oxidase-catalyzed oxidation, *o*-phenylenediamine, 337
laccase-catalyzed oxidation, 2,6-dimethoxyphenol, 337
lipase-catalyzed hydrolysis, 336–7
lipase-catalyzed hydrolysis, 4-nitrophenyl butyrate, 337
capillary electrophoresis, 338
chemical reaction, 337
classifications, 262
colloidal domain
guanidinium ILs, 231
IL-assisted nonaqueous microemulsion, 228
IL-in-oil microemulsion, 228, *228*
lidocaine docusate (LD), 229
silica nanoparticles, 229, *230*
structure-directing ability, 229
components, 263
definition, 262, 325, 375
drug release, 338
features, 253
hemoglobin extraction, 338
IL-in-EG and EG-in-IL, 333
IL-in-IL, 333–4
IL-in-oil (IL/O) and oil-in-IL (O/IL)
additive effects, 331–2
micropolarity and intermolecular interaction, 330
microstructures, 329–30
phase behavior, 326–9

microemulsion(s) (*cont'd*)
solvation dynamics, 331
thermodynamics of formation, 330–331
IL-in-supercritical CO_2 and CO_2-in-IL, 334
IL-in-water and water-in-IL, 332–3
IL surfactants, long-chain anions, 272–6
ionic surfactants and ILs
polar solvents, 256–7
polar solvents and surfactants, 257, *258*
solubilized in water, 254–6, *256*
long-chain ILs
oil-in-water microemulsions, 269–71
water-in-oil microemulsions, 272
material synthesis
bimetallic palladium/gold nanoparticles, 335
gold nanoparticles, 335
mesoporous metal-organic framework, 335–6
polyaniline/silver (PANI/Ag) nanocomposite, 335
silica, 334–5
ZrO_2 nanoparticles, 335
nonmiscible liquids, 253
polymerization
atom transfer radical polymerization, 336
electropolymerization, 336
TiO_2 nanoparticles, 336
water-free IL-based, 304–5
morpholinium cation-based BILs, 414–16
morpholinium ILs
chemical formulae, 153
polarity
coumarin-153 (C-153), $E_T(30)$ values, 153–4, 157
definition, 153
fluorescence probes, 154
normalized steady-state absorption and fluorescence spectra, C-153, *155*
static dielectric constant and refractive index, 153
temperature dependence, n_{max}^{-em}, 155, **156**
structural heterogeneity
excitation wavelength-dependent emission spectra, 157–60
solvation dynamics, 160–163
time-resolved fluorescence anisotropy, 163–70
viscosity, 157, *158*
MSS *see* micelle to solvent stacking (MSS)
NADESs *see* natural deep eutectic solvents (NADESs)
nanotechnology, 92–5
natural deep eutectic solvents (NADESs)
applications, 529
cellular metabolism, 527
description, 518
microbial, mammalian and plant cells, 526–7
physicochemical properties, 527–8
preparation and structure, 527
water, effect of, 528–9
neutron reflectivity (NR)
$C_{12}E_4$, 28, *28*
deuterium labeling, 28
ethylammonium nitrate (EAN) surface, 28
reflected beam intensity, 27
subnanometer resolution, 28
surface layer thickness, 28
surface roughness, 28
NMR *see* nuclear magnetic resonance (NMR)
nonaqueous microemulsions
1-butyl-3-methylimidazolium tetrafluoroborate, 262, 263
catanionic surfactants, 265
$[C_4mim][BF_4]$, 263
$[C_4mim][PF_6]$, 263–4
dynamic light scattering (DLS), 263
IL-in-oil microemulsions, 265
nonionic alkyl oligo (ethylene oxide) surfactants, 265
$[N_{3111}][Tf_2N]$, $[P_{13}][Tf_2N]$ and $[C_2mim][Tf_2N]$, 264
sulfate-based ILs, 264
nonionic surfactants micellization in RTILs, 48, **49**
adsorption species, 48–9
1-butyl-3-methylimidazolium hexafluorophosphate, 48
C_nEO_m in $BmimPF_6$
cmc values, *51,* 51–2
Gibbs–Helmholtz equation, 52
physicochemical parameters estimation, **50**
second-order polynomial, 52
solvophobic interaction, 53
surface tension, *50*
thermodynamics, 52
dilute solution properties, 48
oxyethylene chain length effect, 52
polyoxyethylene alkyl ether surfactants, 48

protic/aprotic nature, 48
static surface tension, 48
NR *see* neutron reflectivity (NR)
nuclear magnetic resonance (NMR)
1-alky-3-methylimidazolium ibuprofenate, 40
$C_{14}E_6$, chemical shift, 39, *39, 40*
diffusion-ordered spectroscopy (DOSY), 40
diffusion, schematic, 40–41, *41*
IL in RMs, 291–2
measured diffusion coefficient, 40, *41*
pulsed-field gradient spin-echo (PGSE), 40
Stokes-Einstein equation, 40
surfactant aggregation, 41

PGSE-NMR *see* pulsed gradient spin-echo nuclear magnetic resonance (PGSE-NMR)
phase behavior study, microemulsions
alkyl side chain length, imidazolium cation, 308–9
RTILs anion, 310–311
phase-forming behaviors, IL-based microemulsions
bicontinuous-type microemulsion, 218, 221, 225, *226*
[bmim][BF_4]/TX-100/triethylamine, 221
clear zones area, 224–5, **226**
CTAB/toluene-pentanol/[emim][$EtOSO_3$], *225*
CTAB/toluene-pentanol/[emim][$HexOSO_3$], *225*
free energy (driving force), microemulsion formation, 219
IL/O-type microemulsion droplets, 221
IL/TX-100/EG ternary system, 221, *222*
microheterogeneous system, microemulsion, 218
phosphonium cation-based ILs, 221
pseudoternary phase diagrams, IL-based microemulsion systems, 222, *223, 224*
[P_{13}][Tf_2N]/TX-100/benzene three-component system, 227, *227*
solvation dynamics, 227
thermodynamic stability and energy diagrams, 220, *220*
varied hydrophobic environment, 220
water-in-[bmim][PF_6] microemulsion system, 224
PLS *see* polymer–layered silicate (PLS) nanocomposites
polarity, DESs, 521
polarized optical microscopy (POM)
hydrogen-bonding, 34–5
mesophases formation, 34
molecular weight role, 35
nonionic surfactants, 34–5
Pluronic P123, amphiphilic triblock copolymer, 35
polybutadiene-*b*-poly(ethylene oxide) (PB*b*-PEO) copolymers, 38
polyester/IL-modified clay nanocomposites, 512
poly(olefins)/IL-modified clay nanocomposites, 511–12
polymerization
atom transfer radical polymerization, 336
electropolymerization, 336
TiO_2 nanoparticles, 336
polymer–layered silicate (PLS) nanocomposites
description, 504
different routes, preparation, 508, 509
in situ intercalative polymerization, 508–9
melt intercalation, 509
organic–inorganic hybrid materials, 508
solution intercalation, 508
POM *see* polarized optical microscopy (POM)
PS/IL-modified clay nanocomposites
ABS/ammonium- and imidazolium-treated MMT, 509
in situ polymerization, 510
nanoparticle suspension, molten state, 509
SAN nanocomposite processing, 510
pulsed gradient spin-echo nuclear magnetic resonance (PGSE-NMR)
advantages, 2
mechanism, 2
molecular exchange, 3
self-diffusion coefficients, 3–4
spin echo, 2
Stokes-Einstein relation, 3
PVC/IL-modified clay nanocomposites, 512
PVDF/IL-modified clay nanocomposites, 510–511
pyrene
polarity scale, 199–200
quenching, fluorescence, 200–201

QCM *see* quartz crystal microbalance (QCM)
Q^* peak, 38
quartz crystal microbalance (QCM), 121
quenching
  bovine serum albumin (BSA), 144
  pyrene fluorescence, 200–201

red edge effect (REE), 157, 158, *159*
reverse micelles (RMs), *284*
  AOT, BHDC, and TX-100 surfactants, 283–4, *284*
  $[bmim]^+$, $[BF_4]^-$, $[Tf_2N]^-$, $[TfO]^-$, and $[TfA]^-$, 285, *285*
  definition, 283
  ILs
    AOT analogues, 287
    $[bmim][BF_4]$, 286
    dynamic light scattering (DLS) measurements, 286
    FT-IR spectroscopy, 292–5
    molecular probe response, 289–91
    NMR spectroscopy, 291–2
    size characteristics, **288,** 288–9, *290*
    stable RMs formation, 286
    steady-state and picosecond time-resolved emission spectroscopy, coumarin 153, 286
  microheterogeneous environments, 283
  nonaqueous, 285
  water properties, 285
RMs *see* reverse micelles (RMs)
room-temperature ionic liquids (RTILs)
  amphiphiles
    critical micelle concentration (CMC), 241–4
    micellar aggregation, 240–241
  binary phase behavior
    BPS-*m* + water systems, 54
    ethylammonium nitrate (EAN), 53–4
    H_1 phase, 54
    imidazolium-type RTILs, 54
    phase states, 54
    self-assembly, 53–4
    temperature-concentration region, 54–5, *55*
  chiral and long-chain, 194
  complexation study, 194
  definition, 47, 375
  imidazolium-type, 194
  lidocaine docusate (LD), 229
  micellization *see* nonionic surfactants micellization in RTILs
  nonionic surfactants, adsorption
    atomic force microscopy (AFM), 56
    BPS-20/EmimTFSI systems, 56, *57*
    detectable layers, 56
    molecular flexibility, 56
    physicochemical properties, 56
    polyethylene oxide (PEO), 56
    polypropylene oxide (PPO), 56
    protic-aprotic nature, 56
    solid/RTIL interfaces, 56
    solvation layers, 56, 58
    sum frequency vibrational spectroscopic (SFVS) data, 58
  physicochemical properties, 193
  RTIL-calixpyrrole inclusion complexes, 194
  volatility, 303
RTILs *see* room-temperature ionic liquids (RTILs)

SAILs *see* surface-active ionic liquids (SAILs)
sample extraction and preconcentration
  food and environmental applications
    biogenic activity/organic contaminants, 498
    conventional cationic surfactant CTAB, 498
    conventional extraction methods, 499
    in situ IL-DLLME method, 498
    in situ preconcentration strategy, 487–9
    methyl-/butylimidazolium cations, 489
    microextraction methods, 487
    MNPs, 498
    organic compounds, 487
    overall extraction scheme, aqueous solutions, 487, 488
    sample preparation schemes, 489–97
    solid-phase extraction (SPE) approach, 489
  partitioning behavior of analytes
    dialysis, ultradialysis and centrifugation, 485
    partition coefficient value ($K_{M,m}$), 484
    solid-phase microextraction (SPME), 485–7
    from water to IL-based surfactant micelle, 484, 485
SANS *see* small-angle neutron scattering (SANS)
SAR *see* structure/activity relationships (SAR)

SAXS *see* small-angle X-ray scattering (SAXS)
scattering, self-assembled amphiphiles
  dynamic light scattering (DLS), 35–6, *37*
  schematic, 35, *36*
  small-angle neutron scattering (SANS), 38–9
  small-angle X-ray scattering (SAXS), 35, 37–8, *38*
self-assembled amphiphiles
  bulk techniques
    fluorescence spectroscopy, 33–4, *34*
    nuclear magnetic resonance (NMR), 39–41
    POM, 34–5
    scattering, 35–9
    transmission electron microscopy (TEM), 42–3
  characterization techniques
    atomic force microscopy (AFM), 30, *30*
    interface techniques, 25
    neutron reflectivity (NR), 27–8
    sum frequency generation (SFG) spectroscopy, 29
    tensiometry, 30–32, *33*
    X-ray photoelectron spectroscopy (XPS), 25–7
  IL cations and anions, 23, *24*
  ionic liquids (ILs) and, 23–4
  structures, 23, **24**
self-assembly, SAILs
  anionic surfactant-like ILs (ASAILs), 182–6
  catanionic surfactant-like ILs (CASAILs), 188–9
  cationic surfactant-like ILs (CSAILs), 177–82
  mixed surfactants, 186–8
small-angle neutron scattering (SANS), 25, 35, 38–9, *39,* 71, 137, 184, 185, 221, 227, **243,** 243–5, 247, 248, 262, 286, 305, 329, 344, 352, 360, 361, 382, 383, 467
small-angle X-ray scattering (SAXS), 25, 35, 37–8, *38, 233,* 257, 264, 267, 272, 273, 329, 330, 352, 383, 413, 435, 467, 469
sodium bis(2-ethylhexyl) sulfosuccinate (NaAOT)
  bilayers in water, 15
  micelles, 16
  surfactant, 8
  ternary systems, 9
  volume fraction, *10,* 11, 19
solid-phase extraction (SPE) approach, 489
solid-phase microextraction (SPME)
  analyte concentrations, 486
  CMC value, 486
  conventional surfactants, 485–6
  $K_{M,m}$ values determination, 486
  monocationics, 487
solid/RTIL interfaces *see* room-temperature ionic liquids (RTILs)
solution intercalation, 508, *509*
spin echo, 2, 40, 286
SPME *see* solid-phase microextraction (SPME)
structural heterogeneity, morpholinium ILs
  excitation wavelength-dependent emission spectra, 157–60
  solvation dynamics, 160–163
  time-resolved fluorescence anisotropy, 163–70
structure/activity relationships (SAR), 85, 87–8
sum frequency generation (SFG) spectroscopy, 24, 25, 29
sum frequency vibrational spectroscopic (SFVS) data, 58
surface-active ionic liquids (SAILs)
  on calixarenes
    dansylated calixarene with [C_{10}mim][Cl], 201–3
    fluorescence, probe pyrene, 199–201
    materials, 195–6
    methods, 196
    micelle formation and critical micelle concentration, [C_{10}mim][Cl], 196–7, *197*
    molecular absorbance, 197–9
  1-decyl-3-methylimidazolium chloride, 194, *195*
  definition, 303
surface tension, DESs, 521
surfactants system
  definition, 1
  structure, 1
  thermodynamics
    AOT micelles, 8
    Gaussian curvature, 7
    microstructure and phase behavior, 7
    percolation theory, 8
    polar and apolar domains, 7
    sodium bis(2-ethylhexyl) sulfosuccinate (NaAOT), 8
    surfactant molecular volume, 7
    water-in-oil (W/O) spherical droplets, 8

TEM *see* transmission electron microscopy (TEM)
tensiometry
  capillary rise method, 31
  cationic C_nTAB surfactant, 32
  C_nE_m surfactants, 32
  fluorinated surfactant, 31–2
  interfacial properties, 31
  isotherm, 31
  neat IL interfaces study, 32
  surface tension, 30–31
  Wilhelmy plate methods, 31
ternary systems
  $bmim^+$ self-diffusion coefficients, *18,* 18–19
  chaotropic ions, 20
  effective cell model (ECM), 13, 15, 16
  interfacial film, effective volume, 12, *12*
  Lindman law, 13
  macroscopic phenomenology, 21
  NaAOT
    $bmim^+$ adsorption isotherms, 19, *19*
    bmimBr aqueous solutions, 9, *9*
    branched bicontinuous micelles, 20
    ^{1}H PGSTENMR experiments, 16–17
    interfacial packing, 9
    self-diffusion coefficients, 10, *10*
    water and AOT- self-diffusion coefficients, *17,* 17–18
  obstruction effects, 11
  small-angle X-ray diffraction (SAXRD), 10
  spherical micelles, 13
  subvolumes, 13
  surfactant, 18
  water diffusion analysis, 14, *14*
  weak binding, 14
  p-xylene, 15–16, *16*
Teubner–Strey model, *248,* 257
thermogravimetric analysis (TGA), 411, 505, *506*
thermostable ILs
  adsorption of, 507
  imidazolium, 507
  polymer nanocomposites, 506
  pyrrolidinium and phosphonium, 506–7
  quaternary alkylammonium, 508
toxicity
  deep eutectic solvents (DESs)
    Hydra sinensis, growth of, 522–3
    phosphonium-based, 522
    vitamin B_4, 522
  ionic liquids (ILs)
    and alkyl chain length, 402
    cell membrane permeability, 402
    halides, 401
    industrial catalytic applications, 401
transmission electron microscopy (TEM)
  advantage, 42
  cryo-TEM, 42–3, *43*
  FC-4 surfactant, 42, *42*
  freeze-fracture TEM (FF-TEM), 42
twisted intramolecular charge transfer (TICT), 201

upper critical solution temperature (UCST), 36, 407

viscosity, DESs, 520–521
Vogel–Fulcher–Tammann (VFT) equation, 157, 208

water/bmim BF_4 binary system
  dimensional arguments, 6
  double logarithmic plot, *5,* 6
  hydrodynamic radius, 6, *6*
  mesoscopic domains, 7
  self-diffusion coefficients, 4, *5*
  Stokes-Einstein equation, 4–5
  structural heterogeneities, 4
  van der Waals radius, 7
  water content, 6
water-free IL-based microemulsions, 304–5

X-ray photoelectron spectroscopy (XPS)
  IL $[C_2C_1IM][EtSO_4]$, 25–7, *26, 27*
  photoelectron spectroscopy, 25
  schematic, 25, *26*
  surface enrichment, 27
  surfactants alkyltrimethylammonium bromide (C_nTAB), 27, *27*
  UHV-based tool, 25

zwitterionic ILBSs, 66, 69